Sodium-Ion Batteries

Sodium-Ion Batteries

Energy Storage Materials and Technologies

Yan Yu

WILEY-VCH

Author

Prof. Yan Yu
University of Science & Technology
China
Materials Science and Engineering
No.96, JinZhai Road
230026 Hefei
China

Cover Image: © tersetki/Shutterstock

Library of Congress Card No.: applied for

British Library Cataloguing-in-Publication Data
A catalogue record for this book is available from the British Library.

Bibliographic information published by the Deutsche Nationalbibliothek
The Deutsche Nationalbibliothek lists this publication in the Deutsche Nationalbibliografie; detailed bibliographic data are available on the Internet at <http://dnb.d-nb.de>.

Print ISBN: 978-3-527-34896-1
ePDF ISBN: 978-3-527-83161-6
ePub ISBN: 978-3-527-83163-0
oBook ISBN: 978-3-527-83162-3

Typesetting Straive, Chennai, India

Printed and bound by CPI Group (UK) Ltd, Croydon, CR0 4YY

C112634_040222

Contents

Foreword

The Paris Agreement is a landmark representing a common decision of all nations to make ambitious efforts to combat the global climate change. At the General Debate of the 75th Session of The United Nations General Assembly, Xi Jinping President of the People's Republic of China expounded China's principled position in alignment with the Paris Agreement to foster the climate resilience with low greenhouse gas emission development. As a long-term goal of development, China will scale up its Intended Nationally Determined Contributions by adopting more vigorous policies and measures. We aim to have CO_2 emissions peak before 2030 and achieve carbon neutrality before 2060.

To achieve the above goals, it is essential to replace fossil fuels with renewable clean energy. However, the temporal and spatial differences in the distribution of typical renewable energies such as solar and wind represent the major challenge for practical applications. Electrochemical energy storage devices are thus particularly important to regulate the energy output of above intermittent energies. At present, lithium-ion batteries (LIBs) are dominating this market by virtue of their long calendar life and high energy density. Over the coming years, the battery manufacturing is expected to grow with the increasing decarbonizing demand. In view of the accelerated expansion of LIBs, concerns have been raised about its short supply given the limited reserve of the global lithium resource. Consequently, a surge in the battery price is almost foreseen, which threatens the large-scale energy storage industry.

Compared with LIBs, sodium-ion batteries (SIBs) are superior with the abundant Na reserves, low cost, and potentially long cycle life. Developing SIB technology would be of great significance for the large-scale energy storage applications. However, because of the larger ionic radius of sodium ions, SIBs normally suffer from slower diffusion kinetics than LIBs, which poses difficulty in seeking suitable electrode materials and electrolytes for SIBs. Through the extensive research over the past 10 years, a diversity of high-performance and promising SIB systems has been developed by employing electrodes such as transition metal oxides, phosphates, and carbon-based materials. In view of the prosperous research progress, Professor Yan Yu, a pioneer in the field of sodium-ion energy storage, wrote the book *"Sodium-Ion Batteries"* to overview the current research status of SIBs inspired by the research works from her own group. Specifically, this book systematically introduces the materials applied as the anodes, cathodes, electrolytes, and binders for SIBs and discusses their key scientific challenges, modification strategies, and future outlook. This book aims to provide both academic and industry community with more comprehensive and cutting-edge knowledge about SIBs.

Finally, I wish that this book will promote the vigorous development of SIB technology for the large-scale energy storage, with more fruitful achievements to facilitate the early accomplishment of CO_2 emissions peak and carbon neutrality.

13 August 2021
Tianjin, China

Professor Jun Chen
Academician of Chinese Academy of Sciences
Nankai University

Preface

Excessive exploitation and use of fossil fuels have caused rapid reduction of the natural resources and a series of environmental issues such as extreme weather and greenhouse effect. With the awakened public awareness of the finiteness of the earth's resources, the renewable energy sources (e.g. solar, wind, tide, etc.) are developed as alternatives. While the intermittent character of these energy sources calls for more efficient energy storage systems. Over the past 30 years, lithium-ion batteries (LIBs) technology has advanced with unparalleled progress and is now dominating the market in portable electronic devices and electric vehicles. While with dwindled reserves of the raw materials, a rise in LIB price is foreseen.

Sodium-ion batteries (SIBs) share similar electrochemistry and fabrication technologies to LIBs while featuring with low cost and better safety due to earth-abundant sodium resources, rendering them a promising alternative to LIBs in large-scale energy storage. Over the past years, a wealth of research works has been dedicated to the development and commercialization of SIBs. There has been gratifying progress achieved and on other hand, urged greater efforts are still needed to address the unresolved troubling issues.

This book intends to provide researchers and graduate/undergraduate students in related fields of materials science and electrochemistry with a snapshot of current efforts in improving the SIB technology, as well as pave the way for the development of a new generation of higher-energy density, rechargeable SIBs through comprehensive optimization in electrode materials, electrolytes, and other key components.

Chapters 1 and 2 provide a basic introduction (including history, operation principle, key materials, etc.) to SIBs, as well as outline some design principles in terms of energy density, power density, cycling life, and safety to build high-performance SIBs.

Chapters 3 to 6 systematically introduce the cathode materials of transition metal oxides, polyanionic compounds, Prussian blue analogues, and organic compounds for SIBs, in terms of structural design, electrochemical properties, and the corresponding reaction mechanisms.

From Chapter 7 to Chapter 10, intercalation-type anode materials (carbon-based materials, titanium-based materials, etc.), alloy- and conversion-type anode materials (metals/metal alloys, metal oxides/metal chalcogenides, phosphorous/phosphides, etc.), and Na metal anodes are described.

Electrolytes including organic liquid electrolytes, ionic liquid electrolytes, and solid electrolytes are mentioned in Chapters 11 to 13. Besides, binder is also an important part of SIBs, and its progress is summarized in Chapter 14.

For practical applications, three types of sodium-ion full batteries (SIFBs) including non-aqueous, aqueous, and all-solid-state SIFBs are discussed in Chapter 15. And, the last Chapter 16 gives perspectives on the opportunities and potential benefits of SIBs for large-scale energy storage.

Finally, I gratefully acknowledge the substantial contribution from Prof. Yanglong Hou, Prof. Xiaogang Zhang, Prof. Xianhong Rui, Prof. Hongfa Xiang, Dr. Lina Zhao, Prof. Yuan-Li Ding, Prof. Changbao Zhu, Prof. Laifa Shen, Prof. Jianmin Ma, Prof. Feixiang Wu, Prof. Xuyong Feng, Dr. Dan Yang, Dr. Xianghua Zhang, and Dr. Wei Luo to this book.

13 August 2021
Hefei, China

Professor Yan Yu
University of Science and Technology of China

1

Introduction to Sodium-Ion Batteries

1.1 Brief Outline

The global energy demand calls for a major transformation in energy sector, driven in large measure by the rapid expansion of power generation from renewables as well as by the increasing focus on electric vehicles. For the former concern, people are looking for possible solutions to shift away from the traditional fossil energy centralized system toward renewable-energy-based power generation. Though renewable resources, such as solar and wind power, are now replacing the roles previously reserved by fossil fuels, fossil fuels are still needed for times when the sun is down or the wind is idle. The intermittency of the electricity generation from renewable energy sources has necessitated the development of energy storage technology that enables the electricity to be delivered and on demand and be able to reach remote rural areas. On the other hand, a profound change in operating norms can be expected with electric vehicles gaining ground as a more environmentally friendly option in the last decade. Therefore, the development of frontier technologies in energy storage is the key that enables them to come into more widespread adoption in the commercial world.

In the past decades, traditional pumped hydrotechnology is once the only commercially viable energy storage that can potentially reshape the energy sector by addressing the limited penetration of renewable energy. However, this technology is eventually losing market share to emerging technologies mainly due to its geographical constraint. Disruptive technologies that come in a combination of flexibility in design, step-change improvement in performance, and good reliability are considered to be critical enabling the transformation in transportation. The emphasis has now been placed on the electrochemical storage that fits in squarely with this focus. As the state of the art in electrochemical energy storage, rechargeable lithium-ion batteries (LIBs) play a significant role in ushering in a swift transition from a fuel-based society to an exclusively all-electric dimension, a fact reflected in the Nobel Prize in Chemistry 2019. They have widely penetrated into the markets owing to the highly sought-after features, including high energy density, good reliability, and lightweight. In fact, LIBs were first commercialized in 1900s by Sony and started out as the power sources of portable electronics and

Sodium-Ion Batteries: Energy Storage Materials and Technologies, First Edition. Yan Yu.

computers. A few decades later, they are increasingly displacing oil and gas to power our vehicles and buildings, thanks to the scientific breakthrough achieved by researchers along the way. Tesla, the pioneering electric car company, stays at the forefront of this change, where the Tesla's first Gigafactory has been a major cornerstone that pushes the development of LIBs to power energy revolution. The company has also announced its intention to unveil LIBs pack for residential storage that will help in the recent move to decentralized energy. Despite the technology's growing prevalence, a year-long focus on the use of the rare and expensive metal lithium has impeded the widespread adoption of this technology in large-scale applications. In addition, the production of lithium is mainly derived from brine or mined from minerals, but they are geographically restricted to Chile, Bolivia, Argentina, and Australia.

Scientific community started to realize that such scarcity makes the price of lithium fluctuates with demand, which is a fatal flaw inherent to LIBs. A question that researchers trying to answer is whether the incumbent LIB technology can be substituted with a more abundant resources without losing its benefits? Sodium-ion batteries (SIBs) could be the answer. In fact, the concept of SIBs was first brought up in 1980s, but its development is unlike that of LIBs that accept wide market adoption mainly because SIBs are not quite up to par with LIBs in all-round performance. Until the last decade, the topic of SIBs has been revisited to look for possible solution over the safety and cost issues posed by LIBs. Moving beyond lithium to sodium is an advantageous step that offers cost-effectiveness and better safety characteristics without a compromise on the incumbent manufacturing setting based on their chemical similarity and components. This is to say, the manufacturers of LIBs will be fully adaptable toward SIB technology, given that the materials being tested from LIBs can also be potentially transplanted to SIBs without major modification on top of their similar manufacturing processes. The working principle of SIBs is highly similar to LIBs, except the use of sodium as anode instead of lithium. Given that sodium resources can be found in sea water and sodium carbonates (the price of sodium carbonate is estimated to be \$135–165 ton^{-1}, while lithium carbonate costs about \$5000 ton^{-1}) that are vastly available on earth, SIBs are recognized as a cost-effective power solution compared to LIBs [1]. Furthermore, Al current collector that are less expensive and lighter can be used in SIBs due to the chemical inertness between sodium and Al in the formation of alloying compounds. With such a replacement with Cu, the battery cost can be reduced by approximately 3%, based on the price of Al (\$0.3 m^{-2}) and Cu (\$1.2 m^{-2}) foils [2]. On top of the cost reduction, safety is another potential benefit after the possible replacement of Cu with Al current collector on the cathode as well as the anode. The sodium-based cells can be stored at zero state of charge without carrying the risk of thermal runaway. Contrarily to the high reactivity between lithium and Cu at low voltage, the chemical stability of SIBs at zero volts is essential for circumventing the increasingly stringent transportation regulations covering LIBs. Furthermore, the use of Al as current collector not only enhances the safety feature of SIBs but also improves the energy density of SIBs with a lighter element.

Being one of the group I neighbors, sodium shares many chemical characteristics with lithium, and hence, the adoption of existing LIB analogs for SIBs has become the logical strategies. While SIBs can be produced on the same manufacturing lines as their lithium counterparts, numerous studies have not proved it to be a straightforward endeavor. Engineering efforts to the process are still needed to tweak the material morphologies and other properties for more seamless adoption in manufacturing. Alongside the obstacle in real practice, sodium technology also comes with specific challenges related to the chemistries of its components. For example, graphite as a commercial anode in LIBs is not energetically favorable for the intercalation of sodium. This anomalous behavior could be due to the larger size of sodium that imposes greater strain to the lattice of the host materials, which leads structural collapse after several cycles of reaction. In addition, SIBs are less powerful in terms of energy density compared to LIBs due to a 0.3-volt lower cell voltage (−2.71 V vs. standard hydrogen electrode (SHE) for sodium and −3.04 V vs. SHE for lithium). In a theoretical study reported by Ceder and coworkers, the lower output voltage of sodium cell can be due to the cathode effect [3]. It is believed that the weaker Na—O bond compared to the Li—O bond is responsible for the significant lower energy gain upon ion insertion into the host structure. This can be further validated by a distinctly lower the formation energy of Na_2O ($-417.98\,kJ\,mol^{-1}$) compared to Li_2O ($-598.73\,kJ\,mol^{-1}$), giving rise to a 1 V difference between Na–O batteries forming Na_2O and Li–O batteries. This study also ruled out the anodic effect to be the predominant factor as Na voltage is supposed to be higher than Li in 0.53 V with the lower cohesive energy of Na.

Despite the drawbacks mentioned earlier, their low cost and abundant nature compared to LIBs may outweigh this concern, which may dictate their wide application in stationary grid storage. The sluggish movement of larger sodium has been recognized as the origin of its slower reaction kinetics. Researcher started to flip their perspective when they found that desolvation energy of ion is likely to play a role as important as the size of ion. Compared to lithium, sodium is a weaker Lewis acid that makes it less stable in organic solvents. The lower desolvation energy exhibited by sodium is indeed beneficial for facile charge transfer at the interface of electrolyte. For ease of comparison, the characteristics of lithium and sodium are summarized in Table 1.1.

The appealing features of SIBs have made their return to the spotlight and achieved astonishing progress in the past decades. For example, the first commercial SIB company, Faradion, announced that its sodium-ion technology that can achieve similar performance to rechargeable batteries powered by conventional chemistries and show exceptional thermal stability and safety. Furthermore, a start-up company, Aquion Energy, has currently experienced commercial success in the development of SIB prototypes that work well at reduced operating temperature and with extended lifespan. If SIBs live up to the promise made by the companies, it could one day accept wider market adoption other than just stationary grid storage.

Table 1.1 Comparison of lithium and sodium properties.

Characteristics	Li^+	Na^+	Reference
Specific capacity ($mAh\,g^{-1}$)	1165	3829	[4]
Ionic radius (Å)	0.76	1.02	[5]
Stokes radius in PC (Å)	4.8	4.6	[6]
E^o (vs. SHE) (V)	-3.04	-2.71	[5]
E^o (A^+_{aq}/A) (vs. Li^+_{aq}/Li) (V)	0	0.326	[7]
Melting point (°C)	97.7	180.5	[4]
E^o (A^+_{PC}/A) (vs. Li^+_{PC}/Li) (V)	0	0.23	[7]
Desolvation energy in PC ($kJ\,mol^{-1}$)	215.8	158.2	[5]
Cohesive energy of pure metal @ 298 K ($kJ\,mol^{-1}$)	−598.73	−417.98	[8]
Formation energy of M—X bond (eV)	-1.55	-1.05	[9]

1.2 Key Materials

On the journey to push the envelope of SIBs performance, researchers are consistently working on the optimization of the main components of SIBs, including electrode materials (anode and cathode), electrolytes, binders, and full cell configuration. Based on our current understanding on the fundamental mechanism of charge storage, the engineering of electrode materials has been successfully achieved through nanostructuring, structural modification, design of hybrid composite, and surface engineering. In short, academic research focus is moving from micron-sized to nano-sized materials to buffer the severe volume change for longer cycle life and shorten the ion diffusion distance for improved reaction kinetics. Current trend dealing with a combination of morphological and architectural controls allows the nanostructuring effect to be preserved in addition to the morphological merits. For example, the construction of three-dimensional (3D) secondary structure enables the advantages derived from the primary and secondary morphology to be maintained. In parallel, structural modification can potentially "repair" the intrinsic properties of electrode materials from the aspect of electronic and ionic conductivity, which leads to an improvement in ion diffusion and electron transfer process. The effect of structural engineering is much more significant in cathode materials where intercalation chemistry is kinetically dependent on the lattice structure. Apart from that, hybrid composite design and surface coating are the strategies putting forward to address the disadvantages of electrode materials by leveraging the synergies between different components (usually highly conductive materials) to improve the electronic conductivity and ion transport rate. At the electrode level, the development of binders and additives has great impact on the performance of SIBs, which is a research direction that worth exploring.

With cathode being the limiting factor in the electrochemical performance of SIBs, the emphasis has now been placed on the development of promising cathode materials as they are usually inferior in specific capacity and rate performance compared to most anodes. Current research on cathode materials for SIBs focuses on modifying existing or direct adoption of LIBs analogs, primarily layered- and tunnel-structured transition metal oxides, polyanion cathodes, Prussian blue analogs (PBAs), and organic compounds. In general, transition metal oxides, including metal-free and Na-based structures, show high specific capacity in working potential of about 2.7–3.0 V vs. Na/Na^+. The research of Na-based transition metal oxides, $NaMO_2$ (M = Ni, Co, Mn, Fe, Cr, V, etc.), was inspired by their lithium counterparts that work well as intercalative cathode for LIBs. The fatal flaw inherent to those transition metal oxides lies in the huge lattice expansion and irreversible multiphase transitions, which leads to a poorer cycling stability and rate capability. Aside from the layered Na-based oxides, the Na-free transition metal oxides MO_x (M = V, Mn, Mo) that can adopt sodium ion in an intercalative manner have also been widely studied for their sodium storage properties. Compared to the Na-based transition metal oxides, the Na-free counterparts are able to deliver higher specific capacity based on their lower molecular weight. At stark different from monoanion compounds, polyanionic compounds, including phosphates, pyrophosphates, and fluorophosphates, are more diverse in structure. Despite their low electronic conductivity, the highly covalent 3D framework of polyanionic compounds allows fast ion diffusion within the structure, in addition to the high structural stability. With these excellent charge storage properties, polyanionic phosphates especially $NaMPO_4$ (Fe, Mn) and sodium super ionic conductor (NASICON)-structured $Na_xM_2(PO_4)_3$ (M = V, Ti) are being researched for SIBs. The great success of $LiFePO_4$ in LIBs has encouraged the investigation of its sodium analogs $NaFePO_4$ that has high theoretical capacity of 154 mAh g^{-1} in the voltage range of 2.9 V (vs. Na^+/Na). As compared to low-dimensional sodium ion transport pathways in $NaFePO_4$, the 3D open framework of NASICON-type $Na_3V_2(PO_4)_3$ seems to allow faster sodium diffusion channels. In a two-phase transition that extracts two sodium ions from the lattice of $Na_3V_2(PO_4)_3$ at 3.4 V vs. Na^+/Na as the result of a V^{3+}/V^{4+} redox reaction, a theoretical capacity of 117 mAh g^{-1} can be obtained. Similarly, the study of pyrophosphates $Na_2MP_2O_7$ (M = Fe, Mn, Co) was inspired by its ultrahigh potential (can reach up to 4.9 V) and the high structural stability of $[P_2O_7]^{4-}$ framework upon accommodation of sodium ion. In another subgroup of phosphate family, the inductive effect of fluoride allows fluorophosphates to operate at higher operating potential when coupled with other transition metals M^{3+}/M^{4+} (M = Ti, Fe, V). For example, the working voltage of vanadium-based fluorophosphates can reach 4.0 V (vs. Na/Na^+) with redox couple V^{3+}/V^{4+}, which reflects that a decent energy density could be potentially obtained in full cell operation. PBAs are a group of cyano-coordination polymers with a generic formula of $AM[M'(CN)_6]\cdot xH_2O$ (A = Li, Na, K; M, M′ = transition metals of Cr, Mn, Fe, Co, Ni, Cu, Zn, etc.; typically M′ = Fe). The metalorganic framework consists of $Fe^{II}—C{\equiv}N—Fe^{3+}$ units all along the three directions of the space, leaving a large zeolitic site at the center that allows the accommodation of guest species (e.g. alkali metals, small organic molecules, water molecules, etc.). With the two different electrochemical active sides in PBAs (M^{2+}/M^{3+} and Fe^{2+}/Fe^{3+} couples), PBAs generally undergo a two-electron transition without breaking their lattice structure apart. What's more is that the partial/complete substitution of Fe^{2+} and Fe^{3+} ions in the cubic lattice of PBAs by other redox-active transition metals (e.g. Co, Ni, and Mn) is kinetically favorable. With such an enhanced flexibility in composition, the charge storage properties can be tuned accordingly to suit certain specifications. For examples, substitution of Fe with Mn and Co pushes the charge/discharge plateaus from around 3.2 V $Na_2Fe[Fe(CN)_6)]$ to higher values of 3.6 V ($Na_2Mn[Fe(CN)_6)]$ [10] and 3.8 V ($Na_2Co[Fe(CN)_6]$) [11], respectively. These features, in addition to their easy preparation through simple and low-cost co-precipitation method, make them highly suitable to be applied in stationary energy storage where cycle life and cost are far more important than energy density. Organic materials are known for their low cost, high tunability, good safety features, and recyclability. With a flexible framework, organic polymers, such as disodium rhodizonate ($Na_2C_6O_6$) and quinone-based tetra-sodium salt, are cathodes that can accommodate large guest ions reversibly with high spatial tolerance. Despite the great progress made in the past decades, the development of SIBs with organic electrode is still very limited. This lies in the fact that conventional polymerization synthesis frequently introduces a large fraction of electrochemical inert residues for the intramolecular linking, which inevitably results in a decrease of the active mass and the gravimetric capacity.

In parallel to the development of cathode, scientists are also working on a number of anode materials to realize high-performance SIBs. The ongoing investigation can be classified into several categories based on their reaction mechanism, including intercalation compounds (carbon-based materials, Ti-based materials), conversion-type compounds (oxides, sulfides, selenides, carbides, etc.), and alloying-type compounds (Si, Ge, Sb, Sn, etc.), as illustrated in Figures 1.1 and 1.2.

Intercalation reactions work exclusively for mostly carbon-based and layered titanium-based materials with large interlayer spacing for the accommodation of guest ions. Typically, the structure of these materials is sufficiently robust to host the occupancy of sodium, but the capacity is limited by the amount of reaction sites. This is to say, the capacities of intercalation compounds are limited by structural stability of the host and stoichiometry. Therefore, it is challenging to achieve specific capacities greater than 200–400 mAh g^{-1} for intercalation compounds.

Carbon-based materials with layered structure are widely used as anode for SIBs, including hard and soft carbons as well as graphene-related carbons. These polymorphs are made of sp^2 carbons arranged in 2D hexagonal covalent form and stacked together via weak Van der Waals interaction. The first anode material that scientists settled on was graphite, but it turned out that its excellent performance in LIBs cannot be duplicated. The Na anomaly has its roots in a general phenomenon: (i) larger size of sodium has made the intercalation reaction kinetically unfavorable; (ii) sodium is intrinsically weaker in binding to carbon (among alkali metals and alkaline earth metals) due to the competition between the ionization of the metal atom and the ion-substrate coupling [14]; (iii) the intercalated carbons, such as NaC_6

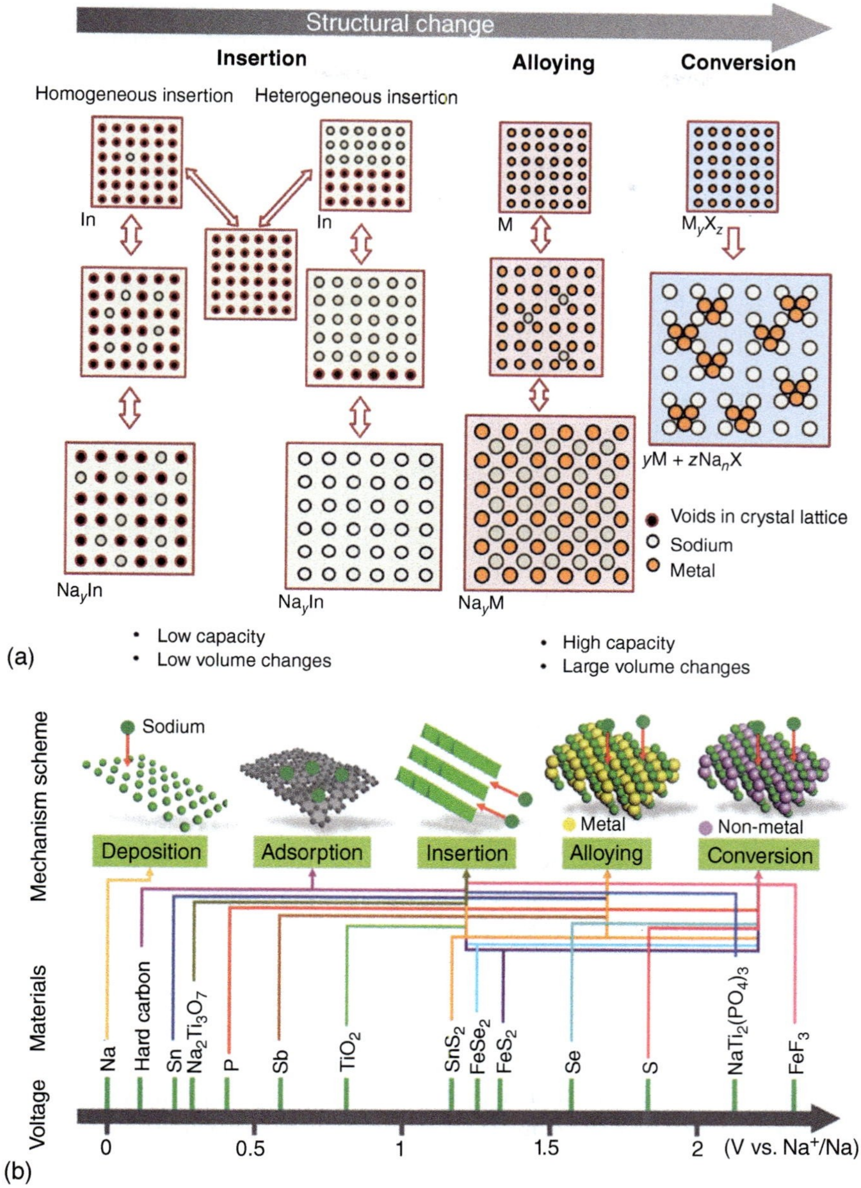

Figure 1.1 (a) Schematic illustration of the generally accepted charge storage mechanism of SIBs. Source: Perveen et al. [4]. Reproduced with permission, 2019, Elsevier. (b) Sodiation voltage and the representative electrode materials for each storage mechanism. Source: Yang and Rogach [12]. Reproduced with permission, 2020, Wiley-VCH.

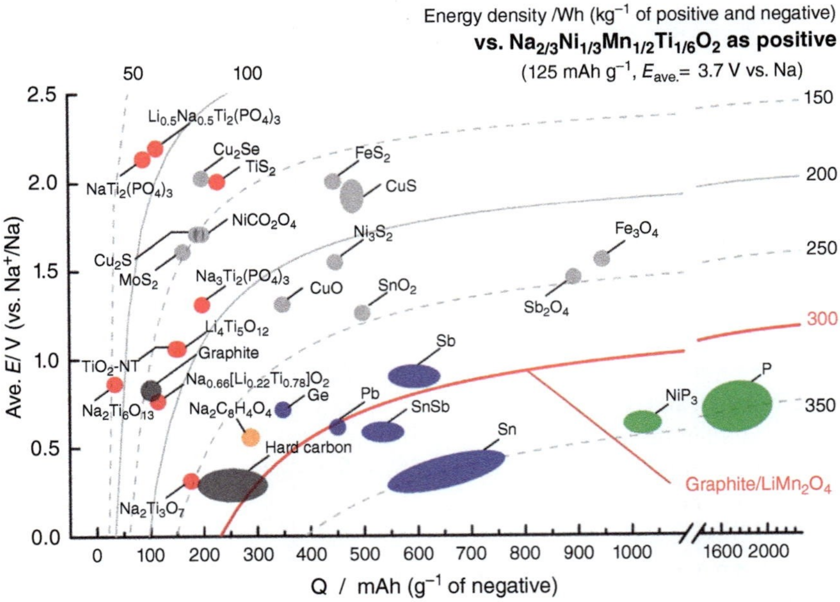

Figure 1.2 Average voltage (V) and energy density (Wh kg^{-1}) vs. gravimetric capacity (mAh g^{-1}) for negative electrode materials for NIBs: (black circles) carbonaceous materials, (red circles) oxides and phosphates as sodium insertion materials, (blue circles) alloy, (green) phosphide/phosphorus, and (gray circles) oxides and sulfides with conversion reaction. Source: Yabuuchi et al. [13]. Reproduced with permission, 2014, American Chemical Society.

and NaC_8, are structurally unstable according to the calculated formation energy [15]. There are several approaches put forward to realize better performance of graphite. The construction of enlarged interlayer spacing in graphite is one of them. Yang and coworkers reported that the lattice distance of graphite was expanded to 4.3 Å through swelling effect of oxygen-containing groups, while the analogous long-range-ordered layered structure of graphite was perfectly retained [16]. Aside from graphite, many other hard carbons have been investigated, especially those that are derived from glucose, cellulose, phenolic resin, etc. [17] Most of these carbons can reach capacity in the range of 250–350 mAh g^{-1}, along with a relatively low oxidation voltages at about 0.3 V. In summary, hard carbons used in SIBs are mostly derived from carbohydrates, while those derived from phenolic resins result in larger energy density. On the other hand, nongraphitizable carbons produced from polymers such as polyaniline (PANI) or polyacrylonitrile (PAN) have also been demonstrated though they usually show relatively low capacities and Coulombic efficiencies. Different from hard carbon, soft carbons usually go on discharge without displaying any voltage plateau at low voltage range, which leads to a lower specific capacity and higher oxidation voltage. Soft carbons can be produced from intrinsic hydrogen-rich precursors, such as polyvinyl chloride (PVC), pitch, petroleum coke, etc. [18] Despite the fact that they have lower energy densities due to aforementioned issues, they outperform hard carbons in rate capability [19].

Overall, the capacities of soft carbons fall in the range of 200–250 mAh g^{-1}, but a slightly higher average oxidation voltages of 0.5 V compared to hard carbons.

As part of the effort to optimize the electrochemical performance of carbonaceous anodes for SIBs, doping strategies have been intensively used to modulate the structure of carbons. Elemental doping, including nitrogen (N), sulfur (S), phosphorus (P), and boron (B), into the lattice of carbons can be easily achieved through the pyrolysis of a heteroatom-containing organic precursors or by mixing the precursor with heteroatom-donor substances. Several improvements can be expected from the incorporation of heteroatoms: (i) larger capacities could be the result of higher fraction of pseudocapacitive contribution from the additional functional groups [20]; (ii) higher ion diffusion rate due to the change in electronic structure of the carbon atoms in the vicinity of heteroatoms [21]; (iii) creates additional reaction sites for sodium [22]; and (iv) induces larger interspacing layers for better reaction kinetics [23]. Different from the N that is electrochemically inert, S itself can serve as the reaction site. In addition, the introduction of larger S atom into the carbon backbone results in significant enlargement in interlayer distance, which further enhances the electrode kinetics. Compared to N- and S-doping, P-doping can be difficult to achieve. Despite the difficulty in synthesis, P-doped carbon has been proven to be excellent in cycling stability and rate capability by Hou et al. [24] For example, a specific capacity of 108 mAh g^{-1} can be achieved at a high current density of 20 A g^{-1}. The long-term cycling investigation shows that the electrode can retain a capacity of 149 mAh g^{-1} after 5000 cycles at a current density of 5 A g^{-1}.

Apart from the intercalation chemistry, reaction mechanism based on redox chemistry is also highly favored due to several reasons: (i) higher specific capacity and energy density arise from the capability to accommodate relatively high stoichiometric ratio of sodium in alloy form; (ii) sodiation takes place at a relatively low voltage range; and (iii) alloy anodes are metallic or metalloid that usually exhibit excellent electrical conductivity. In general, the generic reaction formula of the alloying reaction can be represented as follows:

$$x\text{Na}^+ + x\text{e}^- + \text{M} \leftrightarrow \text{Na}_x\text{M}$$

For alloying reaction, the storage of sodium can be achieved through the formation of binary alloy with different stoichiometry in a stepwise manner, where group IVA and VA elements such as tin (Sn), antimony (Sb), bismuth (Bi), silicon (Si), germanium (Ge), and phosphorus (P) are the materials that fit into this category. The main difference to the conventional intercalation reaction lies in the structural evolution upon alloying reaction. It is believed that continuous increase in the amount sodium induces massive volume change to the alloy compounds, which causes the crumbling and breakdown of alloy-type anodes. As the result of the large volume variation, a series of side effects can be triggered. For example, the solid electrode interface (SEI) could become unstable under large stress and strain, impeding the charge transfer at the interface. On the other hand, the structural integrity of the alloy anodes could be affected or even destroyed, which may lead to the loss of active surface. Over the past decades, researchers have been looking for strategies to overcome the bottlenecks of alloy anodes, mainly through understanding their failure mechanism and identification of the transient intermediate under working states. In general, Sn, Sb, Ge, and Bi exhibit better rate capability thanks to their higher electronic conductivity. As mentioned earlier, their electrochemical performances are limited by mechanical instability due to the large volume variation upon alloying reaction especially in their crystalline state. On the other hand, P and Si are known for their relatively high specific capacity. However, the sluggish reaction kinetics caused by low electronic conductivity has been the major hindrance to their wide adoption in SIBs. Table 1.2 summarizes some of the electrochemical properties of alloying-type anodes that are commonly used in SIBs.

With current alloy anodes reaching almost their performance limits, researchers are scrutinizing every possible strategy to modulate their electrochemical properties for better performance. Therefore, fundamental understanding on how these materials behave in response to sodium (de)insertion from thermodynamic point of view, especially activation energy of sodium ion diffusion, bond breaking, and nucleation energy, is essentially helpful to device effective ways to achieve that goal.

Conversion-type anodes include a diverse group of transition metal oxides, sulfides, and phosphides that are endowed with redox properties. The accommodation sodium in conversion-type anodes involves a series of phase transformation through a change in the valence state of the redox-active centers. The generic equation that represents conversion reactions can be expressed as follows:

$$(b \cdot x)\,\mathrm{Na}^+ + (b \cdot x)\,\mathrm{e}^- + \mathrm{M}_a\mathrm{X}_b \leftrightarrow b\mathrm{Na}_x\mathrm{X} + a\mathrm{M}$$

For typical conversion-type anodes, M is a transition metal element (Fe, Co, Ni, Cu, Mn, etc.), X is a nonmetal (O, N, F, S, Se, P, F, H, etc.), and x is the formal oxidation state of X. Upon sodiation, conversion-type anodes generally convert into metallic M before it nucleates in the form of amorphous or crystalline nanoparticles. Na_xX is the final nucleation product that forms a matrix around the M nanoparticles. It is found that the nonmetal species of X affects the cell potential. As compared to the oxide counterparts (M–O), the conversion reaction of chalcogenides is expected to be more kinetically favorable as the result of weaker M—X bond arising from their larger atomic radius [37]. On top of that, the higher electronic conductivity of the

Table 1.2 Electrochemical properties of alloying-type anodes.

Alloy system	Fully alloyed phase	Theoretical capacity (mAh g^{-1})	Volume expansion (%)	Average voltage (vs. Na$^+$/Na) (V)
Sn	$Na_{15}Sn_4$	847	420 [25]	~0.20 [26]
Sb	Na_3Sb	660	390 [27]	~0.60 [28]
Si	$NaSi/Na_{0.75}Si$	954/725	114 [29]	~0.50 [30]
Ge	NaGe	576	205 [31]	~0.30 [32]
P	Na_3P	2596	>300 [33]	~0.40 [34]
Bi	Na_3Bi	385	250 [35]	~0.55 [36]

discharge products M_2S/M_2Se compared to M_2O in the matrix is also responsible for the better reaction kinetics of the chalcogenides. In the family of transition metal sulfides, they exhibit lower theoretical capacity than the oxides due to the higher molecular weight of sulfur element but higher reaction potential. Compared to the rest, transition metal phosphides have rarely been reported due to the difficulty in material synthesis as well as the low stability. In fact, phosphorus is highly abundant that gives it a relatively low cost. Furthermore, the high theoretical capacity of metal phosphides is their most appealing feature in SIBs, which is arisen from a dual-charge storage mechanism. The transformation from metal phosphides to phosphorus via conversion reaction enables them to proceed with alloying reaction in a lower voltage range. Unfortunately, their SIBs performances are still not up to par with their counterparts in LIBs. Despite the great promise of conversion-type anodes, their feasibility for practical use is still being questioned. For example, oxidation of transition metal sulfides and phosphides under ambient condition has been an obstacle for them to be practically used. As a result, additional measures are required to maintain the material consistency, which might increase the cost for material storage and transportation. In addition, the viability of conversion-type anodes is also challenged by a couple of issues, especially the low electronic conductivity of active materials, dissolution and shuttle effect of chalcogens, irreversibility of conversion reaction, etc.

Sodium metal is the simplest form of anode but is an essential component for room-temperature SIBs. In fact, the early development of SIBs was based on the sodium metal-sulfur molten batteries operating at 300 °C. These systems required an external energy source to maintain the high operating temperature, causing them less promising compared to room-temperature SIBs when the sophisticated system design, high manufacturing costs, and safety issues are taken into consideration as well [38]. In fact, sodium metal has once been regarded incapable as anode because of the safety concern over the dendrite formation, high reactivity against electrolyte, and low melting point (sodium = 97.7 °C and lithium = 180.5 °C). The instability of SEI layer is perhaps the key driver to the remaining associated problems mentioned earlier. For example, the existing explanation of dendrite growth in sodium anode based on sequential growth mechanism is related to the formation of uneven SEI, where dendrite preferentially forms at those protruded spots with concentrated ion flux [39]. On the contrary, researchers might think that SEI is intrinsically heterogeneous concerning its organic–inorganic matrix, which tends to promote the preferential growth of sodium metal at localized region even if it is originally fully isotropic. If dendrites pierce through the separator and short the cell, highly exothermic reaction that can lead to catastrophic thermal runaway such as fire and explosion may result. On the other hand, the preferential dissolution at the base of sodium dendrite causes their delamination from the current collector, which creates so-called "dead" sodium that is no longer electrochemically active but still contributes to the growth of SEI. Hence, a sharp drop in Coulombic efficiency is usually observed accompanied by a rise in impedance [40].

The solutions may be subdivided into the following interrelated taxonomy: (i) tune the compositions of electrolyte and additives; (ii) improve the interfacial reaction between sodium metal and electrolyte; (iii) rational design of the electrode architecture to reduce the current density during plating-stripping and protect the sodium surface from undesired side reactions; and (iv) optimize the properties of sodium metal through alloy design. The power of sodium lies in its energy density compared to other anodes; however, its successful deployment of metal sodium in real practice might take years of research. It is not quite there yet, but this is definitely more than a "pipe dream."

At the electrode level, the overall SIBs performance is contingent upon the choice of electrolytes (e.g. organic liquids, ionic liquids, gel polymers, and solid electrolytes) but is often overlooked in research. Electrolyte generally consists of sodium salt with matching solvents, along with a tiny amount of additives as booster. The electrochemical properties, in particular voltage window, polarity, and wettability, can be optimized through careful tuning the constituents and ratios of the components. The choice of anions in sodium salt affects electrochemical stability window as it is always the first component to be oxidized that set the upper voltage limit for the electrochemical stability window, while the lower limit is more often dictated by solvent reduction. On the other hand, the ionic conductivity of electrolyte is determined by interaction strength between a sodium cation and a number of anions, which affects the amount of available charge carrier. For organic liquid electrolytes, commonly used anions are ClO_4^-, BF_4^-, PF_6^-, $CF_3SO_3^-$ (Tf), and $[N(CF_3SO_2)_2]^-$ bis(trifluoromethylsulfonyl)imide (TFSI). The choice of solvent that changes viscosity of electrolyte has great impact on the ion conductivity. In general, the low viscosity organic solvents give higher ionic conductivity. However, the evaporation of flammable vapors at elevated temperature due to their high vapor pressure pose safety issue for SIBs. The use of polymers or ionic liquids with almost no vapor pressure can bypass this problem. However, the downside of both the latter electrolytes are their high viscosities that impede their widespread adoption in room-temperature applications. The poor interfacial contact has long been a critical issue for polymer electrolytes, which counteracts their safety advantage. Apart from the aforementioned categories, ionic liquid is a promising alternative to organic liquid electrolyte but at the expense of cost. They have intrinsically high ionic conductivities, accompanied by a couple of good safety features such as a large liquidus range, thermal and electrochemical stability, and very low vapor pressure. Solid-state batteries with enhanced safety characteristics but is accessible to high energy and power densities could be answered to addressing the concerns over the organic liquid electrolytes that have limited operating voltage window, flammable nature, and safety hazards caused by leakage.

The research on the optimization of binders has always been overlooked. In fact, binder works complementary with other battery components to ensure good mechanical and electrical integrity of the entire electrode [41]. The binder formula containing polyvinylidene difluoride (PVDF) that has gained great success in LIBs was first explored in SIBs. However, this kind of chain polymeric network fails to provide good cycling stability to the electrode [42]. This has prompted the recent investigation in to the low-cost water-soluble binders such as sodium carboxymethyl cellulose (CMC), poly(acrylic acid) (PAA), and sodium alginate (Na-Alg). The use of these binders with polar functional groups is motivated by the discovery on the unique material–binder interaction that forms a thermally cross-linked 3D interconnection. The formation of this 3D network greatly improves the cycling stability of the electrode by stabilizing the SEI layer and enhancing the adhesion between the active materials and the current collector. As the result, the structural integrity of the active materials can be maintained, along with the intact electrical contact of the electrode. This also works well for electrode materials that undergo high stress and strain upon charge/discharge. Apart from the aforementioned advantages, water-soluble binders (CMC, PAA, and Na-Alg) also show their negligible swellability in organic electrolytes [43]. The promising results of CMC and PAA also prompted the study of their derivatives that contain carboxyl groups or other cross-linked polymers (e.g. chitosan, poly(9,9-dioctylfluorene-*co*-fluorenone-*co*-methylbenzoic ester) [PFM]) [44]. As a rule of thumb, the presence of high density of hydroxyl, carboxyl, or carbonyl groups is responsible for the favorable formation of 3D gel-like structure through hydrogen bonding. Such interaction leads to better mechanical stability and enhanced reaction kinetics of the electrode.

On the other hand, the addition of functional molecules, or so-called additives, is an effective way to stabilize SEI formation process. For example, the use of 1-fluoroethylene carbonate (FEC) and vinylene carbonate (VC) as additives in the PVDF binder formula has successfully suppressed the decomposition of organic electrolytes [44b, 45]. These additives passivate the surface through the decomposition and polymerization of FEC and VC prior to the formation of SEI layer [46]. There are different "sacrificing mechanisms" being reported in the previous studies. In the case of FEC, the protective layer is formed by a reaction associated with the opening of a five-membered ring. On the other hand, the reductive decomposition of VC gives rise to radical anion fragments that are responsible for the suppression of SEI dissolution.

1.3 Toward Future Development

It is recognized that the electrochemical performances examined in a full cell is more conducive to the advancement of SIBs in real setting. There are several ongoing research directions in this field, which can be briefly categorized into nonaqueous liquid sodium-ion full cell, quasi-solid-state sodium-ion full cell, and all-solid-state sodium-ion full cell. Among them, nonaqueous liquid sodium-ion full cell represents the most mature technology that can reach an energy density as high as 300 Wh kg^{-1} by considering the mass of both cathode and anode, despite its inferior cycling stability. Quasi-solid-state sodium-ion full cell that is composed of quasi-solid-state electrolyte composed of a mixture of inorganic ceramics, polymer, and a small amount of solvent exhibits better performance than nonaqueous liquid sodium-ion full cell in terms of safety characteristic, interface stability, wider electrochemical window, and flexibility. Although the liquid content is greatly reduced in the quasi-solid-state electrolytes, the formation of the interface film is still likely to happen, leading to the irreversible consumption of limited sodium supply in the full cell. Therefore, understanding on the pre-sodiation chemistry that can be deployed to offset the irreversible consumption of sodium would be helpful in devising reliable and scalable pre-sodiation strategies. Compared with the above two types of electrolytes, the all-solid-state electrolyte with high voltage stability and safety enables the full redox potential of high-voltage materials to be claimed, which significantly enhances the energy density and power density. Despite these advantages, the full potential of all-solid-state sodium-ion full cell has not been unleashed due to the larger interfacial impedance that gives rise to greater polarization. In general, what is still lack in the research of sodium-ion full cell is the in-depth analysis of the interfacial chemistry of different electrolytes and electrolyte optimization, on top of the development of incumbent strategies to improve the performance of sodium-ion full cells in the aspect of matching capacity and choice of electrode materials.

In recent years, a significant amount of research has been undertaken to develop new battery concept beyond LIBs. SIBs are one of the reliable alternatives that can provide cost-effective solution for future energy storage system. With the development of effective strategies to further optimize different components of SIBs, it is hoped to move this technology past broad commitments toward a durable solution in solving the global energy woes. Before that, there are still a couple of questions need to be answered before SIBs can be deployed on a commercial scale: what are challenges ahead and what are the paths to follow to realize their full potential? Swapping lithium for sodium needs more than just setting sight on tweaking the chemical compositions and lattice structure of the lithium analogs. It requires close partnership between experiment, theory, and simulation to advance fundamental scientific understanding on the charge storage mechanism of those promising electrode materials and to quantify the uncertainty inherent in our assumption. In addition to that, it also requires long-term investment to expedite and facilitate the transformation of scientific research to design tools suitable for the industrial laboratory setting and manufacturing floor. In this book, we aim to provide a comprehensive view on the current development of SIBs by systematically analyzing various important components of SIBs, including anodes, cathodes, electrolytes, binders, and SIBs in full cell configuration. In addition, the design principle is elaborated to provide some guidelines for future development of high-performance SIBs based on those insights gained from the previous studies.

References

1 Slater, M.D., Kim, D., Lee, E., and Johnson, C.S. (2013). *Adv. Funct. Mater.* 23: 947.

2 Vaalma, C., Buchholz, D., Weil, M., and Passerini, S. (2018). *Nat. Rev. Mater.* 3: 18013.

3 Ong, S.P., Chevrier, V.L., Hautier, G. et al. (2011). *Energy Environ. Sci.* 4: 3680.

4 Perveen, T., Siddiq, M., Shahzad, N. et al. (2020). *Renew. Sust. Energy Rev.* 119: 109549.
5 Kubota, K., Dahbi, M., Hosaka, T. et al. (2018). *Chem. Rec.* 18: 459.
6 Zhang, W., Liu, Y., and Guo, Z. (2019). *Sci. Adv.* 5: eaav7412.
7 Komaba, S., Hasegawa, T., Dahbi, M., and Kubota, K. (2015). *Electrochem. Commun.* 60: 172.
8 Poole, R.T. (1980). *Am. J. Phys.* 48: 536.
9 Chase, M.W. (1998). *NIST-JANAF Thermochemical Tables*. New York: American Chemical Society.
10 Wang, L., Lu, Y., Liu, J. et al. (2013). *Angew. Chem. Int. Edit.* 52: 1964.
11 Wu, X., Wu, C., Wei, C. et al. (2016). *ACS Appl. Mater. Interfaces* 8: 5393.
12 Yang, X. and Rogach, A.L. (2020). *Adv. Energy Mater.* 10: 2000288.
13 Yabuuchi, N., Kubota, K., Dahbi, M., and Komaba, S. (2014). *Chem. Rev.* 114 (23): 11636.
14 Liu, Y., Merinov, B.V., and Goddard, W.A. (2016). *Proc. Natl. Acad. Sci. U. S. A.* 113: 3735.
15 Nobuhara, K., Nakayama, H., Nose, M. et al. (2013). *J. Power Sources* 243: 585.
16 Wen, Y., He, K., Zhu, Y. et al. (2014). *Nat. Commun.* 5: 4033.
17 (a) Stevens, D.A. and Dahn, J.R. (2000). *J. Electrochem. Soc.* 147: 1271.
(b) Simone, V., Boulineau, A., de Geyer, A. et al. (2016). *J. Energy Chem.* 25: 761.
(c) Qiu, S., Xiao, L., Sushko, M.L. et al. (2017). *Adv. Energy Mater.* 7: 1700403.
18 (a) Bai, Y., Wang, Z., Wu, C. et al. (2015). *ACS Appl. Mater. Interfaces* 7: 5598.
(b) Li, Y., Mu, L., Hu, Y.-S. et al. (2016). *Energy Storage Mater.* 2: 139.
(c) Alcántara, R., Jiménez Mateos, J.M., and Tirado, J.L. (2002). *J. Electrochem. Soc.* 149: A201.
19 Luo, W., Jian, Z., Xing, Z. et al. (2015). *ACS Cent. Sci.* 1: 516.
20 Zhang, B., Ghimbeu, C.M., Laberty, C. et al. (2016). *Adv. Energy Mater.* 6: 1501588.
21 Li, W., Zhou, M., Li, H. et al. (2015). *Energy Environ. Sci.* 8: 2916.
22 Qie, L., Chen, W., Xiong, X. et al. (2015). *Adv. Sci.* 2: 1500195.
23 Li, Z., Bommier, C., Chong, Z.S. et al. (2017). *Adv. Energy Mater.* 7: 1602894.
24 Hou, H., Shao, L., Zhang, Y. et al. (2017). *Adv. Sci.* 4: 1600243.
25 Wang, J.W., Liu, X.H., Mao, S.X., and Huang, J.Y. (2012). *Nano Lett.* 12: 5897.
26 Ellis, L.D., Hatchard, T.D., and Obrovac, M.N. (2012). *J. Electrochem. Soc.* 159: A1801.
27 Wu, L., Lu, H., Xiao, L. et al. (2015). *J. Mater. Chem. A* 3: 5708.
28 Qian, J., Chen, Y., Wu, L. et al. (2012). *Chem. Commun.* 48: 7070.
29 Jung, S.C., Jung, D.S., Choi, J.W., and Han, Y.-K. (2014). *J. Phys. Chem. Lett.* 5: 1283.
30 Xu, Y., Swaans, E., Basak, S. et al. (2016). *Adv. Energy Mater.* 6: 1501436.
31 Jung, S.C., Kim, H.-J., Kang, Y.-J., and Han, Y.-K. (2016). *J. Alloys Compd.* 688: 158.
32 Kohandehghan, A., Cui, K., Kupsta, M. et al. (2014). *Nano Lett.* 14: 5873.
33 Fu, Y., Wei, Q., Zhang, G., and Sun, S. (2018). *Adv. Energy Mater.* 8: 1703058.
34 Kim, Y., Park, Y., Choi, A. et al. (2013). *Adv. Mater.* 25: 3045.

35 Ellis, L.D., Wilkes, B.N., Hatchard, T.D., and Obrovac, M.N. (2014). *J. Electrochem. Soc.* 161: A416.

36 Mortazavi, M., Ye, Q., Birbilis, N., and Medhekar, N.V. (2015). *J. Power Sources* 285: 29.

37 Zhou, Q., Liu, L., Huang, Z. et al. (2016). *J. Mater. Chem. A* 4: 5505.

38 Dunn, B., Kamath, H., and Tarascon, J.-M. (2011). *Science* 334: 928.

39 Wei, S., Choudhury, S., Xu, J. et al. (2017). *Adv. Mater.* 29: 1605512.

40 Gao, H., Xin, S., Xue, L., and Goodenough, J.B. (2018). *Chem* 4: 833.

41 Bresser, D., Buchholz, D., Moretti, A. et al. (2018). *Energy Environ. Sci.* 11: 3096.

42 (a) Zhang, W., Dahbi, M., and Komaba, S. (2016). *Curr. Opin. Chem. Eng.* 13: 36. (b) Komaba, S., Matsuura, Y., Ishikawa, T. et al. (2012). *Electrochem. Commun.* 21: 65.

43 Kovalenko, I., Zdyrko, B., Magasinski, A. et al. (2011). *Science* 334: 75.

44 (a) Gao, H., Zhou, W., Jang, J.-H., and Goodenough, J.B. (2016). *Adv. Energy Mater.* 6: 1502130. (b) Dai, K., Zhao, H., Wang, Z. et al. (2014). *J. Power Sources* 263: 276.

45 (a) Komaba, S., Ishikawa, T., Yabuuchi, N. et al. (2011). *ACS Appl. Mater. Interfaces* 3: 4165. (b) Zhang, B., Rousse, G., Foix, D. et al. (2016). *Adv. Mater.* 28: 9824.

46 Dahbi, M., Yabuuchi, N., Fukunishi, M. et al. (2016). *Chem. Mater.* 28: 1625.

2

Design Principles for Sodium-Ion Batteries

2.1 Introduction

The working principles of sodium-ion batteries (SIBs) are similar to lithium-ion batteries (LIBs) (Figure 2.1) [1]. For the sodium-containing cathode materials, various materials have been reported including polyanionic compounds, transition metal oxides with tunnel or layered structures, organic cathodes, and Prussian blue analogs. Among them, transition metal oxides ($Na_{0.44}MnO_2$, $Na_4Mn_9O_{18}$, $Na_{1-x}Ni_{0.5}Mn_{0.5}O_2$, $NaNi_{1/3}Mn_{1/3}Co_{1/3}O_2$, Na_xCoO_2, $Na_{0.6}Fe_{0.5}Mn_{0.5}O_2$, etc.) and polyanionic phosphates ($Na_3V_2(PO_4)_3$, $NaFePO_4$, Na_2FePO_4F, $Na_3V_2(PO_4)_2F_3$, etc.) have been extensively investigated [2]. SIB anodes include the following categories: (i) intercalation anodes, such as (Na superionic ion conductor (NASICON)-type $NaTi_2(PO_4)_3$) [3]; (ii) alloy anodes (such as Sb, Sn, etc.) [4]; (iii) conversion anodes, including metal oxides, phosphides, selenides, and sulfides [4c, 5]; (iv) carbon-based materials (i.e. hard carbon) [6]; and (v) nonmetallic elements (i.e. sulfur and phosphorus) [7]. Recent progress of sodium electrodes and electrolytes is summarized in Figure 2.2, displaying the capacities and voltages for typical cathode and anode materials for SIBs.

Although there are various advantages for SIBs, serious disadvantages still exist when compared with lithium. The larger ionic radius (0.76 Å for Li and 1.02 Å for Na) and molecular mass ($7\,g\,mol^{-1}$ for Li and $23\,g\,mol^{-1}$ for Na) result in the inferior electrochemical performance as compared with LIBs: lower theoretical capacity (both volumetric and gravimetric values), lower solubility in the solid state, lower Madelung energies, lower cell voltage, worse transport kinetics, more severe structural damage, and volume change. For example, the storage of Na^+ in the most hosts, such as metal oxides (Na_xCoO_2, $NaNi_{1/3}Mn_{1/3}Co_{1/3}O_2$, V_2O_5, etc.) and sodium phosphates ($NaFePO_4$), is worse than their Li counterparts in terms of various electrochemical indices. Therefore, the effective design principles for SIBs are particularly important.

Sodium-Ion Batteries: Energy Storage Materials and Technologies, First Edition. Yan Yu.

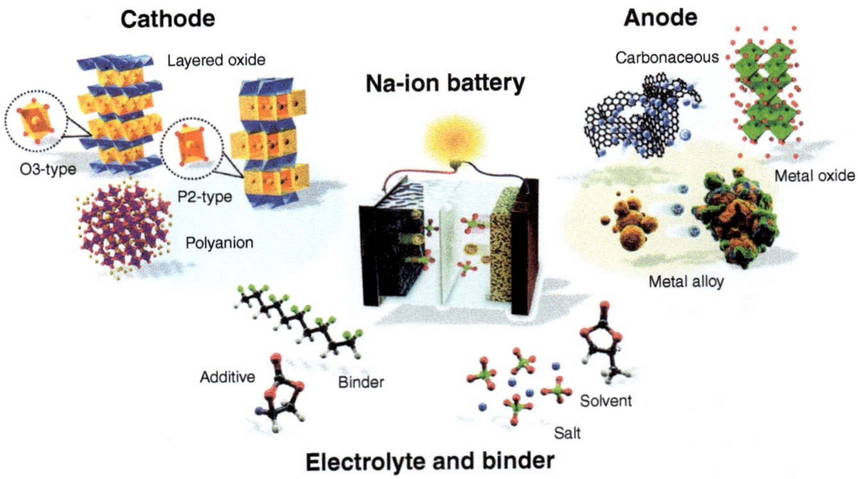

Figure 2.1 Illustration of different components of SIBs. Source: Hwang et al. [1]/with permission from Royal Society of Chemistry.

2.2 Basic Design Principles

The basic design principles for SIBs lie in making the best use of the advantages and bypassing the disadvantages of this potential system. Specifically speaking, we should maintain the strong points such as low cost, high safety of SIBs, and avoid or improve the weak points including energy density, power density, and cycling life in the meantime.

2.2.1 Energy Density

As we all know that the energy density of the battery depends on the reversible capacity and equilibrium voltage. Unfortunately, the larger ionic radius and molecular mass result in the lower energy density of SIBs when compared with LIBs, for instance, both lower theoretical capacity and cell voltage. That is the main reason for the slow growth of SIBs in the past, although SIBs were developed almost in parallel with LIBs. Hence, it is essentially important to unleash the full potential of SIBs with high energy density through rational material/system design.

To increase the reversible capacity, the first step is to make sure that the SIB electrodes are able to fully claim the theoretical capacity, and the key point here is to ensure the effective electronic and ionic transport in the whole electrodes. Secondly, it is believed that breakthrough can be achieved by the conventional electrode materials via the multi-electron transfer reaction of the transition metal elements in the electrode materials. Finally, to develop new SIBs or post SIBs (sodium–sulfur battery or sodium–air battery) is a viable option as well.

To increase the output voltage, we need to increase the voltage of SIB cathodes and/or decrease the voltage of anodes. For the cathodes, according to electrochemical series, we can choose the candidates with suitable voltage for various transition

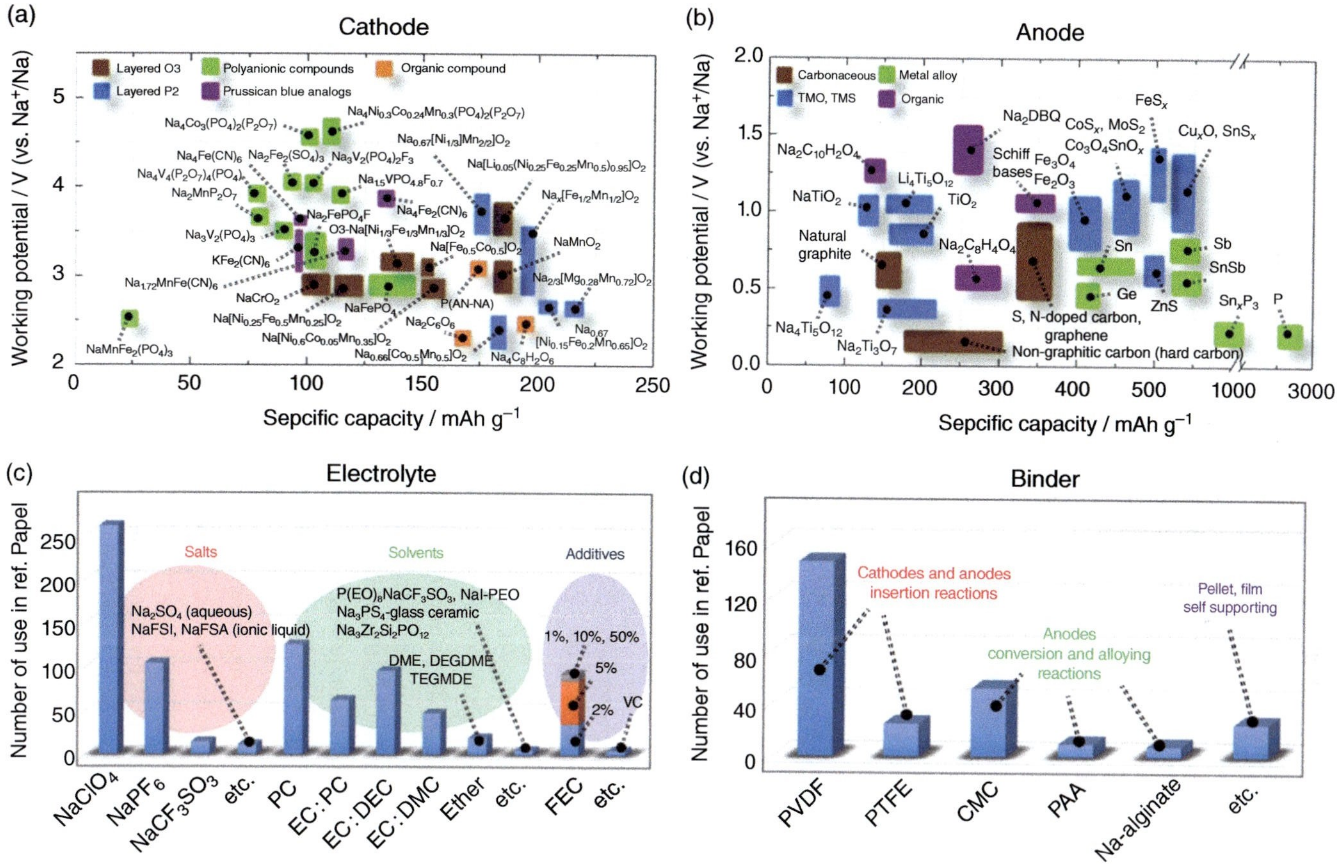

Figure 2.2 Recent progress for SIB electrodes, electrolytes and binders. Source: Hwang et al. [1]. Licensed under CC BY 3.0.

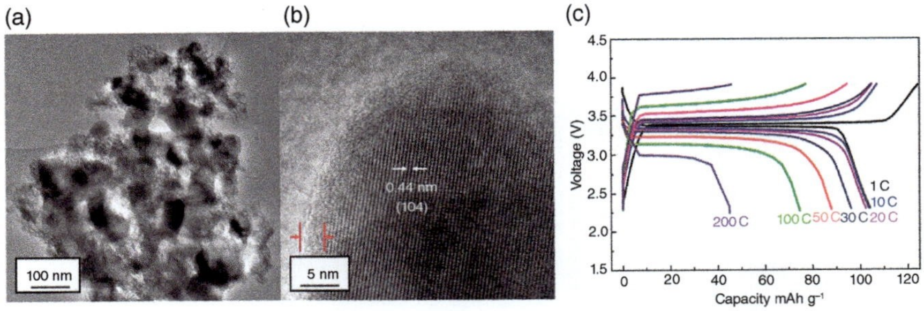

Figure 2.3 Morphology (a, b) and rate performance (c) of $Na_3V_2(PO_4)_3$ prepared by a "double carbon-embedding" approach. Source: Zhu et al. [8]/with permission of American Chemical Society.

metal elements or for various systems, such as oxides and phosphates. Take the phosphate as an example, we can also increase the voltage plateau through inductive effects, which is achieved in $Na_3V_2(PO_4)_3/Na_3V_2(PO_4)_2F_3$ systems. For the anode materials, through controlling the storage mechanisms, the voltage can be reduced. For instance, for the hard carbon anode, by controlling the pore size, sodium ion can intercalate instead of undergoing adsorption, leading to lower voltage plateau. In addition, taking advantages of size effects, a decrease in voltage can be expected as well.

2.2.2 Power Density

In principle, owing to the larger size of Na^+, the transport properties in electrodes are worse compared with its lithium counterparts, leading to the lower rate capability and power density. In order to solve such problem, based on the crystalline structures of electrodes, it is important to construct an efficient electrochemical conducting network, to ensure the effective transport for both electrons and sodium ions, leading to improved rate capability. However, in some cases or for some specific materials (e.g. NASICON-type materials), the sodium transport is even better than that of lithium. Thus, we only need to solve the electronic transport, then a high-rate sodium cathode can be obtained. For instance, we have proposed and realized a powerful electrochemical network for $Na_3V_2(PO_4)_3$ by a "double carbon-embedding" approach (Figure 2.3) [8]. Up to 20 C, polarization is negligible and the capacity almost reaches the theoretical value. Even at 200 C, it still delivers a superior rate performance (44 mAh g^{-1}). Such rate capability even outperforms almost all commonly used lithium cathodes, such as $LiFePO_4$. The excellent rate performance is close or even better than that of a supercapacitor, but energy density is much higher than that of a supercapacitor.

2.2.3 Cycling Life

Due to the larger size of sodium ion, after the repeated insertion/desertion of sodium ions into the hosts, it leads to tremendous volume change of electrode materials, which results in the degradation in the cycling stability performance. Hence, in order to increase the cycling stability, we need to focus on the crystalline structure and morphologies of electrodes and interfaces issues between electrodes and electrolytes. If the sodium ions are more suitable with the transport paths of specific electrodes, e.g. NASICON-type $Na_3V_2(PO_4)_3$, the stable and long cycling can be obtained. In addition, for the alloying anodes or conversion-type anodes, going to construct core–shell or yolk–shell structures can buffer the volume changes, leading to better cycling stability. Finally, choosing suitable sodium electrolytes and electrolyte additives to form the stable solid electrode interface (SEI) layer is also an effective approach to enhance the cycling stability in SIBs.

2.2.4 Safety

As an efficient energy storage device for clean energy systems such as solar energy, wind energy, etc., safety is one of the most important issues for SIBs. In principle, due to lower voltage and energy density of SIBs compared with LIBs, SIBs have better intrinsic safety features. In addition, it is also helpful to choose environmentally friendly elements and materials, decreasing the toxic hazard, to construct green and clean systems. Finally, in terms of electrolyte design, new-type fire-retardant additives or all-solid-state SIBs are effective strategies to further improve the safety.

2.2.5 Cost

As we discussed above, the low cost of SIBs is attributed to not only the abundance of sodium in the earth but also the manufacturing process, such as the usage of low-cost Al as the current collector. When we design the SIBs, cost has been the name of the game with batteries, since it is a prerequisite for large-scale energy storage. For the sodium cathodes, we should focus on the low-cost raw materials, such as Fe- or Mn-based materials rather than Co- and Ni-based materials. For the V-based materials, such as high-rate NASICON-type cathodes, it is more cost-effective to develop the integrated reuse–recycle treatment process. For example, vanadium-based slag can be used as raw materials for sodium cathode materials through extraction and reutilization process, leading to a significant reduction in cost. For the anode materials, low-cost hard carbon, other kinds of carbon-based materials, and cheap alloying system are of great potential. For the electrolytes, cost reduction can be achieved through the design of economical sodium salts and high-performance electrolyte additives. In addition to the low-cost materials systems, we should also focus on the development of processing technology and synthesis process for large-scale production of electrode materials enabling a smooth commercialization in the near future.

From the above discussions, we shed light on the influence of design rules on the energy density, power density, cycling life, safety, and cost of the performance of SIBs. In the following sections, we will discuss the design principles for both electrodes and electrolytes, in order to achieve the successful application of SIBs in the field of large-scale energy storage.

2.3 Design Principles for Electrode Materials

2.3.1 Transport Properties

As we discussed above, in order to increase energy density, power density, and cycling life, transport properties are deemed the crucial parameters. Controlling the transport properties to achieve both effective electronic and ionic transport is the key point to improve the electrochemical performance of SIBs.

Transport properties in SIBs include the transporting behaviors of sodium ions and electrons in the bulk electrodes or interface between the electrodes and electrolytes, which can also be reflected by the chemical diffusion coefficient. The chemical diffusion coefficient determines the comprehensive behaviors of electrons and ions in a gradient of concentration. In principle, diffusion of sodium in the electrode is the rate-determining step among various electrochemical processes in SIBs. Hence, quantitative analysis on the diffusion coefficient provides significance to the intrinsic kinetic properties of the electrodes. In fact, the chemical diffusion coefficient is able to be measured by various electrochemical techniques, such as cyclic voltammetry (CV), electrochemical impedance spectroscopy (EIS), potentiostatic intermittent titration technique (PITT), galvanostatic intermittent titration technique (GITT), etc. [9] under different conditions. One thing needs to be mentioned is that the chemical diffusion coefficient does not directly correlate to the transport properties for both sodium ions and electrons respectively, which can only be obtained through alternating current (AC) impedance together with direct current (DC) polarization coupled with electron-blocking cell or/and ion-blocking cell.

Here we can show an example to demonstrate how to measure and apply transport properties to design high-performance SIB cathode materials. As we known, cathodes of polyanionic-type compounds and sodium layered oxides have been extensively studied. Among them, NASICON $Na_3V_2(PO_4)_3$ has demonstrated its potential as a high-power cathode for SIBs [8]. However, to pursue even higher energy densities, $Na_3V_2(PO_4)_3$ is inferior, owing to its low theoretical energy density (~400 Wh kg^{-1}). On the other hand, $Na_3V_2(PO_4)_2F_3$ is a high-voltage candidate, owing to its higher equilibrium potential and higher ionicity, the theoretical energy density is around ~507 Wh kg^{-1}, which is even closer to commonly used LIB cathodes (e.g. $LiFePO_4$: ~580 Wh kg^{-1}) [10]. There are various methods to fabricate $Na_3V_2(PO_4)_2F_3$, such as the solid-state reaction and the low-temperature solvothermal approach. However, the rate performance and cycling life are still much worse than those of $Na_3V_2(PO_4)_3$. Another shortcoming lies in the lack of understanding of specific transport properties, i.e. chemical diffusion coefficient and electronic/ionic conductivities. In order to solve such problems, we measured and investigated the transport properties experimentally through AC impedance together with DC polarization by the ion-blocking cell: $Ti/Na_3V_2(PO_4)_2F_3/Ti$ (Figure 2.4) [11]. Carbon-free and pure-phase $Na_3V_2(PO_4)_2F_3$ has been obtained through the high-temperature solid-state approaches. The obtained $Na_3V_2(PO_4)_2F_3$ samples were pressed to pellets isostatically (around 500 MPa). Afterward, by

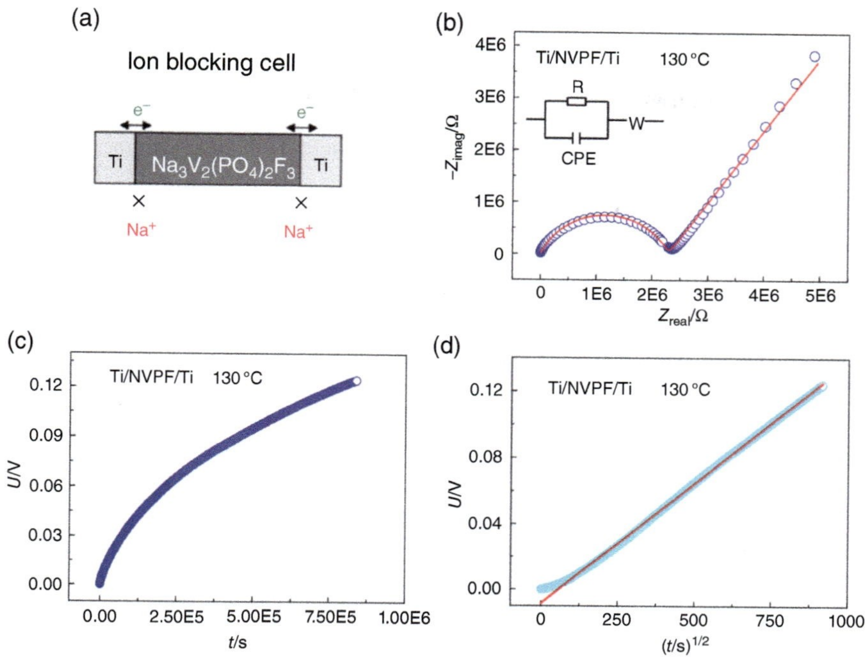

Figure 2.4 Transport investigation on the $Na_3V_2(PO_4)_2F_3$ cathode. (a) Ion-blocking cell, and the corresponding (b) AC impedance spectrum and (c, d) DC polarization curves at 130 °C. Source: Reproduced with permission from Zhu et al. [11]. Copyright 2017, American Chemical Society.

evaporation technique, 400 nm titanium films were deposited on both sides of pellets to prepare ion-blocking cell. The transport properties were measured from 22 to 180 °C, with the waiting times of one day allowing the system to equilibrate. In order to obtain stable conductivity values, at least three heating–cooling cycles were made. Through such testing, we found that the chemical sodium diffusion coefficient is rather small, and the electronic conductivity in $Na_3V_2(PO_4)_2F_3$ is at least 3–4 orders of magnitude lower than its ionic conductivity. Around 0.68 eV of the activation energy is also obtained for ionic conduction in $Na_3V_2(PO_4)_2F_3$. The obtained transport properties provide clear advice on designing SIB electrodes. According to the transport properties, we proposed a low-temperature (without any post heat treatment) and one-step preparation approach to synthesize zero-dimensional (0D)-$Na_3V_2(PO_4)_2F_3$@ three-dimensional (3D)-graphene network, where each nanoparticle embedded uniformly in the graphene network to make sure the good electronic contact. Finally, the electrode showed high energy density and power density and favorable long cycling performance as well. From this case, decisive transport properties are studied, which is beneficial for the optimization and rational design of electrode architecture from this knowledge in the future through the insight gained from the study.

If we have already understood the intrinsic transport properties of the electrodes, optimizing them will be easy. Doping is an effective approach to enhance the transport properties, which can vary chemical diffusion coefficient, the electronic conductivity, or/and ionic conductivity, leading to the improved electrochemical performance. For example, researchers have already investigated $Na_3V_2(PO_4)_3$ doping with Li, K, Mg, Ni, Fe, Cu, Al, Mo, Cr, Ti, Mn, etc. [12], where the crystalline structures, transport properties, and electrochemical performance can be optimized. However, there is still lack of guidance of doping for the SIB electrodes at the moment.

If we would like to obtain the deep understanding of transport properties for SIB electrodes and even to achieve effective prediction for transport properties, building a defect chemistry model is very helpful. Based on defect chemistry reactions and experimental transport data, the nature of charge carriers and the decisive point defects can be confirmed, as well as the dependencies of the concentrations on the control parameters such as temperature, doping content, and Na activity, leading to deep understanding and prediction of transport properties for the electrode materials. Unfortunately, until now, there is still lack of the defect models for typical electrodes, which need further investigation. Hence, we can only draw lessons from LIBs. For LIBs, defect models have been constructed for some important cathode materials, such as $LiFePO_4$ [13], which is quite useful for the rational design of such cathode (Figure 2.5). For $LiFePO_4$, in the native regime, lithium vacancies are main ion conduction charge carriers, while holes are main electronic carriers [14]. Non-intentional impurities or frozen-in native defects dominate. Lithium

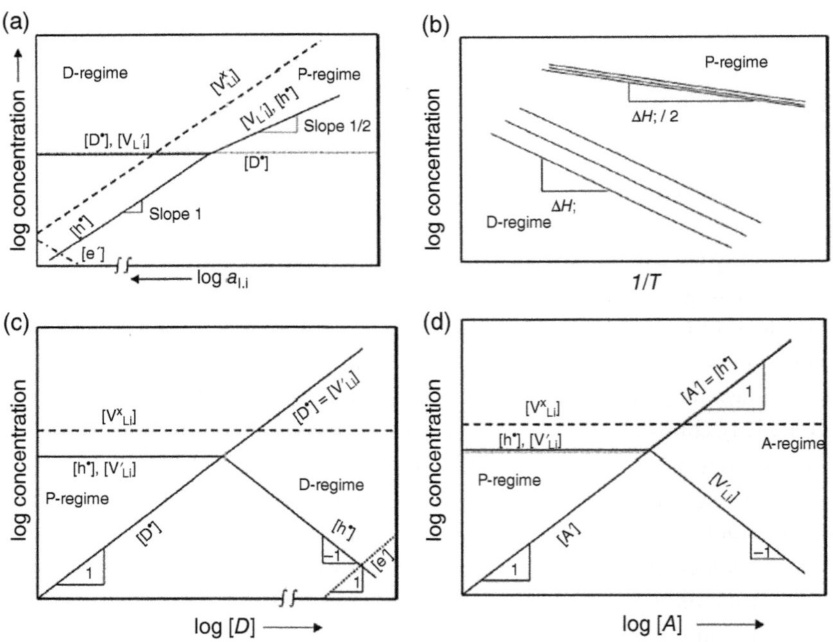

Figure 2.5 Defect chemistry models for $LiFePO_4$. Dependence of defect concentration on lithium activity (a), temperature (b), donor content (c), and acceptor content (d). Source: Maier and Amin [13]. Reproduced with permission.

vacancies are compensated by Fe_{Li}. The same donor effects are obtained if intentionally doping Fe site by Al (Al_{Fe}) or Si (Si_{Fe}) [15]. The log defect concentration vs. log donor/acceptor content or vs. log lithium activity are constructed as shown in the Brouwer diagrams, which can predict the transport properties of $LiFePO_4$ under different doping conditions. Holes trapping by lithium vacancies dominates the effective activation barrier at frozen lithium content, for the temperature dependence of conductivities. For the Al-doped $LiFePO_4$, ionic association is significant. Migration enthalpies and defect reaction are derived for ionic and electronic transport. The process of holes trapping by lithium vacancies and association of lithium vacancies with impurities are keys to understand the chemical diffusion coefficient as a function of the temperature. As a result, it is also necessary to build such defect models for SIB electrodes in order to deeply understand and rationally control the transport properties for specific systems.

In addition to the intrinsic transport properties, for the SIB electrodes with low electronic conductivity, carbon coating is a very efficient method to improve the transport properties and electrochemical performance. A pertinent example here is NASICON-type $Na_3V_2(PO_4)_3$, which has very fast sodium ion conduction but with very poor electronic conductivity. Carbon coating is an effective approach for $Na_3V_2(PO_4)_3$ to further improve the electrochemical performance. In 2012, Hu and coworkers prepared carbon-coated $Na_3V_2(PO_4)_3$, which shows improved cycling stability [16]. Two years later, through "double carbon-embedding" method, we successful synthesized carbon-coated nano-$Na_3V_2(PO_4)_3$ that embedded in a porous carbon, enabling very fast Na storage. Such electrode can discharge quickly at 200 C, which still delivers a high reversible capacity [8]. The excellent rate capability is close to or even better than supercapacitors, whose energy density is much higher than a conventional battery. The outstanding rate performance even outperforms that of the LIB cathodes, such as $LiFePO_4$. However, most of the reported high-rate $Na_3V_2(PO_4)_3$ are powder samples. They require substantial amount of conductive carbon (i.e. ~10–20 wt% Super P) and binder to construct a testing battery. In addition, the binders, e.g. polytetrafluoroethylene (PTFE) and polyvinylidene difluoride (PVDF), are electrically insulating, which will lower the effective electronic transport in the whole electrode. Furthermore, binders and conductive carbons have no electrochemical capacity in the voltage range of investigation, and their existences decrease energy and power density. In addition, the use of powder samples hinders the application of $Na_3V_2(PO_4)_3$ for 3D batteries. Hence, in 2016, we reported a 3D tricontinuous $Na_3V_2(PO_4)_3$/reduced graphene oxide-carbon nanotubes (rGO-CNT), which, without any additional binders and conductive carbons, was grown on the current collector directly [17]. Such composite cathode shows outstanding high rate capability and long cycling performance (Figure 2.6). For instance, at 100 C, the cathode delivers a high reversible capacity of 82 $mAh\,g^{-1}$, which reaches ~70% of the theoretical capacity. The outstanding rate behavior of the composite cathode is comparable to typical supercapacitors but with much higher energy density. A symmetric full sodium battery was also constructed

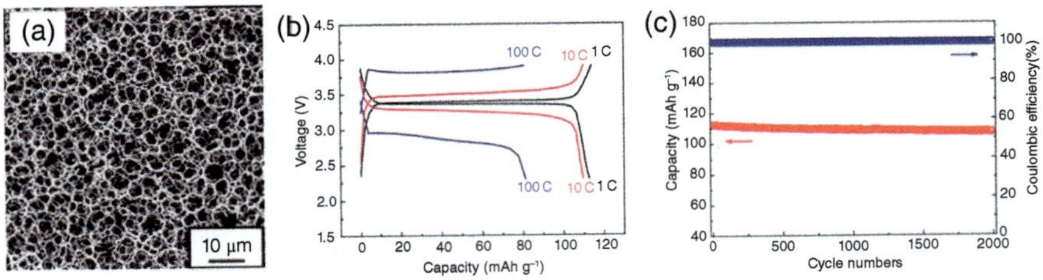

Figure 2.6 Morphology (a) and electrochemical performance (b, c) of interpenetrating 3D tricontinuous $Na_3V_2(PO_4)_3$/rGO-CNT electrodes. Source: Zhu et al. [17]/with permission of Wiley-VCH.

with both cathode and anode based on such interpenetrating 3D tricontinuous $Na_3V_2(PO_4)_3$/rGO-CNT, displaying the potential for real application for 3D batteries.

2.3.2 Size Effects

In addition to controlling the intrinsic transport properties, according to $\tau \sim L^2/D$, we are also able to improve the kinetics through reducing the transport distance of electrons and sodium ions, leading to better electrochemical performance. Hence, at the moment, using size and nanostructuring effect is another effective approach to design and prepare high-performance SIB electrode materials. Up to now, there are a great number of successful examples for both cathodes and anodes, based on various mechanisms such as intercalation reaction, alloying reaction, conversion reaction, and interfacial storage, by using size effects.

In addition, size effects will not only improve the kinetics of electrodes but also have great impact on the thermodynamics, demonstrating a new design strategy for SIBs. With a reduction in sizes, the surface effect becomes more predominant, and the thermodynamic equilibrium shape of electrode's particles will change accordingly, leading to varied preferred orientation of exposed crystalline lattice, which will also influence the transport properties of electrodes.

Moreover, with reducing the particle size, the chemical potential of sodium in the electrodes will change, leading to the variations of the equilibrium voltage, as well as the energy density of SIBs. This concept has been generalized for SIBs and LIBs. Learned from SIBs, the influence of size on the equilibrium potential, involving nanocrystalline and amorphous materials, needs further investigation.

The effect of sizes on the equilibrium has been investigated systematically for nanocrystalline and amorphous RuO_2 [18]. This is the simplest case in terms of thermodynamic complexity. In this case, size effects are for two-phase situation, where one phase is amorphous or nanocrystalline, and the other phase is crystalline and macroscopic. More realistic and complex cases are two-phase simultaneously amorphous/nanocrystalline for two-phase situation. Another case is the single-phase storage in the whole range. If we take $Li_{\zeta}FePO_4$ as a pertinent example, during lithiation, $FePO_4$ keeps intact and the inherent defect chemistry remains invariant as well. Thermodynamically, we can treat it as quasi-binary (LiX).

For nanocrystalline samples, capillary effects lead to changes of the standard chemical potential of defects, components, and compounds, with the excess term of $(2\gamma/r)v$, where r, γ are effective radius of the particles and effective surface tension, and v is the (partial) molar volume of defects, components, and compounds, respectively. For the compound, the chemical potential (μ_{LiX}) increases with a decrease in size, because the volume of a compound is always positive, results in a reduced melting temperature. However, the sign is not definite in terms of the partial molar volume of a component (μ_{Li}) or defect (μ_{defect}). If we take the typical values for γ, r, and v, the excess cell voltages in absolute values are from 1 to 100 mV. If the chemical potential of the compound is taken, for amorphous materials, the excess term is approximated $(2\gamma/r)v$, where r is atomistic.

In terms of excess potentials, in addition to sizes and crystallinity, storage modes (i.e. two-phase or single-phase regimes) also make a great difference. For the two-phase case, the chemical potential of lithium is described by the standard values of the phases. In this case, a positive $(2\gamma/r)v$ is obtained, since the molar volumes are positive. If only one phase is metastable, it is important that which phase has excess term, which determines whether the equilibrium voltage is increased or decreased. If both two phases are metastable, so the difference between the two excess terms plays a decisive role.

For the single phase (α), for the excess lithium potential, we use the partial molar volume of Li in the phase ($v_{Li}{}^{(\alpha)}$); hence it is $\frac{2\gamma_\alpha}{r_\alpha}v_{Li}{}^{(\alpha)}$ instead of $\frac{2\gamma_\alpha}{r_\alpha}v^{(\alpha)}$ as above, a term that may even become negative. If Li intercalated as Li^+ interstitials and compensated by electrons, then we need to consider the respective individual partial molar volumes. The amorphous materials can be regarded as the extreme case of nanocrystalline. In the single-phase case, we address the excess potentials of ionic and electronic defects in the amorphous matrix. The different metastable situations and excess lithium potentials in terms of size, crystallinity, and storage mechanism are summarized in Table 2.1 [19].

Table 2.1 Excess lithium potential in different metastable situations.

Metastable situations	Excess lithium potential
Nano 1 (single phase)	$(2\gamma_1/r_1)v_{Li,1}$
Nano 1: nano 2	$(2\gamma_1/r_1)v_1 - (2\gamma_2/r_2)v_2$
Nano 1: macro 2	$(2\gamma_1/r_1)v_1$
Amorphous 1 (single phase)	$\mu_{Li}^{ex} = \mu_i^{ex} + \mu_n^{ex}$
Amorphous 1: amorphous 2	$\Delta_m G_1 - \Delta_m G_2 \cong (2\gamma_1/r_0)v_1 - (2\gamma_2/r_0)v_2$
Amorphous 1: nano 2	$\Delta_m G_1 - (2\gamma_2/r_2)v_2 \cong (2\gamma_1/r_0)v_1 - (2\gamma_2/r_2)v_2$
Amorphous 1: macro 2	$\Delta_m G_1 \cong (2\gamma_1/r_0)v_1$

Source: Zhu et al. [19]. Reproduced with permission, 2014, American Chemical Society.

As discussed above, size effects influence the equilibrium voltage and energy density of SIBs. However, the effects can be positive or negative depending on whether it is cathode or anode. Size effects also vary the shape of charge–discharge curves, i.e. from plateau to S-type curve, which is also helpful for monitoring the battery and designing the battery management system (BMS).

2.3.3 Morphology and Structure

In addition to improving the transport properties and controlling the sizes, the morphology and structure of electrodes are also quite significant. How to construct an efficient electrochemical circuit is the key point to achieve the high performance of SIBs, which is equally important for both LIBs and SIBs. Hence, the design rules and experience obtained for LIB electrodes can also be transposed to the development of SIB electrodes.

As we know, the development of new materials and the construction of the materials in suitable sizes into a proper morphology and structures are both important tasks to develop battery electrodes. Unfortunately, the design of a composite electrode with optimum phases, suitable sizes, shapes remains challenging due to the following reasons: (i) the transport properties for many electrode materials are unknown or unmeasured; (ii) it is difficult to understand the kinetics in composite electrodes; (iii) it is very complex to include various phases and sizes. To overcome such issues, we have provided some advises in Sections 2.3.1 and 2.3.2 on the transport properties and size effects. From previous studies, the mutual influence can be perceived between the morphology and structural design. In fact, we need to construct an efficient electrochemical circuit to optimize electrode kinetics according to the transport properties and size effects.

Since both LIBs and SIBs follow the same rules, we will discuss in the following based on either lithium or sodium electrodes, which can also be generalized to other batteries, such as K ion batteries, Mg ion batteries, and so on. In the following discussion, the electrochemical active phase is expressed by M (as "mixed ionic/electronic conductor"). The electronic conducting phase is abbreviated by E (as "electronic conductor"). The ionic conducting phase is described by I (as "ionic conductor").

For a single active particle (Figure 2.7), the situation is very straightforward. If E and I are very high electronic and ionic conductivity, respectively, the rate-determining step is solid-state transport in the particles. The transport property is determined by electrical migration and chemical diffusion, involving a local combination of Fick's law and Ohm's law. Typically, M contacts E and I only on part of its surface; hence, the propagation of lithium depends on phase M's electronic and ionic conductivities (σ_{eon} and σ_{ion}). If $\sigma_{\text{ion}} \gg \sigma_{\text{eon}}$, lithium diffuses from the M/E interface. If $\sigma_{\text{eon}} \gg \sigma_{\text{ion}}$, diffusion proceeds inward from the M/I interface. If $\sigma_{\text{ion}} \sim \sigma_{\text{eon}}$, lithium diffuses inward from both interfaces simultaneously. In addition, the relaxation time can be described as $\tau^{\delta} \sim L^2/D^{\delta}_{\text{Li}}$, where L is half-thickness or the particle radius, D^{δ}_{Li} is the effective chemical diffusion coefficient of neutral lithium. Generally, in most cases, the chemical diffusion

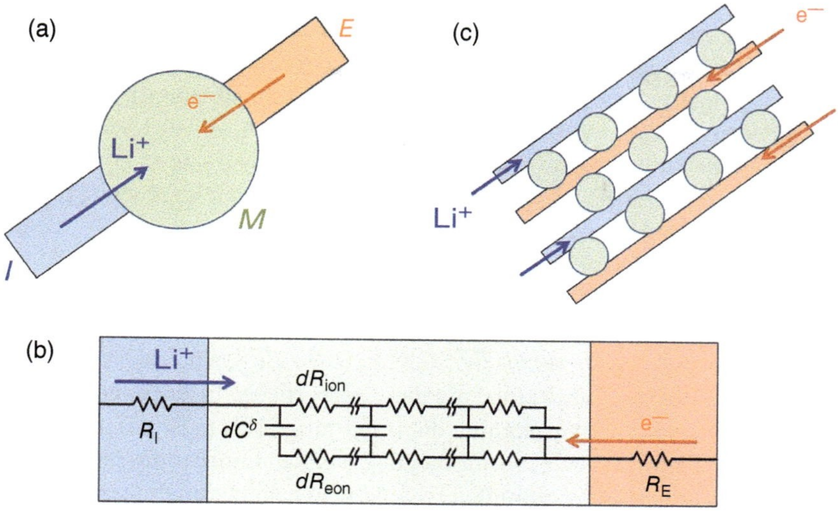

Figure 2.7 Transport properties in Li/Na battery electrodes. (a) Transport in a single particle and (b) the corresponding equivalent circuit. (c) Transport for the network of many particles. Source: Zhu et al. [20]. Reproduced with permission, 2017, American Association for the Advancement of Science.

coefficient for battery electrode materials is fairly low. In principle, the single dense particle electrode structure is only feasible in the devices that need only low power or capacity.

For other more practical cases, the commonly used strategy is to downsize the large dense monolith into many smaller particles (Figure 2.7c). This is also related to the size effects as we discussed in Section 2.3.2. According to $\tau^{\delta} \sim L^2/D^{\delta}_{\text{Li}}$, since it is quadratic dependence of τ^{δ} on L, reducing L can have immense effects. For example, a reduction in size from 1 mm to 10 nm leads to a tremendous decrease of τ^{δ} by 10 orders of magnitude. Hence, size effects seem sufficient in almost all cases. However, since the volume of the particle varies with L^3, when L becomes nanoscale, the number of particles becomes enormous as well. For example, if the size reduces from 1 mm to 10 nm, the number of particles increases by 15 orders of magnitude [20]. However, the limitation is that all of the small particles must maintain good percolation with electrons and ions in order to take advantage of such size effects. In this case, the most difficult task is to construct a good percolating and effective mixed conducting network.

Reducing the sizes of particles will also lead to many other problems. For example, to achieve a percolation network of nanomaterials for both electrons and ions, a great number of inactive E and I phases are needed, decreasing both volumetric and gravimetric energy density. In addition, the higher surface-to-volume ratio will result in an increase in interfacial reactivity, which leads to the formation of SEI, agglomeration, and impurity. Moreover, unnecessary decrease in size tends to increase the level of difficulty in synthesis as well as production cost. Hence, a feasible principle is to prepare particles with the smallest size, which allows full discharging in the fastest rate within the shortest time for the target application. We can refer to the shortest discharge time as τ^*. The corresponding half-thickness or particle radius L^* can be expressed as: $L^* = \sqrt{\alpha D^\delta \tau^*}$, where the constant α depends on charging mode and geometry. Based on Figure 2.7c, the simplified situation for designing the electrochemical circuit is as follows: the electrochemical-active phase M should be incorporated within a bicontinuous network, which constructs an effective and continuous pathway for electronic conducting phases and ionic conducting phases. The M particles should have a half-thickness or radius of approximately L^*, whose percolation is not necessary.

In addition, since σ_{ion} and σ_{eon} for M phase are often different by orders of magnitude, leading to different corresponding optimal "wiring lengths," L^*_{ion} and L^*_{eon}, which are determined by $L^*_{ion} \approx L^* / \sqrt{t_{eon}}$ and $L^*_{eon} \approx L^* / \sqrt{t_{ion}}$, where t_{ion} and t_{eon} are the ionic and electronic transference numbers, respectively. In principle, the general design strategy is taking maximum merit for transport capabilities of particles by optimally using two different wiring lengths, avoiding unnecessary nanoheterogeneity that may cause some unwanted issues due to nanomaterials. As a result, it is not suggested to use wiring lengths below optimal lengths. In short, the most economical way to design the network is as follows: the M particle should be shaped and contacted by the I and E phases and the wiring lengths for ions and electrons should be kept close to L^*_{ion} and L^*_{eon}, respectively. In addition, for electrodes with a phase transformation or conversion reaction (two phases or multiple phases), the optimal values rely on the properties of the new phases in part as well.

Furthermore, the dimensionality will also have effects on volumetric density, percolation, and the wiring length. Taking 3D monolith as an example, if all three dimensions are macroscopic, the diffusion is sluggish; if one parameter goes to nanoscopic, diffusion is much faster. On the other hand, for 0D objects, such as nanoparticles, it is very beneficial for M phases with low D^δ. One-dimensional (1D) objects, such as nanowires, have the mechanical advantage and are very suitable as electronic current collectors. For two-dimensional (2D) objects, such as nanosheets rely on the arrangement and the application. For example, they can serve as M, I, and E phases in the thin-film batteries. In addition, when porous and hierarchical structures are applied, the situation will become much more complicated when other influences are factored in.

Now, we can discuss the general design principles for archetypal architectures based on the transport properties. We can use L^*_{ion} and L^*_{eon}, the optimal length, as classification criteria. For the case 1: if the active phase (i.e. metal or carbon) has a very good electronic conductivity, where $L^*_{eon} \gg L^*_{ion}$. Then the ionic conducting percolation for M should be on a very fine scale, but the E phase without special design. As shown in Figure 2.8a, a typical example is a hierarchical porous carbon monolith [21]. In this case, due to the hierarchical pores, the liquid electrolyte is facile to penetrate into the pore, leading to fast ion transport to the M on a mesoporous length scale. Since M has the high electronic conductivity, it is not necessary for additional mixing electronic conducting phases. For the case 2: as demonstrated in Figure 2.8b, $L^*_{eon} > L^*_{ion}$. In this case, both ion- and electron-conducting phases should be included, while it can be much coarser for the electronic conducting

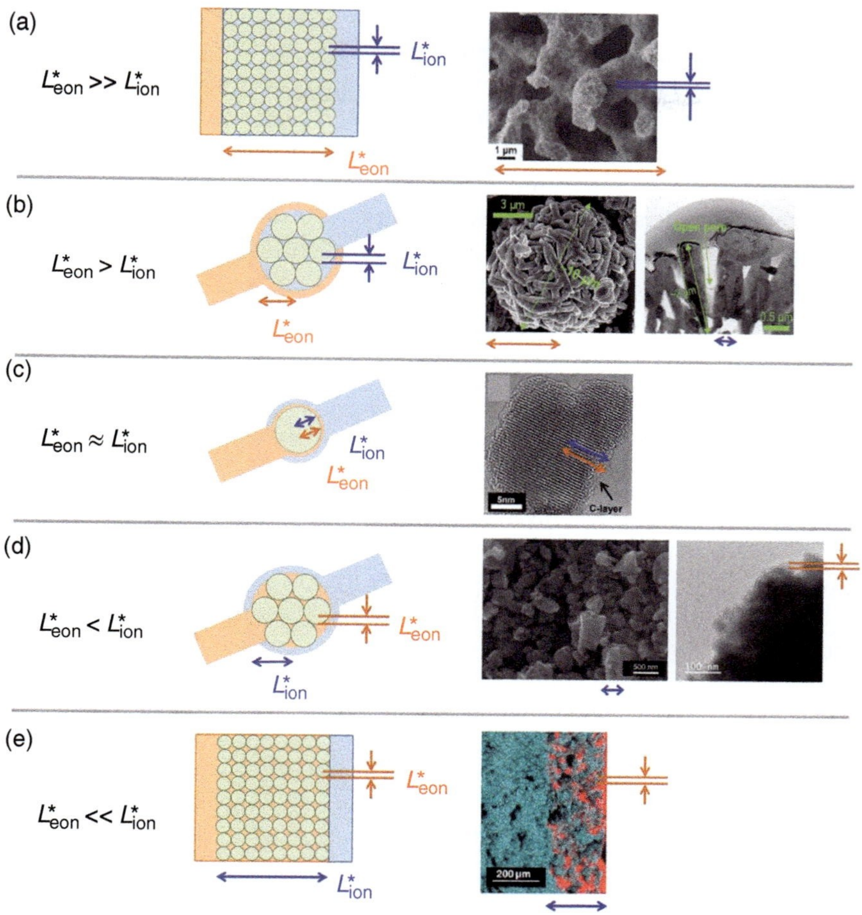

Figure 2.8 Strategies for electrochemical circuit constructing based on the transport properties. (a) $L^*_{eon} \gg L^*_{ion}$, porous carbon monolith anode; (b) $L^*_{eon} > L^*_{ion}$, Li_2MnO_3–$LiNi_{0.5}Mn_{0.5}O_2$–C cathode; (c) $L^*_{eon} \approx L^*_{ion}$, $LiMn_2O_4$–C cathode; (d) $L^*_{eon} < L^*_{ion}$, $Na_3V_2(PO_4)_3$–C sodium cathode; and (e) $L^*_{eon} \ll L^*_{ion}$, $Li_{10}GeP_2S_{12}$–C electrode. Source: Zhu et al. [20]/with permisssion from American Association for the Advancement of Science.

network. Here the typical example is $Li_2MnO_3 : LiNi_{0.5}Mn_{0.5}O_2$ [22], which has the predominantly electronic conductivity. For the electrode structure, it is sufficient to contact agglomerate in about 10 μm with the electronic conducting phase (carbon network), but sub-100 nm pores are necessary for the penetration of electrolytes. For the case 3: as shown in Figure 2.8c, the typical example is $LiMn_2O_4$, where L^*_{ion} and L^*_{eon} are comparable [23]. In this case, the $LiMn_2O_4$ is contacted by the carbon (electronic conducting phase) and liquid electrolyte (ionic conducting phase) on the same length scale of around 20 nm. For the case 4: Figure 2.8d addresses the situation where $L^*_{ion} > L^*_{eon}$. A typical example here is SIB $Na_3V_2(PO_4)_3$ cathode, which is a good NASICON-type Na^+ conductor. As for its design, it is sufficient to contact $Na_3V_2(PO_4)_3$ with polymer electrolyte on a scale of ~1 μm, but coated by carbon at ~10 nm scale [24]. For the case 5: an electroactive phase *M* has much lower electronic conductivity than ionic one, which is the other extreme case. The typical example is $Li_{10}GeP_2S_{12}$ electrodes [25], which is the superionic conductor normally used as a solid electrolyte (higher than 10^{-2} S cm^{-1} for Li^+ conductivity). For such an extreme case, the design strategy is to add electronic conducting phases (Figure 2.8e).

In addition to understanding the design strategies for sodium/lithium electrodes, next step is to look for ways to realize the aforementioned rational electrochemical circuit design. The synthesis approaches can be seen in Table 2.2. Generally,

Table 2.2 Approaches for fabricating nanoscale architectures.

Type of synthesis	Method	Possible dimensionality	Advantages
Solid state	Ball milling	0D, agglomerates	Feasible for large-scale applications
Liquid phase	Hydrothermal	0D, 1D, 2D, 3D porous	Controllable for various nanostructures
	Solvothermal	0D, 1D, 2D, 3D porous	
	Microemulsion	0D, 1D	
	Precipitation	0D, 3D	
	Sol-gel	0D, 3D porous	
	Template	0D, 1D, 2D, 3D porous	
	Spray drying	3D porous	Powerful for 3D porous structures
Field-assisted	Electrospinning	1D	Powerful for 1D structures
	Electrospraying	2D, 3D porous	Powerful for porous nanostructures
	Microwave-assisted synthesis	0D, 1D, 2D, 3D porous	High reaction rates
	Galvanic replacement synthesis	0D, 1D, 2D, 3D porous	Template morphology can be tailored
	Electrodeposition	0D, 2D	Effective for coatings and thin films
Vapor deposition	Chemical vapor deposition	0D, 2D	
	Physical vapor deposition	0D, 2D	
Exfoliation	Various	2D	Powerful for materials with a layered crystal structure

Source: Zhu et al. [20]. Reproduced with permission, 2017, American Association for the Advancement of Science.

solid-state preparation can realize a large-scale production, which is important for battery industry; however, it is not suitable to achieve a nanocircuitry design. Liquid-phase synthesis including hydrothermal and solvothermal methods, as well as template approaches, is powerful for constructing refined nanostructures. By such methods, the size and shape of SIB electrodes can be controlled by changing the parameters such as concentrations, precursors, surfactants, and templates, etc. In addition, some synthesis methods are particularly suitable for target dimensionalities. For example, electrospinning technique is a commonly used method to prepare 1D nanostructures, including nanowires, nanotubes, or core–shell 1D structures [26]. Exfoliation approaches are promising to prepare single- or multilayer sheets structure for the layered crystal structure compounds [27]. Electrostatic spray deposition (ESD) technique is an effective approach to prepare 3D porous structure [28]. In addition, vapor deposition methods are powerful way to synthesize thin-film electrodes, while atomic layer deposition (ALD) is able to grow a conformal coating with atomic accuracy [29].

2.4 Design Principles for Electrolytes

In addition to electrode materials, suitable electrolytes are the key factors to achieve excellent electrochemical performance of SIBs. Concerning the design principles for electrolytes, the following points are quite important, including transport properties, electrochemical stability window, thermal stability, interface issues between electrodes and electrolytes, and battery safety.

2.4.1 Transport Properties

Since the role of SIB electrolyte is to support the sodium ion transport in the electrolyte and between cathodes and anodes, the ionic conductivity is one of the most significant parameters for the electrolyte. There are aqueous liquid electrolyte, nonaqueous liquid electrolyte, and all-solid-state electrolyte for SIBs, whose conductivities are sorted from highest to lowest. In principle, the ionic conductivity is determined by the following equation:

$$\sigma = \sum_i n_i \mu_i z_i e$$

where σ is the conductivity, μ is the ionic migration rate, and z is the valences of charges, n is the free ion number.

The transport properties of electrolyte depend on properties of sodium salts and solvents. As shown in Figure 2.9, the conductivities can be varied for sodium electrolyte with different types of sodium salts dissolved in the same solvent, which is owing to the volume differences of anions and interaction between the cations and anions. In addition, the concentration of sodium salts also has great influence on the ionic conductivity as shown in Figure 2.10. The general trend is that the ionic conductivity starts to increase until it reaches the maximum, then decreases

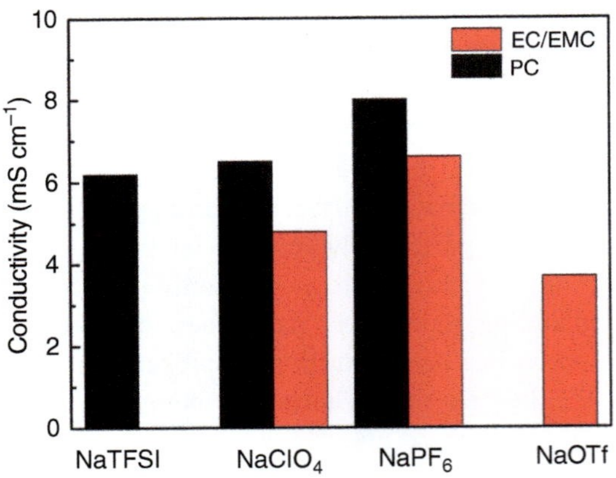

Figure 2.9 Conductivity variations for different sodium salts in EC/ethyl methyl carbonate (EMC) (3 : 7 wt%). Source: Che et al. [30]. Reproduced with permission, 2017, Royal Society of Chemistry.

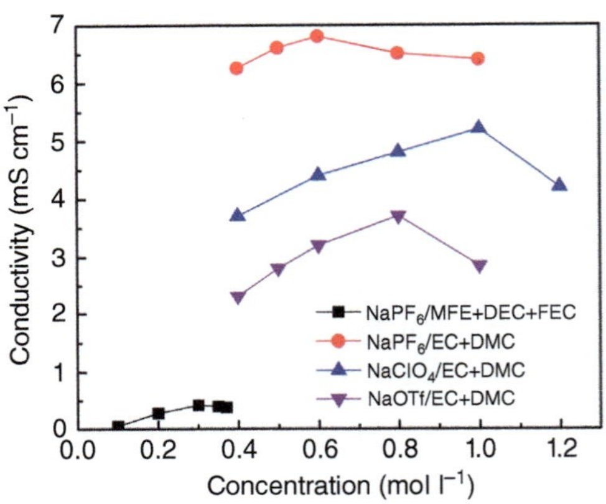

Figure 2.10 Conductivity variations as a function of concentrations. Source: Che et al. [30]. Reproduced with permission, 2017, Royal Society of Chemistry.

with increasing the concentration of sodium slats. For instance, the conductivity of electrolyte system ($NaPF_6$ in ethylene carbonate/dimethyl carbonate [EC/DMC]) increased to the maximum value ($6.8\,mS\,cm^{-1}$), when the concentration of $NaPF_6$ was increased to $0.6\,mol\,l^{-1}$, and then the ionic conductivity decreased slightly if the concentration was further increased. It is because that when the salt concentration is low, the dominant factor is the number of free ions. Therefore, the higher conductivity is obtained by adding more salts. It is generally known that the electrolyte viscosity will be increased accordingly when the high salt concentration increases. Hence, we expected to see a balance between the effect of viscosity and free ions, when the conductivity reaches the maximum value. In the zone of the high salt concentration, the cations and anions in the solvent will associate strongly, leading to a reduced number of free ions, which results in a decrease in electrolyte conductivity after more salts are added.

Solvent is another important factor that affects the ionic conductivity of electrolytes. The viscosity and dielectric constant are the two critical parameters for the transport properties of electrolyte system, which decide ionic mobility and are related to the dissociation of the salts respectively. The viscosity and dielectric constant of commonly used solvents are listed in the Table 2.3. The solvent with

Table 2.3 The dielectric constant and viscosity for different solvents.

Solvent	Dielectric constant	Viscosity (mPa s)
EC	89.6	1.9
PC	66.1	2.5
BC	55.9	3.2
DMC	3.1	0.58
DEC	2.82	0.75
EMC	2.4	0.65
THE	7.58	0.46
DME	7.2	0.455

Source: Che et al. [30]. Reproduced with permission, 2017, Royal Society of Chemistry.

low dielectric constant leads to low viscosity and vice versa. Binary solvent-based electrolytes by mixing a solvent with low viscosity and a solvent with high dielectric constant are applied to achieve proper solvents for SIBs. For SIBs, a systematic investigation shows that the binary solvent electrolytes have much higher conductivity than single-component solvent electrolytes. For example, for the binary solvents, mixing a high dielectric constant solvent (e.g. EC), it dictates the dissociation of the salts; mixing with a low viscosity solvent, e.g. DMC or 1,2-dimethoxyethane (DME), it increases ionic conductivity [31].

2.4.2 Electrochemical Stability Window

Electrochemical stability window of electrolytes in SIBs is another significant factor. A desirable electrolyte should have a wide electrochemical stability window, where reversible redox reactions occur without side reactions. The electrochemical stability window determines the potential energy density, cycling stability, and safety of SIBs. In general, the electrochemical stability window can be measured by commonly used electrochemical characterization techniques such as CV or linear sweep voltammetry (LSV). Working electrodes used in the measurements are usually inert electrode materials, e.g. glassy carbon, stainless steel, Pt, etc., where reactions such as reversible redox processes and decomposition reactions of electrolyte occur simultaneously on the surface of the working electrodes. The choices of different sodium salts, solvents, working electrodes affect the final electrochemical stability windows; however, there are few systematic investigations on the electrochemical stability of electrolytes for SIBs compared with LIBs. Systematic electrochemical stability window tests of some specific SIB electrolyte systems were performed. It indicates that different withstanding voltages are obtained for different solvents. When the solvents were mixed, synergetic effects were observed as well. The electrochemical stability of various solvents can be arranged in the following order: diethyl carbonate (DEC) > propylene carbonate (PC) > triglyme > DME > DMC > tetrahydrofuran

(THF) and EC : PC > EC : DMC > EC : DME > EC : DEC > EC : triglyme, where 1 M $NaClO_4$ was dissolved in the solvents [31]. In addition, the decomposition voltage of the proposed electrolyte of 1 M $NaClO_4$/ethyl methyl carbonate (EMC) + 2% 1-fluoroethylene carbonate (FEC) is around 5.6 V, compared with the PC-based electrolyte of 1 M $NaClO_4$/PC + 2% FEC (decomposition voltage 4.5 V) [32]. Apart from nonaqueous electrolytes mentioned above, ionic liquid-based electrolytes are also quite promising, which exhibit high ionic conductivity at room temperature and excellent safety features. These ionic liquid-based electrolytes can operate at ambient condition in SIBs. It is worth to mention that ionic liquid-based electrolytes exhibit wide electrochemical windows, making them promising for high-voltage electrodes. For instance, the electrochemical stability window of a proposed ionic liquid system NaFSA–C1C3pyrFSA is around 5.2 V [33].

2.4.3 Thermal Stability

Another important parameter for SIBs is the thermal stability, which is associated with the safety of SIBs. As we know, the organic electrolytes used in the conventional battery systems are easy to decompose during (dis)charge process and lead to heat release and an increase in the operation temperature, resulting in safety issues. We should know that both thermal stability of electrolytes and the interactions between electrodes and electrolytes are relevant here. We can apply thermogravimetric analysis (TGA)/differential scanning calorimeter (DSC) techniques to investigate and analyze the thermal stability of various electrolyte systems.

In principle, the thermal stability of the sodium electrolyte has close connection with its composition, such as sodium salts, applied solvents, and various additives. There are some investigations in terms of sodium salts and solvents by DSC techniques on the thermal stability. For the commonly used sodium salts, the thermal stability is as follows: NaTFSI (250 °C) < $NaPF_6$ (280 °C) < $NaClO_4$ (310 °C), while for various solvents, the orders are PC > EC > DEC > DMC > DME (from 235 to 90 °C) [31]. One thing needed to mention is that the ionic liquids-based electrolytes have the best thermal stability among the liquid electrolytes for SIBs.

The interaction between electrolyte and electrodes leads to the complex issues for the thermal stability. For example, if abused conditions such as overcharge or heating occur, it results in explosion if the high internal pressure is created in the battery. The interaction between electrolytes and electrodes, as well as electrolyte decomposition will lead to a series of chemical reactions and result in battery self-heating processes. In this case, it is different for lithium and sodium systems. The investigation on reactivity between sodium solvents and sodiated hard carbon was carried out by accelerating rate calorimetry (ARC). The addition of $NaPF_6$ to EC/DEC solvent does not improve the thermal stability of sodiated hard carbon, because this compound does not thermally decompose to NaF to form a passivated layer and protect the material like LiF from $LiPF_6$ in LIBs. For the stability of sodium cathodes with electrolytes, various cathodes including Na_xCoO_2, Na_xFeO_2, $Na_xNi_{0.5}Mn_{0.5}O_2$, and Na_xCrO_2 have been investigated by DSC and ARC techniques

[34]. They have similar exothermic behavior, which is similar to the corresponding lithium electrodes.

2.4.4 Interfacial Compatibility

The compatibility of sodium electrolytes with sodium anode and sodium cathodes is complicated but very important for the battery performance, which is the key point to optimize the electrolyte.

For the anode part, SEI forms on the surface of anode materials when it contacts with sodium electrolytes. By changing the composition of sodium electrolyte composition, it is possible to construct a stable and compact SEI layer to obtain the optimal performance. Optimizing the proper electrolytes can improve the performance of hard carbons that considered the most promising sodium anodes for SIBs. For instance, Komaba et al. studied various carbonate esters coupled with electrodes of hard carbons. It showed that both the EC + DEC or PC + DEC-based electrolytes were compatible with the hard-carbon anodes by investigating various single or binary solvents systems [35]. As we know, the graphite is commercial lithium anode, which cannot be directly used for sodium anodes. However, by electrolyte optimization, ether-based electrolytes have been proposed to be applied in SIBs, which activates the graphite as sodium anodes through Na^+-solvent co-intercalation mechanism [36]. Further study confirmed that a high rate capability and high cycling stability can be obtained in graphite anode with $NaPF_6$/diethylene glycol dimethyl ether (DEG-DME), which was also evaluated in $Na_{1.5}VPO_{4.8}F_{0.7}$/graphite full cell using this kind of electrolyte [36b].

In addition to the electrolyte compatibility with the anode, the interface between cathode and electrolyte, CEI, is also quite important, which is related to the reversible capacity, charging voltage limit, capacity retention, etc. For example, the Na_2FePO_4F cathode showed higher capacity retention in 1 M $NaPF_6$/EC + DEC than in 1 M $NaPF_6$/PC [37], due to the instability of PC at lower voltage. The interfacial compatibility between sodium electrolytes and electrodes is a complex issue, which is connected with the electrolyte component, electrode structure, morphology, surface area, etc.

2.4.5 Safety Issues

Safety is a significant advantage of SIBs for the large-scale energy storage. Despite a low toxicity of sodium electrolyte, the sodium electrolyte guarantees good safety operation of SIBs. For the organic liquid sodium electrolyte, an effective strategy is to choose a solvent with low flammability and low volatility or apply a high-efficiency, flame-retarding additives. For ionic liquid, which is an intrinsically safe system, it is very promising to replace fully or partially organic solvent with ionic liquid systems. For example, the ionic liquid-based electrolytes $EMIBF_4$–$NaBF_4$ and Pyr14TFSI–NaTFSI have better thermal stability compared with a traditional electrolyte system ($NaPF_6$/$NaClO_4$–EC + DEC) [38]. Last but not least, the application of solid-state electrolytes or gel polymer electrolytes is seen as an all-round solution to safety issues, such as flammability, leakage, and dendrites caused by nonaqueous liquid electrolyte. Nonetheless, the low ionic conductivity of solid-state electrolytes at ambient conditions and solid/solid interface are obstacles to be solved at the moment.

2.5 Conclusions

In summary, SIBs are regarded as a promising alternatives for LIBs, which have a great potential to be applied in the energy storage filed, especially large-scale energy storage. The design strategy of SIBs can be learned from its lithium counterpart; however, there is still a big difference between these two systems. By taking advantages and overcoming the disadvantages of SIBs, through rational design for both electrode materials and electrolytes, we believe that this promising system can live up to its expectation and will also be commercialized in the near future.

References

1 Hwang, J.-Y., Myung, S.-T., and Sun, Y.-K. (2017). *Chem. Soc. Rev.* 46: 3529.

2 (a) Berthelot, R., Carlier, D., and Delmas, C. (2011). *Nat. Mater.* 10: 74. (b) Sauvage, F., Laffont, L., Tarascon, J.M., and Baudrin, E. (2007). *Inorg. Chem.* 46: 3289. (c) Komaba, S., Yabuuchi, N., Nakayama, T. et al. (2012). *Inorg. Chem.* 51: 6211. (d) Oh, S.-M., Myung, S.-T., Hassoun, J. et al. (2012). *Electrochem. Commun.* 22: 149. (e) Kawabe, Y., Yabuuchi, N., Kajiyama, M. et al. (2011). *Electrochem. Commun.* 13: 1225. (f) Yabuuchi, N., Kajiyama, M., Iwatate, J. et al. (2012). *Nat. Mater.* 11: 512. (g) Sathiya, M., Hemalatha, K., Ramesha, K. et al. (2012). *Chem. Mater.* 24: 1846. (h) Ponrouch, A., Dedryvere, R., Monti, D. et al. (2013). *Energy Environ. Sci.* 6: 2361. (i) Guignard, M., Didier, C., Darriet, J. et al. (2013). *Nat. Mater.* 12: 74.

3 Wu, C., Kopold, P., Ding, Y.-L. et al. (2015). *ACS Nano* 9: 6610.

4 (a) Komaba, S., Matsuura, Y., Ishikawa, T. et al. (2012). *Electrochem. Commun.* 21: 65. (b) He, M., Kraychyk, K., Walter, M., and Kovalenko, M.V. (2014). *Nano Lett.* 14: 1255. (c) Farbod, B., Cui, K., Kalisvaart, W.P. et al. (2014). *ACS Nano* 8: 4415. (d) Wu, L., Hu, X., Qian, J. et al. (2014). *Energy Environ. Sci.* 7: 323.

5 (a) Xu, X., Zhao, R., Ai, W. et al. (2018). *Adv. Mater.* 30: 1800658. (b) Zhu, C., Kopold, P., Li, W. et al. (2015). *J. Mater. Chem. A* 3: 20487.

6 (a) Zhu, H., Jia, Z., Chen, Y. et al. (2013). *Nano Lett.* 13: 3093. (b) Cao, Y., Xiao, L., Sushko, M.L. et al. (2012). *Nano Lett.* 12: 3783.

7 (a) Liu, H., Hu, K., Yan, D. et al. (2018). *Adv. Mater.* 30: 1800295. (b) Wang, Y.-X., Lai, W.-H., Chou, S.-L. et al. (2020). *Adv. Mater.* 32: 1903952.

8 Zhu, C., Song, K., van Aken, P.A. et al. (2014). *Nano Lett.* 14: 2175.

9 Tang, K., Yu, X., Sun, J. et al. (2011). *Electrochim. Acta* 56: 4869.

10 Bianchini, M., Fauth, F., Brisset, N. et al. (2015). *Chem. Mater.* 27: 3009.

11 Zhu, C., Wu, C., Chen, C.-C. et al. (2017). *Chem. Mater.* 29: 5207.

12 (a) Li, H., Tang, H., Ma, C. et al. (2018). *Chem. Mater.* 30: 2498. (b) Novikova, S.A., Larkovich, R.V., Chekannikov, A.A. et al. (2018). *Inorg. Mater.* 54: 794. (c) Wang, D., Bie, X., Fu, Q. et al. (2017). *Nat. Commun.* 8: 15888.

13 Maier, J. and Amin, R. (2008). *J. Electrochem. Soc.* 155: A339.

14 (a) Amin, R., Balaya, P., and Maier, J. (2007). *Electrochem. Solid-State Lett.* 10: A13. (b) Amin, R. and Maier, J. (2008). *Solid State Ionics* 178: 1831.

15 (a) Amin, R., Lin, C.T., and Maier, J. (2008). *PCCP* 10: 3519. (b) Amin, R., Lin, C.T., and Maier, J. (2008). *PCCP* 10: 3524. (c) Amin, R., Lin, C.T., Peng, J.B. et al. (2009). *Adv. Funct. Mater.* 19: 1697.

16 Jian, Z., Zhao, L., Pan, H. et al. (2012). *Electrochem. Commun.* 14: 86.

17 Zhu, C., Kopold, P., van Aken, P.A. et al. (2016). *Adv. Mater.* 28: 2409.

18 (a) Delmer, O., Balaya, P., Kienle, L., and Maier, J. (2008). *Adv. Mater.* 20: 501. (b) Delmer, O. and Maier, J. (2009). *PCCP* 11: 6424.

19 Zhu, C., Mu, X., Popovic, J. et al. (2014). *Nano Lett.* 14: 5342.

20 Zhu, C., Usiskin, R.E., Yu, Y., and Maier, J. (2017). *Science* 358: eaao2808.

21 Hu, Y.S., Adelhelm, P., Smarsly, B.M. et al. (2007). *Adv. Funct. Mater.* 17: 1873.

22 Oh, P., Myeong, S., Cho, W. et al. (2014). *Nano Lett.* 14: 5965.

23 Lee, S., Cho, Y., Song, H.-K. et al. (2012). *Angew. Chem. Int. Ed.* 51: 8748.

24 Gao, H., Zhou, W., Park, K., and Goodenough, J.B. (2016). *Adv. Energy Mater.* 6: 1600467.

25 Han, F., Gao, T., Zhu, Y. et al. (2015). *Adv. Mater.* 27: 3473.

26 (a) Zhu, C., Mu, X., van Aken, P.A. et al. (2014). *Angew. Chem. Int. Ed.* 53: 2152. (b) Yu, Y., Gu, L., Zhu, C. et al. (2009). *J. Am. Chem. Soc.* 131: 15984.

27 Chen, K., Song, S., Liu, F., and Xue, D. (2015). *Chem. Soc. Rev.* 44: 6230.

28 Zhu, C., Fu, Y., and Yu, Y. (2019). *Adv. Mater.* 31: 1803408.

29 Wen, L., Zhou, M., Wang, C. et al. (2016). *Adv. Energy Mater.* 6: 1600468.

30 Che, H., Chen, S., Xie, Y. et al. (2017). *Energy Environ. Sci.* 10: 1075.

31 Ponrouch, A., Marchante, E., Courty, M. et al. (2012). *Energy Environ. Sci.* 5: 8572.

32 Oh, S.-M., Myung, S.-T., Yoon, C.S. et al. (2014). *Nano Lett.* 14: 1620.

33 Ding, C., Nohira, T., Kuroda, K. et al. (2013). *J. Power Sources* 238: 296.

34 (a) Zhao, J., Zhao, L., Dimov, N. et al. (2013). *J. Electrochem. Soc.* 160: A3077. (b) Xia, X. and Dahn, J.R. (2012). *J. Electrochem. Soc.* 159: A1048. (c) Xia, X. and Dahn, J.R. (2012). *J. Electrochem. Soc.* 159: A647.

35 Komaba, S., Murata, W., Ishikawa, T. et al. (2011). *Adv. Funct. Mater.* 21: 3859.

36 (a) Jache, B. and Adelhelm, P. (2014). *Angew. Chem. Int. Ed.* 53: 10169. (b) Kim, H., Hong, J., Park, Y.-U. et al. (2015). *Adv. Funct. Mater.* 25: 534.

37 Vidal-Abarca, C., Lavela, P., Tirado, J.L. et al. (2012). *J. Power Sources* 197: 314.

38 Hasa, I., Passerini, S., and Hassoun, J. (2016). *J. Power Sources* 303: 203.

3

Transition Metal Oxide Cathodes for Sodium-Ion Batteries

3.1 Introduction

Since that energy density will be the overriding factor for sodium-ion batteries (SIBs) to be deployed in real practice, the identification of suitable cathode materials is essential to close the performance gap between SIBs and lithium-ion batteries (LIBs) and speed up the commercialization of SIBs. Over the years, a wide spectrum of cathode materials has been investigated to study their viability for SIBs, which primarily include layered- and tunnel-structured transition metal (TM) oxides, polyanion cathodes, Prussian blue analogues (PBAs), and organic compounds. Typical cathode materials for SIBs are based on intercalation chemistry, where sodium ions provided by the cathode shuttle to the anode during charge, and then extract during discharge, with minimal structural change in the host material. General research trend focuses on the tweaking of lithium analogues for SIBs due to their similarity in intercalation chemistry. However, researchers quickly realized the significant difference in their reaction mechanisms. For example, $NaCrO_2$, the analogue of electrochemical inactive $LiCrO_2$, maintains a specific capacity of 90–120 mAh g^{-1} over 50 cycles [1]. On top of that, good performance in cycling stability and capacity retention owned by the lithium analogues do not always pass down to their sodium analogues, which is mainly caused by the sluggish reaction kinetics of the larger sodium ion as well as the lattice environment of the host materials. To provide a glimpse at the current state of SIB cathodes, their electrochemical performances in terms of working voltage and specific capacity are illustrated in Figure 3.1.

In general, TM oxides, including Na-free and Na-based structures, are known for their high specific capacity, despite the lower working potential (≈2.7–3.0 V vs. Na/Na^+). The research of Na-based layered transition oxides, $NaMO_2$ (M = Ni, Co, Mn, Fe, Cr, V, etc.) was inspired by their lithium counterparts that work well as intercalative cathodes for LIBs. Layered Na–metal oxides exist in different crystal structures, which can be denoted as P2, P3, O2, and O3 according to Delmas' notation [3]. O and P indicate the coordination environment of sodium ions, in which O represents that the occupancy at the octahedron sites surrounded by six oxygen

Sodium-Ion Batteries: Energy Storage Materials and Technologies, First Edition. Yan Yu.

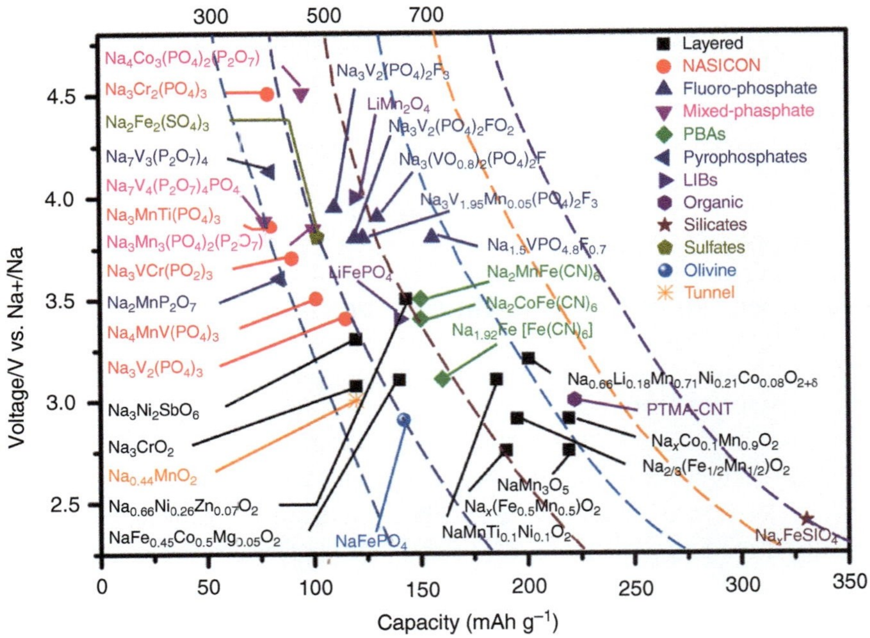

Figure 3.1 A summary on the electrochemical performances of various cathodes reported in literature. Source: Lyu et al. [2]. Reproduced with permission, 2019, Elsevier.

and P represents that the occupancy at the center of prism sites surrounded by six oxygen. The number refers to the oxygen stacking sequence, where 2 and 3 indicate the stacking order of ABBAABBA and ABCABC, respectively. A distorted phase is denoted as prime (′). Among them, O3 phase and P2 phase are widely investigated as cathodes for SIBs. It is generally recognized that the electrochemical behaviors of Na-based oxides are dependent on the stability of oxide layer and the kinetics properties of sodium ions. For example, phase transition is more prevalent in the O3 phase as the structure easily undergoes slab-gliding upon cycling [4]. Despite the relatively high initial specific capacity due to the higher amount of sodium in the pristine state, the capacity of the O3 phase usually fades quickly in the subsequent cycles. On the other hand, P2-type oxides are known for their better overall electrochemistry due to their low diffusion barrier and high ionic conductivity, which facilitates the movement of sodium ions along their layered structure [5].

Aside from the layered Na-based oxides, the Na-free TM oxides MO_x (M = V, Mn, Mo) that can adopt sodium ion in an intercalative manner have also been widely studied for their sodium storage properties. $VO_2(B)$ with layered structure is of particular interest. With an increased edge-sharing connectivity of VO_5 square pyramids, $VO_2(B)$ is endowed with higher resistance to lattice shearing upon cycling. Typical charge storage mechanism of $VO_2(B)$ involves the reversible insertion and withdrawal of sodium ion into the layered structure of $VO_2(B)$. The stable $NaVO_2$ phase is formed upon discharge, but only reversible transition between

$NaVO_2$ and Na_xVO_2 was observed in the subsequent cycles without restoring back to the initial structure [6]. With a single-electron insertion reaction, the theoretical specific capacity for VO_2(B) is estimated to be 323 mAh g^{-1}. V_2O_5, both in the orthorhombic and bilayered form, is electrochemically active for sodium-ion intercalation. Ali et al. has proved the formation of crystalline NaV_2O_5 (majority) and partially amorphous $Na_2V_2O_5$ (trace amount) with the reduction of V^{5+} into V^{4+} upon Na^+ intercalation [7]. However, not all the V_2O_5 are retrieved in the reverse process, where NaV_2O_5 that is irreversibly formed persists in the mixture. Recent trend in the investigation of bilayered V_2O_5 has been inspired by its larger d-spacing than the orthorhombic V_2O_5. The 11.53 Å along the (001) direction allows sodium ion to move freely in the lattice, which leads to an enhancement in SIB performance [8]. In addition, despite the low electronic conductivity, MnO_2 with large open tunnels and interstitial spaces is highly sought after for facile sodium ion storage and transport. Intercalation of sodium in MnO_2 is predicted to be thermodynamically favorable for three sodium insertion at 3.34, 2.84, and 2.19 V (vs. Na^+/Na) [9]. Recently, researchers discovered that β-MnO_2 nanorods with a high density of (1 × 1) tunnel is beneficial for effective sodium-ion insertion and extraction, giving rise to a high initial capacity of 350 mAh g^{-1} and a satisfactory high-rate capability [10].

In this chapter, we introduce the recent development of TM oxides cathode materials for high-performance SIBs, in terms of structural design, electrochemical properties, and the corresponding reaction mechanisms.

3.2 Sodium-free Transition Metal Oxides

3.2.1 Vanadium Oxides

Vanadium oxide is one of the most promising cathode materials for SIBs due to their high capacity. Among them, two polymorphs of VO_2, i.e. VO_2(A) and VO_2(B), are both considered as cathodes for SIBs. They show similar layered structure (Figure 3.2a). The diffusion channels in VO_2(A) are relatively wider and thus VO_2(A) is regarded as more favorable for Na storage. However, the study of VO_2(A) is challenging mainly because of the harsh synthesis condition and capacity reduction during the cycling process. Though Hu et al. managed to synthesize VO_2(A) nanowires by a facile hydrothermal process [11], the nanowires were seriously etched during the sodium-ion intercalation and deintercalation process. On the other hand, metastable VO_2 (B) has been more extensively investigated. It shows a layered structure consisted of two edge-sharing units. The layers connected by the corners further form a three-dimensional structure, allowing rapid diffusion of sodium ions into the VO_2 tunnels [12]. Wang et al. first reported VO_2(B) as the cathode material for SIBs [6]. The discharge capacity maintained at 70 mAh g^{-1} for over 50 cycles. Further investigation into the Na storage mechanism revealed that the reversible charge and discharge process was mainly realized by the conversion between $NaVO_2$ and Na_xVO_2 ($x \approx 0.3$) (Figure 3.2b,c).

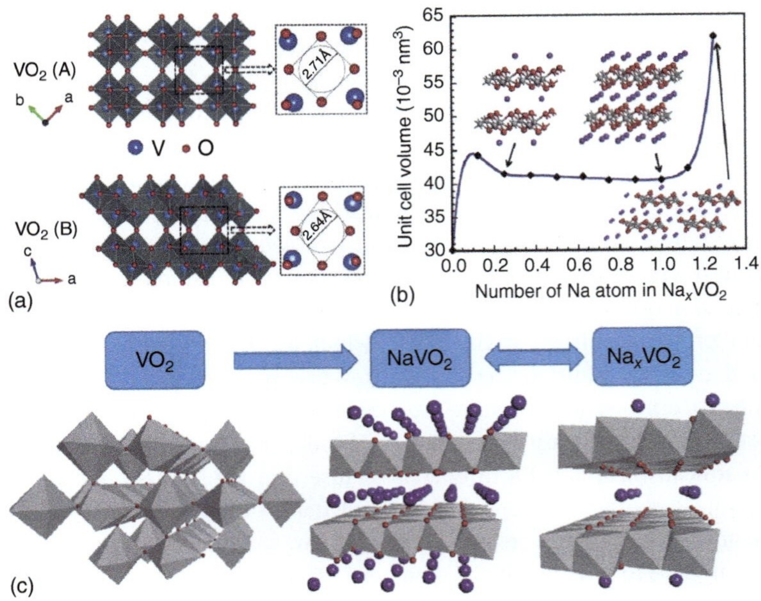

Figure 3.2 (a) The crystal structure of VO_2(A) and VO_2(B). Source: Hu et al. [11]. Reproduced with permission. 2018, Elsevier. (b) The volume variation with different number of Na ions. (c) Illustration of the sodiation/desodiation process. Source: Wang et al. [6]. Reproduced with permission, 2014, Elsevier.

In general, the performance of vanadium oxides is limited by the fast capacity fading and poor high-rate performance, which may be originated from self-aggregation, dissolution, and accumulated charge transfer resistance during cycling. Besides, the insertion of Na ions will result in volume expansion and breakdown of VO_2 tunnels. Nanoscale surface engineering can facilitate the permeation of Na ions between electrolyte and electrodes. Besides, effective surface modification could prevent the agglomeration and dissolution of active materials during the cycling process [13]. For example, the graphene quantum dots (GQDs)-modified VO_2 nanobelts exhibited a high capacity of 306 mAh g^{-1} with superior rate tolerance and good capacity retention (88% of the capacity was retained after 1500 cycles under the current density of 18 A g^{-1}) [13a].

V_2O_5 with higher thermal and chemical stability is also investigated as a cathode material for SIBs. The cathode performance of V_2O_5 has been significantly improved through tailoring of the crystal structure, i.e. configuration of the units, the lattice spacing, and the defects and the ratio of multivalence vanadium ions. Typically, V_2O_5 exists in two crystal structures, which is the bilayer and orthorhombic V_2O_5, respectively. The layered V_2O_5 shows a theoretical capacity of 236 mAh g^{-1} by forming $Na_2V_2O_5$. Practically, the bilayer V_2O_5 with a large interlayer spacing of 11.6 Å delivered a capacity of 220 mAh g^{-1} with good capacity retention after 500 cycles [14]. Tepavcevic et al. investigated the performance of both orthorhombic and bilayered V_2O_5 [15]. They found that the layered structure with short-range ordering was more preferable for the reversible and stable insertion/deinsertion of Na^+ ions.

In particular, the bilayered V_2O_5 electrode delivered a capacity of $250\,mAh\,g^{-1}$ compared to the $150\,mAh\,g^{-1}$ from the orthorhombic V_2O_5 within the voltage range of 1.5–3.8 V. The orthorhombic V_2O_5 starts to lose the crystallinity after 10 cycles, along with the deterioration of the electronic conductivity. In contrast, the bilayer V_2O_5 remained stable after repeated cycling with excellent rate performance. Li et al. compared the electrochemical performance of the orthorhombic and hydrated $V_2O_5 \cdot nH_2O$ cathode for sodium storage. The intercalation reaction of the hydrated V_2O_5 was more favorable owing to the enlarged layer spacing and the presence of more V^{4+}, which can enhance the electrical conductivity of the electrodes [16]. It delivered a specific capacity of $166\,mAh\,g^{-1}$ after 100 cycles at $0.2\,A\,g^{-1}$. However, the orthorhombic V_2O_5 can also exhibit excellent performance after modification with the electrode structure. Raju et al. synthesized the orthorhombic V_2O_5 nanoparticles with the size of around 5–7 nm and encapsulated them in the nanoporous carbon derived from the ambient hydrolysis deposition technology [17]. Benefited from the decreased dimensions and uniform carbon coating, the V_2O_5 cathode achieved a high capacity of $276\,mAh\,g^{-1}$. Córdoba et al. recently reported a new type of β-V_2O_5 cathode from the high-temperature, high-pressure method [18]. It delivered a specific capacity of $147\,mAh\,g^{-1}$ under equilibrium conditions but with moderate stability and an irreversible structural transformation during the initial cycles.

On the other hand, Su et al. synthesized the single-crystalline bilayer V_2O_5 predominantly exposed with the (001) plane [8]. The preferable exposure of the (001) plane with large interlayer spacing was favorable for fast diffusion of Na ions. The V_2O_5 cathode delivered a capacity of $231\,mAh\,g^{-1}$ with excellent rate capability. Pre-insertion of cations or water molecules between the layers is a promising strategy to expand the layer distance for more facile insertion of relatively bulky Na ions. The inserted cations (Na^+ and $NH_4{}^+$) will act as pillars to stabilize the crystal structure and promotes the insertion of more Na ions [19]. But it is worth to note that pre-intercalation of smaller cations such as Mg^{2+}, Fe^{2+}, Mn^{2+}, Al^{3+}, and Sn^{4+} will not change the insertion of Na ions significantly. Layered ammonium vanadium oxide ($NH_4V_4O_{10}$) can realize reversible insertion and deinsertion of more than three Li^+/Na^+ ions during the charge–discharge process without hampering the crystal structure [20]. The $NH_4V_4O_{10}$ nanobelts delivered a specific capacity of $190\,mAh\,g^{-1}$ at $0.2\,A\,g^{-1}$ [21]. With zirconium (Zr) doped into the vanadium sites, the electronic conductivity as well as the electrochemical performance of $NH_4V_4O_{10}$ can be further improved [22]. For example, the Zr-doped $NH_4V_4O_{10}$ with large d-spacing delivered a high capacity of $342\,mAh\,g^{-1}$ at $0.1\,A\,g^{-1}$ with 91% capacity retained after 50 cycles. The SIB full cell assembled with Zr-$NH_4V_4O_{10}$ cathode and H-$Na_2Ti_3O_7$ anode showed a high specific capacity of $145\,mAh\,g^{-1}$ at the current density of $0.5\,A\,g^{-1}$. The cell also demonstrated remarkable cycling stability with 94% capacity retention after 400 cycles. The $H_2V_3O_8$ membrane cathode, which contains mixed V^{5+}/V^{4+} species and the hydrogen bonds as the elastic buffer to accommodate the volume change during the cycling process, was synthesized by a hydrothermal method and exhibited a specific capacity of $168\,mAh\,g^{-1}$ at the current density of $0.01\,A\,g^{-1}$ [23]. Wei et al. synthesized the $V_2O_5 \cdot nH_2O$ with a freeze-drying method and achieved a high capacity of 338 mAh g^{-1} at the current density of 0.05 A g^{-1} [24]. Despite of the large capacity, the voltage range with such hydrated electrodes needs to be compromised to some extent to avoid decomposition of the electrolyte or electrodes.

In contrast to the crystallized material, the amorphous material with a lower entropic energy associated with the ordering of incorporated ions can provide more open framework structure to benefit the diffusion and penetration of the Na ions. The intercalation site energies in a disordered material has been demonstrated by Liu et al., and the Na-ion diffusivity was more independent of the Na-ion concentration [25]. They further investigated the electrochemical behavior of the amorphous and crystalline V_2O_5, respectively, and revealed that the amorphous V_2O_5 showed higher and reversible capacities at low current densities as compared to the crystalline V_2O_5.

As well documented, the deterioration of V_2O_5 is partly induced by the morphology change upon cycling. With properly controlled drying process, a variety of porous structure, i.e. xerogels, aerogels, and cryogels, can be derived and these porous structures are advantageous toward the Na storage. However, some of these porous structures tend to collapse during cycling and affect the stability of the electrode. To better preserve the porous structure, Moretti et al. replaced the solvent with ethanol which has lower vapor pressure and used the supercritical fluid (CO_2) for its further extraction [26]. The resultant porous V_2O_5 electrode showed improved stability for both Li and Na storage. Another issue associated with the V_2O_5 xerogel is the "lattice breathing" which causes the collapse of the electrode structure and rapid capacity fading [24]. To inhibit such breathing phenomenon, Wei et al. pre-inserted some Fe^{3+} ions in between the layers and constrained the lattice expansion during the reaction [27]. The lattice distance was reduced from 3.74 to 0.49 Å, due to strong interaction of the Fe^{3+} ions toward the V_2O_5 bilayers. However, due to the larger size of Na ions, the cathode performance of the optimized electrode toward Na storage was inferior to that for Li storage with lower capacity and poorer stability. In-depth understanding of the reactions between the cathode materials and the Na ions is critical to the design and improvement of the electrode. In a mechanistic study conducted by Ali et al., they elucidated the reaction mechanism between V_2O_5 with the Na ions [7]. The Na ions inserted into the *ab* planes of V_2O_5 with ordered chains of layers and induced a slight expansion along the *c* axis. When fully discharged, V_2O_5 transformed into crystalline NaV_2O_5 with trace amount of amorphous $Na_2V_2O_5$. Besides of tailoring the microstructure of the V_2O_5 electrodes, carbon modification is also effective in improving the rate performance and cycling stability. For example, by forming the nanosized V_2O_5/C composites, electrode stability can be significantly improved and a specific capacity of 255 mAh g^{-1} can be retained after 30 cycles. Modification of the nanosized V_2O_5 with the carbon layer can increase the conductivity and facilitate the penetration of electrolytes. Similarly, the carbon modification strategy was also applied to improve the performance of the NaV_6O_{15} cathode for Na storage [28].

3.2.2 Manganese Dioxides

Manganese dioxide (MnO_2) has many different types of polymorphs, all of which have large open tunnels that can potentially accommodate guest cations like Na^+. Besides, MnO_2 electrodes are advantageous in terms of the structural diversity, low toxicity, and natural abundance [29]. Both α-MnO_2 (Figure 3.3a) and β-MnO_2 (Figure 3.3b) exhibited reversible Na-ion storage capability as cathode materials in SIBs [30]. In particular, the compact tunnels in β-MnO_2 along the [001] direction is favorable for the Na insertion and extraction and can produce higher Na storage capacity (Figure 3.3c). The sodium-ion capacitor with the layered MnO_2/CNTs composite as the cathode delivered a high specific capacitance (322.5 $F\,g^{-1}$ at 0.5 $A\,g^{-1}$) and an excellent cycle stability (5000 cycles with capacitance retention of approximately 90%) [31]. Besides, amorphous MnO_2 prepared by Zhou et al. delivered a specific capacity of 137 $mAh\,g^{-1}$ after 100 cycles at a current rate of 0.1 C [32]. However, the performance of the MnO_2 cathode in LIBs or SIBs is hampered by their intrinsically low electrical conductivity and lack of ion insertion channels, which causes rapid capacity fading and poor cycling stability [33].

Constructing MnO_2 cathode with adjustable channels is highly desirable to optimize their performance for Li or Na storage. Lu et al. sandwiched different

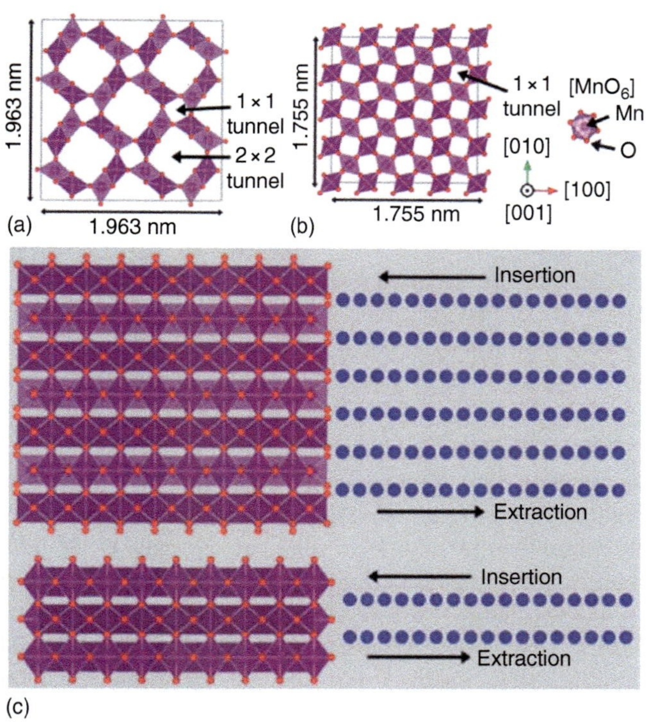

Figure 3.3 The crystal structure of (a) α-MnO_2 and (b) β-MnO_2, and (c) the insertion/extraction of Na ions in β-MnO_2 along different directions. Source: Su et al. [30]. Reproduced with permission, 2013, Royal Society of Chemistry.

types of cations (Li^+, Na^+, K^+, Co^{2+}, and Mg^{2+}) between the 2D MnO_2 layers assembled from the layer-by-layer stacking process [34]. The presence of exchangeable guest cations has endowed the adjustable interlayer spacing between the MnO_2 nanosheets. However, it was revealed that the cathode performance was not solely decided by the interlayer spacing but also was cation-dependent. For the monovalent-cation-pillared MnO_2 framework, their capacity can be enhanced by the enlarged layer spacing. However, the presence of the divalent cations twisted the MnO_2 bilayer and resulted in stronger Coulombic repulsion between these guest cations and Li^+ or Na^+, which led to a lower capacity instead. Similar studies showed that the presence of foreign cations, such as K^+, Ag^+, and Ba^{2+}, in the crystal lattice not only stabilizes the tunnel structure of MnO_2 but also tailors the tunnel size through the interaction between these ions and the host tunnel structure [35]. For example, the presence of potassium (K^+) resulted in a larger tunnel size, while silver (Ag^+) resulted in a smaller tunnel size due to the covalent role in the hollandite structure [35]. Lee et al. showed that more Ag^+ ions induced severe inhomogeneous structural evolution at lower sodiation levels [36].

3.3 Sodium-inserted Layered Metal Oxides

For the sodium-free metal oxides cathodes, it requires an irreversible insertion of Na ions at the initial cycles to couple with the sodium-free anodes (e.g. carbon), while for the sodium-inserted layered metal oxides, the reaction is reversible and stable even in the first two cycles. Study of sodium-containing layered metal oxides (Na_xMO_2, M = Ni, Co, Mn, Fe, Cr, V, etc.) was initiated by Delmas and Hagen muller in the early 1980s [37]. These metal oxides are typically stacked with edge-shared MO_6 octahedral layers and the Na ions. The stacking sequences of the O3, P3, P2, and O2 phases are presented in Figure 3.4 [38]. Based on the coordination environment of the Na^+ ions and the number of oxide layers stacked, they can be classified into two main groups: P2 and O3. In the O3-type electrodes, Na^+ ions have to overcome a relatively higher hopping energy to diffuse from one tetrahedral sites to another, while in the P2-type material Na^+ ions can directly migrate between adjacent prismatic sites, thus enabling fast transport of Na ions [39]. Thus, the P2-type materials have attracted more extensive research interests over the past years. The main concern of using the P2-type electrodes is that the Coulombic efficiency for the initial cycle is abnormally high and requires further elaboration of the fundamental mechanism. Besides, its Na-deficient nature requires additional steps such as the pre-sodiation and use of sacrificial additives to compensate for the insufficient Na ions on the cathode side [40]. NaN_3, Na_3P, $NaCrO_2$, Na_2NiO_2, and $Na_2C_2O_4$ have been implemented as effective sacrificial agents to serve as sodium reservoir for the cathode materials [41]. Guo's group modified the oxidation potential of the $Na_2C_2O_4$ with the conductive additives and used it as a cathode additive to construct the SIB full cell because of the high utilization efficiency (99%) of the high theoretical capacities (e.g. 400 mAh g^{-1}) [42]. The P2-$Na_{2/3}Ni_{1/3}Mn_{1/3}Ti_{1/3}O_2$ cathode with addition of $Na_2C_2O_4$ showed increased capacity retention from 63% to 85% after 200 cycles.

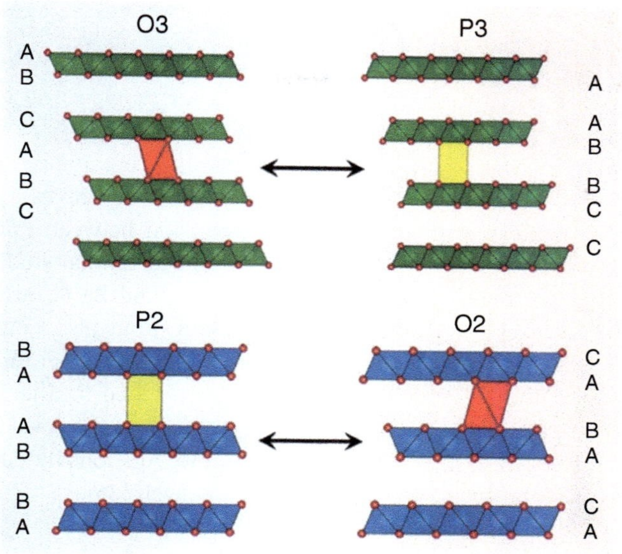

Figure 3.4 The stacking sequence for four typical layered transition metal oxides electrodes. Source: Sun et al. [38]. Reproduced with permission, 2019, Royal Society of Chemistry.

On the other hand, the inferior rate and cycling performance of the P2-type materials are associated with the phase transition from P2 to O2/OP4 and the Na^+–vacancy ordering induced by the bulky Na ions and the small difference in Fermi level of TMs. With substitution of Li^+ ions, it is found that the multiple-phase transition can be eliminated and the electrode can realize solid solution over a wide voltage range with the Na content of 0.85 [43]. The modified electrode could achieve a capacity of 79 mAh g^{-1} at 20 C and 85.4% of the capacity can be retained after cycling at 5 C for 500 cycles.

For the past decades, these layered metal oxides have been extensively investigated as they show promising electrochemical performance for SIBs, including high theoretical capacity, low cost, and facile synthesis. On the other hand, the performance of current SIBs is still insufficient to meet the requirements for real application in large-scale grid storage. One of the main problems is the irreversible and complex phase transition during the electrochemical reaction [44]. Normally, the P2 phase can deliver more stable cycling performance, while the O3 phase can achieve higher capacity due to higher content of the Na ions. The P2 phase can be well preserved during the extraction process of Na ions, while O3 phase always transform into the P3 phase when Na ion is deinserted. It is thus critical to clarify the variation in the electrochemical behavior between P2 and P3 structures. Guo et al. investigated the electrochemical reaction process of the P2 and P3 phases with similar compositions, i.e. P2-$Na_{0.62}Ti_{0.37}Cr_{0.63}O_2$, and P3-$Na_{0.63}Ti_{0.37}Cr_{0.63}O_2$ [45]. Their crystal structures are illustrated in Figure 3.5a,b. They studied the Na-ion conduction behavior in the two compounds and found that the Na-ion diffusion barrier of the

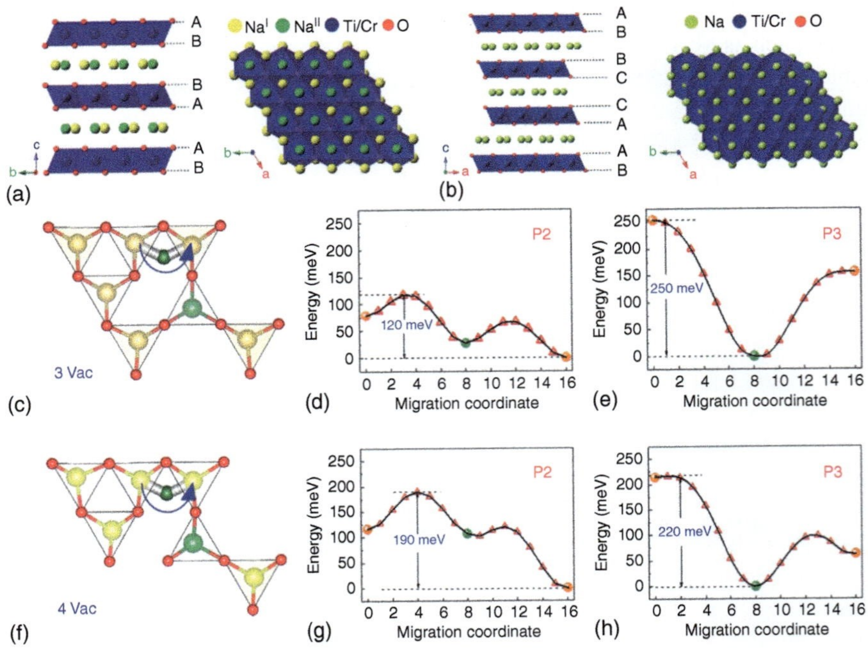

Figure 3.5 The structures of (a) P2-$Na_{0.62}Ti_{0.37}Cr_{0.63}O_2$ and (b) P3-$Na_{0.63}Ti_{0.37}Cr_{0.63}O_2$. (c) The Na-ion diffusion path with three vacancies, and the corresponding Na-ion diffusion barrier for the (d) P2 and (e) P3 structures. (f) The Na-ion diffusion path with four vacancies, and the corresponding Na-ion diffusion barrier for the (g) P2 and (h) P3 structures. Source: Guo et al. [45]. Licensed under CC BY 3.0.

P2 phase was higher than that in the P3 phase if only single vacancy was presented (Figure 3.5c–e); if multiple vacancies were presented, the Na diffusion barrier was largely decreased and even lower than that of the P3 phase (Figure 3.5f–h). Their finding indicates that the Na-ion diffusion kinetics in this type of layered oxide electrodes is mainly determined by the structure instead of the metal compositions.

Another issue regarding these layered metal oxides is their hygroscopic nature which makes them highly unstable upon exposure to air and lead to fast capacity decay of the cells [46]. The electrode structure will be varied due to the reaction with H_2O and the exchange between H^+ and Na^+ will result in weakened shielding effect and increased repulsion between the interlayers [47]. As a result, the interlayer spacing will be increased and results in distortion or collapse of the crystal structure. On the other hand, the layered metal oxides also take up CO_2 in the air and form electrochemically inactive and insulating species, such as NaOH and Na_2CO_3, on the surface. Zuo et al. carried out an in-depth study of the presence of CO_2 and their impact on the structural transformation of layered metal oxides upon exposure to moisture. It was verified that C atoms were mostly on the surface rather than in the bulk [48]. But CO_2 increased the acidity of the oxides and accelerated the Na^+ loss as well as the metal corrosion. Moreover, the Jahn–Teller effects associated with the Mn^{3+} and Fe^{3+} migration also hampers the stability and capacity of the layered metal oxides [49]. Though, it is predicted theoretically that Na-rich layered metal oxides, such as Na_2RhO_3 and Na_2PdO_3, could potentially be stable and serve as new stable high-energy density electrode materials. The preferable metals are expensive and their performance is still awaiting experimental verification [50]. As for the above reasons, it is critical to suppress the complex phase transitions and construct more stable and robust layered metal oxides to improve their electrochemical performance. Zhou's group reported $Na_3Ni_2RuO_6$ with ordered arrangement in the honeycomb superlattice from a solid-state reaction [51]. As compared to the disordered electrodes, it showed an ultralong voltage platform and stable cycling performance. This was attributed to the coexistence of O3 and P3 phase in the firmer superlattice-stabilized structure. The idea to stabilize the layered metal oxides through constructing structures with superlattice is also supported by other researchers [52]. However, as revealed by Zuo et al. [48], to improve the stability of metal oxides, it is required that the Na content should be controlled above the critical value, e.g. 0.38–0.44 for $Na_xNi_{0.33}Mn_{2/3}O_2$ electrode. Besides, other than superlattice stabilization, they tend to believe that the key parameter that determines the stability of TM oxides is the redox couples, and the electrodes with redox couples of $Ni^{2+/4+}$, $Cu^{2+/3+}$, and $O^{2-/n-}$ are regarded as more stable than those with $V^{3+/4+}$, $Mn^{3+/4+}$, $Co^{3+/4+}$, etc. Furthermore, calcinating the electrode at higher temperatures to achieve better crystallinity is also helpful to improve their air stability.

3.3.1 $NaFeO_2$

$NaFeO_2$ has been investigated as a cathode material for SIBs since 1985 [53]. It has two polymorphs, i.e. α- and β-$NaFeO_2$. No electrochemical activity has been reported for the orthorhombic β-$NaFeO_2$ so far. The rhombohedral α-$NaFeO_2$ shows a typical O3-type layered structure with a space group of $R\bar{3}m$. Kikkawa et al. firstly revealed that Na ions can be chemically deintercalated from α-$NaFeO_2$ [53]. Besides, α-$NaFeO_2$ presents a high thermal stability. The thermal stability of α-$NaFeO_2$ in SIBs is similar or even better than that in LIBs with conventional $LiCoO_2$ as the cathode [54], which potentially can solve the safety concerns in large-scale applications such as EVs. The reversible Na storage capability of α-$NaFeO_2$ is based on the Fe^{3+}/Fe^{4+} couple, which can deliver a theoretical capacity as high as 242 mAh g^{-1}. However, the preliminary electrochemical tests of α-$NaFeO_2$ only revealed a capacity of 85 mAh g^{-1} with a potential plateau at 3.4 V. When a higher charging voltage was applied, because of the Jahn–Teller distortion and polarization, the electrode structure was degraded, and the Na-ion diffusion pathways were partly blocked which resulted in the deteriorated rate performance. Upon extraction of Na ions, vacancies were generated at the face-shared tetrahedral sites. The neighboring Fe^{3+} migrated into these tetrahedral sites and blocked the transportation pathway of the Na^+ ions. Though it was reported that the Na-ion migration barrier at the $Fe^{4+}O_6$ octahedral site is much lower, the migration of Fe species is also facilitated when Fe cluster is formed. Other than these issues, Na^+ ion in the α-$NaFeO_2$ is prone to exchange with H^+ and forms the FeOOH and

NaOH when exposed to moisture or air. As mentioned earlier, the uptake of CO_2 also results in the formation of the inactive species like Na_2CO_3 and $NaHCO_3$ on the surface. These insulating and electrochemically inactive species hamper the cycling stability of the electrodes. Therefore, an inert environment is required for the storage and preparation of α-$NaFeO_2$.

When the electrode is charged to high voltages, the performance of the O3-type $NaFeO_2$ is compromised by the migration of Fe ions and the cycling performance is hindered by the as-resulted particle cracks. Partial substitution with Ru not only suppressed the migration of Fe ions but also reduces the particle cracks, thereby has led to an improved electrochemical performance of the O3-type $NaFeO_2$ (e.g. 80% capacity retention after 100 cycles) [55]. So far, various strategies have been developed to suppress the undesirable phase transition and improve the cycling capability of α-$NaFeO_2$, of which reducing the particle size and surface coating are two representative ones. Yabuuchi et al. reported the cutoff voltage dependency on the cathode performance of $Na_{1-x}FeO_2$ in SIBs [56]. It is only in the range of $x = 0$–0.45 that the electrode can show acceptable cycling performance. When x was above 0.5, the electrode underwent irreversible structural changes and drastic capacity decay. Lee et al. systematically studied the structural and chemical quasi-reversibility of the α-$NaFeO_2$ cathode material [57]. They examined the chemical stability of Fe^{4+} by the ex-situ Mössbauer spectroscopy and found that the concentration of Fe^{4+} was reduced in the electrode stored at the charged state due to the spontaneous reduction of partial Fe^{4+} into Fe^{3+} (Figure 3.6a).

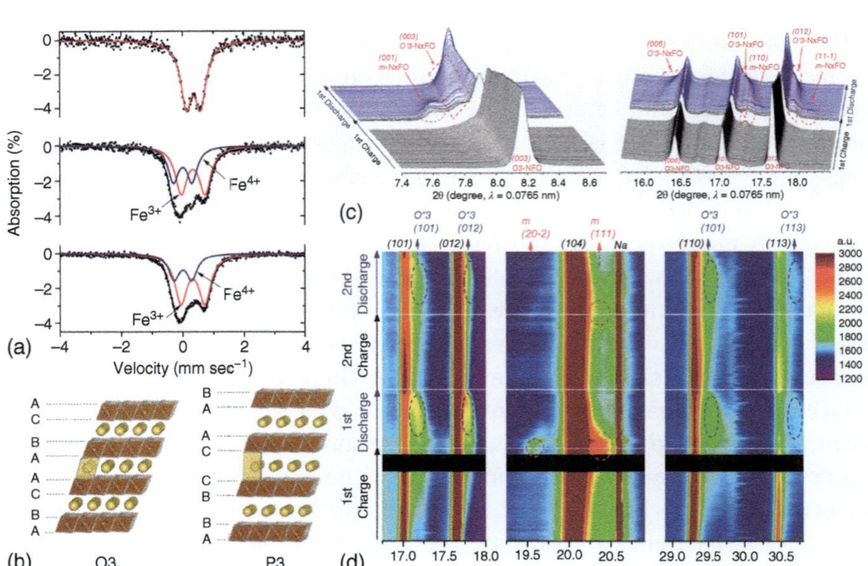

Figure 3.6 (a) Ex situ Mössbauer spectra for $NaFeO_2$ electrodes. (b) Illustration of the O3 and P3 structures that can be reversibly changed. 3D (c) and 2D (d) view of the SXRD pattern of the cathode during the initial cycles. Source: Lee et al. [57]/with permission from American Chemical Society.

In contrast to the general O3-type layered metal oxides, which experiences the phase transition into the P3-type structure by gliding the layers instead of breaking the M—O bonds (Figure 3.6b), desodiation of α-$NaFeO_2$ results in the formation of coherently nucleated O″3-$Na_{1-x}FO$ and m-$Na_{1-x}FO$ phases. The nonequilibrium phase transition was confirmed by the in situ synchrotron X-ray diffraction (SXRD) analysis (Figure 3.6c,d). This unique phase transition is attributed to the presence of tetrahedral Fe^{3+} that resists the layer gliding and blocks the Na-ion diffusion pathway. This results in the concentration gradient of the Na^+ ions and negatively affects the cycling stability of the α-$NaFeO_2$ cathode in SIBs. Silván et al. revealed the mechanisms of such distinctive phase transition behavior of α-$NaFeO_2$ as compared to other O3-type layered metal oxides [58]. They pointed out that Fe migration is responsible for the phase transition. ~40% of the initial Fe migrated into the Na layers when charging the electrode to 4 V and suppressed the O3 to P′3 phase transition. The migration of Fe ions, on the other hand, blocked the diffusion pathway for the Na^+ ions and resulted in the poor performance under high voltage. This highlights the importance of suppressing the unfavorable metal migration as an important measure to optimize the performance of α-$NaFeO_2$ cathode materials.

At the meantime, the highly oxidative Fe^{4+} also results in the decomposition of the electrolyte which not only caused inferior Coulombic efficiency but also increased the impedance by building up the solid–electrolyte interface (SEI) on the electrode surface [56]. Li et al. observed the migration of Fe ions at the atomic scale and before 3.5 V, and the content of Fe^{4+} kept increasing owing to the oxidation of Fe^{3+} ions; however, at higher voltages (~4.4 V), the Fe^{4+} ions decreased [59]. Thus, they proposed that both Fe and O participated in the charge compensation. Susanto et al. also tried to elucidate the origin for the instability of the O3-type $NaFeO_2$ under high voltage [60]. They carried out investigation using in situ X-ray absorption and found that it was the oxygen redox activity instead of Fe^{3+}/Fe^{4+} redox couple that was responsible for the charge compensation. As a result, the oxygen released facilitated the formation of Fe_3O_4 at higher voltage which inhibited later Na insertion and caused the irreversibility of the material.

An effective strategy to improve the electrochemical properties of α-$NaFeO_2$ is to partially replace Fe with other TMs [61]. The electronic structure of Na_xFeO_2 can be systematically tuned without breaking the layered structure when substituting Fe with metals over a wide range of the composition and concentrations. In particular, partial replacement of Fe with other TMs could enhance the migration of electrons and holes, stabilize the structure of the electrode, and smoothen the charge and discharge curves [61d, 62]. Thorne et al. found that Ni^{3+} could suppress the Fe migration more efficiently than the Co^{3+} ions [63]. $NaFeO_2$–$NaCoO_2$ and $NaFeO_2$–$NaNi_{1/2}Mn_{1/2}O_2$ solid solutions with improved electrochemical properties have been proposed [61c, 64]. Compared with $Na_{1-x}FeO_2$, the high-spin Fe^{4+}, which is the main cause for the Jahn–Teller distortion, was significantly decreased owing to the presence of the $Ni^{4+/3+}$ redox couple, thereby suppressing the irreversible migration of Fe^{4+} during the cycling process [61d]. The Ni-doped O3-$NaFe_{0.3}Ni_{0.7}O_2$ delivered an initial discharge capacity of 135 mAh g^{-1} with a high initial Coulombic efficiency of 93% and showed enhanced cycling stability.

Kubota et al. investigated the structural evolution of the $NaFeO_2$–$NaCoO_2$ solid solutions [65]. The $NaFe_{1/2}Co_{1/2}O_2$ cathode delivered a high initial discharge capacity of 160 mAh g^{-1} with good cycling and rate performance. The excellent performance was attributed to the phase transition from O3- to P3-layered structure over a wide capacity range. The partial substitution of Co stabilized the crystal structure upon extraction of Na^+ ions which resulted in more superior rate performance as compared to the $NaFeO_2$ electrodes. Besides, as revealed from the ex situ SXRD analysis and the Monte-Carlo simulation, Na^+–vacancy ordering was observed in the interslab space but not the ordering of Fe or Co in the TM slabs. The P2-type $Na_xFe_{1/2}Mn_{1/2}O_2$ cathode with high purity and good crystallinity was achieved by an eutectic synthesis [66]. By coupling with a pre-sodiated hard carbon anode, the compound delivered a reversible capacity of 180 mAh g^{-1}.

3.3.2 Na_xCoO_2

Na_xCoO_2 are well studied as a promising cathode for SIBs because of the success of its analogy $LiCoO_2$ in LIBs [67]. In general, Na_xCoO_2 can show different crystal forms with variation in the Na content and synthetic environment [68]. The O3-Na_xCoO_2 tends to form at temperatures lower than 650 °C, and in the O3-type structure, Na ions occupy the octahedral holes between the CoO_6 slabs. The structure will transition from O3 to O′3, to P′3, and then to P3 during the desodiation process. The gliding of the CoO_6 slabs results in poor structural stability and inferior conductivity of O3-Na_xCoO_2 as compared to the P2 phase. Most P2-Na_xCoO_2 is produced at temperatures above 650 °C, and in the P2-Na_xCoO_2, Na ions locate in face-shared and edge-shared tetragonal prisms. The P2 phase is more stable and can realize reversible Na insertion/deinsertion without causing the slippage of CoO_6 slabs. P2-type Na_xCoO_2 showed charge–discharge curves complexed with several voltage plateaus caused by Na^+–vacancy ordering for energy minimization. Formation of such domains will result in abrupt change of the crystal structure and hence the instability of the electrode performance. Only limited number of Na^+ ions can be extracted from the octahedral sites without hampering the host structure during the desodiation process, which resulted in a much lower capacity than the theoretical value expected from $NaCoO_2$, i.e. ~235 mAh g^{-1}. Three-component cathode system provides alternative solution to address the undesired Na ordering. Through doping or substitution of TMs, Na^+–vacancy ordering can be controlled, so as the diffusivity and rate capability of the cathode material [69]. Substituting Co with Mn stabilized the crystal structure and eliminated the voltage plateaus in $NaCo_{2/3}Mn_{1/3}O_2$ [70]. The Jahn–Teller distortion can be effectively avoided by the presence of Mn^{4+} instead of Mn^{3+}. But the presence of Mn ions also resulted in higher migration barrier for the Na ions. Theoretical calculation has indicated that addition of Ni dopant could lower the migration barrier. Besides, the introduction of Ni with higher redox potential can enlarge the energy density. As shown in the electrochemical test, the Ni-doped cathode material showed remarkable stability with only 3% loss after 100 cycles. $Na_{0.75}Ni_{0.22}Co_{0.78}O_2$ delivered a specific capacity of 134 mAh g^{-1} within the voltage range between 1.5 and 4.2 V [68b]. Kang et al.

managed to control the Na^+–vacancy ordering in the $Na_{0.67}Co_{1-x}Ti_xO_2$ through Ti doping [71]. As a result, most of the complicated plateaus presented in the pristine Na_xCoO_2 disappeared and the capacity retention was significantly improved, e.g. the capacity maintained at 100 mAh g^{-1} after 150 cycles. Besides, the cutoff voltage can be elevated to 4.5 V due to stronger bonding between oxygen and Ti.

Generally, sodium extraction beyond $x = 0.25$ results in the collapse of the Na layers due to gliding of the MO_2 slab. Therefore, it requires to develop P2-Na_xCoO_2 with more robust crystal structure. Doping of the inactive materials could stabilize the crystal structure as well as maintain stable cycling performance. For example, it was found that presence of Ca^{2+} could suppress the Na ordering effectively and improve the performance of P2-type materials [72]. Since Ca^{2+} is much larger than Co^{3+} and Co^{4+}, it will not substitute them in the crystal structure, but the presence of Ca^{2+} between the CoO_6 slabs could enhance the electrostatic interaction and disrupt the Na^+–vacancy ordering. As a result, the cyclability of electrode materials can be improved. At the meantime, the occupation of large Ca^{2+} will block the transportation of Na ions and eventually compromises the performance of electrode materials. More effective strategies to suppress the Na^+–vacancy ordering as well as more efficient cation doping are still highly desirable in the future exploration.

3.3.3 Na_xMnO_2

The Na_xMnO_2 was firstly reported as a cathode material for SIBs in the 1980s [73]. They were regarded as a more favorable cathode for practical applications as compared to $NaCoO_2$ with a high theoretical capacity of 242 mAh g^{-1} originated from the highly active Mn^{4+}/Mn^{3+} couple. Besides, Mn shows more natural abundance and environmental friendliness than the scare and toxic metal Co. $NaMnO_2$ exists in two polymorphic forms, denoted as α and β, both of which show layered structure composed of alternating MnO_2 and Na sheets [74]. The monoclinic α-$NaMnO_2$ shows flat layers and belongs to the $C2/m$ space group. It is low energy and thermodynamically stable under ambient conditions. The orthorhombic β-$NaMnO_2$ shows corrugated (zigzag-like) layers in the *Pmnm* space group. Mendiboure et al. tested the electrochemical performance of α- and β-$NaMnO_2$ in the 1980s, and only 0.22 and 0.15 Na can be reversibly extracted, respectively [73b]. Ma et al. revisited the performance of α-$NaMnO_2$ and deintercalated 0.85 Na from $NaMnO_2$, corresponding to an initial discharge capacity of 185 mAh g^{-1} [75]. Clément et al. studied the structural changes of the β-$NaMnO_2$ during the charge and discharge process. The results revealed that it consisted of both α-$NaMnO_2$ and β-$NaMnO_2$ domains. At low Na content, the long-range structure collapsed and compromised the reversible Na extraction and insertion processes, i.e. only 70% of the initial reversible capacity is retained after 100 cycles. The solid-state ^{23}Na nuclear magnetic resonance (NMR) and first-principles density functional theory (DFT) study were then carried out to study the evolution of β-$NaMnO_2$ structure during the cycling process [76]. The calculations showed that the formation energy of the α and β-$NaMnO_2$ was quite close and planar defects could form between the nanodomains of the two polymorphs. The Na ions would be firstly extracted from the α-domains, then the β-domains, and finally the twin boundaries. Upon cycling, the Na content in the α-$NaMnO_2$ domain increased and led to the phase separation from the dominant β-$NaMnO_2$. As reported by Jiang et al., the β-$NaMnO_2$ could be stabilized by the addition of a small portion of Cu^{2+} ions [77]. β-$NaCu_{0.1}Mn_{0.9}O_2$ was synthesized from sintering the precursors in air and allowed for cooling to room temperature naturally. The Cu-doped sample showed higher initial Coulombic efficiency and can be reversibly cycled at a high voltage of 3.5 V. After cycling for 100 times, it showed better capacity retention as compared to the undoped samples. This is attributed to the absence of local structure rearrangements in the planar defects that normally occurred in the β-$NaMnO_2$.

Besides of α- and β-$NaMnO_2$, O3 and P2-type Na_xMnO_2 have also been investigated [52b, 78]. The O3-Na_xMnO_2 shows multiple-phase transitions upon charge which vanishes during discharge. Chen et al. identified the "superplanes" in the superstructure ordering under certain Na content [79]. The superplanes contain rich Mn^{3+} and Na, dividing the Mn^{4+}-rich and Na-deficient blocks periodically. The pillaring effects of Na ions in the superplanes allow for larger interlayer distances of the O1 phases and help the diffusion of Na ions, creating an asymmetric phase evolution with strong-ordering-free discharge of high capacity. As compared to the O3 phases, the P2 phases show more robust crystal structure and open prismatic pathway for Na-ion transportation [80]. However, during the electrochemical reaction, the complex phase transitions along with the distortions of the parent Na_xMnO_2 resulted in rapid decay of the electrode performance, especially at high voltages [61a, 78a, 81]. For example, the $Na_{0.6}MnO_2$ cathode delivered a specific capacity of 150 mAh g^{-1} in the first cycle, but in the following cycles the capacity quickly decreased due to the collapse of the host structure during the intercalation–deintercalation process of Na ions. $Na_{0.67}MnO_2$ showed a high initial discharge capacity, while due to structural transitions and the presence of the Jahn–Teller active Mn^{3+}, the cycling performance was unsatisfying [73b, 78a]. The Jahn–Teller distortion with the Na ordering induces the large volume change with more Na content in the crystal. Besides, the disproportion of Mn^{3+} to Mn^{2+} and Mn^{4+} can result in dissolution of active materials in the electrolyte. In order to suppress the complex phase transitions and optimize the stability, the structure of the Na_xMnO_2 compounds were usually carefully tuned by either changing the stoichiometry of the Na and Mn ions or adding hetero-metal species to substitute Mn [81a, 82]. On the other hand, manipulating the crystal structure of the layered metal oxides, e.g. the defects and vacancies, is an effective strategy to achieve higher specific capacities. For example, high-temperature quenching method could eliminate the Mn vacancies and gain more capacity, but this process also generates more Mn^{3+} ions responsible for the Jahn–Teller effect, hampering the long-term stability of the material [83]. Liu et al. thus developed a liquid N_2 quenching strategy that can fully eliminate the Mn vacancies [84]. At the same time, they also doped a small amount of Al and Fe into the compound to tailor the valence of Mn species and suppress the multiple-phase transitions during the charge and discharge process. The Al—O bond enlarged the spacing between the Na layers and stabilized the structure, while the Fe dopants disrupted the cation ordering and enhanced the cycling stability. The $Na_{0.67}Al_{0.1}Fe_{0.05}Mn_{0.9}O_2$ compound delivered a high capacity of 202 mAh g^{-1} in the initial cycle with an outstanding rate performance and excellent capacity retention. But it is worth to mention that the Jahn–Teller distortion is not always disadvantageous to the electrochemical performance of the material. For example, Komaba et al. rationally controlled the stoichiometry ratio of the sodium and manganese in the $Na_{0.67}MnO_2$ and achieved two polymorphs with similar composition but differently distorted structure [83]. By quenching from over 1000 °C, the pure-phase orthorhombic P′2 Na_xMnO_2 was synthesized and denoted as *o*-NMO. It showed a remarkable cooperative Jahn–Teller distortion of six-coordinate Mn^{III} but no Mn vacancy. On the other hand, the undistorted P2-Na_xMnO_2 containing a Mn vacancy (denoted as *h*-NMO) was also investigated. Electrochemical tests revealed that the *o*- and *h*-NMO delivered the first charge capacity of 142 and 94 mAh g^{-1}, corresponding to 0.54 and 0.37 Na, respectively. During cycling, the *o*-NMO displayed a reversible capacity of 216 mAh g^{-1} with an average working potential of 2.74 V. While the *h*-NMO had a capacity of 198 mAh g^{-1} with smoother curves. The cycling stability of the *o*-NMO was also better than the *h*-NMO, which might be resulted from the presence of the 10% Mn defects that induced severe charge disorder in the $Mn_{0.9}O_2$ slab and hampers the diffusivity of the Na ions. Such localized distortion with the Mn defects and the Jahn–Teller Mn^{3+} ions led to the irreversible structural changes and fast capacity decay. In a later work, they further replaced Mn with different metal elements and investigated their electrochemical properties [85]. All the substituted samples showed reversible capacities higher than 200 mAh g^{-1}. Besides, the charge–discharge curves for these doped samples were smoother as compared to the undoped one. This is because the metal substitution disrupted the long-range ordering of both Na vacancy and $Mn^{3+/4+}$ charge, thereby suppressing the phase transitions in the charge–discharge process. The sample doped by Ti and Co with more severe distortion in structure showed higher structural stability when Na was extracted/inserted. The Mg-, Ni-, and Zn-doped samples which were less distorted suffered from the local strain between the distorted and undistorted MO_2 octahedra. Only Cu-doped sample showed both exceptional rate and cycling stability, which might be originated from the high cooperative Jahn–Teller distortion induced by Cu substitution. The positive contribution of Cu^{2+} as a doping cation has also been reported elsewhere. In another work, Li et al. found that the Cu substitution could improve the air stability of the P2-$Na_{2/3}MnO_2$ compound [86]. Wang et al. further demonstrated that the substitution of Cu^{2+} ions could also modulate the water sensitivity of such compound [87]. When Na ions were extracted from the $Na_{2/3}MnO_2$ electrode, the Na^+/vacancies were generated and made it easier for the water to insert into the sodium layer. The water insertion rate was related with the arrangement of the Na^+–vacancy. But no hydrate phase appeared in the Cu-substituted electrode, which affected the arrangement of the Na^+–vacancy ordering and suppressed the water intercalation. Besides, the effect of Jahn–Teller distortion can be alleviated when Mn^{3+} is replaced with Cu^{2+} and Mn^{4+} in the doped sample.

Guo's group reported that partial substitution of Ti for Mn in the compound of $Na_\delta Ni_{1/3}Mn_{2/3}O_2$ could refrain electron delocalization and suppress the charge and Na^+–vacancy ordering, which increased the mobility of Na ions [88]. In the

Ti-substituted tunnel-type $Na_{0.66}[Mn_{0.66}Ti_{0.34}]O_2$, the Na at the Na(1) sites was fully occupied and most of the Mn sites were co-occupied by the Mn^{3+}/Mn^{4+} and Ti^{4+} in a disordered manner [89]. The charge ordering can thus be avoided in such type of compounds [90].

Ni^{3+} was also added to partly replace the Mn^{3+} in the P2-type $Na_{2/3}MnO_2$ [91]. The as-synthesized compound exhibited improved capacity retention and thermal properties. As revealed by the in situ operando SXRD analysis, the stability of the material was improved with the partial substitution of the Ni^{3+} due to suppressed Jahn–Teller distortion. But bi-phasic reaction and the transformation from P2 to O2 phase were observed when charging the electrode to 4.3 V. Figure 3.7a shows the typical charge and discharge curves of the $Na_\delta Ni_{1/3}Mn_{2/3}O_2$, and Figure 3.7b illustrates 2D Na ordering patterns for this compound. For $\delta = 2/3$, a zigzag pattern is formed by the Na_f sites. When δ is reduced to 1/2, the ordering converted into rows where Na orders alternate in one row of Na_f and two rows of Na_e in the plane. With further decrease in δ to 1/3, either Na_f or Na_e arranges in rows within a single layer.

On the other hand, Choi et al. found that the substitution of Fe^{3+} could suppress the Jahn–Teller distortion in the P′2-type $Na_{2/3}MnO_2$ [92]. Owing to the similar sizes of the Fe^{3+} ion and the Mn^{3+} ions, Fe^{3+} preferentially occupied the Mn^{3+} sites and formed the compound $Na_{2/3}[Fe_{0.22}Mn_{0.78}]O_2$. As a result, the oxygen–metal–oxygen (Mn^{3+}–O) length decreased and the multistep phase transition was also inhibited.

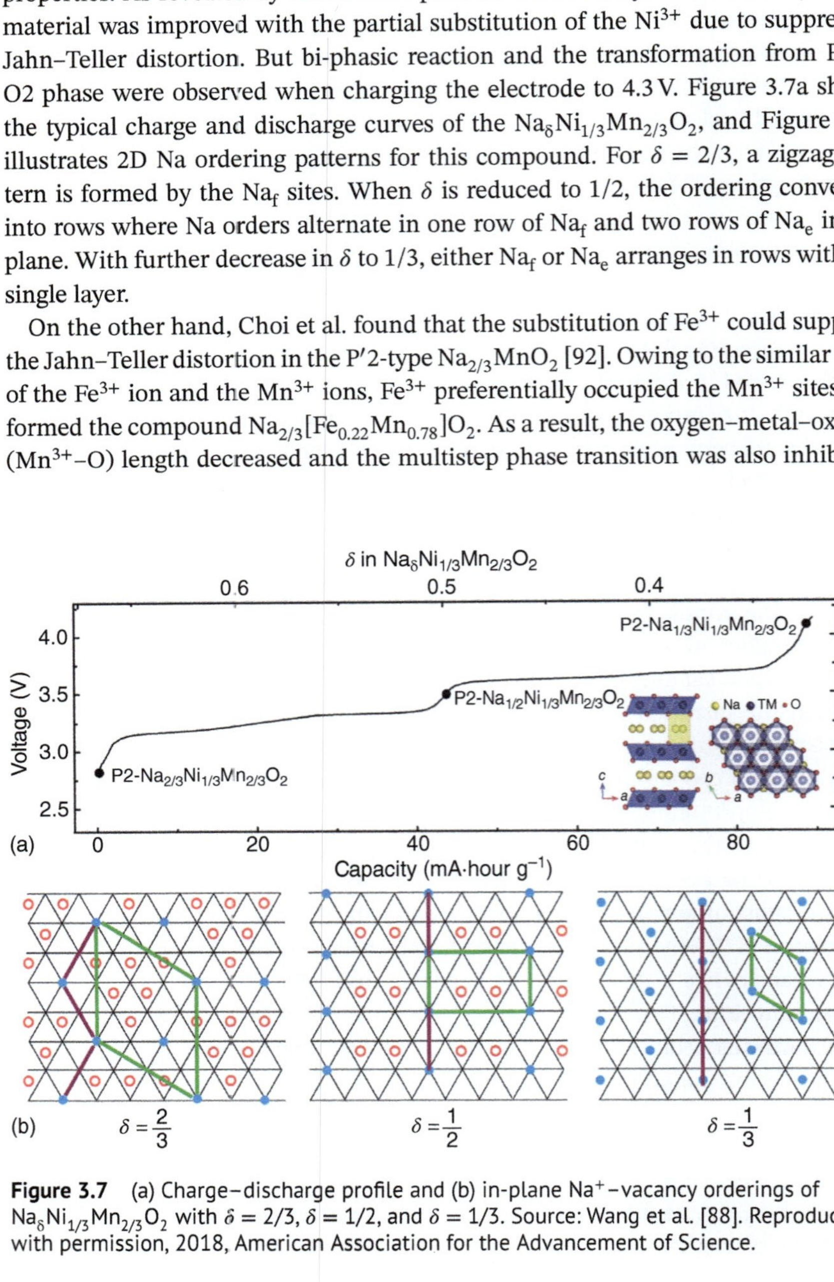

Figure 3.7 (a) Charge–discharge profile and (b) in-plane Na^+–vacancy orderings of $Na_\delta Ni_{1/3}Mn_{2/3}O_2$ with $\delta = 2/3$, $\delta = 1/2$, and $\delta = 1/3$. Source: Wang et al. [88]. Reproduced with permission, 2018, American Association for the Advancement of Science.

With the additional redox couple of Fe^{3+}/Fe^{4+}, the specific capacity, the voltage, along with the cycling stability of the electrodes improved significantly.

Findings show that introduction of Li and Mg can oxidize part of Mn^{3+} into Mn^{4+} and reduce the Jahn–Teller effect. As a result, the charge and discharge curves will be smoothed with better capacity retention [81b, 93]. Sehrawat et al. investigated the phase transitions of P2 $Na_{0.7}MnO_2$ modified with K, Rb, and Cs [94]. The K-modified $Na_{0.7}MnO_2$ formed two P2's and one K-rich phase. One P2 phase was responsible for the solid solution reaction and the other P2 phase was stabilized at the initial stage and remains unchanged in the subsequent charge and discharge cycles (Figure 3.8a). The newly formed P2 phase tended to fuse with the major P2 phase in the Rb-modified P2 $Na_{0.9}MnO_2$ (Figure 3.8b). The P2 phase emerged from the Cs-modified P2 $Na_{0.7}MnO_2$ electrode was highly disordered and multiple phases appeared near the discharge state. On the other hand [95], the Rb-modified $Na_{0.7}Mn_{0.8}Mg_{0.2}O_2$ evolved as a solid solution with multiple-phase region observed (Figure 3.8c). For the Cs-modified $Na_{0.7}Mn_{0.8}Mg_{0.2}O_2$, it also formed a new P2 phase with preexisted P2 phase near the end of the discharge state (Figure 3.8d).

Besides of cation doping, Chae et al. doped both fluorine (F) and aluminum (Al) in the $Na_{0.44}MnO_2$ and achieved a new compound with the chemical composition of $Na_{0.46}Mn_{0.93}Al_{0.07}O_{1.79}F_{0.21}$ (AFMO) [96]. $Na_{0.44}MnO_2$ is orthorhombic with Pbam symmetry. It consists of double and triple linear chains with edge-shared MnO_6 octahedra and single chains of edge-shared MnO_5 square pyramids. For AFMO, it shows a layered structure built with edge-shared MnO_6 octahedra, and Na ions

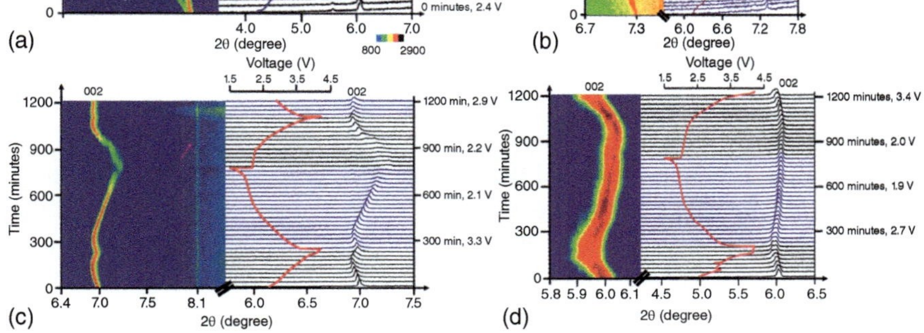

Figure 3.8 In situ synchrotron XRD data show the structural evolution for $K_{0.216}Mn_{0.985}O_2$ (a) and Rb-modified $Na_{0.9}MnO_2$ (b). Source: Sehrawat et al. [94]. Reproduced with permission. 2020, American Chemical Society. (c) Cs-modified $Na_{0.7}Mn_{0.8}Mg_{0.2}O_2$. (d) Rb-modified $Na_{0.7}Mn_{0.8}Mg_{0.2}O_2$. Source: Sehrawat et al. [95]. Reproduced with permission, 2019, American Chemical Society.

occupy two sites active for Na diffusion: Na(1) locates in the 2d crystallographic site, and Na(2) in the 2b site. F/O are positioned at the 4f sites and bonded with the Mn and Al mixed 2a sites. Bond valance energy landscape calculations were applied to study the barrier energies for possible diffusion pathways. With the 2D diffusion path between TM layers, the AFNMO showed a lower energy barrier as compared to the $Na_{0.44}MnO_2$ structure, which is mainly contributed from the fluorine doping. Besides, by substituting the M/O with the Al/F, the phase transition from P2 to O3 at higher voltages was also suppressed. As demonstrated by the mechanistic study, initially a two-phase reaction was observed without any structural transformation. P2 ordering was observed with internal Na diffusion pathways, followed by the subsequent solid solution reactions. The retention of the highly active P2-type ordering in the electrode upon cycling is favorable to the stability and capacity retention of the electrode.

Other than cation or anion doping, surface coating is believed to be able to alleviate the undesirable side reactions and prevent particle exfoliation during the cycling process [97]. Al_2O_3 coating has been applied to the surface of doped and undoped $Na_{2/3}MnO_2$ and enhanced the mechanical stability and suppressed the particle exfoliation [98]. But the nonconductive Al_2O_3 posed increased kinetic barrier and hindered the migration of the Na ions. To address this problem, $Na_2Ti_3O_7$ was applied as the coating material [99]. It is consisted of zigzagging layers of titanium oxygen octahedral with Na ions in the interlayer. The diffusion of Na ions along with the TiO_6 layer can be realized with a relatively lower activation energy based on the vacancy-hopping mechanism. Some Ti^{4+} could be introduced into the bulk structure accompanied by the $Na_2Ti_3O_7$ coating. The synergistic presence of the $Na_2Ti_3O_7$ coating and the Ti^{4+} dopants not only suppressed the Na^+–vacancy ordering but also suppressed the P2–O2 phase transition. As a result, the $Na_2Ti_3O_7$-coated $Na_{2/3}Ni_{1/3}Mn_{2/3}O_2$ showed improved cycle stability and rate capability. As suggested by You et al., surface modification with ZrO_2 could improve the stability of oxide cathodes in the air [100]. The degradation mechanism of O3-type $Na[Ni_xCo_yMn_z]O_2$ was explored by Sun et al. [101] The degradation mechanism was related to the mechanical stress imposed by the phase transitions with the side reaction induced by surface degradations. The electrolyte penetration further accelerated the capacity fade. They improved the stability of $Na[Ni_xCo_yMn_z]O_2$ by designing a nanorod gradient $Na[Ni_{0.65}Co_{0.08}Mn_{0.27}]O_2$ with AlF_3 coating. The AlF_3 coating layer prevented the electrolyte infiltration into the bulk particle and circumvented the side reaction/decomposition of electrolyte with commonly applied Al_2O_3 coating. As a result, the electrode delivered a discharge capacity of 160 mAh g^{-1} at 0.1 C with good capacity retention. Chen's group optimized the coating process and reported a molecular layer deposition (MLD) technology to create an uniform and ultrathin coating layer onto the $Na_{0.66}Mn_{0.9}Mg_{0.1}O_2$ cathode [102]. The alucone-coated cathode exhibited stabilized structure even under a high cutoff voltage of 4.45 V. The cycling stability and rate performance of P2-$Na_{0.66}(Mn_{0.54}Co_{0.13}Ni_{0.13})O_2$ were improved with presence of ultrathin metal oxides (e.g. Al_2O_3, ZrO_2, and TiO_2) deposited by an atomic layer deposition technique on the surface [103].

3.3.4 Na_xNiO_2

Ni behaves like Co when serves as the redox center. But it is much cheaper, more environmentally friendly and natural abundance as compared to Co. Therefore, Na_xNiO_2 cathodes and their analogues are well studied as the Na insertion electrodes. However, it has been found that Ni^{2+}-based Na_2NiO_2 and Ni^{3+}-$NaNiO_2$ show poor stability in terms of both structure and chemical property. The $NaNiO_2$ can stably exist in two polymorphs: the monoclinic and the rhombohedral phases. The monoclinic phase is isostructural to that of $NaMnO_2$ and shows a layered structure in the space group of $C2/m$, in which the Ni-O layer is formed by the edge-shared and Jahn–Teller distorted NiO_6 octahedra [74, 104]. The Na^+ ions that located in between the NiO_2 slabs exhibit a distorted octahedral coordination by oxygen atoms. The rhombohedral phase forms at higher temperature with a similar structure with that of α-$NaFeO_2$ [105]. The electrochemical behavior of $NaNiO_2$ was first reported in 1982 [106]. Only 0.2 Na^+ can be extracted from the pristine material during the first charging process due to the multiple-phase transformation occurred at the voltage range of 1.8–3.4 V. Later on, 0.85 Na^+ can be extracted corresponding to the intercalation of 0.62 Na^+ during the first charge–discharge process at the voltage between 2.0 and 4.5 V [4]. X-ray diffraction (XRD) pattern for the $NaNiO_2$ electrode revealed a new phase at the highly charged state. It was later found from the in situ XRD that there were actually two new phases emerged at the end of discharge and at 3.38 V, respectively [107]. To better clarify the phase evolution process, Wang et al. tried to understand the mechanism of the dissymmetric phase transition and structure evolution of layered $NaNiO_2$ at the initial stage during the cycling process. They demonstrated that at the voltage below 3.0 V and above 4.0 V, the $NaNiO_2$ electrode underwent severe phase transition and deformation which resulted in the irreversible capacity loss during the initial cycle [108]. The sodium diffusion process and the phase transformation kinetics were investigated by the galvanostatic intermittent titration technique, and the sodium diffusion coefficient was three times higher for the electrode cycled at 2–3.8 V as compared to the one at 2–4.5 V. The phase transformation is illustrated in Figure 3.9.

The O3-type $NaTi_{0.5}Ni_{0.5}O_2$ reported by Yu et al. delivered a specific capacity of 121 $mAh\,g^{-1}$ at 0.2 C with an average operation potential of 3.1 V [109]. In general, the degradation of the layered metal oxides is often triggered by the gliding of the metal slabs and the air exposure degradation of the Na layers. Besides, the ordering behavior in the layered metal oxides also results in the poor structural stability and multiple-phase transitions during the reactions. The typical ordering includes the cation ordering, the Na–vacancy ordering, and the charge ordering. The cationic ordering can be achieved with large disparity in the M ionic radii and the M—O bond energies. Yang's group has identified the honeycomb-ordered O3-$Na_3M_2SbO_6$ (M = Ni, Cr, Cu, Co, etc) [110]. It resembles the structure of Na_xMO_2 but with an M^{2+}/Sb^{5+} cationic ordering or a superstructure within each slab, i.e. each SbO_6 octahedron is surrounded by six MO_6 octahedron and forms a honeycomb network. Such kind of structure allows the fast mobility of Na ions with facile changing in

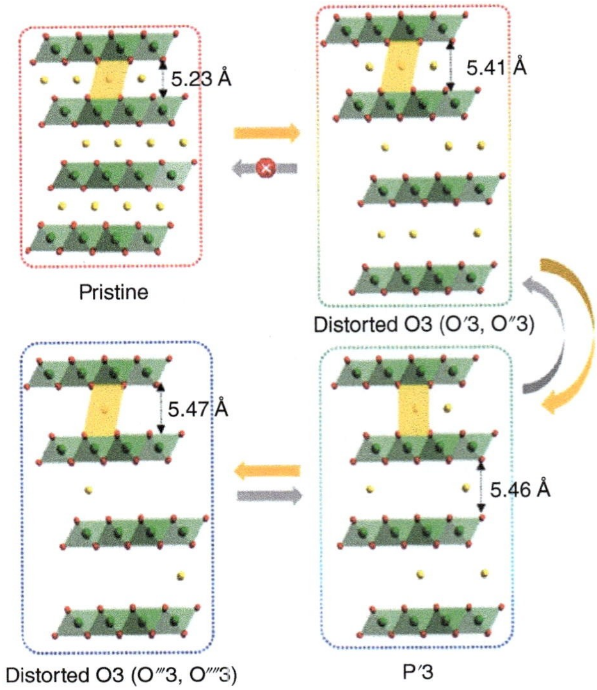

Figure 3.9 Structural evolution of $NaNiO_2$. Source: Wang et al. [108]. Reproduced with permission, 2017, Elsevier.

the oxidation state. In particular, the $Na_3Ni_2SbO_6$ has demonstrated a high capacity and superior stability upon cycling. They further investigated the electrochemical behavior of $Na_3Ni_2SbO_6$ as cathode for SIB. Despite the similar multiphase transitions as shown by the layered metal oxides, the reactions in $Na_3Ni_2SbO_6$ can be completed within a much narrow potential range, rendering an average working potential of 3.3 V. As confirmed by the XRD spectra, the Na-ion extraction/insertion took place via three-phase transformations between O3-$Na_3Ni_2SbO_6$ ↔ P3-$Na_2Ni_2SbO_6$ ↔ O1-$NaNi_2SbO_6$, of which the capacity was mainly contributed by the redox reaction of the Ni^{2+}/Ni^{3+} couple, while SbO_6 octahedron contributed to the structural stability of the honeycomb-ordered Ni_2SbO_6 slabs. Similarly, Wang et al. constructed a highly ordered Ni_6 ring superstructure within the metal layers with two ions that have different M—O bond energies. By substituting 1/3 of Ni with Sb in the $NaNiO_2$, Ni_6 rings are formed. This has rendered the electrode a symmetric atomic configuration, which degenerates the electronic orbital and improves the stability as well as increases the redox potential of the electrode material. Layered s-ordered O′3-$Na_3Ni_{1.5}M_{0.5}BiO_6$ (M = Ni, Cu, Mg, Zn) phases were developed [111]. With the Ni^{2+}/Ni^{3+} and Cu^{2+}/Cu^{3+} as the active redox couples during the cycling, the $Na_3Ni_{1.5}Cu_{0.5}BiO_6$ electrode delivered a discharge capacity of 94 mAh g^{-1} with a high voltage of 3.3 V. The notable difference between M^{2+} and Bi^{5+} resulted in the formation of cationic ordering or superstructures,

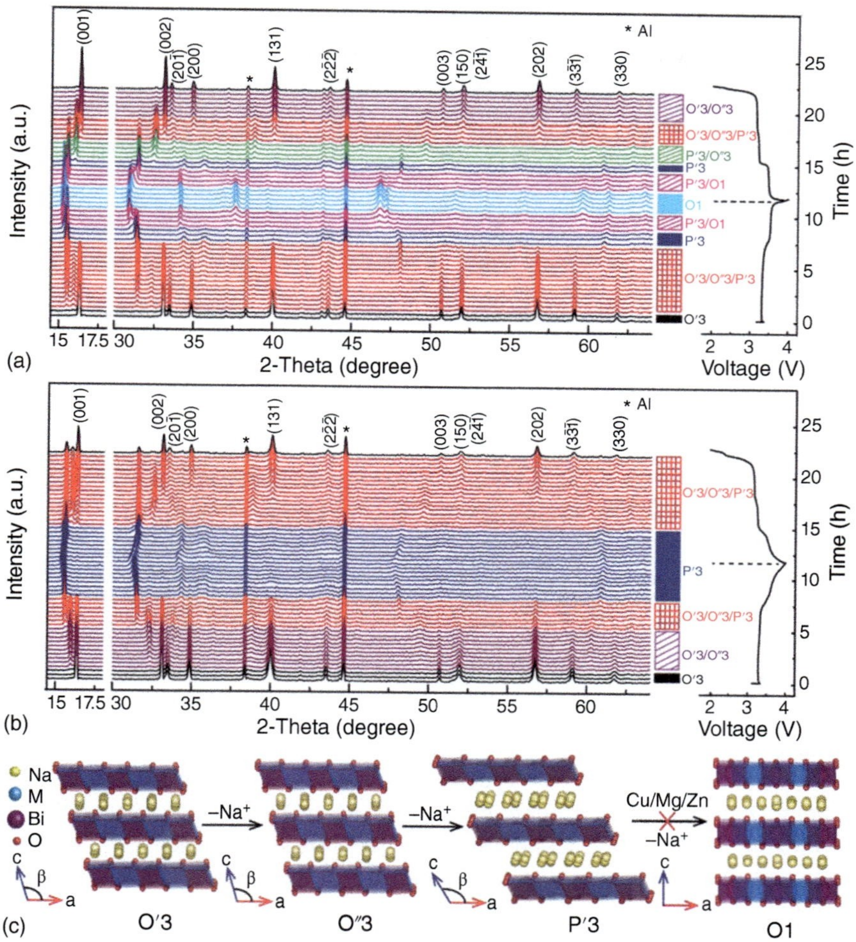

Figure 3.10 In situ XRD patterns of (a) Na/$Na_3Ni_2BiO_6$ and (b) Na/$Na_3Ni_{1.5}Cu_{0.5}BiO_6$ cells at 0.1 C between 2 and 4 V. (c) Phase transition during the sodiation and desodiation process. Source: Wang et al. [111]. Reproduced with permission, 2017, American Chemical Society.

and meanwhile, the substitution of metal in similar ionic radius with Ni^{2+} would suppress the P′3-O1 phase transition and improve the stability. As shown in the in situ XRD patterns (Figure 3.10), the $Na_{3-x}Ni_2BiO_6$ ($x = 0$) showed pristine monoclinic O′3 phase. Upon Na-ion extraction, O′3, O″3, and P′3 coexised in the voltage range of 3.32–3.4 V. Further Na extraction led to the formation of newly formed P′3 phase until the O1 phase emerges at the 3.53 V plateau. And the single O1 phase was retained until 4 V. With partial substitution of Cu, only O″3 structure was observed when charged to 3.28 V without forming any P′3 phase. Then till the end of 3.4 V, O′3, O″3, and P′3 coexisted. The solid solution region can be obtained without any new phase until 4 V. These results indicate that the partial substitution of Cu will extend to solid solution zone over a wider compositional range.

Li_2NiO_2 is a cathode additive in LIBs that can supply extra Li ions to the anode. Based on these previous research, its analogue, Na_2NiO_2 has also been investigated as the cathode additive for SIBs [41d]. Similar to Li_2NiO_2, Na_2NiO_2 shows an orthorhombic structure with Ni occupying the edge-shared square planar sites. If Ni^{2+}/Ni^{4+} redox couple can be fully accessed, the compound can deliver a high capacity of 392.2 mAh g^{-1}. The crystal structures of Li_2NiO_2 and Na_2NiO_2 are illustrated in Figure 3.11. It is also sensitive to air and water, due to the low open circuit voltage at 1.7 V. As reported by Park et al., upon electrochemical reaction, Na_2NiO_2 is transformed into $NaNiO_2$ irreversibly at the initial stage and then the crystalline $NaNiO_2$ can act as the cathode in the following cycles. In this case, the Na_2NiO_2 can also regarded as a cathode additive for SIBs to compensate for the loss of Na ions at the anode side.

To date, the capacity decay of the high-Ni cathode under ambient conditions presents a grand challenge for their further commercialization. The degradation is possibly originated from the water molecules inserted between the TMO_2 slabs, carbonate ions located between the TM layers or oxidation of Na_xTMO_2 [46b, 52a, 100, 112]. Regarding the water stability, recently Zhang et al. theoretically calculated the surface adsorption energy of the water molecules and the results revealed that the Co/Mn and Fe/Mn units exhibit much lower adsorption energy than that of the Li/Mn unit. As a result, the water molecules tended to absorb on the Co/Mn and Fe/Mn units instead of the Li/Mn units. Besides, the Li/Mn unit also suppressed the Na^+–vacancy ordering and improved the Na-ion diffusion kinetics. Benefited from the above reasons, $Na_{2/3}Li_{1/9}Ni_{5/18}Mn_{2/3}O_2$ cathode showed improved structural stability and excellent rate and cycling performance even after water soaking treatment, i.e. 78% capacity retention at 20 C and 87% capacity retention after 1000 cycles [113]. Zheng et al. presented a systematic study of the water stability of the $NaNi_{0.5}Mn_{0.5}O_2$ cathode stored in different atmospheres. Exposure to water resulted in desodiation and material degradation. It was found that ethanol washing can effectively reduce the air sensitivity by removing the side products generated

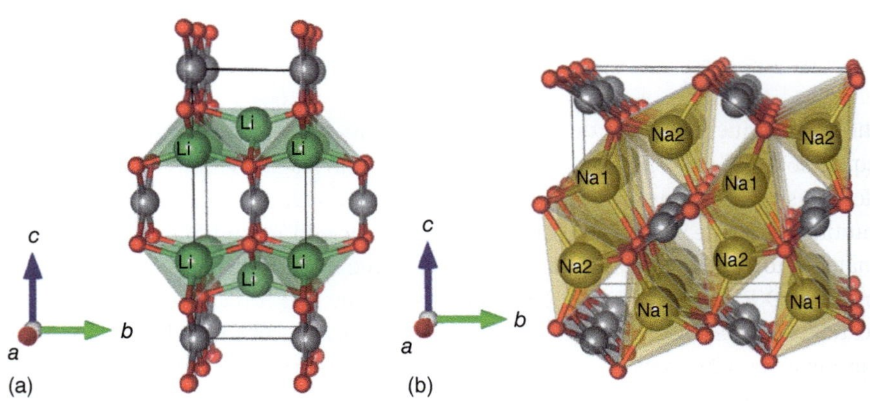

Figure 3.11 Crystallographic structures of Li_2NiO_2 (a) and Na_2NiO_2 (b). Source: Park et al. [41d]. Reproduced with permission, 2015, American Chemical Society.

in air [114]. As indicated by the DFT calculations, Yao et al. found that the air stability can be improved by reducing the interlayer space between the Na layers and increasing the valence state valence of the TMs [52c]. To confirm the DFT prediction, they modified the $NaNi_{0.5}Mn_{0.5}O_2$ structure by Cu/Ti doping. Because Ni ions shows substantial difference in the Fermi level from Mn^{4+}, their presence will prevent charge ordering and decrease electronic localization. Besides, the valence state of Ni ions can be elevated as the electronic delocalization decreases the number of electrons around Ni. Cu^{2+} was further added to deform the TMO_6 octahedrons and rearrange the electrons. As such, the valence of Ni species could be increased. As shown in the electrochemical test results, the pristine NaNM suffered from drastic capacity fading upon exposure to air, while the Cu/Ti co-doped sample showed high-capacity retention after the aging tests. You et al. effectively quantified the concentration of the sodium species in the residue by a chemical titration method and analyzed the spatial distribution and composition of the degradation species at the nanoscale using the sub-nanometer surface-sensitive time-of-flight secondary ion mass spectroscopy (TOF-SIMS) [100]. As revealed by their results, the degradation products were primarily composed of NaOH, Na_2CO_3, $Na_2CO_3{\cdot}H_2O$, and NiO at the surface. Ni dissolved from the electrode and accumulated on the surface. Besides, the side reaction between the alkaline sodium species and the polyvinylidene fluoride (PVDF) binder resulted in the formation of the F-containing species, such as NaF. The products from these side reactions were insulating and led to degradation of the electrodes. Risthaus et al. reported a discharge capacity of 228 mAh g^{-1} within the voltage of 1.5–4.5 V of the P2-$Na_{2/3}Ni_{1/3}Mn_{2/3}O_2$ [115]. The high capacity was attributed to the redox activities of Ni ions, surface Mn ions, and oxygen ions. It was also found that the poor cycling stability of the electrode was partly resulted from the electrolyte decomposition. By using the NaTFSI instead of the ionic liquid, severe drop in capacity drop was suppressed.

3.3.5 Na_xVO_2

$NaVO_2$ resembles the structure of O3-type α-$NaFeO_2$ [116]. Delmas's group isolated the single-phase $Na_{1/2}VO_2$ and $Na_{2/3}VO_2$ from electrochemical deintercalation of Na ions and carried out systematical investigation of their electrochemical properties [117]. During deintercalation of Na ions, the material showed a monoclinic distortion with a O′3 structure. At least 0.5 mol Na^+ can be extracted from $NaVO_2$, corresponding to a capacity of 126.4 mAh g^{-1}. Above 0.5, multiphase single-phase domains were observed. They suggested this material not suitable for battery applications due to the large voltage variation during the cycling and the sensitivity of the material toward the oxygen. Later, they monitored the structure changes of Na_xVO_2 by in situ XRD and identified a superstructure in the $Na_{1/2}VO_2$ [118]. The structure featured the Na–vacancy ordering in the interslab space. Due to repulsions between Na ions, Na ions were displaced from the center of octahedral sites and lowered the crystal symmetry. The structure changed from rhombohedral to monoclinic to allow the deformation of vanadium layers. The vacancies on the $[VO_2]$ slabs were closer to the shortest V–V distances and Na ions faced the longest V–V distances to minimize the $V^{3+/4+}/Na^+$ electrostatic repulsions. Later, Guignard et al. provided the first phase diagram for P2-Na_xVO_2 [119]. This substance can only be obtained at high temperature (850 °C) within a narrow composition range ($x \sim 0.71$). Otherwise, the O3-type $NaVO_2$ or impurities like V_2O_3 and Na_3VO_4 will appear. Due to the competition between the $V^{3+/4+}/Na^+$ repulsions through the face-shared polyhedral (NaO_6 trigonal prisms and VO_6 octahedra) and the Na^+/Na^+ repulsions in the interslab space, some specific properties are expected from this material. When applied as cathode for SIBs, it can be reversibly cycled in the voltage range of 1.4–2.5 V with small polarization and achieves a specific capacity of 110 mAh g^{-1}. When tested at a higher voltage (>2.5 V), the reaction became irreversible. Many potential plateaus emerged and the voltage vs. composition curves showed continuous variation, which is indicative of the formation of the solid solution domains. Single phases only existed at specific Na/V ratios (1/2, 5/8, and 2/3), which suggests a specific ordering of the Na^+ ions between the VO_2 slabs and the V^{3+}/V^{4+} ions within the slabs. During the sodium deintercalation, the average V–V interslab distance decreased owing to the oxidation of vanadium, while the average V–V interslab distances increased. In the $0.5<x<0.9$ range, three ordered phases and two solid solution domains were identified. Besides, this material exhibited pseudo-trimers with very short V–V distances owing to the overlapping of the t_{2g} orbitals through the common edge of the VO_6 octahedra. A detailed thermally or electrochemically driven topotactical transformation pathways in the Na_xVO_3 system were described by the same group [120]. In particular, within a small x-range, the layered structure can accommodate more Na ions to best accommodate the Na ions and minimize the electrostatic repulsions between adjacent alkaline ions.

3.3.6 Na_xCrO_2

It has been generally accepted that $LiCrO_2$ is electrochemically inactive with irreversible Li intercalation behavior in LIBs. However, its analogue, $NaCrO_2$ turns out to be an active electrode material in the organic electrolytes [1]. Since the first report in the 1980s [67], $NaCrO_2$ has drawn considerable attention because of the high specific capacity (~120 mAh g^{-1}) and the thermal stability of $Na_{0.5}CrO_2$ [121]. In terms of the crystal structure, $NaCrO_2$ bears a O3-type layered structure resembled that of α-$NaFeO_2$ [3]. The theoretical capacity for $NaCrO_2$ is 250 mAh g^{-1}, but practically only 0.5 mol Na can be extracted from the crystal structure by cutting off the voltage at around 3.8 V [122]. Otherwise, the structural reversibility will be lost and very small specific capacity can be delivered. Kubota et al. studied the structural change of the $NaCrO_2$ at high voltage and revealed the mechanism of the irreversible loss (Figure 3.12a,b) [123]. They found that in the range of $0.0 \leq x \leq 0.5$, the crystal structure of $Na_{1-x}CrO_2$ evolved from O3 to O′3 and then P′3 reversibly via gliding the CrO_2 slabs. But beyond the voltage of 3.8 V, Cr ions migrated from the slabs to the interslab space when the structure irreversibly changed from P′3 to O3′ type. The interslab space drastically shrunk and caused extremely low discharge capacity (~9 mAh g^{-1} when the cell was charged up to 4.8 V). Therefore, it is only in the range of $0.0 \leq x \leq 0.5$ that the $Na_{1-x}CrO_2$ can be reversibly charged

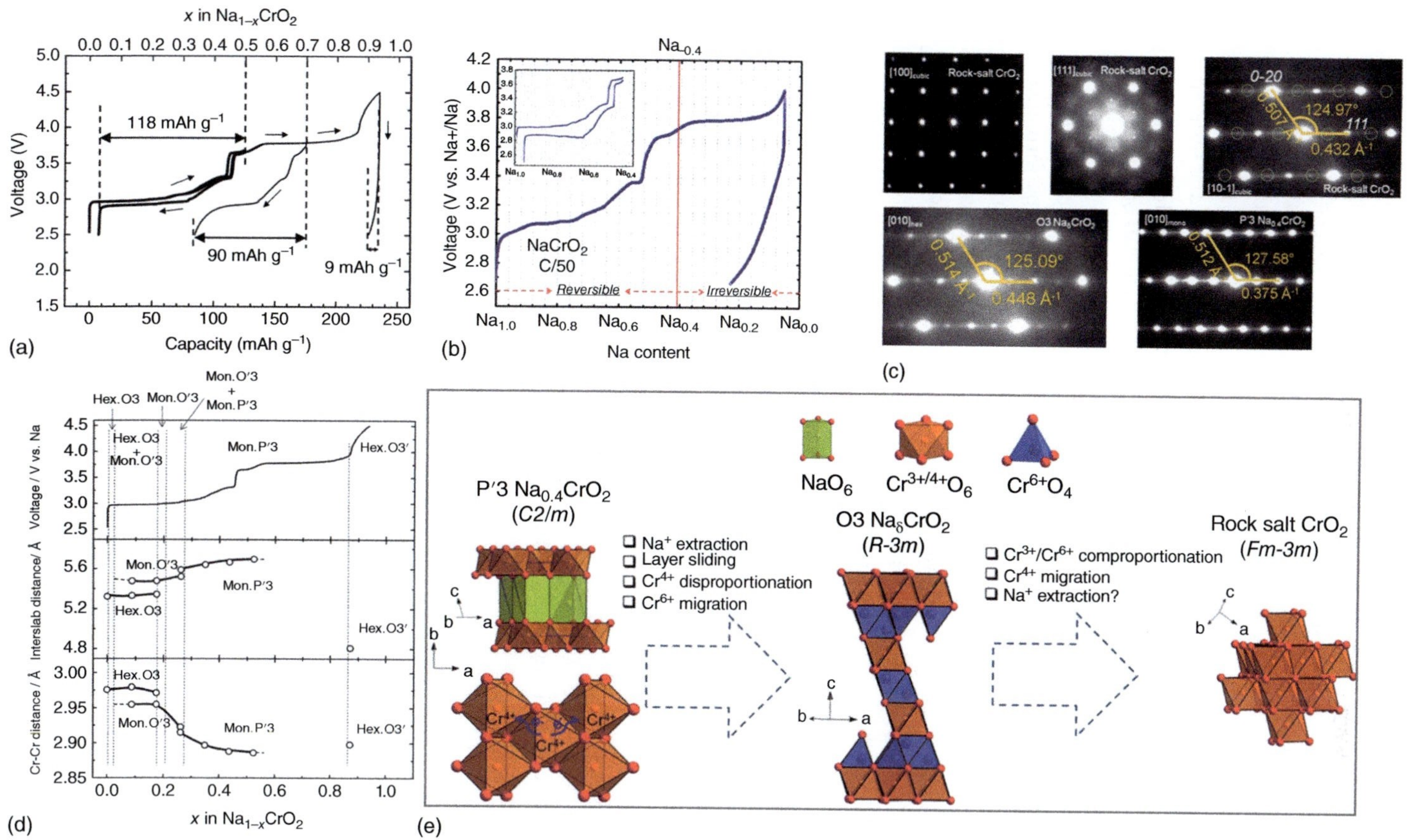

Figure 3.12 (a) The initial charge–discharge curve of the half cell with the $NaCrO_2$ cathode at 0.05 C. (b) Change in structure parameter and phase transition with variation in the Na content (under different Na content). Source: Kubota et al. [123]. Reproduced with permission, 2014, American Chemical Society. (c) Charge–discharge curves of the $NaCrO_2$ at 0.02 C. (d) Electron diffraction patterns of the intermediates formed during the reaction. (e) Schematic illustration of the phase transition from the layered structure to the rock salt structure. Source: Bo et al. [124]. Reproduced with permission, 2016, American Chemical Society.

and discharged with good cycling stability. Ceder's group also studied the mechanism of the capacity fading associated with the O3-type $NaMO_2$ electrodes using combined techniques of SXRD and electron diffraction pattern of the intermediate products [124]. They argued that the O′3-type products identified by Kubota et al. was actually a two-phase mixture comprised of O3 $Na_\delta CrO_2$ and rock salt CrO_2 (Figure 3.12c,d). Na_xCrO_2 ($0<x<1$) can maintain the layered structure framework without Cr migration with the maximum composition of $Na_{0.4}CrO_2$. Desodiation process of this composition triggered the layered-to-rock-salt transition and resulted in poor cycling performance of the electrodes, as illustrated in Figure 3.12e.

Luo et al. reported that the performance of $NaCrO_2$ electrode can be controlled through proper structural defects engineering [125]. The misplacement of Cr^{3+} at the Na-ion sites and the Na—O bond length are two key parameters that determine the capacity and stability of the material. On the other hand, the performance of the $NaCrO_2$ has been improved through nanoengineering and proper surface modification. For example, the nanosized $NaCrO_2$ from heat treatment showed suppressed phase transition and MO_2 layer gliding due to the presence of rich grain boundaries [126]. The high-performance $NaCrO_2$ synthesized via a novel emulsion-drying method was reported by Yu et al. [127]. In contrary to the conventional solid-state synthetic process, such a "water-in-oil"-based method enabled the modification of electrode material surface with uniform carbon coating. Liang et al. synthesized the ultralong $NaCrO_2$ nanowires and used them as the cathode material for SIBs that can survive a wide operating temperature ranged from −15 to 55 °C [128]. The atomic-scale-controlled carbon coating has brought about a significant improvement in the electrical conductivity (e.g. from 8×10^{-5} S cm^{-1} of the bare $NaCrO_2$ to 0.47 S cm^{-1} of the C-coated $NaCrO_2$). At the same time, the carbon coating also retarded the exothermal decomposition of the $Na_{0.5}CrO_2$ particles at high temperature by generating less heat. The C-$NaCrO_2$ cathodes exhibited excellent rate capability at the rate of up to 150 C, holding great promise for the manufacturing of safe and high-performance SIBs. Even coated with the inactive Cr_2O_3, the rate performance of the $NaCrO_2$ cathode was significantly improved as the side reactions between the active material and the electrolyte were effectively addressed [129]. As a result, the Cr_2O_3-coated $NaCrO_2$ sample achieved a capacity of 108 mAh g^{-1} at a high current rate of 60 C.

On the other hand, partial substitution of Cr^{3+} with Ti^{4+} induces a strong inductive effect and endows the $Na_{2/3-x}Cr_{2/3}Ti_{1/3}O_2$ electrode with higher operating voltage, though x is limited to $<1/3$ due to the concerns over the oxidation and migration of Cr^{6+} into the tetrahedral sites upon charging [130]. Similar to most layered metal oxides, another challenging issue of this type of cathode material is their hygroscopic and vulnerable nature upon long-term exposure to moisture or air. With the presence of Ru, the irreversible migration of Cr ions could be suppressed. As a result, the Ru-doped $Na_{0.88}Cr_{0.88}Ru_{0.12}O_2$ delivered a high capacity of 156 mAh g^{-1} with an elevated operating voltage at 3.8 V [131]. The stability of the Ru-doped electrode was also significantly improved with 80.7% capacity retained after cycling for 1100 times. With minor ratio of Mn doped in the $NaCr_{0.8}Mn_{0.2}O_2$, the operating voltage was increased and as the oxidation of Mn^{3+} took place ahead of the oxidation of Cr^{3+}, the formation of the toxic Cr^{6+} ions was also suppressed [132]. The stability of the $NaCrO_2$ could also be improved with the presence of minor Ca^{2+} ions in the Na sites [133].

3.3.7 Mixed Cation Oxides

As a popular strategy in optimizing the property and electrochemical performance of the TM oxides cathodes, integration of various metals in one system not only endows the layered oxides with improved electrochemical activity but also improves the structural stability. The ternary, quaternary, and even more complexed metal oxide have showed more remarkable performance as cathodes for SIBs when compared to the one composed of single-TMs [134]. For example, the capacity and cycling stability of the P2-type $Na_{2/3}[Ni_{1/3}Mn_{2/3}]O_2$ consisted of active Ni^{2+} ions and inactive Mn^{4+} ions were significantly enhanced [135]. The three types of ordering in the layered metal oxides, i.e. TM ordering, charge ordering, and Na^+–vacancy ordering, are detrimental to the ion transportation and will lower the Na-ion diffusion coefficient. By forming compounds from different TMs, the Jahn–Teller-induced structural transition and the Na^+–vacancy ordering could be effectively suppressed and the Na^+ diffusion kinetics were also enhanced. Wang et al. selected two TMs, Cr^{3+} and Ti^{4+}, which are similar in ionic radii but different in redox potential to suppress the Na^+–vacancy ordering [136]. The Na^+–vacancy disordered $Na_x[Cr_{0.6}Ti_{0.4}]O_2$ (x = 0.33–1.0) showed high ionic conductivity and rate capability. On the other hand, Cao et al. realized the multi-electron transfer of $Cr^{2+\sim3+}/Cr^{4.4+}$ redox in the layered $Na_{0.66}Fe_{1/3}Cr_{1/3}Ti_{1/3}O_2$ [137]. The substitution of Fe^{3+} and Ti^{4+} suppressed the disproportion reaction of Cr^{4+} and improved the cyclability of the compounds. The presence of the electrochemically inactive tetravalent Ti^{4+} also improved the cycling stability of oxide cathodes [109]. P2 phase $Na_{2/3}Mn_{0.8}Fe_{0.1}Ti_{0.1}O_2$ cathode delivered a capacity of 146.5 mAh g^{-1} in the second cycle with good capacity retention (95.9% of the initial capacity was retained after 50 cycles at 0.1 C) [90]. Sometimes, the presence of the electrochemically inactive metal ions such as Sb^{3+}, Mg^{2+}, and Ca^{2+} could also bring about improved performance by effectively suppress the phase separation during the charge and discharge process [138]. Mariyappan et al. reported that by partial substituting Ni^{2+} with inactive cations such as Zn^{2+}, Mg^{2+}, and Ca^{2+}, the potential capacity of the layered $NaMO_2$ could be fully exerted, as the undesired phase transition and structural degradation were effectively suppressed upon the cycling. The $Na_{0.2}Ni_{0.45}Zn_{0.05}Mn_{0.4}Ti_{0.1}O_2$ formed the nanodomains composed of intergrown P3–O1 phases and showed more robust crystal structure upon long-term cycling [138a]. Besides, the introduction of Li^+ ion eased the O3–P3 phase transformations and was believed to be favorable to optimize the electrochemical performance of the O3-type cathode in SIBs. Xu et al. introduced a new type of potassium-containing layered oxides $K_{0.5}Mn_{0.7}Fe_{0.2}Ti_{0.1}O_2$ as the cathode material for SIBs [139]. The cathode containing 0.5 K^+ showed an interlayer spacing of 0.63 nm which expanded with the subsequent extraction of K^+ ions. The lattice expansion enabled the exchange between K^+ and Na^+ ions. With a small portion of K^+ left in the structure and acted as the pillar to maintain the stability of the electrodes, Na^+ ions could insert the structure and decrease the interlayer spacing without disturbing the P2 structure of the material. Such type of cathode could achieve a high initial capacity of 147.1 mAh g^{-1} and 71.5% of the capacity was retained after 1000 cycles, indicating its good stability.

Though O3- and P2-type electrode materials suffer from different issues, forming the hybrid phases of O3 and P2 structures in turn brings about combined advantages while at the meantime can eliminate some undesired features. For example, the O3-type $NaNi_{0.5}Mn_{0.5}O_2$ materials with minor P2 phases showed smoother profile [140]. The P2/O3 $Na_{0.66}Li_{0.18}Mn_{0.71}Ni_{0.21}Co_{0.08}O_2$ compound delivered a specific capacity of 200 mAh g^{-1} [141]. The O3 phase in the $Na_{0.67}Mn_{0.55}Ni_{0.25}Ti_{0.2-x}Li_xO_2$ could optimize the stability of the electrode [142]. Qi et al. systematically studied the hybrid P2/O3 phase of $Na_x[Ni_{0.2}Fe_{x-0.4}Mn_{1.2-x}]O_2$ ($x = 0.7-1.0$) [143]. The in situ XRD pattern revealed a reduced polarization when the cutoff voltage was below 4 V, which can be attributed to the co-presence of P2 phase.

High entropy oxides (HEOs) is an another new concept material [144]. It is formed with multiple metals that can crystallize in a single phase. Different elements share equiatomic sites in HEO, and owing to the complex compositions, they exhibit better electrochemical performance. The reason for the superior performance of HEOs may lie in the oxygen vacancies created by the metals and increase percolating channels which enable diffusion of Na ions. A nine-component HEO, $NaNi_{0.12}Cu_{0.12}Mg_{0.12}Fe_{0.15}Co_{0.15}Mn_{0.1}Ti_{0.1}Sn_{0.1}Sb_{0.04}O_2$, was reported [145]. The presence of these different elements causes variation in the local interactions between elements, TMO_2 slabs, and Na in NaO_2 slabs, and some of the elements are responsible for the charge compensation during the charge–discharge process. The O3 to P3 phase transition was delayed in the HEO cathode, and the entropy stabilization could facilitate the long-term stability and rate performance of the electrode structure.

3.3.8 Other Emerging Metal Oxides

In recent years, introducing the anionic redox reaction (ARR) becomes an effective strategy to boost the specific capacity and energy density of the battery [146]. ARR normally occurs in TM oxides partly substituted by Li, Na, Mg, and Zn. By triggering ARRs, it is viable to achieve higher voltage and thus higher specific capacity. For example, small amount of Mg^+ in the P2-type electrode could stabilize the structure and improve the cycling performance. However, as the Mg species are electrochemically inactive, concerns were raised that their presence would decrease the overall capacity. But a latter finding revealed that the large quantity of Mg will suppress the oxygen evolution reaction and contribute to the extra capacity of the electrode materials. For example, the Mg-doped P2 electrode can better maintain the original crystal structure at high-voltage region, i.e. 2–4.5 V [147]. However, a trade-off was observed between the voltage and the specific capacity in such type of material. That is, the higher voltage brings about higher specific capacity, while at the same time, it also induces the electrolyte decomposition and multiple-phase transitions during the charge–discharge process, causing obvious voltage hysteresis and rapid capacity decay of the electrodes [148].

The TM deficiency is believed to the key to activate the oxygen anion redox. Doping of metal ions to form the ionic bond, e.g. Li—O, Na—O, and Mg—O bond, creates O(2p) nonbonding orbitals so that the electrons are fully localized on the oxygen anions. $Na_{0.78}Ni_{0.23}Mn_{0.69}O_2$ as a cathode delivered a specific capacity of 138 mAh g^{-1} with an average potential of 3.25 V vs. Na^+/Na [149]. Oxidation of Ni^{2+} contributed to the charge compensation, while lattice oxygen participated in part of the reaction and contributed to the excessive capacity of the electrode.

Besides of boosting the capacity of layered metal oxides, Rong et al. demonstrated that ARR is also favorable to stabilize the structure of the electrode [150]. They reported a P2-type $Na_{0.72}[Li_{0.24}Mn_{0.76}]O_2$ with a high initial discharge capacity of 270 mAh g^{-1} and a high energy density of 700 Wh kg^{-1}. The electrode maintained the P2 structure even when 0.93 Na was extracted. The change in the oxygen radii and charges carried by the oxygen ions resulted in a decrease in oxygen repulsion around the empty Na layer and hence stabilized the structure. Moreover, triggering the oxygen redox activity has also been highlighted as a potential strategy to increase the intrinsically low-redox potential of SIBs. Kim et al. investigated the thermodynamic mixing enthalpy value, the bonding length, and electronic structure of the $Na[Mn_{1-x}Ni_x]O_2$, and they predicted that the thermodynamically stabilized $Na[Mn_{0.5}Ni_{0.5}]O_2$ would show a double-redox reaction from both Ni^{2+}/Ni^{4+} and O^{2-}/O^{n-}, which should achieve a higher-redox potential than that from single-redox reaction in $NaMnO_2$ [151].

But materials based on ARR often suffer from structural degradation associated with the complex phase transitions, migration of cations, and loss of oxygen during the cycling. Zn ion shows higher Pauling electronegativity than other alkaline ions but no orbital and is expected to be covalently bond with oxygen. P2-$Na_{2/3}Mn_{1-y}Zn_yO_2$ was synthesized by Bai et al. and exhibited oxygen redox activity associated with nonbonding O(2p) orbitals [148]. Wang et al. reported that doping of Zn ions could not only trigger the ARR but also enhance the air stability of the electrode material [152]. The high Zn content suppressed the reduction from Mn^{4+} to Mn^{3+}, thereby alleviating the strain stress and enhanced the structural stability. As a result, the Zn-doped electrode showed a volume change as low as 1% and no phase transformation was observed during the initial cycles. Even after exposure to air for 30 days, the single P2 phase could be well preserved in the Zn-doped samples. But the overall performance of Zn-based cathode was inferior to those based on alkaline metals due to cation-migration-induced phase transition.

It has also been discovered that some metal oxides exhibit the intrinsic ARR activity even without doping of heteroatoms. For example, Su et al. investigated the $NaVO_3$ as a cathode material for SIBs and the electrode delivered a high capacity of 245 mAh g^{-1} without phase transition [153]. $NaVO_3$ is consisted of O–V–O slabs made of VO_4 chains. Na ions are sandwiched between the VO_2 slabs and occupy the half distorted octahedral sites. The O atoms adopt three different configurations surrounded with different numbers of Na and V with vacancies nearby. The VO_4 tetrahedral framework suppresses the migration of metal ions and endows the electrode with superior rate and cycling performance. The cathodic (V^{4+}/V^{5+}) and anionic (O^{2-}/O^{-}) redox couples during discharge contribute to the high capacity. Boisse et al. identified the self-repairing phenomenon of stacking defaults driven by the strong long-range cooperative Coulombic interactions between MO_3 slabs intermediated by the ordered vacancies in the Na_2RuO_3 [146a]. Qiao et al. demonstrated that the Na storage capacity of Na_3RuO_4 was dominated solely by ARR. With the highly covalent Ru—O bond and the existence of the high Na/Ru ratio, pure nonbonding O2p states can be generated during Na extraction, facilitate the formation of sable peroxo O—O bond upon charging, and contribute to the high-redox reversible capacity [154].

As layered oxides are mostly air-sensitive, tunnel-type oxides are also widely investigated. However, the specific capacity achieved by the tunnel-type material is lower. For example, the tunnel-type $Na_{0.44}MnO_2$ only delivered a specific capacity of 80 mAh g^{-1} at 0.1 C. Besides, the capacity retention after long-term cycling was even poor. Cation doping could improve the stability of tunnel-type oxides. For example, it has been widely demonstrated that the Ti-doped $Na_{0.44}MnO_2$ compounds could show better capacity retention [89, 155]. Shi et al. found that the Ti substitution could relieve the strain in the tunnel structure and they have achieved a higher capacity retention, e.g. 88 mAh g^{-1} at 1 C after 600 cycles [156].

3.4 Concluding Remarks

As above, cathode materials based on TM oxides have been extensively investigated because of their low cost and high specific capacity. However, the relatively low working potential and complex phase variation have severely limited their application. With the tremendous efforts devoted by the researchers, a variety of effective strategies have been developed. For example, for the Na-free metal oxide, such as MnO_2 and V_2O_5, the stability and specific capacity can be significantly improved by tailoring the particle size, crystal structure (crystallinity, defects, and exposed crystal plane), and inserting suitable ions or molecules to serve as pillars to enlarge the channels and enhance the structural stability. In comparison, the Na-containing cathode materials are more intriguing in terms of practical application as presodiation process can be excluded. The main challenge lies in the complicated phase transition which not only results in unstable electrochemical performance but also poses difficulties in clarifying the mechanisms for the sodium storage. Introducing single or multiple dopants into these layered metal oxides has proved to be effective in suppressing the phase transition, increasing the specific capacity, and widening the operation voltage window. Besides of the materials based on the intercalation mechanism, several new concept cathodes have emerged in recent years. For example, triggering the ARR activity by partial substitution of layered TM oxides with Li, Na, Mg, and Zn can boost the specific capacity and promise the high energy density. Tunnel-structured oxides show better stability as compared to layered oxides, but less capacity. The HEOs have also been proposed and showed superior electrochemical performance.

In general, the Na-containing layered metal oxides are the most promising candidate that might be suitable for the future commercialization. As they are most extensively studied, technologies in controlling their composition and crystal structure are relatively mature, and successful attempts have been made in suppressing the phase transitions and realizing stable output. To realize cathode materials with higher capacity and high voltage and seek further breakthrough in achieving high-performance SIBs, more research attention should be paid to the new concept materials as well, given the intriguing performances they have demonstrated and the insufficient fundamental understanding in their sodium storage mechanism.

References

1 Komaba, S., Takei, C., Nakayama, T. et al. (2010). *Electrochem. Commun.* 12: 355.

2 Lyu, Y., Liu, Y., Yu, Z.-E. et al. (2019). *Sustain. Mater. Technol.* 21: e00098.

3 Delmas, C., Fouassier, C., and Hagenmuller, P. (1980). *Physica B+C* 99: 81.

4 Vassilaras, P., Ma, X., Li, X., and Ceder, G. (2012). *J. Electrochem. Soc.* 160: A207.

5 Han, M.H., Gonzalo, E., Singh, G., and Rojo, T. (2015). *Energy Environ. Sci.* 8: 81.

6 Wang, W., Jiang, B., Hu, L. et al. (2014). *J. Power Sources* 250: 181.

7 Ali, G., Lee, J.H., Oh, S.H. et al. (2016). *ACS Appl. Mater. Interfaces* 8: 6032.

8 Su, D. and Wang, G. (2013). *ACS Nano* 7: 11218.

9 Tompsett, D.A. and Islam, M.S. (2013). *Chem. Mater.* 25: 2515.

10 Su, D., Ahn, H.-J., and Wang, G. (2013). *NPG Asia Mater.* 5: e70.

11 Hu, X., Zhao, Z., Wang, L. et al. (2019). *Electrochim. Acta* 293: 97.

12 (a) Murphy, D. and Christian, P. (1979). *Science* 205: 651. (b) Popuri, S.R., Miclau, M., Artemenko, A. et al. (2013). *Chin. Phys. Lett.* 52: 4780.

13 (a) Chao, D., Zhu, C., Xia, X. et al. (2015). *Nano Lett.* 15: 565. (b) Balogun, M.-S., Luo, Y., Lyu, F. et al. (2016). *ACS Appl. Mater. Interfaces* 8: 9733.

14 Li, H.-Y., Yang, C.-H., Tseng, C.-M. et al. (2015). *J. Power Sources* 285: 418.

15 Tepavcevic, S., Xiong, H., Stamenkovic, V.R. et al. (2012). *ACS Nano* 6: 530.

16 Li, S., Li, X., Li, Y. et al. (2017). *Electrochim. Acta* 244: 77.

17 Raju, V., Rains, J., Gates, C. et al. (2014). *Nano Lett.* 14: 4119.

18 Córdoba, R., Kuhn, A., Pérez-Flores, J.C. et al. (2019). *J. Power Sources* 422: 42.

19 (a) Dong, Y., Li, S., Zhao, K. et al. (2015). *Energy Environ. Sci.* 8: 1267. (b) Kang, H., Liu, Y., Shang, M. et al. (2015). *Nanoscale* 7: 9261. (c) Lee, Y., Oh, S.M., Park, B. et al. (2017). *CrystEngComm* 19: 5028.

20 Fei, H., Liu, X., Lin, Y., and Wei, M. (2014). *J. Colloid Interface Sci.* 428: 73.

21 Sarkar, A., Sarkar, S., Sarkar, T. et al. (2015). *ACS Appl. Mater. Interfaces* 7: 17044.

22 Sarkar, A., Sarkar, S., and Mitra, S. (2017). *J. Mater. Chem. A* 5: 24929.

23 Wang, D., Wei, Q., Sheng, J. et al. (2016). *Phys. Chem. Chem. Phys.* 18: 12074.

24 Wei, Q., Liu, J., Feng, W. et al. (2015). *J. Mater. Chem. A* 3: 8070.

25 Liu, S., Tong, Z., Zhao, J. et al. (2016). *Phys. Chem. Chem. Phys.* 18: 25645.
26 Moretti, A., Maroni, F., Osada, I. et al. (2015). *ChemElectroChem* 2: 529.
27 Wei, Q., Jiang, Z., Tan, S. et al. (2015). *ACS Appl. Mater. Interfaces* 7: 18211.
28 Song, X., Li, J., Li, Z. et al. (2019). *ACS Appl. Mater. Interfaces* 11: 10631.
29 Qian, J., Jin, H., Chen, B. et al. (2015). *Angew. Chem. Int. Ed.* 54: 6800.
30 Su, D., Ahn, H.-J., and Wang, G. (2013). *J. Mater. Chem. A* 1: 4845.
31 Zhang, Y., An, Y., Jiang, J. et al. (2018). *Energy Technol.* 6: 2146.
32 Zhou, Y., Chen, T., Zhang, J. et al. (2017). *Chin. J. Chem.* 35: 1294.
33 Guo, C.X., Wang, M., Chen, T. et al. (2011). *Adv. Energy Mater.* 1: 736.
34 Lu, K., Hu, Z., Xiang, Z. et al. (2016). *Angew. Chem.* 128: 10604.
35 Brady, A.B., Huang, J., Durham, J.L. et al. (2018). *MRS Adv.* 3: 547.
36 Lee, S.-Y., Housel, L.M., Huang, J. et al. (2019). *J. Power Sources* 435: 226779.
37 Braconnier, J.-J., Delmas, C., Fouassier, C., and Hagenmuller, P. (1980). *Mater. Res. Bull.* 15: 1797.
38 Sun, Y., Guo, S., and Zhou, H. (2019). *Energy Environ. Sci.* 12: 825.
39 Xu, G.L., Amine, R., Abouimrane, A. et al. (2018). *Adv. Energy Mater.* 8: 1702403.
40 Sun, Y.-K. (2020). *ACS Energy Lett.* 5: 1278.
41 (a) De Ilarduya, J.M., Otaegui, L., del Amo, J.M.L. et al. (2017). *J. Power Sources* 337: 197. (b) Zhang, B., Dugas, R., Rousse, G. et al. (2016). *Nat. Commun.* 7: 1. (c) Shen, B., Zhan, R., Dai, C. et al. (2019). *J. Colloid Interface Sci.* 553: 524. (d) Park, K., Yu, B.-C., and Goodenough, J.B. (2015). *Chem. Mater.* 27: 6682. (e) Shanmukaraj, D., Kretschmer, K., Sahu, T. et al. (2018). *ChemSusChem* 11: 3286.
42 Niu, Y.B., Guo, Y.J., Yin, Y.X. et al. (2020). *Adv. Mater.* 32: 2001419.
43 Jin, T., Wang, P.-F., Wang, Q.-C. et al. (2020). *Angew. Chem.* 59: 14511.
44 Vinckevičiūtė, J., Radin, M.D., and Van der Ven, A. (2016). *Chem. Mater.* 28: 8640.
45 Guo, S., Sun, Y., Yi, J. et al. (2016). *NPG Asia Mater.* 8: e266.
46 (a) Nam, K.W., Kim, S., Yang, E. et al. (2015). *Chem. Mater.* 27: 3721. (b) Buchholz, D., Chagas, L.G., Vaalma, C. et al. (2014). *J. Mater. Chem. A* 2: 13415.
47 Yang, X., Takada, K., Itose, M. et al. (2008). *Chem. Mater.* 20: 479.
48 Zuo, W., Qiu, J., Liu, X. et al. (2020). *Nat. Commun.* 11: 1.
49 Talaie, E., Duffort, V., Smith, H.L. et al. (2015). *Energy Environ. Sci.* 8: 2512.
50 Do, J., Kim, I., Kim, H., and Jung, Y. (2020). *Energy Storage Mater.* 25: 62.
51 Li, Q., Xu, S., Guo, S. et al. (2020). *Adv. Mater.* 32: 1907936.
52 (a) Lu, Z. and Dahn, J. (2001). *Chem. Mater.* 13: 1252. (b) Wang, P.F., You, Y., Yin, Y.X., and Guo, Y.G. (2018). *Adv. Energy Mater.* 8: 1701912. (c) Yao, H.-R., Wang, P.-F., Gong, Y. et al. (2017). *J. Am. Chem. Soc.* 139: 8440.
53 Kikkawa, S., Miyazaki, S., and Koizumi, M. (1985). *Mater. Res. Bull.* 20: 373.
54 Zhao, J., Zhao, L., Dimov, N. et al. (2013). *J. Electrochem. Soc.* 160: A3077.
55 Xu, J., Han, Z., Jiang, K. et al. (2020). *Small* 16: 1904388.
56 Yabuuchi, N., Yoshida, H., and Komaba, S. (2012). *Electrochemistry* 80: 716.
57 Lee, E., Brown, D.E., Alp, E.E. et al. (2015). *Chem. Mater.* 27: 6755.

58 Silván, B., Gonzalo, E., Djuandhi, L. et al. (2018). *J. Mater. Chem. A* 6: 15132.
59 Li, Y., Gao, Y., Wang, X. et al. (2018). *Nano Energy* 47: 519.
60 Susanto, D., Cho, M.K., Ali, G. et al. (2019). *Chem. Mater.* 31: 3644.
61 (a) Yabuuchi, N., Kajiyama, M., Iwatate, J. et al. (2012). *Nat. Mater.* 11: 512. (b) Dose, W.M., Sharma, N., Pramudita, J.C. et al. (2017). *Chem. Mater.* 29: 7416. (c) Yoshida, H., Yabuuchi, N., and Komaba, S. (2013). *Electrochem. Commun.* 34: 60. (d) Wang, X., Liu, G., Iwao, T. et al. (2014). *J. Phys. Chem. C* 118: 2970.
62 (a) de Boisse, B.M., Cheng, J.-H., Carlier, D. et al. (2015). *J. Mater. Chem. A* 3: 10976. (b) Mu, L., Xu, S., Li, Y. et al. (2015). *Adv. Mater.* 27: 6928. (c) Li, X., Wu, D., Zhou, Y.-N. et al. (2014). *Electrochem. Commun.* 49: 51.
63 Thorne, J., Zheng, L., Lee, C.L. et al. (2018). *ACS Appl. Mater. Interfaces* 10: 22013.
64 Yabuuchi, N., Yano, M., Yoshida, H. et al. (2013). *J. Electrochem. Soc.* 160: A3131.
65 Kubota, K., Asari, T., Yoshida, H. et al. (2016). *Adv. Funct. Mater.* 26: 6047.
66 Li, M., Wood, D.L., Bai, Y. et al. (2020). *ACS Appl. Mater. Interfaces* 12: 23951.
67 Delmas, C., Braconnier, J.-J., Fouassier, C., and Hagenmuller, P. (1981). *Solid State Ionics* 3: 165.
68 (a) Kaufman, J.L. and Van der Ven, A. (2019). *Phys. Rev. Mater.* 3: 015402. (b) Bianchini, M., Wang, J., Clément, R., and Ceder, G. (2018). *Adv. Energy Mater.* 8: 1801446.
69 Doubaji, S., Ma, L., Asfaw, H.D. et al. (2018). *ACS Appl. Mater. Interfaces* 10: 488.
70 Carlier, D., Cheng, J., Berthelot, R. et al. (2011). *Dalton Trans.* 40: 9306.
71 Kang, S.M., Park, J.H., Jin, A. et al. (2018). *ACS Appl. Mater. Interfaces* 10: 3562.
72 Han, S.C., Lim, H., Jeong, J. et al. (2015). *J. Power Sources* 277: 9.
73 (a) Molenda, J., Stokłosa, A., and Than, D. (1987). *Solid State Ionics* 24: 33. (b) Mendiboure, A., Delmas, C., and Hagenmuller, P. (1985). *J. Solid State Chem.* 57: 323.
74 Parant, J.-P., Olazcuaga, R., Devalette, M. et al. (1971). *J. Solid State Chem.* 3: 1.
75 Ma, X., Chen, H., and Ceder, G. (2011). *J. Electrochem. Soc.* 158: A1307.
76 Clément, R.J., Middlemiss, D.S., Seymour, I.D. et al. (2016). *Chem. Mater.* 28: 8228.
77 Jiang, L.-W., Lu, Y.-X., Wang, Y.-S. et al. (2018). *Chin. Phys. Lett.* 35: 048801.
78 (a) Clément, R.J., Bruce, P.G., and Grey, C.P. (2015). *J. Electrochem. Soc.* 162: A2589. (b) Wang, S., Sun, C., Wang, N., and Zhang, Q. (2019). *J. Mater. Chem. A* 7: 10138. (c) You, Y. and Manthiram, A. (2018). *Adv. Energy Mater.* 8: 1701785.
79 Chen, X., Wang, Y., Wiaderek, K. et al. (2018). *Adv. Funct. Mater.* 28: 1805105.
80 Sauvage, F., Laffont, L., Tarascon, J.-M., and Baudrin, E. (2007). *Inorg. Chem.* 46: 3289.
81 (a) Billaud, J., Singh, G., Armstrong, A.R. et al. (2014). *Energy Environ. Sci.* 7: 1387. (b) Hemalatha, K., Jayakumar, M., Bera, P., and Prakash, A. (2015).

J. Mater. Chem. A 3: 20908. (c) Xiao, Y., Wang, P.F., Yin, Y.X. et al. (2018). *Adv. Energy Mater.* 8: 1800492.
82 (a) Xu, X., Ji, S., Gao, R., and Liu, J. (2015). *RSC Adv.* 5: 51454. (b) Kang, W., Zhang, Z., Lee, P.-K. et al. (2015). *J. Mater. Chem. A* 3: 22846.
83 Kumakura, S., Tahara, Y., Kubota, K. et al. (2016). *Angew. Chem.* 128: 12952.
84 Liu, X., Zhong, G., Xiao, Z. et al. (2020). *Nano Energy* 76: 104997.
85 Kumakura, S., Tahara, Y., Sato, S. et al. (2017). *Chem. Mater.* 29: 8958.
86 Li, Y., Yang, Z., Xu, S. et al. (2015). *Adv. Sci.* 2: 1500031.
87 Wang, J., Liu, H., Yang, Q. et al. (2020). *ACS Appl. Mater. Interfaces* 12: 34848.
88 Wang, P.-F., Yao, H.-R., Liu, X.-Y. et al. (2018). *Sci. Adv.* 4: eaar6018.
89 Wang, Y., Mu, L., Liu, J. et al. (2015). *Adv. Energy Mater.* 5: 1501005.
90 Han, M.H., Gonzalo, E., Sharma, N. et al. (2016). *Chem. Mater.* 28: 106.
91 Konarov, A., Choi, J.U., Bakenov, Z., and Myung, S.-T. (2018). *J. Mater. Chem. A* 6: 8558.
92 Choi, J.U., Park, Y.J., Jo, J.H. et al. (2018). *ACS Appl. Mater. Interfaces* 10: 40978.
93 Yabuuchi, N., Hara, R., Kajiyama, M. et al. (2014). *Adv. Energy Mater.* 4: 1301453.
94 Sehrawat, D., Rawal, A., Cheong, S. et al. (2020). *Inorg. Chem.* 59: 12143.
95 Sehrawat, D., Rawal, A., Cheong, S., and Sharma, N. (2019). *ACS Omega* 4: 5784.
96 Chae, M.S., Kim, H.J., Lyoo, J. et al. (2020). *Adv. Energy Mater.* 10: 2002205.
97 (a) Li, N., Ren, J., Dang, R. et al. (2019). *J. Power Sources* 429: 38. (b) Li, X., Liu, W., Kou, H. et al. (2019). *Mater. Today Energy* 14: 100353. (c) Yu, F., Du, L., Zhang, G. et al. (2020). *Adv. Funct. Mater.* 30: 1906890.
98 (a) Liu, Y., Fang, X., Zhang, A. et al. (2016). *Nano Energy* 27: 27. (b) Alvarado, J., Ma, C., Wang, S. et al. (2017). *ACS Appl. Mater. Interfaces* 9: 26518.
99 Dang, R., Chen, M., Li, Q. et al. (2018). *ACS Appl. Mater. Interfaces* 11: 856.
100 You, Y., Dolocan, A., Li, W., and Manthiram, A. (2018). *Nano Lett.* 19: 182.
101 Sun, H.-H., Hwang, J.-Y., Yoon, C.S. et al. (2018). *ACS Nano* 12: 12912.
102 Kaliyappan, K., Or, T., Deng, Y.P. et al. (2020). *Adv. Funct. Mater.* 30: 1910251.
103 Kaliyappan, K., Liu, J., Xiao, B. et al. (2017). *Adv. Funct. Mater.* 27: 1701870.
104 Dick, S., Müller, M., Preissinger, F., and Zeiske, T. (2013). *Powder Diffract.* 12: 239.
105 (a) Dyer, L.D., Borie, B.S. Jr., and Smith, G.P. (1954). *J. Am. Chem. Soc.* 76: 1499. (b) Takeda, Y., Nakahara, K., Nishijima, M. et al. (1994). *Mater. Res. Bull.* 29: 659.
106 Braconnier, J., Delmas, C., and Hagenmuller, P. (1982). *Mater. Res. Bull.* 17: 993.
107 Han, M.H., Gonzalo, E., Casas-Cabanas, M., and Rojo, T. (2014). *J. Power Sources* 258: 266.
108 Wang, L., Wang, J., Zhang, X. et al. (2017). *Nano Energy* 34: 215.
109 Yu, H., Guo, S., Zhu, Y. et al. (2014). *Chem. Commun.* 50: 457.
110 Yuan, D., Liang, X., Wu, L. et al. (2014). *Adv. Mater.* 26: 6301.
111 Wang, P.-F., Guo, Y.-J., Duan, H. et al. (2017). *ACS Energy Lett.* 2: 2715.

112 (a) Duffort, V., Talaie, E., Black, R., and Nazar, L.F. (2015). *Chem. Mater.* 27: 2515. (b) Kubota, K. and Komaba, S. (2015). *J. Electrochem. Soc.* 162: A2538.
113 Zhang, Y., Wu, M., Teng, W. et al. (2020). *ACS Appl. Mater. Interfaces* 12: 15220.
114 Zheng, L., Li, L., Shunmugasundaram, R., and Obrovac, M. (2018). *ACS Appl. Mater. Interfaces* 10: 38246.
115 Risthaus, T., Zhou, D., Cao, X. et al. (2018). *J. Power Sources* 395: 16.
116 (a) Barker, M.G. and Hooper, A.J. (1973). *J. Chem. Soc. Dalton Trans.* 15: 1517. (b) Hamani, D., Ati, M., Tarascon, J.-M., and Rozier, P. (2011). *Electrochem. Commun.* 13: 938.
117 Didier, C., Guignard, M., Denage, C. et al. (2011). *Electrochem. Solid-State Lett.* 14: A75.
118 Didier, C., Guignard, M., Darriet, J., and Delmas, C. (2012). *Inorg. Chem.* 51: 11007.
119 Guignard, M., Didier, C., Darriet, J. et al. (2013). *Nat. Mater.* 12: 74.
120 Didier, C., Guignard, M., Suchomel, M.R. et al. (2016). *Chem. Mater.* 28: 1462.
121 Xia, X. and Dahn, J. (2011). *Electrochem. Solid-State Lett.* 15: A1.
122 Komaba, S., Nakayama, T., Ogata, A. et al. (2009). *ECS Trans.* 16: 43.
123 Kubota, K., Ikeuchi, I., Nakayama, T. et al. (2015). *J. Phys. Chem. C* 119: 166.
124 Bo, S.-H., Li, X., Toumar, A.J., and Ceder, G. (2016). *Chem. Mater.* 28: 1419.
125 (a) Luo, M., Ortiz, A.L., and Shaw, L.L. (2020). *ACS Appl. Energy Mater.* 3: 7216. (b) Sawicki, M., Ortiz, A.L., Luo, M., and Shaw, L.L. (2017). *ChemElectroChem* 4: 3222.
126 Tsuchiya, Y., Glushenkov, A.M., and Yabuuchi, N. (2017). *ACS Appl. Nano Mater.* 1: 364.
127 Yu, C.-Y., Park, J.-S., Jung, H.-G. et al. (2015). *Energy Environ. Sci.* 8: 2019.
128 (a) Liang, L., Sun, X., Denis, D.K. et al. (2019). *ACS Appl. Mater. Interfaces* 11: 4037. (b) Liang, L., Zhang, W., Denis, D.K. et al. (2019). *J. Mater. Chem. A* 7: 11915.
129 Wang, S., Chen, F., Zhu, T.-y. et al. (2020). *ACS Appl. Mater. Interfaces* 12: 44671.
130 Tsuchiya, Y., Takanashi, K., Nishinobo, T. et al. (2016). *Chem. Mater.* 28: 7006.
131 Xi, K., Chu, S., Zhang, X. et al. (2020). *Nano Energy* 67: 104215.
132 Wang, Y., Cui, P., Zhu, W. et al. (2019). *J. Power Sources* 435: 226760.
133 Zheng, L., Bennett, J.C., and Obrovac, M. (2019). *J. Electrochem. Soc.* 166: A2058.
134 (a) Hasa, I., Buchholz, D., Passerini, S. et al. (2014). *Adv. Energy Mater.* 4: 1400083. (b) Lv, T., Guan, L., Xiao, P. et al. (2019). *J. Mater. Sci.* 54: 5584. (c) Zhu, Y.-E., Qi, X., Chen, X. et al. (2016). *J. Mater. Chem. A* 4: 11103.
135 Wang, H., Yang, B., Liao, X.-Z. et al. (2013). *Electrochim. Acta* 113: 200.
136 Wang, Y., Xiao, R., Hu, Y.-S. et al. (2015). *Nat. Commun.* 6: 1.
137 Cao, M., Wang, T., Shadike, Z. et al. (2018). *J. Electrochem. Soc.* 165: A565.
138 (a) Mariyappan, S., Marchandier, T., Rabuel, F. et al. (2020). *Chem. Mater.* 32: 1657. (b) Li, Z.Y., Wang, H., Chen, D. et al. (2018). *ChemSusChem* 11: 1223.
139 Xu, Y.-S., Gao, J.-C., Tao, X.-S. et al. (2020). *ACS Appl. Mater. Interfaces* 12: 15313.

140 Lee, E., Lu, J., Ren, Y. et al. (2014). *Adv. Energy Mater.* 4: 1400458.
141 Guo, S., Liu, P., Yu, H. et al. (2015). *Angew. Chem. Int. Ed.* 54: 5894.
142 Li, Z.-Y., Zhang, J., Gao, R. et al. (2016). *J. Phys. Chem. C* 120: 9007.
143 Qi, X., Liu, L., Song, N. et al. (2017). *ACS Appl. Mater. Interfaces* 9: 40215.
144 Bérardan, D., Franger, S., Meena, A., and Dragoe, N. (2016). *J. Mater. Chem. A* 4: 9536.
145 Zhao, C., Ding, F., Lu, Y. et al. (2020). *Angew. Chem. Int. Ed.* 59: 264.
146 (a) de Boisse, B.M., Reynaud, M., Ma, J. et al. (2019). *Nat. Commun.* 10: 1. (b) Maitra, U., House, R.A., Somerville, J.W. et al. (2018). *Nat. Chem.* 10: 288. (c) Dai, K., Wu, J., Zhuo, Z. et al. (2019). *Joule* 3: 518. (d) Sathiya, M., Rousse, G., Ramesha, K. et al. (2013). *Nat. Mater.* 12: 827.
147 Or, T., Kaliyappan, K., Bai, Z., and Chen, Z. (2020). *ChemElectroChem* 7: 3284.
148 Bai, X., Sathiya, M., Mendoza-Sánchez, B. et al. (2018). *Adv. Energy Mater.* 8: 1802379.
149 Ma, C., Alvarado, J., Xu, J. et al. (2017). *J. Am. Chem. Soc.* 139: 4835.
150 Rong, X., Hu, E., Lu, Y. et al. (2019). *Joule* 3: 503.
151 Kim, D. and Lee, J. (2020). *Chem. Mater.* 32: 5541.
152 Wang, Y., Wang, L., Zhu, H. et al. (2020). *Adv. Funct. Mater.* 30: 1910327.
153 Su, B., Wu, S., Liang, H. et al. (2020). *Chem. Mater.* 32: 8836.
154 Qiao, Y., Guo, S., Zhu, K. et al. (2018). *Energy Environ. Sci.* 11: 299.
155 Wang, Y., Liu, J., Lee, B. et al. (2015). *Nat. Commun.* 6: 1.
156 Shi, W.-J., Zhang, D., Meng, X.-M. et al. (2020). *ACS Appl. Mater. Interfaces* 12: 14174.

4

Polyanion-Type Cathodes for Sodium-Ion Batteries

4.1 Introduction

In stark contrast to the transition metal (TM) oxide cathodes, polyanionic compounds are more stable in structure and display good cycling stability. $NaFePO_4$ in olivine structure is highly pursued and optimized as a sodium-ion battery (SIB) cathode to duplicate the success of its counterpart $LiFePO_4$ in lithium-ion batteries (LIBs). The most appealing feature of $NaFePO_4$ lies in its high theoretical capacity of 154 mAh g^{-1} and working voltage plateau of 2.9 V (vs. Na^+/Na), along with rigid lattice framework that guarantees facile sodium-ion migration within the host structure. Moreover, as compared to low-dimensional sodium-ion transport pathways in $NaFePO_4$, the three-dimensional (3D) open framework of NASICON (Na Super Ionic Conductor)-type $Na_3V_2(PO_4)_3$ seems to allow faster sodium diffusion channels. Two sodium ions can be extracted from the lattice via a two-phase transition ($Na_3V_2(PO_4)_3 \leftrightarrow NaV_2(PO_4)_3$, 3.4 V vs. Na^+/Na) based on the V^{3+}/V^{4+} redox reaction, which yields a theoretical capacity of 117 mAh g^{-1} [1].

Higher operating potential can be achieved by pyrophosphates $Na_2MP_2O_7$ (M = Fe, Mn, and Co). Despite the low capacities of ~90 mAh g^{-1}, a high energy density of around 350 Wh kg^{-1} can be delivered. However, the ultrahigh potential (even up to 4.9 V) of pyrophosphates exceeds the stable voltage window of commercial electrolytes. In fact, the pyrophosphate anion $[P_2O_7]^{4-}$ is more thermodynamically stable than phosphates, which is beneficial for the accommodation of sodium ions without undergoing severe distortion. Research findings also show that the relatively high working voltage of $Na_2MP_2O_7$ arisen from its abundance crystal configuration can manage to hold sodium ions over a wide potential range [2].

Furthermore, the inductive effect of fluoride allows fluorophosphates to operate at higher operating potential when coupled with other TMs M^{3+}/M^{4+} (M = Ti, Fe, and V). For example, the working voltage of vanadium-based fluorophosphates can reach 4.0 V (vs. Na/Na^+) with redox couple V^{3+}/V^{4+}, which reflects that a decent energy density could be potentially obtained in full cell operation [3]. In fact, there are three polymorphic forms of sodium–vanadium fluorophosphates that have been studied for SIBs, which are $NaVPO_4F$, $Na_3(VO)_2(PO_4)_2F$, and $Na_3V_2(PO_4)_2F_3$. $NaVPO_4F$ has been demonstrated to be electrochemically active in the voltage range of 2.5–4.2 V (vs. Na/Na^+) with the successive reduction of V^{5+} to V^{3+} at 3.3/3.5 V

Sodium-Ion Batteries: Energy Storage Materials and Technologies, First Edition. Yan Yu.

and 3.65/3.75 V, respectively, giving rise to a theoretical capacity of 143 mAh g^{-1} [4]. NASICON-type $Na_3(VO)_2(PO_4)_2F$ shows slightly higher reaction plateaus at 3.68 and 4.15 V (vs. Na/Na^+) upon extraction of two sodium ions per formula, along with the oxidation of V^{3+} to V^{4+} [5]. Therefore, a specific capacity of 120 mAh g^{-1} can be obtained if the theoretical value is fully claimed. For $Na_3V_2(PO_4)_2F_3$, it outperforms the $Na_3(VO)_2(PO_4)_2F$ counterpart in terms of structural stability thanks to the stable V^{4+} phase that remains in the structure under oxidizing environment, despite of relatively lower theoretical capacity (120 mAh g^{-1}) [6].

Unfortunately, the absence of direct electron delocalization of polyanionic units results in inferior intrinsic electron conductivity. To solve this problem, strategies such as coating with a variety of conductive materials, the construction of unique micro-nano structures, and lattice doping are useful to achieve excellent energy storage performance for polyanionic cathode materials. Based on the knowledge and developments accumulated from fundamental research and application practices of polyanionic compound cathode materials, it is considered that this type of material will have great potential in large-scale energy storage systems.

In this chapter, we focus on the recent developments of advanced polyanionic compound cathode materials (i.e. olivine-type, NASICON-type, pyrophosphates, fluorophosphates, sulfates, silicates, other polyanion-type compounds, etc.) for high-performance SIBs.

4.2 Phosphates

4.2.1 $NaMPO_4$ (M = Fe and Mn)

Olivine-type $LiFePO_4$ cathode with advantages of high energy density, high safety, and low cost has been commercialized in LIB systems. Its sodium counterpart $NaFePO_4$ has also been widely studied, which is considered to be one of the earliest discussed cathode materials for polyanion-based SIBs [7]. The $NaFePO_4$ has a orthorhombic (space group: *Pnma*) crystal structure, where the hexagons are tightly arranged in the oxygen lattice and P atoms are confined to 1/8 of the tetrahedral center, while Na and Fe atoms occupy 1/2 of the octahedral center [8]. The NaO_6 octahedra and FeO_6 octahedra are distinguished owing to the difference in ionic radius and charge between sodium and iron ions [9]. The NaO_6 octahedra are arranged in a straight chain along the [010] direction through a common edge, while FeO_6 units are connected with each other through angle sharing to form a linear 1D zigzag chain parallel to the *c*-axis. Phosphate-ion PO_4 tetrahedron and FeO_6 octahedron share one edge and four corners, which constitute the main skeleton structure of $NaFePO_4$ [10]. The co-bordered NaO_6 octahedra provide a 1D Na^+ transfer channel following the direction of *b*-axis in the $NaFePO_4$ phase, which allows a relative low diffusion barrier (Figure 4.1a) [14]. The $Fe^{3+/2+}$ redox couple is able to retain a relatively stable polyanion skeleton structure; thus, a high working potential (~2.8 V vs. Na^+/Na, corresponding to Fe^{3+}/Fe^{2+} redox couple) together

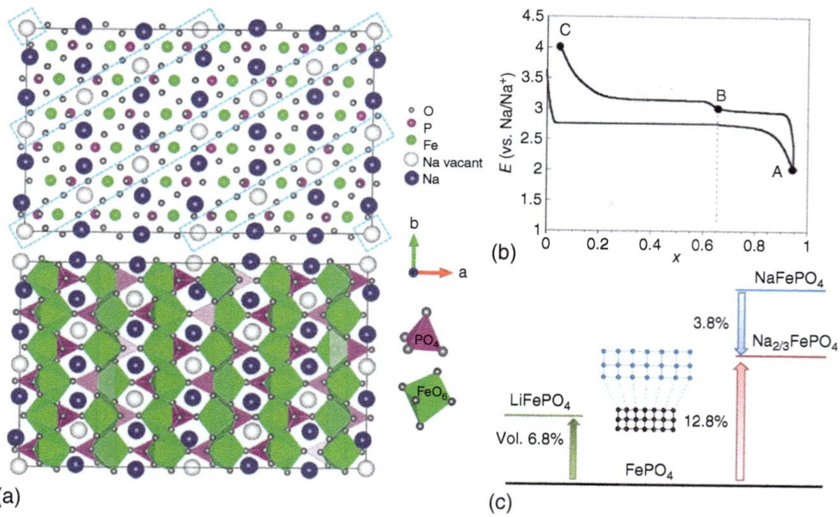

Figure 4.1 (a) Scheme of the refined crystal structure of $NaFePO_4$ including Na^+ vacancies and polyhedrons. Source: Galceran et al. [11a]. Reproduced with permission, 2014, American Chemical Society. (b) Typical electrochemical curve for the formation of $NaFePO_4$ and $Na_{0.7}FePO_4$ in potentiostatic intermittent titration technique (PITT) mode. Source: Moreau et al. [12]. Reproduced with permission, 2012, Elsevier. (c) Schematic diagram comparing the volume expansion of the $FePO_4$ upon sodium insertions. Source: Lu et al. [13]. Reproduced with permission, 2013, American Chemical Society.

with a high theoretical capacity (154 mAh g^{-1}) for $NaFePO_4$ cathode materials can be achieved [15].

Olivine-type $NaFePO_4$ was commonly derived from homologue $LiFePO_4$ via chemical/electrochemical ion-exchange, instead of conventional solid-phase transition reactions [10]. In one unit mole of $NaFePO_4$ material, one Na^+ ion participates in the redox reaction, corresponding to the Fe^{3+}/Fe^{2+} redox couple [16]. The electrochemical properties of olivine-type $NaFePO_4$ prepared by electrochemical Li–Na ion exchange route were studied for the first time in 2014 [17]. The material had a high initial capacity of 125 mAh g^{-1} with a voltage platform of ~2.8 V vs. Na^+/Na, posing itself a promising cathode material for SIB applications [18]. During the reversible reaction with Na^+, a transition stage occurs to buffer the mechanical strain that caused by large volumetric mismatch between iron phosphate ($FePO_4$) and $NaFePO_4$ (difference in unit volume of 17.58%). Olivine $NaFePO_4$ cathode usually has two independent charging platforms and one discharge platform (Figure 4.1b) [12]. The asymmetric (de)insertion of sodium ions into olivine $NaFePO_4$ cathode under different mechanisms was reported. For example, there is a stable intermediate phase composed of $Na_{2/3}FePO_4$ in the (de)intercalation process in Moreau and Casas-Cabanas's reports [12, 15, 19]. In the range of Na_xFePO_4 ($0 < x < 2/3$), two-phase transformation reaction is occurred, and Na_xFePO_4 ($2/3 < x < 1$) is in the solid solution phase. The existence of $Na_{2/3}FePO_4$ phase is beneficial to alleviate the large volume expansion in electrochemical reaction (Figure 4.1c) [13]. The Na ordering, electron mobility, and local magnetic environment of olivine-type $NaFePO_4$ were studied [11]. The average occupancy rate is 0.7, and there is no restriction on the modulation component.

The electrochemical performance of the $NaFePO_4$ cathodes in Na-ion cells is also investigated. The $NaFePO_4$-based SIBs could deliver a theoretical capacity of 125 mAh g^{-1} with a working voltage of 2.7 V (vs. Na/Na^+) [18]. However, a high volumetric mismatch of 17.58% between olivine-type $NaFePO_4$ and desodiated $FePO_4$ is highly detrimental to structural stability. Thus, reducing the volume change and improving the stability are the key matters to improve their Na^+ storage performance. The amorphous $NaFePO_4$ nanoengineering is able to relieve volume expansion and boost Na^+/e^- transport by providing large surface area and short diffusion distance. Li et al. employed an in situ hard template method to synthesize hollow amorphous $NaFePO_4$ nanospheres, whose hollow amorphous structure significantly improved the electronic transport and Na^+ (de)intercalation kinetics. As a result, it showed a high sodium storage capacity of 152 mAh g^{-1} at 0.1 C and long-term cycling stability with a retained capacity of 144 mAh g^{-1} after 300 cycles, implying superior performance to those bare $NaFePO_4$ materials [20]. Other than amorphous modification, improving the volumetric mismatch between the $NaFePO_4$ and the desodiated $FePO_4$ in the electrochemical reaction process is considered as a fundamental starting point to essentially strengthen the structure stability during cycling. Lately, a new type of $Na_{0.71}Fe_{1.07}PO_4$ cathode was also studied owing to its small volume change (~1%) upon desodiation, which not only promoted the rapid transport of ions and electrons but also ensured the structural stability during the repeated (de)sodiation. The $Na_{0.71}Fe_{1.07}PO_4$ cathode delivered a high discharge capacity of 140 mAh g^{-1} at a 0.1 C rate and an excellent rate capability exceeding 50 C as well as exceptional cycling stability over 5000 cycles at a high charging/discharging rate of 20 C [21].

Besides olivine-type $NaFePO_4$, $NaMnPO_4$ has also been considered as one of the most popular cathode materials for SIBs owing to its high theoretical capacity of 160 mAh g^{-1}. Unfortunately, the low electronic conductivity, poor sodium-ion diffusivity, and possible Mn dissolution in the electrolyte prevent $NaMnPO_4$ from getting satisfying electrochemical performance. Effective strategies such as nanocrystallization, uniform carbon coating, and cationic doping were implemented to gain excellent electrochemical performance of $NaMnPO_4$ [22]. Boyadzhieva et al. adopted uniform carbon coating and Mg-doping to improve the kinetics and thermal stability of the crystals during chemical and electrochemical desodiation. The Mg-substituted $NaMnPO_4$/C (Mg-NMP/C) displayed a more stable plateau at 3.9 V and higher discharging capacity (78 mAh g^{-1}) than that of the NMP phase [23]. Zou et al. also adopted a combination of the above strategies to improve the kinetics and thermal stability of the crystals during electrochemical (de)sodiation. The carbon-coated Mg-doped $Na(Mn,Fe)PO_4$ (NMMFP) nanoparticles (~50 nm) delivered a reversible specific capacity of 127 mAh g^{-1} at 0.1 C with an average discharge voltage of 3.1 V [24].

4.2.2 NASICON-Type Phosphates

The typical NASCION (Na Super Ionic Conductor) compounds were first studied as solid electrolytes by Goodenough 20 years ago [25]. Since then, Delmas reported that NASICON is a suitable and attractive framework for electrode materials based on the study of NASICON-structured $NaTi_2(PO_4)_3$ [26]. Generally, the basic formula of NASICON cathode materials can be defined as $Na_xMM'(PO_4)_3$ (M = V, Mn, Ti, Fe, Ni, Cr, etc.; $x = 0$–4). And, the NASICON-type structure can be divided into two MO_6 octahedrons separated by three XO_4 tetrahedrons, sharing the same angular oxygen. The hexagonal NASICON has a total of four possible sodium positions in each recipe unit. While only one is a sixfold coordination 6b between the eight edges of MO_6, and the other three are 10-fold coordination 18e between these bands on the *c*-axis. Moreover, there are no shared edges or faces in the matrix, which result in all these large sodium sites being interconnected, thus forming an open, flexible 3D framework. It is unrealistic to expect high electronic conductivity in this structure because of the isolated metal octahedrons [27]. Necessary modifications are desirable.

4.2.2.1 NASICLON-type $Na_3V_2(PO_4)_3$

Many NASICON materials, such as $Na_3V_2(PO_4)_3$, $Na_3MnTi(PO_4)_3$, $Na_3Ni_2Cr(PO_4)_3$, $Na_3MnZr(PO_4)_3$, $Na_4MnCr(PO_4)_3$, $Na_2Co_2Fe(PO_4)_3$, $Na_2Ni_2Fe(PO_4)_3$, and $Na_3CoFe_2(PO_4)_3$, have been particularly well investigated as cathode materials of SIBs [28]. Among them, $Na_3V_2(PO_4)_3$ (NVP) with better thermal stability, longer cycle life, and better safety features has been considered as a popular cathode in SIBs [29]. The rhombohedral NVP with the space group of $R3\bar{c}$ offers an open framework, in which VO_6 octahedra and PO_4 tetrahedra interlinked each other by corner-sharing O, forming a conceptual framework: $[V_2(PO_4)_3]^{3-}$ "lantern" unit (Figure 4.2a) [30]. The unit cell shows the lattice parameters of $a = b = 8.728$ Å and $c = 21.804$ Å via Visualization for Electronic and Structural Analysis (VESTA). Moreover, there are two kinds of isolated Na ions locating in the passages with two distinct oxygen environments: Na1 (M(1), sixfold coordination) locating in a $V_2(PO_4)_3$ environment along the *c*-axis and Na2 (M(2), eightfold coordination) coordinated analogously but along *b*-axis. When Na^+ is extracted from NVP, two Na2-sited ions (2/3 occupancy) are easily extracted and the remaining one Na^+ anchored in Na1 site (1 occupancy) [32]. When applied as cathode, NVP based on the V^{3+}/V^{4+} redox conversion demonstrates a typical two-phase transition reaction via the equation: $Na_3V_2(PO_4)_3 \leftrightarrow NaV_2(PO_4)_3 + 2Na^+ + 2e^-$. Specifically, an impressive theoretical capacity of 117.6 mAh g^{-1} with a flat voltage platform at ~3.4 V (vs. Na^+/Na) is achieved (Figure 4.2b) [31]. The detail Na-extraction mechanism from Na2 sites was studied by applying first-principles calculations. The three possible migration routines for sodium ions were listed to evaluate their internal migration energies (*E*). Diffusion pathway 1, carrier Na^+ diffuses in the channels between two PO_4 tetrahedral along the a-axis direction. Gaps between PO_4 tetrahedron and VO_6 octahedron in the *b* orientation are defined as diffusion pathway 2. The migration energies (*E* values) of the above two pathways were

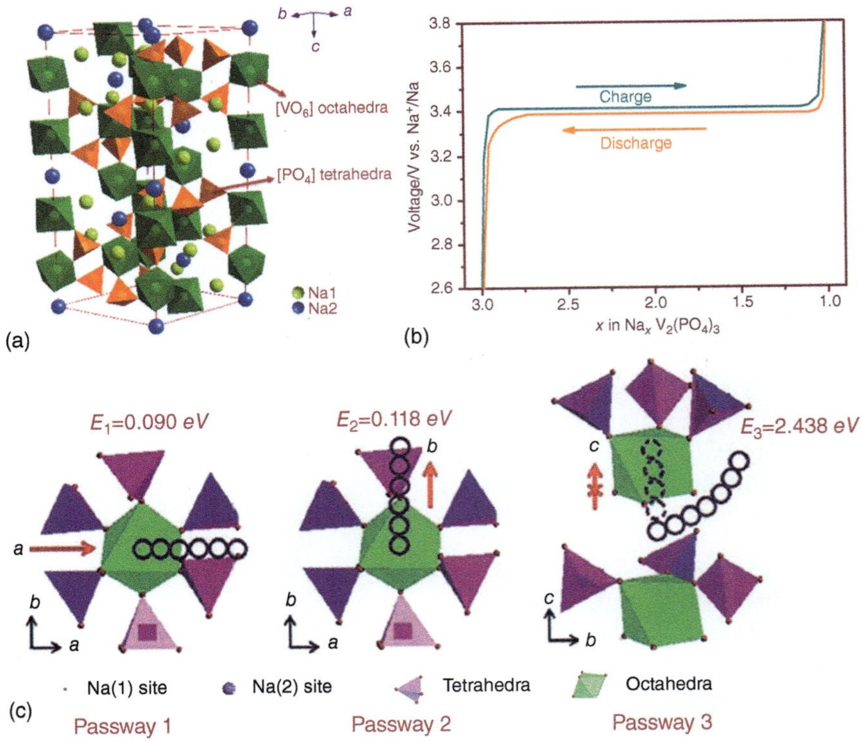

Figure 4.2 (a) The crystal structure of rhombohedra NVP. Source: Rui et al. [30]. Reproduced with permission, 2017, Wiley-VCH. (b) Electrochemical voltage – composition curve of the NVP cathode. Source: Zhang et al. [31]. Reproduced with permission, 2019, Royal Society of Chemistry. (c) Migration pathways of Na ions in NVP along *a*, *b*, and curved *c* axis. Source: Song et al. [32d]. Licensed under CC BY 3.0.

calculated to be very low (0.0904 eV for pathway 1 and 0.11774 eV for pathway 2), showing the most likely migration ways for Na^+ diffusion. While Na ions migrating in channels between adjacent $[VO_6]$ octahedra along the c direction consume large energy ($E > 200$ eV), which is considered as an unachievable pathway. Yet, Na^+ transportation along a curved *c* direction (noted as pathway 3) significantly decreased *E* value to reach 2.438 eV, demonstrating another possible migration pathway (Figure 4.2c) [32d, 33].

Though NVP shows good thermal stability and safety, the inferior intrinsic electronic conductivity causes obstacles in achieving good electrochemical performance and practical applications [34]. The traditional modification strategies are mainly focusing on surface conductivity modification (e.g. conductive carbon coating), micro-nano structure designing, elemental doping, and so on. Thus, conductive media and carbon materials are usually utilized including amorphous carbon, hollow carbon nanospheres, carbon nanotubes, graphene, mesoporous-ordered carbon frameworks, and interconnected carbon frameworks, which not only enhance the electrical conductivity of grains but also prevent particle agglomeration and electrode polarization during long-period cycling at a high current density [32c, 35]. Moreover, doped carbon materials by nitrogen and boron seem to be particularly advantageous [36]. Figure 4.3 summarizes the various ways to improve the overall conductivity of NVP [37].

Many organic polymers, sugars, organic acids, and surfactants can be pyrolyzed into amorphous carbon in inert atmosphere under high-temperature calcination. For example, the carbon-coating NVP nanoparticles were embedded in a porous carbon matrix originated from vanadium(III) 2,4-pentanedionate, which generated a double-carbon-embedding structured (C@NVP)@pC nanocomposite. The obtained (C@NVP)@pC composite showed exceptional electrochemical performance [35c]. In addition, Mai and co-workers evaluated the different electrochemical properties of the NVP nanoparticles coated with various carbon materials such as amorphous carbon, carbon nanotube, and graphite [34]. The result showed that the uniform coated NVP/amorphous carbon composite exhibited the highest energy storage properties among all tested samples.

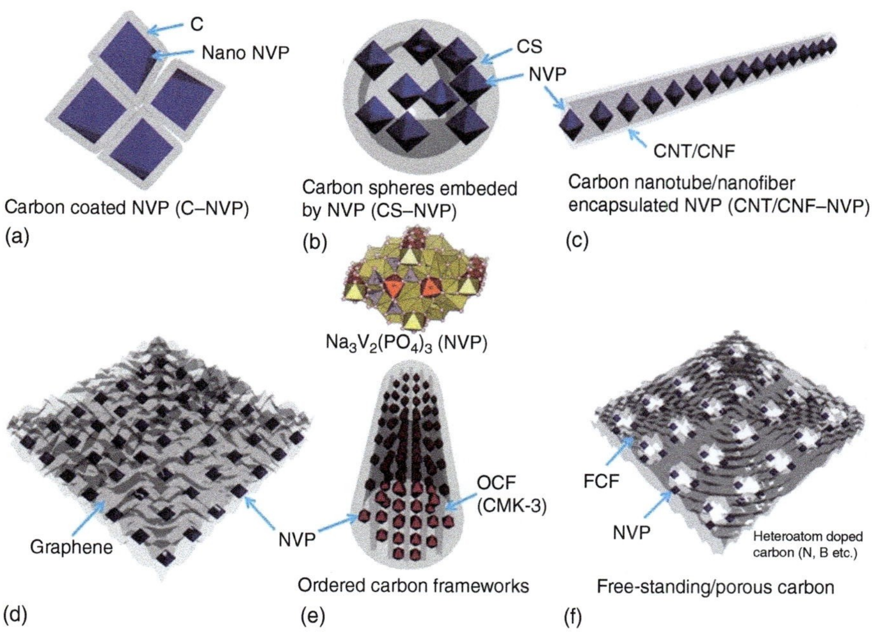

Figure 4.3 The scheme of modifying NVP particles via different approaches to enhance the conductivity of composites, including minimizing particles size (a), porous carbon spheres embedded NVP (b), carbon nanotube/nanofiber-encapsulated NVP nanoparticles (c), graphene film-supported NVP nanoparticles (d), ordered carbon framework (CMK-3)-encapsulated NVP nanoparticles (e), and free-standing/porous carbon-supported NVP nanoparticles or heteroatoms (N, B, etc.) doped carbon frameworks carried NVP nanoparticles (f). Note that electrolyte access to the electroactive mass must be guaranteed in all these cases. Source: Chen et al. [37]. Reproduced with permission, 2017, Wiley-VCH.

Carbon nanotube is a carbon allotrope with superior conductivity, high surface area, and excellent chemical stability, which is considered as an excellent current collecting phase. The commercial multi-walled carbon nanotubes (MWCNTs, 2 wt%) and NVP precursors were mixed and ball-milled to achieve an uniform composite [35d]. Due to the incorporation of MWCNT as a charge transfer medium, the MWCNT-modified NVP/C (MWCNT-NVP/C) composites exhibited favorable electrochemical behavior such as high sodium storage capacity of 146.5 mAh g^{-1} and high capacity retention of 94.3% after 50 cycles compared with the carbon-coated NVP.

Graphene, a 2D layered sp^2-bonded carbon atoms, has been applied in many research fields because of its particularly high electronic conductivity, large specific surface area, good thermal and mechanical performances as well as steady chemical stability [38]. For example, the NVP nanoparticles coated with the graphene sheets (amount of 5 wt%) achieved a superior high-rate capability (e.g. 75 and 60 mAh g^{-1} at 20 and 30 C, respectively) and exhibited good structure stability at both 1 C and 10 C over 300 cycles [39]. Besides, the layer-by-layer structured NVP@reduced graphene oxide (NVP@rGO) nanocomposites prepared by interfacial coupling of NVP/acetylacetone gel and reduced graphene oxide (rGO) nanosheets could provide high electronic and ionic conductivity and mechanical stability in the sodium insertion/extraction processes and exhibited remarkable high rate capacity (e.g. 73 and 41 mAh g^{-1} at 100 and 200 C, respectively) and ultra-long cyclic life with the capacity of 63 mAh g^{-1} at 50 C after 15 000 cycles [40].

Ordered porous carbon frameworks are widely used in many fields because of their high specific surface area, uniform porosity, good skeletal connectivity, and high electrical conductivity [41]. For example, a NVP@C nanoparticle set in mesoporous carbon material (CMK-3) chambers (abbreviation for NVP@C@CMK-3) was studied by Yu and co-workers [35a] Thin carbon coatings and mesoporous carbon CMK-3 matrix allowed rapid Na^+/electron transfer paths for NVP particles during electrochemical reactions. As a result, the unique architecture stimulated a super rate performance (81 mAh g^{-1} at 30 C) and stable cyclability (78 mAh g^{-1} at 5 C over 2000 cycles). A novel "honeycomb-graded porous microsphere" was prepared via a simple sol–gel method by Wang et al. The surfactant of cetyltrimethylammonium bromide (CTAB) played an important role in the preparation process, which acted as both a carbon source and a soft template to form a multistage porous structure. The prepared composites demonstrated a macro/mesopore structure with interconnected ion diffusion channels on the surface of the spherical particles, which facilitated the penetration of electrolytes and efficient Na^+ transport, obtaining high sodium storage capacities of 97 and 80 mAh g^{-1} at the 5 and 20 C, respectively [42].

Interconnected carbon frameworks are free-standing and binder-free substrates, which are easily acquired via varied techniques such as electrostatic spray deposition (ESD), filtration-freezing-drying, and spray-drying methods. For instance, a self-supported interpenetrating 3D tricontinuous cathode (NVP: reduced graphene oxide–carbon nanotube, abbreviated as NVP:rGO-CNT) was synthesized by the ESD technique, in which no additives or binders were used [32c]. The homogenously distributed NVP nanoparticles displayed outstanding electrochemical performance owing to the high porosity and interconnectivity. Specifically, the NVP:rGO-CNT composite demonstrated an impressive capacity (e.g. 82 mAh g^{-1} at the extremely high current density of 100 C rate) and excellent cycling performance with 96% of its initial capacity at 10 C after 2000 cycles. Besides, carbon-coated hollow-structured NVP/C spheres were prepared by using the ultrasonic spray pyrolysis procedure [43]. NVP particles uniformly distributed on the surfaces of the porous hollow carbon spheres accelerated the Na^+ transmission, exhibiting an enhanced energy storage property for SIB systems.

Beyond the pure carbon materials, nonmetallic elements such as nitrogen- and boron-doped carbon exhibit significantly higher electron transfer kinetics [44]. For nitrogen-doped carbon materials, three types of binding configurations, i.e. pyridine nitrogen, pyrrolidine nitrogen, and quaternary ammonium (or graphite nitrogen), are generally studied. (i) Pyridine nitrogen is interwoved with two adjacent carbon atoms at the edge or vacancy, which provides a p electron for the π system, thus showing a sp^2 hybrid structure. (ii) Pyrrolidine nitrogen provides two p electrons in the π system with a sp^3 hybrid structure, forming a five-membered ring. (iii) The quaternary nitrogen is formed by substituting n-type dopant for C in hexagonal ring, resulting in sp^2 hybridization [44f, 45]. Studies have shown that all these three types of N doping can affect the conductivity of carbon coating on host materials. Specifically, both pyridinic N and pyrrolic N create atomic defects and active sites that are beneficial to sodium storage [44f]. Different nitrogen/carbon ratio-coated NVP (NVP–C–N) nanoparticles were prepared, and their electrochemical properties were discussed by Wang and co-workers [36a] The N-doped NVP–C–N142 (molar ratio of N to C was 14.2%) sample with a thin carbon shell (~5 nm) released a high specific capacity of 102 mAh g^{-1} at 0.2 C, excellent rate capacities, and stable cycling performance (e.g. 92.6% of initial capacity after 100 cycles at 5 C). The good electrochemical performance is mainly profited from the preferentially sodium traveling through vacancies caused by N-doping on carbon shells. As a contrast, the sodium-ion conductivity was poor in the undoped carbon matrix. Furthermore, recent reports demonstrated that the N-doped carbon shell can improve sodium-ion diffusion and electron transfer simultaneously during charging and discharging processes [46]. Later, Wang and co-workers applied the similar strategy to compose the double-nanocarbon synergistically modified NVP composite, in which the N-doped carbon nanotubes encapsulated amorphous carbon-coated NVP nanoparticles (NVP@C + N@CNT) in the form of self-linking network [36c]. Benefiting from the extrinsic defects produced by the N-doping, the NVP@C-N@CNT composite exhibited exceptional rate capabilities from 0.2 C to 70 C.

Although N-doped carbon shells improved the electrochemical performance of NVP materials greatly, they are relatively sensitive to water and oxygen, resulting in adverse effects on its conductivity. Recently, boron-doped carbon materials derived from boron trichloride (BCl_3), borane tetrahydrofuran adduct (BH_3–THF), or boron oxide (B_2O_3) were not affected by the environment and had high electrical conductivity [47]. For instance, Shen et al. adopted sodium tetraphenylborate ($NaBC_{24}H_2O$) as boron source to modify a carbon-coated NVP (NVP–C–B) by a facile sol–gel method [36b]. The obtained B-doped carbon layer displayed quick charge transfer and fast Na^+ ion diffusion. Thus, with a proper molar ratio of B to C (~0.38%), the NVP–C–B(0.38%) exhibited lower electrochemical impedance and better rate capabilities than those of NVP–C–B(0.24%), NVP–C–B(0.51%), and NVP–C composites. Moreover, Jiang et al. designed a nitrogen (N), boron (B) co-doped carbon-coated 3D flower-like NVP composite (NVP@C-BN) using a wet chemical method [48]. This special micro-nano architecture with N, B co-doping not only improved surface wettability to relieve the volume expansion, shortened the distance of electron/Na-ion diffusion but also provided numerous active sites for Na-ion storage. As a result, the NVP@C–BN showed improved sodium storage capability, especially in cycling stability (e.g. maintained 79 mAh g^{-1} over 1000 cycles at 100 C).

In addition to the incorporating carbon materials into NVP and micro-nano structure design, crystal lattice doping with metal element is proven to be another effective method to improve electronic/ionic conductivity of NVP and many cations, including Mg^{2+} [49], Mn^{2+} [50], Al^{3+} [51], Ca^{2+} [52], Ni^{2+} [53], Ce^{3+} [54], Fe^{3+} [55], Ti^{4+} [56], Mo^{6+} [57], K^+ [58], Li^+ [59], etc., have employed as dopants in NVP cathodes. For instance, Li et al. composed different amounts of Mg^{2+}-doped NVP/C cathodes (i.e. $Na_3V_{2\text{-}x}Mg_x(PO_4)_3/C$) with the different values of $x = 0$, 0.01, 0.03, 0.05, 0.07, and 0.1 via a simple sol–gel method [49a]. Because the radius of Mg^{2+} ($r_{Mg^{2+}} = 0.65$ Å) is smaller than that of V^{3+} (i.e. $r_{Mg^{2+}} = 0.65$ Å $< r_{V^{3+}} = 0.74$ Å), the mixing of the Mg^{2+} dopant could cut down the average lengths of V—O and P—O bonds, providing broader Na^+ diffusion passages. The electrochemical testing revealed that the modified composites had remarkable enhancement in terms of the rate capability and cycle stability performance. The optimized $Na_3V_{1.95}Mg_{0.05}(PO_4)_3/C$ composites displayed a high capacity of 94 mAh g^{-1} at a high rate of 30 C. Besides, another cathode of Mg^{2+}-doped NVP/C nanocrystals embedded in graphene sheets was also studied, which showed an optimum performance of 85 mAh g^{-1} at 20 C [49c]. Klee et al. employed a citric acid as carbon source to prepare Mn^{3+}-doped NVP/C composites, that is, $Na_3V_{2\text{-}x}Mn_x(PO_4)_3/C$ ($0 \leq x \leq 0.7$) [50b]. Moderate amount of Mn^{3+} played a considerable important role in the sample morphology. And, a homogeneous composite was controlled in the range of $x = 0.3$–0.5 for the Mn substitution. It was noted that the low Mn^{3+} content ($x < 0.5$) would not alter the unit cell parameters of NVP. And, the $Na_3V_{1.7}Mn_{0.3}(PO_4)_3/C$ composite as the cathode for SIBs exhibited the excellent properties, e.g. showing discharge capacity of 104 and 92 mAh g^{-1} at 0.5 and 2 C, respectively. Aragón et al. employed iron to partially substitute the vanadium site in NVP ($Na_3V_{2-x}Fe_x(PO_4)_3/C$, $0 \leq x \leq 0.5$) [55b]. It was found that the unit cell volume was slightly increased with the increasing of iron content. The main reason is the larger radius of Fe^{3+} ($r_{Fe^{3+}} = 0.785$ Å) than that of V^{3+}. Galvanostatic cycling implied a significant raise of capacity for the composites with $x = 0.1$, 0.2, and 0.3. Later, Aragón et al. selected Al^{3+} cation to dope NVP and improved its electrochemical performance [51c]. In this work, a simple sol–gel method was employed to prepare four $Na_3V_{2-x}Al_x(PO_4)_3/C$ ($0 \leq x \leq 0.3$) samples. And, the $Na_3V_{1.8}Al_{0.2}(PO_4)_3/C$ showed the perfect capacity of 97 mAh g^{-1} at 6 C. Additionally, the high-valance

Mo^{6+} ($r_{Mo^{6+}}$ = 0.62 Å) was introduced to substitute the vanadium sites of NVP (i.e. $Na_{3-5x}V_{2-x}Mo_x(PO_4)_3/C$, $0 < x < 0.04$) by Li et al. [59] It was found that the Mo^{6+}-doped NVP showed an enhanced conductivity and Na^+-ion diffusion kinetics and displayed a superior rate capability of 94 mAh g^{-1} at 10 C and a stable cycle life with a capacity retention of 93.5% over 500 cycles.

Besides, doping monovalent cations such as Li^+ and K^+ at the Na sites in the NVP system was also discussed. It was concluded that there was a positive effect on the utilization of active substances and capacity attenuation by univalent cations doping. For example, the large K^+ as the doping ions can broaden the Na^+ transfer channels, making the NASICON framework more robust during long-term cycling [58]. Depending on the different K^+ doping contents in the crystal structure of NVP (i.e. $Na_{3-x}K_xV_2(PO_4)_3/C$, $x = 0.04$, 0.09, and 0.12), the electrochemical activity of the as-obtained composites was also inequable. The $Na_{2.91}K_{0.09}V_2(PO_4)_3/C$ owned a lower charge transfer resistance and a higher Na^+ diffusivity than those of other controlled composites, exhibiting the best rate capability. The Li^+ ions doped $Na_{3-x}Li_xV_2(PO_4)_3/C$ ($x = 0$, 0.01, 0.05, 0.1, 0.5, 0.7, and 1.0) composites were synthesized and investigated by Zhang and co-workers [59] When $x = 0.01$, 0.05, and 0.1, Li^+ preferred occupied the Na2 sites. Yet, Li^+ continues to occupy the Na1 site as $x > 0.1$. The electrochemical testing verified a favorable rate capability with a high discharge capacity of 102 mAh g^{-1} at 20 C for the $Na_{2.5}Li_{0.5}V_2(PO_4)_3/C$ compound.

Apart from the cation doping, anion doping was also one of the effective approaches to improve the cell performance of NVP. For instance, Muruganantham et al. selected the F^- as the dopant to prepare $Na_{3-x}V_2(PO_{4-x}F_x)_3$ ($x = 0$, 0.1, 0.15, and 0.3) composites via applying a simple sol-gel method [60]. As the radius of F^- is very close to that of O^{2-} and much larger than that of V^{3+} and Na^+ ions (i.e. $r_{O^{2-}}$ = 1.40 Å, r_{F^-} = 1.33 Å, $r_{V^{3+}}$ = 0.64 Å, r_{Na^+} = 1.02 Å), it is easy to replace O^{2-} with F^- and increase the operating voltage by induction effect. And, the ionic conductivity of the host NVP material is also significantly improved. Specifically, the pre-obtained $Na_{2.85}V_2(PO_{3.95}F_{0.05})_3$ compound exhibited good cyclic stability and brilliant rate performance.

4.2.2.2 NASICON-type $Na_3MnTi(PO_4)_3$

The NASICON-structured $Na_3MnTi(PO_4)_3$ (NMTP) material has drawn increasing attention as an immensely promising high energy substitute for the cathode material of SIBs due to its features of robust crystal framework, low toxicity, low cost, as well as high performance. The NMTP crystallizes in a rhombohedral structure ($R-3c$) with the cell parameters of $a = b = 8.8343$ Å and $c = 21.6654$ Å. In the internal crystal structure of NMTP, the isolated TiO_6 and MnO_6 octahedra shared all the corners with PO_4 tetrahedra and formed [$MnTi(PO_4)_3$] "lantern" units, which further construct a 3D open framework with excellent structural stability and intrinsic safety [61]. When NMTP is applied as the cathode for SIB system, similar Na^+ extraction mechanism as NVP is demonstrated. A single interstitial site (Na1 site, sixfold coordination) per formula is occupied by a less mobile Na^+ ion (occupancy factor: 0.981), and three equivalent sites (Na2 sites, eightfold coordination) per formula unit are occupied by two mobile Na^+ ions (occupancy factor: 0.673) [62]. Specifically, the NMTP electrode can display a high theoretical capacity of 176 mAh g^{-1} based on the reversible three-electron redox reactions (i.e. $Ti^{3+/4+}$ (≈2.1 V), $Mn^{2+/3+}$ (≈3.6 V), and $Mn^{3+/4+}$ (≈4.1 V vs. Na^+/Na) redox couples) [63].

Despite the favorable characteristics mentioned above, the NMTP cathode still suffers from inferior electronic conductivity, leading to disappointing cycling stability and unsatisfactory rate performance. For example, the NMTP only displayed a discharge capacity of 80 mAh g^{-1} at the current density of 0.1 C and obvious capacity fading (from 80 to 62 mAh g^{-1}) during the first 20 cycles [64]. Some efficient methods aimed at improving the electron/ion transportation dynamics and enhancing the electrochemical performance of SIB system have been explored and applied. For instance, a stable structured NMTP/carbon hollow microspheres were synthesized by a spray-drying-assisted process, and they provided a superior cyclability with 92% capacity retention after 500 cycles at 2 C [61]. Zhu et al. employed a dual carbon decoration strategy to optimize the sodium storage performance. Semi-graphitic carbon and rGO were selected as carbon sources to co-functionalize NMTP microspheres (NMTP/C@rGO). The prepared NMTP/C@rGO manifested a high energy density of 410 Wh kg^{-1} and long-term durability with a capacity of 76% after 800 cycles [63]. And, Li et al. also carried out the structural optimization by interpenetrating graphene coating and carbon-shell design for NMTP (rGO@NMTP-C) cathode to overcome its drawback of inferior electron/ion diffusion dynamics. Consequently, the rGO@NMTP-C showed outstanding cycling performance (e.g. 74.5% of capacity retention after 3500 cycles at 20 C) and ultra-high rate capability (92.4 mAh g^{-1} at 50 C) [65].

In addition to the above representative cathode materials, other NASICON-type cathode materials, such as $Na_3Ni_2Cr(PO_4)_3$ [66], $Na_3MnZr(PO_4)_3$ [67], $Na_4MnCr(PO_4)_3$ [68], $Na_2Co_2Fe(PO_4)_3$ [69], $Na_2Ni_2Fe(PO_4)_3$ [70], and $Na_3CoFe_2(PO_4)_3$ [71], also showed their superiority in distinctive open framework geometry, excellent structural flexibility, facile scalable synthesis, and outstanding electrochemical performance, which are expected to become promising sodium insertion cathode contenders for large-scale energy storage application.

4.3 Pyrophosphates

Encouraged by the application of $Li_2FeP_2O_7$ as high-performance cathode material for LIBs [72], Na-based pyrophosphate has emerged as an insertion host of SIBs in the past few years [73]. There are several major types of pyrophosphates: sodium monometallic pyrophosphates $NaMP_2O_7$ (M = Fe, V, and Ti) [74], sodium bimetallic pyrophosphates $Na_2MP_2O_7$ (M = Fe, Co, and Mn) [75], and the mixed metal sodium pyrophosphates $Na_4M_3(PO_4)_2P_2O_7$ (M = Fe, Co, Mn, and Ni) [55a, 76]. Pyrophosphate, which is easily obtained from the thermal decomposition and oxygen evolution of phosphate, has thermodynamic stability and is comparable to other polyanionic compounds. The high intrinsic stability pyrophosphate $P_2O_7^{4-}$

anions offer many clear advantages for Na-based pyrophosphate materials in the large-scale SIB applications.

4.3.1 $NaMP_2O_7$ (M = Fe, V, and Ti)

Sodium pyrophosphate compounds are usually composed of TM octahedral MO_6 and P_2O_7 units, which are interconnected to form a solid framework [74b]. The study of Na-pyrophosphate can be dated back to 1969 when $NaFeP_2O_7$ was first unveiled [77]. $NaFeP_2O_7$ has two different crystal structures: $NaFeP_2O_7$-I is isostructural with $KAlP_2O_7$ and $KFeP_2O_7$; $NaFeP_2O_7$-II is isostructural with $NaAlP_2O_7$. As shown in Figure 4.4, $NaFeP_2O_7$ (monoclinic crystals, space group: $P2_1/c$) can be divided into two distinct crystal structures (I-type and II-type) owing

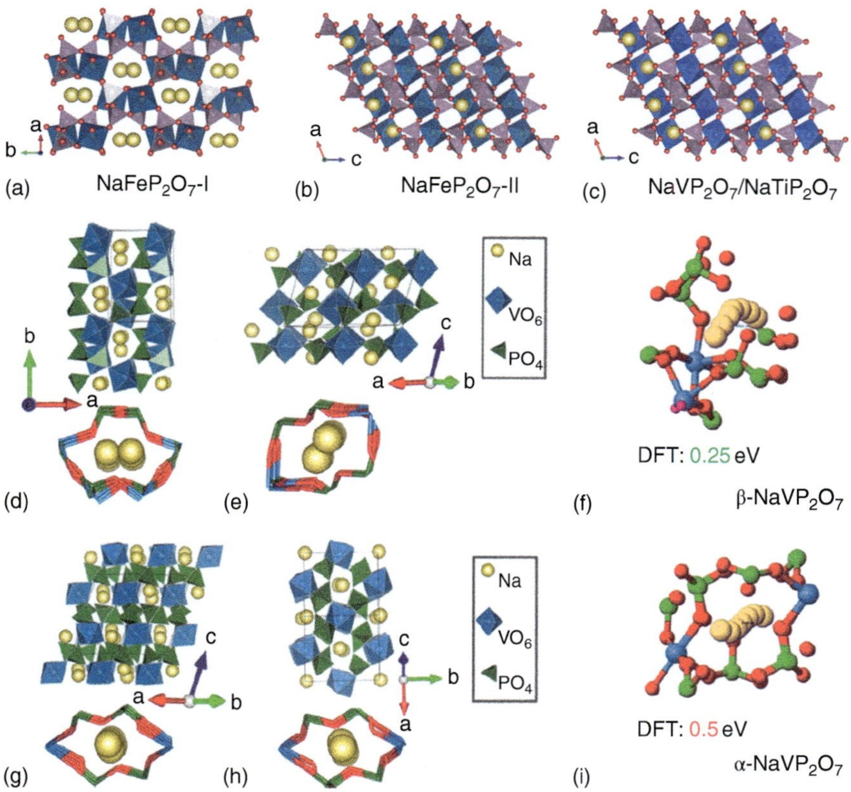

Figure 4.4 Crystal structures of (a) $NaFeP_2O_7$-I, (b) $NaFeP_2O_7$-II, and (c) $NaVP_2O_7$/$NaTiP_2O_7$. Source: Ng and Calvo [78]. Reproduced with permission, 1973, National Research Council. Crystal structure of β-$NaVP_2O_7$ viewed along the [001] and [110] directions (d, e) and α-$NaVP_2O_7$ along the [110] and [101] directions (g, h). Schematic representation illustrates Na^+ local coordination in the β- and α-phases (red, green, and blue "semi-bonds" correspond to O, P, and V atoms, respectively). Overlapped Na^+ positions from every image obtained from optimization with the NEB method for the shortest migration pathways in (f) β-$NaVP_2O_7$ and (i) α-$NaVP_2O_7$ structures. Source: Drozhzhin et al. [79]. Reproduced with permission, 2019, American Chemical Society.

to the irreversible phase transition at different temperatures [78]. $NaFeP_2O_7$-I is composed of FeO_6 octahedrons and PO_4 tetrahedrons sharing angles, and its structure has channels for sodium-ion migration in the [001] direction (Figure 4.4a). Furthermore, $NaFeP_2O_7$-I can easily transform into the thermodynamically stable $NaFeP_2O_7$-II phase under high temperature conditions (≥750 °C). For example, Robert and Hagenmuller's groups studied the structure of $NaFeP_2O_7$-II, which was synthesized by annealing the mixed precursor at 900 °C, then melting at 1100 °C for one hour, and finally quenching the crystal to ambient temperature [80]. On the other hand, $NaFeP_2O_7$-II is a cage structure composed of corner-sharing FeO_6 octahedrons and PO_4 tetrahedrons, where the P–O–P angle reaches a high angle of 132.86° (Figure 4.4b). Although this solid frame structure is very dense, it can also provide Na^+ migration channels along the [101] direction. However, even though the above two structures can provide open channels for sodium (de)insertion and are electrochemically active in at least one direction, they have not been reported as electrodes for SIBs.

In addition to $NaFeP_2O_7$, there are many substitutes studied (e.g. $NaTiP_2O_7$ and $NaVP_2O_7$) with good electrochemical activities for SIBs. As shown in Figure 4.4c, $NaVP_2O_7$/$NaTiP_2O_7$ was isostructural to $NaFeP_2O_7$-II polymorphs. In 1988, Leclair et al. determined two-type crystalline forms and cell parameters of α- and β-$NaTiP_2O_7$, in which α-$NaTiP_2O_7$ was closed to β-cristobalite structure with intersecting hexagonal tunnels and β-$NaTiP_2O_7$ was isostructural to $NaFeP_2O_7$ and $NaMoP_2O_7$ [74a]. Both α- and β-$NaTiP_2O_7$, belong to $P2_1/c$ space group, constructed by TiO_6 octahedra that share their corners with diphosphate P_2O_7 groups, forming an alternated stacking of octahedral layers and phosphate layers parallel to [001] plane. In literature, the differences between α- and β-$NaTiP_2O_7$ originate from the fact that the diphosphate P_2O_7 groups share one or two corners with the same TiO_6 octahedron, respectively. Despite they pose stable NASICON structure that known for good ion-exchange capability and ionic conductivity, their application in energy storage is still lacking.

Centrosymmetric $NaVP_2O_7$ (space group: $P2_1/c$ (Z = 4), a = 7.3169(2) Å, b = 7.9350(2) Å, c = 9.567(2) Å) consists of a slightly twisted VO_6 octahedron connected with five unique P_2O_7 groups, first reported by Okada and coworkers in 2015 [74c]. And, $NaVP_2O_7$ can deliver a high theoretical specific capacity of 108 mAh g^{-1} with the average V^{4+}/V^{3+} redox potential centered at 3.4 V (vs. Na^+/Na) when it was employed as cathode materials for SIBs. However, it just exhibited an inferior initial discharge capacity of 38.4 mAh g^{-1} at 1/20 C (1 C = 108 mA g^{-1}) in the voltage window of 2.5–4.0 V [74c]. In this case, the electrochemical impedance spectroscopy (EIS) results indicated that the increased internal resistance limited the phase change dynamics between $NaVP_2O_7$ ($P2_1/c$) and $Na_{1-x}VP_2O_7$ ($P2_1$) during the charging–discharging processes. Additionally, on the basis of density functional theory (DFT) calculation, X-ray photoelectron spectroscopy (XPS), Fourier-transform infrared spectroscopy (FT–IR), and *ex situ* X-ray diffraction (XRD) analyses, it was further confirmed that the deintercalation of Na ions would cause slight deformation of the pyrophosphate group and reduce the symmetry of the pyrophosphate group, resulting in the limited electrochemical activity of $NaVP_2O_7$.

Based on the thermal dehydration of hydrogen phosphate intermediate ($NaV(HPO_4)_2 \rightarrow NaVP_2O_7 + H_2O$), Drozhzhin's group first reported a new type of β-polymorphic sodium vanadium pyrophosphate (β-$NaVP_2O_7$) [79]. The obtained β-$NaVP_2O_7$ is isomorphic with $KAlP_2O_7$-type structure, and its 3D skeleton is formed by VO_6 octahedron and P_2O_7 bi-tetrahedral pyrophosphate group. In this work, they also found that the different structural configurations of α- and β-$NaVP_2O_7$ would lead to significant differences in the Na^+ coordination environment (Figure 4.4d,e,g,h). For instance, the sodium channels of β-$NaVP_2O_7$ are along the [001] and [110] directions. The penetration and diffusion of Na vacancies in the direction of [110] requires two different hops with distances of 4.54 and 8.51 Å. In addition, the authors further focused on the direction of [001], where the penetration and diffusion requires two hops with almost the same distances of 4.12 Å. Furthermore, based on the result of DFT calculations, the corresponding migration barriers of β-$NaVP_2O_7$ were calculated to be 0.25 and 0.17 eV, and these low diffusion barriers were resulted from the small displacement of oxygen around the Na vacancy (Figure 4.4f). More importantly, the barrier of Na migration along [001] is even reduced to 0.2 eV after Na is completely released from β-$NaVP_2O_7$.

For α-$NaVP_2O_7$, its diffusion channels are along the [110] and [101] directions. The permeation diffusion along [110] requires two different hops with distances of 5.37 and 5.72 Å, and the corresponding migration barrier in this direction is 0.85 eV, which prevents the effective diffusion. For the other direction of [101], two hops of 4.12 and 5.68 Å are required with the corresponding migration barriers of 0.5 and 2.6 eV, respectively, making this direction completely independent of diffusion (Figure 4.4i). For SIBs, β-$NaVP_2O_7$ could provide a reversible capacity of 104 $mAh\,g^{-1}$ at 10 $mA\,g^{-1}$, which corresponded to 96% of its theoretical capacity. Remarkably, it also displayed two high-voltage platforms of 4.1 and 3.8 V and a high-rate capacity of 77 $mAh\,g^{-1}$ even at 50 C (1 C = 108 $mA\,g^{-1}$). Besides, it was worth nothing that the β-$NaVP_2O_7$ could also act as the anode, which provided a reversible capacity of 93 $mAh\,g^{-1}$ with an average potential of 1.5 V at a low rate of 10 $mA\,g^{-1}$. Finally, the full battery system with the β-$NaVP_2O_7$ electrodes delivered an overall sodium storage capacity of more than 200 $mAh\,g^{-1}$ within the voltage range between 1.0 and 4.4 V.

4.3.2 $Na_2MP_2O_7$ (M = Co, Fe, Mn, Cu, and Zn)

The $Na_2MP_2O_7$ based on the possible tetrahedral MO_4 or octahedral MO_6 coordination with metal M can be classified into three distinct polymorphs: (i) orthorhombic [$P2_1/cn$], (ii) triclinic [$P\bar{1}$], and (iii) tetragonal [$P4_2/mnm$] [75a, 81]. Among them, I-type pyrophosphate has the highest thermodynamics stability, while that of the III-type is the weakest. The polymorphism of the two-sodium metal pyrophosphates present the layered and 3D structure with tunnels containing Na atoms, promoting the transporting of the sodium ions, and showing fine sodium storage performance.

$Na_2CoP_2O_7$ is one of the typical two-sodium pyrophosphates with the three different normal polymorphs [82]. In Yamada's study, a layered orthorhombic $Na_2CoP_2O_7$ belonging to $Pna2_1$ space group was determined [2]. As shown in Figure 4.5a, this layered structure was constructed by CoO_4 tetrahedra and PO_4 tetrahedra, in which both tetrahedral units forming $[Co(P_2O_7)]^{2-}$ slabs are parallel to (001) plane and filled with Na atoms. Inspired by these structural features, Yamada and co-workers have explored an orthorhombic $Na_2CoP_2O_7$ for SIBs through a combination of solid-state and solution combustion process [85], which exhibited a reversible discharge capacity of ~80 mAh g^{-1} at C/20, which corresponded to 80% of the theoretical capacity. Furthermore, the average Co^{3+}/Co^{2+} redox potential was centered at 3.0 V (vs. Na/Na^+) during the (de)sodiation reaction process.

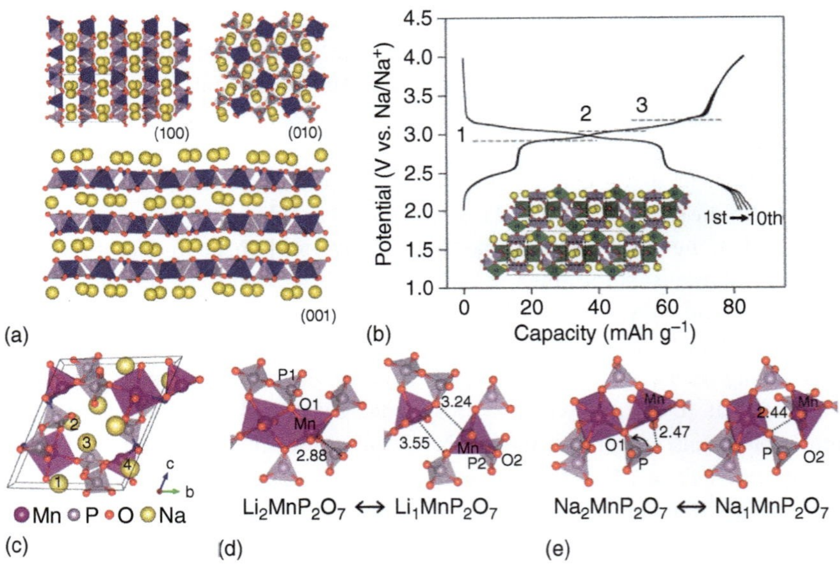

Figure 4.5 (a) Schematic presentation of layer-structured orthorhombic polymorph of $Na_2CoP_2O_7$ viewed along the (100), (010), and (001) planes. The CoO_4 tetrahedra (blue), PO_4 tetrahedra (light gray), and Na atoms (yellow) are illustrated. Source: Barpanda et al. [2]. Reproduced with permission, 2013, American Chemical Society. (b) Galvanostatic charge–discharge curves of $Na_2FeP_2O_7$ at a rate of C/20 showing highly reversible Na (de)insertion activity with the average Fe^{3+}/Fe^{2+} redox potential centered at ~3 V. Inset: a representative illustration of the crystal structure of $Na_2FeP_2O_7$ viewed normal to the *bc* plane. The tunnels for Na^+ migration (black dotted rectangles), FeO_6 octahedra (green), PO_4 tetrahedra (light purple), and Na atoms (yellow) are showed. Source: Barpanda et al. [83]. Reproduced with permission, 2012, Elsevier. (c) A representative illustration of the crystal structure of $Na_2MnP_2O_7$. In this periodic image, the octahedral MnO_6 and pyramidal MnO_5 moieties share one oxygen atom (blue) to make Mn_2O_{10} unit. The rearrangements of atomic positions and their bond lengths between the charged and discharged states for (d) $Li_2MnP_2O_7$ and (e) $Na_2MnP_2O_7$. Source: Park et al. [84]. Reproduced with permission, 2013, American Chemical Society.

In literature, the triclinic polymorph in disodium metal pyrophosphates ($Na_2MP_2O_7$) is referred to as the "rose" form, while the orthorhombic polymorph in $Na_2MP_2O_7$ calls the "blue" form. In this context, Jung et al. proposed that the "blue" form is more stable than "rose" type [86]. By introducing Na deficiency in $Na_{2-x}CoP_2O_7$, they further found that the "rose"-type $Na_2CoP_2O_7$ trended to become systematically stabilized with the increase of Na deficiency. Consequently, the authors have induced defects in triclinic $Na_2CoP_2O_7$ via controlling the reaction temperature (600–800 °C), which provided an increase in energy density of more than 40% and a high average voltage of 4.3 V (vs. Na/Na^+). Jung et al. further devised a computational strategy for selecting the best substituents to generate the desired polymorphs $Na_2Co_{1-x}M_xP_2O_7$ (M = Ca, Ni, or Mn) [87].

Besides, $Na_2FeP_2O_7$ is also one popular cathode material for SIBs in the family of pyrophosphates [83]. In 2010, Yamada's research group first demonstrated a decent electrochemical property of $Li_2FeP_2O_7$ particles [88]. Inspired by this, for the first time, Yamada and co-workers explored the application of pyrophosphates $Na_2FeP_2O_7$ in SIBs [83]. As shown in the inset of Figure 4.5b, the crystal framework of the as-prepared $Na_2FeP_2O_7$ was assumed to own a triclinic structure with the cell parameters of a = 6.43382(16) Å, b = 9.4158(3) Å, c = 11.0180(3) Å, α = 64.4086(15)°, β = 85.4794(19)°, γ = 72.8073(17)°, and V = 574.16(3) Å^3. $Na_2FeP_2O_7$ is composed of FeO_6 octahedrons and PO_4 tetrahedrons that connect in a staggered manner, forming an open tunnel for the accommodation of Na^+ ions along [100], [$\bar{1}$10], and [01$\bar{1}$] directions [75a]. In this context, its interesting structure offers the possibility to store metal ions. When used as the cathode for SIBs in the voltage window of 2.0–4.0 V, $Na_2FeP_2O_7$ displayed a high reversible capacity (82 mAh g^{-1} at C/20), stable average voltage platform (~3.0 V vs. Na/Na^+), and good rate performance (10 C). Meanwhile, it is worth noting that the Na ions (de)insertion process was divided into three small steps. This phenomenon can be ascribed to the structural rearrangement or Na^+ ordering upon cycling, which can be also observed in other Na-based cathodes (such as P2-$Na_{2/3}Co_{2/3}Mn_{1/3}O_2$) [89].

In addition to the advantages of low cost, stable and high-voltage platform, and promising electrochemical activity, the superior thermal stability of $Na_2FeP_2O_7$ further promotes its application in large-scale energy storage [90]. For example, Chen et al. studied the cycling lifespan of batteries at high temperature condition [91]. When tested at 363 K, the $NaFeP_2O_7$ electrode realized a high reversible capacity (91 mAh g^{-1} at 10 mA g^{-1}), excellent rate performance (59 mAh g^{-1} at 2000 mA g^{-1}), and good cycling stability (a capacity retention of 91% after 1000 cycles at 100 mA g^{-1}). In this study, based on the X-ray absorption spectroscopy analysis, the authors indicated that the increase of average oxidation state of Fe was accompanied by the shorten of Fe—O bonds upon charging process. Along this line, the $NaFeP_2O_7$ cathode-based batteries assembled by Chen and colleagues also manifested an impressive electrochemical performance over a wide temperature range (253–363 K) [75c]. In Yamada's study, they found that the charged state (i.e. $NaFeP_2O_7$) of $Na_2FeP_2O_7$ had good stability at ultra-high temperatures (up to

600 °C) and would not suffer thermal decomposition that generates oxygen from the active material [90a]. Nevertheless, owing to the temperature-induced phase transition, $NaFeP_2O_7$ will generate an irreversible polymorphic transition above 560 °C (i.e. from triclinic $P{-}1$ to monoclinic $P2_1/c$). The high operational safety can prevent the occurrence of chemical/thermal breakdown in battery system, which is beneficial to drive its application in SIBs [90a].

While the actual capacity of $NaFeP_2O_7$ is inferior to its theoretical capacity of 97 mAh g^{-1}. Aiming to improve the achievable capacity, cycling stability and rate capability of $NaFeP_2O_7$, large amounts of efforts have been directed to this material over the past few years [92]. The modification strategies of $NaFeP_2O_7$ mainly involve the hybridization with carbon materials and introduction of cations [93]. Owing to the intrinsically poor conductivity of $NaFeP_2O_7$, the researchers have paid attention to constructing $NaFeP_2O_7$-carbon material composites for the enhancement of electrochemical performance. For example, through aqueous synthesis followed by thermal treatment, Ruffo and co-workers have prepared a $Na_2FeP_2O_7$/multiple-walled carbon nanotube ($NaFeP_2O_7$/MWCNT) composite that featured a decent electrochemical property (68 mAh g^{-1} at 10 C) [93b]. Additionally, Zhang et al. reported an *in situ* carbon-coated layer and rGO matrix-modified $Na_2FeP_2O_7$ composite (NFPO@C@rGO) through a one-step urea-nitrate combustion process [93e]. As a result, the NFPO@C@rGO cathode achieved improved electrochemical performance compared to the NFPO@C and pure NFPO electrodes, retaining a capacity of ~100% after 300 cycles at 1 C. Very recently, to further modify $Na_2FeP_2O_7$ (NFPO), Liu and co-workers adopted a scalable sol–gel method combined with the ball-milling treatment to prepare a $Na_2FeP_2O_7$@C/expanded graphite (NFPO@C/EG) cathode for SIBs [93f]. With this unique multiscale carbon-modified structure, the electronic conductivity of active material can be ameliorated and the diffusion pathway of sodium ions can be shortened. Impressively, ultrahigh reversible capacities of 95 (almost close to theoretical capacity) and 60 mAh g^{-1} can be obtained at 11.6 and 2320 mA g^{-1}, respectively.

To raise the operational voltage, many research groups have focused on the Mn-based pyrophosphates ($A_2MnP_2O_7$) due to the electrochemical activity of the Mn^{2+}/Mn^{3+} redox couple. Owing to the sluggish kinetics of above redox couple, Whittingham's and Yamada's groups proposed the inactive nature of $Li_2MnP_2O_7$ at room temperature in LIBs [94]. But fortunately, Choi and co-workers have demonstrated a Mn-based pyrophosphates ($Na_2MnP_2O_7$) cathode with effective Na^+ storage at room temperature. Based on the XRD pattern and Rietveld refinement result, $Na_2MnP_2O_7$ was assumed to adopt a triclinic structure with the cell parameters of $a = 6.5373$ Å, $b = 9.5404$ Å, $c = 11.0735$ Å, $\alpha = 64.5664°$, $\beta = 79.9632°$, and $\gamma = 73.4167°$. As shown in Figure 4.5c, the octahedral MnO_6 and neighboring pyramidal MnO_5 in triclinic $Na_2MnP_2O_7$ form a Mn_2O_{10} unit by corner-sharing of one oxygen atom, which is different from the type of edge-sharing in monoclinic $Li_2MnP_2O_7$ [95]. Based on DFT calcgulations, they further found that the two edge-sharing Mn–O bonds in $Li_2MnP_2O_7$ were completely ruptured after charging. Furthermore, one new bond was formed to accommodate Jahn–Teller distortion and to minimize the electrostatic repulsion between Mn^{3+} ions. As a result, a pair of octahedral and pyramidal TM polyhedrons were disjointed into two pyramidal-like polyhedrons (Figure 4.5d). This transition would produce substantial kinetic barrier which made the deintercalation reaction unlikely to occur at room temperature. In contrast, benefiting from the corner-sharing type in $Na_2MnP_2O_7$ as shown in Figure 4.5e, the Mn—O bonds only rearranged. Therefore, the authors concluded that the significantly enhanced kinetics of $Na_2MnP_2O_7$ is mainly due to the corner-sharing crystal structure assisting the local flexible adjustment of Jahn–Teller distortion. Besides, the low degree of atomic rearrangement is beneficial for reducing the barrier energy of the electronic conduction and ionic migration. Consequently, the $Na_2MnP_2O_7$ cathode featured an impressive electrochemical activity for SIBs.

Moreover, Yamada's study unveiled a novel $Na_2MnP_2O_7$ phase, β-$Na_2MnP_2O_7$, registering a high Mn^{3+}/Mn^{2+} redox potential (ca. 3.6 V vs. Na/Na^+). The β-$Na_2MnP_2O_7$ was assumed to adopt a triclinic structure with the cell parameters of $a = 9.922(3)$ Å, $b = 11.083(4)$ Å, $c = 12.473(2)$ Å, $\alpha = 148.39(1)°$, $\beta = 121.94(5)°$, $\gamma = 68.42(3)°$, and $V = 599.4(1)$ Å^3. It is worth nothing that the lattice constants of the above two types of $Na_2MnP_2O_7$ turn out to be larger in all three directions compared to those of the Fe and Co counterpart perhaps because the ionic radius of Mn (0.97 Å) is larger than that of Fe (0.92 Å) and Co (0.89 Å). This structure consists of twisted MnO_6 octahedrons and PO_4 tetrahedrons and has 3D channels for ion (de)intercalation along [−110], [111] and [100] directions. By virtue of this open framework, the β-$Na_2MnP_2O_7$ was found to be electrochemically active in SIBs, showing a good reversible capacity of 80 mAh g^{-1} at C/20 and high average voltage platform of 3.6 V (vs. Na/Na^+) [75d].

Even though $Na_2MnP_2O_7$ has been reported as a promising cathode for SIBs due to its high potential, low cost and non-toxicity, its low initial CE, poor rate and cycling performance caused by the inferior electronic conductivity and manganese dissolution seriously hinder its practical applications. In this regard, through a feasible high-energy vibrating activation process, Jiao et al. proposed a robust graphene layer (GL)-modified $Na_2MnP_2O_7$ (NMP@GL) that can effectively improve its electrochemical performance [96]. In the ultrahigh-frequency vibrating activation process, the bulk NMP material was pulverized into sub-micron size and the rGO wrapped on the surface prevented these smaller particles from aggregating during the re-annealing process. The size reduction of particles not only shortened the migration pathway of sodium ions but also increased the specific surface area, which was beneficial for the contact between the electrolyte and the electrode, thus improving the rate capability of the active material. As a result, the NMP@GL electrode could display a reversible capacity of 93 mAh g^{-1} at 0.1 C with an initial CE of ~90%, and capacity retention of 83% over 600 loops at 2 C.

4.3.3 $Na_4M_3(PO_4)_2P_2O_7$ (M = Fe, Co, Mn, Ni, and Mg)

Currently, a unique mixed pyrophosphate $Na_4M_3(PO_4)_2P_2O_7$ (M = Fe, Co, Mn, Ni, and Mg) with a strong structural stability during the sodiation/desodiation reaction is considered as an interesting new type of polyanionic cathodes for SIBs [76a, 97]. Many researchers found the phosphate PO_4^{3-} and pyrophosphate $P_2O_7^{4-}$ exhibited good structural stability during reversible electrochemical reactions, promoting excellent electrochemical properties in sodium storage [9, 55a, 98].

In terms of electrode materials, the earth-abounded and low-cost elements are instrumental for large-scale energy density storage application. Iron-based $Na_4Fe_3(PO_4)_2P_2O_7$, as a representative of this category, was confirmed to be one of the most popular candidates for cathode materials of SIBs [55a, 99]. $Na_4Fe_3(PO_4)_2P_2O_7$ was crystallized into orthorhombic structure (space group: $Pn2_1a$) with the structural parameters of a = 18.0165(16) Å, b = 6.5510(6) Å, c = 10.6972(9) Å, and unit cell volume = 1262.54(24) Å^3 [100]. The structure of $Na_4Fe_3(PO_4)_2P_2O_7$ can be defined as a 3D network of $Fe_3P_2O_{13}$ sheets blocking parallel to the bc plane built up by FeO_6 octahedra and PO_4 tetrahedra. The blocks are connected along the a-axis by diphosphate groups. Along the [100], [010], and [001] directions, large tunnels for Na^+ diffusion are produced by FeO_6 octahedra with PO_4 tetrahedra and P_2O_7 diphosphates [101]. On the basis of first-principles calculation and the solid-state nuclear magnetic resonance (NMR) spectroscopy, there are four symmetrically distinguishable Na sites in the structure of $Na_4Fe_3(PO_4)_2P_2O_7$. Two Na sites (Na1 and Na4) formed by NaO_6 octahedra are shown on the b–c plane (Figure 4.6a). And, the other two Na sites (Na2 and Na3) formed by seven coordinated NaO_7 polyhedra and NaO_6 octahedra are located along the a-axis orientation [99a]. The extraction sequence of Na is described as follows: Na2 site (five-coordinated) is extracted first, followed by 1/2 Na in Na1 and 1/2 Na in Na4 in the meantime. And then, 1/2 Na in Na3 and the remaining Na in Na1 are extracted simultaneously, leaving the rest of Na in the Na3 and Na3 sites in $NaFe_3(PO_4)_2P_2O_7$ occupied [99a, 101b]. The (de)sodiation mechanism of $NaFe_3(PO_4)_2P_2O_7$ was further analyzed by electrochemical potentiostatic intermittent titration technique (PITT) measurements to monitor the redox reaction. And, the "Cottrel-type" curves represented the one-phase reaction with a reversible Fe^{2+}/Fe^{3+} redox reaction (Figure 4.6b,c) [55a].

Inspired by the feasible Na diffusion in the iron-based phosphate crystal, the electrochemical activity of $NaFe_3(PO_4)_2P_2O_7$ was evaluated. $NaFe_3(PO_4)_2P_2O_7$ exhibited a theoretical capacity of 129 mAh g^{-1} with a high average discharge voltage of ~3.1 V (vs. Na/Na$^+$) and low volume change (<4%) [102]. While, the major obstacles of $NaFe_3(PO_4)_2P_2O_7$ as cathode material for SIBs are the low electronic and ionic conductivities which hinder its rate capability and cyclability. Surface modification (e.g. highly conductive carbon coating), fine granulationp treatments, and functionalization treatments (e.g. acidification) can obviously improve the electronic and ionic conductivities of electrodes for SIBs. Profiting from

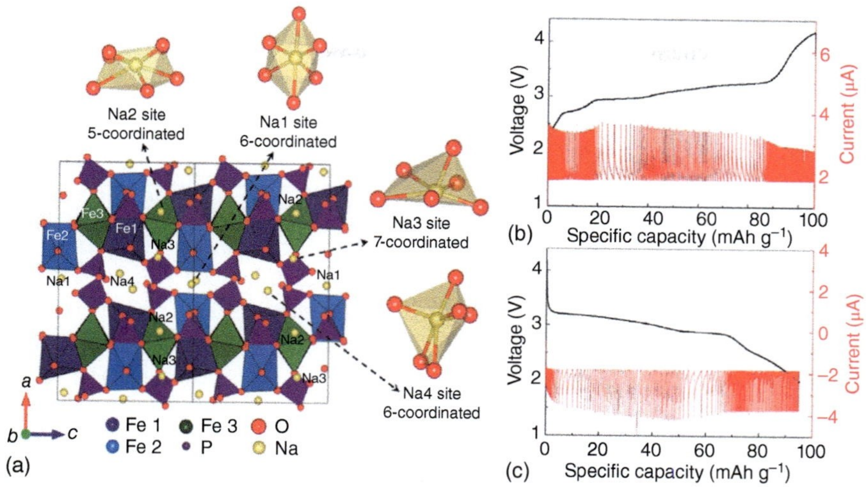

Figure 4.6 Schematic representations (a) and PITT measurements of $Na_4Fe_3(PO_4)_2P_2O_7$ during (b) charging and (c) discharging processes. Source: Kim et al. [99a]. Reproduced with permission, 2013, American Chemical Society.

the presence of an in situ formed 3D carbon network, the electronic conductivity of $Na_4Fe_3(PO_4)_2P_2O_7$/C nanocomposite prepared via the sol–gel process was improved markedly, which displayed a high initial capacity of 110 mAh g^{-1} at 0.05 C with the average discharge voltage of ~3.1 V, excellent rate capability of 78 mAh g^{-1} at 10 C, and stable cycling stability (e.g. 89% of capacity retention over 300 cycles) [103]. Cao et al. succeeded in scalable synthesizing the nanospherical $Na_4Fe_3(PO_4)_2P_2O_7$ growing on the MWCNTs ($Na_4Fe_3(PO_4)_2P_2O_7$@MCNTs) by using a simple acidification method and then a quickly spray drying method. The obtained composite as a cathode material for SIBs exhibited high electronic conductivity and revealed a high discharge capacity of 116 mAh g^{-1} at 0.1 C, an excellent rate capability with a capacity of 63 mAh g^{-1} at 20 C, as well as an outstanding cycle stability with a capacity retention of 95% after 1200 cycles at 2 C rate [104]. Yuan et al. studied the Na^+ storage property of the 3D graphene-decorated $Na_4Co_3(PO_4)_2P_2O_7$ microspheres (NFPP@rGO) prepared by applying a simple spray drying method. Combing the superiorities of $Na_4Co_3(PO_4)_2P_2O_7$ nanoparticles and the flexible 3D graphene network, the electronic conductivity and thermal stability of the as-prepared NFPP@rGO composite have been significantly improved, leading to the enhanced electrochemical properties in terms of high reversible capacity (e.g. 128 mAh g^{-1} at 0.1 C), super rate performance up to 200 C, and ultra-long cycling lifetime over 6000 cycles at 10 C [105]. Besides, $Na_4Fe_3(PO_4)_2P_2O_7$/C (NFPP/C) nanospheres with tunable size (~30 nm) and carbon layer thickness (~3 nm) synthesized by a template approach delivered a high discharge capacity of 128.5 mAh g^{-1} at 0.2 C and 79 mAh g^{-1} at 100 C [106]. Moreover, designing and fabricating structurally optimized $Na_4Fe_3(PO_4)_2P_2O_7$ cathodes also realized excellent electrochemical performance for SIBs. A core-double shell-structured $Na_4Fe_3(PO_4)_2P_2O_7$@$NaFePO_4$@C nanoparticles grown on a carbon cloth (CC) substrate (NFPP@NFP@C-CC) was designed to study the enhanced electrochemical activation. Combined with the superiorities of high voltage of $Na_4Fe_3(PO_4)_2P_2O_7$, high capacity of $NaFePO_4$, and high conductive CC, the as-obtained composite exhibited outstanding rate capability (68 mAh g^{-1} at 100 C) and cycling stability (without capacity decay over 3000 cycles at 10 C) [102]. Later, $Na_4Fe_3(PO_4)_2P_2O_7$ plates (NFPP-E) were also investigated to exploit the enhanced electrochemical performance by structural designing. Specifically, the NFPP-E delivered high capacity retention of 69% after 4400 cycles, which was much higher than that of the microporous $Na_4Fe_3(PO_4)_2P_2O_7$ particles (NFPP-C, capacity retention: 57%) under the same operating conditions [107]. $Na_4Fe_3(PO_4)_2P_2O_7$ thin films with uniform grain size (~220 nm) and surface morphology delivered a reasonable electrochemical performance with a highly reversible Na^+ storage capacity of ~120 mAh g^{-1} and good reversibility of over 500 cycles [108]. Additionally, the selection of electrolyte is also important to ameliorate the Na^+ storage performance of $Na_4Fe_3(PO_4)_2P_2O_7$ cathodes. Recently, Jang et al. substantially improved the electrochemical performance of $Na_4Fe_3(PO_4)_2P_2O_7$ cathodes by employing an ethylene carbonate (EC)/propylene carbonate (PC)-based $NaClO_4$ salt electrolyte. The $Na_4Fe_3(PO_4)_2P_2O_7$||Na cell with EC/PC/1 M $NaClO_4$ delivered a high discharge capacity of 128 mAh g^{-1}, closing to the theoretical capacity of 129 mAh g^{-1} [109]. And, Lee et al. used the linear carbonate-containing electrolytes with fluoroethylene carbonate (FEC) additive (i.e. EC/PC/DEC (5/3/2, v/v/v)/0.5 M $NaClO_4$ with FEC additive) to assist the exploration of sodium storage of $Na_4Fe_3(PO_4)_2P_2O_7$ cathodes. The investigation revealed that a remarkable enhancement in electrochemical performance of this cathode material with high-rate and long-cycle characteristics (e.g. 98% capacity retention after 300 cycles at 30 C) [110].

Following the research of $Na_4Fe_3(PO_4)_2P_2O_7$, homologue $Na_4Co_3(PO_4)_2P_2O_7$ has also shown its superiority in sodium storage property [98b, 111]. Recently, Nose et al. reported high voltages and high rate behavior for SIBs using $Na_4Co_3(PO_4)_2P_2O_7$ as the cathode material [76c]. $Na_4Co_3(PO_4)_2P_2O_7$ shows an orthorhombic structure with a $Pn2_1a$ space group [112]. There are four crystallographically distinct Na sites. Na1 and Na4 sites are sixfold coordinated to oxygen, while Na2 and Na3 sites have sevenfold coordination. The large channels enable a long-range Na-ion migration to occur with a low activation energy, making $Na_4Co_3(PO_4)_2P_2O_7$ a good 3D ion conductor [76a, 113]. This is advantageous over 1D conductors because even if some channels are blocked (e.g. by crystalline defects), other pathways would still be available to allow Na insertion into (and extraction from) the cathodes.

Despite having a high-voltage plateau at 4.5 V (vs. Na^+/Na) and a good theoretical capacity of 129 mAh g^{-1}, the rate capabilities of $Na_4Co_3(PO_4)_2P_2O_7$ cathode are limited due to unsatisfactory electronic/ionic conductivity and structural instability at high voltage [76c]. Highly conductive carbon modification is considered as the common strategy to solve this problem. Recently, the carbon-coated $Na_4Co_3(PO_4)_2P_2O_7$ (NCP–C) synthesized by a sol–gel method exhibited an enhanced electrochemical property. Specifically, the NCP–C cathode material delivered a stable reversible capacity of 70 mAh g^{-1} after 100 cycles with an excellent capacity retention of above 95% [113]. Additionally, the MWCNT-decorated $Na_4Co_3(PO_4)_2P_2O_7$ cathode (NCPP–CNT) was investigated by using suitable high-voltage electrolytes, which delivered a high initial discharge capacity of 92 mAh g^{-1} at 0.1 C and negligible capacity fading after 50 cycles [114]. Such excellent properties make $Na_4Co_3(PO_4)_2P_2O_7$-based composites one of the most promising cathode materials for SIBs. However, another drawback of $Na_4Co_3(PO_4)_2P_2O_7$ cathode material is its high cobalt content. Then, research efforts to reduce the cobalt amount in this material (i.e. metallic ionic doping) have been performed [115]. Metal doping (e.g. Mn, Ni, and Al) not only enhances its ionic conductivity but also improves the structural stability, which is beneficial to achieve excellent sodium storage properties [115a, 116]. For example, Nose et al. explored the electrochemical properties of Mn and Ni co-doped $Na_4Co_3(PO_4)_2P_2O_7$ material. It showed a high redox potential of 4.5 V and excellent rate capability with the reversible capacity of 103 mAh g^{-1} at 5 C rate [115a]. And, the effect of the Al dopant content on the sodium storage property of the $Na_4Co_3(PO_4)_2P_2O_7$ cathode was also investigated in Liu's work, from which concluded that the $Na_{3.85}Co_{2.85}Al_{0.15}(PO_4)_2P_2O_7$ ($Al_{0.15}$-NCPP) delivered a superior rate capability (80 and 73 mAh g^{-1} at 20 C and 50 C, respectively) and better ultralong cycling stability (96% capacity retention after 900 cycles at 10 C and 83% capacity retention after 8000 cycles at 30 C) than that of the NCPP, $Al_{0.05}$–NCPP, $Al_{0.1}$–NCPP, and $Al_{0.2}$–NCPP (e.g. only 80% capacity retention during 900 cycles at 10 C for NCPP) [116].

Ni-based polyanionic material $Na_4Ni_3(PO_4)_2P_2O_7$ as a potential high-voltage cathode for SIBs is also explored recently. Similar to $Na_4Fe_3(PO_4)_2(P_2O_7)$, $Na_4Ni_3(PO_4)_2P_2O_7$ is crystallized in the orthorhombic $Pn2_1a$ space group with $a = 18.0006(3)$ Å, $b = 6.4933(1)$ Å, and $c = 10.4115(2)$ Å [117]. It offers a stable 3D $(P_2O_7)^{4-}$ framework with a low migration barrier for diffusion of multiple Na sites [90b]. Zhang et al. first experimentally demonstrated the $Ni^{2+}/^{3+}$ redox activity in the mixed polyanionic framework for SIBs. $Na_4Ni_3(PO_4)_2P_2O_7$ enabled a reversible (de-)insertion of 1.3 Na with a theoretical capacity of 127.2 mAh g^{-1} under a high operating potential of 4.8 V vs. Na^+/Na [118]. Kumar et al. proposed the investigation of $Na_4Ni_3(PO_4)_2P_2O_7$ with rGO and carbon to increase the electronic conductivity. $Na_4Ni_3(PO_4)_2(P_2O_7)$-rGO and $Na_4Ni_3(PO_4)_2(P_2O_7)$-C delivered the discharge capacities of 72 mAh g^{-1}, exceeding the data reported in the recent literatures [119].

The Mn-based mixed-phosphate $Na_4Mn_3(PO_4)_2P_2O_7$ cathode material is very popular in SIB systems due to its high stable capacity, which would not be jeopardized by the structural volume changes induced by Jahn–Teller distortion (Mn^{3+}). The orthorhombic $Na_4Mn_3(PO_4)_2P_2O_7$ (group space: $Pn2_1a$; lattice parameters: $a = 18.02651(7)$ Å, $b = 6.65673(2)$ Å, and $c = 10.76886(4)$ Å) demonstrates a high energy density of 416 W h kg^{-1} under the high working potential of 3.84 V vs. Na^+/Na based on Mn^{2+}/Mn^{3+} redox couple [120]. Ryu et al. adopted a Co-substituted procedure and incorporated reduced graphene oxide (NMCMP/rGO) to study the sodium storage property of the $Na_4Mn_3(PO_4)_2P_2O_7$ materials and compared with no Co adjunction (NMMP/rGO). The NMCMP/rGO cathode exhibited a better electrochemical property with a high-voltage platform (4.5 V, corresponding to $Co^{2+/3+}$ redox couple) and excellent cyclic stability (88% capacity retention at 0.1 C) [121].

4.3.4 Other Pyrophosphates

Apart from the above-mentioned pyrophosphates, new type mixed-polyanion compounds such as sodium vanadium pyrophosphate $Na_7V_3(P_2O_7)_4$ and vanadium-based *ortho*-diphosphate $Na_7V_4(P_2O_7)_4(PO_4)$ are also considered as potential cathode candidates for SIBs owing to their merits of rich crystal chemistry, upstanding framework, and good sodium ionic mobility.

Sodium vanadium pyrophosphate $Na_7V_3(P_2O_7)_4$ shows a structural characteristic of a space group of $C2/c$ and lattice parameters of $a = 9.7219(4)$ Å, $b = 8.3185(4)$ Å, $c = 27.6051(11)$ Å, which is isostructural with the well-known $Na_7Fe_3(P_2O_7)_4$ composite [122]. The schematic crystal structure of $Na_7V_3(P_2O_7)_4$ is mainly composed of two consecutive layers (i.e. quasi-layer 1: $V(1)[P(1)P(2)O_7]_2$, and quasi-layer 2:$[V(2)P(3)P(4)O_7]_2$). Here, the three distinct sites Na2, Na3, and Na4 are interlinked in quasi-layer 1. While Nal ions locate in quasi-layer 2, connecting the Na2 and Na4 ions in quasi-layer 1. The octahedra VO_6 shares every corner with P_2O_7, which creates large tunnels and a stable 3D skeleton for the rapid Na^+ diffusion (Figure 4.7a). The redox potential of $Na_7V_3(P_2O_7)_4$ is significantly affected by the interaction between TMs and polyanions, and will increase through the induction effect. Therefore, the $Na_7V_3(P_2O_7)_4$ is expected to display a higher theoretical redox potential than other cathode materials. The (de)sodiation mechanism in a high-voltage zone manifests almost four Na ions per formula unit are deintercalated from the structure with the following order: two Na^+ in Na2 site → one Na^+ from Na3 site → one Na^+ in Na4 site (>4.5 V vs. Na/Na^+), remaining one Na^+ in Na4 site and all Na^+ in Nal site in the $Na_xV_3(P_2O_7)_4$ structure (Figure 4.7b). The overall electrochemical reaction within a practical voltage window (e.g. 2.5–4.35 V) mainly occurs between $Na_7V_3(P_2O_7)_4$ and $Na_4V_3(P_2O_7)_4$ based on the V^{3+}/V^{4+} redox reaction. Notably, a high redox potential at ≈4.13 V (vs. Na^+/Na) and a satisfactory theoretical capacity (≈80 mAh g^{-1}) can be obtained (Figure 4.7c) [123]. While Kovrugin et al. studied the V^{5+}/V^{4+} redox activity without achieving fully reversible reaction at high voltage, which delivered a capacity of about 106 mAh g^{-1}. Moreover, based on three redox couples (V^{3+}/V^{2+}, V^{4+}/V^{5+}, and V^{3+}/V^{4+}) in a wide voltage range of 1.3–4.8 V vs. Na^+/Na, $Na_7V_3(P_2O_7)_4$ was able to release a reversible capacity of 118 mAh g^{-1} [124]. In addition to broadening the voltage range, micro-nano structure design and surface conductive carbon decoration are always applied to facilitate fast electron/ion transport and enhance the electrochemical activities [125]. Ke et al. introduced a biochemical-enabled strategy with the hexahedral natural fern spore as the biotemplate. The hierarchical structure bio-$Na_7V_3(P_2O_7)_4$ composites owned small nanoparticles and high surface area, greatly facilitating the rapid ion transport. As a result, it provided excellent rate capability with a high capacity retention of 93% after 500 cycles at 20 C [126]. Then, freestanding 3D hybrid foams of $Na_7V_3(P_2O_7)_4$@biomass-derived porous carbon were also synthesized, and

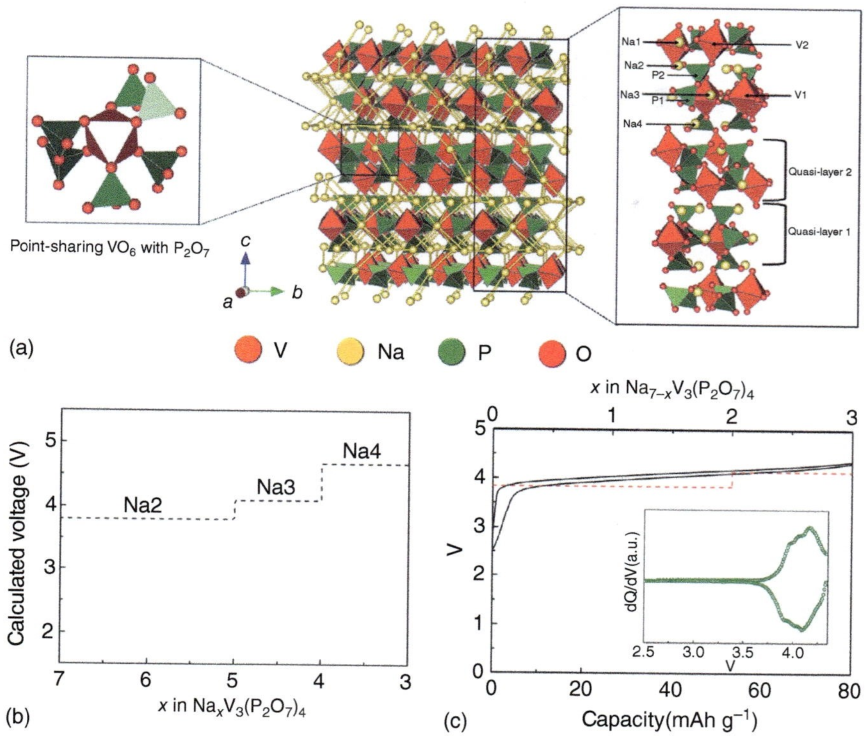

Figure 4.7 (a) Schematic of $Na_7V_3(P_2O_7)_4$ structure with 3D Na diffusion path. (b) Calculated average redox potential of $Na_xV_3(P_2O_7)_4$ ($3 \leq x \leq 7$). (c) A charge/discharge profile of $Na_7V_3(P_2O_7)_4$ with the calculated voltage (inset: the corresponding differential capacity plots of $Na_7V_3(P_2O_7)_4$). Source: Kim et al. [123]. Reproduced with permission, 2016, Wiley-VCH.

its sodium storage performance was studied. In this work, a graphene-like frame is extracted from the fungus and applied to realize the construction of a highly conductive 3D porous framework, which closely wrapped nanocrystals together to form a hierarchical hybrid structure. The self-assembled 3D hybrid foam achieved fast Na^+ (de)insertion kinetics and excellent cycling stability (e.g. 91% of initial capacity retained after 800 cycles at 3 C) thanks to the advantages of hierarchical porosity, high electrical conductivity, and good stability [127].

The $Na_7V_4(P_2O_7)_4(PO_4)$ compound is also a popular alternative cathode for high-performance SIBs, which demonstrates the tetragonal $P{-}42_1c$ space group with well-defined lattice parameters (i.e. a = 14.225(3) Å and c = 6.364 (17) Å). The crystal structure of $Na_7V_4(P_2O_7)_4(PO_4)$ cathode is assembled with the repetition of a basic unit: $(VP_2O_7)_4PO_4$ (Figure 4.8a), which interconnect together to make up a 3D skeleton structure with explicit Na^+ diffusion channels [128]. The sodium (de)intercalation mechanism of $Na_7V_4(P_2O_7)_4(PO_4)$ was evaluated using galvanostatic intermittent titration technique (GITT) measurement (Figure 4.8b). Based on the quasi-open-circuit potential (QOCP)

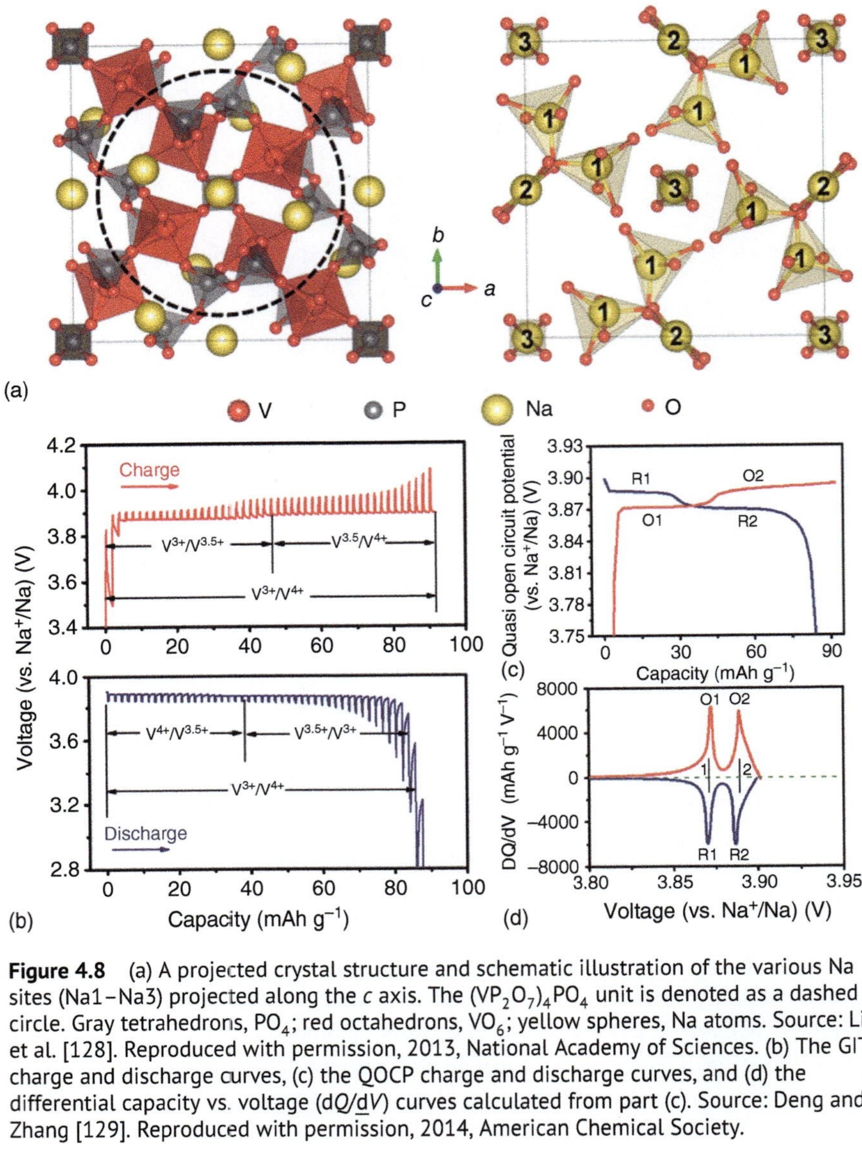

Figure 4.8 (a) A projected crystal structure and schematic illustration of the various Na sites (Na1–Na3) projected along the *c* axis. The $(VP_2O_7)_4PO_4$ unit is denoted as a dashed circle. Gray tetrahedrons, PO_4; red octahedrons, VO_6; yellow spheres, Na atoms. Source: Lim et al. [128]. Reproduced with permission, 2013, National Academy of Sciences. (b) The GITT charge and discharge curves, (c) the QOCP charge and discharge curves, and (d) the differential capacity vs. voltage (d*Q*/d*V*) curves calculated from part (c). Source: Deng and Zhang [129]. Reproduced with permission, 2014, American Chemical Society.

charge and discharge curves obtained from the GITT curves and the differential capacity (dQ/dV) curves calculated from Figure 4.8c, an intermediate phase $Na_5V_4^{3.5+}(P_2O_7)_4(PO_4)$ during the discharge and charge process was found, which derived two equilibrium potential plateaus at 3.8713 and 3.8879 V, corresponding to the $V^{3+}/V^{3.5-}$ and $V^{3.5+}/V^{4+}$ redox reactions, respectively (Figure 4.8d). The evaluation during sodium extraction process can be described in two steps: $Na_7V_4^{3+}(P_2O_7)_4(PO_4) \xrightarrow{\text{charge}} Na_5V_2^{3+}V_2^{4+}(P_2O_7)_4(PO_4) \xrightarrow{\text{charge}} Na_3V_4^{4+}(P_2O_7)_4(PO_4)$ [129]. Thanks to the robust host framework and favorable charge–discharge mechanism, the $Na_7V_4(P_2O_7)_4(PO_4)$ held exceptional electrochemical performances with long cycling capability over 1000 cycles. Additionally, the impurity of $Na_7V_4(P_2O_7)_4(PO_4)$ samples synthesized via conventional solid-state methods and their intrinsic low electronic conductivity limit their further development. Nanoengineering with a feasible synthetic method is needed to improve the purity and facilitate the electron/ionic diffusion, leading to a distinctly improved electrochemical performance. Recently, a 1D nanostructured $Na_7V_4(P_2O_7)_4(PO_4)$ with high purity prepared by a sol–gel method and subsequent high-temperature calcination process displayed favorable Na^+ intercalation kinetics and excellent electrochemical property with a large reversible capacity of 92.1 mAh g^{-1} at 0.05 C (nearly close to the theoretical capacity: 92.8 mAh g^{-1}) [129]. Subsequently, a freestanding 3D $Na_7V_4(P_2O_7)_4(PO_4)$-based hybrid foam also achieved favorable Na^+/electron diffusion kinetics and a ultralong-term cycling capability with 94% capacity retention following 800 cycles at a high rate of 20 C [127]. Coating conductive carbon is also an efficient strategy to promote a fast electron transfer. Prepared by a hydrothermal-assisted method with the introduction of templates and carbon additive, the $Na_7V_4(P_2O_7)_4(PO_4)$/C nanorods exhibited enhanced electronic dynamics and a good sodium storage capability [130].

4.4 Fluorinated Phosphate Cathodes

Fluorophosphates, produced by the addition of electronegative fluorine anions (F^-) into Na-based phosphate cathodes, are capable to maintain the charge balance of the structure in relation to phosphates. Na_2VPO_4F first caught the attention of researchers in 2003 [131], which not only stimulated the exploration of various fluorophosphates as SIB cathode materials but also promoted the development and extension of fluorophosphate to other TMs. The anionic F^- in the strong covalent anion lattice structure delivers strong inductive effect, resulting an enhanced redox voltage. There are three types of promising fluorophosphate-based cathodes for SIBs, which are $NaVPO_4F$, Na_2MPO_4F (M = Fe, Mn, and Ni), and $Na_3(VO_{1-x}PO_4)_2F_{1+2x}$ ($0 \leq x \leq 1$).

4.4.1 $NaVPO_4F$

$NaVPO_4F$ is the first type that been studied as cathode material for SIBs in the fluorophosphate group, which is crystallized into two polymorphs: tetragonal symmetric phase (space group: *I*4/*mmm*) and monoclinic phase (space group: *C*2/*c*) [132]. The tetragonal $NaVPO_4F$ showed an extended 3D framework consisting of $[VO_4F_2]$ octahedron and $[PO_4]$ tetrahedron for sodium-ion diffusion. Due to the synergistic effect of strong P—O bond and the inductive effect of F-anion, a high working platform of ~3.7 V can be achieved for sodium-ion full cell coupled with hard carbon as anode. The cycling stability of the tetragonal $NaVPO_4F$, however, was inferior, whose discharge capacity faded to less than half of the initial capacity only after 30 cycles [131]. In the monoclinic $NaVPO_4F$, two PO_4 tetrahedra are

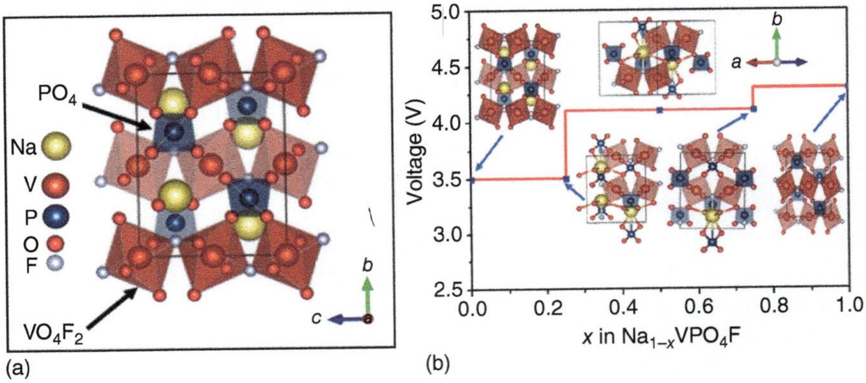

Figure 4.9 (a) Crystal structure of $NaVPO_4F$. (b) Theoretical voltage profile for $Na_{1-x}VPO_4F$. Source: Mamoor et al. [133]. Reproduced with permission, 2019, Royal Society of Chemistry.

connected with two different VO_4F_2 octahedra by the two angular oxygen atoms (Figure 4.9a), which provides a two-dimensional pathway along the (010) plane for the rapid Na^+ diffusion [134]. Moreover, monoclinic-Na_2VPO_4F realized a reversible (de)insertion of one Na^+ per $NaVPO_4F$ formula unit on the basis of V^{4+}/V^{3+} redox reaction, showing a high theoretical capacity of 143 mAh g^{-1}. The desodiation process can be analyzed by removing Na^+ from $Na_{1-x}VPO_4F$ with $x = 0, 0.25, 0.5, 0.75$, and 1 (Figure 4.9b). The evolution process can be divided into the following processes: $Na_1VPO_4F(C2/c) \rightarrow Na_{0.75}VPO_4F(C2/c) \rightarrow Na_{0.5}VPO_4F(P2/c) \rightarrow Na_{0.25}VPO_4F(C2/c) \rightarrow VPO_4F(C2/c)$. It was found that the structural symmetry of $NaVPO_4F$ shifted from the space group of $C2/c$ to $P2/c$ when 0.5 Na^+ was extracted. The relevant reason behind the transition may be the significant difference in bond length between $C2/c$ and $P2/c$ phases [133].

Although the $NaVPO_4F$ material showed high capacity, its cycling performance needed to be enhanced [135]. Carbon coating, doping with other elements, and designing porous networks with different synthesis methods are effective approaches to increase the overall electrochemical performance of the $NaVPO_4F$ cathode material [136]. Zhao et al. developed a novel sol–gel route to synthesize $NaVPO_4F$, which showed 83% of capacity retention after 100 cycles at 2 C [137]. Ling et al. employed a molecular level blending technique, followed by a facile ball-mining to prepare a $NaVPO_4F$@C, which delivered a high initial specific capacity of 135 mAh g^{-1} at the current density of 0.2 C and ultrahigh rate performance of 112.1 mAh g^{-1} at 30 C [134a]. Researchers also doped Cr into $NaVPO_4F$ to improve its cyclability (e.g. a reversible capacity retention of 91.4% during the 20th cycles) [138]. Xu et al. reported a novel approach to synthesize $NaVPO_4F/C$ network consisting of nano/microparticles with open interconnected 3D porous microstructure, which achieved an enhanced discharge capacity of 121 mAh g^{-1} and long cycle life (76% capacity retention after 65 cycles) [139]. Ge et al. employed a solution method to prepare the interface conductivity enhanced $NaVPO_4F/C$, in which PVP acted as both the particle size regulator to restrict particles growth and the carbon source to generate a uniform carbon coating layer on the surface of $NaVPO_4F$. The obtained $NaVPO_4F/C$ composite showed much enhanced electrochemical performance, delivering a highly reversible capacity of 111 mAh g^{-1} at 0.1 C [140]. In addition, modifying morphology of the $NaVPO_4F$ material is also an effective approach to enhance their sodium storage properties. Jin et al. successfully prepared novel $NaVPO_4F/C$ nanofibers via a feasible electrospinning method as self-supporting cathode. The 3D conductive network formed by the interconnection of 1D $NaVPO_4F/C$ nanofibers improved the ionic and electronic transport of $NaVPO_4F$ and effectively restrained the aggregation of $NaVPO_4F$ particles during charge/discharge process, leading to the high electrochemical performance of the $NaVPO_4F/C$ [141]. Cheng et al. designed nano-$NaVPO_4F$ particles (<100 nm) enwrapped in rGO sheets. Even at an ultrahigh rate of 100 C, a reversible capacity of 86.5 mAh g^{-1} can be achieved [134b]. A 3D coral-like structured $NaVPO_4F/C$ was synthesized by Feng et al. using sol–gel as the main strategy. The $NaVPO_4F$ particles were anchored by effective fluorine–hydrogen bonds in the sol–gel template route, resulting in a final product with an interconnected porous 3D coral-like structure consisting of an oriented stack of uniform nanoparticles (15–20 nm). The prepared NVPF/C exhibited enhanced electronic conduction and demonstrated excellent sodium storage performance based on SIBs electrochemical testing [142].

4.4.2 Na_2MPO_4F (M = Fe, Mn, and Ni)

Na_2FePO_4F shows a layered orthorhombic structure with a space group of *Pbcn* ($a = 5.2352(2)$ Å, $b = 13.8365(4)$ Å, and $c = 11.7728(3)$ Å) [143]. Layered Na_2FePO_4F consists of 2D framework with PO_4 tetrahedra connected by $Fe_2O_6F_3$ biotetrahedra, which are formed by sharing the faces of two iron-based octahedra and interconnected by F atoms. The long chain of $Fe_2O_6F_3$ extends along the *a*-axis and is connected to the PO_4 tetrahedron on the *ac* plane. There are dense large channels between $[Fe_2O_6F_3\text{–}PO_4]_\infty$ chains, where Na^+ ions are mobile upon extraction/insertion reactions (Figure 4.10a,b). The facile 2D Na^+ pathways lead to good structural stability with only 3.7% volume variation during Na^+ insertion/extraction processes, satisfying theoretical capacity (~124.2 mAh g^{-1}), and decent operating voltage (~3.0 V vs. Na^+/Na) [144]. However, the electrochemical performance of Na_2FePO_4F also suffered from poor intrinsic electronic conductivity [144a, d].

Wang et al. reported an electrospinning method to homogeneously encapsulate small Na_2FePO_4F nanoparticles (≈3.8 nm) into porous N-doped carbon nanofibers [145]. The unique network structure of carbon nanofibers surrounding the ultrafine Na_2FePO_4F nanoparticles facilitated the enhancement of the electrochemical activity, electron/ion transport capacity, and structural stability of the cathode material. The Na_2FePO_4F@NC nanofibers, as expected, exhibited an excellent long cycling life with 85% capacity retention after 2000 cycles. Ko et al. used a simple low-temperature process to prepare Na_2FePO_4F surrounded by nanospheres of conductive polymer, poly(3,4-ethylenedioxythiophene) (Na_2FePO_4F–PEDOT). The

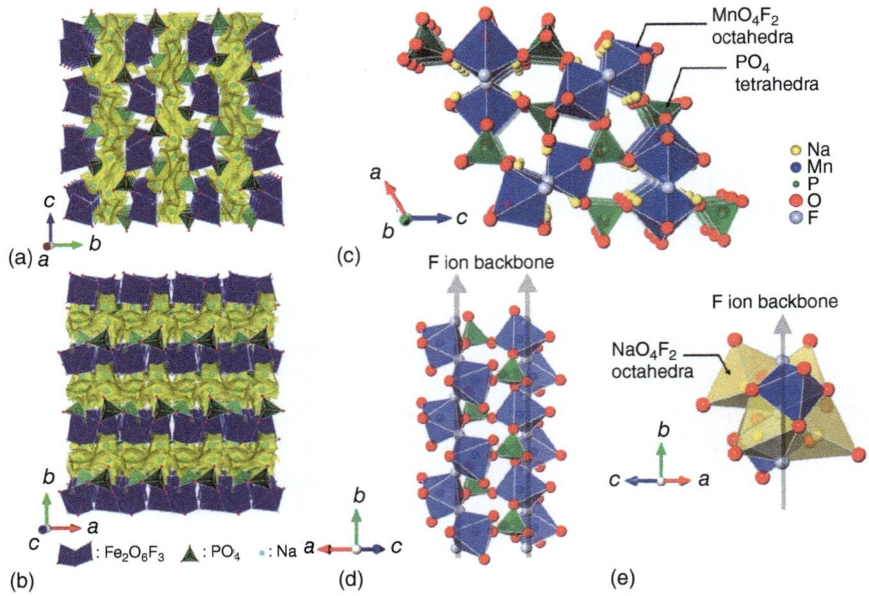

Figure 4.10 Bond-valence site energy maps of Na_2FePO_4F in (a) *bc* plane and (b) *ab* plane. Source: Ko et al. [144b]. Reproduced with permission, 2019, Elsevier. Crystal structure of Na_2MnPO_4F with *P2₁/n* space group: (c) *b*-direction projected image, (d) arrangement of MnO_4F_2 octahedra and PO_4 tetrahedra along the *b*-direction, and (e) arrangement of NaO_4F_2 and MnO_4F_2 octahedra surrounding F ion backbone along the *b*-direction. Source: Kim et al. [143]. Reproduced with permission, 2012, Royal Society of Chemistry.

prepared Na_2FePO_4F–PEDOT materials delivered a high rate capacity at 10 C, which was ~10 times higher than that of the pristine Na_2FePO_4F [144b].

Zhang and co-workers designed a mesoporous Na_2FePO_4F@C (M-NFPF@C) composite, which was synthesized by a facile ball-milling method [144d]. Thanks to the mesoporous structure and high conductive carbon, the M-NFPF@C material had good SIB performance of high discharge capacity (e.g. 114 mAh g^{-1} at 0.1 C), good rate performance as high as 10 C, and excellent cycling property over 600 cycles at 5 C. Analogously, Hu et al. also used ball-mining as a synthetic method to prepare Na_2FePO_4F, and the as-prepared Na_2FePO_4F/C nanocomposites delivered 108.6 mAh g^{-1} at 0.1 C [144c]. This good performance can be ascribed to uniformly dispersed Na_2FePO_4F nanoparticles and mixed-carbon sources derived unique carbon network with hierarchical pore structure. Additionally, Yan et al. reported a multi-layered Na_2FePO_4F/CNT material synthesized via employing layer-by-layer nanoassembly method, which easily generated a 3D porous structure in which the highly interwoven CNT layers and abundant pores act as conductive carbon network and ion transport paths, respectively. Expectedly, the Na_2FePO_4F/CNT material provided a capacity of 103.5 mAh g^{-1} at 0.4 C [146].

Na_2MnPO_4F possesses a 3D $P2_1/n$ structure, as shown in Figure 4.10c. The F-ions are distributed around the MnO_4F_2 and NaO_4F_2 octahedra extending along the *b*-axis. The two MnO_4F_2 octahedra alternate with F-ions along the main chain angles to form single chains. And, the MnO_4F_2 chains are interconnected with the PO_4 tetrahedra (Figure 4.10d). The sodium ions, and every four sodium ions located in the gaps of the framework, are surrounded by one F^- ion along the backbone (Figure 4.10e). The open skeleton of Na^+ along the MnO_4F_2 octahedral backbone suggested that the Na^+ ions may conduct along the *b*-axis direction [143]. Due to the strong induction effect of $[PO_4]^{3-}$ and the strong electronegativity of F^-, Na_2MnPO_4F had high working potential (3.6 V), high theoretical capacity (124 mAh g^{-1}), and good thermal stability [144a, 146]. Unfortunately, the poor electrochemical kinetics, including intrinsic low electron conductivity and sluggish diffusion of Na^+, hinder its further application in SIBs [144a, 147, 148]. And, many scholars have made their efforts to solve this problem.

For example, a micro-nano structured Na_2MnPO_4F/C composite prepared via wet ball-milling was reported by Zhong et al. [149] The micrometer-scale secondary aggregates were composed of small primary nanoparticles (10–30 nm) wrapped in a carbon network and presented a mesoporous structure. The material served as the cathode in SIBs exhibited excellent capacity capability (121 mAh g^{-1} at 10 mA g^{-1}) and cycling behavior. And, a 3D hybrid conductive network composed of interconnecting nanofibers was fabricated by the electrospinning method [144a]. After high-temperature calcination under 650 °C, the cathode materials with good crystallinity exhibited good electrochemical performances with a high initial discharge specific capacity of 122.4 mAh g^{-1} at 0.05 C, which was very close to its theoretical capacity of 124.7 mAh g^{-1}. Ling et al. applied a hydrothermal method to synthesize a dual-carbon confined Na_2MnPO_4F nanoparticles, in which core–shell structured nanoparticles (Na_2MnPO_4F nanoparticles coated by carbon coating layers) were uniformly anchored on the surface of well-dispersed reduced graphite oxide nanosheets. Evidently, it delivered an improved discharge capacity of 122 mAh g^{-1} at a current density of 0.05 C [150]. Besides, an efficient synthesis method is also important to improve the electrochemical performance of cathode materials in SIBs. For example, spray drying is a simple method to produce solid powders from liquids by fast drying of hot gases. Its advantages include the following: (i) the raw materials are well mixed and homogeneous in the liquid phase; (ii) spherical precursors with uniform particle size distribution; and (iii) *in situ* uniform carbon thin layers can be obtained after sintering. Therefore, the spray drying method would be the ideal preparation method based mainly on the construction of small particles and well-distributed carbon coatings [147, 148]. Hu et al. designed carbon-coated Na_2MnPO_4F hollow spheres via a spray drying method. The obtained hollow spherical shell Na_2MnPO_4F/C consisted of nanosized primary Na_2MnPO_4F particles coated with the uniform amorphous carbon layer, which released an enhanced discharge capacity of 122.4 mAh g^{-1} at 0.05 C [147]. Analogously, Lin et al. successfully synthesized a Na_2MnPO_4F/C by spray drying. The highly phase-pure Na_2MnPO_4F was uniformly mosaiced in the carbon framework, which was particularly important for building efficient electron transfer channels and improving the electrical conductivity of the nanocomposites.

Unsurprisingly, an excellent initial discharge capacity of 140 mAh g^{-1} at 0.05 C for the Na_2MnPO_4F/C could be achieved [148].

Na_2NiPO_4F has a unit cell volume of 823.4 A^3 which is 4% and 1.5% smaller than that for Na_2FePO_4F and Na_2CoPO_4F respectively, owing to the smaller size of the nickel ion compared to the other TM ions. The SIB with Na_2NiPO_4F as the cathode material did not show any electrochemical activity below 5 V. It was expected that this compound has a Ni^{2+}/Ni^{3+} redox couple above 5 V, similar to other nickel phosphates such as $LiNiPO_4$ and Li_2NiPO_4F [151].

4.4.3 $Na_3(VO_{1-x}PO_4)_2F_{1+2x}$ ($0 \leq x \leq 1$)

Sodium vanadium fluorophosphate ($Na_3(VO_{1-x}PO_4)_2F_{1+2x}$ ($0 \leq x \leq 1$)) belongs to polyanionic compound type, which contains a variety of molecular structures due to the variation of O and F. It can be divided into three categories: $Na_3V_2(PO_4)_2F_3$ ($x = 1$), $Na_3V_2O_2(PO_4)_2F$ ($x = 0$), and $Na_3(VO_{1-x}PO_4)_2F_{1+2x}$ ($0 < x < 1$) with mixing valence states of vanadium ions. They all belong to the tetragonal system but different space groups, which is attributed to the regulation of the lattice parameters and unit cell volume by different oxygen content.

The synthesis and crystal structure of $Na_3V_2(PO_4)_2F_3$ (NVPF) were first reported by Meins et al. in 1999 [152]. When one phosphate group in NVP is replaced by three fluorine atoms, a rhombohedral structure $Na_3V_2(PO_4)_2F_3$ (space group: $P4_2/mnm$) is formed, in which $[V_2O_8F_3]$ bi-octahedra and $[PO_4]$ tetrahedra are interconnected by sharing angles (Figure 4.11a,b) [155]. The lattice parameters are calculated to be $a = b = 9.047$ Å and $c = 10.749$ Å. In the open framework structure, the occupancy ratio of Na1 sites and Na2 sites is 2 : 1. While when three F^- ions replace one $(PO_4)^{3-}$ anion, additional interstitial spaces for fast sodium-ion diffusion are generated. On account of the extreme electronegativity of fluorine, NVPF displays high redox potentials (3.9 and 4.28 V for desodiation and sodiation processes, respectively) and good thermal stability [156]. The full electrochemical reaction can be described as the following equation [154]:

$$Na_3V_2(PO_4)_2F_3 \leftrightarrow NaV_2(PO_4)_2F_3 + 2Na^+ + 2e^- \quad (4.1)$$

The GITT technique and ^{23}Na NMR were employed to study the charge/discharge process [154a]. The Na^+ in NVPF was removed non-selectively from two different Na sites (i.e. the fully occupied Na1 site, and the partially occupied Na2 site) with three separate phases: Stage I (3.6–3.7 V) and Stages II and III (4.05–4.5 V) at the earlier stage of charging process (Figure 4.11c). And, a high theoretical capacity of 128 mAh g^{-1} can be achieved for NVPF cathode material. The *ex situ* ^{23}Na NMR spectrum showed no obvious peak shift; however, the peak intensity and line width (after $x = 0.3$) decreased significantly in the first stage. For the second and third stages, there were obvious peak shift and intensity decrease with new peaks appearing at lower resonance (Figure 4.11d). There is no obvious tendency of Na transfer in Na1 and Na2 sites at early stage of sodium extraction or late stage of sodium insertion, which is in contradiction with previous reports [153, 157]. As a result,

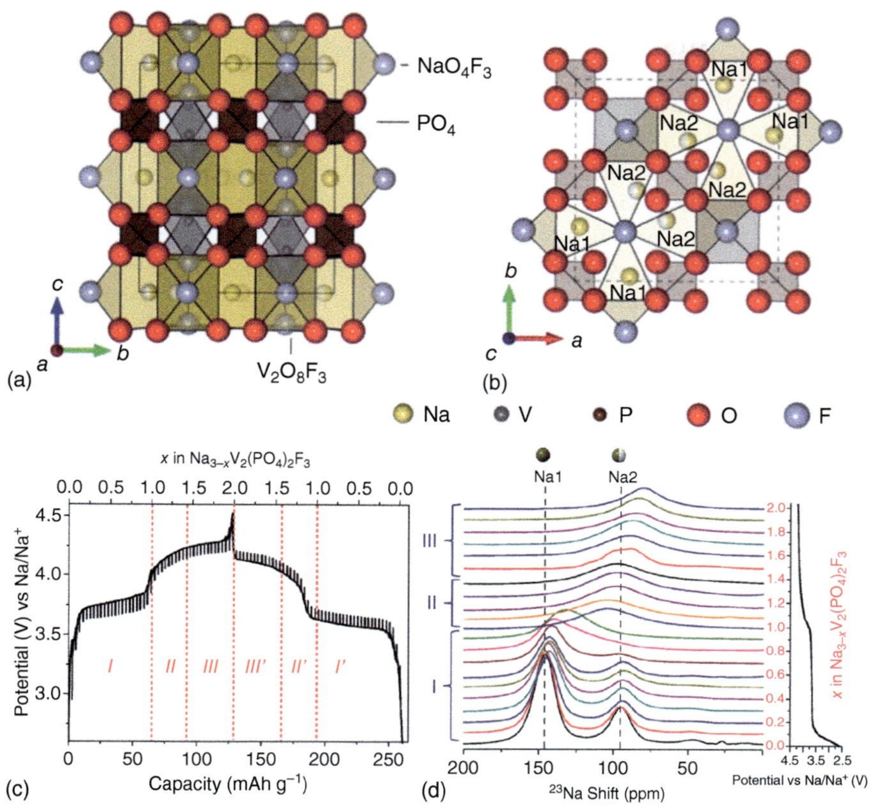

Figure 4.11 Illustration of $Na_3V_2(PO_4)_2F_3$ projected on *bc* plane (a) and on *ab* plane (b). Source: Shakoor et al. [153]. Reproduced with permission, 2012, Royal Society of Chemistry. (c) Galvanostatic intermittent titration technique plot, and (d) *ex situ* ^{23}Na NMR spectra of $Na_3V_2(PO_4)_2F_3$ at different charge states. Source: Liu et al. [154a]. Reproduced with permission, 2014, American Chemical Society.

the $Na_3V_2(PO_4)_2F_3$ cathode exhibits high-voltage platforms located at 3.66 and 4.1 V (vs. Na/Na^+) [158], a very low volume change (1.79%) and exceptional thermodynamic stability with only 5% weight loss at 550 °C, manifesting the rapid Na^+ diffusion and minimal structural collapse. These all indicate that the NVPF material shows good potential for safe SIB application.

Similar to other types of polyanionic compounds, NVPF also suffers from low electron conductivity. Therefore, modification strategies such as carbon coating, nanoparticles granularity controlling, morphology construction, and metallic ion doping have been excavated by scholars to improve the conductivity of NVPF compounds. For example, Song et al. reported the sodium storage performance of amorphous carbon-coated NVPF nanoparticles (NVPF@C) synthesized by the carbothermal reduction [159]. Compared with the pure NVPF, the NVPF@C nanoparticles exhibited superior electrochemical properties with a high discharge capacity of 114.6 mAh g^{-1} at 0.045 C, and a high capacity retention ratio of 97.6% (Coulombic efficiency: ~98%) after 50 cycles. And, multifunctional dual NVPF@C composites with a thin carbon shell (thickness: ~7 nm) derived from citric acid were proposed by Du and co-workers [160] These novel NVPF@C composites yielded high capacity of about 130 mAh g^{-1} and a good rate capacity up to 30 C [156]. Noked and co-workers used a solvothermal method to construct $Na_3V_2(PO_4)_2F_3$ hollow sphere as the cathode material for SIB, which released a pronounced specific capacity of 197 mAh g^{-1} in a wide voltage window (1.0–4.8 V (vs. Na/Na^+)) at 10 mA g^{-1}. But stability was still a key problem [161]. In addition, the researchers have also constructed microcubes [162], nanofibers [163], and flower shapes [164] to optimize the transport path and stability of Na^+ ions through morphology modification. In the aspect of metallic ion doping, researchers have also spent a lot of effort. Zheng and co-workers first proposed doping rare earth element yttrium (Y) into the $Na_3V_2(PO_4)_2F_3$ system to improve the intrinsic conductivity. After partially replacing the V site, the electrochemical performance of the material was improved and the discharge capacity (121 mAh g^{-1}) at low rate of 0.5 C was very close to the theoretical capacity [165]. Then, the effects of different titanium ions doping on the electronic conductivity and Na^+ diffusion of $Na_3V_{2-x}Ti_x(PO_4)_2F_3$ materials were explored. In this doping routine, the valence state of the titanium ions (i.e. Ti^{2+}, Ti^{3+}, and Ti^{4+}) changed during the synthesis and thus the $Na_3V_{2-x}Ti_x(PO_4)_2F_3$ compounds were abbreviated to $NVPF - Ti_y^{x+}$samples ($x = 2, 3, 4$; $y = 0, 0.01, 0.05, 0.1, 0.2$). As a result, the optimized $NVPF - Ti_{0.1}^{2+}$ material effectively enhanced the rate capability, which released 125 mAh g^{-1} reversible capacity at 0.2 C, and 41 mAh g^{-1} capacity even at high rate of 200 C [166]. Other researchers have also explored the effects of K [167], Cr [168], Zr [169], and Mn doping [170] on the electrochemical activity of NVPF in SIB application. These element doping experiments can improve the conductivity and enhance the sodium storage capacity to a certain extent.

Additionally, another new type of sodium vanadyl (IV) fluoride phosphate $Na_3V_2O_2(PO_4)_2F$ was discovered by Massa's team in 2002 using hydrothermal synthesis. $Na_3V_2O_2(PO_4)_2F$ (space group: I_4/mmm; cell parameters: a = 6.3856(2) Å, c = 10.6119(9) Å) is formed by replacing fluorine atom with oxygen atom in NVPF (Figure 4.12a,b) [171]. The crystal framework consists of alternating [VO_5F] octahedra and [PO_4] tetrahedra sharing O atoms. Along the c direction, the [VO_5F] octahedra pairs are inversely connected with fluorine ions as the common vertex. And, the disordered sodium ions locate in the space formed by octahedron and tetrahedron [174]. Sauvage et al. investigated the structure of stoichiometric $Na_3V_2O_2(PO_4)_2F$ and found that its structural framework was the same as that of $NaVPO_4F$, but there are two different Na sites (8h and 8j), one F site (2a), and a octahedral V^{4+} site [VO_5F]. Moreover, Sauvage et al. also measured that $Na_3V_2O_2(PO_4)_2F$ had a lower ionic conductivity of 1.8×10^{-7} S cm^{-1} at room temperature [175]. Subsequently, Sharma et al. studied the sodium storage mechanism of the $Na_3V_2O_2(PO_4)_2F$ cathode, and revealed that there were two-phase reaction and solid solution reaction in the process of charge and discharge, as well as the

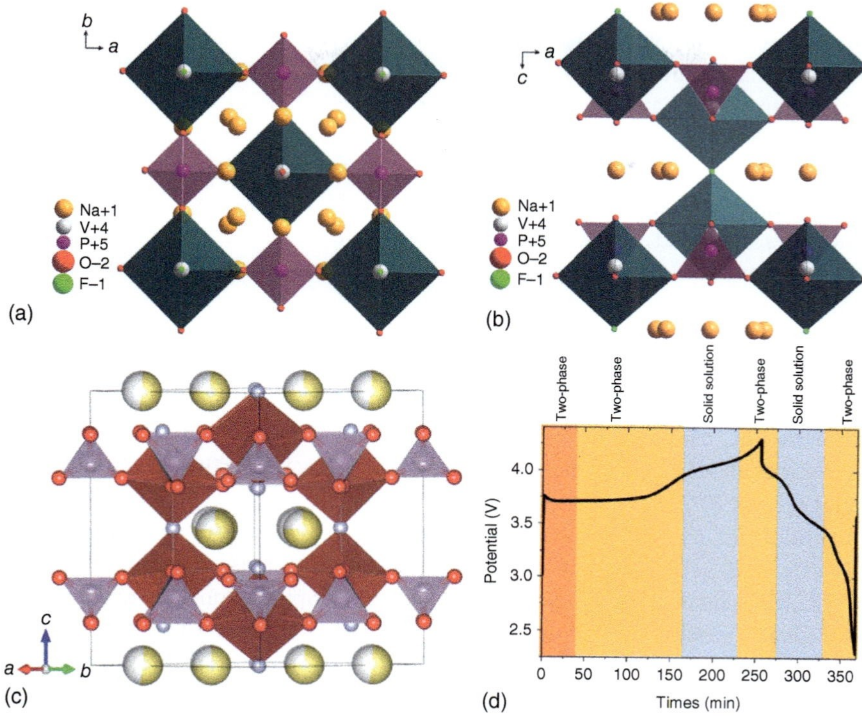

Figure 4.12 Schematic representation of the crystal structure of $Na_3V_2O_2(PO_4)_2F$ projected (a) down the *c* axis and (b) down the *b* axis. Source: Li et al. [171]. Reproduced with permission, 2015, Wiley-VCH. (c) Crystal structure of $Na_3(VO_{1-x}PO_4)_2F_{1+2x}$ with PO_4 shown in purple and VO_4F_2 in red. Oxygen is red, fluorine is light blue, and sodium is yellow with the shading indicating occupancy. Source: Serras et al. [172]. Licensed under CC BY 3. (d) The reaction mechanism evolution during charge/discharge with brown/orange shading indicating two-phase behavior (first order transition) and blue shading indicating solid solution behavior (second order transition). Source: Palomares et al. [173]. Reproduced with permission, 2018, Royal Society of Chemistry.

structural order of the cathode remained unchanged during the whole process [176]. Goodenough and co-workers determined through theoretical calculation that there should be three chemical potential values and three voltage platforms in the phase transition process of $Na_3V_2O_2(PO_4)_2F$ [177]. However, due to the limitation of electrolyte decomposition under high voltage, the third sodium extraction which can only be occurred at 5.3 V has never been successful.

With the further investigation of $Na_3V_2O_2(PO_4)_2F$ in SIB, the existing problem is also exposed: an inferior electronic conductivity (1.8×10^{-7} S cm^{-1}). The main solution is to coat or composite with different carbon materials, including 3D graphene [178], graphene quantum dots [179], MWCNTs [6], rGO [180], amorphous carbon [181], and so on. Moreover, controlling the morphology is also a good way to improve the electrochemical performance. Hu and co-workers used a one-step room temperature strategy to design in situ bubble templates to generate nanocrystals assembled $Na_3(VOPO_4)_2F$ microspheres, where the unique structure design provided a large surface area and a short electron diffusion path with enhanced electrochemical performance. The obtained $Na_3(VOPO_4)_2F$ microspheres exhibited a capacity of 81 mAh g^{-1} at a high rate of 15 C and maintained excellent sodium storage performance of 70% after 3000 cycles [182]. Other researchers also designed hierarchical mulberry-shaped $Na_3V_2(PO_4)_2O_2F$@C [183], $Na_3(VO)_2(PO_4)_2F$@C nanocubes [179], and other structures to shorten the ion/electron transport pathway and improve the electrochemical performance. In addition to these conventional methods, Ru doping [184], RuO_2 coating [185], Fe substitution for O [186], and defect-containing TiO_{2-x} coating [187] were proposed to improve the electrochemical performance. To combine the advantages of high conductivity of RuO_2 and nanostructure, Xia and co-workers prepared a RuO_2-coated $Na_3V_2O_2(PO_4)_2F$ nanowires by microemulsion regulated hydrothermal reaction method with a diameter of about 50 nm, which exhibited a high capacity of 80 mAh g^{-1} after 1000 cycles even at 20 C [185].

In addition to the above materials, researchers have also synthesized $Na_3(VO_{1-x}PO_4)_2F_{1+2x}$ with $0<x<1$ (Figure 4.12c). This material is particularly fascinating and worth exploring for cathode application due to their high theoretical capacity (130 mAh g^{-1}), high operating voltage (3.6 and 4.1 V vs. Na/Na^+), and high structural stability (less than 2% volume change) [172, 188]. Palomares et al. [173] found there were two phases in the synthesized material: a mixed valence phase with an intermediate V oxidation state (~66%, $P4_2/mnm$ space group) and another phase with a V oxidation state closing to V^{3+} (~33%, *Amam* space group). At the beginning of charging, the *Amam* phase was transformed to the $P4_2/mnm$ phase, which is similar to low potential plateau reactions (e.g. $V^{3.8+}$ and V^{4+}), while this transformation during discharge process is not reversible (Figure 4.12d) [173].

The most common methods of preparing $Na_3(VO_{1-x}PO_4)_2F_{1+2x}$ ($0\leq x\leq 1$) include solvothermal and hydrothermal reactions [189]. Park et al. adjusted the ratio of VPO_4 to $VOPO_4$ precursors to synthesize the $Na_3(VO_{1-x}PO_4)_2F_{1+2x}$ ($0\leq x\leq 1$) compounds, which manifested continuous change of the lattice parameters with the increase of fluorine content in $Na_3(VO_{1-x}PO_4)_2F_{1+2x}$ ($0\leq x\leq 1$). While their internal mechanism of sodium storage needed to be further explored [190]. Zhao et al. used a phase-transfer assisted solvothermal strategy to synthesize $Na_3(VO_{1-x}PO_4)_2F_{1+2x}$ ($x=0$, 0.5, 1) at a rather low temperature (80–140 °C) [191]. The as-prepared samples exhibited high Na storage capability and extraordinary cycling stability. The preparation of phase-pure $Na_3(VO_{1-x}PO_4)_2F_{1+2x}$ ($0\leq x\leq 1$) was investigated by Qi et al. via a low-temperature hydrothermal strategy. It was shown that the pH value of the reaction system played an important role in the successful synthesis of micro-structured $Na_3(VO_{1-x}PO_4)_2F_{1+2x}$ ($0\leq x\leq 1$) [188a]. Serras et al. employed a hydrothermal method to synthesize $Na_3(VO_{1-x}PO_4)_2F_{1+2x}$ ($0\leq x\leq 1$) containing a medium percentage of electrochemical grade carbon (i.e. Ketjen Black) as an additive, which showed a high specific capacity of approximately 100 mAh g^{-1} at 1 C [192].

Adding carbon material to form a composite is proven to be an effective modification strategy. Kumar et al. investigated the effects of super P carbon, MWCNTs, and rGO on gain size and electrochemical activity of $Na_3(VO_{1-x}PO_4)_2F_{1+2x}$ cathode material [193]. Among them, the composites using MWCNT as carbon substrate can obtain the smallest particle size, which exhibited significantly improved electrochemical properties with a high sodium storage capacity of 98 $mAh\,g^{-1}$ at 0.1 C. Kumar et al. investigated the cell performance through regulating the ratio of $Na_3(VO_{1-x}PO_4)_2F_{1+2x}$–rGO composite synthesized by a hydrothermal method with the hydrophilic carboxymethyl cellulose sodium (CMC-Na) as the binder. Here, in addition to the introduction of rGO to increase the electronic conductivity, the hydrophilic CMC-Na binders also facilitated the electron conduction by maintaining the integrity of the electrodes during charge/discharge cycles. Specifically, the as-obtained $Na_3(VO_{1-x}PO_4)_2F_{1+2x}$–rGO composite showed enhanced sodium storage capacity (e.g. 108 $mAh\,g^{-1}$ at 0.1 C rate) and highly stable cyclability with 98% capacity retention over 250 cycles [188b]. Nano-$Na_3(VO_{0.5})_2(PO_4)_2F_2$ embedded in porous graphene was confirmed as an excellent high-performance SIB cathode material by Xiang et al., which displayed a high reversible capacity of 100 $mAh\,g^{-1}$ at 1 C and remained capacity of 77 $mAh\,g^{-1}$ with a capacity retention of 73% after 1000 cycles at 50 C [194].

$Na_3V_2(PO_4)_2O_{1.6}F_{1.4}$ with space group of $P4_2/mnm$ presented an average vanadium valence value of +3.8, which can release a high theoretical capacity of 156 $mAh\,g^{-1}$ based on 2.4 electrons transferring per formula unit during the sodiation/desodiation process. Palomares et al. studied the electrochemical behavior and structural evolution of $Na_3V_2O_{1.6}(PO_4)_2F_{1.4}$, which corresponds to a vanadium oxidation state of +3.8 for the as-prepared material [195]. It is further found that the $V^{3.8+}$-based $Na_3V_2(PO_4)_2O_{1.6}F_{1.4}$ allowed a maximum sodium extraction as overcharge voltage achieved up to 4.8 V. Li et al. synthesized a nanostructured $Na_3V_2(PO_4)_2O_{1.6}F_{1.4}$ via microwave-assisted solvothermal method. The as-prepared $Na_3V_2(PO_4)_2O_{1.6}F_{1.4}$ showed the improved Na^+ transport kinetics and manifested impressive rate capability of 67 $mAh\,g^{-1}$ at 30 C and long-term cycling performance with the capacity of 61 $mAh\,g^{-1}$ after 1000 cycles at 10 C [196].

Another representative sodium vanadium fluorophosphates for advance cathode materials are $Na_{1+y}VPO_4F_{1+y}$ ($0 \leq y \leq 0.5$), which is able to exhibit a variable reversible discharge capacity of 103 and 87 $mAh\,g^{-1}$ for $y = 0.25$ and 0, respectively. And, the tetragonal $Na_{1+y}VPO_4F_{1+y}$ with the group space of $P4_2/mnm$ can achieve a high discharge capacity of 116 $mAh\,g^{-1}$ only when the value of y is 0.5 [197].

Except for the materials mentioned above, a new cathode material of $Na_{1.5}VPO_{4.8}F_{0.7}$ is also developed. The new $Na_{1.5}VPO_{4.8}F_{0.7}$ provides a high energy density (~600 $Wh\,kg^{-1}$), which is mainly attributed to the high potential (3.8 V vs Na^+/Na) corresponding to the multivalent vanadium redox couple ($V^{3.8+}/V^{5+}$) [198]. Furthermore, the volume change of the $Na_{1.5}VPO_{4.8}F_{0.7}$ cathode material is small (~2.9%) during the Na^+ insertion/extraction processes, whose structural rigidity enables its excellent cycle. For example, 95% capacity retention after 100 cycles and 84% capacity retention after 500 extended cycles can be achieved for this new cathode material. Additionally, the $Na_{1.5}VPO_{4.8}F_{0.7}$ cathode material possesses an open crystal framework with a 2D Na^+ diffusion path leading to a low activation potential barrier for Na^+ transport, resulting in excellent rate performance.

4.5 Sulfates

4.5.1 $Na_xFe_y(SO_4)_z$

The M—O bond in sulfates is more ionic than those in silicates or phosphates, posing strong inductivity potential of $[SO_4]$ group, as a result sulfates are considered potential candidates for high-voltage cathodes in SIBs. Various sulfate-based sodium hosts including $Na_2Fe_2(SO_4)_3$, $Na_2Fe(SO_4)_2$, and $Na_2(SO_4)_2{\cdot}nH_2O$ have been studied because of the high natural abundance and the favorable sodium intercalation capability. In particular, alluaudite-structured $Na_2Fe_2(SO_4)_3$ with a high Fe^{3+}/Fe^{2+} redox potential (~3.8 V vs. Na/Na^+) and considerable energy density (>450 Wh kg^{-1}) is regarded as the most pragmatic sodium host material. However, the low conductivity and thermal instability of $Na_2Fe_2(SO_4)_3$ hinder its application in SIBs. To solve this problem, some conventional strategies such as structural modulation, metal ion doping, and conductive material combination have been adopted.

The alluaudite $Na_2Fe_2(SO_4)_3$ was first reported by Barpanda et al. [199]. After replacing the phosphate PO_4^{3-} with sulfate SO_4^{2-}, which has higher electronegativity, the $Na_2Fe_2(SO_4)_3$ reached a high potential of 3.8 V vs. Na^+/Na with a capacity of 100 mAh g^{-1}. In the alluaudite-type framework, Fe^{2+} ions occupy octahedral sites and form Fe_2O_{10} dimer units. Those Fe_2O_{10} dimers are in turn bridged together by SO_4 units and form a 3D framework with large tunnels along *c* axis (Figure 4.13a). Na^+ occupies three different crystal sites, one fully occupied Na1 site and two partially occupied Na2 and Na3 sites. The Na2 and Na3 sites along the *c* axis form one-dimensional Na^+ transportation channels. Na^+ ions in Na1 can be extracted through the Na3 sites and then took part in reaction, which was shown the low activation energies for the Na ions diffusion, such as the value of the activation energy (~0.55 eV) for the Na2 channel along the *c* axis, 0.68 eV for the transfer between Na1 and Na2 sites, 0.54 eV for the migration between Na1 and Na3 sites, and only 0.28 eV for the Na3 site along the *c* axis (Figure 4.13b). Thus, all the Na ions are able to diffuse through the channels and inhere excellent kinetics.

The first reported alluaudite was synthesized by the solid-state method, whereas the $Na_2Fe_2(SO_4)_3$ would be thermal decomposed at high temperature and react with moisture [201]. Plewa et al. reported a facile aqueous synthesis of stoichiometry $Na_2FeM(SO_4)_3$ with M = 3d TM (M = Fe, Mn, Ni) alluaudites. Based on the DFT computations, the replaced 3d elements were not directly involved in the electrochemical reaction; they instead took effect on pushing p-type oxygen orbitals to the Fermi level and then activating the anionic redox. Compared with other non-stoichiometry materials such as $Na_{2.56}Fe_{1.72}(SO_4)_3$ [202] and $Na_{2.5}Fe_{1.75}(SO_4)_3$

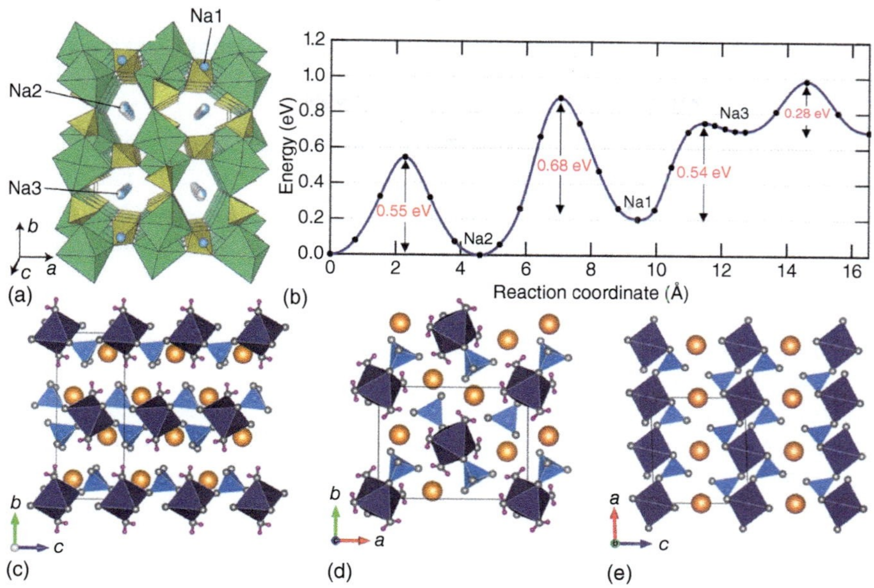

Figure 4.13 (a) The structure of $Na_2Fe_2(SO_4)_3$ projected along the *c* axis. Green octahedra, yellow tetrahedral, and blue spheres show FeO_6, SO_4, and Na, respectively. (b) Migration activation energy of Na^+ calculated with DFT, showing the migration values (from left to right) for migrations along the *c* axis for the Na2 sites, between Na2 and Na1 sites, between Na1 and Na3 sites, and along the c axis for the Na3 sites, respectively. Source: Barpanda et al. [199]. Licensed under CC BY 3.0. Structures of (c) $Na_2Fe(SO_4)_2{\cdot}2H_2O$, (d) $Na_2Fe(SO_4)_2{\cdot}4H_2O$, and (e) eldfellite $NaFe(SO_4)_2$. FeO_6 and SO_4 polyhedra are shown in blue and turquoise, respectively. Na, O, and H are orange, grey, and pink, respectively. Source: Lander et al. [200]. Reproduced with permission, 2018, Wiley-VCH.

[203], the strict stoichiometry composite $Na_2Fe_2(SO_4)_3$ raised the concentration of 3d metals and was able to deliver its specific capacity. As a result, the as-synthesized $Na_2Fe_2(SO_4)_3$ composite showed good Na-ion diffusion kinetics and delivered a capacity of 110 mAh g^{-1} with a high voltage of 3.6 V vs. Na^+/Na [201]. Thomas et al. studied the properties of stoichiometric $Na_2Fe_2(SO_4)_3$ and the sodium-rich composite $Na_{2+2x}Fe_{2-x}(SO_4)_3$ [204]. Both composites had the same alluaudite structure, and no distinct differences were found. Due to the lower amount of redox-active Fe ions, Na-rich composite has inferior capacity of 106 mAh g^{-1}. However, the more intensive repulsion between Fe^{2+} in stoichiometric $Na_2Fe_2(SO_4)_3$ makes it more unstable, and thus, the reversible capacity of Na-rich composite was higher than that of $Na_2Fe_2(SO_4)_3$.

Dwibedi et al. reported an ionothermal synthesis at low temperature to produce pure products [205]. The as-synthesized $Na_{2.44}Fe_{1.78}(SO_4)_3$ delivered a reversible discharge capacity of 75 mAh g^{-1} (Coulombic efficiency: 98%). To further improve the conductivity of $Na_{2+2x}Fe_{2-x}(SO_4)_3$, Wang et al. proposed a nitrogen-doped graphene coating strategy to improve electrochemical activity. By surface modification using N-doping graphene, the $Na_{2+2x}Fe_{2-x}(SO_4)_3$@N-rGO displayed a high reversible capacity of 93 mAh g^{-1} at 0.05 C and an excellent cycling life over 400 cycles with a capacity retention of 83% even though the discharge rate rose to 10 C [206]. Nam et al. synthesized $Na_2Fe_2(SO_4)_3$@C nanofibers by electrospinning method [207]. The Na^+ conductivity of $Na_2Fe_2(SO_4)_3$@C nanofibers was 1.48×10^{-6} S cm^{-1}, one order of magnitude higher than $Na_2Fe_2(SO_4)_3$ nanoparticles. Other 3d TM-based alluaudite materials such as Na_2Mn_2 $(SO_4)_3$ and Na_2Co_2 $(SO_4)_3$ have also been investigated [208]. The Mn, Co-based alluaudites show higher redox potentials of 4.1 and 4.8 V, indicating excellent application in batteries. However, there are not any suitable electrolytes to match with those materials at such high voltages.

$Na_2Fe(SO_4)_2{\cdot}2H_2O$ crystallizes in the monoclinic $P2_1/c$ unit cell and consists of isolated $MO_4(OH_2)_2$ octahedra, which are interconnected via four SO_4 tetrahedra through their oxygen corners forming linear chains along the *c*-axis (Figure 4.13c). The H atom of the H_2O group and the free oxygen of the SO_4 group interact through hydrogen bonds. The Fe^{3+}/Fe^{2+} redox process is located at an average potential of 3.25 V vs. Na^+/Na and delivers a reversible capacity of 70 mAh g^{-1}. The tetrahydrate $Na_2Fe(SO_4)_2{\cdot}4H_2O$ crystallizes in the bloedite-type mineral with typical monoclinic space group $P2_1/c$. The structure is built of isolated $FeO_2(OH_2)_4$ octahedra connected to two SO_4 tetrahedra (Figure 4.13d). The eldfellite-type $NaFe(SO_4)_2$ was reported as Na^+ insertion material, which could be described in a monoclinic $C2/m$ unit cell. The framework consists of edge-sharing FeO_6 forming layers, which are bridged via SO_4 tetrahedra (Figure 4.13e).

For synthesis and modification, the alluaudite $Na_2Fe_2(SO_4)_3$ will convert into bisulfates $Na_2Fe(SO_4)_2{\cdot}4H_2O$ after it is exposed to moisture atmosphere. Thus, it was not suitable to synthesize $Na_2Fe_2(SO_4)_3$ via a simple dissolution and precipitation process, whereas the $Na_2Fe(SO_4)_2{\cdot}2H_2O$ or $Na_2Fe(SO_4)_2{\cdot}4H_2O$ can also be introduced into SIBs and show moderate performance. The $Na_2Fe(SO_4)_2{\cdot}4H_2O$ could be obtained by using hydrophobic alcohol, while the $Na_2Fe(SO_4)_2{\cdot}2H_2O$ was synthesized via a heating process. Compared to the solid-state method, low-temperature solution synthesis is more feasible. Note that the crystal water in the material alters the crystal structure; thus, the $Na_2Fe(SO_4)_2{\cdot}2H_2O$ is denoted as kroehnkite, whereas the $Na_2Fe(SO_4)_2{\cdot}4H_2O$ is bloedite. Barpanda et al. synthesized Kröhnkite-type $Na_2Fe(SO_4)_2{\cdot}2H_2O$ via a dissolution-precipitation route and studied its sodium storage properties [209]. The Kröhnkite-type $Na_2Fe(SO_4)_2{\cdot}2H_2O$ compound had a pseudolayered monoclinic structure, where Na^+ ions occupied the interstitial positions and paved a pathway for the Na^+ transportation along the *b*-axis. The product exhibited 70 mAh g^{-1} at a rate of C/20 and reached an average potential of 3.25 V vs. Na/Na^+. Meng et al. synthesized different dimensional carbon matrix incorporated $Na_2Fe(SO_4)_2{\cdot}2H_2O$/C and explored the architecture-properties relationship [210]. Based on the results, the zero-dimensional activated-carbon can increase the porosity and enlarge the surface, thus inducing the best ion diffusion capability and rate capacity.

In addition, the incorporation of two-dimensional graphene is able to provide a robust structure with high compactness, improving the structure stability. Yao et al. synthesized a hierarchical porous $Na_2Fe(SO_4)_2$@reduced graphene oxide/carbon dot ($Na_2Fe(SO_4)_2$@rGO/C) materials [211]. The cross-linked carbon matrix improved the reaction kinetics; thus, the composite delivered 85 mAh g^{-1} at 0.05 C with a high Na^+ diffusion coefficient of 1.19×10^{-12} cm^2 s^{-1}. Pan et al. reported an anhydrous carbon coating $Na_2Fe(SO_4)_2$ with a good thermal stability and moisture resistance [212]. The as-synthesized composite exhibited 82 mAh g^{-1} at 0.1 C with a high Fe^{2+}/Fe^{3+} redox potential of ~3.6 V (vs. Na^+/Na). The product showed a relatively high Na^+ diffusion coefficient of 1.73×10^{-12} cm^2 s^{-1}, indicating the enhanced reaction kinetics. The carbon coating strategy increased the deposition temperature of sulfate radicals from 450 to 580 °C and enhanced its moisture resistivity.

4.5.2 Fluorosulfates

$NaFeSO_4F$ has attracted attention due to the high operating voltage (3.7 V vs. Na^+/Na) and 138 mAh g^{-1} of theoretical capacity. Tripathi studied the Na^+ diffusion paths in monoclinic $NaFeSO_4F$ and found that the most favorable migration path was along the [110] direction, which was six orders of magnitude higher than other directions, suggesting that $NaFeSO_4F$ was a one-dimensional Na-ion conductor [213]. Yuan et al. then proposed an effective benzene – water azeotrope route to synthesize $NaFeSO_4F$ and $NaFeSO_4F{\cdot}2H_2O$. In this work, $NaFeSO_4F$ illustrated apparently electrochemical activity in 2.5–5.0 V (vs. Na/Na^+) at a low current density, but $NaFeSO_4F{\cdot}2H_2O$ showed little electrochemical activity [214]. Additionally, this technique can also be extended to synthesize other $NaMSO_4F{\cdot}2H_2O$ (M = Co, Ni) [215]. Though $NaFeSO_4F$ displayed a high theoretical capacity (138 mAh g^{-1}), the practical capacity hardly achieved ~7% of the theoretical values. To improve the inherent low electrochemical properties of $NaFeSO_4$, all sorts of $NaFeSO_4F$@C composite have been studied. Rajagopalan et al. reported a new carbon nanotube-lined sodium iron fluro-sulfate ($NaFeSO_4F$-CNT) system with excellent electrochemical properties [216]. The introduction of CNT facilitated the Na ion transportation and released the stress from volume expansion, leading to the composite delivered a high discharge capacity of 110 mAh g^{-1} with 91% capacity retention after 200 cycles.

4.6 Silicates

The TM silicates Na_2MSiO_4 (M = Fe, Co, Mn) are regarded as ideal cathode materials for advanced SIB applications on the merits of environmental friendliness, cost-effectiveness, and multi-electron reaction per unit molecular weight [217]. The structural feature of monoclinic Na_2CoSiO_4 with the space group of *Pn* is characterized by CoO_4 and SiO_4 tetrahedra forming a cristobalite-type 3D skeleton by corner-sharing (Figure 4.14a). This crystal framework is composed of corner-sharing alternating cobalt and silicate tetrahedral units, in which Na^+ ions fill in the tetrahedral vacancies. Two polycrystalline types are found: one is the monoclinic Si/Co ordered in *Pc* space group, and the other is orthogonal Co/Si disordered in *Pbca* space group. The Na_2CoSiO_4 cathode exhibited fast Na^+ transfer kinetics in the 3D network with low activation barrier [220]. The theoretical specific capacity of Na_2CoSiO_4 in a two-electron process is 272.06 mAh g^{-1}. However, the

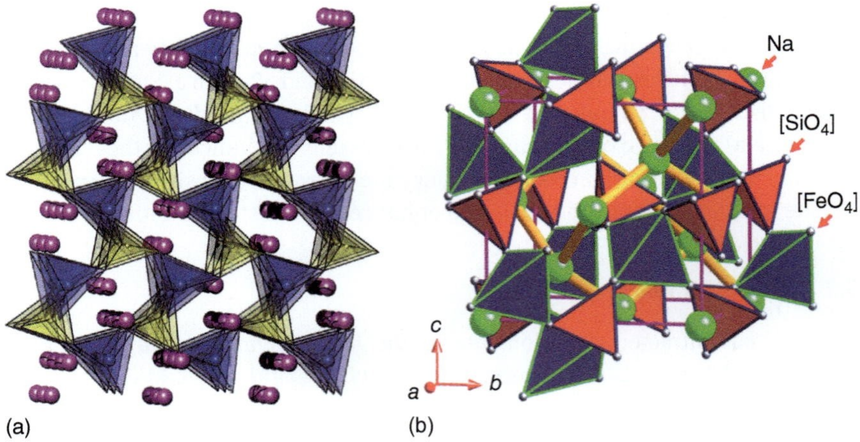

Figure 4.14 (a) Crystal structure of monoclinic Na_2CoSiO_4, Na^+ ions, CoO_4, and SiO_4 tetrahedra in Na_2CoSiO_4 highlighted by pink, blue, and yellow colors, respectively. Source: Treacher et al. [218]. Reproduced with permission, 2016, Royal Society of Chemistry. (b) Crystal structure of cubic Na_2FeSiO_4, Na^+ ions, FeO_4 and SiO_4 tetrahedra in Na_2CoSiO_4 highlighted by green, blue, and red colors, respectively. Source: Li et al. [219]. Reproduced with permission, 2016, American Chemical Society.

specific capacities of sodium silicates reported so far were based on single-electron process. For instance, Treacher et al. employed the co-precipitation and solid-state method to prepare the Na_2CoSiO_4 cathode material, which displayed a reversible capacity of 100 mAh g^{-1} under the operation potential of 3.3 V (vs. Na/ Na^+) at a current density of 5 mA g^{-1} [218].

The cubic Na_2FeSiO_4 (space group: *Pbca*) also provides a 3D skeleton structure, in which FeO_4 groups are linked to the SiO_4 tetrahedron by sharing corners (Figure 4.14b). While, there is another new cubic Na_2FeSiO_4 polymorph with the space group of $F{-}43m$, which was first synthesized through a sol–gel method. Electrochemical studies showed that this new cubic Na_2FeSiO_4 presented strong structure stability with almost no obvious volume change in a long-term sodiation/desodiation processes. Surprisingly, the new cubic Na_2FeSiO_4 cathode exhibited a reversible capacity of 106 mAh g^{-1} within the voltage window of 1.5–4.0 V [219].

Manganese-based Na_2MnSiO_4 is characterized with a monoclinic *Pn* structure of a polyanionic skeleton composed of corner-sharing MnO_4 and SiO_4 tetrahedra [221]. Na_2FeSiO_4 was first obtained as a precursor to prepare Li_2MnSiO_4 through ion-exchange, which showed a capacity of 330 mAh g^{-1} based on the capacity of Li_2FeSiO_4 within the voltage window of 1.2–4.5 V [222]. Although Na_2MnSiO_4 showed good sodium storage performance and a higher cell potential (>0.5 V) than their Fe-counterparts, the Jahn–Teller distortion of Mn^{3+} ($3d^4$ valence electron configuration) caused the poor electronic conductivity and sluggish structural conversion between the (de)sodiated phases [223]. Generally, nano-sizing of the active particles and the use of electrical conducting coatings were the common approaches to resolve the potential challenge [224]. Chen and Chiu synthesized Na_2MnSiO_4/C by a sol–gel method and achieved a high reversible capacity (e.g. 125 mAh g^{-1} at 0.1 C) and acceptable rate performance [225]. And the carbon-coated Na_2FeSiO_4 nanoparticle (20–50 nm) exhibited the enhanced sodium storage capability in terms of high discharge capacity (e.g. 181 mAh g^{-1} at 27.6 mA g^{-1}) and stable cycling performance (e.g. 88% capacity retention over 100 cycles) [226].

4.7 Other Polyanion-Type Compounds

Beyond the aforementioned polyanion compounds which are widely applied as cathode materials of SIBs, some other polyanion-type compounds such as carbonophosphates [227], anti-NASICON type compounds, and amorphous polyanion compounds [228] have also been concerned and studied.

Carbonophosphates, with a general formula of $Na_3M(CO_3)(PO_4)$ (M = Co, Mn, Fe, or Ni) are regarded as a new class of potential cathode materials owing to their superior security and high theoretical specific energy, which are nearly half higher than that of iron-based phosphate [229]. $Na_3M(CO_3)(PO_4)$ is crystallized in the monoclinic $P2_1/m$ space group. As highlighted in black in Figure 4.15, the crystal

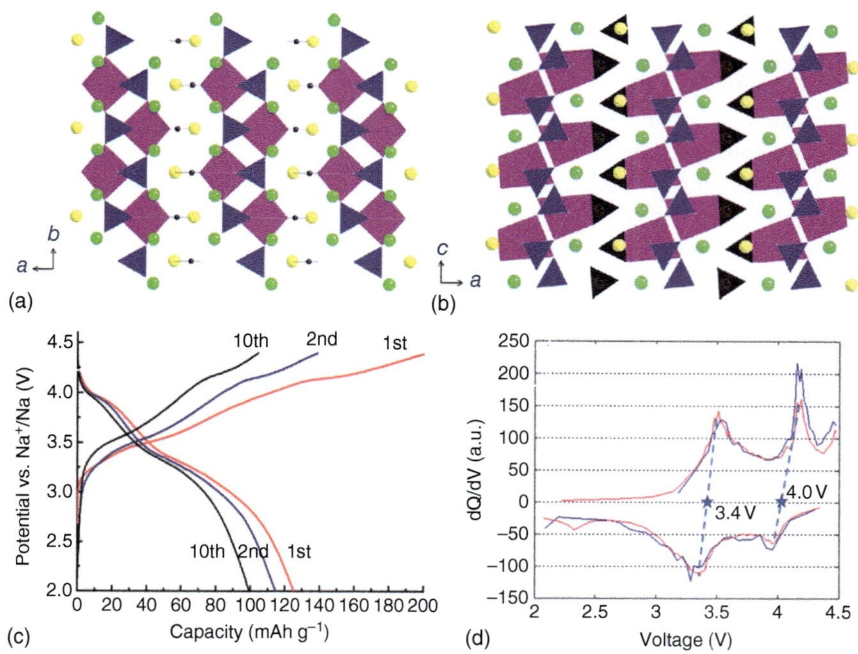

Figure 4.15 Structure of $Na_3MPO_4CO_3$ (M = Mn, Fe, etc.) viewed along [001] direction (a) and [010] direction (b). Transition metal octahedra, purple; PO_4 tetrahedra, blue; CO_3 triangle planar, black; Na, green (Na1) and yellow balls (Na2). Source: Chen et al. [227a]. Reproduced with permission, 2012, American Chemical Society. (c) The voltage curves of sidorenkite during the first, second, and 10th cycles at a C/100 rate. (d) Differential capacity plots of the first (red line) and second (blue line) cycles. Source: Chen et al. [227c]. Reproduced with permission, 2013, American Chemical Society.

structure is defined in a 2×2×2 super cell. Each MO_6 octahedron is connected with four vertices of the PO_4 tetrahedron and shares one edge with a CO_3 group. The connected MO_6 octahedron, PO_4 tetrahedron, and CO_3 group form double layers extending along the (100) plane, in which the occupied two different Na sites coordinate with six (Na1, multiplicity = 4) and seven (Na2, multiplicity = 2) oxygen atoms, respectively (Figure 4.15a,b) [227c]. Na2 is preferentially extracted from the lattice before Na1, and then 50% Na1 is released, corresponding a two-electron intercalation reaction based on the M^{2+}/M^{3+} and M^{3+}/M^{4+} redox reactions. The sodium versions of Mn compound, sidorenkite ($Na_3MnCO_3PO_4$), is known as the common compounds in this class for SIB cathode application [227a, b, g, h]. $Na_3MnCO_3PO_4$ exhibited a two-electron redox behavior with a theoretical capacity of 191 mAh g^{-1}. And, two obvious plateaus with the average value of 3.4 and 4.0 V were observed in both galvanostatic charge/discharge curves and the related differential capacity plots, which were attributed to Mn^{2+}/Mn^{3+} and Mn^{3+}/Mn^{4+} redox reactions, respectively (Figure 4.15c,d). This result is also consistent with the voltages computed by the ab initio calculations [227c]. $Na_3MnCO_3PO_4$ prepared by an oil path process in inert atmosphere displayed a high discharge capacity of 125 mAh g^{-1} under the specific energy of 374 Wh kg^{-1}. While the $Na_3MnCO_3PO_4$ with high conductive carbon content of 60 vol% delivered an enhanced specific capacity of 176 mAh g^{-1}, nearly 92.5% of its theoretical capacity. Further study on carbonophosphate cathodes showed that ionic conductivity was positively correlated with their specific capacity [227i]. In addition, particle size and structural defects (such as vacancy) also have marked impact on relative discharge capacity. Thus, strategies such as micro-nanofabrication, carbon coating, and ionic doping should be introduced to enhance the electronic conductivity of carbonophosphate cathodes.

Besides, one polyanion-based insertion host, anti-NASICON-type $Fe_2(MO_4)_3$ (M = Mo or W), has also been identified as an interesting intercalation cathode material owing to their open 3D framework capable of intercalating alkali metal ions such as Li^+, Na^-, and K^+ [230]. The monoclinic $Fe_2(MoO_4)_3$ with the group space of $P2_1/c$ is built by FeO_6 octahedra and MoO_4 tetrahedra, which interconnect together via the corner sharing oxygen atoms (Figure 4.16a). And, two FeO_6 octahedra connecting three MoO_4 tetrahedra to constitute the basic "lantern unit" (Figure 4.16b). These basic units are stacked in an antiparallel manner in the ($2b+c$)-direction in the anti-NASICON structure, while the equivalent units are stacked in a parallel way along the c-axis direction in the NASICON structure [232]. Moreover, Na^+-intercalation into monoclinic $Fe_2(MoO_4)_3$ crystal structure is divided into two-stage solid-solution behavior, which is signified by two distinct potential plateau centered at 2.7 and 2.58 V (inset of Figure 4.16c), respectively. The reversible capacity of $Fe_2(MoO_4)_3$ reached approximately 79 mAh g^{-1} for the Na/$Fe_2(MoO_4)_3$ cell (Figure 4.16c). Apart from $Fe_2(MoO_4)_3$, homologous $Fe_2(WO_4)_3$ has also been studied in aspects of synthetic methods, reaction mechanism and structural evolution with electrochemical characterizations. It also exhibited favorable electrochemical behavior in the application of rechargeable batteries [233].

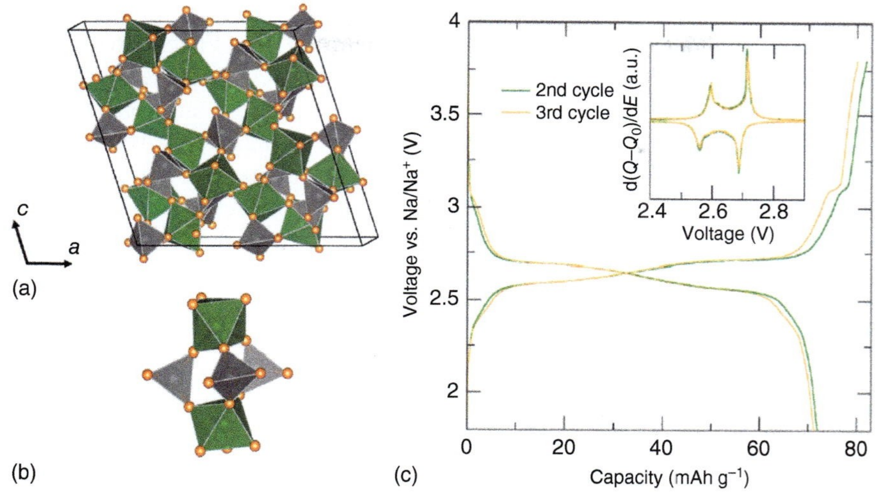

Figure 4.16 (a) Illustration of the unit cell of monoclinic $Fe_2(MoO_4)_3$. (b) The "lantern unit" that consists of three MoO_4 tetrahedra (grey) connecting two FeO_6 octahedra (green). (c) Galvanostatic electrochemical cycling of the $Na/Fe_2(MoO_4)_3$ cell and its derivative (shown as inset). Source: Zhou et al. [231]. Reproduced with permission, 2016, American Chemical Society.

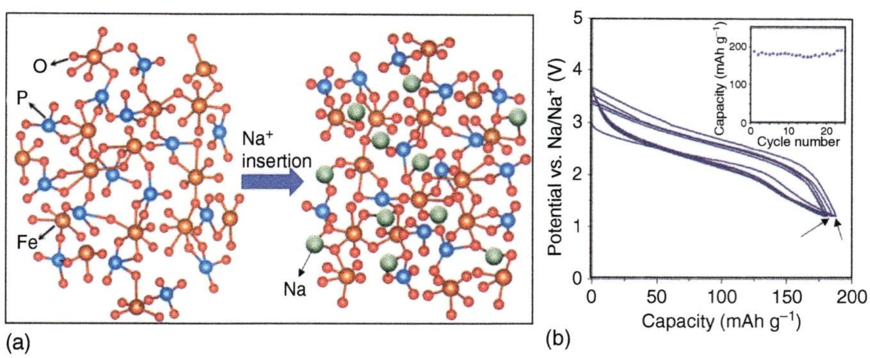

Figure 4.17 (a) Schematic representation of Na-ion insertion in amorphous $FePO_4$ electrode host. (b) The initial five voltage composition curves of the amorphous $FePO_4$ cathode in sodium cells cycled between 3.7–1.2 V at current rate of 10 mA g^{-1} (inset: the cycle performance). Source: Mathew et al. [228d]. Licensed under CC BY 4.0.

Compared to crystalline electrodes, amorphous polyanion compounds demonstrate isotropic structure with no lattice limitations, which can provide more facial channels for fast Na^+ diffusion, leading to good structural stability and redox kinetics during the repeated Na^+ insertion/extraction processes. Typical representative compound, $FePO_4$, has attracted the research interest of a large number of scholars. Amorphous $FePO_4$ host with short-range ordering may facilitate the insertion of guest ions such as Na^+ ions (Figure 4.17a). The electrochemical capabilities of the amorphous $FePO_4$ cathode in sodium test cells were evaluated within the potential windows of 3.7–1.2 V (Figure 4.17b). When tested for its Na^+ insertion properties, the $FePO_4$ cathode delivered a high discharge capacity of 179 mAh g^{-1} (corresponding to 100% theoretical capacity) and retained 97% of its initial capacity at the 25th cycle (Figure 4.17b, inset) [228d]. However, amorphous $FePO_4$ cathode is also restricted by poor electrical conductivity and inferior diffusion coefficient. Some approaches such as conductive carbon material modification, nanoengineering design, morphology, and control initiatives have been investigated to overcome those above-mentioned shortcomings. Recently, Fang et al., reported the cell performance of the mesoporous amorphous $FePO_4$ nanospheres prepared by a simple chemically induced precipitation method. It was found that the unique mesoporous structure was favorable to a full wetting of electrolyte, shortening the Na^+ migration distance, and alleviating the large change of electrode volume during the (de)sodiumation process. The as-prepared $FePO_4$/C composite was expected to greatly enhance the sodium storage kinetics and structural stability, giving a high reversible sodium storage capacity of 151 mAh g^{-1} at 20 mA g^{-1} and steady cyclability with 94% capacity retention ratio over 160 cycles [234]. Apart from that, hydrothermally synthesized porous amorphous $FePO_4$ nanoparticles connected by single-wall carbon nanotubes ($FePO_4$–SWNTs) were also studied, which showed markedly improved electronic conductivity and excellent cell performance with high-rate long cycling capability (e.g. 50 mAh g^{-1} retained after 300 cycles at 100 mA g^{-1}) [235]. Subsequently, Xu et al. invented a maize-like $FePO_4$@MWCNTs core–shell nanowire composite through a microemulsion processing route. The nanowire structure provided fast pathways for Na^+ transportation, and MWCNTs facilitated electron-transfer property. The resulting $FePO_4$@MWCNT nanowires showed improved dynamics and excellent electrochemical performance with significantly elevated specific capacities of 155 and 75 mAh g^{-1} at 0.1 and 1 C rate, respectively [236]. Additionally, a graphene-amorphous $FePO_4$ porous nanowire hybrid as an advanced cathode material for SIBs also illustrated an enhanced kinetics, reversibility for Na^+ storage, and superior rate capability of 42 mAh g^{-1} at high 20 C rate, owing to the high electrical conductivity of graphene, large surface area, and excellent chemical stability of the nanowire structure [237]. In addition to the exploration of 1D nanostructured amorphous $FePO_4$ cathode for sodium storage applications, 2D amorphous $FePO_4$ nanocomposites with surface modifications have also been studied accordingly. For example, Liu et al. synthesized mono/bi-layer amorphous $FePO_4$ 2D nanosheets as a cathode for SIB application. The 2D nanosheets showed short pathways and large implantation surface areas for rapid electron transfer and fast Na^+ diffusion. This in turn improved the electrochemical performance of SIBs, e.g. a long cycle life with 92.3% capacity retention after 1000 cycles [238]. And, single-layer $FePO_4$/reduced graphene oxide ($FePO_4$/rGO) nanosheets composite prepared by a micro-emulsion technique was also investigated, which demonstrated greatly enhanced electronic and ionic conductivity, enabling high specific capacities and reversibility (e.g. 155 mAh g^{-1} at 0.1 C and 152 mAh g^{-1} at 0.2 C) for this $FePO_4$ cathode material [239].

Furthermore, amorphous $FeSO_4$ had also been systematically studied as a promising electrode for SIBs. Kitajou et al. improved the electrochemical performance by increasing the sodium concentration per unit volume to accelerate the transmission of sodium ions. In this work, an amorphous xNaF–$FeSO_4$ systems ($1 \leq x \leq 2$) with excellent sodium storage properties were illustrated [240]. Specifically, the 1.3NaF–$FeSO_4$ showed an initial discharge capacity of 115 mAh g^{-1} and a capacity retention of 64% at 0.05 C. In addition, the cathode properties of crystalline NaF–$FeSO_4$ have been investigated and showed an inferior capacity of 66 mAh g^{-1}. Given the volume change during (de)insertion process, amorphous samples had low density local structure for Na ion movement. Thus, it was believed that the lower Na diffusion energy and higher Na concentration were responsible for the improved cathode properties of amorphous NaF–$FeSO_4$.

4.8 Concluding Remarks

Polyanionic compounds have the advantages of good structural stability, high ionic conductivity, and good thermal stability, which provide a suitable and feasible system for reliable and well-function cathode materials. The electrochemical performance of the polyanionic system can be significantly enhanced through the established routes such as surface engineering with highly conductive carbon compounds, the design and optimization of specific micro-nano structures and crystal lattice doping. Although the currently designed polyanion-based cathodes have achieved fairly good results, their energy/power densities are still unsatisfactory, which seriously hinder their practical application. To develop high-performance cathode materials and maximize the overall electrochemical activity for the entire battery system, rational modification strategies must be employed. Therefore, for accelerating the practical application of polyanionic cathode materials, the following problems need to be put on the agenda: (i) developing and optimizing suitable Na^+ intercalation structures with high reversible capacity and working voltage via the theoretical simulation and calculation methods; (ii) controlling microstructure parameters with increased active material loading to achieve high energy density; (iii) exploring rational electrode modification strategy to produce polyanionic cathodes with high specific capacity and power density; and (iv) inventing high-performance electrolyte with good safety to match the high potential cathode material. On the basis of these efforts on the fundamental research, the realization of industrial-scale production is just around the corner. Polyanionic materials are expected to play an important role in realizing high-energy large-scale energy storage applications.

References

1 Saravanan, K., Mason, C.W., Rudola, A. et al. (2013). *Adv. Energy Mater.* 3: 444.
2 Barpanda, P., Avdeev, M., Ling, C.D. et al. (2013). *Chinese Phys. Lett.* 52: 395.
3 Xu, M., Wang, L., Zhao, X. et al. (2013). *Phys. Chem. Chem. Phys.* 15: 13032.
4 Lu, Y., Zhang, S., Li, Y. et al. (2014). *J. Power Sources* 247: 770.
5 Ponrouch, A., Dedryvère, R., Monti, D. et al. (2013). *Energy Environ. Sci.* 6: 2361.
6 Kumar, P.R., Jung, Y.H., Wang, J.E., and Kim, D.K. (2016). *J. Power Sources* 324: 421.
7 Islam, M.S., Heath, J., and Tealdi, C. (2016). *J. Mater. Chem. A* 4: 6998.
8 Lin, L., Jing, W., and Ya-Ru, Z. (2019). *J. Power Sources* 438: 227016.
9 Padhi, K.A. (1997). *J. Electrochem. Soc.* 144: 1188.
10 Zhu, Y., Xu, Y., Liu, Y. et al. (2013). *Nanoscale* 5: 780.
11 (a) Galceran, M., Roddatis, V., Zuniga, F.J. et al. (2014). *Chem. Mater.* 26: 3289. (b) Tracy, S.J., Mauger, L., Smith, H.L. et al. (2016). *Chem. Mater.* 28: 3051. (c) Sugiyama, J., Nozaki, H., Harada, M. et al. (2014). *Phys. Rev. B* 90: 014426.
12 Moreau, P., Guyomard, D., Gaubicher, J., and Boucher, F. (2010). *Chem. Mater.* 22: 4126.
13 Lu, J., Chung, S.C., Nishimura, S.-i., and Yamada, A. (2013). *Chem. Mater.* 25: 4557.
14 Padhi, A.K., Nanjundaswamy, K.S., Masquelier, C. et al. (1997). *J. Electrochem. Soc.* 144: 1609.
15 Casas-Cabanas, M., Roddatis, V.V., Saurel, D. et al. (2012). *J. Mater. Chem.* 22: 17421.
16 Le Poul, N., Baudrin, E., Morcrette, M. et al. (2003). *Solid State Ionics* 159: 149.
17 Wongittharom, N., Lee, T.-C., Wang, C.-H. et al. (2014). *J. Mater. Chem. A* 2: 5655.
18 Oh, S.-M., Myung, S.-T., Hassoun, J. et al. (2012). *Electrochem. Commun.* 22: 149.
19 (a) Galceran, M., Saurel, D., Acebedo, B. et al. (2014). *Phys. Chem. Chem. Phys.* 16: 8837. (b) Boucher, F., Gaubicher, J., Cuisinier, M. et al. (2014). *J. Am. Chem. Soc.* 136: 9144. (c) Fang, Y., Liu, Q., Xiao, L. et al. (2015). *ACS Appl. Mater. Interfaces* 7: 17977.
20 Li, C., Miao, X., Chu, W. et al. (2015). *J. Mater. Chem. A* 3: 8265.
21 Zhu, X., Mochiku, T., Fujii, H. et al. (2018). *Nano Res.* 11: 6197.
22 Boyadzhieva, T., Koleva, V., Zhecheva, E. et al. (2015). *RSC Adv.* 5: 87694.
23 Boyadzhieva, T., Koleva, V., Kukeva, R. et al. (2020). *RSC Adv.* 10: 29051.
24 Zou, B.K., Shao, Y., Qiang, Z.Y. et al. (2016). *J. Power Sources* 336: 231.
25 Goodenough, J.B. (1983). *Solid State Ionics* 9–10: 793.
26 Delmas, C., Cherkaoui, F., Nadiri, A., and Hagenmuller, P. (1987). *Mater. Res. Bull.* 22: 631.
27 Zhao, L., Zhang, T., Zhao, H., and Hou, Y. (2020). *Mater. Today Nano* 10: 100072.

28 (a) Anantharamulu, N., Koteswara Rao, K., Rambabu, G. et al. (2011). *J. Mater. Sci.* 46: 2821. (b) Patoux, S., Rousse, G., Leriche, J.-B., and Masquelier, C. (2003). *Chem. Mater.* 15: 2084. (c) Sun, Q., Ren, Q.-Q., and Fu, Z.-W. (2012). *Electrochem. Commun.* 23: 145.

29 (a) Duan, W., Zhu, Z., Li, H. et al. (2014). *J. Mater. Chem. A* 2: 8668. (b) Li, H., Bai, Y., Wu, F. et al. (2015). *Solid State Ionics* 278: 281.

30 Rui, X., Sun, W., Wu, C. et al. (2015). *Adv. Mater.* 27: 6670–6676.

31 Zhang, X.H., Rui, X.H., Chen, D. et al. (2019). *Nanoscale* 11: 2556.

32 (a) Lim, S.Y., Kim, H., Shakoor, R.A. et al. (2012). *J. Electrochem. Soc.* 159: A1393. (b) Jian, Z., Yuan, C., Han, W. et al. (2014). *Adv. Funct. Mater.* 24: 4265. (c) Zhu, C., Kopold, P., van Aken, P.A. et al. (2016). *Adv. Mater.* 28: 2409. (d) Song, W., Ji, X., Wu, Z. et al. (2014). *J. Mater. Chem. A* 2: 5358. (e) Song, W., Cao, X., Wu, Z. et al. (2014). *Phys. Chem. Chem. Phys. PCCP* 16: 17681. (f) Jian, Z., Han, W., Lu, X. et al. (2013). *Adv. Energy Mater.* 3: 156.

33 Bui, K.M., Dinh, V.A., Okada, S., and Ohno, T. (2015). *Phys. Chem. Chem. Phys.* 17: 30433.

34 Li, S., Dong, Y., Xu, L. et al. (2014). *Adv. Mater.* 26: 3545.

35 (a) Jiang, Y., Yang, Z., Li, W. et al. (2015). *Adv. Energy Mater.* 5: 1402104. (b) Ren, W., Zheng, Z., Xu, C. et al. (2016). *Nano Energy* 25: 145. (c) Zhu, C., Song, K., van Aken, P.A. et al. (2014). *Nano Lett.* 14: 2175. (d) Chen, M., Kou, K., Tu, M. et al. (2015). *Solid State Ionics* 274: 24. (e) Liu, J., Tang, K., Song, K. et al. (2014). *Nanoscale* 6: 5081. (f) Fang, J., Wang, S., Li, Z. et al. (2016). *J. Mater. Chem. A* 4: 1180. (g) Zhang, Q., Wang, W., Wang, Y. et al. (2016). *Nano Energy* 20: 11. (h) Jian, Z., Zhao, L., Pan, H. et al. (2012). *Electrochem. Commun.* 14: 86. (i) Zhai, Y., Dou, Y., Zhao, D. et al. (2011). *Adv. Mater.* 23: 4828. (j) Wang, D.-W., Zeng, Q., Zhou, G. et al. (2013). *J. Mater. Chem. A* 1: 9382.

36 (a) Shen, W., Wang, C., Xu, Q. et al. (2015). *Adv. Energy Mater.* 5: 1400982. (b) Shen, W., Li, H., Wang, C. et al. (2015). *J. Mater. Chem. A* 3: 15190. (c) Shen, W., Li, H., Guo, Z. et al. (2016). *ACS Appl. Mater. Interfaces* 8: 15341.

37 Chen, S., Wu, C., Shen, L. et al. (2017). *Adv. Mater.* 29: 1700431.

38 (a) Elias, D.C., Nair, R.R., Mohiuddin, T.M.G. et al. (2009). *Science* 323: 610. (b) Chen, S., Yeoh, W., Liu, Q., and Wang, G. (2012). *Carbon* 50: 4557. (c) Chen, S., Wang, Y., Ahn, H., and Wang, G. (2012). *J. Power Sources* 216: 22. (d) Zhan, B., Li, C., Yang, J. et al. (2014). *Small* 10: 4042. (e) Liao, L., Zhu, J., Bian, X. et al. (2013). *Adv. Funct. Mater.* 23: 5326. (f) Raccichini, R., Varzi, A., Passerini, S., and Scrosati, B. (2015). *Nat. Mater* 14: 271. (g) Raccichini, R., Varzi, A., Wei, D., and Passerini, S. (2017). *Adv. Mater.* 29: 1603421.

39 Jung, Y.H., Lim, C.H., and Kim, D.K. (2013). *J. Mater. Chem. A* 1: 11350.

40 Xu, Y., Wei, Q., Xu, C. et al. (2016). *Adv. Energy Mater.* 6: 1600389.

41 (a) Gu, D., Bongard, H., Deng, Y. et al. (2010). *Adv. Mater.* 22: 833. (b) Wang, J., Li, W., Wang, F. et al. (2014). *Nanoscale* 6: 3217. (c) Chen, S., Huang, X., Sun, B. et al. (2014). *J. Mater. Chem. A* 2: 16199. (d) Schuster, J., He, G., Mandlmeier, B. et al. (2012). *Angew. Chem. Int. Ed.* 51: 3591.

42 Wang, Q., Zhao, B., Zhang, S. et al. (2015). *J. Mater. Chem. A* 3: 7732.

43 Mao, J., Luo, C., Gao, T. et al. (2015). *J. Mater. Chem. A* 3: 10378.

44 (a) Sforzini, J., Hapala, P., Franke, M. et al. (2016). *Phys. Rev. Lett.* 116: 126805. (b) Lin, Y., Wu, S., Shi, W. et al. (2015). *Chem. Commun.* 51: 13086. (c) Xie, X., Su, D., Zhang, J. et al. (2015). *Nanoscale* 7: 3164. (d) Cui, K., Hu, S., and Li, Y. (2016). *Electrochim. Acta* 210: 45. (e) Guo, J.-Z., Wu, X.-L., Wan, F. et al. (2015). *Chem. Eur. J.* 21: 17371. (f) Wang, H., Maiyalagan, T., and Wang, X. (2012). *ACS Catal.* 2: 781.

45 (a) Fei, H., Dong, J., Arellano-Jiménez, M.J. et al. (2015). *Nat. Commun.* 6: 8668. (b) Joucken, F., Tison, Y., Le Fèvre, P. et al. (2015). *Sci. Rep.* 5: 14564. (c) Shao, Y., Zhang, S., Engelhard, M.H. et al. (2010). *J. Mater. Chem.* 20: 7491.

46 (a) Liang, H.-W., Wei, W., Wu, Z.-S. et al. (2013). *J. Am. Chem. Soc.* 135: 16002. (b) Chen, T., Pan, L., Loh, T.A.J. et al. (2014). *Dalton Trans.* 43: 14931. (c) Bulusheva, L.G., Okotrub, A.V., Kurenya, A.G. et al. (2011). *Carbon* 49: 4013.

47 (a) Kuchta, B., Firlej, L., Roszak, S., and Pfeifer, P. (2010). *Adsorption* 16: 413. (b) Wu, Z.-S., Ren, W., Xu, L. et al. (2011). *ACS Nano* 5: 5463. (c) Han, J., Zhang, L.L., Lee, S. et al. (2013). *ACS Nano* 7: 19. (d) Yeom, D.-Y., Jeon, W., Tu, N.D.K. et al. (2015). *Sci. Rep.* 5: 9817. (e) Xia, K., Zhan, H., Wei, Y., and Gu, Y. (2014). *Beilstein J. Nanotechnol.* 5: 329. (f) Lin, Y., Zhu, Y., Zhang, B. et al. (2015). *J. Mater. Chem. A* 3: 21805.

48 Jiang, Y., Zhou, X., Li, D. et al. (2018). *Adv. Energy Mater.* 8: 1800068.

49 (a) Li, H., Yu, X., Bai, Y. et al. (2015). *J. Mater. Chem. A* 3: 9578. (b) Inoishi, A., Yoshioka, Y., Zhao, L. et al. (2017). *ChemElectroChem* 4: 2755. (c) Xu, G. and Sun, G. (2016). *Ceram. Int.* 42: 14744.

50 (a) Shen, W., Li, H., Guo, Z.Y. et al. (2016). *RSC Adv.* 6: 71581. (b) Klee, R., Lavela, P., Aragón, M.J. et al. (2016). *J. Power Sources* 313: 73.

51 (a) Chen, Y., Xu, Y., Sun, X., and Wang, C. (2018). *J. Power Sources* 375: 82. (b) Lalère, F., Seznec, V., Courty, M. et al. (2015). *J. Mater. Chem. A* 3: 16198. (c) Aragón, M.J., Lavela, P., Alcántara, R., and Tirado, J.L. (2015). *Electrochim. Acta* 180: 824.

52 Zhu, Q., Cheng, H., Zhang, X. et al. (2018). *Electrochim. Acta* 281: 208.

53 Li, H., Bai, Y., Wu, F. et al. (2016). *ACS Appl. Mater. Interfaces* 8: 27779.

54 Zheng, Q., Yi, H., Liu, W. et al. (2017). *Electrochim. Acta* 238: 288.

55 (a) Kim, H., Park, I., Seo, D.H. et al. (2012). *J. Am. Chem. Soc.* 134: 10369. (b) Aragon, M.J., Lavela, P., Ortiz, G.F., and Tirado, J.L. (2015). *J. Electrochem. Soc.* 162: 257.

56 Zhang, B., Zeng, T., Liu, Y., and Zhang, J. (2018). *RSC Adv.* 8: 5523.

57 Li, X., Huang, Y., Wang, J. et al. (2018). *J. Mater. Chem. A* 6: 1390.

58 Lim, S.J., Han, D.W., Nam, D.H. et al. (2014). *J. Mater. Chem. A* 2: 19623.

59 Zheng, Q., Ni, X., Lin, L. et al. (2018). *J. Mater. Chem. A* 6: 4209.

60 Muruganantham, R., Chiu, Y.T., Yang, C.C. et al. (2017). *Sci. Rep.* 7: 1.

61 Zhu, T., Hu, P., Wang, X. et al. (2019). *Adv. Energy Mater.* 9: 1803436.

62 Gao, H. and Goodenough, J.B. (2016). *Angew. Chem. Int. Ed.* 55: 12768.

63 Zhu, T., Hu, P., Cai, C. et al. (2020). *Nano Energy* 70: 104548.

64 Gao, H., Li, Y., Park, K., and Goodenough, J.B. (2016). *Chem. Mater.* 28: 6553.

65 Li, H., Xu, M., Gao, C. et al. (2020). *Energy Storage Mater.* 26: 325.

66 Yahia, H.B., Essehli, R., Avdeev, M. et al. (2016). *J. Solid State Chem.* 238: 103.
67 Gao, H., Seymour, I.D., Xin, S. et al. (2018). *J. Am. Chem. Soc.* 140: 18192.
68 (a) Zhang, J., Liu, Y., Zhao, X. et al. (2020). *Adv. Mater.* 32: 1906348. (b) Zhang, W., Li, H., Zhang, Z. et al. (2020). *Small* 16: 2001524.
69 Essehli, R., Belharouak, I., Ben Yahia, H. et al. (2015). *Dalton Trans.* 44: 7881.
70 Essehli, R., Belharouak, I., Ben Yahia, H. et al. (2015). *Dalton Trans.* 44: 4526.
71 Dwibedi, D., Gond, R., and Barpanda, P. (2019). *Chem. Mater.* 31: 7501.
72 (a) Bando, X.W.M.-S.W.Y. and Golberg, D. (2010). *J. Am. Chem. Soc.* 132: 13592. (b) Wang, X., Niu, C., Meng, J. et al. (2015). *Adv. Energy Mater.* 5: 1500716.
73 Yabuuchi, N., Kubota, K., Dahbi, M., and Komaba, S. (2014). *Chem. Rev.* 114: 11636.
74 (a) Leclaire, A., Benmoussa, A., Borel, M.M. et al. (1988). *J. Solid State Chem.* 77: 299. (b) Gabelica-Robert, M., Goreaud, M., Labbe, P., and Raveau, B. (1982). *J. Solid State Chem.* 45: 389. (c) Kee, Y., Dimov, N., Staikov, A. et al. (2015). *RSC Adv.* 5: 64991. (d) Leclaire, A., Borel, M.M., Grandin, A., and Raveau, B. (1988). *J. Solid State Chem.* 76: 131.
75 (a) Erragh, F., Boukhari, A., Elouadi, B., and Holt, E.M. (1991). *J. Crystallogr. Spectrosc. Res.* 21: 321. (b) Clark, J.M., Barpanda, P., Yamada, A., and Islam, M.S. (2014). *J. Mater. Chem. A* 2: 11807. (c) Chen, C.Y., Matsumoto, K., Nohira, T. et al. (2014). *Electrochim. Acta* 133: 583. (d) Barpanda, P., Ye, T., Avdeev, M. et al. (2013). *J. Mater. Chem. A* 1: 4194. (e) Lokanath, N.K., Sridhar, M.A., Prasad, J.S. et al. (2000). *Bull. Mater. Sci.* 23: 175.
76 (a) Wood, S.M., Eames, C., Kendrick, E., and Islam, M.S. (2015). *J. Phys. Chem. C* 119: 15935. (b) Aksyonov, D.A., Fedotov, S.S., Stevenson, K.J., and Zhugayevych, A. (2018). *Comput. Mater. Sci.* 154: 449. (c) Nose, M., Nakayama, H., Nobuhara, K. et al. (2013). *J. Power Sources* 234: 175.
77 Barpanda, P., Nishimura, S., and Yamada, A. (2012). *Adv. Energy Mater.* 2: 841.
78 Ng, H.N. and Calvo, C. (1973). *Can. J. Chem.* 51: 2613.
79 Drozhzhin, O.A., Tertov, I.V., Alekseeva, A.M. et al. (2019). *Chem. Mater.* 31: 7463.
80 Moya-Pizarro, T., Salmon, R., Fournes, L. et al. (1984). *J. Solid State Chem.* 53: 387.
81 (a) Sanz, F., Parada, C., Rojo, J.M. et al. (1999). *J. Solid State Chem.* 145: 604. (b) Barpanda, P., Avdeev, M., Ling, C.D. et al. (2013). *Inorg. Chem.* 52: 395.
82 Beaury, L., Derouet, J., Binet, L. et al. (2004). *J. Solid State Chem.* 177: 1437.
83 Barpanda, P., Ye, T., Nishimura, S.-i. et al. (2012). *Electrochem. Commun.* 24: 116.
84 Park, C.S., Kim, H., Shakoor, R.A. et al. (2013). *J. Am. Chem. Soc.* 135: 2787.
85 Barpanda, P., Lu, J., Ye, T. et al. (2013). *RSC Adv.* 3: 3857.
86 Kim, H., Park, C.S., Choi, J.W., and Jung, Y. (2016). *Angew. Chem. Int. Ed.* 55: 6662.
87 Kim, H., Park, C.S., Choi, J.W., and Jung, Y. (2016). *Chem. Mater.* 28: 6724.

88 Nishimura, S.-i., Nakamura, M., Natsui, R., and Yamada, A. (2010). *J. Am. Chem. Soc.* 132: 13596.
89 Carlier, D., Cheng, J.H., Berthelot, R. et al. (2011). *Dalton Trans.* 40: 9306.
90 (a) Barpanda, P., Liu, G., Ling, C.D. et al. (2013). *Chem. Mater.* 25: 3480. (b) Kim, H., Shakoor, R.A., Park, C. et al. (2013). *Adv. Funct. Mater.* 23: 1147.
91 Chen, C.-Y., Matsumoto, K., Nohira, T. et al. (2014). *J. Power Sources* 246: 783.
92 (a) Hwang, J., Matsumoto, K., and Hagiwara, R. (2018). *J. Phys. Chem. C* 122: 26857. (b) Makhlooghiazad, F., Sharma, M., Zhang, Z. et al. (2020). *J. Phys. Chem. Lett.* 11: 2092.
93 (a) Tealdi, C., Ricci, M., Ferrara, C. et al. (2016). *RSC Adv.* 6: 99735. (b) Longoni, G., Wang, J.E., Jung, Y.H. et al. (2016). *J. Power Sources* 302: 61. (c) Cao, J., Ni, L., Qin, C. et al. (2017). *Ionics* 23: 1783. (d) Song, H.J., Kim, D.-S., Kim, J.-C. et al. (2017). *J. Mater. Chem. A* 5: 5502. (e) Chen, X., Du, K., Lai, Y. et al. (2017). *J. Power Sources* 357: 164. (f) Zeng, L., Li, F., Xu, X. et al. (2020). *ChemElectroChem* 7: 3874. (g) Shakoor, R.A., Park, C.S., Raja, A.A. et al. (2016). *Phys. Chem. Chem. Phys.* 18: 3929.
94 Zhou, H., Upreti, S., Chernova, N.A. et al. (2011). *Chem. Mater.* 23: 293.
95 Shakoor, R.A., Kim, H., Cho, W. et al. (2012). *J. Am. Chem. Soc.* 134: 11740.
96 Li, H., Chen, X., Jin, T. et al. (2019). *Energy Storage Mater.* 16: 383.
97 Essehli, R., El Bali, B., Benmokhtar, S. et al. (2010). *J. Alloys Compd.* 493: 654.
98 (a) Padhi, A.K., Manivannan, V., and Goodenough, J.B. (1998). *ChemInform* 29: 514. (b) Sanz, F., Parada, C., Rojo, J.M., and Ruiz-Valero, C. (2001). *Chem. Mater.* 13: 1334.
99 (a) Kim, H., Park, I., Lee, S. et al. (2013). *Chem. Mater.* 25: 3614. (b) Kosova, N.V. and Shindrov, A.A. (2019). *Batteries-Basel* 5: 39.
100 (a) Fernadez-Ropero, A.J., Zarrabeitia, M., Reynaud, M. et al. (2018). *J. Phys. Chem. C* 122: 133. (b) Lee, M.H., Kim, S.J., Chang, D. et al. (2019). *Mater. Today* 29: 26.
101 (a) Baskar, S., Angalakuthi, R., Murugesan, C. et al. (2018). *J. Electrochem. Soc.* 85: 227. (b) Kosova, N.V. and Belotserkovsky, V.A. (2018). *Electrochim. Acta* 278: 182.
102 Ma, X., Pan, Z., Wu, X., and Shen, P.K. (2019). *Chem. Eng. J.* 365: 132.
103 Wu, X., Zhong, G., and Yang, Y. (2016). *J. Power Sources* 327: 666.
104 Cao, Y., Xia, X., Liu, Y. et al. (2020). *J. Power Sources* 461: 228130.
105 Yuan, T., Wang, Y., Zhang, J. et al. (2019). *Nano Energy* 56: 160.
106 Pu, X., Wang, H., Yuan, T. et al. (2019). *Energy Storage Mater.* 22: 330.
107 Chen, M., Hua, W., Xiao, J. et al. (2019). *Nat. Commun.* 10: 1480.
108 Senthilkumar, B., Rambabu, A., Murugesan, C. et al. (2020). *ACS Omega* 5: 7219.
109 Jang, J.Y., Kim, H., Lee, Y. et al. (2014). *Electrochem. Commun.* 44: 74.
110 Lee, Y., Lee, J., Kim, H. et al. (2016). *J. Power Sources* 320: 49.
111 Moriwake, H., Kuwabara, A., Fisher, C.A. et al. (2016). *J. Power Sources* 326: 220.
112 Sanz, F., Parada, C., Amador, U. et al. (1996). *J. Solid State Chem.* 123: 129.
113 Zarrabeitia, M., Jauregui, M., Sharma, N. et al. (2019). *Chem. Mater.* 31: 5152.

114 Kumar, P.R., Essehli, R., Yahia, H.B. et al. (2020). *RSC Adv.* 10: 15983.
115 (a) Nose, M., Shiotani, S., Nakayama, H. et al. (2013). *Electrochem. Commun.* 34: 266. (b) Li, H., Cormier, M., Zhang, N. et al. (2019). *J. Electrochem. Soc.* 166: A429.
116 Liu, X., Tang, L., Li, Z. et al. (2019). *J. Mater. Chem. A* 7: 18940.
117 Livitska, O.V., Strutynska, N.Y., Zatovsky, I.V., and Slobodyanik, N.S. (2015). *Cryst. Res. Technol.* 50: 626.
118 Zhang, H., Hasa, I., Buchholz, D. et al. (2017). *NPG Asia Mater.* 9: 370.
119 Kumar, P.R., Yahia, H.B., Belharouak, I. et al. (2020). *J. Solid State Electrochem.* 24: 17.
120 Kim, H., Yoon, G., Park, I. et al. (2015). *Energy Environ. Sci.* 8: 3325.
121 Ryu, S., Wang, J.E., Kim, J.-H. et al. (2019). *J. Power Sources* 444: 227274.
122 Masquelier, C., Yvoire, F.D., and Rodier, N. (1991). *J. Solid State Chem.* 95: 156.
123 Kim, J., Park, I., Kim, H. et al. (2016). *Adv. Energy Mater.* 6: 1502147.
124 Kovrugin, V.M., Chotard, J.-N., Fauth, F., and Masquelier, C. (2020). *J. Mater. Chem. A* 8: 21110.
125 Deng, C., Zhang, S., and Zhao, B. (2016). *Energy Storage Mater.* 4: 71.
126 Ke, L., Yu, T., Lin, B. et al. (2016). *Nanoscale* 8: 19120.
127 Li, Q., Lin, B., Zhang, S., and Deng, C. (2016). *J. Mater. Chem. A* 4: 5719.
128 Lim, S.Y., Kim, H., Chung, J. et al. (2014). *Proc. Natl. Acad. Sci. USA* 111: 599.
129 Deng, C. and Zhang, S. (2014). *ACS Appl. Mater. Interfaces* 6: 9111.
130 Deng, C., Zhang, S., and Wu, Y. (2015). *Nanoscale* 7: 487.
131 Barker, J., Saidi, M.Y., and Swoyer, J.L. (2003). *Electrochem. Solid State Lett.* 6: A1.
132 Barker, J., Saidi, M.Y., and Swoyer, J.L. (2004). *J. Electrochem. Soc.* 151: 1662.
133 Mamoor, M., Lian, R., Wang, D. et al. (2019). *Phys. Chem. Chem. Phys.* 21: 14612.
134 (a) Ling, M., Li, F., Yi, H. et al. (2018). *J. Mater. Chem. A* 6: 24201. (b) Cheng, B., Zhang, S., Zou, F. et al. (2019). *J. Alloys Compd.* 811: 151828.
135 Chang, C., Li, Y., He, W. et al. (2017). *Mater. Lett.* 209: 82.
136 Law, M. and Balaya, P. (2018). *Energy Storage Mater.* 10: 102.
137 Zhao, J., He, J., Ding, X. et al. (2010). *J. Power Sources* 195: 6854.
138 Zhuo, H., Wang, X., and Tang, A. (2006). *J. Power Sources* 160: 698.
139 Xu, M., Cheng, C.J., Sun, Q.Q. et al. (2015). *RSC Adv.* 5: 56686.
140 Ge, X., Li, X., Wang, Z. et al. (2019). *Chem. Eng. J.* 357: 458.
141 Jin, T., Liu, Y., Li, Y. et al. (2017). *Adv. Energy Mater.* 7: 1700087.
142 Feng, P., Wang, W., Hou, J. et al. (2018). *Chem. Eng. J.* 353: 25.
143 Kim, S.W., Seo, D.H., Kim, H. et al. (2012). *Phys. Chem. Chem. Phys.* 14: 3299.
144 (a) Wu, L., Hu, Y., Zhang, X. et al. (2018). *J. Power Sources* 374: 40. (b) Ko, W., Yoo, J.-K., Park, H. et al. (2019). *J. Power Sources* 432: 1. (c) Hu, H., Bai, Y., Miao, C. et al. (2020). *J. Electroanal. Chem.*: 867. d Zhang, J., Zhou, X., Wang, Y. et al. (2019). *Small* 15: e1903723.
145 Wang, F., Zhang, N., Zhao, X. et al. (2019). *Adv Sci.* 6: 1900649.
146 Yan, J., Liu, X., and Li, B. (2015). *Electrochem. Commun.* 56: 46.
147 Hu, Y., Wu, L., Liao, G. et al. (2018). *Ceram. Int.* 44: 17577.

148 Lin, X., Hou, X., Wu, X. et al. (2014). *RSC Adv.* 4: 40985.

149 Zhong, Y., Wu, Z., Tang, Y. et al. (2015). *Mater. Lett.* 145: 269.

150 Ling, R., Cai, S., Shen, K. et al. (2019). *Ceram. Int.* 45: 19799.

151 Ellis, B.L., Makahnouk, W.R.M., Rowan-Weetaluktuk, W.N. et al. (2010). *Chem. Mater.* 22: 1059.

152 Le Meins, J.M., Crosnier-Lopez, M.P., Hemon-Ribaud, A., and Courbion, G. (1999). *J. Solid State Chem.* 148: 260.

153 Shakoor, R.A., Seo, D.-H., Kim, H. et al. (2012). *J. Mater. Chem.* 22: 20535.

154 (a) Liu, Z., Hu, Y.-Y., Dunstan, M.T. et al. (2014). *Chem. Mater.* 26: 2513. (b) Bianchini, M., Fauth, F., Brisset, N. et al. (2015). *Chem. Mater.* 27: 3009.

155 Barker, J., Gover, R.K.B., Burns, P., and Bryan, A.J. (2007). *J. Electrochem. Soc.* 154: A882.

156 Song, W., Ji, X., Wu, Z. et al. (2014). *RSC Adv.* 4: 11375.

157 Chihara, K., Kitajou, A., Gocheva, I.D. et al. (2013). *J. Power Sources* 227: 80.

158 Song, W., Wu, Z., Chen, J. et al. (2014). *Electrochim Acta* 146: 142.

159 Song, W., Ji, X., Wu, Z. et al. (2014). *J. Power Sources* 256: 258.

160 Liu, Q., Wang, D., Yang, X. et al. (2015). *J. Mater. Chem. A* 3: 21478.

161 Mukherjee, A., Sharabani, T., Perelshtein, I., and Noked, M. (2020). *Batter. Supercaps* 3: 52.

162 Guo, H., Hu, Y., Zhang, X. et al. (2019). *Front. Chem.* 7: 689.

163 Li, Y., Liang, X., Zhong, G. et al. (2020). *ACS Appl. Mater. Interfaces* 12: 25920.

164 Mukherjee, A., Sharabani, T., Sharma, R. et al. (2020). *Batter. Supercaps* 3: 510.

165 Liu, W., Yi, H., Zheng, Q. et al. (2017). *J. Mater. Chem. A* 5: 10928.

166 Yi, H., Ling, M., Xu, W. et al. (2018). *Nano Energy* 47: 340.

167 Li, L., Liu, X., Tang, L. et al. (2019). *J. Alloys Compd.* 790: 203.

168 Criado, A., Lavela, P., Perez-Vicente, C. et al. (2020). *J. Electroanal. Chem.* 856: 113694.

169 Wang, M., Wang, K., Huang, X. et al. (2020). *Ceram. Int.* 46: 28490.

170 Zhang, Y., Guo, S., and Xu, H. (2018). *J. Mater. Chem. A* 6: 4525.

171 Li, J., Du, G., Huang, T. et al. (2020). *Energy Technol.* 8: 2000494.

172 Serras, P., Palomares, V., Rojo, T. et al. (2014). *J. Mater. Chem. A* 2: 7766.

173 Palomares, V., Blas, M., Setien, S. et al. (2018). *Dalton Trans.* 47: 2610.

174 Massa, W., Yakubovich, O.V., and Dimitrova, O.V. (2002). *Solid State Sci.* 4: 495.

175 Sauvage, F., Quarez, E., Tarascon, J.M., and Baudrin, E. (2006). *Solid State Sci.* 8: 1215.

176 Sharma, N., Serras, P., Palomares, V. et al. (2014). *Chem. Mater.* 26: 3391.

177 Xu, M., Xiao, P., Stauffer, S. et al. (2014). *Chem. Mater.* 26: 3089.

178 (a) Jin, H., Dong, J., Uchaker, E. et al. (2015). *J. Mater. Chem. A* 3: 17563. (b) Yin, Y., Xiong, F., Pei, C. et al. (2017). *Nano Energy* 41: 452. (c) Liu, Y., Tai, Z., Zhang, J. et al. (2018). *Nat. Commun.* 9: 3645. (d) Chang, W., Zhang, X.-Y., Qu, J. et al. (2020). *ACS Appl. Mater. Interfaces* 12: 41419.

179 Deng, G., Chao, D., Guo, Y. et al. (2016). *Energy Storage Mater.* 5: 198.

180 (a) Chen, H., Mi, H., Sun, L. et al. (2018). *ChemElectroChem* 5: 2593. (b) Zhang, Z., Chen, Z., Mai, Z. et al. (2019). *Small* 15: 1900356.

181 (a) Serras, P., Palomares, V., Kubiak, P. et al. (2013). *Electrochem. Commun.* 34: 344. (b) Jin, H., Liu, M., Uchaker, E. et al. (2017). *CrystEngComm* 19: 4287. (c) Zhang, L.-L., Liu, J., Wei, C. et al. (2020). *ACS Appl. Mater. Interfaces* 12: 3670.

182 Qi, Y., Tong, Z., Zhao, J. et al. (2018). *Joule* 2: 2348.

183 Hou, Y., Chang, K., Wang, Z. et al. (2019). *Sci. China Mater.* 62: 474.

184 Peng, M., Zhang, D., Zheng, L. et al. (2017). *Nano Energy* 31: 64.

185 Peng, M., Li, B., Yan, H. et al. (2015). *Angew. Chem. Int. Ed.* 54: 6452.

186 Nguyen, L.H.B., Olchowka, J., Belin, S. et al. (2019). *ACS Appl. Mater. Interfaces* 11: 38808.

187 Du, P., Li, T., Jiang, X.L. et al. (2020). *J. Alloys Compd.* 814: 10.

188 (a) Qi, Y., Zhao, J., Yang, C. et al. (2018). *Small Meth.* 3: 1800111. (b) Kumar, P.R., Jung, Y.H., Ahad, S.A., and Kim, D.K. (2017). *RSC Adv.* 7: 21820.

189 Qi, Y., Mu, L., Zhao, J. et al. (2015). *Angew. Chem. Int. Ed.* 54: 9911.

190 Park, Y.-U., Seo, D.-H., Kim, H. et al. (2014). *Adv. Funct. Mater.* 24: 4603.

191 Zhao, J., Mu, L., Qi, Y. et al. (2015). *Chem. Commun. (Camb)* 51: 7160.

192 Serras, P., Palomares, V., Goñi, A. et al. (2012). *J. Mater. Chem.* 22: 22301.

193 Kumar, P.R., Jung, Y.H., and Kim, D.K. (2016). *J. Solid State Electrochem.* 21: 223.

194 Xiang, X., Lu, Q., Han, M., and Chen, J. (2016). *Chem. Commun. (Camb)* 52: 3653.

195 Palomares, V., Serras, P., Brand, H.E.A. et al. (2015). *J. Mater. Chem. A* 3: 23017.

196 Li, C., Shen, M., Hu, B. et al. (2018). *J. Mater. Chem. A* 6: 8340.

197 Kosova, N. and Rezepova, D. (2017). *Inorganics* 5: 19.

198 Park, Y.U., Seo, D.H., Kwon, H.S. et al. (2013). *J. Am. Chem. Soc.* 135: 13870.

199 Barpanda, P., Oyama, G., Nishimura, S.-i. et al. (2014). *Nat. Commun.* 5: 4358.

200 Lander, L., Tarascon, J.-M., and Yamada, A. (2018). *Chem. Rec.* 18: 1394.

201 Plewa, A., Kulka, A., Hanc, E. et al. (2020). *J. Mater. Chem. A* 8: 2728.

202 Oyama, G., Nishimura, S.i., Suzuki, Y. et al. (2015). *ChemElectroChem* 2: 1019.

203 Wei, S., Mortemard de Boisse, B., Oyama, G. et al. (2016). *ChemElectroChem* 3: 209.

204 Jungers, T., Mahmoud, A., Malherbe, C. et al. (2019). *J. Mater. Chem. A* 7: 8226.

205 Dwibedi, D., Baskar, S., and Barpanda, P. (2017). *ECS Trans.* 80: 337.

206 Wang, W., Liu, X., Xu, Q. et al. (2018). *J. Mater. Chem. A* 6: 4354.

207 Nam, Y.C., Lee, S.J., and Son, J.T. (2020). *J. Korean Phys. Soc.* 77: 836.

208 Araujo, R.B., Chakraborty, S., Barpanda, P., and Ahuja, R. (2016). *Phys. Chem. Chem. Phys.* 18: 9658.

209 Barpanda, P., Oyama, G., Ling, C.D., and Yamada, A. (2014). *Chem. Mater.* 26: 1297.

210 Meng, Y., Li, Q., Yu, T. et al. (2016). *CrystEngComm* 18: 1645.

211 Yao, G., Zhang, X., Yan, Y. et al. (2020). *J. Energy Chem.* 50: 387.

212 Pan, W., Guan, W., Liu, S. et al. (2019). *J. Mater. Chem. A* 7: 13197.

213 (a) Tripathi, R., Ramesh, T.N., Ellis, B.L., and Nazar, L.F. (2010). *Angew. Chem. Int. Ed.* 49: 8738. (b) Tripathi, R., Gardiner, G.R., Islam, M.S., and Nazar, L.F. (2011). *Chem. Mater.* 23: 2278.

214 Yuan, Z., Wei, D., Wang, Y. et al. (2012). *CrystEngComm* 14: 4251.

215 Melot, B., Chotard, J., Rousse, G. et al. (2011). *Chinese Phys. Lett.* 50: 7662.

216 Rajagopalan, R., Wu, Z.G., Liu, Y.M. et al. (2018). *J. Power Sources* 398: 175.

217 Guo, S.-P., Li, J.-C., Xu, Q.-T. et al. (2017). *J. Power Sources* 361: 285.

218 Treacher, J.C., Wood, S.M., Islam, M.S., and Kendrick, E. (2016). *Phys. Chem. Chem. Phys.* 18: 32744.

219 Li, S., Guo, J., Ye, Z. et al. (2016). *ACS Appl. Mater. Interfaces* 8: 17233.

220 Rangasamy, V.S., Thayumanasundaram, S., and Locquet, J.-P. (2018). *Electrochim. Acta* 276: 102.

221 Duncan, H., Kondamreddy, A., Mercier, P.H.J. et al. (2011). *Chem. Mater.* 23: 5446.

222 Zhang, Y., Yu, H., and Zhou, H. (2014). *J. Mater. Chem. A* 2: 11574.

223 Chen, G., Shukla, A.K., Song, X., and Richardson, T.J. (2011). *J. Mater. Chem.* 21: 10126.

224 Nordh, T., Younesi, R., Hahlin, M. et al. (2016). *J. Phys. Chem. C* 120: 3206.

225 Chen, C.-M. and Chiu, H.-H. (2014). *Int. J. Hosp. Manag.* 41: 63.

226 Guan, W., Pan, B., Zhou, P. et al. (2017). *ACS Appl. Mater. Interfaces* 9: 22369.

227 (a) Chen, H., Hautier, G., and Ceder, G. (2012). *J. Am. Chem. Soc.* 134: 19619. (b) Chen, H., Hautier, G., Jain, A. et al. (2012). *Chem. Mater.* 24: 2009. (c) Chen, H., Hao, Q., Zivkovic, O. et al. (2013). *Chem. Mater.* 25: 2777. (d) Xie, B., Sakamoto, R., Kitajou, A. et al. (2020). *Sci. Rep.* 10: 3278. (e) Kosova, N.V., Shindrov, A.A., Slobodyuk, A.B., and Kellerman, D.G. (2019). *Electrochim. Acta* 302: 119. (f) Kosova, N.V. and Shindrov, A.A. (2019). *Ionics* 25: 5829. (g) Gao, J., Huang, W., Wu, X. et al. (2015). *Phys. Chem. Miner.* 42: 191. (h) Amaral, I.R., Alencar, R.S., Paraguassu, W. et al. (2019). *Spectrochim. Acta Part A* 223: 117298. (i) Wang, C., Sawicki, M., Kaduk, J.A., and Shaw, L.L. (2015). *J. Electrochem. Soc.* 162: A1601.

228 (a) Lee, J., Urban, A., Li, X. et al. (2014). *Science* 343: 519. (b) Uchaker, E., Zheng, Y.Z., Li, S. et al. (2014). *J. Mater. Chem. A* 2: 18208. (c) Shiratsuchi, T., Okada, S., Yamaki, J., and Nishida, T. (2006). *J. Power Sources* 159: 268. (d) Mathew, V., Kim, S., Kang, J. et al. (2014). *NPG Asia Mater.* 6: 138.

229 Hautier, G., Jain, A., Chen, H. et al. (2011). *J. Mater. Chem.* 21: 17147.

230 (a) Manthiram, A. and Goodenough, J.B. (1987). *J. Solid State Chem.* 71: 349. (b) Nadiri, A., Delmas, C., Salmon, R., and Hagenmuller, P. (1984). *ChemInform* 21: 537. (c) Bruce, P.G. and Miln, G. (1990). *J. Solid State Chem.* 89: 162. (d) Senthilkumar, B., Selvan, R.K., and Barpanda, P. (2019). *Electrochem. Commun.* 110: 106617.

231 Zhou, S., Barim, G., Morgan, B.J. et al. (2016). *Chem. Mater.* 28: 4492.

232 Masquelier, C. and Croguennec, L. (2013). *Chem. Rev.* 113: 6552.

233 Barim, G., Cottingham, P., Zhou, S. et al. (2017). *ACS Appl. Mater. Interfaces* 9: 10813.

234 Fang, Y., Xiao, L., Qian, J. et al. (2014). *Nano Lett.* 14: 3539.

235 Liu, Y., Xu, Y., Han, X. et al. (2012). *Nano Lett.* 12: 5664.

236 Xu, S., Zhang, S., Zhang, J. et al. (2014). *J. Mater. Chem. A* 2: 7221.

237 Yang, G., Ding, B., Wang, J. et al. (2016). *Nanoscale* 8: 8495.
238 Liu, T., Duan, Y., Zhang, G. et al. (2016). *J. Mater. Chem. A* 4: 4479.
239 Liu, Y., Xu, S., Zhang, S. et al. (2015). *J. Mater. Chem. A* 3: 5501.
240 Kitajou, A., Momida, H., Yamashita, T. et al. (2019). *ACS Appl. Energy Mater.* 2: 5968.

5

Prussian Blue Analogue Cathodes for Sodium-Ion Batteries

5.1 Introduction

In addition to the transition metal oxide cathodes (Chapter 3) and polyanion-type cathodes (Chapter 4), Prussian Blue Analogues (PBAs) with large three dimensional (3D) channels are also favorable for the sodium insertion/extraction. The history of Prussian Blue (PB) dates back to the discovery of distinct blue hue from an old pigment recipe by the paint maker Diesbach in Berlin around 1706. Their striking color of PB had once been the myth that scientists were highly curious about. The eye-catching optical transition property of PB was first correlated to mixed-valence chemistry of a specific group of elements. With the hindsight of quantum mechanics, people eventually understand that the source of coloration is tied with the transfer of an electron from one metal ion to another under the excitation of light. Some questions arise following this: Where do we find mixed valency in the periodic table and in what kind of coordination compound? Apparently, variable oxidation state is the prerequisite for electron transition, where most d-block (e.g. Cr, Mn, Fe, Co, Ni, and Cu) elements are those elements that fulfill this requirement. Following the first identification of PB as ferric ferrocyanide consisted of two different Fe redox centers, it was then developed into a plethora of mixed-valence polynuclear transition metal cyanide complex by substituting Fe with other transition metals as well as by playing around with the counteranions. In the cubic crystal structure of PB, N and C atoms of the bridging cyanide ligands are both possible coordination sites, which allow the occupancy of high-spin Fe^{3+} and low-spin Fe^{II} respectively. It is noteworthy that the bonding interaction between the C atom and low-spin Fe^{II} is stronger than that at the N-coordinated site, which can be reflected in the much difficult substitution in the $[Fe(CN)_6]^{4-}$ unit even under very acidic condition due to the strong covalent interaction at this site.

The highly tunable composition and geometry of PB have resulted in a family of cyanide derivatives denominated as PBAs with a generic formula of $AM[M'(CN)_6]{\cdot}xH_2O$ (A = Li, Na, K; M, M′ = transition metals Cr, Mn, Fe, Co, Ni, Cu, Zn etc.; typically M′ = Fe). PBA is also named hexacyanoferrates (when M′ = Fe) or hexacyanometallates (when M′ = other transition metal), which correspond to chemical acronyms of hexacyanoferrate (HCF) and hexacyanometallate (HCM), respectively [1]. The large 3D channel framework of most PBAs is also structurally

Sodium-Ion Batteries: Energy Storage Materials and Technologies, First Edition. Yan Yu.

and dimensionally robust to tolerance to the insertion/extraction processes of gust ions [2]. Furthermore, there are two different electrochemical active sites in PBAs, where the M^{2+}/M^{3+} and Fe^{2+}/Fe^{3+} couples allow reversible intercalation/insertion of more than one guest ion without breaking the structure apart. On top of that, the adjustment in the composition of PBAs can be done through partial/complete substitution of Fe^{2+} and Fe^{3+} ions in the cubic lattice of PB by other redox-active transition metals (e.g. Co, Ni, and Mn) to satisfy certain applications. For examples, substitution of Fe with Mn and Co pushes the charge/discharge plateaus from around 3.2 V ($Na_2Fe[Fe(CN)_6]$) to higher values of 3.6 V ($Na_2Mn[Fe(CN)_6]$) [3] and 3.8 V ($Na_2Co[Fe(CN)_5]$) [4] respectively. The elevation in charge/discharge voltage paves a way to enhance the energy density of PBA. In addition, several reports have evidenced an enhancement in capacity after the elemental substitution. In the case of $Na_2Mn[Fe(CN)_6]$, an exceptionally high capacity of >200 mAh g^{-1} can be achieved through additional Na^+ ion insertion on top of the conventional two-sodium-ion insertion mechanism [5]. This value is higher than most of the transition metal oxides (100–150 mAh g^{-1}) and phosphates (≈120 mAh g^{-1}). These features, in addition to their easy preparation through simple and low-cost coprecipitation method, make them highly suitable to be applied in stationary, grid-scale energy storage where cost, cycle life, and rate capability are of greater importance than energy density.

5.2 Crystal Structure

The first successful identification of crystal structure was achieved in 1936 by Keggin and Miles. It was satisfactorily explained by using powder X-ray diffraction (XRD) analysis as a 3D cubic face-centered unit cell of 5.1 Å edges with ferrous (Fe^{II}) and ferric (Fe^{3+}) ions arranged alternately at the corner and linked by bidentate cynides (—C≡N—) ligands in the edges [6]. The metalorganic framework can be depicted as repeated units of Fe^{II}—C≡N—Fe^{3+} in the 3D cubic structure, leaving a large zeolitic site at the center that allows the accommodation of guest species (e.g. alkali metals, small organic molecules, water molecules, etc.) with ionic radius of ~1.6 Å. As shown in Figure 5.1, a unit cell of PB with *Fm*3*m* space group symmetry contains eight cubes, giving a lattice parameter of ~10.2 Å. More specifically, high-spin Fe^{II} and low-spin Fe^{3+} fill the 4*a* and 4*b* Wyckoff positions respectively. C, N, and O of coordinated water reside at 24*e* positions. With zeolitic cavities in the lattice (represented as 8*c* or a slightly offset 32*f* position), alkali ions A^+ and/or water molecules can accommodate the interstitial space by compensating charge of $Fe^{3+}[Fe^{II}(CN)_6]^-$.

In a typical PB, Fe coordinated to C and N in different electronic configuration despite the same Fe element in PBAs, which is determined by the strength of ligands around Fe (Figure 5.1c). For example, low-spin Fe^{II} atoms show a d^6 electronic configuration, which is hexacoordinated to six C atoms via strong covalent interaction (strong C-coordinated crystal field). In contrast, high-spin Fe^{3+} is in d^5 electronic configuration that leads to a relatively weak ionic bonding with N atoms (weak N-coordinated crystal field).

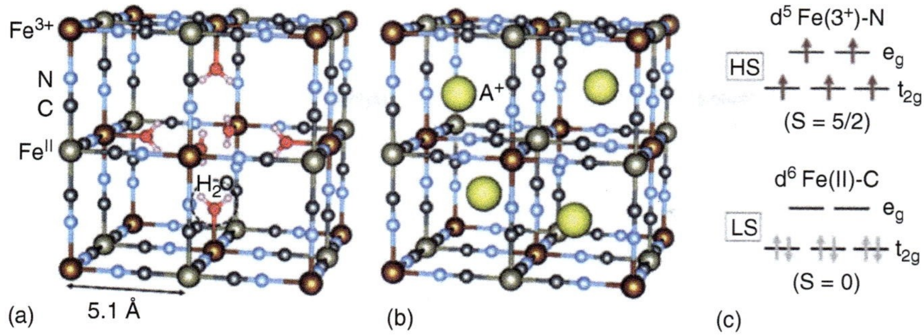

Figure 5.1 Crystal structure of (a) insoluble PB, $Fe_4[Fe(CN)_6]_3 \cdot xH_2O$ and (b) soluble PB, $AFe^{3+}[Fe^{II}(CN)_6] \cdot xH_2O$. (c) Spin state of Fe element in a typical PB compound. Source: Piernas Muñoz and Castillo Martínez [7]. Reproduced with permission, 2018, Royal Society of Chemistry.

Despite the primitive crystal structure of HCF being cubic ($a = b = c = 10.22$ Å), the successive occupation of Na could eventually alter its crystal lattice of HCM. Continuously increasing the amount of Na expands the lattice until it reaches the limit and turns into a slightly distorted monoclinic phase. This new phase is elongated in a (10.45 Å) but shrinks in b (7.52 Å) and c (7.27 Å), along with a distorted β of 92.35° [8]. This has given rise to an increase in volume of the unit cell in about 6.5%. Furthermore, the removal of interstitial water from the monoclinic phase induces a transition to rhombohedral structure, where the primitive cell shows distinct elongation in c (18.88 Å) and further reduction in a and b (6.54 Å) with $\gamma = 120°$. Therefore, a shrinkage in unit cell is usually seen in rhombohedral structure after the removal of water, which is 18% and 12.5% smaller than the monoclinic and cubic structure, respectively.

In fact, the changes in crystal structure of various HCF have been documented in literature, where NaMnHCFs (Figure 5.2a) and NaFeHCFs (Figure 5.2b) are those good examples that show different characteristic in XRD analysis at their different states. Due to the apparent structure difference between the rhombohedral and the other two structures, they are reported to behave differently in redox reaction. As illustrated in Figure 5.2c, the voltage profile of the dehydrated rhombohedral NaFeHCF shows two distinct plateaus in the scanning range. They can be assigned to the two first-order phase transitions of low-spin Fe and high-spin Fe redox reactions, which occur at 3.31/3.28 V and 3.11/2.98 V, respectively [12]. It was revealed that the severe distortion from monoclinic to rhombohedral phase has triggered a large polarization at the low-voltage plateaus due to the large volume variation. In addition, the low conductivity of the Na-rich phase is also responsible for large polarization. On the contrary, the voltage profiles of hydrated cubic and monoclinic NaFeHCF resemble to each other, which features a second-order phase transitions with a significant inclined curve [13]. It is perceived that the rhombohedral structure displays a much higher specific capacity and discharge voltage in sodium-ion batteries (SIBs), where similar trend is also observed for rhombohedral NaMnHCF (Figure 5.2d). In the case of rhombohedral NaMnHCF,

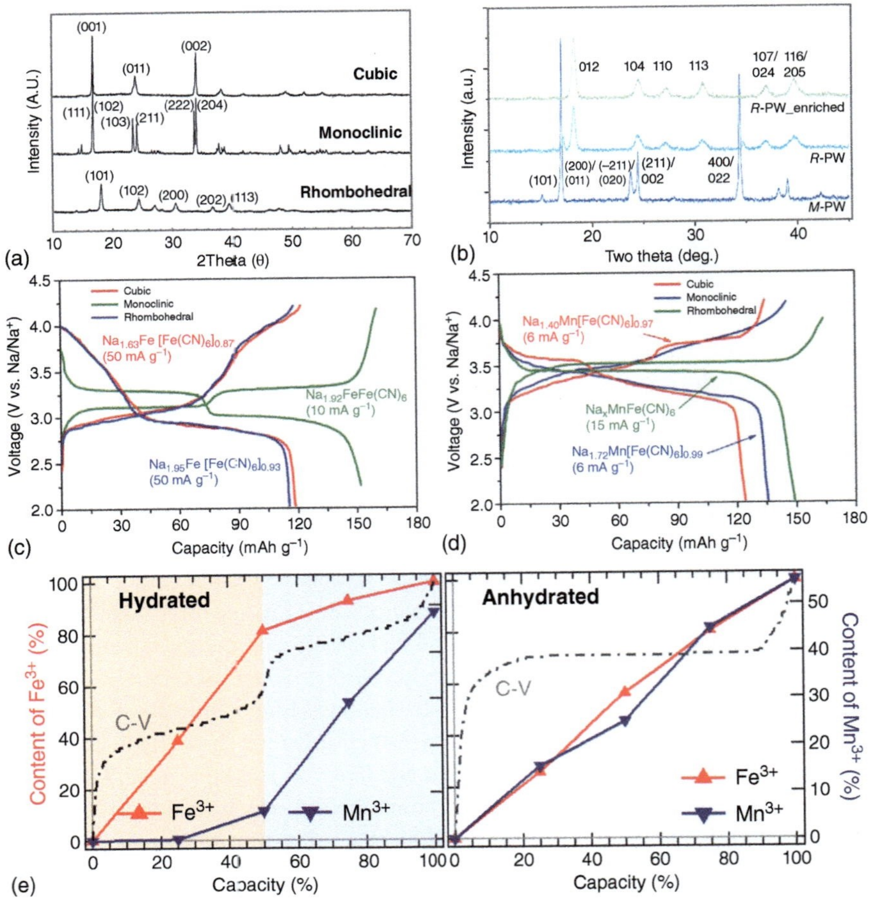

Figure 5.2 XRD patterns of (a) NaMnHCFs [8] and (b) NaFeHCFs [9]. Voltage profiles of (c) NaMnHCFs and (d) NaFeHCFs in nonaqueous SIBs [10]. Plot that illustrates the change in the oxidation state upon charging: (e) hydrated and dehydrated NaMnHCFs [11]. Source: (a) Jo et al. [8]. Reproduced with permission, from Elsevier, (b) Brant et al. [9]. Reproduced with permission, from American Chemical Society, (c, d) Zhou et al. [10]. Reproduced with permission, from Wiley-VCH, (e) Wu et al. [11]. Reproduced with permission, from American Chemical Society.

the two redox reaction associated with the $Mn^{III/II}$ and $Fe^{III/II}$ couples cannot be resolved in the voltage profile. Researchers rationalized this phenomenon to be the effect arisen from energy coupling of high-spin Mn and low-spin Fe [14]. From the result obtained from synchrotron-based soft X-ray absorption spectroscopy (sXAS) and theoretical calculations, it was proposed that the water content has great impact on the electrochemical profile of NaMnHCF [11]. As illustrated in Figure 5.2e, the transition metal redox reaction in hydrated system takes place at different potentials that leads to separated reaction plateaus while the mixed Fe and Mn redox occurs simultaneously in anhydrated system. In fact, it was explained that this elusive redox behavior of rhombohedral NaMnHCF comes from the balance between standard ionization energy (SIE) and ligand field stabilization energy (LFSE). Despite the intrinsically lower redox potential of $Fe^{3+/2+}$ compared with $Mn^{3+/2+}$ from the consideration of SIE, the LFSE of low-spin $Fe^{III/II}$ is calculated to be higher than that of high-spin $Mn^{III/II}$ due to stronger interaction of the Fe–C than Mn–N coordination. These two forces counterbalance each other, which causes the low-spin $Fe^{III/II}$ and high-spin $Mn^{III/II}$ redox potentials to be overlapped when the interstitial water effect is absent, resulting in only one reaction plateau. In contrast, the ligand field between the Fe–C and Mn–N coordination in the hydrated form is weakened by an abundant amount of interstitial water, which changes the spin states after the original structure is disturbed. In this case, the LFSE effect losses its control over SIE. Therefore, the potentials of low-spin $Fe^{III/II}$ and high-spin $Mn^{III/II}$ are primarily determined solely by SIE, exhibiting two separated plateaus.

Apart from FeHCF and MnHCF, the electrochemical study on NaCoHCF and NaNiHCF mainly focuses on the monoclinic symmetry. Owing to the smaller ionic radius of Co^{2+} (0.75 Å) and Ni^{2+} (0.69 Å), NaCoHCF and NaNiHCF exhibit smaller lattice volume than that of the former ones in monoclinic phase. For NaCoHCF, the redox potential of low-spin $Co^{III/II}$ (3.4 V) is slightly lower than that of the high-spin $Fe^{III/II}$ (3.8 V), which is attributed to the high energetic stability of low-spin Co^{III} [15]. The use of Zn in HCF has been rare. In most cases, Zn^{2+} preferentially coordinates to tetrahedral site with N to form ZnN_4 rather than occupying the octahedral site. Therefore, unlike other HCFs, it is more common to find hydrated ZnHCFs in rhombohedral symmetry rather than in cubic or monoclinic [16]. In terms of energy density, the higher discharge voltage of NaMnHCF endows it with higher value compared with NaFeHCF (Figure 5.2c,d). Summarized from the previous studies, the energy density of monoclinic and rhombohedral NaMnHCFs ranges from 446 to 504 Wh kg^{-1}, while NaFeHCFs can only reach approximately 360–472 Wh kg^{-1}. For NaCoHCF, the energy density of its monoclinic phase is about 505 Wh kg^{-1}.

When sodium is replaced by potassium in HCF, the structure of KHCFs can be less predictable than its sodium counterpart. For example, cubic structure is basically retained when M = Co, Ni, and Cu but both cubic and monoclinic structures are available for KFeHCFs depending on the amount of potassium [10, 17]. On the contrary, KMnHCFs have always been reported in monoclinic symmetry [18]. Compared with NaHCFs, the formation of KHCFs is more thermodynamically favorable. In addition to the smaller Stokes radius of K^+, the lower Gibbs free energy for the intercalation of K^+ in the zeolitic site is responsible for the preferential formation of KHCFs. Studies also found that the intercalation of crystal water is greatly reduced in KHCFs as compared with NaHCFs, which can be attributed to the steric hindrance posed by K^+ to repeal the water molecules from residing at 8c sites. Under the same synthesis condition, the radius of M^{2+} also has substantial impact on the amount of crystal water in the lattice of KHCFs, where decreasing trend is usually observed when M^{2+} is getting larger [17b]. From the crystallographic point of view, the lack of interstitial water and larger size of K^+ are two main factors that drive a more severe local deformation in the structure of monocline KMnHCF compared with NaMnHCF.

5.3 Electrochemistry Mechanisms

Before PBA was deployed as battery electrode, the electrochemical behavior of PB was first explored as thin films in aqueous electrolyte more than 30 years ago by Neff [19]. The PB thin films were electrochemically prepared by using $FeCl_2{\cdot}6H_2O$ and $K_3Fe(CN)_6$ as precursors in HCl solution. This important discovery of its electrochemical reversibility opens up new avenues for its future application in batteries. PBAs can exit in different forms, which are given the notation such as PB, Prussian white (PW), Prussian green (PG), Berlin green (BG), and Prussian yellow (PY). Their correlations are illustrated in Figure 5.3. In fact, a color transition reflects the changes in the bandgap and —C≡N— stretching frequency of PBAs. When the PB is being reduced, the reaction involves a reduction of Fe^{3+} into Fe^{2+} along with K^+ insertion, in addition to a color change from blue to transparent (fully reduced Everitt's salt or PW) as described below:

$$KFe^{3+}[Fe^{II}(CN)_6] + K^+ + e^- \rightarrow K_2Fe^{2+}[Fe^{II}(CN)_6] \tag{5.1}$$

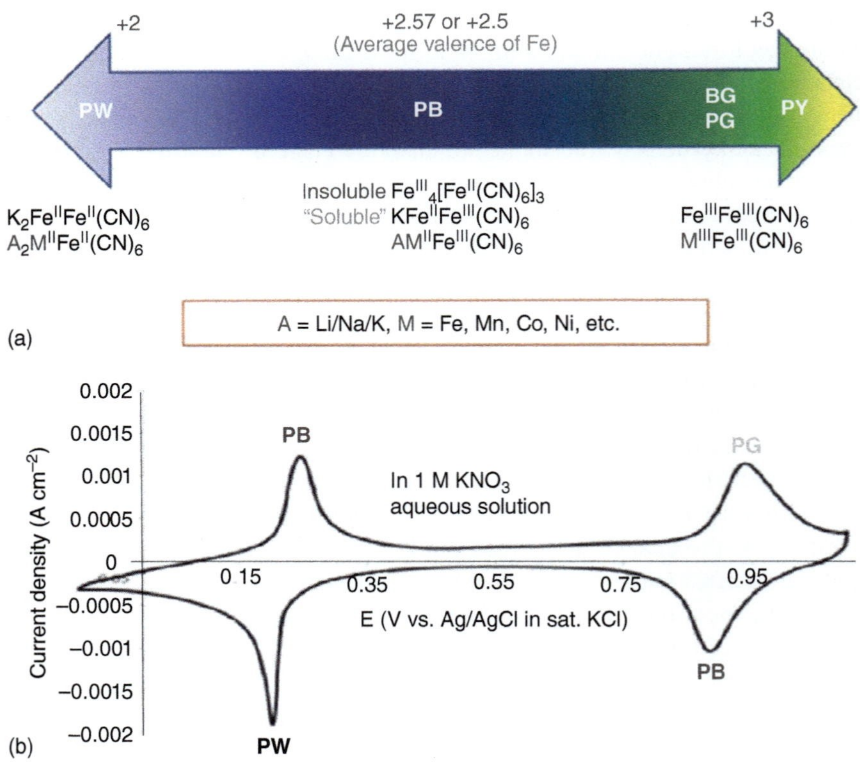

Figure 5.3 (a) Illustration of PBAs in different compositions. Source: Zhou et al. [10]. Reproduced with permission, from Wiley-VCH. (b) Cyclic voltammogram of $Fe^{III}Fe^{III}(CN)_6$ showing electrochemical conversions between PW, PB, and PG. Source: Padigi et al. [20]. Reproduced with permission, from Elsevier.

Upon the extraction of K^+ from PB, oxidation of Fe^{II} occurs in multiple steps to give a color transition from green (BG) to yellow (PY), which corresponds to the following equations, respectively:

$$KFe^{3+}[Fe^{II}(CN)_6] \rightarrow K_{1/3}Fe^{3+}[Fe^{II}_{1/3}Fe^{III}_{2/3}(CN)_6] + 2/3K^+ + 2/3e^- \quad (5.2)$$

$$K_{1/3}Fe^{3+}[Fe^{II}_{1/3}Fe^{III}_{2/3}(CN)_6] \rightarrow Fe^{3+}[Fe^{III}(CN)_6] + 1/3K^+ + 1/3e^- \quad (5.3)$$

The oxidation of Everitt's salt to PB differs from BG. The former involves the transition of high-spin Fe^{II} in an N coordination site to high-spin Fe^{3+}. It was evidenced that unnoticeable change in the Fourier transform infrared spectroscopy (FTIR) spectrum can be detected from this oxidation reaction. This could be due to the weak back-bonding effects arisen between the Fe atoms and the N end of the CN^- ligand. On the contrary, the shift becomes more apparent when PB is oxidized to BG due to the stronger metal-ligand t_{2g}-π^* back-bonding between low-spin Fe and the C end of the CN^- ligand. This holds true since the π^* CN^- orbital is more localized on the carbon, leading to a stronger π back-bonding for the metal electrons in the t_{2g} orbitals in the low-spin configuration [21].

The electrochemical potential of HCFs can be controlled from the following aspects: (i) polarizing effect of the Mn^{2+} coordinated to the N atom; (ii) hydration free energy of A^+ to be intercalated into the zeolitic cavities when $[M(CN)]^-$ units upon the reduction of $[M(CN)]^-$ units. In general, the polarization power of Mn^{2+} influences the electrochemical potential of HCFs with the same intercalating cation. An almost linear dependency between the effective radii of the Mn^{2+} and electrochemical potential can be explained by the fact that the stronger polarizing effect weakens the σ bond between the Fe atom and the CN^-, which has driven a stronger π back bonding. As a result, the oxidation of Fe^{2+} becomes more energetically favorable with a decrease in the absolute energy of the fully filled t_{2g} orbitals (low-spin Fe^{2+} ions) [22].

On the other hand, the electrochemical potential of HCM also correlates to the nature of the A^+ ions. Alkali metal ions with larger radius migrate more quickly from aqueous to HCM phase. In general, the diameter of naked cation can be sorted in the following order: Li (1.48 Å) < Na (2.04 Å) < K (2.76 Å) < Rb (2.98 Å) < Cs (3.40 Å). We can expect that the number of diameters of the solvation sphere decreases from Li to Cs, and hence the free energy of solvation decreases in the same way in the channel of HCM, which makes the electrochemical insertion and migration of alkali metal ion with larger atomic radii much more facile. The change in formal potentials of the hexacyanoferrates, hexacyanochromates, and hexacyanomanganates also fits this general trend when they are coordinated to Mn, Fe, and Cr ions. The formal potential decreases with increasing lattice constant. This linear correlation revealed that the enhanced acid–base interaction from Cr to Fe can potentially strengthen the M′–C σ back bonding on the N end of CN^-.

One of the most appealing features that makes them suitable to be used in energy storage system is the open framework and nanoporous structure with wide channels. For example, PB lattices have large interstitial sites (4.6 Å in diameter) and open framework along the (100) direction (3.2 Å in diameter) that translate into a high diffusion coefficient of 10^{-9}–10^{-8} cm^2 s^{-1} [1]. On top of that, the cubic geometry shows high tolerance to the mechanical stress imposed during ion insertion, which is deemed an important criterion for long cycle life. Resembling Ti-based intercalation-type materials, the large 3D channel framework of most PBAs is also structurally and dimensionally robust to tolerance to the volume variation upon insertion/extraction process of gust ions. Minimal lattice strain is seen in highly crystallized PBAs, such as $Na_2Cu[Fe(CN)_6]$ [2a] and $Na_2Co[Fe(CN)_6]$ [2b]. With proper and rational tailoring of their lattice parameters, they can maintain good capacity retention over prolonged cycles. Furthermore, PBAs contain two different electrochemical active sites that allow redox reaction of M^{2+}/M^{3+} and Fe^{2+}/Fe^{3+} couples, both of which can undergo transitions in the oxidation states for reversible intercalation of more than one guest ion (usually two-electron transfer) without breaking the structure apart. The partial/complete substitution of Fe^{2+} and Fe^{3+} ions in the cubic lattice of PBAs by other redox-active transition metals (e.g. Co, Ni, and Mn) is kinetically favorable. With such an enhanced flexibility in composition, the charge storage properties can be tuned accordingly to suit certain specifications. For examples, substitution of Fe with Mn and Co pushes the charge/discharge plateaus from around 3.2 V $Na_2Fe[Fe(CN)_6)]$ to higher values of 3.6 V ($Na_2Mn[Fe(CN)_6)]$ [3] and 3.8 V ($Na_2Co[Fe(CN)_6]$) [4] respectively. Previous findings on the calibration of the redox potentials have explained anomalous behavior in redox potential compared with the standard redox potential (E_0 = +0.16 V vs. Ag/AgCl) of soluble $Fe(CN)_6^{4-}/Fe(CN)_6^{3-}$ couple in aqueous solutions for $Na_2Co[Fe(CN)_6)]$ (+0.90 V) and $Na_2Ni[Fe(CN)_6)]$ (+0.45 V) and $Na_2Cu[Fe(CN)_6)]$ (+0.58 V) to be the effect of strong charge–spin–lattice coupling in the PBA lattice [2b]. The elevation in charge/discharge voltage paths a way to enhance the energy density of PBA. In addition, several reports have evidenced an enhancement in capacity after an elemental substitution. For the case of $Na_2Mn[Fe(CN)_6)]$, an exceptionally high capacity of >200 mAh g^{-1} can be achieved through the contribution of an additional Na^+ ion insertion steps on top of the conventional two Na^+ insertion [5]. This value is higher than most of the transition metal oxides (100–150 mAh g^{-1}) and phosphates (≈120 mAh g^{-1}). These features, in addition to their easy preparation through simple and low-cost coprecipitation method, make them highly suitable to be applied in large-scale stationary energy storage where sustainability and cost are of greater importance than energy density.

5.4 Preparation Approaches

The synthesis of PBAs can be achieved through two chemical routes, which are the coprecipitation and self-decomposition of precursor. It is widely recognized that the former method has been extensively studied due to its versatility in achieving PBAs with diverse structure while the latter one is only restricted to FeHCFs owing to the limited selection of precursors that are available to be used. Despite the coprecipitation synthesis being more feasible to mass-produce PBAs, the precise control over the quality of PBAs could be problematic. On the other hand, self-decomposition synthesis usually exhibits slower nucleation and grain growth rate that makes the particle size and morphology more homogeneous.

5.4.1 Coprecipitation

In addition to the advantages of being low-cost, less toxic, and highly scalable, the simplicity in coprecipitation approach to mixed-valence chemistry that leads to the precipitation of PBAs at ambient gives rise to a virtually unlimited number of PBAs with relatively little effort. However, whenever precipitation reaction is concerned, solubility product constant (Ksp) is always the governing factor over the crystallization of the final products. In the case of HCFs, their relatively small Ksp value causes the nucleation and grain growth processes take place almost instantly after the precursor solutions are mixed. As a result, irregular and aggregated HCFs in nanoscale tend to form, along with the generation of abundant vacancies and crystal water in the lattice. In addition, the ambient reaction affects the crystallinity of HCFs, which is also one of the factors that leads to degraded electrochemical performance. In the past decades, researchers have manipulated the mixed-valence chemistry to "separate" these two rate-determining steps further. The use of chelating agents with nucleophilic property is deemed effective to slow down the precipitation rate, thus improving the homogeneity of HCFs. The underlying mechanism can be explained from the ability of the chelating agent to coordinate with the transition metal ions to prevent them from complexing with the $Fe(CN)_6^{4-}$ ions, which significantly delays the nucleation and precipitation of HCFs. On top of that, the negatively charged chelating molecules that stay on the surface of the initial nuclei suppress the growth rate and aggregation through the force of electrostatic repulsion. Given a much slower reaction rate in the solution by having transition metal chelating agent complex that acts as a reservoir to slowly release the transition metal ion, HCFc nucleus grows slowly into a well-defined shape with more monodispersed grains and less defect [23]. So far, in addition to the widely reported sodium and potassium citrates (Na_3CA, K_3CA), chelates such as $EDTA^{4-}$, oxalate, and pyrophosphoric salts have been reported for the precipitation of HCFs. Other roles played by the chelating agents are to promote homogeneous particle size and morphology and more importantly to reduce the water inclusion in the crystals through occupation of excess Na^+ or K^+ in the lattice. For example, Liu et al. reported the successful preparation of sodium-rich NaFeHCFs through the Na_3CA–assisted precipitation [24]. With an increasing amount of sodium being incorporated into the lattice of the sodium-rich NaFeHCFs, the water content is seen to decrease from 18 to 13.2 wt%. On the other hand, the morphology of the NaFe-HCFs changed from irregular nanoparticles to homogeneous nanocubes after the addition of Na_3CA, along with an expansion in the lattice when the concentration of Na_3CA increases. The improved properties of the sodium-rich NaFeHCFs enable high capacities of 120 and 71 mAh g^{-1} to be delivered at current densities of 0.2 and 2 A g^{-1} respectively. In another chelate-assisted precipitation study reported by Peng et al. [25] bidendate ethylene diamine tetraacetic acid (EDTA) was also employed for stronger binding to slow down the precipitation rate. In this case, low-defect $Na_{1.92}Mn[Fe(CN)_6]_{0.98}$ can be successfully synthesized through a self-chelate Mn source $MnC_{16}H_{12}N_2Na_2O_8$ (EDTA-$MnNa_2$). In fact, the strong ligancy of EDTA with $MnNa_2$ precursor suppresses the precipitation of MnHCF unless a certain amount of ascorbic acid is added to facilitate the dissociation of EDTA-$MnNa_2$ complex. When the monoclinic NaMnHCF was tested in nonaqueous SIBs, it exhibits a high-rate performance (153 and 110 mAh g^{-1} at 0.01 and 1 A g^{-1} respectively), not to mention the excellent cycling stability that shows a capacity retention of 81.7% after 500 cycles at 0.1 A g^{-1}. Goodenough's group also studied the effect of cations on the particle size, water content, and defects of MnHCFs. A reaction solution containing high concentration of K/Na-citrate was used to slacken the nucleation process in order to achieve homogeneous and near-stoichiometric crystal growth [18a]. In this study, a co-insertion/extraction of K^+ and Na^+ was proposed as the possible sodium charge storage mechanism of the KMHCF, which is driven by the high concentration of Na^+ and the lower energy barrier of K^+ insertion. It was also validated by observing the stepwise K^+ insertion into the monoclinic KMHCF to form intermediate KNaMHCF before it reaches the cubic NaMHCF final phase. The KMHCF electrode shows excellent electrochemical performance as cathode for SIBs, where 100% and 80% capacity retention can be achieved after 200 and 1000 cycles at a current density of 1 C, respectively, along with a stable capacity of 95 mAh g^{-1} at 20 C. The precipitation of CoHCFs, NiHCFs, and even NaNiCoHCF can be achieved through this simple precipitation method in the presence of Na_3CA chelating agent. For instances, monodispersed monoclinic $Na_{1.85}Co[Fe(CN)_6]_{0.99}$ nanocubes were successfully fabricated using Na_3CA but not the case in the absence of this chelating agent. Aggregated $Na_{1.32}Co[Fe(CN)_6]_{0.88}$ nanoparticles with cubic structure were produced due to the uncontrolled precipitation that leads to the undesired composition and morphology [2b]. Moreover, Li et al. studied the precipitation of NaCoHCFs in a mixed NaCl/Na_3CA solution [26]. In contrast to the NaCoHCFs ($Na_{1.47}Co[Fe(CN)_6]_{0.90}$) nanoparticle obtained in the absence of chelating agent, the use of NaCl/Na_3CA enables the structure to accommodate higher amount of Na($Na_{1.19}Co[Fe(CN)_6]_{0.87}$). In addition to the more regular cuboid morphology, the grain size of $Na_{1.19}Co[Fe(CN)_6]_{0.87}$ (200–300 nm) is also larger than that of the $Na_{1.47}Co[Fe(CN)_6]_{0.90}$ (30–50 nm). Moreover, it is found that an increase in the concentration of NaCl (Na_3CA concentration remains constant) can further improve the stoichiometry to $Na_{1.87}Co[Fe(CN)_6]_{0.98}$ without affecting the grain size. Based on the aforementioned studies, there is some golden rules to follow: high crystallinity can potentially give better cycling stability and high capacity relies on stoichiometry.

Apart from the use of chelates, researchers are also playing around with other reaction parameters to optimize the electrochemical performances. The most fundamental one lies at the tuning of precipitation temperature and concentration of reagents. It is reasonable to apply lower reaction temperature in the attempt to slow down the precipitation rate. However, it has been proved not to be a direct endeavor as very little effect is shown in the effort to control over the morphology of HCFs, though temperature has great impact on the crystallinity and stoichiometry of HCFs.

In a study reported by Yang et al. a decrease in reaction temperature to 0 °C does help with maintaining the stoichiometry of the NaFeHCF, which is at start contrast to the Na-poor compositions of $Na_{1.27}Fe[Fe(CN)_6]_{0.94}$ and $Na_{1.12}Fe[Fe(CN)_6]_{0.92}$ obtained at 40 and 80 °C, respectively [27]. On the other hand, the use of concentrated Na^+ solution helps to improve the structure and stoichiometry of HCF. For example, Goodenough's group found that water content can be reduced from 17 to 12 wt% by changing the stoichiometry after they purposely increased the amount of Na in NaMnHCF [3]. Similar trend was also observed by Jo et al. by increasing the precipitation temperature [8]. It is found that higher reaction temperature at 60 °C results in monoclinic Na-rich composition ($Na_{1.90}Mn[Fe(CN)_6]_{0.99}$), which is in contrast to the cubic Na-poor compounds ($Na_{1.38}Mn[Fe(CN)_6]_{0.91}$) that precipitate at room temperature. In addition to these parameters, the precursor concentration and mixing rate are also crucial to determine the particle size of the HCFs. An increase in the concentration of $NiCl_2$ and $K_3Fe(CN)_6$ precursors or in mixing rate significantly brings down the particle size from hundreds to tens of nanometers [28]. Apart from those conventional way of synthesizing HCFs, electrostatic spraying-assisted coprecipitation (ESAC) was also developed for the preparation of high-quality NaCoHCF and NaNiHCF [29]. This technique can potentially slow down the precipitation process that successfully eliminates the formation of abundant crystal water and vacancy, without a compromise on the crystallinity and morphology of the final products as compared with the chelate-assisted precipitation.

The simplicity in coprecipitation approach to mixed-valence chemistry that leads to the precipitation of PBAs at ambient gives rise to a virtually unlimited number of PBAs with relatively little effort. However, this holds true only if ligand-bridged complex can be formed via a possible Lewis acid–base interaction among the components. Classic $Fe_4[Fe(CN)_6]_3$ is a notable example, where the formation is favorable as $Fe(H_2O)_6{}^{3+}$ (Lewis acid) is susceptible to be substituted by N of the CN^- ligands in ferrocyanide (Lewis base). In the presence of CN^- ligand that bridges Fe^{+2} and Fe^{+3}, intervalence transfer is allowed between these two sites. Nevertheless, coincident reactions are likely to take place along with the precipitation reaction due to the complexity of the equilibria. As a result, the formation of PBAs is accompanied by other impurities that are difficult to be isolated from the solution, which makes the work on predicting and controlling the quality of final products even harder. In general, there are several parameters need to be taken into account in the design of synthetic routes and the possible competing reactions that are likely to occur, which primarily are the redox potential differences between the redox centers, bonding nature of the precursors, and oxidizing/reducing agents.

5.4.2 Self-decomposition of Precursors

Self-decomposition of precursor in the synthesis of HCFs is relatively straightforward compared with the coprecipitation reaction. Given the slower intrinsic reaction rate of the self-decomposition method, the particle size and morphology of the synthesized HCFs can be more precisely tuned from tens of nanometers to several microns, despite its relatively low production yield. In fact, the reaction mechanism can be seen as the pyrolysis of the precursors, preferentially $Na_4Fe(CN)_6$ and $K_4Fe(CN)_6$, which leads to the dissociation of Fe^{2+} in the $Fe(CN)_6{}^{4-}$ and its partial oxidation to Fe^{3+} in the subsequent step [30]. When the concentration of Fe^{2+}/Fe^{3+} reaches a certain limit, nucleation will take place between Fe^{2+}/Fe^{3+} and $Fe(CN)_6{}^{4-}$ ions until the initial nuclei grow into larger particles to solid precipitates. It is worth noting that the reaction does not occur spontaneously, where a certain amount of acid is needed to serve as nucleophiles to dissociate the Fe^{2+} from the CN^- ligand. In the presence of divalent Fe, a higher amount of A^+ ions can be accommodated in the lattice. In summary, the final structure, composition, and morphology of HCFs can be carefully tuned by several important reaction parameters, especially the concentration of intercalating cations, reaction temperature, pH of the solution, and annealing environment.

The research in this area was pioneered by Guo's group. In their early study, high-quality $Na_{0.61}Fe[Fe(CN)_6]_{0.94}$ with cuboid morphology can be achieved with the aid of hydrochloric acid by slow decomposition of $Na_4Fe(CN)_6$ in air (60 °C) [31]. They also proved that the NaFeHCF synthesized by this route results in lesser amount of vacancies and lattice water than the one that was synthesized through coprecipitation between $Na_4Fe(CN)_6$ and $FeCl_3$. However, only $x = 0.61$ of Na content was found in the lattice, which is equivalent to 30% occupation Na in the framework. Further effort was also made by the same group to achieve higher sodium content. The x value in NaFeHCFs was further increased to 1.63 when N_2 protection was implemented and ascorbic acid was used as the reductant to prevent oxidation of Fe^{2+} [32]. On top of that, an increase in lattice parameters inevitably induced structural change to the NaFeHCF from cubic (Na-poor) to monoclinic (Na-rich). In addition to the effect of acidic reagent, the use of surfactants such as polyvinyl pyrrolidone (PVP) [33] polyallylamine hydrochloride (PAH) [34], cetyltrimethylammonium bromide (CTAB) [35], and sodium dodecylbenzene-sulfonate (SDBS) [36] is also reported for their ability to modulate the properties of HCFs in terms of homogeneity, orientation, grain size, and shape. Similarly, KFeHCFs can also be prepared via the self-decomposition route, where the features of the final products depend strongly on the precursor. For example, KFeHCFs synthesized by two different sources ($K_4Fe(CN)_6$ and $K_3Fe(CN)_6$) result in a change in crystallinity and electrochemical performances [37].

5.5 Optimizing Electrochemical Performance

PBA was first employed as the cathode for aqueous-based materials due to their excellent compatibility and based on previous understanding on their electrochemistry in aqueous electrolytes. For example, NiFe–PBA has been demonstrated to be electrochemically toward aqueous-based SIBs and suitable to be in grid-scale energy storage by Cui and coworkers [38]. The $K_{0.6}Ni_{1.2}Fe(CN)_6{\cdot}3.6H_2O$ exhibits stable long-term cycling performance over 5000 cycles at 8.3 C due to the isotropic lattice strain as little as 0.18% upon insertion and removal of sodium ion. However, only a moderate specific capacity of 60 mAh g^{-1} can be achieved at 0.83 C in aqueous electrolyte. In a similar study conducted by Cui and coworkers, the CuHCF retains ≈77% of its original capacity at a rate of 8.3 C over 500 cycles [39]. In another study, the redox potential has been successfully extended from 0.6 to 1.0 V in aqueous electrolyte by tuning the composition of the $Cu_xNi_{1-x}HCF$ electrode, which follows a trend that the redox potential increases with the Cu content (0.6–1.0 V) but without compromising on the reversible capacity. In addition, the $Cu_{0.56}Ni_{0.44}HCF$ also exhibits almost 100% capacity retention after 2000 cycles [40].

Until recently, Goodenough and coworkers pioneered the research of PBAs for SIBs. They disclosed a series of compounds ($KMFe(CN)_6$, M = Fe, Mn, Ni, Cu, Co, Zn) that are promising insertion host for Na^+ ion, though Na^+ ion might not be necessarily incorporated in the organic framework [41]. These findings started flipping the perspective on PBAs, which sparked a surge of research on modified PBAs to optimize their electrochemical performance. Despite the potential advantages of PBAs being signified in thermodynamic expressions, the theoretical value or even appreciable capacity is difficult to be fully claimed in the early studies investigating the insertion behaviors of PBAs. This puzzling phenomenon was later understood to be mainly associated with structural irregularity and coordinated water in the lattices as well as the electronic conductivity of PBAs. The discovery of these subtle relationships enables researchers to understand and engineer the crystallographic feature of PBAs in a more precise way, which is of great importance to accelerate the future development of PBAs in SIBs.

5.5.1 Effect of Lattice Architecture on Electrochemistry

5.5.1.1 Substitution of Cation

Taking Fe^{II}—C≡N—Fe^{3+} unit as an example, Fe coordinated to C and N differ from each other in electronic configuration despite the same Fe element in PBAs. This is to say, the substitution of Fe coordinated to the C site that is located at the center of the hexacynometallate complex is much more difficult due to the bonding nature, which has a determining effect on the lattice parameter and subsequently the reaction potential of PBAs. For example, substituted elements with smaller radius reduce the lattice parameter. Counterintuitively, chemical reduction due to the insertion of alkali ion into the PBA lattice also reduces the lattice parameter, given that an increase in electron density causes the lattice to shrink accordingly. For example, the volume of the unit cell in Mn-PBA decreases 3.5% on lithiation [42], and other compositions also show similar trend upon insertion of different alkali ions [2a].

On the other hand, the substitution at N-coordinated site enjoys a broader variety of choice as the geometry does not alter much with the substitution. However, the reaction potential at the C-coordinated site has great influence on N-coordinated site. The underlying cause could be due to the polarization of the s bond in the CN^- ligand that results in stronger π-back-bonding t_{2g} orbitals on N-coordinated site [22]. The substituents can be either electrochemically inactive or active. For example, Zn, Cu, and Ni are electrochemically inert but help to push the reaction potential to higher value and optimize the lattice parameter for more rapid ion conduction.

On the contrary, Fe, Mn, and Co participate in the electrochemical reaction, where additional electron involved in the redox reaction contributes to higher capacity. However, cycle lives of these types of cathodes are compromised at elevated potentials due to the reaction of structural water. In addition, potential change in crystal structure arising from the change of oxidation state at the N-coordinated site is also a concern, especially those elements that are unstable at certain oxidation state. For example, Mn^{3+} are susceptible to an irreversible disproportionation reaction associated with the Jahn–Teller distortion [43]. Aside from Mn, energetic stability of the LS d6 electron configuration Co induces a favorable HS–LS Co^{II}/Co^{III} spin-state transition, along with a lattice contraction arising from a significant reduction in ionic radius from 7.45 to 5.45 nm [44]. Despite the clear advantages of N site substitution, these cathodes usually suffer from slower kinetics and irreversibility, which is attributed to significant crystallographic modification of the open framework structure when Co or Mn is cycled and hence, hinders the solid-state diffusion of ions in the framework. To obtain a compelling balance between these few criteria, exciting research path has been set on the preparation of single-phase solid-solution PBAs with more than one element at N site. The lattice parameter in a solid-solution (Mn, Co) hexacyanoferrate has been shown to be simply the average of the lattice parameters of their individual compound, which is weighted by atomic fraction [45]. Through this mixing strategy, it is believed that phase transition upon successive insertion of alkali ion can be greatly ameliorated.

In addition, there is also a possibility to incorporate certain amount of dopant without breaking down the lattice structure. For example, the introduction of Ni in MnFe-PBA$(CN)_6$ has been proved to be able to stabilize the lattice distortion caused by the Jahn–Teller effect of Mn^{2+}/Mn^{3+} redox reaction. With the alleviation of the Jahn–Teller effect, the Ni-doped MnFe-PBA shows improvement in capacity retention over long-term cycling [46]. In another investigation on the doping effect of Ni in FeFe-PBA, Ni plays an essential role in activating the low-spin Fe^{2+}/Fe^{3+} couple for Na storage. It was proposed that the electron atmosphere of C-coordinated Fe is changed by Ni, which makes the Na^+ intercalation/extraction more favorable under the influence of low-spin Fe^{2+}/Fe^{3+} couples [47].

5.5.1.2 Inserting Cation

The open framework of PBAs allows the insertion of a wide spectrum of guest ions, where alkali, alkali earth, and transition metals, including Li, Na, K, Ca, Mg, Al, and Zn, can accommodate the subcubes of the lattice with a change in the oxidation state of the coordinated transition metal ion. Thermodynamically, the dimension of the guest ion determines the insertion potential of the reaction based on the correlation between the hydration energy and the Gibbs free energy of ion insertion. This holds true from monovalent to multivalent ions [48]. In addition, it was also theorized that the stronger steric hindrance of larger ions is responsible for the higher reaction potential despite their slower migration kinetics [49]. Density functional theory (DFT) calculations also suggest that the geometrical stability within the large interstitial sites in the lattice is influenced by the size of cations. Larger cations (K^+, Rb^+, and Cs^+) preferentially reside the body-centered sites that are more spacious rather than the face-centered sites that are more favored by small cations (Li^+, Na^+, and all alkaline earth ions) due to shorter cation–anion bonding distance. Despite the fact that PBAs are structurally stable with almost zero lattice strain toward the insertion/extraction processes, the introduction of high amount of inserting ions displaces toward the corners in the <111> direction, resulting in a slight distortion from a cubic to a less symmetric rhombohedral geometry [50]. As a result, the ionic conductivity of the PBAs can be lowered due to a change in structure, in addition to the reaction reversibility. Kinetically, the migration of the inserting ion along the nanoporous channels is strongly affected by the lattice parameter, where ionic conductivity increases with the lattice parameter. For example, the Na diffusion coefficient improves from 2.3×10^{-10} to 7.7×10^{-10} $cm^2\ s^{-1}$ with an increase in the lattice parameter [51].

5.5.1.3 Vacancy

PBAs are easily prepared through simple chemical precipitation method, but their lattice frameworks are usually filled with structural defects such as vacancy and structural water. They have been identified as a detrimental factor that prevents the full potential of PBAs to be unlocked. Vacancy, as a positively charged defect, significantly reduces the specific capacity especially at high content. To tackle this issue, a common way for vacancy reduction falls on the modification of synthetic method. Generally, the use of excess alkali ion can suppress the formation of vacancy. It has been proved that the vacancy content drops from 19% to 7% when the Na^+ per formula unit increases from 1.24 to 1.72 with higher concentration of precursor [52]. On top of that, the deployment of chelating agent is able to modulate the rate of crystallization process that is critical to minimize the formation of vacancy. For example, citrate-based reagent with enhanced affinity with transition metal ion precursor was used by Yang and coworkers to achieve low content of vacancies of 1% in a cobalt hexacyanoferrate for Na^+. The reversible specific capacity of 150 mAh g^{-1} is close to the theoretical value, along with an excellent long-term cyclability of ~90% over 200 cycles [4]. Despite their adverse impacts on the electrochemical performance, several studies have also revealed the roles of vacancy within the lattice framework in improving the rate capability and cycle life of batteries. First, vacancy has long been tied with the ionic conductivity, where these connected cavities offer an alternative transport pathway for inserting ions [53]. Such an exclusive feature particularly facilitates the conduction of larger ions along within the lattice. On the other hand, vacancy is known for its ability to suppress the <111> cooperative displacement of the crystal structure, especially in the lattice with higher concentration of Li and Na [54].

5.5.1.4 Water Molecules

Water molecules can occupy the lattice framework of PBAs in two structurally distinct positions, which are located at zeolitic site and coordinated to transition metal ion at vacancies. For the zeolitic water molecules, they compete with inserting ion to accommodate the center of the lattice's subcube. The other type of water molecule coordinates to the dangling transition metal ions at vacancies. Due to this special coordination that shields the positive charge of the exposed ions, the migration of inserting ion in the lattice framework can be retarded. Compared with coordinated water, zeolitic water is easier to be removed as they only bond weakly at the interstitial space [55]. There are a couple of role water molecules can play in the thermodynamic stability of the inserting ions. For example, the solvation of Li and Na is energetically favorable, which might lead to a reduction in ionic potential. In addition, these thermodynamically favored phases manage to mitigate a cubic to rhombohedral transformation in the crystal structure. The downside is that the water molecules can trigger a cascade of side reactions that can potentially deteriorate the performance of the PBAs. Hence, there is a need to remove excess water especially when organic electrolytes are employed for high-voltage applications. In addition, Xie et al. also discovered that the reduced defects coordinated H_2O in the framework of MnCoNi-PBA allows the reversible transition between cubic and rhombohedral phase upon Na^+ insertion/extraction [56]. To obtain an optimum performance, researchers are required to consider the trade-off of using water molecules by striking a good balance between a severe structural distortion that is caused by high concentration of inserting ion after complete removal of water molecules and a degradation in performance arisen from the side reaction.

5.5.2 Effect of Morphological Optimizations on Electrochemistry

Along with the optimization and expansion of the PBAs family, various strategies such as nanostructure design, surface modification, and morphological control have been actively developed in recent years. There are some distinct advantages of nanostructuring: (i) enhancing the interfacial interaction by increasing the surface to volume ratio; (ii) shortening the ion diffusion pathway to accelerate the reaction kinetics; (iii) providing buffering effect for volume change. More specifically, the increasing surface area of nanostructured PBAs allows higher flux of ion moving across the interface. With the enhanced contact between the electrode and electrolyte, the penetration of electrolyte ion becomes more effective. In addition, the ion diffusion becomes more kinetically favorable when PBAs are made into smaller size. For example, a dual-textured KFeFe-PB (porous PB@nonporous PB) in nano-size was synthesized by Kim et al. [57] The structural features of the porous PB@ nonporous PB allow it to exhibit reversible capacities of 85, 65, and 52 $mAh\,g^{-1}$ at current densities of 10, 240, and 480 $mA\,g^{-1}$ respectively. Going small might not be a universal antidote, several other strategies have to be coupled simultaneously to address the battery problems fundamentally. Dai and coworkers designed a meso/macroporous NiFe-PBA to validate the effect of porous structure on the electrochemical performance of PBAs. In this study, the authors claimed that the transportation of Na^+ from the external to interior of the PBA framework is enhanced in pores that are larger in size [58]. In short, significant improvement has been seen in nanosized PBAs, but most of them remain in the lab-scale production, in addition to the long validation cycles before the new materials can make their way into the market.

Morphological optimization is commonly used for property modulation of PBAs. This can be done through the preparation of porous or three-dimensional structures that show better reaction kinetics through a continuous ion replenishment and rapid ion movement in the PBA framework. The common way to achieve mesoporous morphology relies on the instability of PBAs under acidic condition, where a stepwise crystal growth-etching strategy is frequently employed to generate porous architecture [59]. Despite a great success being reported to generate porous/hollow structure in various acid mediums, the tunability of this method is relatively limited as the etching rate is less controllable and the pore formation is not uniform, which render the collapse of original structure somewhat more likely to happen. On top of that, an aggregation-driven formation mechanism was also demonstrated to be viable for assessing mesoporous PBAs. This concept has been realized in the synthesis of NiHCF, where the NiHCF aggregates joined together to become a single-grained particle [58]. Furthermore, the interfacial contact area can be enhanced substantially with construction of border-rich structure, giving rise to better wettability and ions transport resistance between the electrode–electrolyte interfaces. The PBAs that are rich in boundaries remain 95 $mAh\,g^{-1}$ after 280 cycles and 60 $mAh\,g^{-1}$ at a high rate of 1.6 $A\,g^{-1}$ [60]. Furthermore, juxtaposition of disparate materials in a hybrid structure is effective to leverage synergies between different components to provide solution to the battery problems associated with their limitation in electronic and ionic transport rate. In the case of PBAs, the effect of core–shell design is particularly significant when it comes to addressing the cycling stability problem of most PBAs encountered. For example, the combination of core–shell CuFe-PBA@NiFe-PBA makes full utilization of the high capacity CuFe-PBA core and highly stable NiFe-PBA protective shell. In spite of the low capacity contribution, the interfacial coupling of CuFe-PBA core to the NiFe-PBA shell suppresses possible structural phase transition in the CuFe-PBA [61]. This concept also stands for dual-textured PBA, where Kim et al. converted the shell into porous structure to further improve the penetration of electrolyte for better reactivity at the core [57].

Aside from that, the PBAs are frequently composited with carbons (i.e. graphenes [62], carbon nanotubes (CNTs) [63], Ketjen black [64], etc.) as surface coating or deposition metrics, thanks to their excellent chemical and physical properties. For example, the full potential of FeFe-PB prepared by Dou and coworkers can be unleashed with fast electron transfer within the electrode and shorter ion diffusion length after coupling with Ketjen black. Therefore, a capacity of 77.5 $mAh\,g^{-1}$ can be retained at high rate of 90 C, along with a capacity of 90 $mAh\,g^{-1}$ at 20 C after 2000 cycles [64]. Similarly, uniform PB nanocrystals threaded by a CNT intertwined mat prepared by Guo and coworkers retain good electric contact between the PB particles and the current collector [65]. The PBAs/CNT electrode can deliver a discharge capacity of 142 $mAh\,g^{-1}$ with a specific energy density of 408 $Wh\,kg^{-1}$ and retain a 86% capacity retention after 1000 cycles.

On the other hand, surface coating with conductive polymer or small organic molecules not only enhances the electronic conductivity of the composite, but also provides better accommodation of the lattice strain upon insertion reaction.

In the case when Mn species is present, the polymer coating can potentially suppress severe dissolution problem of Mn. In addition, researchers also found that redox-active polymers exhibit capacitive contribution, albeit this could probably come at the expense of degradation in structural stability as a protective layer. In this regard, Dou and coworkers have demonstrated the viability of polymer coating that renders the MnFe-PBA@PPy composite enhanced rate capability and cycling stability [66]. We also witnessed significant improvement after the surface of PBAs was modified with specially designed organic molecules. For example, effective linkage that bridges the NiHCF and 7,7,8,8-tetracyanoquinododimethane leads to electronic coupling that allows further enhancement in conductivity due to more facile electron movement between these two species [67].

5.5.3 Na_xMFe-PBAs with Two Na^+ Insertion Sites

The high tunability in composition of PBAs is one of their most significant advantages, where the redox centers in Na_xMFe-PBAs are determined by the nature of M element. As mentioned earlier, electrochemically active elements, such as Fe, Mn, and Co with multiple valency, contribute to another single-electron transition reaction, aside from the reaction occurring at the N-coordinated site. This type of PBAs can serve a theoretical two Na^+ insertion within a suitable potential range and therefore, exhibit higher specific capacity, in principle. This advantage, however, has not been conferred unconditionally to all PBAs with the same crystal structure, where their expected potential in capacity and cycling stability is not fully claimed. Early studies revealed the contradiction in experimental results, but these puzzling questions were left unsolved.

For example, Na_2FeFe-PBAs are classical compounds in this category. Early attempt in the deployment of Na_2FeFe-PBAs for SIBs was not smooth sailing until researchers found out the critical roles played by lattice vacancy and water molecules. With the ability to access vacancy-free FeFe-PB lattice, a capacity of 120 mAh g^{-1} was successfully achieved, along with a remarkable stable capacity retention of 87% over 500 cycles [68]. Further insight into the investigation revealed that lattice vacancies that are filled with coordinating and zeolitic water cause lattice distortion, which blocks the redox active site as well as the ion conduction pathway for inserting ion. This finding has provided essential guidelines in the development of synthetic strategies for high-performance PBAs. Given that high degree of structural regularity is a critical criterion that brought significant electrochemical improvement, Guo and coworkers synthesized a FeFe-PB electrode with low contents of vacancy (6%) and water molecules (15.7%), which make the active sites more accessible by Na^+ [31]. A capacity of 170 mAh g^{-1} was obtained, which is nearly equivalent to the theoretical value of two Na^+ insertion. The deployment of defect-free PBAs in aqueous-based SIBs also shows similar effect, which further validates the critical roles of lattice architecture in electrochemistry of PBAs [69].

Alternatively, a Na-rich lattice structure can effectively reduce the number of vacancies and enhance the structural integrity upon the insertion/desertion of ions [24, 70]. In numerous reports, the Na content can be perfectly tuned by using high concentration of Na source coupled with suitable chelating agents. For example, a Na content can be increased to 1.63 Na per PB unit with the use of ascorbic acid as reducing agent [32]. This Na-rich $Na_{1.63}Fe_{1.89}(CN)_6$ compound can deliver a reversible capacity of 150 mAh g^{-1} in the first cycle and maintain 90% of the initial capacity after 200 cycles.

Recent study on the effect of lattice water suggested a possible way to achieve improved electrochemical performance. A $Na_{1.92}$FeFe(CN) with low water content synthesized by Goodenough and coworkers exhibits two redox active sites in response to Na^+, which was evidenced from two distinct voltage plateaus in the soft X-ray spectroscopic results [12]. With the elimination of interstitial water in the lattice in a rhombohedral $Na_{1.92}$FeFe(CN), a capacity retention of 80% can be extended up to 800 cycles, along with a capacity of as high as 100 mAh g^{-1} achieved at 15 C.

To suit the specification of different purposes, one of Fe redox centers can be substituted with other elements without negligible disruption to the lattice structure. For example, Mn^{2+}/Mn^{3+} couple that is known for its high reaction voltage (3.6 V) can potentially improve the energy density for practical use. For example, vacancy-free MnFe-PBA was proposed as a way to achieve high capacity based on a three Na^+ insertion mechanism [5]. A value of 209 mAh g^{-1} apparently exceeds the typical 172 mAh g^{-1} expected for two Na^+ ion reaction. Despite the fact that the discharge capacity has been significantly improved after the substitution of Mn, coprecipitated and electrodeposited Na_xMnFe-PBAs still suffer from low cycling stability [71]. Such a phenomenon is primarily caused by the Jahn–Teller effect of Mn^{3+} and hence, lattice distortion and dissolution of Mn element into electrolyte are often the problems encountered by them [55, 72]. Several strategies have been put forward to suppress the effect, where the preparation of water- and vacancy-free MnFe-PBAs structure gives the more prominent result. This was exemplified in literature through the removal of interstitial water under vacuum drying by Goodenough and coworkers [55] The crystal structure of the dehydrated MnFe-PBAs undergoes a transition from monoclinic to rhombohedral, which exhibits higher tolerance to structural distortion upon (de)insertion of Na^+ ion. The change in electrochemical response also results in a single-platform charge/discharge profiles arising from a stable phase transition between tetragonal and rhombohedral. Given a water-free lattice, the MnFe-PBA manages to maintain 75% capacity retention after 500 cycles at 0.7 C. Aside from Mn, metallic Co can serve as a stabilizer and conductor in the lattice of PBAs. Upon substitution with Co, two distinct plateaus are shown at 3.4 and 3.8 V [44]. In a representative literature, the authors worked through the structural instability of PBAs by using a low-defect Na_2CoFe-PBA with low content of vacancies and coordinated water molecules. In addition to the two Na^+ insertion reaction, it also retains 90% of its initial capacity after 200 cycles [4].

5.5.4 Na_xMFe-PBAs with One Na^+ Insertion Sites

At variance to the aforementioned Na_xMFe-PBAs that are endowed with two redox centers, replacement of Fe at electrochemical inert elements, especially Zn, Cu, and

Ni, has different roles to play. For example, the incorporation of those elements in the lattice counteracts the irreversible volume change of the PBA framework. Furthermore, the reduced interaction between the anion lattice and the incoming Na^+ ion significantly alleviates the mechanical stress of the PBA framework. As evidence, the NaNiFe-PBA ($Na_{0.84}Ni[Fe(CN)_6]_{0.71}$) synthesized by Guo and coworkers exhibits a zero-strain insertion of Na^+ ion that results in a negligible volume change of <1% [73]. The electrode is not susceptible to capacity fading and retained 99.7% of its original capacity after 200 cycles. Along this line, the substitution of Fe in the lattice of PBA with electrochemically inert elements has been found to be successful to enlarge the ionic channel for Na^+ ion insertion. Choi and coworkers have evidenced the modified tunnels Na_2ZnFe-PBA can prevent rupture of lattice by better accommodating the volume variation upon ion insertion/removal [74]. Similar explanations were made to elucidate the poorer cycling stability and rate capability, such as the impact of coordinated water and lattice vacancy in the diffusion of Na^+ ion, where their presence significantly blocks the ion movement. Conclusion and insight gained from this study can also be extended to other members in this family.

5.6 Concluding Remarks

Although the PBA-based SIBs suffer from low specific energy and energy density caused by their low crystal density, their inexpensive raw materials, long cycling stability, and high-rate capabilities make them economically competitive to complement the state-of-the-art lithium-ion batteries in grid-scale energy-storage applications. The topic of PBAs was revitalized in battery applications mainly due to the advancement in characterization techniques and technologies that endow researchers with capability to unlock the potential PBAs through deeper understanding on the reaction mechanism and modulation of the material properties. Despite their great progress in the past decades, we still see knowledge gaps in certain areas of research. The observation of several phenomena remains speculative as they have not been fully supported by experimental results and theoretical calculations. For example, the contribution of water in enhancing the ionic conduction generally accepted to be associated with the "paddlewheel mechanism" that was proposed to elucidate the ion movement through channels and vacancies. The migration of inserting ion can be accelerated through the help of coordinated water that rotates its bond axis with a P-site ion at a vacancy. However, it has not been fully understood due to complexity of the reaction environment. Similarly, the elimination of zeolitic water has dramatically improved the electronic conductivity of PBAs, but the possible explanations to this phenomenon are still debatable. Hence, previous computational insight is suggested to couple with other characterization techniques (e.g. isotopic labeling study) that can distinguish roles of zeolitic and coordinated water in ion conduction of PBAs. Deeper understanding of PBAs from fundamental aspects is likely to further underpin their future development as promising electrode for high-performance SIBs.

Notwithstanding the simplicity of the synthesis and highly scalable potential, PBAs, however, are particularly difficult to be isolated from other impurities and get rid of adsorbed water. Later studies have revealed that the abundant defects and water content during synthesis are responsible for their inferior performance. In this regard, machine learning in material design can be useful to predict the possible changes in the properties of PBAs by careful control over the reaction parameters such as concentration, temperature, and mixing rate. On top of that, the possible improvement is expected from the elaboration of functional architectures such as core–shell, interconnected porous structure, and hybrid structure with multiple components. Recently, the use of mirocfluidic reactors has demonstrated the better control over the preparation of PBAs, but what is still lack is the fundamental understanding of the reaction mechanism of synthesis process, which might enhance our ability to tune the lattice architecture and electrochemistry of PBAs. In addition, the major safety concern posed by PBAs is the release of cyanide as potent toxicant. Toxicological studies have claimed that PB is nontoxic unless the cyanide ligand is detached from the framework. The use of PBAs needs to be extra careful as the stability of cyanide ligands is highly susceptible to pH of the electrolyte, where ligand exchange with hydroxide group usually takes place under basic condition. This has necessitated an extra effort in handling the waste of PBAs so that the CN^- can be completely oxidized into environmentally friendly products, such as CO_2, N_2, etc. before it is channeled back to the earth.

References

1 Shibata, T. and Moritomo, Y. (2014). *Chem. Commun.* 50: 12941.

2 (a) Wessells, C.D., Huggins, R.A., and Cui, Y. (2011). *Nat. Commun.* 2: 550. (b) Wu, X., Sun, M., Guo, S. et al. (2015). *ChemNanoMat* 1: 188.

3 Wang, L., Lu, Y., Liu, J. et al. (2013). *Angew. Chem. Int. Ed.* 52: 1964.

4 Wu, X., Wu, C., Wei, C. et al. (2016). *ACS Appl. Mater. Interfaces* 8: 5393.

5 Lee, H.-W., Wang, R.Y., Pasta, M. et al. (2014). *Nat. Commun.* 5: 5280.

6 Paolella, A., Faure, C., Timoshevskii, V. et al. (2017). *J. Mater. Chem. A* 5: 18919.

7 Piernas Muñoz, M.J. and Castillo Martínez, E. (2018). *Prussian Blue and Its Analogues. Structure, Characterization and Applications*, SpringerBriefs in Applied Sciences and Technology, 9–12. Switzerland: Springer.

8 Jo, I.-H., Lee, S.-M., Kim, H.-S., and Jin, B.-S. (2017). *J. Alloys Compd.* 729: 590.

9 Brant, W.R., Mogensen, R., Colbin, S. et al. (2019). *Chem. Mater.* 31: 7203.

10 Zhou, A., Cheng, W., Wang, W. et al. (2021). *Adv. Energy Mater.* 11: 2000943.

11 Wu, J., Song, J., Dai, K. et al. (2017). *J. Am. Chem. Soc.* 139: 18358.

12 Wang, L., Song, J., Qiao, R. et al. (2015). *J. Am. Chem. Soc.* 137: 2548.

13 Chen, S., Wu, C., Shen, L. et al. (2017). *Adv. Mater.* 29: 1700431.

14 Xiao, P., Song, J., Wang, L. et al. (2015). *Chem. Mater.* 27: 3763.

15 Takachi, M., Matsuda, T., and Moritomo, Y. (2013). *Jap. J. Appl. Phys.* 52: 090202.

16 (a) Heo, J.W., Chae, M.S., Hyoung, J., and Hong, S.-T. (2019). *Chin. Phys. Lett.* 58: 3065. (b) Islas-Vargas, C., Guevara-García, A., Oliver-Tolentino, M. et al. (2018). *J. Electrochem. Soc.* 166: A5139.
17 (a) Wang, J.-G., Zhang, Z., Zhang, X. et al. (2017). *Nano Energy* 39: 647. (b) Wu, X., Jian, Z., Li, Z., and Ji, X. (2017). *Electrochem. Commun.* 77: 54. (c) Ren, W., Chen, X., and Zhao, C. (2018). *Adv. Energy Mater.* 8: 1801413.
18 (a) Zhou, A., Xu, Z., Gao, H. et al. (2019). *Small* 15: 1902420. (b) Bie, X., Kubota, K., Hosaka, T. et al. (2017). *J. Mater. Chem. A* 5: 4325.
19 Piernas Muñoz, M.J. and Castillo Martínez, E. (2018). *Prussian Blue Based Batteries*. Cham, Switzerland: Springer International Publishing.
20 Padigi, P., Thiebes, J., Swan, M. et al. (2015). *Electrochim. Acta* 166: 32.
21 Crumbliss, A.L., Lugg, P.S., Childers, J.W., and Palmer, R.A. (1985). *J. Phys. Chem.* 89: 482.
22 Scholz, F. and Dostal, A. (1996). *Angew. Chem. Int. Ed.* 34: 2685.
23 Wu, X., Zhong, G., and Yang, Y. (2016). *J. Power Sources* 327: 666.
24 Liu, Y., Qiao, Y., Zhang, W. et al. (2015). *Nano Energy* 12: 386.
25 Peng, F.W., Yu, L., Gao, P.Y. et al. (2019). *J. Mater. Chem. A* 7: 22248.
26 Li, C., Zang, R., Li, P. et al. (2018). *Chem. Asian J.* 13: 342.
27 Yang, Y., Liu, E., Yan, X. et al. (2016). *J. Electrochem. Soc.* 163: A2117.
28 Li, C.H., Nanba, Y., Asakura, D. et al. (2014). *RSC Adv.* 4: 24955.
29 (a) Ma, X.-H., Wei, Y.-Y., Zhou, J.-F. et al. (2018). *Electrochim. Acta* 272: 44. (b) Ma, X.-H., Wei, Y.-Y., Wu, Y.-D. et al. (2019). *Electrochim. Acta* 297: 392.
30 Azhar, A., Li, Y., Cai, Z. et al. (2019). *Bull. Chem. Soc. Jpn.* 92: 875.
31 You, Y., Wu, X.-L., Yin, Y.-X., and Guo, Y.-G. (2014). *Energy Environ. Sci.* 7: 1643.
32 You, Y., Yu, X., Yin, Y. et al. (2015). *Nano Res.* 8: 117.
33 Liu, Y., Wei, G., Ma, M., and Qiao, Y. (2017). *Eur. J. Chem.* 23: 15991.
34 Meng, Q., Zhang, W., Hu, M., and Jiang, J.-S. (2016). *Chem. Commun.* 52: 1957.
35 Wang, H., Wang, L., Chen, S. et al. (2017). *J. Mater. Chem. A* 5: 3569.
36 Chen, M., Wang, E., Liu, Q. et al. (2019). *Energy Storage Mater.* 19: 163.
37 Cai, D., Yang, X., Qu, B., and Wang, T. (2017). *Chem. Commun.* 53: 6780.
38 Wessells, C.D., Peddada, S.V., Huggins, R.A., and Cui, Y. (2011). *Nano Lett.* 11: 5421.
39 Wessells, C.D., Peddada, S.V., McDowell, M.T. et al. (2011). *J. Electrochem. Soc.* 159: A98.
40 Wessells, C.D., McDowell, M.T., Peddada, S.V. et al. (2012). *ACS Nano* 6: 1688.
41 Lu, Y., Wang, L., Cheng, J., and Goodenough, J.B. (2012). *Chem. Commun.* 48: 6544.
42 Okubo, M., Asakura, D., Mizuno, Y. et al. (2010). *J. Phys. Chem. Lett.* 1: 2063.
43 Moritomo, Y., Wakaume, K., Takachi, M. et al. (2013). *J. Phys. Soc. Jpn.* 82: 094710.
44 Takachi, M., Matsuda, T., and Moritomo, Y. (2013). *Appl. Phys. Express* 6: 025802.
45 Pasta, M., Wang, R.Y., Ruffo, R. et al. (2016). *J. Mater. Chem. A* 4: 4211.
46 Yang, D., Xu, J., Liao, X.-Z. et al. (2014). *Chem. Commun.* 50: 13377.
47 Yu, S., Li, Y., Lu, Y. et al. (2015). *J. Power Sources* 275: 45.
48 Wang, R.Y., Shyam, B., Stone, K.H. et al. (2015). *Adv. Energy Mater.* 5: 1401869.

49 Ling, C., Chen, J., and Mizuno, F. (2013). *J. Phys. Chem. C* 117: 21158.

50 Moritomo, Y., Matsuda, T., Kurihara, Y., and Kim, J. (2011). *J. Phys. Soc. Jpn.* 80: 074608.

51 Takachi, M., Fukuzumi, Y., and Moritomo, Y. (2016). *Dalton Trans.* 45: 458.

52 Moritomo, Y., Kurihara, Y., Matsuda, T., and Kim, J. (2011). *J. Phys. Soc. Jpn.* 80: 103601.

53 Moritomo, Y., Igarashi, K., Kim, J., and Tanaka, H. (2009). *Appl. Phys. Express* 2: 085001.

54 Kurihara, Y., Matsuda, T., and Moritomo, Y. (2013). *Jap. J. Appl. Phys.* 52: 017301.

55 Song, J., Wang, L., Lu, Y. et al. (2015). *J. Am. Chem. Soc.* 137: 2658.

56 Xie, B., Zuo, P., Wang, L. et al. (2019). *Nano Energy* 61: 201.

57 Kim, D.S., Zakaria, M.B., Park, M.-S. et al. (2017). *Electrochim. Acta* 240: 300.

58 Yue, Y., Binder, A.J., Guo, B. et al. (2014). *Angew. Chem. Int. Ed.* 53: 3134.

59 (a) Hu, M., Furukawa, S., Ohtani, R. et al. (2012). *Angew. Chem. Int. Ed.* 51: 984. (b) Hu, M., Belik, A.A., Imura, M., and Yamauchi, Y. (2013). *J. Am. Chem. Soc.* 135: 384.

60 Huang, Y., Xie, M., Zhang, J. et al. (2017). *Nano Energy* 39: 273.

61 Asakura, D., Li, C.H., Mizuno, Y. et al. (2013). *J. Am. Chem. Soc.* 135: 2793.

62 (a) Prabakar, S.J.R., Jeong, J., and Pyo, M. (2015). *RSC Adv.* 5: 37545. (b) Bu, F., Feng, X., Jiang, T. et al. (2017). *Chem–Eur. J.* 23: 8358.

63 Xu, Y., Wei, Q., Xu, C. et al. (2016). *Adv. Energy Mater.* 6: 1600389.

64 Jiang, Y., Yu, S., Wang, B. et al. (2016). *Adv. Funct. Mater.* 26: 5315.

65 You, Y., Yao, H.R., Xin, S. et al. (2016). *Adv. Mater.* 28: 7243.

66 Li, W.-J., Chou, S.-L., Wang, J.-Z. et al. (2015). *Nano Energy* 13: 200.

67 Nie, P., Yuan, J., Wang, J. et al. (2017). *ACS Appl. Mater. Inter.* 9: 20306.

68 Wu, X., Deng, W., Qian, J. et al. (2013). *J. Mater. Chem. A* 1: 10130.

69 Wu, X., Luo, Y., Sun, M. et al. (2015). *Nano Energy* 13: 117.

70 Li, W.-J., Chou, S.-L., Wang, J.-Z. et al. (2015). *Chem. Mater.* 27: 1997.

71 (a) Matsuda, T., Takachi, M., and Moritomo, Y. (2013). *Chem. Commun.* 49: 2750. (b) Sottmann, J., Bernal, F.L.M., Yusenko, K.V. et al. (2016). *Electrochim. Acta* 200: 305.

72 Zhang, W., Liu, Y., Chen, C. et al. (2015). *Small* 11: 3822.

73 You, Y., Wu, X.-L., Yin, Y.-X., and Guo, Y.-G. (2013). *J. Mater. Chem. A* 1: 14061.

74 Lee, H., Kim, Y.-I., Park, J.-K., and Choi, J.W. (2012). *Chem. Commun.* 48: 8416.

6

Organic Cathodes for Sodium-Ion Batteries

6.1 Introduction

Compared to the inorganic cathodes in previous chapters, organic cathode materials (OCMs) have the benefits of being renewable, environmentally friendly, low cost, and high capacity because they are rich in carbon, hydrogen, oxygen, and other elements and have received extensive attention in recent years [1]. OCMs can be directly extracted from plants (such as crops, fruits, and vegetables) or prepared by simple methods using biomass as raw materials. The carbon dioxide produced in the process of extraction and preparation of organic materials, battery assembly, and recycling can be absorbed and utilized by plants, thus reflecting good recyclability and reproducibility, as shown in Figure 6.1 [2].

The development of OCMs can be looked back upon the late 1960s, when carbonyl compounds were initially reported for lithium primary batteries [3, 4]. In the past decade, a growing number of OCMs have been developed. There are several types of OCMs, which can be categorized based on the nature of their active centers (functional groups). In addition to carbonyl compounds, currently reported OCMs mainly include conductive polymers (CPs), organic sulfides, organic free radicals, imine compounds, nitrile compounds, and azo compounds [2c, 5]. Figure 6.1b shows the development history of OCMs. In general, OCMs can be roughly divided into three types, namely n-type, p-type, and bipolar type, based on the change in charge.

The pioneer OCM being reported was dichloroisocyanuric acid in lithium-ion batteries (LIBs) by Williams' group in 1969 [4]. Although the electrochemical reaction of this material is irreversible because of its high solubility in electrolyte, it lays the foundation for the development of OCMs in batteries. The superiorities of OCMs include high theoretical specific capacity, abundant sources and raw materials, structural flexibility, and controllable preparation methods [6]. Thus, organic materials are considered to be potential electrode materials for the next-generation sodium-ion batteries (SIBs), which have the following advantages as compared with the inorganic materials: (i) organic electrodes are mostly composed of elements with abundant natural reserves (C, H, O, N, S, etc.) and are relatively light in weight; (ii) the structure of OCMs is more diverse, and by adjusting their substituents or functional groups, the redox potential and specific capacity of the battery can be tuned [7]; and (iii) OCMs can be synthesized at low temperatures (<200 °C) because

Sodium-Ion Batteries: Energy Storage Materials and Technologies, First Edition. Yan Yu.

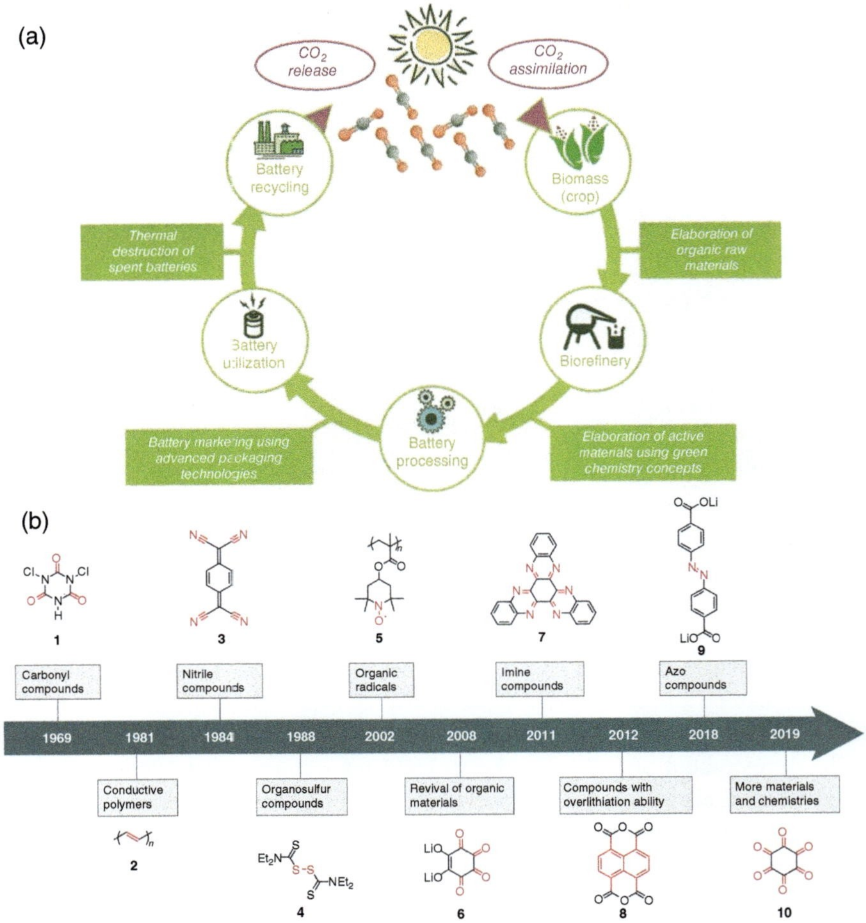

Figure 6.1 (a) A diagram for green cycle of OCMs. Source: Chen et al. [2c]. Reproduced with permission, 2008, Wiley-VCH. (b) Development history of OCMs. Source: Lu and Chen [3]. Reproduced with permission, 2020, Springer Nature.

they do not contain metal elements that usually require higher temperature for reaction, which is environmentally friendly and inexpensive. OCMs can be divided into small organic molecular electrodes and polymer electrodes according to the size of the molecular structure. Most small organic compounds dissolve quickly in the organic electrolytes, resulting in a short cycle life. Polymerization of small molecules is a simple and effective way to address this issue [5e, 8]. Compared with small organic molecules, polymer skeletons with a large number of repeating units have lower solubility in the electrolyte and thus have better cycle performance.

According to the electrochemical reaction mechanisms, the as-known OCMs for SIBs can be divided into three types of C=O bond reaction, doping reaction, and C=N bond reaction (Figure 6.2) [9], and their theoretical capacities and operation voltages are illustrated in Figure 6.3. The details will be introduced in the following sections.

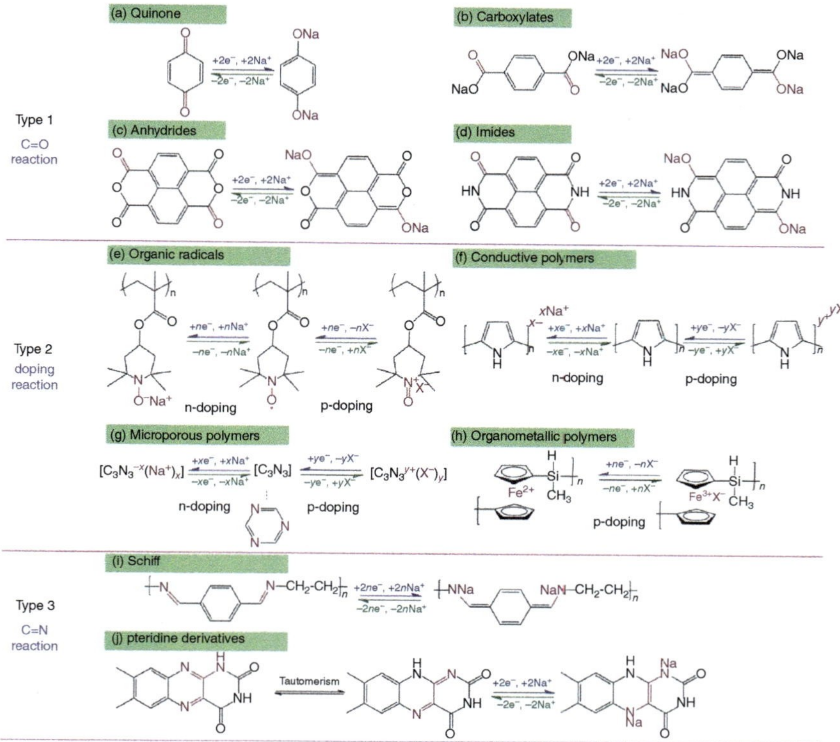

Figure 6.2 Three typical reaction mechanisms of OCMs for SIBs. Type 1 is C=O reaction: (a) quinone compounds with benzoquinone (BQ), (b) carboxylates with disodium terephthalate (Na_2TP), (c) anhydrides with 1,4,5,8-naphthalenetetracarboxylic dianhydride (NTCDA), and (d) imides with 1,4,5,8-naphthalenetetracarboxylicdimide (HNDI). Type 2 is doping reaction: (e) organic radical compounds of poly(2,2,6,6tetramethylpiperidinyloxy-4-vinylmethacrylate), (f) conductive polymers of polypyrrole, (g) an active site of microporous polymer, and (h) organometallic ferrocene-based polymer. Type 3 is C=N reaction: (i) Schiff base compound and (j) pteridine derivatives. Source: Zhao et al. [2b]. Reproduced with permission, 2017, Wiley-VCH.

6.2 C=O Reaction

As early as 40 years ago, carbonyl groups have been proven to be reversible redox centers in batteries [6a, 7b, 10]. Due to the fact that oxygen atom has a small mass and strong oxidizability, the theoretical specific capacity and redox potential of the carbonyl compound cathodes are relatively high. In addition, organic carbonyl compounds have faster redox reaction kinetics, making these electrode-like materials have good charge and discharge reversibility, thereby obtaining higher Coulombic efficiency. Based on these advantages, carbonyl compounds are the first choice for SIB OCMs. A carbonyl group undergoes reversible single-electron reduction to constitute a monovalent anion. This anion is equilibrated by a sodium ion and stabilized by structural substituents. On the basis of the reaction mechanism of carbonyl and sodium ions, carbonyl compounds can be divided into three types, as shown in

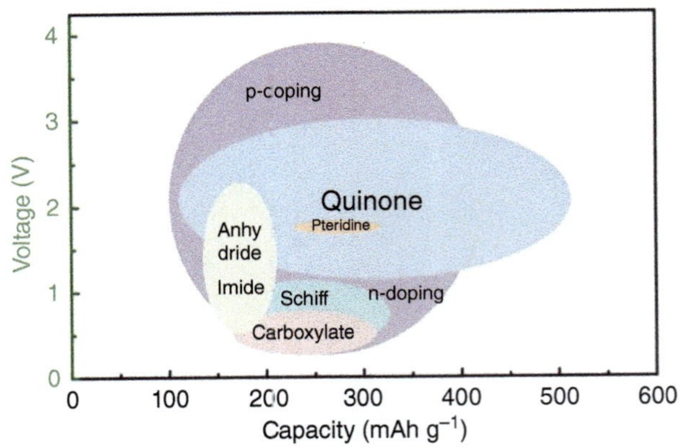

Figure 6.3 Typical theoretical capacities and operation voltages of selected OCMs for SIBs. Source: Zhao et al. [2b]. Reproduced with permission, 2017, Wiley-VCH.

Figure 6.4 [11]. Type I compounds use ortho-carbonyl groups to form stable enol acids for the reversible insertion and extraction of ions [12]. In the Type II compounds, the carbonyl group is straightly connected to the aromatic nucleus, and the aromatic nucleus disperses negative charges by delocalization to take up and release sodium ions. Such compounds include aromatic carboxylates, imines, and anhydrides [13]. Type III compounds are mostly quinones and ketones, which possess some of the characteristics of Type I and II compounds; however, the main stabilizing power of Type III comes from the additional aromatic system formed after reduction to take up and release sodium ions [14]. In addition, according to the structure, it can be divided into four categories: quinone compounds, carboxylate compounds, acid anhydride compounds, and imide (including polyimide) compounds. These four kinds of compounds are described as below.

6.2.1 Quinones

Quinones are kinds of aromatic organic compounds with two double bonds and a six-carbon atom cyclic diketone structure. Common quinone electrode materials include benzoquinone (BQ), anthraquinone (AQ), and naphthoquinone (NQ) [15].

The electrochemical oxidation–reduction mechanism of the quinone group is the enolization reaction of the conjugated carbonyl group C=O and its reverse reaction. At present, it is generally believed that the electrochemical reaction mechanism of conjugated quinone-based materials is a two-electron or multiple-electron process. Taking the sodium salt of 2,5-dihydroxy-1,4-benzoquinone (Na_2DBQ) as an example [16], its electrochemical oxidation–reduction reaction mechanism is as below. When the first sodium ion is intercalated, one of the carbonyl groups (C=O) accepts an electron to convert the C=O double bond into a C—O single bond, and oxygen becomes a negative ion state. At the same time, there is a Coulombic electrostatic interaction between the positively charged sodium ion and the oxygen anion, and sodium is occupied around the oxygen to form a sodium salt. With the intercalation of the second sodium ion, another carbonyl group (C=O) receives

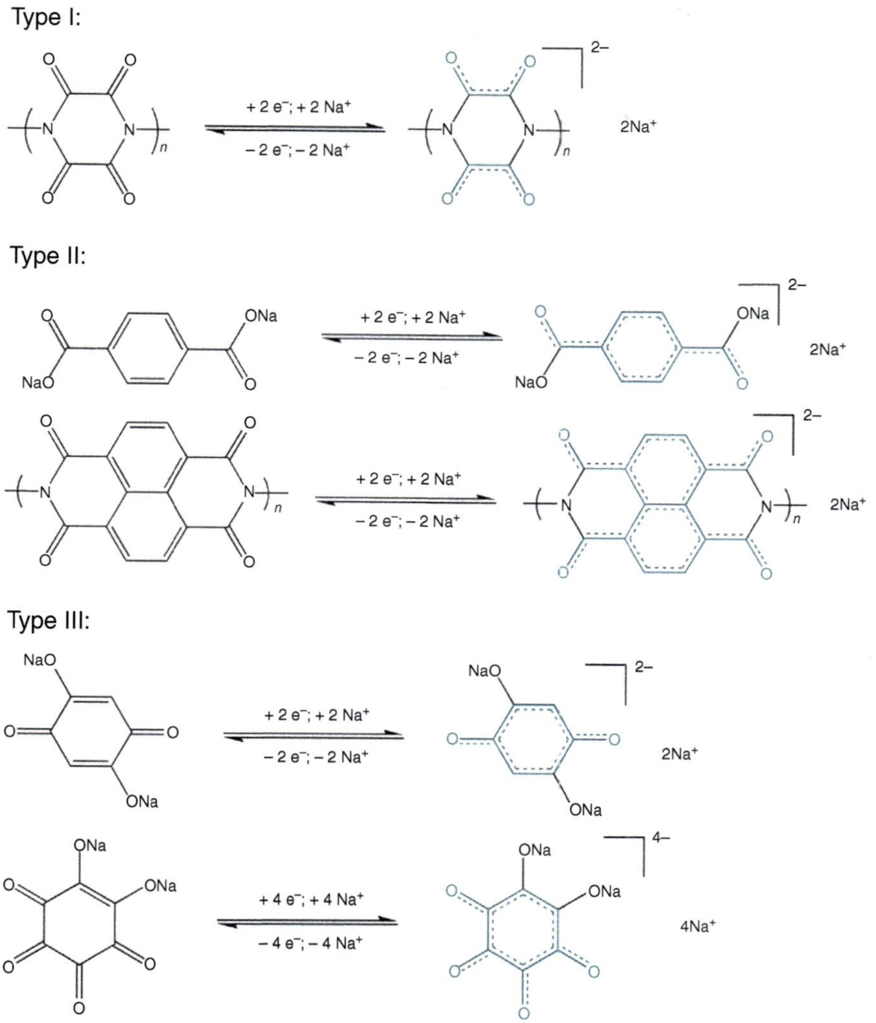

Figure 6.4 Representative of three types of carbonyl compounds for SIBs. Source: Xu et al. [11]. Reproduced with permission, 2018, Elsevier.

another electron. The intercalation of two electrons not only changes the C=O but also rebuilds the bond length of the six-membered ring. The reverse process corresponds to loss of two electrons to restore its carbonyl structure.

Quinone compounds are considered to be the most important class of carbonyl electrodes for rechargeable batteries because they have higher redox potential and theoretical capacity than other imide and carboxylate electrodes. Quinone compounds can be divided into two different groups according to their structures, namely small-molecule and polymer quinone electrodes. Small molecular quinones include simple quinones, polycarbonyl quinones, fused heteroaromatic quinones, substituted quinones, carbon oxide salts, and carboxylate quinones. BQ, AQ, and phenanthrenequinone (PAQ) can produce a reversible multi-electron transfer reaction, where the multiple sodium ion reaction sites endow them with high theoretical capacity. For fused heteroaromatic quinones, the introduction of O, S, and N into the heteroaromatic quinone can significantly increase the discharge potential of the quinone electrode, and the discharge plateau of some materials can arrive >3.0 V. For quinones including alkoxy, hydroxyl, and halogen groups, the addition of substituents can increase the discharge potential of the quinone electrode. Quinone salt is composed of carbon oxide salt and quinone carboxylate. The introduction of functional groups into this type of quinone salt can reduce the solubility of electrodes in organic electrolytes [10a, 17]. Other types of quinone polymer, including simple conjugated polymers, alkyl-coupled polymers, nitrogen-coupled polymers, and sulfur-coupled polymers, are also the important OCMs.

Quinone electrode materials generally have higher theoretical specific capacity [18]. For example, a simple small-molecule quinone compound such as BQ can reach the theoretical capacity of 500 mAh g^{-1}, and the theoretical capacity of AQ and phenanthraquinone (PQ) can arrive 257 mAh g^{-1}. Some organic small molecular quinone compounds and their theoretical specific capacities are summarized in Table 6.1.

The application of quinone materials for batteries was firstly displayed in LIBs. In 1972, Alt et al. firstly synthesized a tetrachloro-*p*-benzoquinone as the cathode for a lithium-organic battery [10a]. The battery has an open-circuit voltage of 3.3 V at a current density of 3 mA cm^{-1}. When discharged to 2.6 V, the actual specific capacity is 85% of the theoretical value (C_{th} = 218 mAh g^{-1}). Another example, Wang and coworkers reported the electrochemical properties of the 1,4,5,8-tetrahydroxy-9,10-anthraquinone (THAQ) and its oxidation 1,4,5,8-tetrahydroxy-9,10-anthraquinone product (O-THAQ), which showed that when the four hydroxyl groups of THAQ are oxidized to carbonyl groups, the resulting oxidation product has better cycling performance and higher discharge capacity (first discharge capacity: 250 mAh g^{-1}) [19]. The successful application of these types of quinone compounds in LIBs shows their potential application in SIBs. In recent years, researchers have used quinone cathode materials for SIBs. Xiong et al. investigated the electrochemical performance of pillar[5]quinone (P5Q) with a theoretical capacity of 446 mAh g^{-1} as cathode in SIBs [20]. This quinone cathode showed an initial capacity of 418 mAh g^{-1} and maintained 290 mAh g^{-1} after 300 cycles at 0.1 C. To avoid the dissolution of P5Q, it was needed to encapsulate P5Q into CMK-3 to form a composite material, and then integrated with single-walled carbon nanotubes (SWCNTs). Calix[4]quinone (C4Q) with a similar structure to P5Q has also been explored as the cathode of SIBs [21]. The C4Q shows a high theoretical capacity of 446 mAh g^{-1}. But similarly, C4Q also needs to combine with CMK-3 and SWCNT to avoid the dissolution.

It can be seen from the above examples, small-molecule quinone compounds have the benefits of large theoretical capacity and high utilization rate. However, some major problems, such as solubility of quinone electrode in organic electrolyte and poor conductivity, need to be addressed. In order to alleviate the solubility of small-molecule quinone compounds in organic electrolytes, researchers have tried

Table 6.1 A summary of the theoretical specific capacities of some organic small-molecule-conjugated carbonyl compounds.

Molecular quinone compound	Structure	Theoretical capacity ($mAh\,g^{-1}$)
BQ	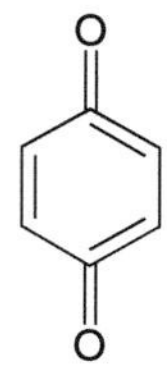	496
NQ	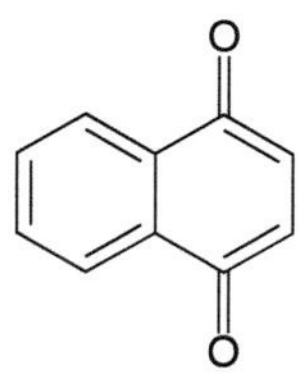	339
AQ	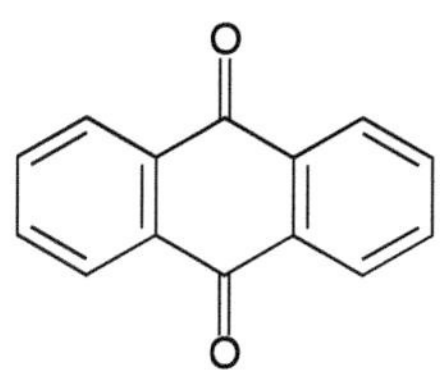	257
PQ	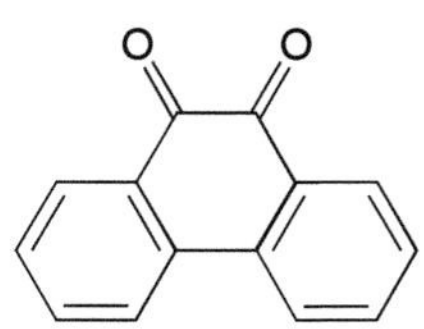	257
Chloranil	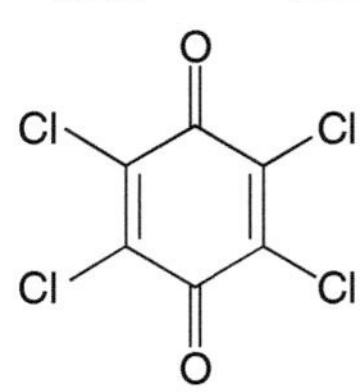	218
DDQ	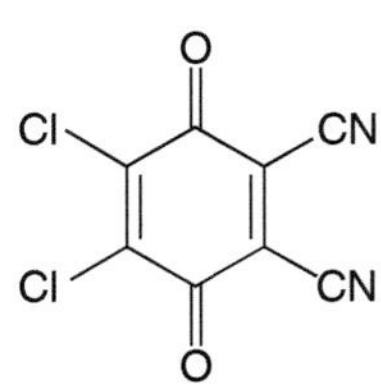	236
THBQ	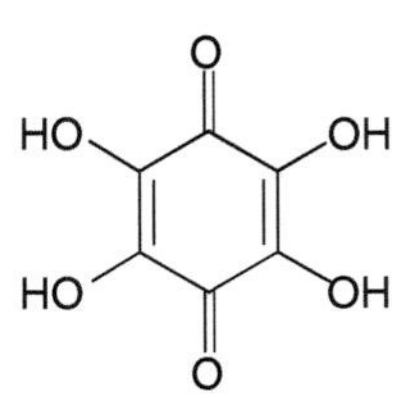	311

(continued)

Table 6.1 (Continued)

Molecular quinone compound	Structure	Theoretical capacity ($mAh\,g^{-1}$)
DHBQ	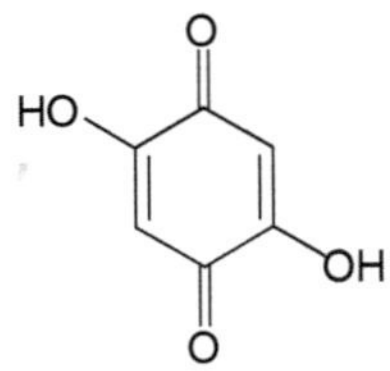	382
DMBQ	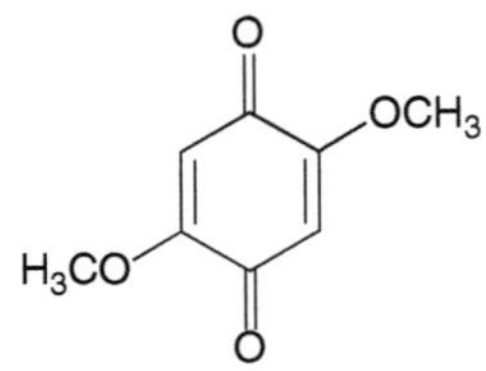	319
PT	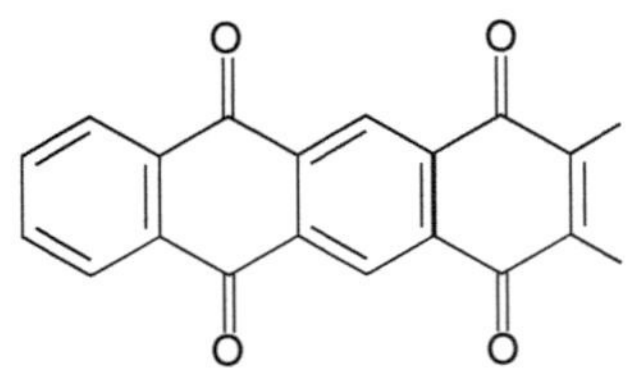	317
C4Q	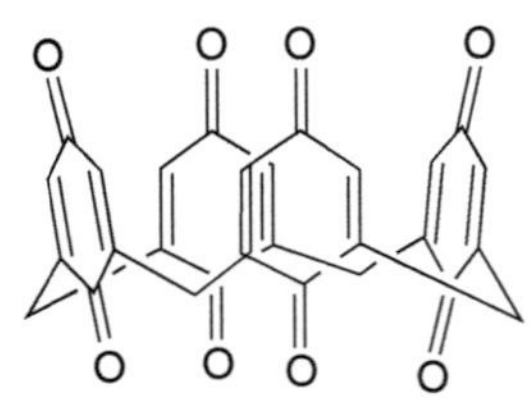	446

various methods in recent years. The most commonly used method is to increase the conductive carbon content during the electrode preparation process, which not only inhibits the dissolution of the active materials but also can enhance the conductivity of the electrodes. Along this line of thinking, porous mesoporous carbon (MC) with a large specific surface area is used to reinforce the active materials, reduce its dissolution, and improve the cycle performance [22]. For example, Zhou et al. prepared an economical and practical bio-carbon-PPL with a porous tubular structure by pyrolyzing physalis (*Peralis peruviana* L. Calyx) and compounded it with C4Q to obtain C4Q/PPL cathode material for SIBs [23]. Its initial capacity can reach 435 $mAh\,g^{-1}$, and it has a stable cycle capacity with 195 $mAh\,g^{-1}$ after 100 cycles at 0.1 C. In addition, Kim et al. also improved the cycle stability of the quinone derivative $C_6Cl_4O_2$ by combining $C_6Cl_4O_2$ with the porous carbon template nanocomposite material and prevented the active molecules from dissolving into the electrolyte [24]. Recently, some new methods have been proposed to tackle this problem, including loading advanced carbon materials

[6b, 25], polymerizing small-molecule carbonyl compounds, and constructing multifunctional membranes [26].

Graphene has many advantages, including large specific surface area, remarkable electronic conductivity, advanced mechanical strength, and chemical stability, which have been widely used to enhance the performance of quinone compound electrodes. For instance, Wang et al. recombined the walnut molecule and reduced graphene oxide (rGO) through a strong π–π interaction between the aromatic structure and the carbon scaffold to obtain a composite material with a stable reversible capacity of 305 mAh g^{-1} [27]. The capacity can still maintain 91% after 100 cycles. Song et al. used in situ polymerization to mix graphene and anthraquinone sulfide (poly(anthraquinone-based sulfide) [PAQS]) to prepare nanocomposite electrode materials [28]. 5 wt% of the mixed amount of graphene can greatly increase the electronic conductivity of the PAQS composite electrode by more than six orders of magnitude. When the graphene content in the anode increased to 26 wt%, the composite electrode shows a more than of 100 mAh g^{-1} capacity at the ultra-high rate of 100 C, and the utilization rate of the active material reaches 95%. In 2016, Wang and coworker developed graphene-supported poly(1,5-diaminoanthraquinone) nanocomposites (GNS@PDAA) by chemical oxidation polymerization [29]. When the mass ratio of DAA/GNS is 6 : 1, the electrochemical performance of this composite material is the best. The GNS@PDAA composite material also has better cycle stability and electron transport ability. Li group employed graphene as a substrate to grow porous sodium salts of poly(2,5-dihydroxy-*p*-benzoquinone sulfide) (Figure 6.5a) [30]. The composite material can reversibly insert two sodium ions, and the corresponding theoretical capacity can reach 250 mAh g^{-1}. After 150 cycles, the composite can still exhibit a reversible capacity of 179 mAh g^{-1} at 100 mA g^{-1}, which is better than the single polymer (NaPDHBQS), monomer (Na_2Cl), and rGO. In addition, the composite material also exhibits a higher rate capacity (147 mAh g^{-1}, Figure 6.5b) than that of the NaPDHBQS. Moreover, this composite material can be used as a cathode material in an organic full battery with the coupling of a disodium terephthalate anode material, where an initial reversible

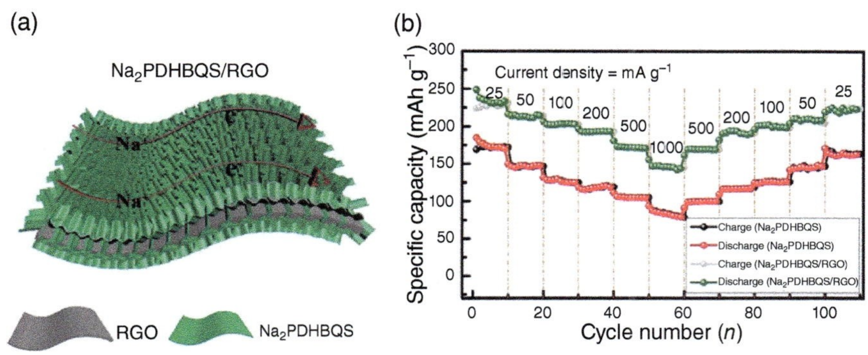

Figure 6.5 (a) Schematic diagram of the synthesis process of $Na_2PDHBQS/RGO$ and (b) rate performance at various current densities. Source: Li et al. [30]. Reproduced with permission, 2017, Elsevier.

capacity of 210 mAh g^{-1} is displayed at 25 mA g^{-1}. The improved performance is profited from the successful polymerization, incorporation of graphene, and utilization of ether-based electrolyte.

MC and carbon nanotube (CNT) with interconnected nanochannels are in favor of electrolyte penetration, and the highly conductive carbon matrix helps to maintain smooth electron flow. These two types of carbon materials are also commonly used in the modification of quinone compound electrodes. Chen and coworkers coated an AQ molecule by MC material of CMK-3 [17]. The obtained composite material can reach 205 mAh g^{-1} in the initial cycle. The capacity retention rate is as high as 84% after 100 charge–discharge cycles. Wu et al. prepared composite electrode materials by introducing CNTs into Na_2DBQ, which reduced the dissolution of Na_2DBQ [31]. This composite material provides capacities of 161 and 142 mAh g^{-1} at 1.45 and 2.03 A g^{-1}, respectively.

Polymerization of small-molecule quinone compounds is also an effective method to prevent the dissolution of small-molecule carbonyl compounds, thereby improving battery cycling performance [32]. Figure 6.6 shows the structure of several polyquinones for SIBs. In 2013, Yang team reported the polyquinone

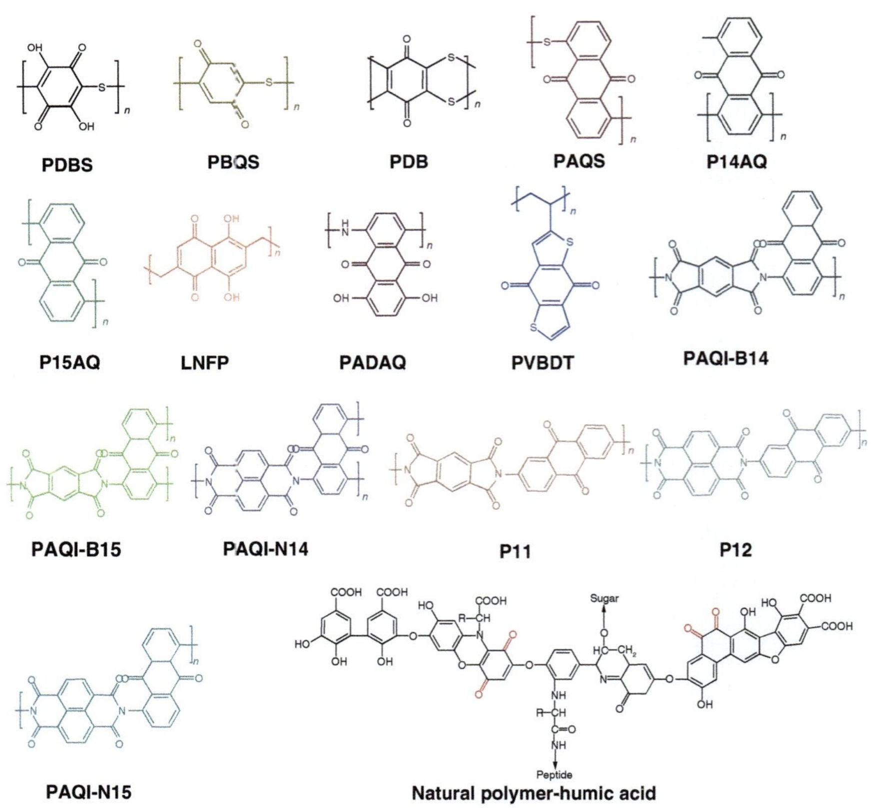

Figure 6.6 Structures of ployquinone compounds for SIBs.

cathode for SIBs, namely PAQS [5b]. These PAQS show two charging and discharging platforms at 1.5 and 2.0 V, respectively, with a capacity of 220 mAh g^{-1} that corresponds to 98% of the theoretical storage capacity. Even at a high current density of 6400 mA g^{-1}, the quinone polymer still has a capacity of 160 mAh g^{-1}. Due to the excellent reactivity of the BQ monomers, Zhou and coworkers prepared poly(benzoquinonyl sulfide) as the SIB cathode by an oxidative polymerization technique [9a]. Poly(benzquinoline sulfide) provides a large capacity of 268 mAh g^{-1} and an energy density of 557 Wh kg^{-1}. However, its cycling stability and multiplier performance are moderate (maintaining a capacity retention of 68% after 100 cycles). The authors indicated that the poor compatibility between sodium and electrolyte is the reason for the inferior cycling stability and rate performance.

Using solid/ionic liquid electrolytes can also effectively prevent the dissolution of small carbonyl compounds. Solid electrolytes cannot dissolve quinone compounds, while ionic liquid electrolytes can inhibit the dissolution of quinone compounds due to polarity. Wang et al. combined quinone cathode and ionic liquid electrolyte for organic SIBs [33]. Through density functional theory (DFT) and spectroscopy studies, the authors found that ionic liquids have an inhibitory effect on the dissolution of quinones and improve the stability of the quinone compounds' cathodes. Similarly, Casado et al. used lignin which contains quinone/hydroquinone redox moieties as cathode of SIBs [34]. To avoid the dissolution of lignin, they also used ionic liquid as electrolyte. The application of solid electrolyte can more directly prevent the dissolution of quinone compound cathode materials. This concept is initially applied in the LIBs. For example, Zhu and coworkers used P5Q as the positive electrode material and gel polymer of polymethacrylate/polyethylene glycol as the electrolyte to assembly a quasi-solid LIB, which not only delivers the first specific discharge capacity of 410 mAh g^{-1} but also shows an improved cycle life [35]. Chen et al. utilized calix quinone (C4Q) cathode and a polymer gel electrolyte to assemble a superior LIB [36]. At the current rate of 0.2 C, the C4Q shows the initial discharge capacity of 422 mAh g^{-1}, and the capacity retention rate can reach 90% after 100 cycles. Compared with the liquid organic electrolyte, this battery has higher initial capacity and preferable cycling performance. Inspired by this, the application of solid electrolytes in SIBs has also begun to develop.

What's more, the preparation of organic salts with small polar molecules can also reduce their dissolution in organic electrolytes. The functional groups with strong hydrophilic in the quinone salt, including –ONa, –COONa, and $-SO_3Na$, can reduce the solubility of the quinone salt in the organic electrolyte [37]. As an example, Chen's research group fabricated a cathode material of Na_2DBQ for SIBs [18b]. At a current rate of 0.1 C, the capacity of Na_2DBQ can arrive 265 mAh g^{-1}. In addition, the Na_2DBQ exhibits an outstanding cycling stability, and a capacity of 231 mAh g^{-1} is retained after 50 cycles. Besides, an organic 2-dihydroxy tetra-sodium terephthalate (Na_4DHTPA: $Na_4C_8H_2O_6$) with two –COONa groups was developed [38]. Figure 6.7a shows the electrochemical redox reaction mechanism of sodium ions in $Na_4C_8H_2O_6/Na_6C_8H_2O_6$ and $Na_2C_8H_2O_6/Na_4C_8H_2O_6$ [5d]. Combining with Na_4DHTPA, the as-prepared SIB has an initial capacity of 183 mAh g^{-1} at a current

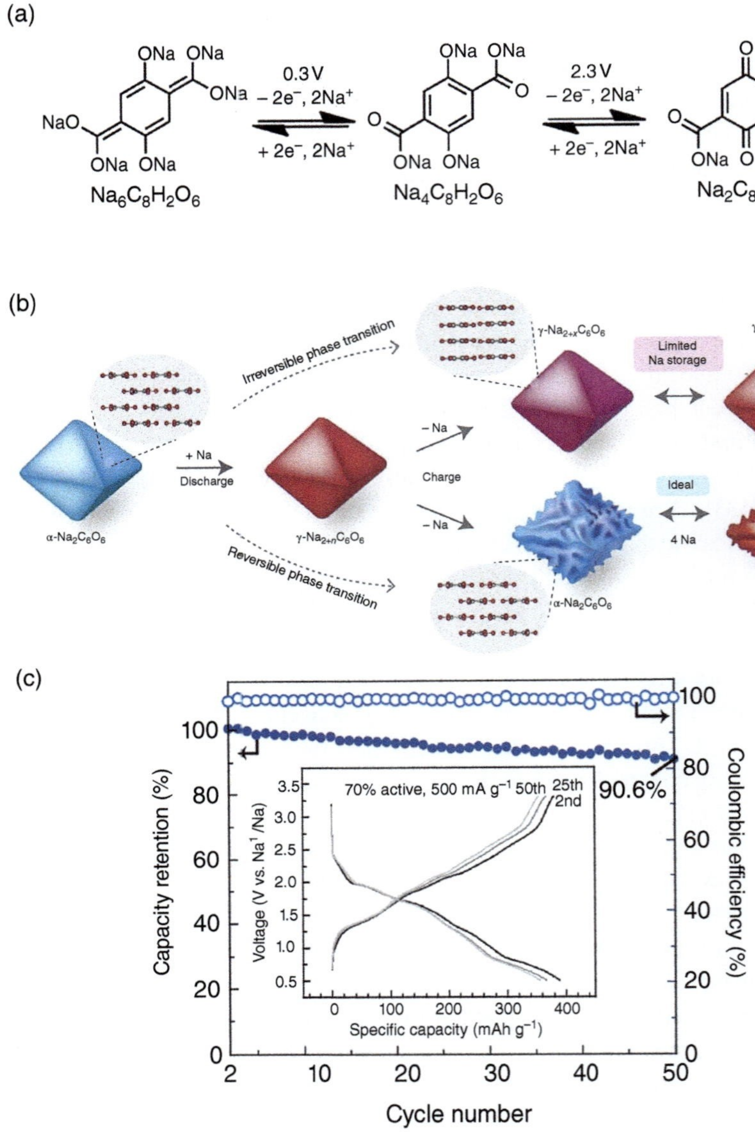

Figure 6.7 (a) Electrochemical redox reaction mechanism of sodium ions in $Na_4C_8H_2O_6$/$Na_6C_8H_2O_6$ and $Na_2C_8H_2O_6$/$Na_4C_8H_2O_6$ at potentials of 0.3 and 2.3 V, respectively. Source: Wang et al. [5d]. Reproduced with permission, 2014, Wiley-VCH. (b) A schematic of the proposed (de)sodiation mechanism of the $Na_2C_6O_6$. (c) Cycling retention of the $Na_2C_6O_6$ electrode at 500 mA g^{-1}. Source: Lee et al. [39]. Reproduced with permission, 2017, Springer Nature.

rate of 0.1 C and maintains at 154 mAh g^{-1} after 100 cycles. At a high rate of 5 C, its capacity of about 80 mAh g^{-1} is still obtained, manifesting a good rate performance. For quinone salts ($Na_2C_6O_6$) cathode, the reversible storage of four sodium ions can be effectively achieved by reducing the size of the active material and optimization of electrolyte (Figure 6.7b) [39]. The results show that this type of quinone salts can release an actual capacity approaching the theoretical value (498 mAh g^{-1}) with a capacity retention of 87% after 50 cycles (Figure 6.7c). Most importantly, this quinone salt exhibits better performance (the energy density of 410 Wh kg^{-1} and the power density of 3151 W kg^{-1}) than the reported inorganic cathode materials.

Quinone salts containing –SONa functional groups also refrain from the dissolution of quinones in organic electrolytes. For a disodium anthraquinone-1,5-disulfonante (AQDS) with –SONa functional groups [18a], the inversible capacity of the AQDS cathode is 100 mAh g^{-1} at 0.1 C, and a inversible capacity of 120 mAh g^{-1} can still be retained after 100 cycles.

The introduction of diverse functional groups into the structure of quinone compounds can effectively adjust the working voltage of OCMs. Halogen atoms have strong electron-withdrawing properties, which can effectively reduce the energy of the lowest unoccupied molecular orbital (LUMO). Generally, the energy of LUMO is tied with the electrochemical stable window, where lower LUMO energy can effectively increase the reduction voltage. Based on the DFT calculations, the redox voltage of various materials can be arranged in the following order: $C_6F_4O_2 > C_6Cl_4O_2 > C_6Br_4O_2 > C_6H_4O_2$. Except for halogen atoms, there are other usual and strong electron-withdrawing groups such as cyano (CN) and sulfonate ($-SO_3Na$).

Natural organic compounds can also be served as positive electrodes for SIBs, such as humic acid with a large number of carbonyl groups (Figure 6.6), which provides some active sites for sodium ion insertion, and the theoretical specific capacity is 244 mAh g^{-1} [9c]. However, due to the existence of some inactive groups, its theoretical specific capacity can not be fully claimed.

6.2.2 Carboxylates

Carboxylic acid-based organic materials have been widely reported as electrode materials for SIBs. The structure of the carboxylates includes an aromatic ring connected to two or more sodium carboxylates. Figure 6.8 shows some cathode materials of carboxylate organic compounds for SIBs [7b, 40]. Since the electron-donating group (–ONa) is usually linked to the carboxyl group, the voltage of carboxylate intercalating sodium ions is <1.0 V, so it is often selected as the anode material. For example, disodium terephthalate (Na_2TP, $Na_2C_8H_4O_4$) and sodium 4,4-diphthalate ($Na_2C_{14}H_8O_4$) display the discharge platform of 0.3 and 0.4 V, respectively [41]. Each molecule of $Na_2C_8H_4O_4$ and $Na_2C_{14}H_8O_4$ can reversibly insert two Na ions.

When the two functional groups of carboxylate (–OONa) and sodium enolquinone are simultaneously connected to the structure of an organic compound, it may serve as both a cathode material and an anode material. A research team designed an

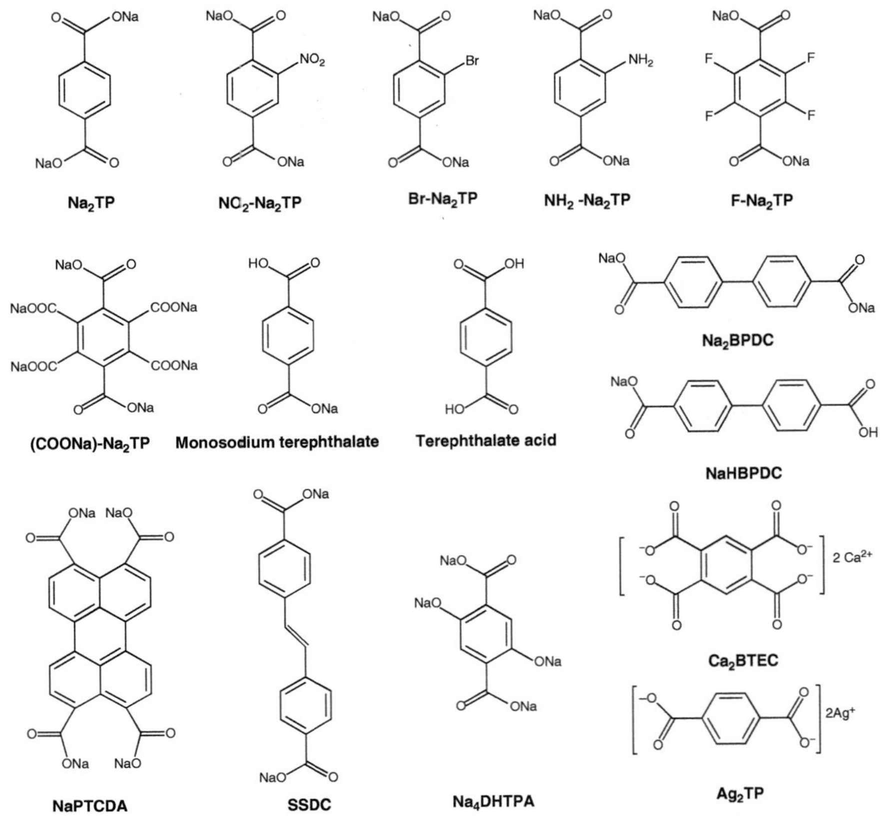

Figure 6.8 Structures of carboxylate-based organic compounds for SIBs.

organic compound of 2.5-dihydroxyterephthalate ($Na_4C_8H_2O_6$). When it acts as a cathode material, the redox potential of $Na_4C_8H_2O_6/Na_6C_8H_2O_6$ occurs at 2.3 V. When it is used as an anode material, the redox reaction of $Na_4C_8H_2O_6/Na_6C_8H_2O_6$ occurs at 0.3 V.

Like quinone compounds, the solubility of carboxylate compounds in electrolyte is also one of the key issues. The solubility of these compounds in the electrolyte is related to the symmetry and conjugation degree of the molecules. Generally speaking, the higher the symmetry and conjugation degree, the better the stability of the organic-conjugated carboxylic acid sodium salt in the electrolyte. However, the disadvantage of this type of material is their poor performance at high charge–discharge rates. To address this problem, preparing composite electrodes or using advanced carbon materials as conductive additives to optimize the active materials can increase the electrochemical performance. Similar with quinone electrode material, forming inorganic cation salts is also one of the methods to reduce carboxylate dissolution. For instance, organic calcium salt of 1,2,4,5-benzenetetracarboxylic acid and silver salt of terephthalate exhibit a stable cycling life, up to hundreds of cycles [42].

The electrochemical properties of cathode materials can be improved by modifying the surface of electrode materials. The morphology and structure of the material are the important factors affecting the electrochemical performance. Based on the molecular design perspective, the extension of the π-conjugated system in materials is an effective method to enhance the rate performance. The extension of the π-conjugated system can both increase the charge traffic and stabilize the charged/discharged states and enhance the intermolecular interaction to promote the insertion and de-intercalation of sodium ions. On the side, compared with oxygen atoms, the incorporation of sulfur atoms which have larger atomic radius and electron density is another effective method to increase electrical conductivity, resulting in high-rate performance. The 4,4′-stilbene-dicarboxylate sodium prepared by Wang et al. has excellent rate performance, where a high capacity of 72 mAh g^{-1} can be achieved at a high current density of 10 A g^{-1} [5f]. The reason for these achievements can be ascribed to the existence of π-conjugated system that can both enhance the charge transport and the intermolecular interaction and promote the insertion/extraction of sodium ions. Modification of carboxylate functional group by the addition of sulfur atoms with larger atomic radius and electron density can increase the conductivity of the material, thereby obtaining high-rate performance. In addition, the substitution of oxygen by sulfur in carboxylic acid group can increase the electron delocalization and sodium absorption capacity.

6.2.3 Anhydrides

OCMs based on anhydride, which generally include aromatic core and two anhydride groups, have been widely used for SIBs due to their multi-electron reactions and large specific capacity [13b, 43]. The acid anhydride compounds used as SIB cathodes usually have a large conjugated planar structure, and the number of sodium ion insertion active sites can be adjusted by regulating the discharge voltage. Pyromellitic dianhydride (PMDA), 1,4,5,8-naphthalenetetracarboxylic dianhydride (NTCDA), and 3,4,9,10-perylene tetracarboxylic dianhydride (PTCDA) are the commonly used SIB cathodes [43a, 43c].

Especially, PTCDA has received considerable attention. The conjugated acid anhydride PTCDA (Figure 6.9a) can accept two sodium ions at 1.0–3.0 V and has a inversible specific capacity of about 150 mAh g^{-1} [40b, 43a]. When the discharge voltage decreased, 15 sodium ions can be inserted, and the first discharge capacity of 1017 mAh g^{-1} is displayed. This extra capacity is believed to be caused by the appearance of a solid electrolyte mesophase and the insertion of sodium ions into the aromatic ring structure. The mechanism of the chemical redox reaction is shown in Figure 6.9b [42b]. The aromatic nucleus in the PTCDA structure can make up a conjugated structure equipped with the carbonyl group to facilitate the redox enolization reaction. Aromatic carbonyl groups can form sodium enolate with sodium ions, thereby promoting the intercalation of sodium ions. However, in the low potential range, the cycling performance of PTCDA is relatively poor, and the capacity frequently drops down to 227 mAh g^{-1} after only five cycles of charge and discharge, because its crystal structure is destroyed. Similarly, Ji and coworkers

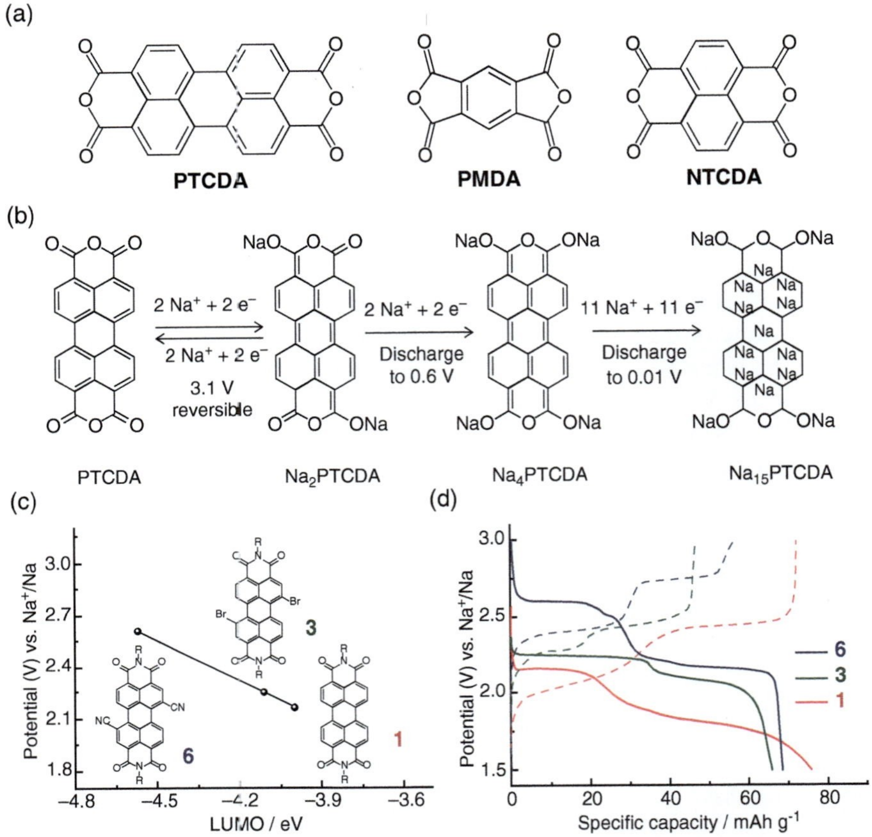

Figure 6.9 (a) Structures of anhydrides. (b) The proposed reaction mechanism of PTCDA. Source: Luo et al. [42b]. Reproduced with permission, 2014, Wiley-VCH. (c) First reduction potential vs. calculated LUMO energy levels for various perylene diimide derivatives, and (d) the corresponding charge–discharge voltage profiles at a current rate of C/4. Source: Banda et al. [44]. Reproduced with permission, 2017, Wiley-VCH.

studied the sodium storage performance of perylene tetracarboxylic anhydride dye [42b]. The material has a voltage plateau of about 2.3 V with a reversible capacity of 145 mAh g^{-1}. At the same time, there is still a reversible capacity of 91 mAh g^{-1} at 1000 mA g^{-1}, showing excellent rate performance.

The combination with carbon materials is an available way to enhance the electrochemical performance of acid anhydride-based electrode materials. Recently, Yuan et al. synthesized nano-ultralight materials (polymer/graphene composite [PGC]) based on organic compounds/graphene aerogels, in which aromatic carbonyl compounds are confined in a three-dimensional hybrid structure [43c]. PGC forms a stable composite structure through the interaction of the π=π bond between the aromatic nucleus and the graphene, and the conjugated acid anhydride group shows electrochemical activity toward sodium ions. Experimental results demonstrated that PGC shows a considerable initial capacity of about 98 mAh g^{-1}

at 25 mA g^{-1}, and the 90% capacity retention rate after 100 cycles is remained. The restricted growth in porous graphene not only enhances the electrochemically active PTCDA molecule's accessibility to sodium ions but also facilitates electron transfer and shortens the diffusion pathway of sodium ions.

The energy levels including the highest occupied molecular orbital (HOMO) and the LUMO can determine the redox potential of organic molecules. Based on the molecular orbital theory, low LUMO energy will make electron affinity (EA) greater, thereby increasing the oxidizability and ultimately increasing the reduction potential [9d, 13b, 45]. In order to reduce the LUMO energy and increase the oxidation–reduction potential, two methods are viable here. One is expanding the conjugated system of the aromatic ring, which can both increases the amount of sodium ions to be inserted to increase the final capacity and reduces the HOMO–LUMO gap and the polarization. The second strategy is to incorporate the electron-withdrawing groups [46]. For example, Banda et al. used applicable electron-withdrawing groups, such as Br and CN, to form the substituents of perylene diimides, which significantly enhanced the discharge potential to 2.6 V vs. Na^+/Na, as shown in Figure 6.9c,d [44].

6.2.4 Amides

The general formula of amide electrode material is R–C(O)–N(R)–C(O)–R, and the nitrogen atom in its molecular structure is connected to two carbonyl groups. At present, the amide compounds used in SIBs are mainly small-molecule imides and polyimides. The redox mechanism of imines is electrochemical enolization of carbonyl groups, and sodium ion binds and separates with oxygen atoms. In the process, the aromatic nuclei are used to stabilize negative charge. Half of the carbonyl group in the imide compound can be used to accept sodium ions, thereby improving the structural stability of the electrode material. The sodium hydride group in the imide salt enables the exchange reaction with protons. A battery assembled with disodium naphthalimide as the cathode and Prussian blue as the anode has a working voltage of 1.1 V.

Unfortunately, small-molecule imide compounds are highly soluble in aprotic electrolytes [13b]. For instance, the phthalimide disodium salt prepared by Brandell and coworkers has a specific capacity of 150 mAh g^{-1} at 100 mA g^{-1} and an efficiency of 67.4% in the first cycle [47]. However, after 100 cycles, the capacity retention rate drops to 53%. This phenomenon has seriously degraded the viability of this kind of compounds. To solve this, the "polar inversion" strategy has been used to make small polar common yoke carbonyl compounds into salts, thereby increasing the polarity of imide compounds and reducing their solubility in organic electrolytes. For example, the diimide salt (Figure 6.10a,b) can be used as a cathode material in both aqueous and aprotic electrolytes [49]. As the same with acid anhydride compounds, a large proportion of the carbonyl groups in the diimide can be used to accept sodium ions [48]. The electrochemical redox reaction mechanism is shown in Figure 6.10c. Although the energy density is much lower than that of commercial LIBs, the low-cost of organic positive electrode outweighs this concern

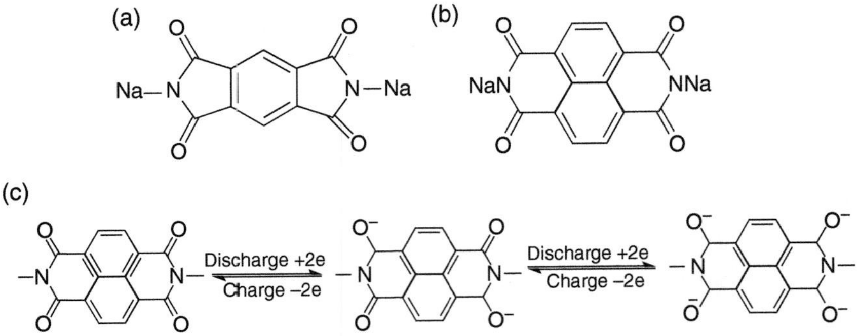

Figure 6.10 (a, b) Structures of diimide salts. (c) Mechanism for reversible redox reaction of imide salt. Source: Deng et al. [48]. Reproduced with permission, 2015, American Chemical Society.

and makes it still one of good choices for SIBs. Establishing compounds with large π-conjugated structure and intermolecular hydrogen bonds is beneficial to avoid the dissolvability of electrode materials in organic electrolytes. Deng et al. [48] studied the electrochemical performance of 3,4,9,10-perylene tetracarboximide (PTCDI) as a cathode in SIBs, which has large π-conjugated structure and intermolecular hydrogen bonds [50]. As the results, the PTCDI has a specific capacity of 140 mAh g^{-1} at 5 C, and after 300 cycles the capacity remains at 103 mAh g^{-1}.

Polymerizing small-molecular imide salts to form polyimides is an effective method to prevent them from dissolving in the electrolyte. Polyimide has lower solubility in organic electrolyte than imide, and the cycling performance is relatively good. Polyimide is generally formed by high-temperature polymerization of dicarboxylic anhydride and diimine. The synthesis of aromatic polyimide can mainly be divided into two steps in the traditional synthesis method. First, the dianhydride and diamine react in a polar aprotic solvent to form polyxamic acid. Immediately afterward, the polyxamic acid ring is dehydrated to form polyimide by heating or chemical dehydrating agent treatment. Polyimides can also be synthesized through one-step solvothermal approach, and here the condensation polymerization and imidization reactions proceed spontaneously. According to the synthesis method, the formation of the selected dihydride and diamine determines the structure and properties of the as-synthesized polyimide aromatic skeleton. Thus, a variety of molecular structures of polyimides can be achieved by judiciously selecting the dihydride and diamine used in synthesis. Figure 6.11 shows the structure of polyimides synthesized from different diamines and dianhydrides [51]. Due to its higher molecular weight, polyimide has better stability in the electrolyte. Polyimide has a high redox potential (>1.7 V vs. Na^+/Na) and is generally acted as a cathode material in organic SIBs. Wu et al. reported polyimide-based OCM-polyperylene dianhydride ethylene diamine/carbon nanotubes (PTCDA-EDA/CNT) and polyperylene dianhydride *p*-phenylenediamine/carbon nanotubes (PTCDA-*p*PDA/CNT) composite materials and studied the electrochemical performance [52]. The results showed that they all have better rate performance and cycling stability.

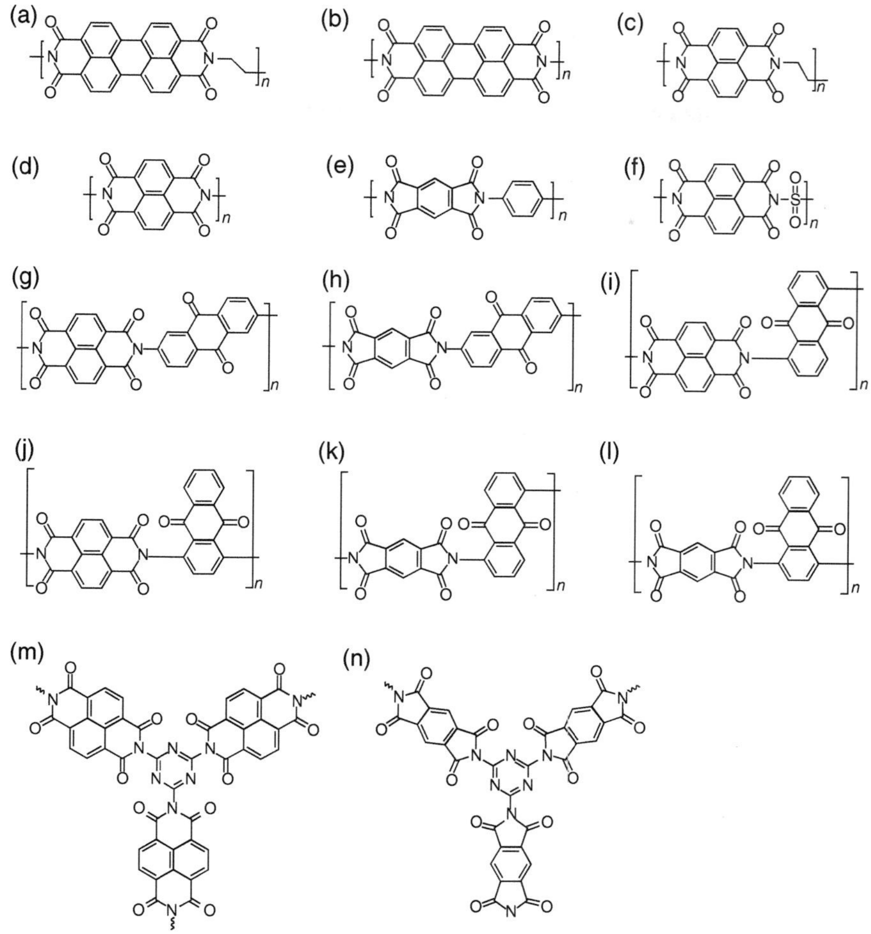

Figure 6.11 The structures of typical polyimides applied in SIBs.

The polyimides reported for SIBs are mainly derived from pyromellitic acid diimide (PMDI, Figure 6.12a), 1,4,5,8 naphthalene tetracarboxylic acid diimide 1,4,5,8 naphthalene tetracarboxylic acid diimide (NTCDI, Figure 6.12b–e), and 3,4,9,10-perylenetetraformodiimide (PTCDI, Figure 6.12f–i) [53]. Zhan et al. used different dianhydride and diamine monomers to synthesize a series of polyimide compounds (Figure 6.12j–l). The polyimide based on the dicarboxylic anhydride PTCDA has the highest redox voltage of 2.2 V vs. Na^+/Na. And after 5000 cycles, its capacity retention rate is 98%, showing exceptionally high stability. Wang et al. also manufactured a series of polyimides and investigated the correlation between the electrochemical properties of polyimides and dihydride [9d]. By changing the active core, the discharge voltage platform increases from 1.73 to 1.89/1.94 V, and the circulation stability is improved due to the decrease in dissolvability. The reduction of alkyl chains in polyimides leads to a decrease in molecular weight and thus an increase in capacity. It has a capacity of 137.6 $mAh\,g^{-1}$ at 25 $mA\,g^{-1}$ after 400

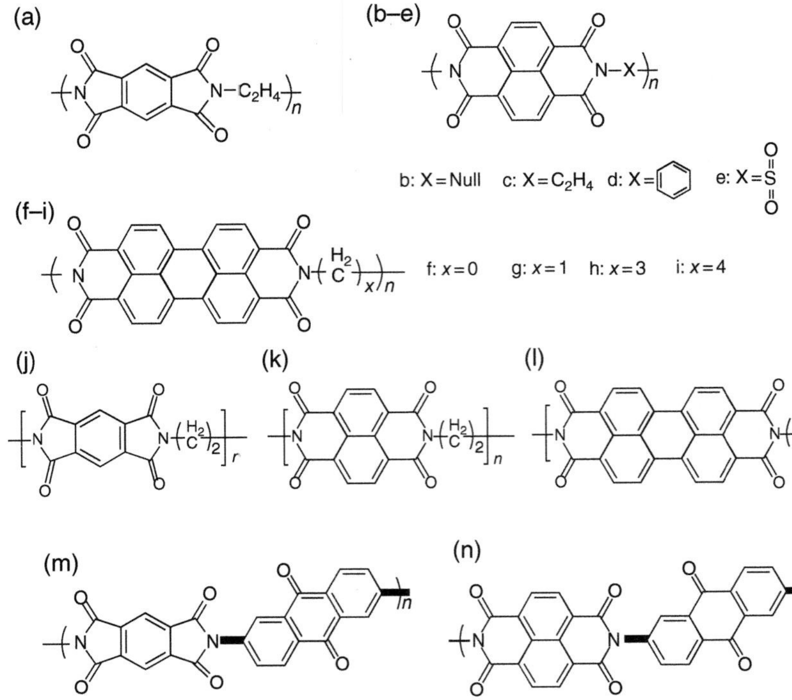

Figure 6.12 Structures of polyimide compounds.

cycles, and a capacity of 110.8 mAh g^{-1} at 200 mA g^{-1} after 5000 cycles. It also has a brilliant rate capacity with a capacity of 50 mAh g^{-1} at 10 A g^{-1}.

Polyimide has a very low solubility in the electrolyte because of the stable conjugated structure; hence, it has more stable cycling performance. The current challenge is that the theoretical specific capacity of polyimide is limited, which is usually below 150 mAh g^{-1}. Xu et al. copolymerized PMDA with a high-capacity AQ group as a linking unit to synthesize two new polyimides (Figure 6.11l,m), each with theoretical specific capacities of 383 and 342 mAh g^{-1} [51]. Experimental results expressed that the capacities of these two cathodes are higher than that of pure polyimide. Similar, a series of other poly(mide-quinone) complexes were also synthesized to increase capacity, as shown in Figure 6.11i–l. In addition, the capacity of polyimide can also be improved by introducing electron groups into the polyimide chain. For example, Li et al. introduced the triazine ring into polyimide (Figure 6.11m,n) to increase the storage capacity of sodium ions through the synergistic effect of the amide group and the triazine ring [54]. Besides, Xu et al. used the sulfonyl group as the connecting unit of the polyimide, which not only improves the cycling performance of the material but also increases its discharge voltage without affecting the capacity [53c]. The experimental results expressed that the Coulombic efficiency of this material is close to 100%, indicating that its good stability is achieved at the interface. The sulfonyl group is an electron-withdrawing group that can reduce the electron density of the redox-active carbonyl group through inductive effect, so that the average charge–discharge voltage of the connected polyimide can reach 2.0 V, which is higher than that of the general polyamide with vinyl connection unit [55].

6.3 Doping Reaction

Organic batteries that rely on doping reactions can be divided into p-type and n-type doping reactions according to the different charges carried by the doped ions of the electrodes. When the organic electrode undergoes an electrochemical oxidation–reduction reaction, the segment structure loses its electrons and becomes positively or negatively charged. At this time, in order to maintain the charge balance inside the electrode, the anions/cations in the electrolyte are transferred to the electrode which are called p-type and n-type doping, respectively. When the reverse reaction proceeds, the corresponding ions are extracted from the electrode and then re-enter the electrolyte (dedoping), and the polymer segment returns to its original state.

Taking polymer P as the cathode and metallic sodium as the anode, the doping reactions for the organic batteries are briefly introduced as below.

$$\text{Anode}: \text{Na}^+ + \text{e}^- \rightleftharpoons \text{Na} \tag{6.1}$$

$$\text{Cathode}: \text{P} + \text{A}^- - \text{e}^- \rightleftharpoons \text{P}^+\text{A}^- \tag{6.2}$$

$$\text{Cell reaction}: \text{P} + \text{A}^- + \text{Na}^+ \rightleftharpoons \text{P}^+\text{A}^- + \text{Na} \tag{6.3}$$

where P is the polymer and A^- is the negative ion in the electrolyte. In the charging process, the cathode polymer P is oxidized to P^+, and the negative ions in the electrolyte travel to the cathode and dope the cathode to balance the charge to obtain P^+A^-. The electrons flow from the external circuit to the anode, and sodium ion in the electrolyte is reduced at the anode to produce metallic sodium. During discharge, the negative metal element is oxidized to form sodium ion, which is then dissolved in the electrolyte. At the same time, electrons flow to the cathode through the external circuit. The cathode of doped polymer obtains electrons to generate polymer P, and release A^- into the electrolyte.

This type of battery has advantages unmatched by other batteries:

(i) During the charging and discharging process, the conversion of electrical energy and chemical energy is achieved by doping and dedoping ions. The capacity of the battery depends on the doping degree. Taking poly batch as an example, if each cell can be doped with an anion, then the theoretical capacity can reach up to 412 mAh g^{-1}, which is much larger than the current inorganic cathode materials.

(ii) This type of battery has nothing to do with the cation of the electrolyte, then in theory, we can also replace metallic lithium with resource-rich metallic sodium or other inorganic materials, or even with polymers.

(iii) This type of battery does not work the way as the conventional battery system where the inorganic material is used as the positive electrode. The sluggish solid-state ion diffusion problem in the lattice can be bypassed, and the material itself has high conductivity, so it can be quickly charged and discharged. It also has excellent cycling stability, not to mention their rich resources and environmental friendliness.

6.3.1 Conductive Polymers

Since 1977, Hideki and Heeger have discovered that doped polyacetylene has the property of conducting metals or semiconductors, and since then research in the field of CPs has been initiated [56]. The CP is characterized by a large conjugated π bond in the structure, that is, highly delocalized π electrons. In the intrinsic state, it is a semiconducting state or an insulating state, and after doping, it behaves similar to metal that exhibits a certain degree of conductivity. CPs are widely concerned because of their excellent conductivity and insolubility [57]. At present, the widely studied CPs include polypyrrole (PPy) and polyaniline (PANI). Table 6.2 shows the common CPs and their conductivity after doping [56, 58]. Doping usually includes p-type and n-type, which can be done either chemically or electrochemically.

In fact, the first type of organic materials used in SIBs is the CPs. As early as 1985, Shacklette et al. proved the storage capacity of the reduced polyacetylene and polyparaphenylene for sodium ions [59]. In the past few decades, scientists have conducted a lot of research on CPs as electrode materials for sodium-ion secondary

Table 6.2 Common conductive polymers and their conductivity after doping.

Conductive polymer	Time of first appearance	Structure	Conductivity after doping at room temperature (S cm^{-1})
Polyacetylene (PAc)	1977	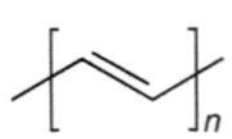	10^3 [56]
Polypyrrole (PPy)	1978	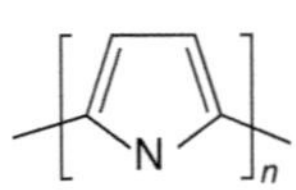	10^3 [58a]
Polythiophene (PTh)	1981	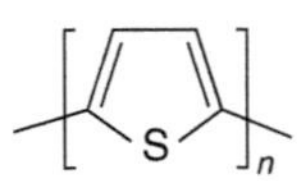	10^3 [58b]
Polyparaphenylene (PPP)	1979		10^3 [58c]
Polystyrene (PPV)	1979	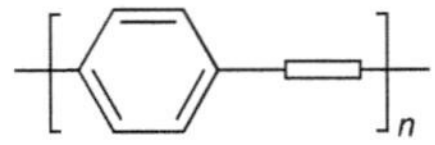	10^3 [58d]
Polyaniline (PANI)	1980	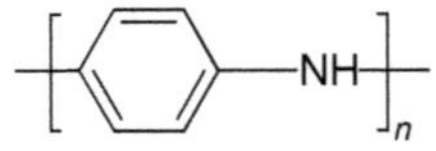	10^2 [58e]

batteries, and the most studied polymers are shown in Table 6.2. The redox reaction of these polymers is the transition between their doped and intrinsic states, which is usually accompanied by ion doping and dedoping. When they are used as electrode materials, the actual capacity is not only related to the molecular weight of the unit itself, but also to the degree of doping. Unfortunately, the doping degree of most of these polymers is not high, so the specific capacity is limited to less than 150 mAh g^{-1} that is much lower than their theoretical values. The CPs in Table 6.2 are described in detail below.

Shirakawa et al. first synthesized polyacetylene in the 1970s, but its intrinsic and doped states are unstable in air and will react with oxygen [60]. Since then, researchers have proposed improvements on its synthesis method. In 1981, Nigrey et al. reported the performance of polyacetylene as the battery cathode material [61]. The cathode is 6% $LiClO_4$-doped polyacetylene and the anode is metallic lithium. Since the charge storage mechanism of CP is only related to the doping or dedoping of anion/ionic groups in the electrolyte during charge and discharge, it is reasonable to believe that the reaction mechanism of CP in SIBs is similar to that of LIBs. Here, the discharge reaction is used as an example to introduce the energy storage mechanism of a CP. The reaction formula is shown as below.

$$[CH^{0.06+}(ClO_4)_{0.06}{}^{-}]_x + 0.06x\text{Na} \rightarrow (CH)_x + 0.06x\text{NaClO}_4 \tag{6.4}$$

During charging, the positive polyacetylene is oxidized to carry a positive charge, and then ClO_4^- in the electrolyte is doped into the polyacetylene chain to balance the charge. In the external circuit, the electrons flow to the negative electrode. Sodium ion in the electrolyte obtains electrons at the negative electrode and is reduced to metallic sodium. During discharge, the metal sodium is oxidized and loses electrons to form sodium ion. At the same time, sodium ion enters the electrolyte, and electrons flow from the external circuit to the polyacetylene positive electrode. The polyacetylene then accepts electrons and undergoes a reduction reaction to become electrically neutral, while ClO_4^- moves to the electrolyte.

PPy as a traditional CP has been fully studied in SIBs [62]. PPy is generally prepared by electrochemical oxidizing or chemical oxidizing the pyrrole monomer. The pyrrole monomer is safe and nontoxic and is an organic substance that can be used in organisms. PPy is the polymer with the lowest oxidation potential among the CPs. Common oxidants ($FeCl_3$, $(NH_4)_2S_2O_8$, etc.) can be used for the oxidative polymerization of pyrrole. PPy with high stability in the air, high doping conductivity, excellent redox reversibility, high operation potential of 2.5–4.0 V, and no n-type doping properties, has engaged wide attention as potential cathode materials for rechargeable batteries. Machida et al. developed PPy that polymerized with anhydrous $FeCl_3$ in methanol [63], showing highly reversible redox properties and excellent conductivity (190 S cm^{-1}). In the subsequent study, PPy was reported as a positive electrode material. If each pyrrole unit gains or loses one electron, the theoretical specific capacity can be calculated to be 412 mAh g^{-1}. However, due to its low utilization rate of polymer chains, its capacity is relatively low. Methods such as doping anions, copolymerization, and nanocrystallization of the electrodes can effectively enhance the capacity of the PPy.

Generally, PPy can be doped with small molecules with redox activity during the polymerization process, such as sodium indigo disulfonate, 2,2-diazobis(3-ethylbenzothiazolin-6-sulfonic acid) diammonium salt, lignin, etc., to greatly improve its electrochemical performance. Both $Fe(CN)_6^{4-}$ and diphenylamine-4-sulfonate have been proven as effective dopants for PPy, and the doped PPy (polypyrrole/$Fe(CN)_6^{4-}$ [PPy/FC], polypyrrole/diphenylamine-4-sulfonate [PPy/DS]) has higher capacity than PPy. The large capacity of the PPy/FC electrode is attributed to the insertion/extraction of sodium ion, the PF_6-doping/dedoping process, and the pseudocapacitive manifestation of $Fe(CN)_6^{4-}$ anion. Due to the natural redox characteristics of DS, the doping of DS in the PPy/DS electrode increases the utilization of PPy chains and the capacity. As an example, Zhou et al. doped $Fe(CN)_6^{4-}$ into the PPy chain, and the specific capacity of the material can reach up to 135 mAh g^{-1}. After 100 cycles, the capacity retention rate is 85% [64]. Similarly, Yang and coworkers used CPs including PANI and PPy to dope with iron cyanide as the positive electrode materials for SIBs [65]. The theoretical specific capacity can reach 296 mAh g^{-1}. On the other hand, PPy can be made into a nano-electrode with a hollow spherical shape, which can also increase its utilization rate [66]. For example, Su et al. made PPy into the shape of nano-hollow spheres, thereby increasing the utilization rate of PPy chains and increasing the specific capacity [62b]. Liu et al. used the dual template method to control the pore size to prepare two-dimensional mesoporous nanosheets (mesoporous polypyrrole [mPPy]) [67]. Figure 6.13 demonstrates the synthesis process of mPPy. Being a cathode material for SIBs, the nanosheet mPPy exhibits a large inversible discharge capacity of 123 mAh g^{-1} at 50 mA g^{-1}. Its large surface area provides active sites for storing

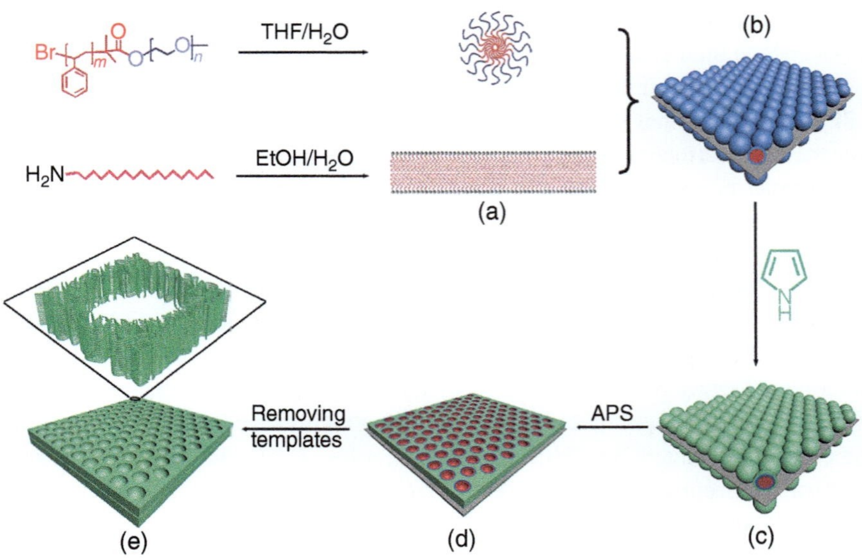

Figure 6.13 Schematic illustration of the fabrication of 2D mesoporous PPy nanosheets. Source: Liu et al. [67]. Reproduced with permission, 2016, Wiley-VCH.

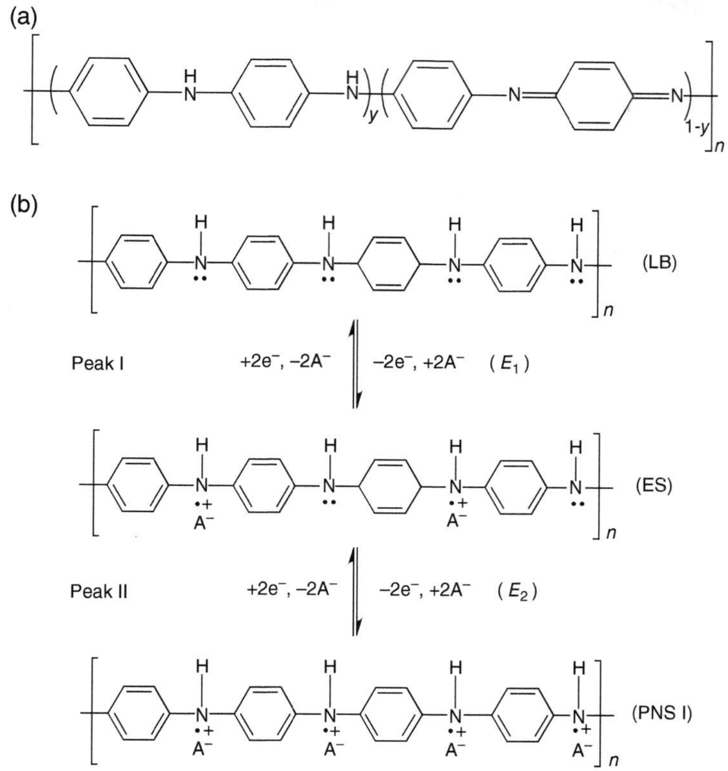

Figure 6.14 (a) The structure of the PANI. (b) Schematic illustration of the PANI doping mechanism.

more sodium ions. The mPPy nanosheets show excellent cycling stability, maintaining the discharge capacity of 83 mAh g^{-1} at 300 mA g^{-1} after 120 cycles. Zhou et al. explored the electrochemical properties of PPy doped with diphenylamine sulfonate, which can give a invertible specific capacity of 115 mAh g^{-1}, as well as better cycling stability and rate performance [68].

The research on the modification of PANI CP is also underway. Although PANI was discovered long ago, its structural model was not fully resolved until 1987 by MacDiarmid et al. [69]. As shown in Figure 6.14a, when $y = 0$, it is a fully oxidized structure; when $y = 0.5$, it is an intrinsic state structure; when $y = 1$, it is a fully reduced structure. The intrinsic PANI is basically nonconductive, but its conductivity can be increased by 12 orders of magnitude after doping. Generally, PANI can undergo a transition between conductor and insulator by proton acid doping and ammonia dedoping. The synthesis method of PANI is similar to that of PPy, but the polymerization reaction is usually carried out in an acidic aqueous solution to increase the length of the conjugated chain segment and improve the stability of the polymer. Compared with other CPs, PANI has a high redox potential, the highest degree of doping and the stability of the doped state in the air.

The charging and discharging mechanism of the organic battery with PANI as the positive electrode is similar to that of polyethylene block. As shown in Figure 6.14b, when it is charged, PANI is positively charged by the oxidized polymer chain, and the anions in the electrolyte are doped into the polymer to balance the charge. During discharge, PANI is reduced by the polymer chain to obtain electrons and release anions. According to the specific capacity calculation formula, each aniline unit can gain or lose one electron, so the theoretical capacity of the fully oxidized PANI is calculated to be 296 mAh g^{-1}. However, the fully oxidized state is unstable and the polymer main chain is easily broken by oxidation. Therefore, when PANI is used as a battery positive electrode, it usually only releases about half of its theoretical capacity.

There have been a lot of reports on PANI as an electrode material in the literatures. For example, Manuel et al. prepared PANI nanofibers using interfacial polymerization method [70]. As the positive electrode of SIBs, the discharge capacity can reach 112 mAh g^{-1}, but the cycling performance is not so good. After 100 cycles, the specific capacity decreases by about 20%. Poor cycling performance is a common problem towards PANI.

Several improvements have been reported in the literatures: (i) pre-doping or synthesizing self-doped PANI; (ii) replacing large anions with the movement of small positive ions; (iii) adding conductive carbon to improve the utilization of electrode materials; and (iv) adding appropriate dopant to improve the morphology of PANI (e.g. by adding and orienting acid-containing naphthalene ring to obtain fibrous PANI with better performance). These methods can improve the electrochemical performances of PANI to a certain extent. Shen et al. prepared sodium-rich poly(sodium dibenzenesulfonic acid)/PANI polymers with a invertible capacity of 100 mAh g^{-1} [71]. Zhou et al. designed a new type of cathode material for SIBs, namely PANI-*co*-aminobenzene sodium sulfonate [72]. This cathode provides a high inversible capacity of 133 mAh g^{-1}. Moreover, it has great cycling stability that after 200 cycles, the capacity can still remain of 96.8%. The structure and preparation method of PANI are also important factors affecting its electrochemical performance. For example, nano-structured polystyrene/PANI/graphene electrodes and aniline/*o*-nitroaniline copolymer electrodes show increased capacity compared to pure PANI electrodes [5e]. Coupling PANI with carbon materials is also an important strategy to improve the electrochemical performance. Li et al. synthesized PANI-graphene oxide (GO) polymer by grafting (5-amino-1,4-dihydrobenzo[d]-1, 2-dithiadiene)/aniline copolymer [P(DTAN-*co*-AN)] on the GO surface [73]. The fabrication theory of the polymer is based on the affinity of $-NH_2$ group and –NCO group to GO, which greatly improves the cycling stability of P(DTAN-*co*-AN).

As a derivative of PANI, polytriphenylamine (PTPAn) has the characteristics of large energy density and high power density, mainly because its structure has a polyparaphenylene main chain structure with good conductivity and a nitrogen radical-active center. Moreover, PTPAn-type CPs show highly reversible redox behavior that makes them the material of choice for SIBs. The typical PTPAn cathode material can be charged and discharged in the voltage range of 3.0–4.2 V, and the discharge voltage platform is at 3.8 V. The charging and discharging mechanism of PTPAn is similar to that of PANI. Deng et al. synthesized a PTPAn compound and assembled an all-organic battery by coupling a polysulfide AQ negative electrode

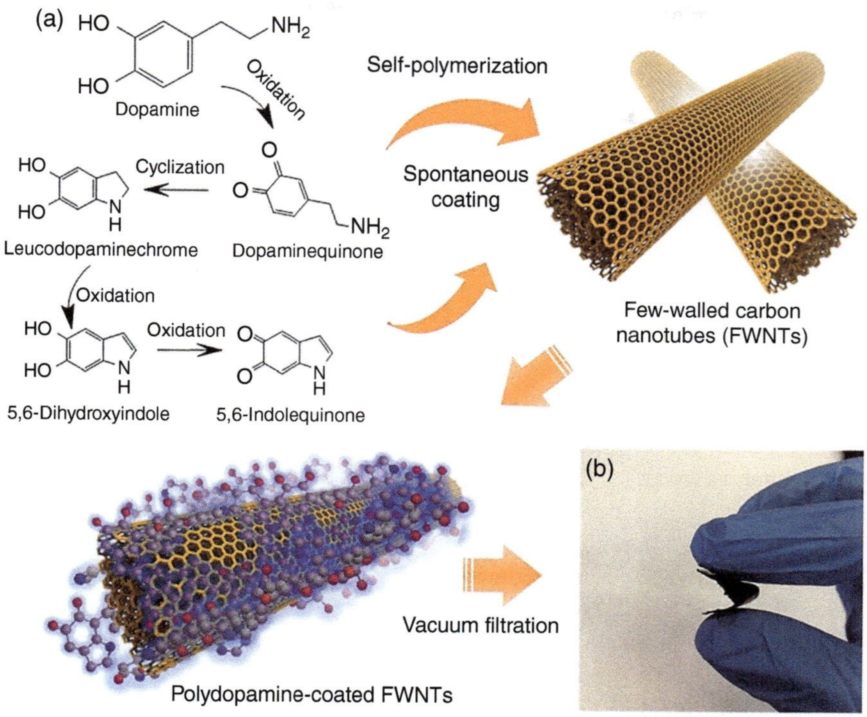

Figure 6.15 (a) Oxidative self-polymerization reactions of dopamine and a continuous coating process of polydopamine on the surface of carbon nanotubes, and (b) digital image of a flexible polydopamine carbon nanotube hybrid film. Source: Liu et al. [74]/with permisssion from Royal Society of Chemistry.

[5b]. The full battery has an energy density of 92 Wh kg^{-1} with a voltage range of 1.8 V. At the same time, after 500 cycles, the capacity retention rate is 85%, which proves the viability of all organic SIBs.

Recently, polydopamine (PDA) has also been extensively studied as a CP. Interestingly, the carbonyl group in the PDA molecular structure exhibits n-type reaction, and the secondary amine (RNH-R) has p-doped/dedoped properties. The PDA prepared by Zhang and coworkers has a good viscosity, and the electrode can be prepared without addition of binder [6a]. This cathode can maintain a capacity of 500 mAh g^{-1} after 1024 charge and discharge cycles. Liu et al. proved that dopamine can self-polymerize on the surface of CNTs in an alkaline solution, and then a self-supporting flexible composite membrane was prepared by vacuum filtration (Figure 6.15a) [74]. The composite film can be directly cut into electrodes without current collectors, conductive agents, and adhesives, avoiding complicated preparation processes (Figure 6.15b). The results showed that the flexible electrode exhibits a remarkable capacity of 213 mAh g^{-1}, outstanding rate performance, and good cycling stability.

The oxidation potential of thiophene is high (3.0–4.2 V) and the solubility of the polymer is poor, so the polymerization reaction is generally carried out in a polar

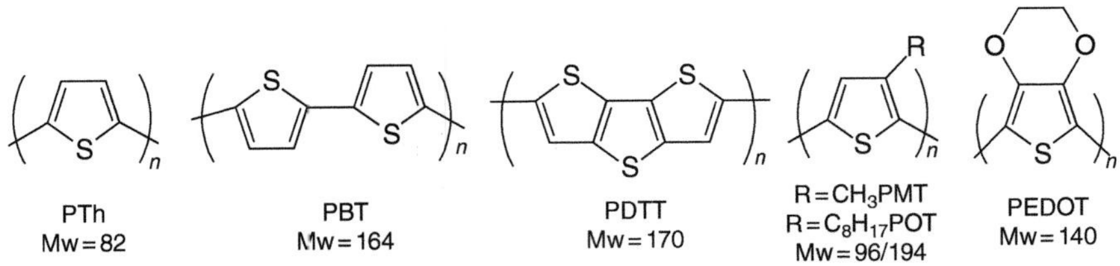

Figure 6.16 Several derivatives of polythiophene.

solvent (chloroform, acetonitrile, carbonate, etc.). Like poly(*p*-phenylene), polythiophene has a strong inter-chain interaction force, which is insoluble in any organic solvents and difficult to process. Usually long-chain groups are used to replace hydrogen at the position to reduce the force between molecular chains and improve their processing properties. Figure 6.16 lists the structures of several representative polythiophene derivatives. The performance of polythiophene as a secondary battery cathode material is not satisfactory. The highest doping degree reported in the literature is 0.24–0.3, and the corresponding specific capacity is about 100 mAh g^{-1} [75]. The capacity of electrochemical polymerization poly(thiophene) (PTh) in most literatures is about 33 mAh g^{-1}. Higher degree of doping can be achieved by polybutylene terephthalate (PBT) and polymethoxythiophene (PMT), and their specific capacities will be slightly increased [76]. Within a certain temperature range, the performance of CPs improves with an increase of temperature [77].

6.3.2 Organic Radical Compounds

Traditional CPs generally have a low degree of electrochemical doping, and when used as secondary battery electrode materials, the capacity is often far lower than their theoretical capacity and decay faster. In recent years, free radical polymers have come to the fore and exhibited excellent electrochemical performance. A free radical polymer is a kind of polymer with free radical groups attached to the main chain or side chain [78]. Organic radical polymers have the characteristics of rapid kinetics and doping reaction with sodium ions under high pressure. From a kinetic point of view, free radical redox reactions are the fastest electron transfer reactions, and stable free radical polymers have reversible oxidation and reduction reactions. The oxidation process of free radical compounds has little structural change and exhibits fast kinetics. Moreover, the redox process does not contain the rupture and constitution of chemical bonds, and the electron rearrangement is also almost nonexistent which is suitable for using as a cathode material for high power density SIBs. Free radical polymers not only can perform p-type doping and dedoping reactions with anions but also can have n-type doping and dedoping reactions with sodium ions to achieve the insertion and de-intercalation of sodium ions at a relatively low voltage. The electrochemical redox reaction mechanism is shown in Figure 6.17a.

The redox reaction process of radical polymer is the essence of the reaction mechanism of radical polymer battery. As shown in Figure 6.17a, nitroxide radicals can undergo reversible oxidation reactions to form free radical cations, or they

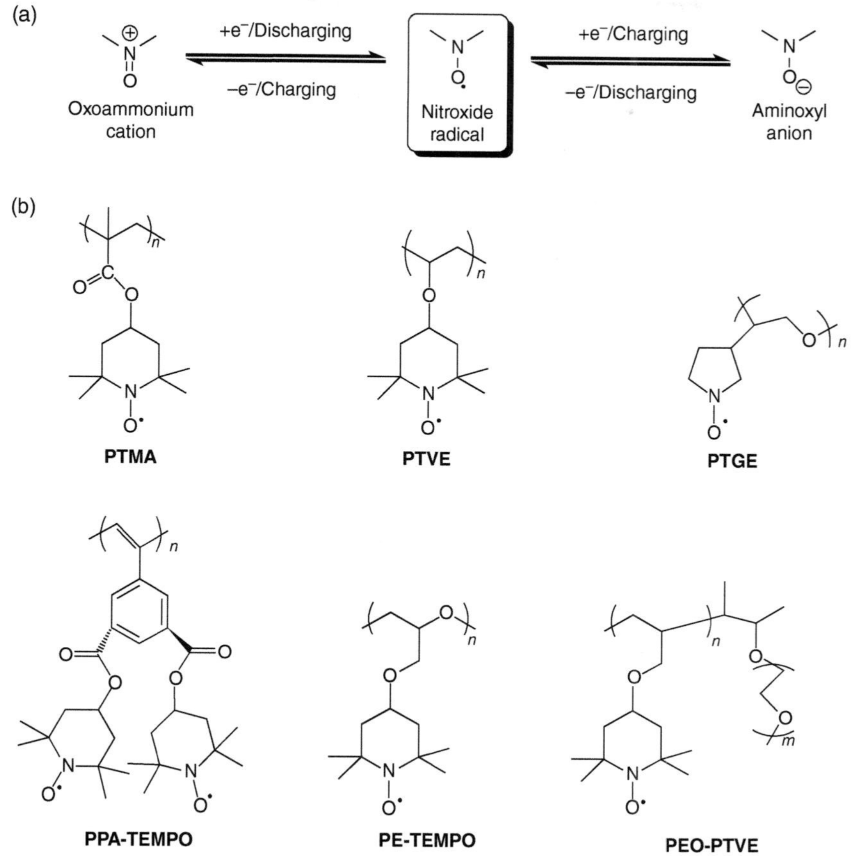

Figure 6.17 (a) Mechanisms for reversible redox reaction of organic radical compounds (nitroxide radicals). (b) Structures of nitroxide radical compounds commonly used as cathode materials.

can undergo reversible reduction reactions to form free radical anions. The p-type radical polymer generally performs the oxidation reaction part, and the n-type radical polymer generally takes part in the reduction reaction. Generally, the positive electrode material of an ideal radical polymer battery requires a higher potential p-type free radical polymer, while the negative electrode material needs a lower potential n-type free radical polymer.

During the charging process, free radicals undergo an oxidation reaction on the negative electrode to generate free radical cations, and at the same time, free radicals undergo a reduction reaction on the positive electrode to generate free radical anions. The discharge process is the opposite. The free radical polymer battery does not break and form chemical bonds during the charge and discharge processes but realizes the transfer of electrons through the gains and losses of the unpaired electrons of the free radicals to complete the conversion between electrical energy and chemical energy.

The energy density of radical polymer batteries is determined by two factors: voltage and specific capacity. The voltage plateau of the radical polymer battery is closely related to the redox reaction type of the radical polymer. The free radical polymers can undergo p-type redox reactions between free radicals and cations, or n-type redox reactions between free radicals and anions. Among them, the reaction voltage of the p-type redox reaction is higher than that of the n-type reaction. Moreover, the theoretical specific capacity obtained by the p-type reaction is usually higher than that of the n-type reaction. Under the same situation, the lower molecular weight of the structural unit of the radical polymer, the higher theoretical specific capacity of the radical polymer is achieved. At present, p-type radical polymer is considered to be one of the most prospective new organic sodium-ion battery electrode materials.

Common free radical polymers are shown in Figure 6.17b, and their most prominent advantages are illustrated as below. (i) The free radical polymer has a high degree of doping and a high specific capacity. Among the electrode materials that have been reported so far, poly(4-vinyloxy-2,2,6,6-tetramethylpiperidine-*N*-oxyl) (PTVE) has the remarkable specific capacity of 135 mAh g^{-1} [79]. (ii) Nitrogen–oxygen free radical has fast electron transfer rate, so it has high-rate charge and discharge performance. Different from other organic reactions, since the electrons of the electrochemically active sites of the organic radical polymer are moderately delocalized, only the shell layer electrons take part in the electrode reaction without the breakage and formation of chemical bonds. It thus has an extremely fast charge transfer rate of 10^{-1} cm s^{-1}. (iii) The superior chemical stability of the structure displays a cycle life of over 1000 times. The main chain of organic radical polymer basically has structures such as polystyrene, polymethyl diethyl ester, cross-linked polynorbornene, polyvinyl alcohol, and polyethylene oxide. The electrochemical window of these main chain structures is wide, and the structure is stable in common organic electrolytes or water systems. The reaction process does not involve the rupture and formation of chemical bonds. Hence, organic radical polymer electrode materials mostly have prominent cycling performance.

In the early twenty-first century, Nakahara et al. synthesized a stable nitroxyl poly(2,2,6,6tetramethylpiperidinyloxy-4-vinylmethacrylate) (PTMA) for the first time, and it was suitable to be used as cathode for rechargeable batteries [80]. In the past few decades, although organic radical polymers have been extensively researched as electrodes in LIBs, there have been few studies on their application in SIBs. The first free radical polymer used in SIBs was the organic radical polymer of poly[Norbornene-2,3-*endo*, *exo*-(COO-4-tempo)$_2$] prepared by Dai et al. in 2010. It has a discharge capacity of 75 mAh g^{-1} at 50 mA g^{-1}, with a discharge capacity retention rate of 64.5% after 50 cycles [5c].

Organic radical polymer batteries certainly have many advantages, including their high power density and stable cycle performance, diversified structure, designability, sustainable development, and the ability to prepare flexible batteries. However, there are still many challenges in the research of organic radical batteries, which are mainly related to their poor conductivity and low specific capacity. On one hand, since most of the organic radical polymer materials themselves are electrically insulating materials, a large number of conductive agents must be added to improve their conductivity. Relevant researches have demonstrated that the mass fraction of the conductive agent in the electrode material can be as high as 60–80 wt% of the entire electrode, and this will lead to a decrease in the mass fraction of the active material of the electrode material, thereby affecting the energy density of the electrode material. On the other hand, compared with the current commercialized inorganic cathode materials, the specific capacity of organic radical polymer materials is relatively low.

The current methods for improving the performance of organic radical batteries mainly include reducing the molecular weight of organic radical polymer structural units or increasing their redox-active sites to increase their theoretical specific capacity, adding electrode materials with high-conductivity and high-ratio new-type nano-conductive materials with high surface area, such as CNTs, graphene, etc., through in situ compounding and other chemical methods.

Recently, Kim et al. have contributed to improve the electrochemical capacity of PTMA [81]. The authors used the ultrasonic perfusion technology to compress PTMA into CNTs which have a diameter of 100–300 nm, as shown in Figure 6.18. This method effectively solves the self-discharge phenomenon of PTMA, and the open-circuit voltage is almost unvaried in last 20 days. They then carried out other electrochemical experiments at the voltage between 1.5 and 4.0 V. During the initial charging process, only the p-type doping reaction is occurred, and the capacity is

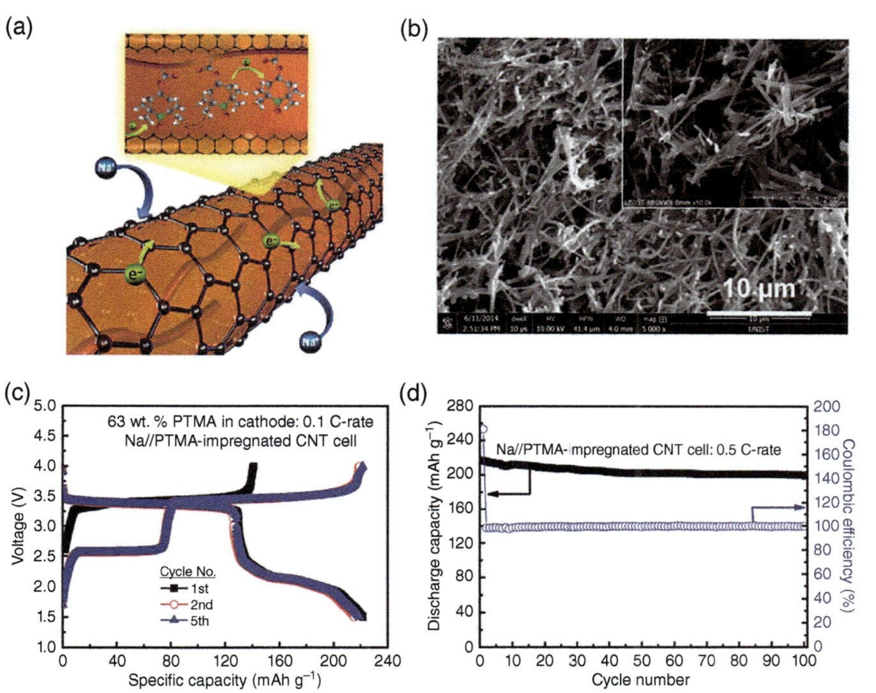

Figure 6.18 (a) Illustration of the composite of PTMA and CNT, (b) SEM images of PTMA-impregnated CNT electrodes, (c) charge–discharge curve, and (d) cycling performance of the composite electrode. Source: Kim et al. [81]/with permission from Royal Society of Chemistry.

only 130 mAh g^{-1}. However, the p-type dedoping reaction and the n-type doping reaction (sodium ion insertion) can proceed reversibly at the same time during the subsequent discharge process, and the capacity exceeds 220 mAh g^{-1}. After 100 cycles, the capacity retention rate is 93%, and the rate performance is also significantly enhanced. Under a high current density of 5 C, the capacity can still maintain at 190 mAh g^{-1}.

The above researches show that nitroxide radicals have been widely used as cathodes for rechargeable energy storage batteries due to their excellent electrical conductivity, cycle ability, and rate performance. Polymers containing nitroxyl molecules have been used in nontoxic organic batteries due to their fast electrochemical reaction kinetics. Besides, nitroxide radical compounds are renewable and environmentally benign, along with their high flexibility in structural design. All in all, organic radical battery is a new type of green rechargeable battery.

6.3.3 Microporous Polymers

In recent years, new crystalline porous materials have been attracted great interest. Different linkers and nodes are combined to form millions of structures, e.g., metal–organic frameworks (MOFs), metal–organic macrocycles (MOMs), covalent organic frameworks (COFs), etc., with different morphologies, pore shapes, and chemical characteristics [82]. According to their structural characteristics, organic porous materials can be divided into covalent organic networks (COFs), polymers of intrinsic microporosity (PIMs), covalent triazine-based frameworks (CTFs), porous aromatic frameworks (PAFs), hyper-cross-linked polymers (HCPs), and conjugated microporous polymers (CMPs).

For the past few years, microporous polymers have gradually come into people's field of vision. Their synthesis methods are simple and their production processes are safe. Due to their micropores and large specific surface area, microporous polymers are expected to be used as a catalysts, gas adsorption and storage, chemical sensing, and other applications. The size and shape of the micropores can be controlled at the molecular level through the design of the functional groups. CMPs were first prepared by Cooper's research group in 2007, which has a rigid conjugated structure formed by the regular and alternating connection of aromatic structural units. The specific surface area and pore properties of this type of polymer can be adjusted by changing the structure of the unit.

CMP is a promising cathode material for SIBs. The large surface area and microporous structure of this type of material can facilitate the rapid transmission of sodium ions. The structure of the polymer relieves its dissolution problem in the organic electrolyte. At the same time, the large number of redox-active sites afford an appreciable theoretical capacity.

In the 2010s, Sakaushi et al. manufactured a bipolar porous organic polymer battery with a capacity of 230 mAh g^{-1} [83]. As shown in Figure 6.19a, the structure of the active material is composed of benzene and triazine rings. Its special porous structure is conducive to electrolyte infiltration, and its large molecular weight inhibits its dissolution in organic electrolytes. These characteristics are conducive to

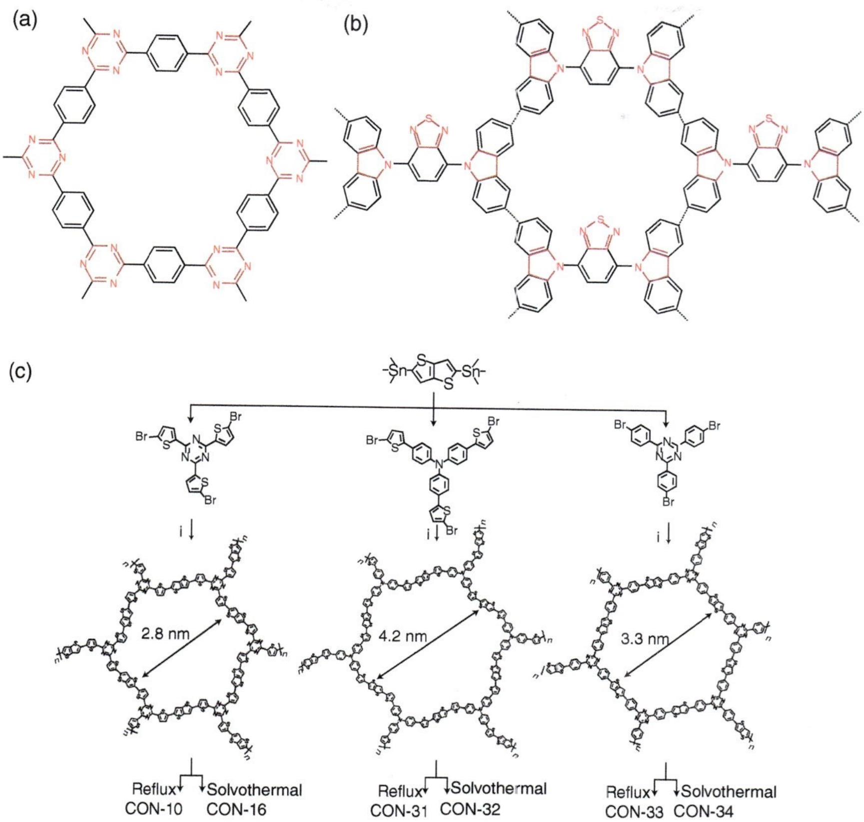

Figure 6.19 (a, b) Structures of conjugated microporous polymers. (c) General synthesis procedure and molecular structures of CONs along with their precursor monomers. Source: Kim et al. [84]. Reproduced with permission. 2018, American Chemical Society.

the stability of its cycling performance. However, most of the microporous polymers have low electrical conductivity. Using S heterocycles, the conductivity of CMPs can increase to $6.83 \times 10^{-4}\,S\,cm^{-1}$ [85]. On the other hand, in order to enhance the ionic conductivity of the CMP, Zhang et al. synthesized the polymer 4,7-dicarbazole [2,1,3]-benzothiadiazole with ferric chloride as an oxidative coupling agent (polymer 4,7-dicarbazyl-[2,1,3]-benzothiadiazole [PDCzBT]) (Figure 6.19b) [85]. The prepared PDCzBT has an ultra-high specific surface area and a large number of redox-active units. These structural advantages can provide plentiful active sites for ion storage and have a uniform microporous structure, which are favorable for the rapid transmission of electrons and ions. When used as a cathode in a SIB, the PDCzBT has high specific capacity and excellent cycling stability. After 100 cycles, an inversible capacity of $145\,mAh\,g^{-1}$ is obtained at $20\,mA\,g^{-1}$.

Self-polymerizing microporous polymers (PIMs) are special organic microporous polymers with both rigid and twisted structures. During the polymerization process, the polymer chains connected by rigid structural units cannot be bent and twisted freely, which avoids the accumulation of molecular chains and thus produces a large number of micropores. At the same time, the molecular chains of PIMs are difficult to rotate and bend so that PIMs can maintain a stable structure for a long time. Mckeown et al. prepared PIMs for the first time using bis-indene in 2002 [86]. The two types of PIMs contain phthalocyanine and porphyrin structures, respectively. The rigidity of the porphyrin-conjugated macrocycle prevents molecular chain accumulation, and the surface area can reach up to 950 m g^{-1} [2]. PIMs not only produce stable and abundant pore structures due to the twisted accumulation of molecular chains but also have a relatively simple preparation process, so they have promising potential application in secondary battery electrode materials.

COFs are a kind of organic microporous polymer with long-range order and crystalline structure. The monomers can exhibit a periodic and orderly arrangement structure through self-polymerization and have relatively uniform micropores, so the pore structure can easily controlled. In the preparation of COFs, polycondensation reactions are usually used. Such reaction processes are controlled by thermodynamics, and the resulting network structure is two-dimensional or three-dimensional. Yaghi and coworkers firstly used self-condensation reaction to prepare COF-1 with high specific surface area in 2005 [87]. The study found that this kind of thermodynamically controlled reaction can get a very stable polymer structure. COFs have the peculiarities of crystal structure, uniform pore size, low density, and high specific surface area. They have promising applications in the fields of adsorption, catalysis, semiconductor, and energy storage. At the same time, the large specific surface area and microporous structure prefer the speedy transference of sodium ions; the polymer structure avoids its solubility; and a great number of redox-active centers can provide a larger theoretical capacity. Hence, COF can also be employed as a new OCM for batteries. Kim et al. prepared a series of network polymers (covalent organic nanosheets, CONs) by cross-coupling under conventional reflux and solvothermal conditions, as shown in Figure 6.19c, and studied the relationship between their structure and energy storage performance [84]. It shows that the specific surface area and self-assembly morphology of CON can be effectively controlled by carefully selecting the synthesis route and monomer combination. By enhancing the planarity of the polymer backbone, the charge-carrier conductivity of the polymer can be improved, thereby increasing the sodium ion storage capacity. The results show that the CON-16-based electrode exhibits the best cycling and rate performance and can maintain a reversible discharge capacity of about 250 mAh g^{-1} after 30 cycles at 100 mA g^{-1}.

6.4 C=N Reaction

6.4.1 Schiff Base Organic Compounds

Schiff base organic compounds mainly refer to a class of organic compounds containing imine or methylimine groups (—RC=N—). Schiff base is usually formed by condensation of amine and active carbonyl group. Through the simple polycondensation reaction of diamine and dialdehyde or diketone, Schiff bases and

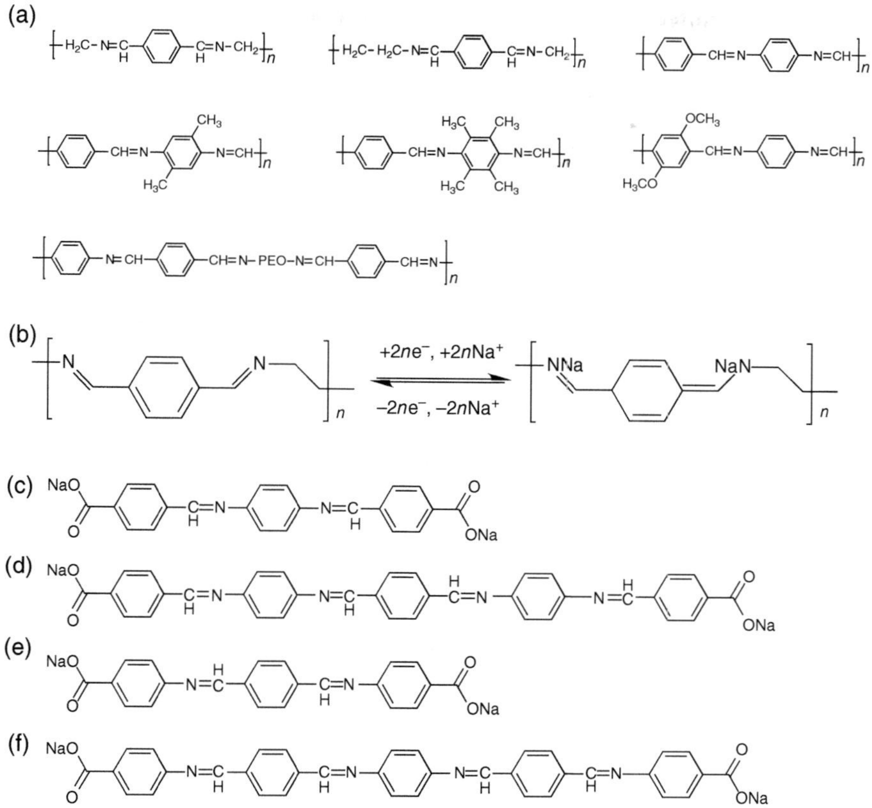

Figure 6.20 (a) Structures of a series of Schiff base polymers. (b) The sodium storage mechanism of Schiff base polymer. Source: Zhao et al. [88]. Licensed under CC BY 4.0. (c–f) Structures of Schiff base organic compounds.

polymers with various structures can be synthesized. The multiplicity of Schiff base polymers depends on the structure of diamine and dialdehyde or diketone precursors. Figure 6.20a shows the structure of a series of Schiff base polymers consisting of different diamines and dialdehydes or diketones. The Schiff base group can coordinate with the metal atom (or ion) through the nitrogen (N) atom on the carbon–nitrogen double bond and the adjacent oxygen (O), sulfur (S), and phosphorus (P) atoms with a lone pair of electrons. Schiff base compounds generally have a low redox potential. The reported anode materials for SIBs are mainly aromatic Schiff base compounds. They have repetitive conjugated units —N=CH—Ar—CH—=N— which are also electrochemically active centers. However, the isomer —CH=N—Ar—N=CH— is not a conjugated structure and has no electrochemical activity. The sodium storage mechanism of Schiff base polymer is shown in Figure 6.20b [88].

Introducing carboxylate end groups can effectively increase the capacity and operating voltage of Schiff base organic compound electrodes. DFT calculations confirmed that the active Hückel coplanar groups (—OOC—Ar—C=N— and —N=C—Ar—C=N—) provide storage locations for sodium ions. However, the isoelectronic groups (—OOC—Ar—N=C— and —C=N—Ar—N=C—) can not act as active centers because of their π–π interaction or loss of flatness, but are able to stabilize the discharge–charge process and weaken N–N repulsion. Similarly, the inactive aromatic group also acts as a crucial role for the stabilization of the Schiff base structure. Less aromatic groups containing compounds show lacking stability and capacity. The insertion of conjugated and planar structures is an effective method to enhance the electrochemical activity of oligomeric Schiff base materials.

Lopez-Herraiz et al. prepared four kinds of oligomeric Schiff base compounds with carboxylate groups [89], as shown in Figure 6.20c–f, and compounds with —OOC—Ar—CH=N— terminal groups have a inversible specific capacity of 250 mAh g^{-1}. In comparison, compounds with nonconjugated structure of —OOC—A—N=CH— end groups only have specific capacities of 120–170 mAh g^{-1}. This indicates that the conjugated structure plays an important part in the electrochemical activity of Schiff base organic materials.

Sheng et al. used ethanol as solvent, salicylaldehyde, ethylenediamine, and cobalt acetate as raw materials to synthesize small-molecule salicylaldehyde-based Schiff base cobalt complex (SSB(Co)(II)) and coupled it with rGO to obtain SSB(Co)(II)/rGO complex [90]. The composite exhibits excellent electrochemical performance. At 0.2 C, the specific discharge capacity can reach 183 mAh g^{-1}, and the specific discharge capacity remains at 136 mAh g^{-1} after 100 cycles.

The advantage of Schiff base polymer organic electrodes lies in their low solubility in most organic solvents; however, this makes them difficult to process. By using a suitable linking agent between the azomethine units to modify the polymer chain, the equilibrium between solubility and processability can be controlled. In addition, the electrical conductivity of Schiff base materials can be enhanced to semiconductor levels through doping with iodine. This crucial property makes Schiff base polymer a great promising electrode material for SIBs.

6.4.2 Pteridine Derivatives

Pteridine derivatives are an important class of heterocyclic compounds that exist widely in organisms. Pteridine is also called 1,3,5,8-tetraazanaphthalene, which is formed by the fusion of pyrimidine heterocycle and pyrazine heterocycle [91]. Pteridine derivatives with biological redox centers have great application potential in SIB electrode materials.

Hong et al. manufactured a series of pteridine derivatives as OCMs for SIBs [6c], as shown in Figure 6.21. The theoretical capacities are 270 mAh g^{-1} for lumichrome (LC), 250 mAh g^{-1} for alloxazine (ALX), and 327 mAh g^{-1} for lumazine (LMZ). However, the volume expansion of this type of electrode materials during charging and discharging causes the rupture of electrode. At 10 mA g^{-1}, the capacity is only 138 mAh g^{-1} (LC), 168 mAh g^{-1} (ALX), and 70 mAh g^{-1} (LMZ), respectively. In response to this problem, Kang and coworkers combined such compounds with CNTs through strong π–π interactions to immobilize molecules [6c]. Their specific capacities eventually increase to 255 mAh g^{-1} (LC), 225 mAh g^{-1} (ALX), and 220 mAh g^{-1} (LMZ).

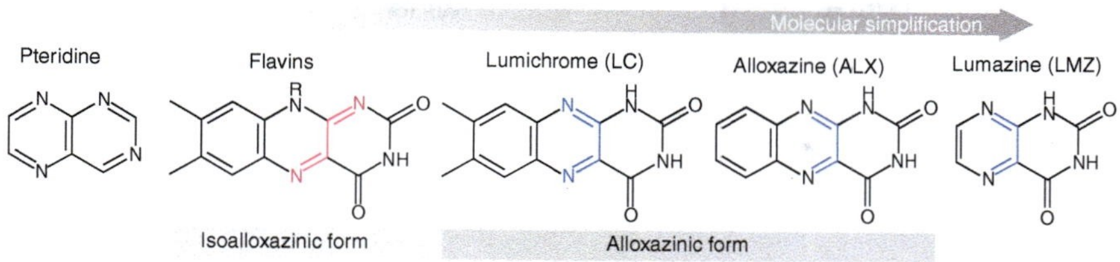

Figure 6.21 Structures of pteridine derivatives. Source: Hong et al. [6c]. Reproduced with permission, 2014, Springer Nature.

In addition to the C=N double bond, tetracyanoethylene (TCNE) and tetracyanoaminodimethane (TCNQ) with C≡N bonds have already been studied as the cathodes for SIBs. In the 1990s, Ratnakumar et al. prepared a Na/β″-alumina/TCNE half-cell with an operating temperature of 230 °C [92]. Based on the Fourier Transform infrared spectroscopy (FTIR) spectrum analysis, the redox center of TCNE is composed of C≡N bonds and C=C bonds. Newly, Manzhos et al. employed the DFT method to study sodium ion storage in TCNE and its composites. They demonstrated that up to four sodium ions can be stored in the TCNE molecule.

6.5 Concluding Remarks

The organic compounds acquired from natural biomass are green environmental protection and price-moderate materials for SIBs. Although the application of OCMs in rechargeable SIBs is still in its infancy, the appealing features and design of such materials have made them a research focus for the past few years. Among the above-mentioned types of organic materials, simple small-molecule quinone compounds have large theoretical specific capacity, sodium carboxylate compounds have suitable working voltage, and polyimides have excellent cycling performance. Most of the organic free radical compounds and conductive polymers in the heterogeneous reaction have high power density and fast kinetics. Schiff base organic materials and pteridine derivatives are new types of compounds used as OCMs. After a deeper understanding on their reaction mechanisms, rational design of their structure can be further developed. Although rechargeable SIBs using organic materials have been studied, the problems such as high dissolvability of organic cathodes in organic liquid electrolytes, weak thermal stability, and lacking electron conductivity have hindered their large-scale practical application. Hence, finding a suitable electrolyte to avoid the dissolvability of electrode materials and designing a reasonable structure to enhance the conductivity are crucial for realizing its practical application as the next-generation electrode material for SIBs. The improvement of OCMs and future development focus can be expanded from the following four directions. (i) Polymerizing small-molecule organic electrodes to design and synthesize new polymers, which reduces solubility on the one hand and increases specific capacity on the other. (ii) Combining various OCMs with reasonable molecular design, such as forming a conjugated structure to improve ionic conductivity, synthesizing salts to reduce dissolution, and combining appropriate functional groups to increase or decrease operating voltage. (iii) Combining low-conductivity organic electrodes with carbon materials to increase the conductivity of the electrode materials for higher capacity. (iv) Improving other components of the battery, such as modification of electrolyte or separator. OCMs can be developed and prepared from renewable biomass which have abundant resources and low price, showing a broad development prospect. Considering the abundant sodium resources and a wide range of organic compounds available in the earth, OCMs have great prospects for large-scale electrochemical energy storage.

References

1 (a) Lee, S., Kwon, G., Ku, K. et al. (2018). *Adv. Mater.* 30: 1704682. (b) Shea, J.J. and Luo, C. (2020). *ACS Appl. Mater. Interfaces* 12: 5361. (c) Liang, Y., Tao, Z., and Chen, J. (2012). *Adv. Energy Mater.* 2: 742.

2 (a) Lin, Z., Xia, Q., Wang, W. et al. (2019). *InfoMat* 1: 376. (b) Zhao, Q., Lu, Y., and Chen, J. (2017). *Adv. Energy Mater.* 7: 1601792. (c) Chen, H., Armand, M., Demailly, G. et al. (2008). *ChemSusChem* 1: 348.

3 Lu, Y. and Chen, J. (2020). *Nat. Rev. Chem.* 4: 127.

4 Williams, D.L., Byrne, J.J., and Driscoll, J.S. (1969). *J. Electrochem. Soc.* 116: 2.

5 (a) Bhargav, A., Bell, M.E., Karty, J. et al. (2018). *ACS Appl. Mater. Interfaces* 10: 21084. (b) Deng, W., Liang, X., Wu, X. et al. (2013). *Sci. Rep.* 3: 2671. (c) Dai, Y., Zhang, Y., Gao, L. et al. (2010). *Electrochem. Solid-State Lett.* 13: A22. (d) Wang, S., Wang, L., Zhu, Z. et al. (2014). *Angew. Chem.* 53: 5892. (e) Zhao, R., Zhu, L., Cao, Y. et al. (2012). *Electrochem. Commun.* 21: 36. (f) Wang, C., Xu, Y., Fang, Y. et al. (2015). *J. Am. Chem. Soc.* 137: 3124. (g) Mukhopadhyay, A., Jiao, Y., Katahira, R. et al. (2017). *Nano Lett.* 17: 7897.

6 (a) Sun, T., Li, Z.-J., Wang, H.-G. et al. (2016). *Angew. Chem.* 128: 10820. (b) Janoschka, T., Martin, N., Martin, U. et al. (2015). *Nature* 527: 78. (c) Hong, J., Lee, M., Lee, B. et al. (2014). *Nat. Commun.* 5: 5335.

7 (a) Kai, H.A., Jawahar, N., Komba, T. et al. (2018). *ACS Appl. Mater. Interfaces* 10: 7982. (b) Renault, S., Oltean, V.A., Araujo, C.M. et al. (2016). *Chem. Mater.* 28: 1920. (c) Emanuelsson, R., Sjödin, M., Sterby, M., and Strømme, M. (2017). *J. Am. Chem. Soc.* 139: 4828.

8 (a) Zhang, Y., Wang, J., and Riduan, S.N. (2016). *J. Mater. Chem. A* 4: 14902. (b) Banda, H. (2015). *J. Mater. Chem. A* 3: 10453. (c) Dugas, R., Zhang, B., Rozier, P., and Tarascon, J.M. (2016). *J. Electrochem. Soc.* 163: A867.

9 (a) Song, Z., Qian, Y., Zhang, T. et al. (2015). *Adv. Sci.* 2: 1500124. (b) Zhao, L., Zhao, J., Hu, Y.-S. et al. (2012). *Adv. Energy Mater.* 2: 962. (c) Zhu, H., Yin, J., Zhao, X. et al. (2015). *Chem. Commun.* 51: 14708. (d) Wang, H.-G., Yuan, S., Ma, D.-L. et al. (2014). *Adv. Energy Mater.* 4: 1301651.

10 (a) Alt, H., Binder, H., Köhling, A., and Sandstede, G. (1972). *Electrochim. Acta* 17: 873. (b) Lu, Y., Hou, X., Miao, L. et al. (2019). *Angew. Chem. Int. Ed.* 58: 7020. (c) Wu, X., Jin, S., Zhang, Z. et al. (2015). *Sci. Adv.* 1: e1500330.

11 Xu, Y., Zhou, M., and Lei, Y. (2018). *Mater. Today* 21: 60.

12 Geng, J., Bonnet, J.-P., Renault, S. et al. (2010). *Energy Environ. Sci.* 3: 1929.

13 (a) Armand, M., Grugeon, S., Vezin, H. et al. (2009). *Nat. Mater.* 8: 120. (b) Song, Z., Zhan, H., and Zhou, Y. (2010). *Angew. Chem. Int. Ed.* 49: 8444.

14 Bostjan, G., Klemen, P., Romana, C.-K. et al. (2010). *Angew. Chem.* 122: 7380.

15 Wu, Y., Zeng, R., Nan, J. et al. (2017). *Adv. Energy Mater.* 7: 1700278.

16 Gall, T.L., Reiman, K.H., Grossel, M.C., and Owen, J.R. (2003). *J. Power Sources* 119: 316.

17 Guo, C., Zhang, K., Zhao, Q. et al. (2015). *Chem. Commun.* 51: 10244.

18 (a) Wan, W., Lee, H., Yu, X. et al. (2014). *RSC Adv.* 4: 19878. (b) Zhu, Z., Li, H., Liang, J. et al. (2015). *Chem. Commun.* 51: 1446.

19 Zhang, Y.-Y., Wang, W.-K., Wang, A.-B. et al. (2010). *Acta Phys. -Chim. Sin.* 26: 47.

20 Xiong, W., Huang, W., Zhang, M. et al. (2019). *Chem. Mater.* 31: 8069.

21 Yan, B., Wang, L., Huang, W. et al. (2019). *Chin. Phys. Lett. Front.* 6: 1977.

22 Zhao, L., Wang, W.-K., Wang, A.-B. et al. (2011). *J. Electrochem. Soc.* 158: A991.

23 Zhou, W., Zhang, X., Zhang, W. et al. (2020). *Org. Electron.* 82: 105702.

24 Kim, H., Kwon, J.E., Lee, B. et al. (2015). *Chem. Mater.* 27: 7258.

25 Wei, X., Xu, W., Huang, J. et al. (2015). *Angew. Chem. Int. Ed.* 54: 8684.

26 Renault, S., Mihali, V.A., Edström, K., and Brandell, D. (2014). *Electrochem. Commun.* 45: 52.

27 Wang, H., Hu, P., Yang, J. et al. (2015). *Adv. Mater.* 27: 2348.

28 Song, Z., Xu, T., Gordin, M.L. et al. (2012). *Nano Lett.* 12: 2205.

29 Zhou, S.M.-Q.X. and Wang, G.C. (2016). *Acta Phys. -Chim. Sin.* 32: 975.

30 Li, A., Feng, Z., Sun, Y. et al. (2017). *J. Power Sources* 343: 424.

31 Wu, X., Ma, J., Ma, Q. et al. (2015). *J. Mater. Chem. A* 3: 13193.

32 (a) Hanyu, Y., Ganbe, Y., and Honma, I. (2013). *J. Power Sources* 221: 186. (b) Senoh, H., Yao, M., Sakaebe, H. et al. (2011). *Electrochim. Acta* 56: 10145.

33 Wang, X., Shang, Z., Yang, A. et al. (2019). *Chem* 5: 364.

34 Casado, N., Hilder, M., Pozo-Gonzalo, C. et al. (2017). *ChemSusChem* 10: 1783.

35 Zhu, Z., Hong, M., Guo, D. et al. (2014). *J. Am. Chem. Soc.* 136: 16461.

36 Huang, W., Zhu, Z., Wang, L. et al. (2013). *Angew. Chem. Int. Ed.* 52: 9162.

37 (a) Zhao, Q., Guo, C., Lu, Y. et al. (2016). *Ind. Eng. Chem. Res.* 55: 5795. (b) Luo, C., Zhu, Y., Xu, Y. et al. (2014). *J. Power Sources* 250: 372. (c) Li, S., Guo, J., Ye, Z. et al. (2016). *ACS Appl. Mater. Interfaces* 8: 17233.

38 Wang, S., Wang, L., Zhu, Z. et al. (2014). *Angew. Chem. Int. Ed.* 53: 5892.

39 Lee, M., Hong, J., Lopez, J. et al. (2017). *Nat. Energy* 2: 861.

40 (a) Wang, H.-G., Yuan, S., Si, Z., and Zhang, X.-B. (2015). *Energy Environ. Sci.* 8: 3160. (b) Iordache, A., Delhorbe, V., Bardet, M. et al. (2016). *ACS Appl. Mater. Interfaces* 8: 22762. (c) Abouimrane, A., Weng, W., Eltayeb, H. et al. (2012). *Energy Environ. Sci.* 5: 9632. (d) Huang, X., Liu, W., Hooker, J.M., and Groves,

J.T. (2015). *Angew. Chem. Int. Ed.* 54: 5241. (e) Zhao, L., Ren, Y., Li, M., and Sakurai, K. (2012). *J. Netw. Comput. Appl.* 35: 962.
41 (a) Park, Y., Shin, D.-S., Woo, S.H. et al. (2012). *Adv. Mater.* 24: 3562. (b) Zhang, Y., Niu, Y., Wang, M.-Q. et al. (2016). *Chem. Commun.* 52: 9969.
42 (a) Xue, J., Fan, C., Deng, Q. et al. (2016). *Electrochim. Acta* 219: 418. (b) Luo, W., Allen, M., Raju, V., and Ji, X. (2014). *Adv. Energy Mater.* 4: 1400554.
43 (a) Qin, H., Song, Z.P., Zhan, H., and Zhou, Y.H. (2014). *J. Power Sources* 249: 367. (b) Manuel, J., Zhao, X., Cho, K.-K. et al. (2018). *ACS Sustain. Chem. Eng.* 6: 8159. (c) Yuan, C., Wu, Q., Li, Q. et al. (2018). *ACS Sustain. Chem. Eng.* 6: 8392.
44 Banda, H., Damien, D., Nagarajan, K. et al. (2017). *Adv. Energy Mater.* 7: 1701316.
45 Liang, Y., Peng, Z., and Chen, J. (2013). *Chem. Sci.* 4: 1330.
46 (a) Xu, Y., Chen, J., Zhu, C. et al. (2018). *Sci. China Mater.* 61: 707. (b) Yanliang, L., Peng, Z., Siqi, Y. et al. (2013). *Adv. Energy Mater.* 3: 600. (c) Vadehra, G.S., Maloney, R.P., Garcia-Garibay, M.A., and Dunn, B. (2014). *Chem. Mater.* 26: 7151. (d) Iordache, A., Maurel, V., Mouesca, J.M. et al. (2014). *J. Power Sources* 267: 553. (e) Hernández-Burgos, K., Burkhardt, S.E., Rodríguez-Calero, G.G. et al. (2014). *J. Phys. Chem. C* 118: 6046.
47 Mihali, V.A., Renault, S., Nyholm, L., and Brandell, D. (2014). *RSC Adv.* 4: 38004.
48 Deng, W., Shen, Y., Qian, J. et al. (2015). *ACS Appl. Mater. Interfaces* 7: 21095.
49 Kim, D.J., Jung, Y.H., Bharathi, K.K. et al. (2014). *Adv. Energy Mater.* 4: 1400133.
50 Jouhara, A., Dupré, N., Guyomard, D. et al. (2020). *J. Electrochem. Soc.* 167: 070540.
51 Xu, F., Xia, J., and Shi, W. (2015). *Electrochem. Commun.* 60: 117.
52 Wu, H., Wang, K., Meng, Y. et al. (2013). *J. Mater. Chem. A* 1: 6366.
53 (a) Dong, X., Chen, L., Liu, J. et al. (2016). *Sci. Adv.*: 2. (b) Gu, T. (2016). *RSC Adv.* 2: 1501038. (c) Xu, F., Xia, J., Shi, W., and Cao, S.-A. (2016). *Mater. Chem. Phys.* 169: 192.
54 Li, Z., Zhou, J., Xu, R. et al. (2016). *Chem. Eng. J.* 287: 516.
55 Chen, L., Li, W., Wang, Y. et al. (2014). *RSC Adv.* 4: 25369.
56 Shirakawa, E.J.L.H., Macdiarmid, A.G., Chiang, C.K., and Heeger, A.J. (1977). *J. Chem. Soc. Chem. Commun.* 16: 578.
57 Novák, P., Müller, K., Santhanam, K.S.V., and Haas, O. (1997). *Chem. Rev.* 28: 207.
58 (a) Liu, Y.-C. and Chung, K.-C. (2003). *Synth. Met.* 139: 277. (b) Udum, Y.A., Pekmez, K., and Yıldız, A. (2004). *Synth. Met.* 142: 7. (c) Ahlskog, M., Reghu, M., Noguchi, T., and Ohnishi, T. (1997). *Synth. Met.* 89: 11. (d) Yang, B.D., Yoon, K.H., and Chung, K.W. (2004). *Synth. Met.* 143: 25. (e) Wei, X. and Epstein, A.J. (1995). *Synth. Met.* 74: 123.
59 Shacklette, L.W., Murthy, N.S., and Baughman, R.H. (1985). *Mol. Cryst. Liq. Cryst.* 121: 201.
60 Shirakawa, H., Louis, E.J., MacDiarmid, A.G. et al. (1977). *J. Chem. Soc. Chem. Commun.*: 578.
61 Nigrey, J. and Paul, J. (1981). *Electrochem. Soc.* 128: 1651.

62 (a) Chen, X., Liu, L., Yan, Z. et al. (2015). *RSC Adv.* 6: 2345. (b) Su, D., Zhang, J., Dou, S., and Wang, G. (2015). *Chem. Commun.* 51: 16092.

63 Machida, S., Miyata, S., and Techagumpuch, A. (1989). *Synth. Met.* 31: 311.

64 Zhou, M., Zhu, L., Cao, Y. et al. (2012). *RSC Adv.* 2: 5495.

65 Zhu, L., Shen, Y., Sun, M. et al. (2013). *Chem. Commun.* 49: 11370.

66 Chen, J., Liu, Y., Li, W. et al. (2015). *J. Mater. Sci.* 50: 5466.

67 Liu, S., Wang, F., Dong, R. et al. (2016). *Adv. Mater.* 28: 8365.

68 Zhou, M., Xiong, Y., Cao, Y. et al. (2013). *J. Polym. Sci. B* 51: 114.

69 MacDiarmid, A.G., Chiang, J.C., and Richter, A.F. (1987). *Synth. Met.* 18: 285.

70 Manuel, J., Ahn, J.H., Kim, D.S. et al. (2012). *J. Nanosci. Nanotechnol.* 12: 3534.

71 Shen, Y.F., Yuan, D.D., Ai, X.P. et al. (2014). *Electrochem. Commun.* 49: 5.

72 Zhou, M., Li, W., Gu, T. et al. (2015). *Chem. Commun.* 51: 14354.

73 Li, L., Sun, Y., Bao, H. et al. (2011). *Eur. Polym. J.* 47: 1630.

74 Liu, T., Kim, K.C., Lee, B. et al. (2017). *Energy Environ. Sci.* 10: 205.

75 Lu, X., Xia, G., Lemmon, J.P., and Yang, Z. (2010). *J. Power Sources* 195: 2431.

76 Panero, S., Prosperi, P., Zane, D., and Scrosati, B. (1992). *J. Appl. Electrochem.* 22: 189.

77 Arbizzani, C. and Mastragostino, M. (1990). *Electrochim. Acta* 35: 251.

78 Oyaizu, K. and Nishide, H. (2010). *Adv. Mater.* 21: 2339.

79 Suguro, M., Iwasa, S., and Nakahara, K. (2008). *Macromol. Rapid Commun.* 29: 1635.

80 Nakahara, K., Iwasa, S., Satoh, M. et al. (2002). *Chem. Phys. Lett.* 359: 351.

81 Kim, J.-K., Kim, Y., Park, S. et al. (2016). *Energy Environ. Sci.* 9: 1264.

82 (a) Sakaushi, K., Nickerl, G., Wisser, F.M. et al. (2012). *Angew. Chem. Int. Ed.* 51: 7850. (b) Ding, S.-Y. and Wang, W. (2013). *ChemInform* 42: 548.

83 Sakaushi, K., Hosono, E., Nickerl, G. et al. (2013). *Nat. Commun.* 4: 1485.

84 Kim, M.-S., Lee, W.-J., Paek, S.-M., and Park, J.K. (2018). *ACS Appl. Mater. Interfaces* 10: 32102.

85 Zhang, S., Huang, W., Hu, P. et al. (2015). *J. Mater. Chem. A* 3: 1896.

86 McKeown, N.B., Makhseed, S., and Budd, P.M. (2002). *Chem. Commun.* 23: 2780.

87 Côté, A.P., Benin, A.I., Ockwig, N.W. et al. (2005). *Science* 310: 1166.

88 Zhao, Q., Whittaker, A., and Zhao, X. (2018). *Materials* 11: 2567.

89 Lopez-Herraiz, M., Castillo-Martınez, E., Carretero-Gonzalez, J. et al. (2015). *Energy Environ. Sci.* 8: 3233.

90 Q. Sheng (2016), *Master Thesis*, Nanjing University of Aeronautics and Astronautics.

91 Quideau, S., Deffieux, D., Douat-Casassus, C., and Pouysegu, L. (2011). *Angew. Chem. Int. Ed.* 50: 586.

92 Ratnakumar, B.V., Sistefano, S., Williams, R.M. et al. (1990). *J. App. Electrochem.* 20: 357.

7

Intercalation-Type Anode Materials for Sodium-Ion Batteries

7.1 Introduction

Intercalation-type anode materials (e.g. carbon-based materials [1], titanium-based materials [2], etc.) show a great potential for practical applications of sodium-ion batteries (SIBs) owing to highly stable structure and cycling stability, low cost, and relatively high safety. For example, titanium-based oxides including TiO_2, $Li_4Ti_5O_{12}$, and $Na_2Ti_3O_7$ possess favorable structural dynamics owing to stable crystalline structure and two-dimensional (2D) diffusion channel for sodium ions, which is very beneficial for the whole structure stability upon repeated Na^+ intercalation and deintercalation for SIBs [3]. Moreover, such anodes show relatively high operating working voltage, thus avoiding the formation of sodium dendrites and enhancing battery safety for full cells. On the other hand, relatively high working potential can efficiently prevent some side reactions of electrolyte, exhibiting higher electrode/electrolyte interface stability and better rate capability. However, the intrinsic poor electron transfer capability and low Na^+ diffusion coefficient seriously limit their applications for SIBs. In this regard, various strategies have been widely developed for enhancing sodium storage performance of titanium-based anode materials in recent years.

In this chapter, the currently mainstream intercalation-type anode materials would be introduced and discussed in detail for their potential applications for sodium storage in the future. To date, there are mainly carbon-based materials (graphite, hard carbon, soft carbon, etc.) and titanium-based materials (TiO_2, $Li_4Ti_5O_{12}$, $Na_2Ti_3O_7$, $TiNb_2O_7$, $NaTi_2(PO_4)_3$, etc.) as intercalation-type anodes for SIBs [4]. The related fundamental issues, challenges, and advances of these anode materials would be summarized and discussed.

7.2 Carbon-Based Anode Materials

Currently, there are mainly three types of carbon-based materials for SIBs, such as graphite, hard carbon, and soft carbon. The Na^+ storage performance of such carbon-based anode materials is mainly based on the intercalation process owing to

Sodium-Ion Batteries: Energy Storage Materials and Technologies, First Edition. Yan Yu.

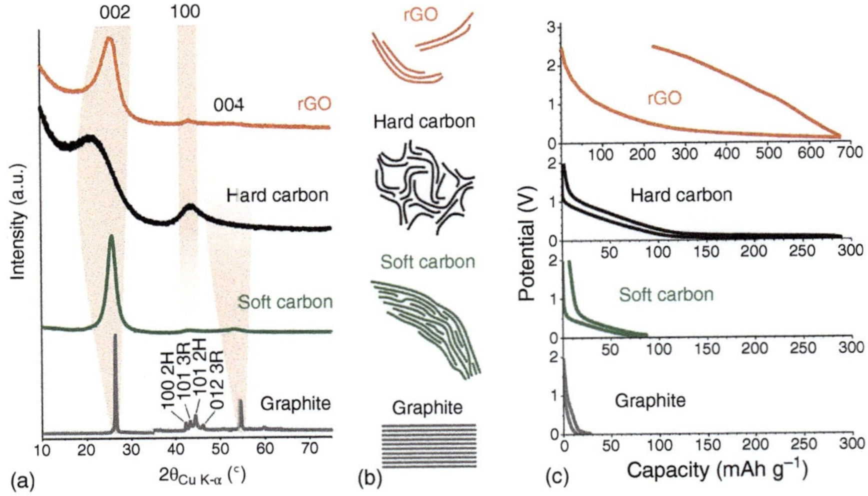

Figure 7.1 (a) Typical XRD patterns of various carbon-based anode materials including graphite, soft carbon, hard carbon, and reduced graphene oxides (rGO); (b) structural schematic illustration and (c) charge and discharge profiles of these carbon materials. Source: Saurel et al. [7]. Reproduced with permission, 2018, Wiley-VCH.

their layered structure [5]. Graphite, soft carbon, hard carbon, and graphene-related carbon are all based on polymorphs of sp^2 carbons, thus tending to form layered structure by hexagonal covalent carbon lattices (graphene layers) based on the weak Van der Waals interactions [6]. As displayed in Figure 7.1a, graphite possesses a crystalline structure with a high graphitic degree, while other carbon materials including soft carbon, hard carbon, and graphene-related carbon materials exhibit relatively low crystallinity and low graphitic degree [7]. When they are employed as anode for sodium storage, these four carbon materials show different electrochemical activities (Figure 7.1b). Graphite anode shows the lowest sodium storage properties, while soft carbon, hard carbon, and rGO deliver better sodium storage capabilities. The detailed analyses and discussion are as follows.

7.2.1 Graphite Anode

Graphite shows the crystalline form, while other carbon materials exhibit disordered structure nature and increased interlayer distance from soft carbon to hard carbon (Figure 7.1a). Layered graphite is a widely commercial anode material for lithium-ion batteries (LIBs) with a high specific capacity of about 350 mAh g^{-1} (theoretical capacity: 372 mAh g^{-1}). However, it only delivers a low reversible capacity of 40 mAh g^{-1} for SIBs because thermodynamic issue is not favorable for sodium intercalation. Theoretical studies indicate that the sodium–graphite intercalation compounds in SIBs are more unstable in comparison to the corresponding graphite–lithium or graphite–potassium intercalation compounds in LIBs or K-ion batteries (KIBs), respectively. To enhance sodium storage capability of graphite, one effective solution is to utilize co-intercalation strategy by using ether-based

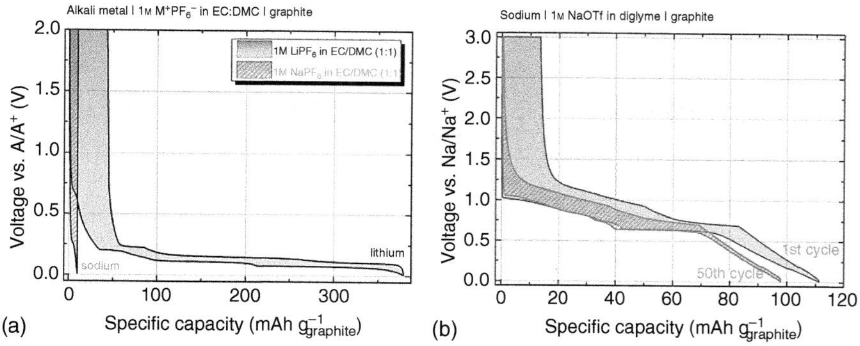

Figure 7.2 Charge and discharge profiles of graphite anode in different electrolytes: (a) 1 M $LiPF_6$ in EC/DMC (1 : 1) and 1 M $NaPF_6$ in EC/DMC (1 : 1), and (b) 1 M NaOTf in diglyme. Source: Jache and Adelhelm [8]. Reproduced with permission, 2014, Wiley-VCH.

electrolyte, which plays an important role in enhancing Na^+ intercalation kinetics. As shown in Figure 7.2, graphite anode in sodium triflate (NaOTf)/diglyme electrolyte delivers a specific capacity of 100 mAh g^{-1}, five times higher than that in $NaPF_6$/ethylene carbonate/dimethyl carbonate (EC/DMC) electrolyte (~20 mAh g^{-1}) [8]. Na^+ ions can form Na^+-solvent complex in the ether solvent and then intercalated into the interlayer structure of graphite. In this electrolyte system, graphite anode can exhibit high-efficiency Na^+ intercalation kinetics and cycling performance. The theoretical and experimental results have shown that the solvent with long-chain molecules can reduce energy barriers and facilitate Na^+ intercalation into the interlayer of graphite due to its shielding effect, leading to a higher specific capacity and cycling stability.

Another effective approach to enhance Na^+ storage properties of graphite is to engineering expanded graphite by increasing interlayer distance [9]. Pristine graphite is oxidized to graphite oxide through an oxidation reaction and then a reduction process (Figure 7.3a). Expanded graphite shows a larger interlay distance (0.43 nm) compared to raw graphite and still maintains a layered structure with the long-range order of the raw graphite crystal. As shown in Figure 7.3b, such expanded graphite undergoes the reversible expansion/shrinkage and exhibits a high reversible capacity of 284 mAh g^{-1} at 20 mA g^{-1}, which is much higher than that of raw graphite (~35 mAh g^{-1}). Moreover, the expanded graphite also presents a high capacity retention of 73% over 2000 cycles (Figure 7.3c) [9].

7.2.2 Hard Carbon Anode

As for SIBs, hard carbon is the most promising candidate as anode owing to the total balanced performance, working voltage, cycle lifetime, raw material, and manufacturing costs [10]. Furthermore, hard carbon is composed of turbostratic carbon layers with abundant internal nanopores and defects (Figure 7.4a), which both act as active sites for sodium storage.

Compared to soft carbon, hard carbon possesses more disordered structure with a high amount of structural microporosity [11], as shown in Figure 7.4b. Besides, it can

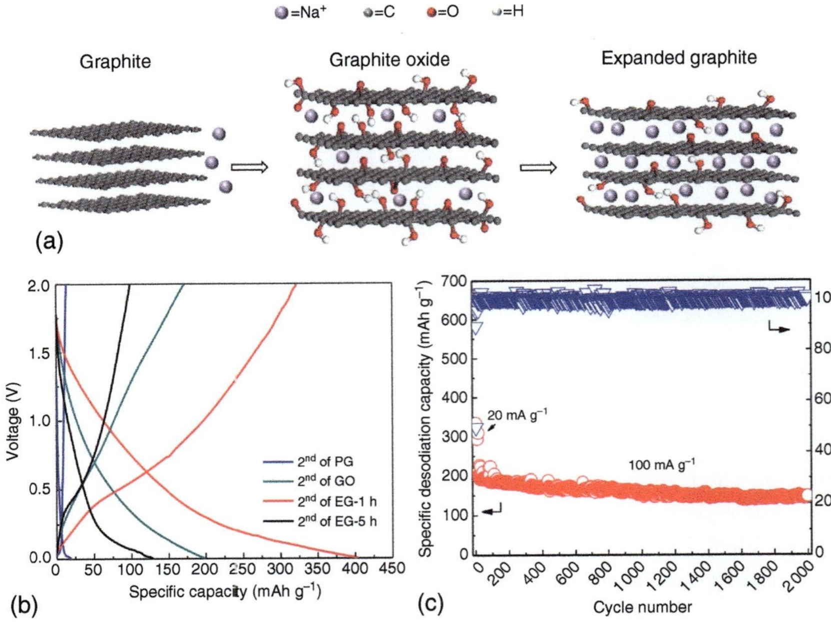

Figure 7.3 (a) Schematic diagram of Na^+ storage in expanded graphite. Left panel (graphite): Na^+ is hard to be electrochemically intercalated owing to the small interlayer distance. Middle panel (graphite oxide): Na^+ can be electrochemically intercalated into the interlayer due to the enlarged interlayer distance by an oxidation process. The intercalation process, however, is still hindered by a large amount of oxygen-containing groups in the graphite oxides. Right panel (expanded graphite): a large amount of Na^+ can be reversible intercalated and extracted owing to reduced oxygen-containing groups in the interlayers and expanded interlayer distance. (b) Charge/discharge profiles of the expanded graphite compared to pristine graphite (PG). (c) Cycling stability test of the resulting expanded graphite electrode. Source: Wen et al. [9]. Reproduced with permission, 2014, Springer Nature.

also provide more active sites to store Na^+ ions due to the cross-linked micropores with layered structure of graphite carbon. Up to now, hard carbon is the first carbon material with a reversible capacity for Na^+ storage compared with graphite anode for LIBs. As displayed in Figure 7.4c, hard carbon shows the higher specific capacity than those of graphite and soft carbon. During the sodiation/desodiation process, hard carbon presents a sloping curve at high-voltage range and a plateau at low-voltage range. To better explain the Na^+ storage mechanism of hard carbon, various advanced characteristics and technologies have been used and developed including small-angle XRD, DFT calculations, in situ TEM, and NMR. In general, there are two main explanations for Na storage mechanisms. One is based on intercalation and adsorption processes. Namely, the slope profile is attributed to Na^+ intercalation into graphene layers and adsorption effects as a result of the existence of adequate defect sites, and the plateau at low-voltage range can be ascribed to adsorption of cross-linked micropores. Another is based on adsorption mechanism

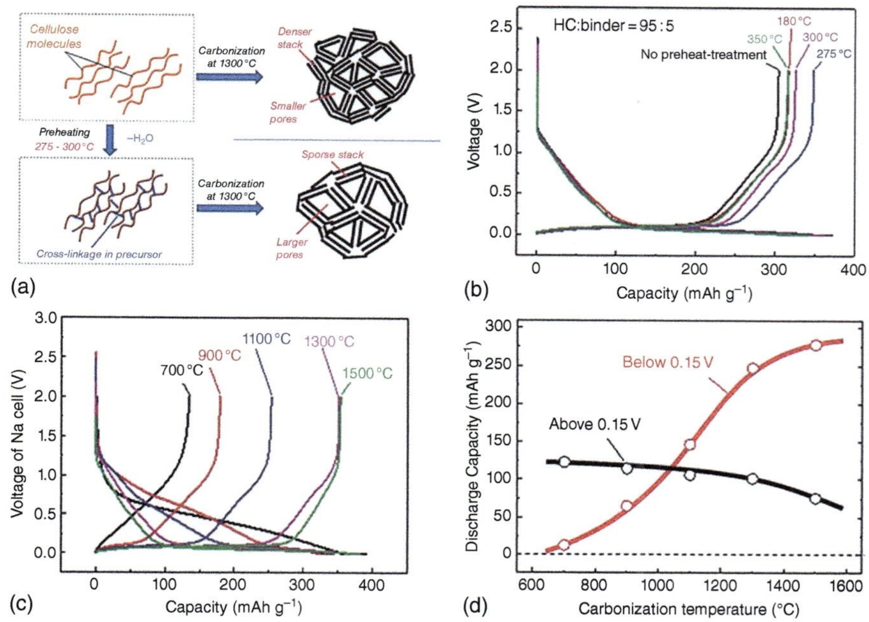

Figure 7.4 (a) Schematic diagram of the molecular structure and the resulting hard carbon nanostructure from cellulose upon different pretreatment conditions; (b) the first sodiation and desodiation profiles of cellulose derived hard carbon at different pretreatment temperatures; (c) the discharge and charge curves of the cellulose derived hard carbon at different annealing temperatures; and (d) the reversible capacities of hard carbon in the different potential regions below and above 0.15 V. Source: Yamamoto et al. [11]. Reproduced with permission, 2018, Royal Society of Chemistry.

in the defect sites (sloping profile) and intercalation and adsorption processes in the graphitic layers and porous structures corresponding to the low potential plateau area. Despite the Na^+ storage mechanism, more defect sites and porous structures for hard carbon materials can offer better Na storage performance.

Hard carbon is usually prepared by pyrolysis process of biomass precursors, such as sugars (e.g. glucose [12], pectin [13], sucrose [14], etc.), plants (e.g. algal [15], wood [16], cotton [17], etc.), fruit peels (e.g. pomegranate shell [18], coconut shells [19], apples [20], peanuts [21], etc.), and polymers (e.g. cellulose [11], lignin [22], etc.). Moreover, sodium storage properties are strongly dependent on the choice of precursors and related pretreatment conditions (temperature, time, etc.). For example, Komaba et al. compared various non-graphitizable hard carbons by using different carbon precursors such as sucrose, glucose, maltose, cellulose, glycogen, and amylopectin [11]. It was found that cellulose-derived carbon exhibits higher Na^+ ion storage capacity. Furthermore, the pretreatment (275 °C) in air plays an important role in regulating dehydration and cross-linkage degrees of cellulose, thus affecting nanoscale structure of hard carbon, as displayed in Figure 7.4b. The higher calcination temperature provides higher sodium storage capability and higher specific capacity below 0.15 V (Figure 7.4d).

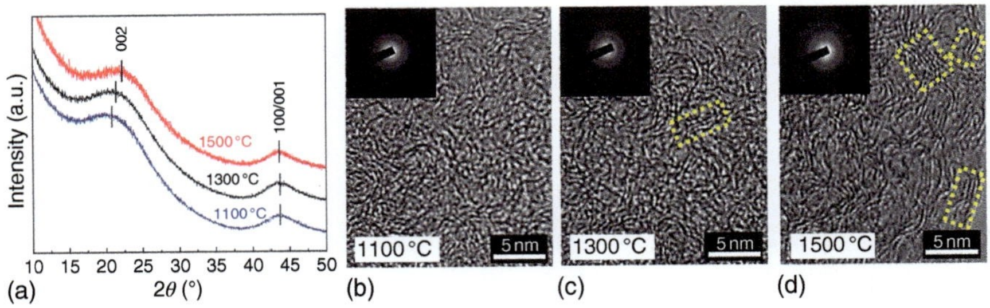

Figure 7.5 (a) XRD patterns of hard carbon derived from phenolic resin at different temperatures (1100, 1300, and 1500 °C); and (b–d) HRTEM images of hard carbon derived from phenolic resin at different temperatures: (b) 1100, (c) 1300, and (d) 1500 °C. The inset images in (b–d) show the corresponding SAED patterns. Source: Kamiyama et al. [23]/with permission from American Chemical Society.

Annealing temperature also plays an important role in influencing structure and electrochemical performance of hard carbon (Figure 7.5a) [23]. Two broad peaks at about 22° and 43° can be observed on the XRD patterns of hard carbon derived from phenolic resin at 1100, 1300, and 1500 °C, corresponding to the graphite crystal of (002) and (100)/(101), respectively. Besides, the XRD pattern at the higher temperature obviously shifts toward the higher angle of 2θ. However, these samples still show the non-graphitizable behaviors. The corresponding interlayer distances of (002) plane can be calculated to be about 4.008 Å (1100 °C), 3.940 Å (1300 °C), and 3.832 Å (1500 °C). It was found that the interlayer distance decreases with increasing annealing temperature, which is further confirmed by high-resolution transmission electron microscopy (HRTEM) analyses (Figure 7.5b–d). The sample at 1100 °C is disordered and consisted of curving sp^2 carbon, and the samples obtained at higher temperatures exhibit relatively ordering and higher graphitic degree structure (Figure 7.5c,d). In addition, the selected area electron diffraction (SAED) rings are clearer with the rising of annealing temperature, which is in good agreement with XRD patterns.

For industrialization, the yield of hard carbon from carbonization is a crucial factor to realize large-scale applications. In general, the yield of hard carbon from saccharides (<10%) is lower than that from phenolic resin (~50%) [24]. Thus, choosing phenolic resin could be the most suitable strategy for synthesizing high-performance and cost-competitive hard carbon materials in the future. The electrochemical behaviors of hard carbon for SIBs derived from phenolic resin are displayed in Figure 7.6a. The discharge and charge curves show a slope and plateau region with the voltage boundary of 0.15 V. As anode materials for SIBs, they can present a discharge specific capacity of 337 mAh g^{-1} (1100 °C), 358 mAh g^{-1} (1300 °C), and 386 mAh g^{-1} (1500 °C). Besides, all the three hard carbon anodes deliver superior cycling performance and show a high capacity retention over 93% (Figure 7.6b). This work exhibits a high specific capacity of 386 mAh g^{-1}, which is higher than those (350 mAh g^{-1}) of most reported hard carbon. Such performance is strongly dependent on the raw material types, pretreatment and carbonization processes. The unique mesoporous structure inherited from phenolic

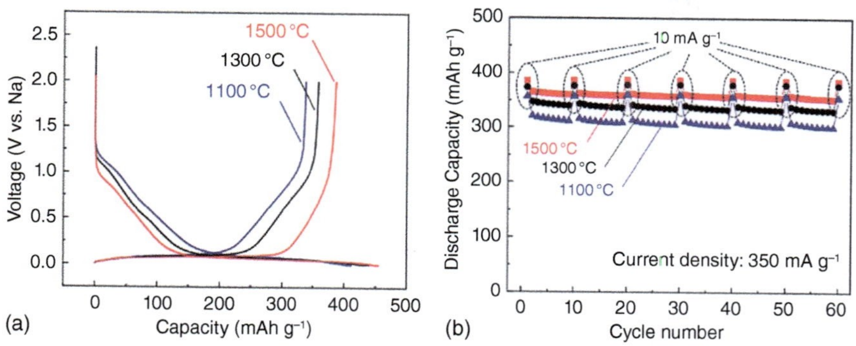

Figure 7.6 (a) The initial discharge and charge curves of the obtained hard carbon anode at different carbonization temperatures (1100, 1300, and 1500 °C) at a current density of $10\,mA\,g^{-1}$, and (b) cyclability tests. Source: Kamiyama et al. [23]/with permission from American Chemical Society.

resin and appropriate annealing temperature have both effect on the sodium storage performance of the obtained hard carbon. With the increase of annealing temperature (1100–1500 °C), the size of nanovoid increases with the decrease of interlayer distance between the stacked carbon sheets. Meanwhile, the electrochemical performance of the obtained hard carbon was examined, and the total specific capacity is improved at a higher annealing temperature, especially when the voltage is <0.15 V.

Besides annealing temperature, heteroatom doping (e.g. N, B, O, S, P, etc.) is an efficient strategy to optimize its electronic structures or introduce more defects to enhance the Na^+ storage performance of hard carbon. Very recently, Li et al. reported P-doped hard carbon via a pyrolysis process from phosphoric acid-solidified epoxy resin [25]. To deeply understand the correlation between groups containing phosphorus element and micro/nanostructure, and sodium storage properties, the authors prepared highly active functional groups containing phosphorus element and microstructure (large interlayer distance) at different annealing temperatures, realizing high capacity and excellent rate capability. Among various samples, the as-prepared PHC-700 (700 denotes annealing temperature) electrode delivers a high specific capacity ($379.3\,mAh\,g^{-1}$ at $0.1\,A\,g^{-1}$) and it can also maintain $158.1\,mAh\,g^{-1}$ after 6500 cycles even at a high current density of $5\,A\,g^{-1}$. It is worth noting that P–O/P–C bonds transform into P–P bonds with the increase of annealing temperature, and the electrochemical activity of hard carbon is enhanced with the reduction of layer spacing. The realized high capacity is attributed to its enlarged interlayer distance and high electrochemical activity. It is believed that the annealing temperature has a great influence on optimizing the group types containing phosphorus element and micro/nanostructure, directly determining sodium storage performance. This work also offers a new strategy to engineering highly active hard carbon anode for SIBs. Besides P doping, N or S or N, S co-doping have also been widely employed as the efficient strategies to enhance sodium storage performance. On the other hand, engineering defects in hard carbon is also an efficient method to promote sodium storage performance. Ji group recently reported hard carbon with abundant defects by a microwave and pyrolysis process of cellulose [26]. Only after 6-second treatment of microwave, its specific capacity improved greatly from 204 to 308 mAh g^{-1}, much higher than that (274 mAh g^{-1}) with a high temperature treatment of 1100 °C. The proposed microwave approach to prepare hard carbon with energy efficiency can generate abundant defects and further confirmed by the results of the associated pair distribution function and neutron total scattering. This work offers a facile and feasible treatment way to remarkably enhance sodium storage performance of hard carbon.

7.2.3 Soft Carbon Anode

Compared to graphite, soft carbon shows a relatively disordered graphitic structure. Similar to hard carbon, soft carbon shows a long sloping region with high reversibility and a voltage plateau during the first discharge process due to the excessive defects in the carbon matrix, as shown in Figure 7.7a. Moreover, an irreversible

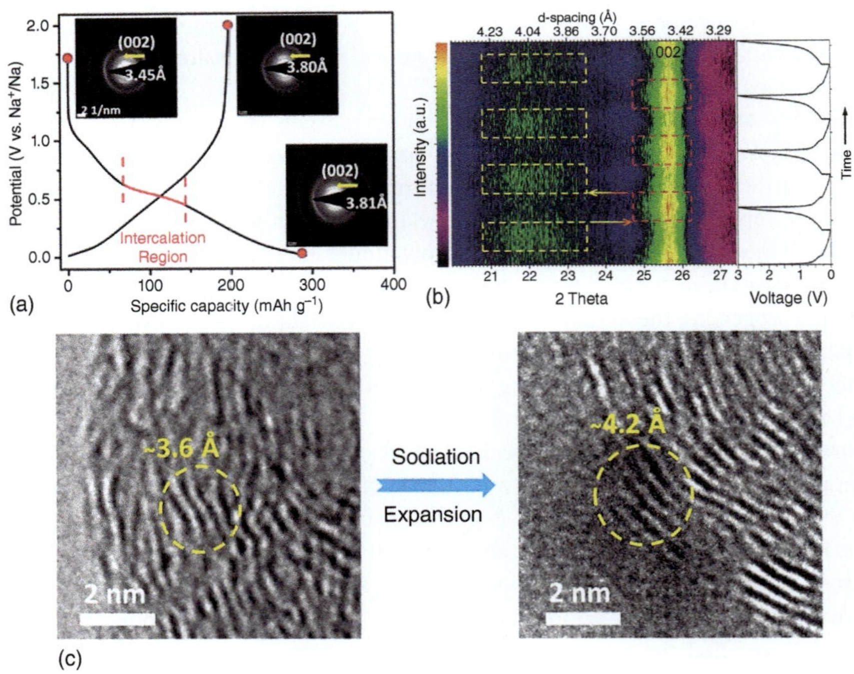

Figure 7.7 (a) Charge and discharge curves of soft carbon. Source: Jian et al. [27a]/with permission from American Chemical Society. (b) The in situ XRD patterns with the time-potential curve under voltage window of 0.01–3 V. Source: Yao et al. [27b]/with permission from Wiley-VCH. (c) Ex situ HRTEM images of C-900 electrode before and after sodiation. Source: Luo et al. [28]/with permission from American Chemical Society.

voltage plateau at about 0.5 V (vs. Na^+/Na) can be observed owing to the irreversible expansion of crystal lattice when the sodium ions intercalate into graphite layers (Figure 7.7b) [27]. The high potential plateau is related to the local defect structure stacked with carbon sheets and the high binding energy of sodium ions, which indicates that it is a trapping mechanism [27a]. Compared to hard carbon, soft carbon exhibits lower specific capacity and higher working potential, thus leading to lower energy density. Fortunately, some soft carbon anode delivers good rate capability, which could become a partial substitute for hard carbon in the high-power applications.

During the insertion process of sodium ions, electrochemically expandable of the turbostratic lattice from 3.6 to 4.2 Å in soft carbon was found by Luo et al., as shown in Figure 7.7c [28]. These expanded lattice shows high reversibility, suggesting its excellent cycling performance for SIBs. After phosphorus doping, the layer spacing and defects can be increased and its electrochemical performance is also improved significantly [29]. The P-doped soft carbon electrode presents a reversible capacity of 251 $mAh\,g^{-1}$, and its specific capacity can also retain 80.1% after 200 cycles. Cao et al. prepared mesoporous soft carbon from mesophase pitch with the template of nano-$CaCO_3$, and the path of ion diffusion is shortened, which is benefit to the penetration of electrolytes [30]. The electrode material can remain a reversible specific capacity of 103 $mAh\,g^{-1}$ at the current density of 500 $mA\,g^{-1}$ after long-term cycles (3000 times). Yao group reported a soft carbon with a microporous structure and lots of defects by the microwave exfoliation method [27b]. With the edges synergistically of micropores and defects, kinetics and extra sodium-ion storage sites are enhanced with capacitance-dominated sodium-ion storage mechanism and the capacity increases to 232 $mAh\,g^{-1}$, which is more than twice as much as before (103 $mAh\,g^{-1}$). Furthermore, soft carbon presents an excellent rate performance at the current density of 1 $A\,g^{-1}$ for SIBs. In this work, sodium ions intercalate into graphitic layers was observed by in situ XRD analyses.

7.3 Titanium-Based Anode Materials

Ti-based anodes such as $Li_4Ti_5O_{12}$ and TiO_2 in LIBs have been widely studied, owing to their unique advantages: (i) stable crystal structure during lithium interaction and deintercalation processes; (ii) high safety working voltage (1.2–1.7 V vs. Li^+/Li); (iii) better rate capability compared to carbon anode; and (iv) earth-abundant Ti resource and low cost [31]. Taking $Li_4Ti_5O_{12}$ as a case, it possesses a high safety working voltage of 1.5 V with almost zero structure change during repeated lithiation and delithiation cycling [32]. To date, it has been successfully commercialized as anode for LIBs, especially for fast-charging LIBs [33]. Considering the above mentioned merits and favorable lithium storage properties of Ti-based anode for LIBs, they have recently got considerable attention and shown attractive sodium storage performance for SIBs such as TiO_2, $Li_4Ti_5O_{12}$, $Na_2Ti_3O_7$, $TiNb_2O_7$, $NaTi_2(PO_4)_3$, etc. [34] In this section, the recent development of Ti-based anode materials for application in SIBs would be summarized and discussed.

7.3.1 TiO_2

TiO_2 shows n-type semiconductor properties with a wide bandgap and has been widely studied in various areas such as semiconductor [35], optical devices [36], photovoltaic cells [37], photocatalysis [38], solar cell [39], and rechargeable batteries [4a, 40]. TiO_2 has been proved as a promising anode with excellent Li^+ storage performance and a relatively high working voltage (1.5–2.0 V vs. Li^+/Li). As for sodium storage, TiO_2 anode shows a relatively low working potential (0–2.0 V vs. Na^+/Na), thus providing a higher energy density for a full cell. To date, various TiO_2 including rutile, amorphous phase, TiO_2-B, and anatase TiO_2 have been developed for sodium storage (Figure 7.8) [41]. It is found that these TiO_2 phases show various redox characteristics and Na^+ storage mechanism in comparison with the corresponding properties for LIBs. In this section, various TiO_2 would be analyzed and discussed for sodium storage.

7.3.1.1 Amorphous TiO_2

Rajh and coworkers first reported the sodium storage behaviors of amorphous TiO_2 [42]. The as-prepared TiO_2 presented a tube-like nanostructure grown on the Ti foil, and the compound material can be used as the anode directly for SIBs. The electrochemical results show that such anode undergoes a gradually increased specific capacity. As shown in Figure 7.9, it shows a specific capacity of only 75 mAh g^{-1} in the first cycle and then increases to 150 mAh g^{-1} after 15 cycles. Based on the

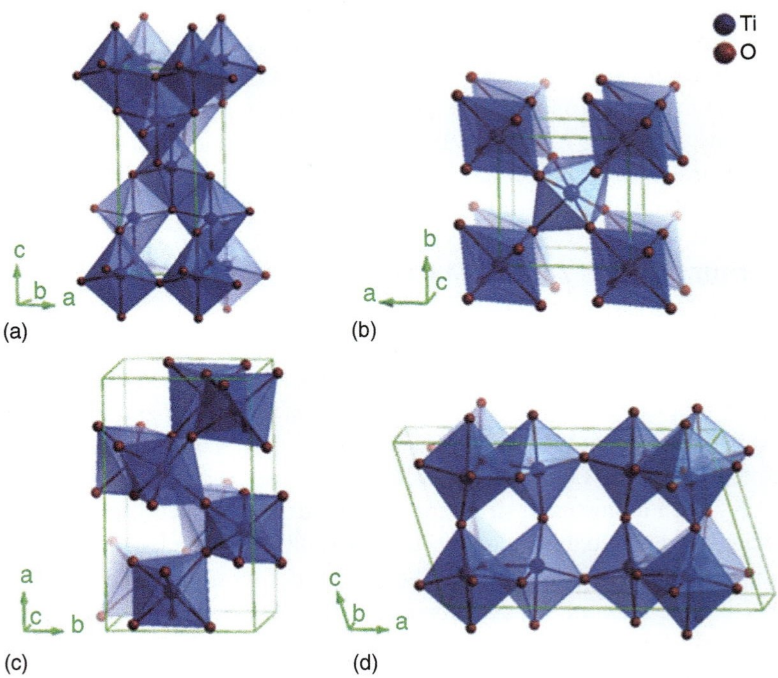

Figure 7.8 The crystalline structures of TiO_2: (a) anatase, (b) rutile, (c) brookite, and (d) TiO_2-B.

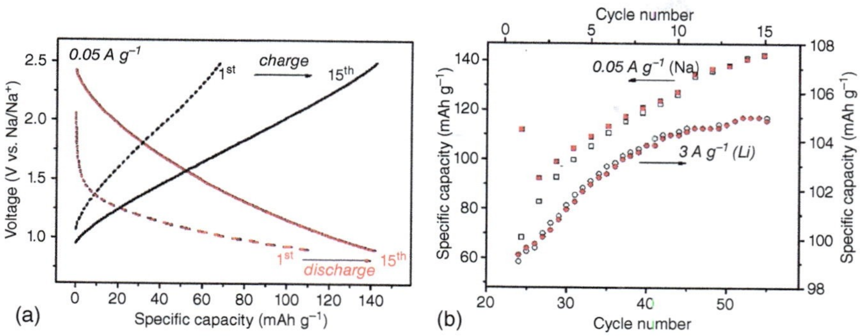

Figure 7.9 (a) The charge and discharge curves, and (b) cycling tests of amorphous TiO_2. Source: Xiong et al. [42]. Reproduced with permission, 2011, American Chemical Society.

calculated results, the amorphous TiO_2 electrode shows two kinds of Na^+ storage mechanisms with capacitance-limiting mechanism (1.4–0.9 V), diffusion and surface pseudocapacitive mechanism. This finding was also confirmed by Paranthaman and coworkers [43] The resulting amorphous TiO_2 nanotubes deliver the similar sodium storage behaviors. Such Na^+ storage phenomena were probably attributed to a significant active process during initial charge and discharge cycles.

In recent years, Gerbaldi and coworkers investigated the crystallinity change of amorphous TiO_2 upon different State-of-Charge (SOC) states [44]. The XRD patterns of TiO_2 anode materials after 300 cycles remain unchanged, revealing a superior stability of amorphous phase. The Na^+ storage mechanism is attributed to the combination of pseudocapacitive and solid-solution reaction process.

7.3.1.2 Anatase TiO_2

Anatase TiO_2 presents a tetragonal body-centered crystal structure (space group: *I*41/amd), which is made up of edge-shared TiO_6 octahedra (Figure 7.10a) [46]. As anode for SIBs, sodium ions can intercalate into interstitial sites and de-intercalate out of the crystal structure reversibly [47]. In addition, the diffusion pathway is three-dimensional (3D), thus showing favorable ion diffusion kinetics. Compared to amorphous or rutile TiO_2, anatase TiO_2 presents a 3D diffusion network structure stacked by zigzag chains with edge-shared TiO_6 octahedra, leading to better Na^+ storage capability. Such stacked zigzag structure can generate more diffusion channels and accommodation sites for sodium ions. In addition, theoretical calculation also shows the lower energy barrier for Na^+ intercalation in the anatase TiO_2.

To enhance the sodium storage performance of anatase TiO_2, Wu and coworkers designed and constructed oxygen vacancy-rich anatase TiO_2, showing greatly enhanced sodium intercalation behaviors [45]. First of all, theoretical calculation was used to analyze the effect of introducing oxygen vacancies (OVs) into anatase TiO_2. The DFT method was employed to optimize the structure of the supercell. As shown in Figure 7.10b, the electronic energy bandgap of F-TiO_{2-x} was reduced from 1.63 to 1.472 eV, revealing the enhanced electronic conductivity by introducing oxygen vacancies into bulk TiO_2. When the anatase TiO_2 with oxygen vacancies and

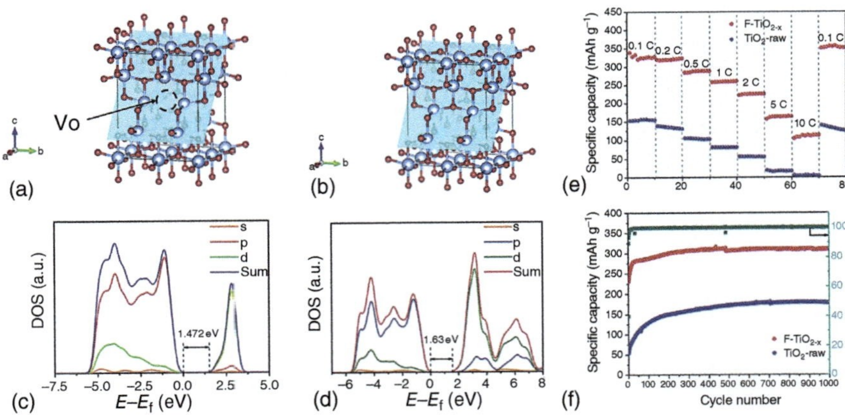

Figure 7.10 Crystal structure of (a) F-TiO_{2-x} and (b) TiO_2-raw. The blue plane denotes the (101) plane of TiO_2 crystal. V_o is the oxygen vacancy between the two Ti atoms. Calculated density of sate (DOSs) of (c) F-TiO_{2-x} and (d) TiO_2-raw. Electrochemical performance of F-TiO_{2-x} and TiO_2-raw: (e) rate capability and (f) cycling performance. Source: Ni et al. [45]. Reproduced with permission, 2020, Elsevier.

bulk counterpart were used as anode for SIBs, the samples with OVs present better Na^+ storage performance. As displayed in Figure 7.10c, the F-TiO_{2-x} electrode shows a reversible capacity of 122 mAh g^{-1} (10 C). In contrast, $TiO_{2\text{-ras}}$ exhibits only 25 mAh g^{-1} at 5 C. Moreover, the oxidization potentials of two samples are obviously different. The overpotential of F-TiO_{2-x} is much smaller than that of TiO_2-bulk. When they were examined at the rate capability tests (Figure 7.10d), it was found that the F-TiO_{2-x} shows superior sodium storage performance. As displayed in Figure 7.10e, the sample with OVs delivers a specific capacity of 330 mAh g^{-1} (0.1 C) and 114 mAh g^{-1} (10 C), respectively, which are much higher than those of TiO_2-raw electrodes. Moreover, on decreasing the current density to 0.1 C, the F-TiO_{2-x} presents a specific capacity of 350 mAh g^{-1}, showing superior Na^+ storage reversibility in the sample with OVs. When such electrode was further investigated as anode for long-term cycling tests (1000 cycles), it shows a high reversible capacity of 331 mAh g^{-1}, while only 178 mAh g^{-1} was obtained for TiO_2-raw electrode (Figure 7.10f). These results fully indicate that the introduction of OVs into anatase TiO_2 can efficiently tune electronic structure and greatly improve sodium storage performance.

7.3.1.3 TiO_2-B

TiO_2-B was first prepared with an approximate ReO_3-type layered structure by an ion-exchange and calcination process [48]. Layered TiO_2-B belongs to the monoclinic system with a space group of $C2/m$, which can be considered to originate from the ABO_3 perovskite structure without A atoms. The shear planes of TiO_2-B consist of TiO_6 octahedra, which are arranged along the [010] direction sharing with edge or corners. Compared to other TiO_2 phases, TiO_2-B possesses lower tap density and more abundant channels and is thus the best candidate for Na^+ intercalation. Dawson and Robertson found that TiO_2-B shows highly reversible Na^+

storage capability [49]. However, the electrochemical performance of TiO_2-B is still not satisfactory owing to limited Na^+ ion diffusion and poor electronic conductivity.

Many works have been devoted on the investigation of sodium storage mechanism in the TiO_2-B. In recent years, Gao group first examined the electrochemical performance in TiO_2-B for Na^+ storage. It was found that TiO_2-B can well accommodate Na^+ ions reversibly owing to the large interlayer distance of 0.56 nm (001) [50]. Owing to open tunnel structure, TiO_2-B shows a solid solution mechanism. Passerini et al. indicated that crystalline TiO_2-B became fully amorphous upon the first sodiation process [51]. The related XRD characteristics further support such findings that the XRD patterns during the subsequent sodiation reaction still remain amorphous, thus confirming the solid solution mechanism.

To enhance electrochemical performance of TiO_2-B, Ji group facilitates Na^+ storage properties by engineering oxygen vacancies in the TiO_2-B, generating blue TiO_2 [52]. The DFT calculation was used to reveal the intercalation energies of TiO_2-B (C site) and TiO_2-B with oxygen vacancies (O_{3f}), as displayed in Figure 7.11. It was computationally found that TiO_2-B with oxygen vacancies possesses lower Na^+ intercalation energy barriers, revealing that sodium ions have a strong disposition to intercalate into TiO_2-B with oxygen vacancies (TiO_2-B-OVs) in comparison to pristine TiO_2-B. When evaluated as anode for SIBs, TiO_2-B with oxygen vacancies exhibits

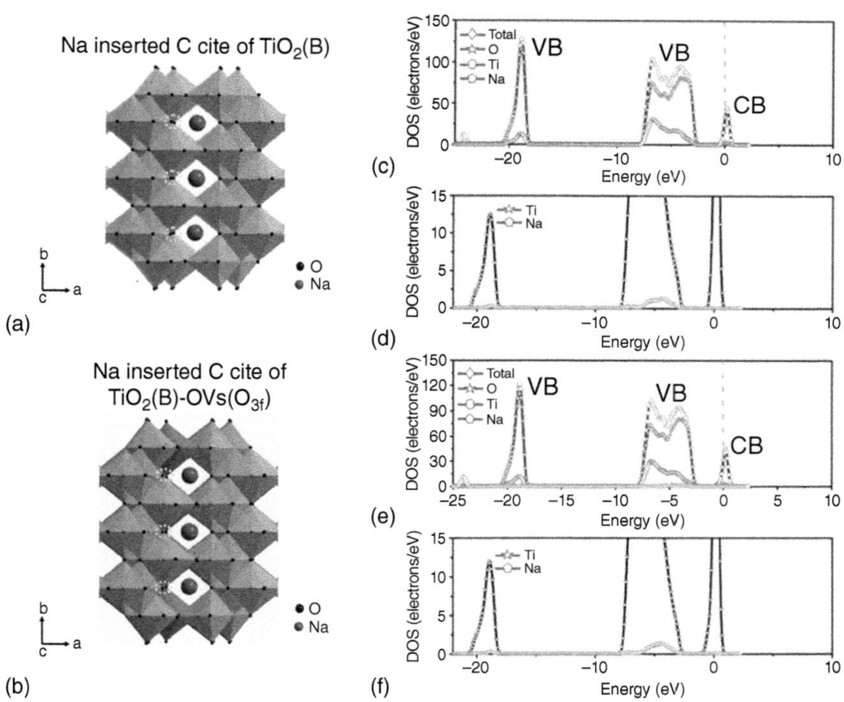

Figure 7.11 The structure models of Na inserted in (a) TiO_2-B and (b) TiO_2-B-OVs; and the corresponding calculated density of states (DOSs) and partial density states (PDOSs): (c, d) Na-inserted TiO_2-B and (e, f) Na-inserted TiO_2-B-OVs. OV denotes oxygen vacancy. Source: Zhang et al. [52]. Reproduced with permission, 2017, Wiley-VCH.

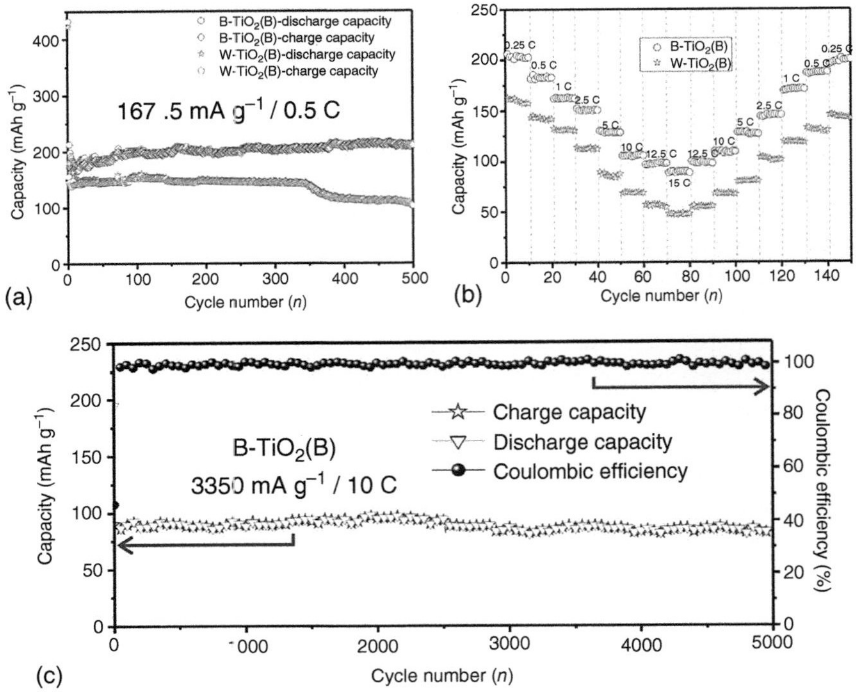

Figure 7.12 Electrochemical performance of blue-TiO_2-B (B-TiO_2-B) and white TiO_2-B (W-TiO_2-B): (a) cycling performance, (b) rate capability and (c) long-term cycling performance at 10 C. Source: Zhang et al. [52]. Reproduced with permission, 2017, Wiley-VCH.

a specific capacity of 200 mAh g^{-1} (0.5 C). Furthermore, it still can maintain a high specific capacity of 210 mAh g^{-1} after 500 cycles. Moreover, the rate capabilities of TiO_2-B-OVs and white-TiO_2-B were also compared, as displayed in Figure 7.12a. On increasing the current rate, TiO_2-B-OVs show reversible capacities of 204.6, 182.5, 162.9, 150.9, 134.7, 114.9, and 106.8 mAh g^{-1} at rates of 0.25, 0.5, 1.0, 2.5, 5.0, 10.0, and 12.5 C, respectively (Figure 7.12b). In contrast, W-TiO_2-B underwent capacity decay to 55.6 mAh g^{-1} at the current rate of 12.5 C, which is much lower than that at the current rate of 0.25 C (160.3 mAh g^{-1}). Even at a higher current rate of 15 C, the TiO_2-B-OV sample still can retain a high reversible capacity of 89.8 mAh g^{-1}, double higher than that of W-TiO_2-B sample. Furthermore, the excellent electrochemical performance of TiO_2-B-OV electrode was then verified by long-term cycling tests at 10 C. It was found that the as-prepared electrode still exhibits a high specific capacity of 80.9 mAh g^{-1} after 5000 cycles (capacity retention: 94.4%), as shown in Figure 7.12c.

7.3.1.4 Rutile TiO_2

Rutile TiO_2 is seldom employed as anode materials for LIBs owing to lower electrochemical activity than anatase TiO_2. It possesses polycrystal properties with unique diffusion characteristics of lithium ions. Moreover, it is worth noting that the diffusion coefficient (*D*) along the *c*-axis (10^{-6} cm^2 s^{-1}) is much higher than that in the *ab* plane (10^{-15} cm^2 s^{-1}) [53]. Owing to the high degree of anisotropy, 3D diffusion in TiO_2 particles with micro/nanoscale for lithium ions is limited by dynamics, thus leading to poor electrochemical performance. In this regard, Maier and Tarascon groups greatly enhanced lithium storage electrochemical activity of rutile TiO_2 by using a nanosized strategy [40d, 54]. However, the sodium storage study of rutile TiO_2 is still seldom reported.

In recent years, the investigation of rutile TiO_2 for SIBs has been proposed and studied. For example, Sakaguchi and coworkers reported that Nb-doped rutile TiO_2 has been employed for sodium storage for the first time [55], showing better electrochemical performance than that of anatase counterparts. Figure 7.13a shows the XRD patterns of Nb-doped rutile TiO_2 with different Nb amounts x ($x = 0 - 0.25$) via sol–gel method. When the amount of Nb is below 0.18, rutile TiO_2 with single phase can be obtained, while there exists an impurity phase ($TiNb_2O_7$) when $x = 0.25$. It was accepted that grain-enlarging behaviors of TiO_2 can be effectively suppressed by introducing Nb dopant and the ionic oxygen mobility is also reduced. As shown in Figure 7.13b, there is no obvious change in *c*-axis of $Ti_{1-x}Nb_xO_2$ with

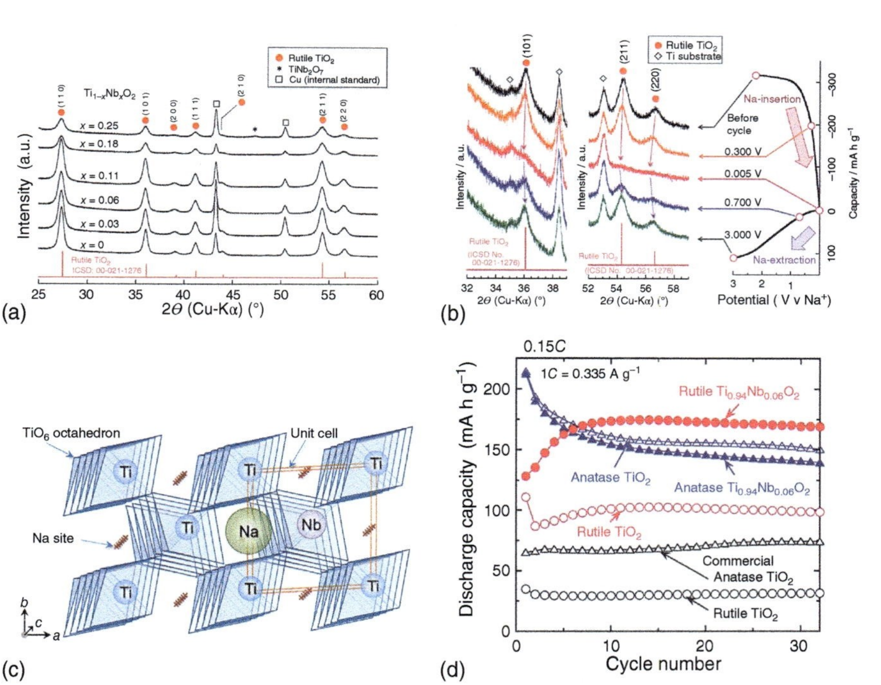

Figure 7.13 (a) XRD patterns of $Ti_{1-x}Nb_xO_2$. (b) Ex-situ XRD patterns of the rutile TiO_2 electrode during the first sodiation and desodiation cycles. (c) Possible occupation sites of Na in the Nb-doped rutile TiO_2 structure. Na occupies probably in the octahedral oxygen-coordinated interstitial sites. (d) Sodium storage performance of Nb-doped rutile TiO_2, Nb-doped anatase TiO_2, pristine rutile and anatase TiO_2, and commercial rutile and anatase TiO_2. Source: Usui et al. [55]. Reproduced with permission, 2015, American Chemical Society.

an increase of Nb dopant. In contrast, the lattice parameter shows a linear increase with Nb amount up to 0.06, which is mainly attributed to a substitution of Ti sites by a larger Nb^{5+}(ionic radius: 64 pm) compared to Ti^{4+} (ionic radius: 60.5 pm). With further increase of Nb dopant amount, the lattice parameter keeps no change. Thus, the impurity phase will be produced in the grain boundary due to the existence of excess Nb amount.

To identify the phase change of TiO_2 after Nb doping upon different charge and discharge states, the authors carried out ex-situ XRD measurements for Nb-doped TiO_2 electrode with different sodiation and desodiation conditions, as shown in Figure 7.13c. Its diffraction peaks gradually shift to lower angles with the increase of diffraction intensities during the sodiation process. When discharged to 0.005 V, the intensity of diffraction peaks became weak. However, the peak intensities of (101) and (211) planes are retained. On the contrary, the peaks gradually shift to higher angles during the subsequent desodiation process. Such findings confirm the reversible Na^+ storage of rutile TiO_2 during sodiation and desodiation reactions, although its crystallinity decreases to some extent. It is probably because the partially disordered state of lattice arrangement is caused by sodium-ion intercalation. However, such partially disordered state shows an irreversible change because the XRD diffraction peak intensities were well recovered upon desodiation process. These results demonstrate the structural stability upon repeated Na^+ insertion and extraction reactions.

When evaluated as anode for SIBs, the Nb-doped rutile TiO_2 shows better electrochemical performance than that of anatase counterparts. For a comparison, Nb-doped rutile and anatase TiO_2, pristine rutile and anatase TiO_2, and commercial rutile and anatase TiO_2 were prepared and tested as anodes for sodium storage, as displayed in Figure 7.13d. Commercial anatase TiO_2 shows better sodium storage performance than corresponding rutile phase. After introducing Nb dopant into TiO_2 phase, it was found that rutile TiO_2 exhibits better electrochemical performance than Nb-doped anatase phase with the same Nb doping amount. Such findings demonstrate that Nb doping is an efficient strategy to improve sodium storage properties of TiO_2 electrode.

7.3.2 $Li_4Ti_5O_{12}$

Spinel $Li_4Ti_5O_{12}$ (LTO) shows little volume change with a trifling change of lattice axis (0.2%) during repeated charging/discharging process between $Li_4Ti_5O_{12}$ and $Li_7Ti_5O_{12}$, which has been considered as one of the "zero strain materials". LTO exhibits a high safety working voltage (1.5 V) [56], efficiently avoiding the formation of SEI and electrolyte decomposition. Therefore, it can enable fast Li^+ transportation across the electrode/electrolyte interface, leading to superior rate capability. To date, LTO has been successfully used as high-rate anode for LIBs. Owing to its unique structural and lithium storage properties, LTO has recently received much attention for sodium storage and also shows attractive electrochemical performance for SIBs.

Considering the unique advantages of LTO, Chen and coworkers first reported its sodium storage behaviors of LTO in 2012 and demonstrated reversible Na^+

intercalation and deintercalation reactions with a working voltage of 1.0 V corresponding to a specific capacity of 145 mAh g^{-1} [57]. The Na^+ storage mechanism of LTO for SIBs resembles Li^+ storage behavior but different from Li^+ intercalation mechanism in LTO. Furthermore, the Na^+ diffusion coefficient of LTO (~10^{-16} cm^2 s^{-1}) is much lower than that of lithium ions in LTO (~10^{-11} cm^2 s^{-1}), revealing a lower Na^+ diffusion kinetics in spinel LTO structure. As anode for sodium storage, LTO shows typical discharge and charge profiles with voltage plateaus of 0.75 and 1.0 V (vs. Na^+/Na), respectively. And, its average working voltage (0.91 V) for SIBs is much lower (~0.64 V) than that of LTO for LIBs, which is exactly beneficial for enhancing its energy density when assembled into a full cell.

The discharge product of LTO could be a mixture of $LiNa_6Ti_5O_{12}$ and $Li_7Ti_5O_{12}$ after three Na^+ insertions, forming $Li_4Na_3Ti_5O_{12}$ single phase. Hu and coworkers confirmed the existence of $LiNa_6Ti_5O_{12}$ and $Li_7Ti_5O_{12}$, and the corresponding mechanism can be regarded as a three-phase reaction (V: vacancy, Eq. (7.1)) [58].

$$2[Li_3]^{8a}V^{16c}[Ti_3Li]^{16d}O_{12} + 6Na^+ + 6e^- \leftrightarrow V^{8a}[Li_6]^{16c}[Ti_5Li]^{16d}O_{12} + V^{8a}[Na_6]^{16c}[Ti_5Li]^{16d}O_{12} \tag{7.1}$$

During the discharge process (Figure 7.14a), Na^+ occupies the vacancy of 16c sites to form $V^{8a}[Na_6]^{16c}[Ti_5Li]^{16d}O_{12}$ phase (Na_6Li). In the meanwhile, the ions of Li^{8a}

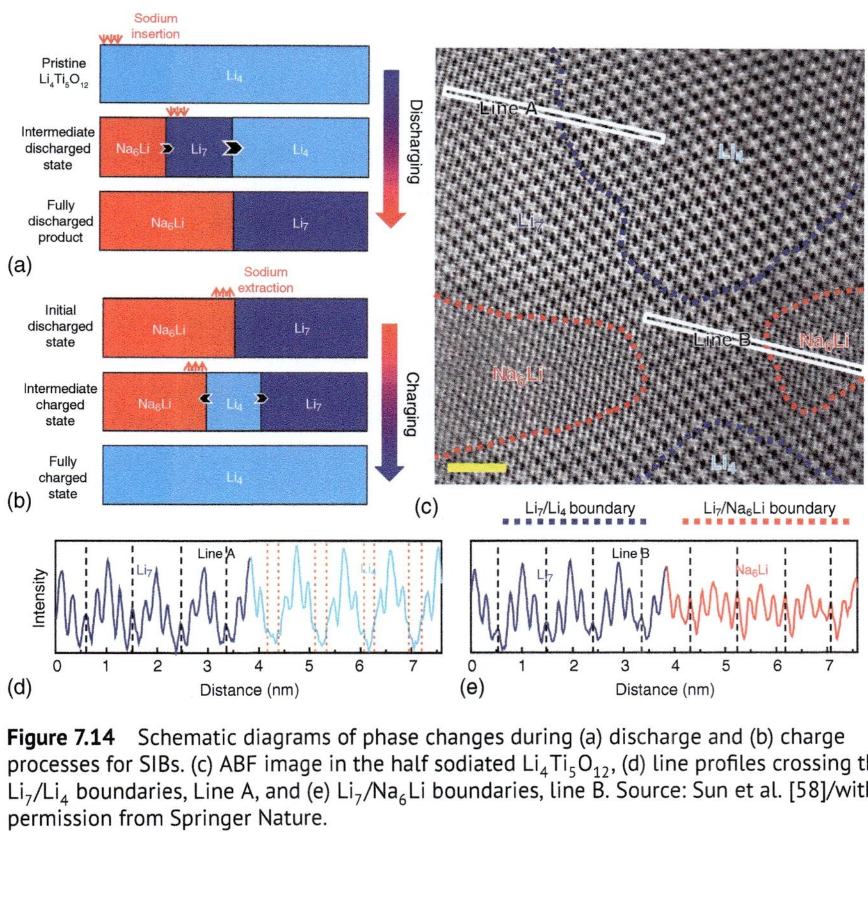

Figure 7.14 Schematic diagrams of phase changes during (a) discharge and (b) charge processes for SIBs. (c) ABF image in the half sodiated $Li_4Ti_5O_{12}$, (d) line profiles crossing the Li_7/Li_4 boundaries, Line A, and (e) Li_7/Na_6Li boundaries, line B. Source: Sun et al. [58]/with permission from Springer Nature.

are forced to transport to near active sites of Li_4 in the $Li_4Ti_5O_{12}$ phase to generate $V^{8a}[Li_6]^{16c}[Ti_5Li]^{16d}O_{12}$ phase (Li_7), forming the phases of Na_6Li and Li_7 simultaneously. With further Na^+ intercalation, Li_7 transforms to Na_6Li gradually and pushes the Na_6Li/Li_7 boundary forward.

Simultaneously, the insertion mechanism of Na^+ in SIBs differs from that of Li^+ in LIBs. As for LIBs, a bi-phase reaction occurs with a flat potential plateau (1.55 V) and efficiently avoids the formation of lithium dendrite. In contrast, sodium storage of LTO shows a three-phase reaction mechanism proposed by Hu group [58]. Moreover, the three-phase Na^+ storage was further confirmed through DFT calculations, in situ synchrotron XRD, and STEM. During Na^+ insertion in LTO, sodium ions occupy the 16c sites to form Na_6Li phase and the instant migration of Li^{8a} into Li_4 ($Li_4Ti_5O_{12}$) to form Li_7 ($Li_7Ti_5O_{12}$) phase. The so-called three-phase mechanism can be presented by the following Eq. (7.2):

$$2Li_4Ti_5O_{12} + 6Na^+ + 6e^- \leftrightarrow Li_7Ti_5O_{12} + Na_6Ti_5O_{12} \tag{7.2}$$

If Na_6Li does not form any nucleation, Na^+ storage will continuously intercalate in the boundary between Na_6Li and Li_7, leading to the phase transition from Li_7 to Na_6Li. The formation of Na_6Li would incur limited kinetics behavior and electrochemical properties for sodium storage because of larger size of Na^+ ions and larger volume change upon the whole Na^+ intercalation process.

The proposed three-phase intercalation mechanism for SIBs was then examined by spherical aberration STEM, and three-phase coexistence area has been identified during half sodiated process (Figure 7.14c). The sharp boundaries of Li_7/Li_4 and Li_7/Na_6Li without dislocation can be observed, which is in good agreement with lithiated $Li_4Ti_5O_{12}$ (Figure 7.14d,e). Different from $Li_4Ti_5O_{12}$ for lithium storage (two-phase mechanism), the above three-phase intercalation mechanism for sodium storage was first found in the intercalation-type anode materials.

Despite the existence of three-phase Na^+ storage mechanism, LTO still delivers excellent Na^+ intercalation and deintercalation reversibility, which is probably attributed to the slight change of lattice parameters, thus ensuring the structure integrity and cycling stability even of the bulk materials. However, sodium storage of LTO shows lower ion diffusion kinetics and electronic conductivity, thus limiting its practical applications. To addressing these issues, various strategies have been developed including nanostructure design, surface modifications, structure modifications, and engineering composites.

To improve sodium storage capacity, many nanostructured LTO including nanoparticles [59], nanofibers [60], nanorods [61], nanosheets [62], and nanowires [63] have been developed. Although nanostructure designing can shorten the transport pathways of electron and ions, the rate capability is still restricted owing to its sluggish Na^+ diffusion coefficient and low electron conductivity. Recently, doping and surface coating are proposed to realize the enhanced sodium storage performance including Cu doping [64], lanthanide doping [65], Na doping [66], and B-doped carbon coating [67]. For example, Yu et al. constructed LTO@C nanocomposite by embedding LTO nanoparticles inside carbon nanofibers (CNFs) by electrospinning [68]. When evaluated as anode for SIBs, the resulting LTO@CNFs show a reversible capacity of 133 mAh g^{-1} at 0.5 C and remain a specific capacity of 163 mAh g^{-1} after 100 cycles at 0.2 C. Sun group prepared boron-doped carbon coating micro-size LTO by a facile wet-chemical method [67]. It was found that the introduction of boron dopant into carbon can efficiently enhance electron transport. Three different boron species (BC_3, BC_2O, and BCO_2) were used for coating LTO. Such heteroatoms can deliver extrinsic defects in the carbon layers, thus leading to enhanced Na^+ diffusion kinetics. To identify the uniform distributions of the carbon layers, TEM and electron energy loss spectroscopy (EELS) were carried out, as shown in Figure 7.15a–c. The uniform carbon layer with 2 nm in thickness was observed in the HRTEM image. The so-formed carbon layers enhance electron conductivity and are also favorable for speeding up ion ionic transport by shortening the pathway from carbon layer to active components. In addition, the presence of carbon protection layer can efficiently prevent LTO corrosion from the electrolyte during repeated cycling. As displayed in Figure 7.15d–f, boron and carbon are distributed uniformly on the surface of LTO (BC-LTO), revealing the complete and uniform protection of boron-doped carbon materials. When evaluated as anode for SIBs, the BC-LTO exhibits higher specific capacity than that of P-LTO (pristine LTO) and C-LTO (C coated LTO), as shown in Figure 7.15g. In addition, BC-LTO also delivers better rate capability and cycling stability (Figure 7.15h,i). Such findings confirm that B-doped carbon coated LTO not only increases electronic conductivity but also offers an efficient protection layer from electrode corrosion in the electrolyte.

Considering the great potential of LTO as anode for SIBs, it has been used as anode by assembling full cell (cathode materials: $Na_3V_2(PO_4)_3$) [69]. It delivers an initial discharge capacity of 138.5 mAh g^{-1} with an average potential of 2.3 V. After 50 cycles, it also retains a reversible specific capacity of 114.7 mAh g^{-1} with a capacity retention of 82.8%. Furthermore, the related energy density can also be calculated around 165 Wh kg^{-1} with the total weight of cathode and anode. Such findings confirm that LTO would be very promising as anode materials for SIBs in the future.

7.3.3 $Na_2Ti_3O_7$

$Na_2Ti_3O_7$ is promising as anode material for SIBs owing to its low intercalation voltage (~0.3 V) and suitable specific capacity (theoretical capacity: 177 mAh g^{-1}), which is favorable to increase the full cell energy density [70]. As shown in Figure 7.16a, $Na_2Ti_3O_7$ possesses layered crystal structure consisted of TiO_6 octahedra with shared edges and simultaneously shared corners, leading to a formation of $(Ti_3O_7)^{2-}$ layers. In general, two Na^+ can be intercalated into $Na_2Ti_3O_7$ framework to generate $Na_4Ti_3O_7$ phase based on the following Eq. (7.3):

$$Na_2Ti_3O_7 + 2Na^+ + 2e^- \leftrightarrow Na_4Ti_3O_7 \tag{7.3}$$

The reversible phase transformation between $Na_2Ti_3O_7$ and $Na_4Ti_3O_7$ was clearly observed in the Na^+ intercalation and deintercalation processes with a plateau potential at ~0.3 V, as displayed in Figure 7.16b. Such sodium storage is also confirmed by in situ XRD characteristics. The two-phase transformation between $Na_2Ti_3O_7$ and $Na_4Ti_3O_7$ is highly reversible. Moreover, DFT theoretical calculation

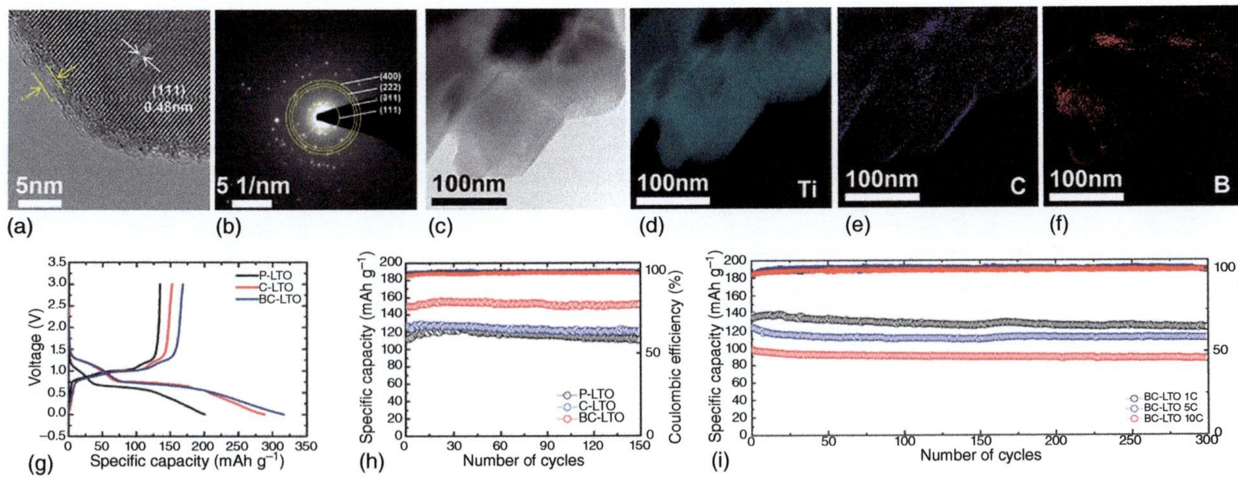

Figure 7.15 (a) High-resolution TEM image, (b) SAED pattern, (c) TEM image, (d–f) Ti, C, and B elemental mapping images by EELS for BC-LTO. (g) Charge and discharge profiles and cycling performance of BC-LTO, P-LTO, and C-LTO at 0.5 C. (i) Cycling performance of BC-LTO at different current rates (1, 5, and 10 C). Source: Yun et al. [67]. Reproduced with permission, 2013, Royal Society of Chemistry.

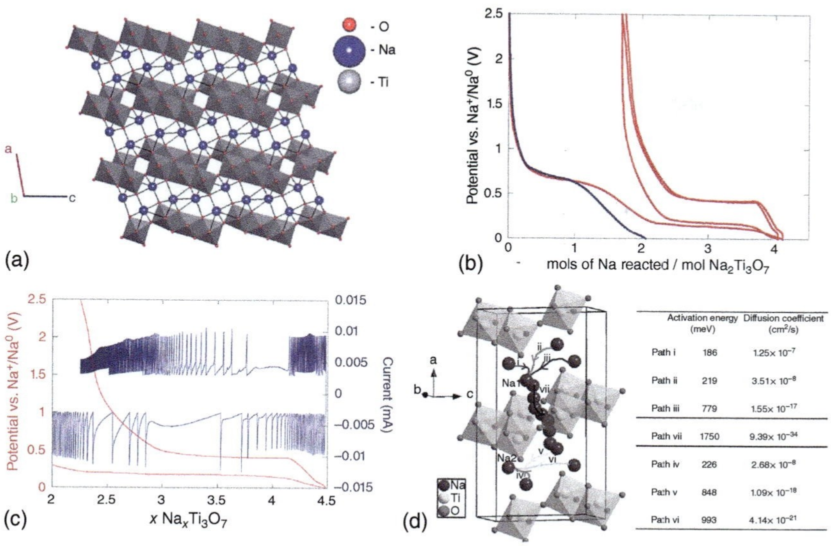

Figure 7.16 (a) Schematic diagram of layered structure of $Na_2Ti_3O_7$ viewed along the b-axis. Source: Rudola et al. [70c]. Reproduced with permission, 2013, Royal Society of Chemistry. (b) Charge and discharge curves carbon black (blue) and $Na_2Ti_3O_7$ composite electrode including 30% carbon black (red). (c) Stepwise potentiodynamic experiment. Source: Senguttuvan et al. [70a]. Reproduced with permission, 2011, American Chemical Society. (d) The possible sodium diffusion pathways (path i–vii), and the calculated Na^+ diffusion coefficients in $Na_2Ti_3O_7$ along different paths. Source: Pan et al. [70b]. Reproduced with permission, 2013, Wiley-VCH.

also demonstrates that the $Na_4Ti_3O_7$ phase is dynamically and mechanically stable, which is very beneficial for long-term cycling stability of $Na_2Ti_3O_7$ for SIBs. Based on the previous report, two Na^+ ions could insert into $Na_2Ti_3O_7$ at different sites including Na1 and Na2 (Figure 7.16c). It has been further confirmed the lower formation energies of vacancies (Na1: 3.03 eV; Na2: 3.0 eV) in $Na_2Ti_3O_7$ than those in other oxides such as $SrTiO_3$ (6 eV) and $LaMnO_3$ (4 eV). Meanwhile, Na^+ ion transport behaviors are also a key. Na^+ ion can transfer through the vacancies (Na1/Na2) in $Na_2Ti_3O_7$ between TiO_6 octahedron layers, forming a quasi-3D diffusion mechanism of sodium ions (Figure 7.16d).

When $Na_2Ti_3O_7$ was employed as anode for sodium storage, there are still some issues to be addressed such as electrode/electrolyte side reactions, low electronic conductivity, and sluggish Na^+ diffusion kinetics. For example, $Na_2Ti_3O_7$ anode usually suffers from the formation of unstable SEI layer. The SEI layer in the interface of $Na_2Ti_3O_7$/electrolyte consists of chemisorbed oxygen, alkyl carbonates and PEO, NaF, and NaCl. In contrast to the SEI layer in LIBs, the SEI layer in the $Na_2Ti_3O_7$ is thinner and exhibits more inorganic components. Some SEI components are partially soluble, thus leading to repeated formation of SEI in the subsequent cycles. The instability of SEI layer directly results in electrolyte degradation and capacity decaying, as well as low Coulombic efficiency. Hence, guaranteeing the interface stability between $Na_2Ti_3O_7$ and electrolyte is very crucial. On the other hand, enabling intrinsic and external electron transfer capability of $Na_2Ti_3O_7$ is also a key for enhancing sodium storage performance. In this regard, various strategies have been developed and used for realizing high-performance $Na_2Ti_3O_7$ anode for SIBs.

7.3.3.1 Surface Modifications

Carbon coating is always an efficient strategy to enhance electronic conductivity and the interface stability for electrode materials in the field of batteries. Wang group reported a simple approach to engineering carbon-coated $Na_2Ti_3O_7$ and demonstrated its superior sodium storage properties [71]. As shown in Figure 7.17a, the authors successfully prepared carbon-coated $Na_2Ti_3O_7$ by a two-step method. Compared to pristine $Na_2Ti_3O_7$, $Na_2Ti_3O_7/C$ exhibits a lower XRD intensity mainly owing to the existence of carbon. The resulting $Na_2Ti_3O_7/C$ composite is spherical and relatively uniform. The so-formed carbon protection layers not only guarantee electron transfer between active component and carbon but also facilitate rapid electron transport between each individual particle, thus leading to enhanced Na^+ storage properties. The $Na_2Ti_3O_7/C$ nanocomposites exhibit a specific capacity of 72 mAh g^{-1}, around twice higher than that of pristine $Na_2Ti_3O_7$ (Figure 7.17b). When it was evaluated for rate capability tests, the obtained $Na_2Ti_3O_7/C$ composites also deliver a higher specific capacity and better rate capability. Such findings confirm that the introduction of carbon can efficiently improve sodium storage properties of $Na_2Ti_3O_7$.

7.3.3.2 Micro-Nano Structure Design

Engineering micro-/nanostructure is an efficient approach to not only realize highly efficient electron/ion transport efficiency but also guarantee high structure stability for repeating Na^+ insertion and extraction processes. Ding and

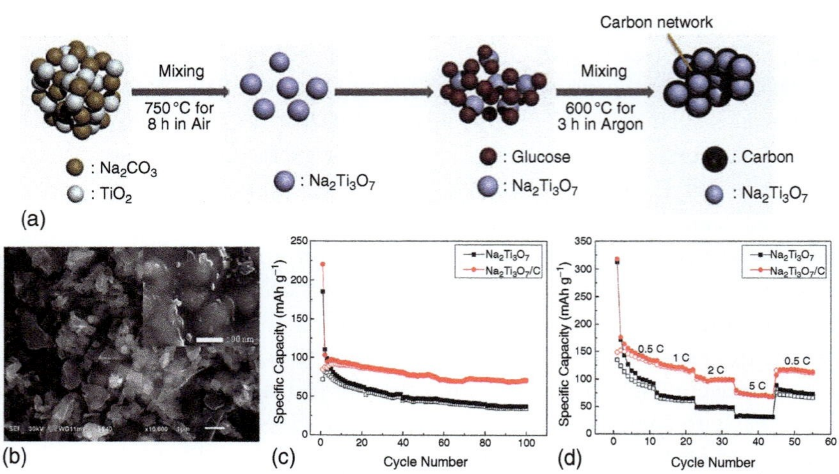

Figure 7.17 (a) The illustration diagram of the synthesis of $Na_2Ti_3O_7/C$ nanocomposite, (b) SEM image of the resulting $Na_2Ti_3O_7/C$, (c) cycling performance, and (d) rate capability of the $Na_2Ti_3O_7/C$ nanocomposite and pristine $Na_2Ti_3O_7$. Source: Yan et al. [71]/with permission from Elsevier.

coworkers designed and constructed $Na_2Ti_3O_7$ nanosheets based on 3D architectures assembled from ultrathin $Na_2Ti_3O_7$ nanosheets on the red blood cell-like hollow carbon spheres [72]. Figure 7.18a illustrates the synthesis process of $Na_2Ti_3O_7$ nanosheets@RHCS, and the corresponding SEM, TEM, and HRTEM images are shown in Figure 7.18b–d. The $Na_2Ti_3O_7$ nanosheets are well anchored on the surface of RH-PS via the hydrothermal reaction. Subsequently, the obtained precursors were further annealed at high temperature, thus generating highly crystalline $Na_2Ti_3O_7$ with carbon scaffolds. The unique 3D nanoarchitectures show superior sodium storage performance (specific capacity: 110.5 mAh g^{-1} at 5 C). In general, pure $Na_2Ti_3O_7$ suffers from poor rate capability owing to inferior electronic conductivity. In this work, the unique 3D micro-/nanohybrids show better performance than $Na_2Ti_3O_7$@HCS and pure $Na_2Ti_3O_7$ at various current rates. As shown in Figure 7.18h, $NaTi_3O_7$@RHCS nanocomposites exhibit a high reversible capacity of 175.5 mAh g^{-1} (1 C). Even at a high current density of 50 C, such 3D composites still can retain a specific capacity of 45.7 mAh g^{-1}, much higher than those of $Na_2Ti_3O_7$@HCS and pure counterpart. In addition, the $Na_2Ti_3O_7$@RHCS can retain reversible capacities of 110.5 mAh g^{-1} (5 C) and 70.6 mAh g^{-1} (20 C) after 1000 cycles, respectively, as shown in Figure 7.18j. Such findings fully confirm that the unique red blood cell-like hollow carbon hybrid structure not only can facilitate rapid transport of electrons/ions but also can guarantee the whole structural stability upon Na^+ insertion and extraction. This is probably attributed to the following factors: (i) the unique hollow micro-/nanocomposites provide a higher compacted density, shorter electron and ion transport pathways, and sufficient active sties for sodium storage; (ii) the as-prepared carbon nanostructures not only facilitate rapid electron transfer capability but also ensure the entire structure stability; (iii) ultrathin $Na_2Ti_3O_7$ nanosheets are beneficial for rapid transport and electrolyte penetration and are also capable to well accommodate the structural strain caused by repeated Na^+ insertion and extraction, especially at a high current rate.

7.3.3.3 Self-Supported Electrode Design

In recent years, the self-supported electrode directly grown on conductive substrates or forming free-standing films has attracted more and more attentions for rechargeable batteries, supercapacitors, and fuel cell owing to their merits such as lightweight, bendable, rugged, portable, and potentially foldable [73]. Importantly, self-supported electrode design can allow higher energy density and power density without necessity to use binder and conductive agent. As for $Na_2Ti_3O_7$, Yang group reported a freestanding vertically aligned $Na_2Ti_3O_7$ nanofiber arrays/flexible carbon matrix@N-doped graphene quantum dots (QDs) coating as a binder-free electrode directly for SIBs by a hydrothermal treatment and subsequent calcination treatment (Figure 7.19a) [74]. Each nanofiber array is composed of ultrathin nanosheets (length: 1–2 mm; thickness: <10 nm). Subsequently, the obtained $Na_2Ti_3O_7$ nanofiber was coated by N-doped graphene QDs. Such uniformly N-doped graphene QD coating can remarkably enhance electronic conductivity and offer adequate active sites for sodium ions accommodation between a large number of graphene nanoflakes. Moreover, such electrode can be used directly with

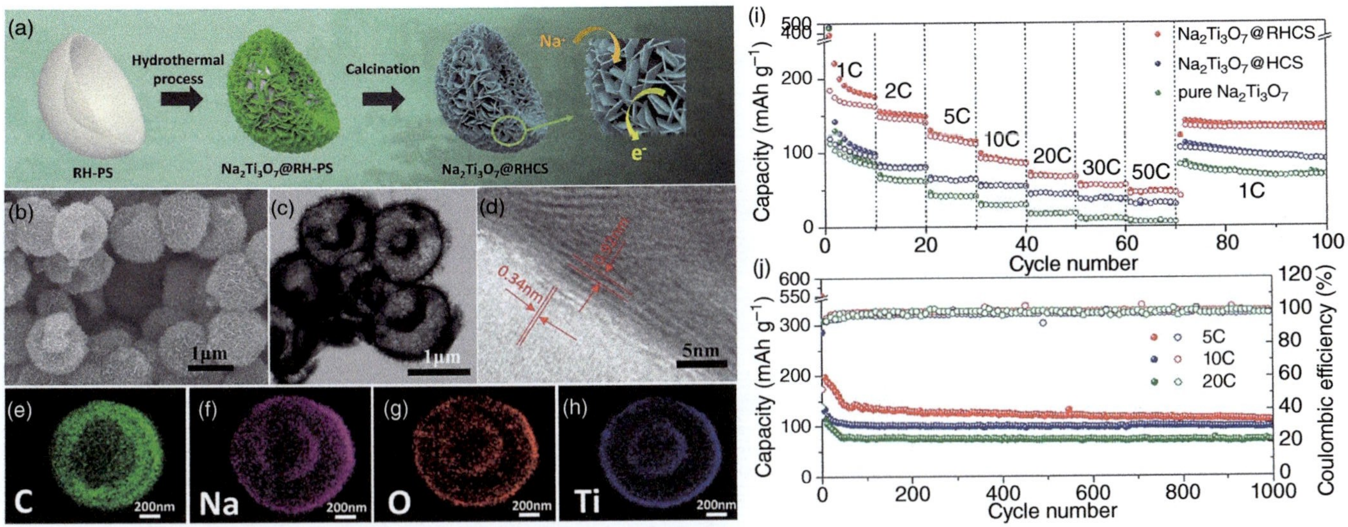

Figure 7.18 (a) Schematic diagram of the synthesis of $Na_2Ti_3O_7$ nanosheets@RHCS; (b) SEM and (c) TEM images of $Na_2Ti_3O_7$@RHCS; (d) HRTEM image of $Na_2Ti_3O_7$@RHCS and the corresponding elemental mapping of (e) C, (f) Na, (g) O, and (h) Ti; (i) rate tests of $Na_2Ti_3O_7$@RHCS, $Na_2Ti_3O_7$@HCS, and pure $Na_2Ti_3O_7$; and (j) cycling stability tests of the as-prepared $Na_2Ti_3O_7$@RHCS at different current rates of 5, 10, and 20 C. Source: Chen et al. [72]/with permission from Royal Society of Chemistry.

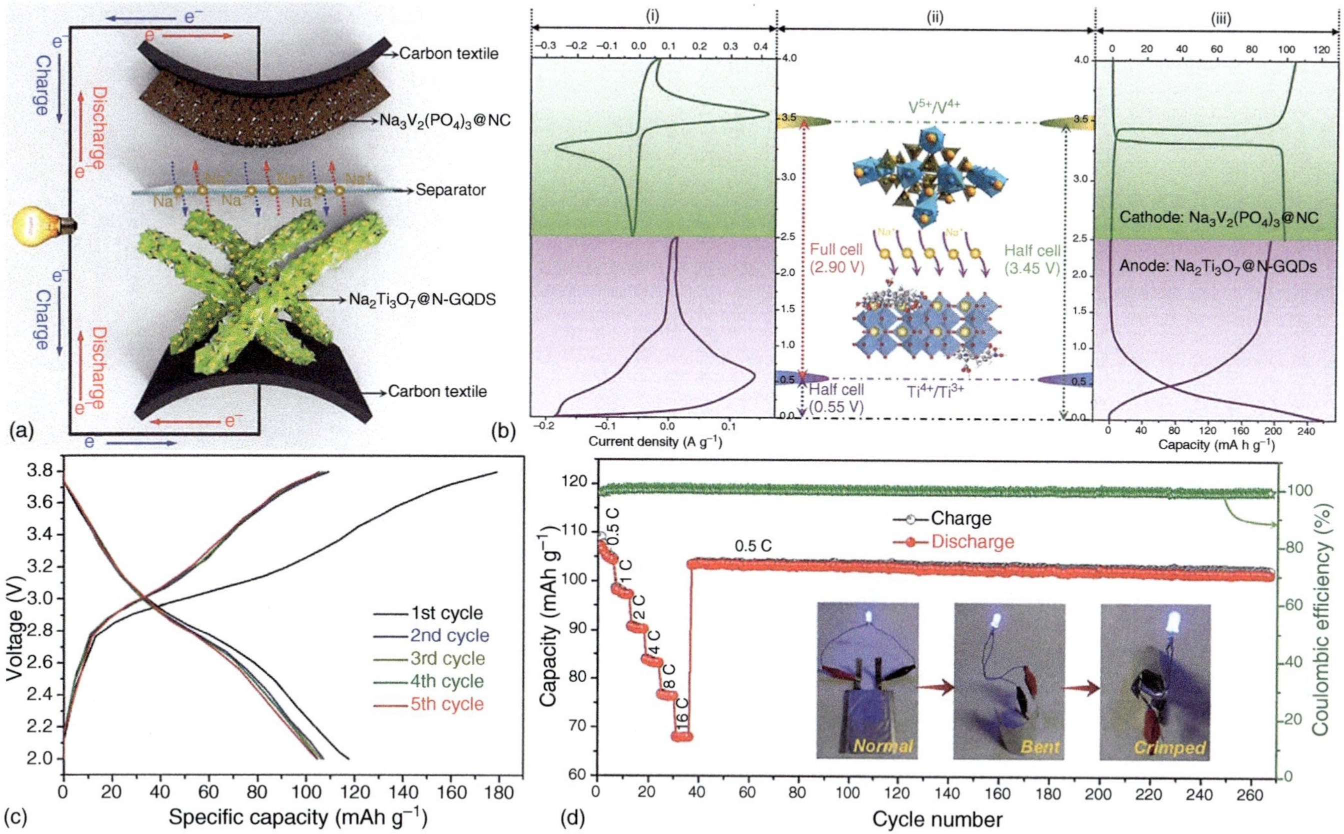

Figure 7.19 (a) Schematic diagram of the assembled $Na_3V_2(PO_4)_3$//$Na_2Ti_3O_7$@N-GQDs full cell, (b) CV and charge/discharge profiles of the $Na_3V_2(PO_4)_3$ and $Na_2Ti_3O_7$@N-GQDs half cells, (c) charge and discharge profiles of $Na_3V_2(PO_4)_3$//$Na_2Ti_3O_7$@N-GQDs full cell at 0.5 C, and (d) rate capability and cycling performance of $Na_3V_2(PO_4)_3$//$Na_2Ti_3O_7$@N-GQDs full cell (the insets: the optical images with different bending states lighting a light-emitting diode (LED)). Source: Kong et al. [74]/with permission from Royal Society of Chemistry.

no necessity to use binder and conductive agent. When evaluated as sodium storage anode, the as-prepared self-supported electrode exhibits excellent electrochemical performance. High reversible capacities of 164 mAh g^{-1} (4 C) and 58 mAh g^{-1} (64 C) were obtained, respectively. Moreover, such electrode delivers an ultra-stable cycling performance with capacity loss (~0.0075%) per cycle over 4000 cycles. Finally, they also demonstrated the full cell performance (cathode: $Na_3V_2(PO_4)_3$), as shown in Figure 7.19b, and it presented high flexibility, good rate capability, and excellent cyclic stability. The potential plateau of full cells is around 2.9 V.

Figure 7.19c,d shows the charge/discharge curves of the full cell with two voltage plateaus at around 2.9 V, revealing a two-step Na^+ intercalation/deintercalation mechanism. Compared to reported full cell, the developed nitrogen-doped graphene quantum dots (NGQDs)-doped $Na_2Ti_3O_7$//$Na_3V_2(PO_4)_3$ flexible full batteries show a higher energy density (273.5 Wh kg^{-1}) and power density (5.09 KW kg^{-1}). Xia and coworkers prepared ultra-long $Na_2Ti_3O_7$ nanowires@carbon cloth as anode for SIBs [75]. These nanowires of $Na_2Ti_3O_7$ distribute on the carbon matrix uniformly, thus generating one dimensional (1D)@3D porous micro-/nanostructure. The resulting pores can facilitate rapid electrolyte penetration, and sufficient contact between active nanowires and electrolyte, thus leading to shortened ion and electron transport pathways. This is very beneficial to sodium storage, especially for high-rate sodium insertion and extraction processes. These results confirm that the self-supported electrode design of $Na_2Ti_3O_7$ not only ensures rapid transport of electrons and ions but also reduces the use of non-electrochemically active materials, thus enabling higher power density and higher energy density. In addition, such design also shows a great potential for high-performance flexible electronic devices in the future.

7.3.3.4 Anion Doping

Although the previously carbon coating strategy can efficiently enhance rate capability of $Na_2Ti_3O_7$, it just can ensure rapid electron transfer ability on the surface of active materials, while intrinsic electron conductivity of $Na_2Ti_3O_7$ is still not substantially enhanced. Owing to the semiconductive nature of $Na_2Ti_3O_7$ with a large bandgap (3.7 eV) [76], it is highly desirable for enhancing intrinsic charge transfer capability for SIB applications. Dou and coworkers designed and constructed S-doped $Na_2Ti_3O_7$ hollow microspheres for tuning and increasing intrinsic electron conductivity of $Na_2Ti_3O_7$ [77], since sulfurization is capable to narrow the bandgap and improve electronic conductivity of $Na_2Ti_3O_7$ (Figure 7.20a–d). Compared to pristine $Na_2Ti_3O_7$ (NTO, bandgap: 3.7 eV) and calcinated $Na_2Ti_3O_7$ (bandgap of C-NTO: 3.49 eV), S-doped NTO exhibits lower bandgap (2.12 eV). The deduction of the bandgap is mainly attributed to the substation of sulfur atoms into lattice of $Na_2Ti_3O_7$ which can lower the electron transition energy from valence band to conduction band and consequently induce a shift in the absorption to lower energy. To identify the effect of sulfur into $Na_2Ti_3O_7$, ultraviolet–visible (UV–vis) diffuse reflectance spectra measurement was carried out, as displayed in Figure 7.20e,f. In contrast to NTO, the absorption edges of C-NTO and S-NTO show a shift toward higher wavelength. The trailing absorption of S-NTO shifts

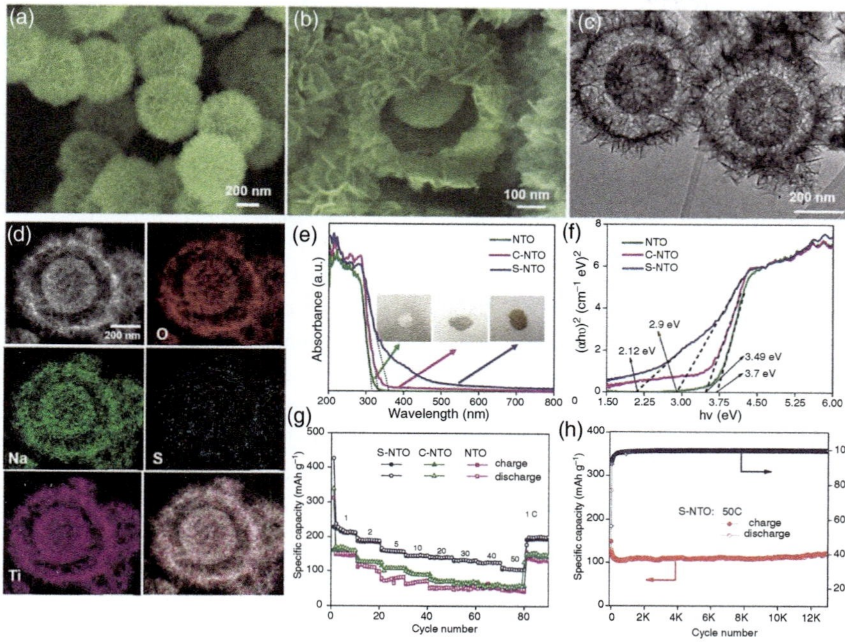

Figure 7.20 (a, b) SEM, (c) TEM, and (d) EDS mapping images of S-doped $Na_2Ti_3O_7$. (e–h) Electrochemical performance of the proposed S-doped S-NTO electrode. Source: Wang et al. [77]/with permission from Wiley-VCH.

to 320–480 nm (Figure 7.20e). In addition, NTO and C-NTO show white and light grey, while S-NTO exhibits yellow (Figure 7.20f). When used as anode material for SIBs, the proposed S-doped S-NTO electrode presents good rate capability and cycle stability, as shown in Figure 7.20g,h. Such findings reveal that sulfur doping could be an efficient strategy to develop advanced anode materials for SIBs.

Besides sulfur doping, phosphorus doping is also an efficient approach to enhance electron transport capability and electrochemical properties. Li and coworkers prepared P-doped NTO by introducing phosphorus into NTO for tuning the surface electron configuration of NTO [78]. P-doped NTO was designed and prepared in the Ti substrate by a low-temperature phosphorous strategy using NaH_2PO_2 as phosphorous source. The introduction of phosphorous can induce the formation of vacancies to improve the electronic conductivity and enhanced sodium storage kinetics. P-doped 3D $Na_2Ti_3O_7$ nanowire arrays exhibit a high reversible capacity of 290 mAh g^{-1}, superior rate performance (50 mAh g^{-1} at 20 C), and ultrastable cycle stability (capacity retention: 98% after 3100 cycles). The excellent sodium storage capability is probably due to the synergistic functions of phosphorus doping and nanostructure design. Among them, the introduction of phosphorous can induce the formation of surface vacancies and offer much more active sites and remarkably enhance electron conductivity of $Na_2Ti_3O_7$, which can enable NTO better rate capability. On the other hand, the construction of $Na_2Ti_3O_7$ nanowire array is beneficial to improve electron and ion rapid transport.

7.3.3.5 Cation Doping

Besides anion doping modifications, cation doping into $Na_2Ti_3O_7$ is also an attractive strategy to enhance Na^+ storage performance of $Na_2Ti_3O_7$. Chen et al. synthesized Nb-doped $Na_2Ti_3O_7$ via sol–gel process and demonstrated better Na^+ storage capability [79]. By introducing Nb element into cation on Ti sites, the trace Ti^{4+} ion was transformed into Ti^{3+} owing to the charge compensation and Ti^{4+} ions still retain the original state, leading to Na-deficient $Na_{2-x}Ti_{2.97}Nb_{0.03}O_7$. Meanwhile, the electron density and electronic conductivity were also enhanced. Owing to Nb doping, the unit cell is slightly expanded owing to the increased ionic radius of Nb^{5+} (0.64 Å) and Ti^{3+} (0.67 Å) in contrast to Ti^{4+} (0.60 Å). When employed as anode for SIBs, the Nb-doped NTO shows better sodium storage performance than pristine NTO. Du group reported lanthanide-doped $Na_2Ti_3O_7$ for enhancing sodium storage properties of SIBs [80]. As shown in Figure 7.21a, various lanthanide elements including La, Ce, Nd, Sm, Gd, Er, and Yb have been investigated and evaluated as anode for SIBs. Based on the first-principles calculations, the electronic conductivity of NTO changes by introducing Yb dopant. As shown in Figure 7.21b,c, the pristine NTO shows a direct bandgap of 3.07 eV, while Yb-doped NTO delivers a lower bandgap (0.58 eV), revealing that the polar migration energy barrier along the constrained pathway is remarkably reduced, thus resulting in an increased electronic conductivity of Yb-NTO. The introduction of lanthanide element into NTO usually induces a slight lattice distortion and thus leads to the formation of oxygen vacancies, which can significantly enhance donor density and electronic conductivity of NTO and facilitate rapid electron transport, leading to superior rate performance and cycling stability. All of lanthanide-doped NTO samples exhibit better sodium storage performance in Figure 7.21d,e, especially for Yb-doped NTO. After 1600 cycles at 5 C,

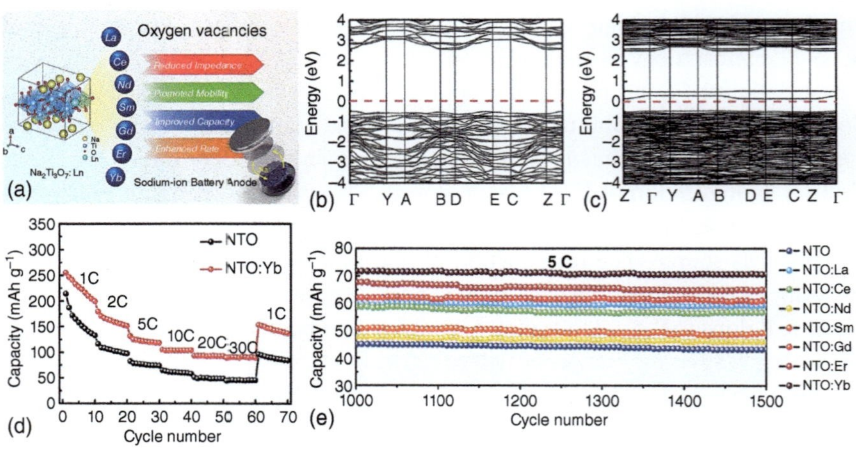

Figure 7.21 (a) Schematic diagram of the lanthanide-doped $Na_2Ti_3O_7$ for SIBs. The calculated band structure in the irreducible Brillouin zone of (b) $Na_2Ti_3O_7$ and (c) Yb doped $Na_2Ti_3O_7$. (d) Rate capability of $Na_2Ti_2O_7$ and Yb-doped $Na_2Ti_3O_7$, and (e) cycling performance of various lanthanide doped $Na_2Ti_3O_7$ electrodes from 1000 to 1500 cycles at 5 C. Source: Xia et al. [80]. Licensed under CC BY 3.0.

Yb-doped NTO still delivers a specific capacity of 71.6 mAh g^{-1}, almost twice that of pristine NTO. The as-prepared samples are micro-sized by a solid-state method without any complicated synthesis, hence showing a great potential for practical application in the future. Such findings are probably attributed to a slight lattice distortion and oxygen vacancies caused by introducing lanthanide dopants. This work also shows a facial and feasible strategy to realize high-performance $Na_2Ti_3O_7$ anode for SIBs.

7.3.4 $NaTi_2(PO_4)_3$

7.3.4.1 Structure and Properties of $NaTi_2(PO_4)_3$

The crystal structure of $NaTi_2(PO_4)_3$ is rhombohedron with space group of *R-3c* (167) composed of TiO_6 octahedrons and PO_4 tetrahedrons (Figure 7.22) [82]. The cell parameters are as follows: $a = 8.4913$ Å, $b = 8.4913$ Å, $c = 21.7858$ Å [83]. In this crystal, three PO_4 tetrahedrons separate two TiO_6 octahedrons and the skeleton unit of $[Ti_2(PO_4)_3]^-$ is constructed along with the *c*-axis through the oxygen atoms sharing the vertex angle. There are two types of gap positions in the skeleton of $[Ti_2(PO_4)_3]^-$: M_1 is located at the intersection of three conduction channels in the environment of elongated octahedral oxygen; and M_2 is located at each inflection point of the conduction channel in the environment of three eight coordination sites, which is close to M_1. It has an open 3D frame structure, which can be used as a fast transmission channel for Na^+ [34f, 81]. Furthermore, the crystal structure can keep stable even at high charge state owing to its short sodium ion conduction path and strong P—O bond [81d].

The volume expansion of $NaTi_2(PO_4)_3$ is only 7% after the insertion of sodium ions with the voltage platform of 2.1 V, which is belonging to "zero strain" material and shows high safety [83, 84]. As shown in Eq. (7.4), two phases convert to each other between $NaTi_2(PO_4)_3$ and $Na_3Ti_2(PO_4)_3$ with the insertion and desorption of two moles of sodium ions during the charging/discharging processes, causing a high theoretical capacity of 132.8 mAh g^{-1} [81b, 84]. During the discharging process, Na^+ ions are inserted into $NaTi_2(PO_4)_3$ crystal with the reduction of Ti^{3+} from Ti^{4+}

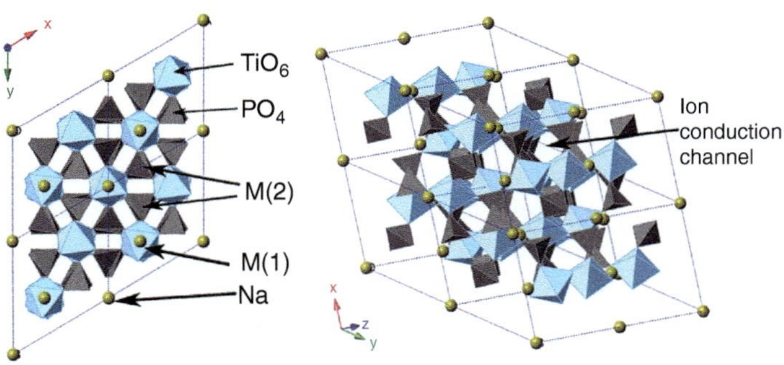

Figure 7.22 Structure diagram of $NaTi_2(PO_4)_3$. Source: Li et al. [81b]. Reproduced with permission, 2017, Elsevier.

and electrons reach the positive electrode through the external circuit. The charging process is the reverse of this process.

$$NaTi_2(PO_4)_3 + 2Na^+ + 2e^- \leftrightarrow Na_3Ti_2(PO_4)_3 \quad (7.4)$$

7.3.4.2 Modification Strategies of $NaTi_2(PO_4)_3$

Even if $NaTi_2(PO_4)_3$ shows many advantages used as anode materials for SIBs, but there are still great challenges in the practical application of large-scale energy storage. On account of the redox reaction between Ti^{3+} and Ti^{4+}, the electrons need to be transferred through the transmission channel of Ti–O–P–O–Ti owing to the separation of TiO_6 in the 3D frame structure. The insulation of phosphate results in the poor conductivity of $NaTi_2(PO_4)_3$, and it cannot provide enough charge transfer power between active particles, which affects its rate capability and cycle stability [85]. Meanwhile, sodium ions show high resistance during intercalation/deintercalation processes into the crystal lattice of $NaTi_2(PO_4)_3$ due to its large radius, leading to poor reversibility and large initial irreversible capacity loss [86]. In addition, it is difficult to prepare small size $NaTi_2(PO_4)_3$ due to its large molecular configuration, which is not conducive to the rapid diffusion of sodium ions.

The ideal $NaTi_2(PO_4)_3$ electrode material should offer a short transport distance of ions and electron, and stable material structure in high rate and long cycle is also needed. So far, three methods have been reported to solve these problems: (i) design of micro-nanostructure, which can shorten the transmission distance of ions and electrons and prevent particle agglomeration, facilitating to increase the contact area between $NaTi_2(PO_4)_3$ and electrolyte; (ii) combined with good conductive materials to construct an effective conductive network, and electrons can transferred between active materials and conductive materials quickly (such as carbon), so as to increase the surface conductivity of $NaTi_2(PO_4)_3$; and (iii) doping with other ions can increase the electronic conductivity and structural stability, and the carrier concentration and Hall mobility of $NaTi_2(PO_4)_3$ can also be improved, thus improving the sodium storage performance.

7.3.4.2.1 Micro-nanostructure Design Wei et al. reported a porous $NaTi_2(PO_4)_3$/CNFs composite materials prepared by the electrospinning method (Figure 7.23a) [85a]. The presence of carbon nanofibers provides 1D continuous conductive element, and the 1D nanofibers interweave into a conductive network, causing the high electronic conductivity of composite materials. Furthermore, the structural stability can also be improved owing to the existence of $NaTi_2(PO_4)_3$ nanoparticles inside of carbon nanofibers. The porous $NaTi_2(PO_4)_3$/CNFs electrode presents a high specific capacity (120 mAh g^{-1}) for SIBs, and it shows good cycling stability after 700 cycles at 2 C (capacity retention: 93%). Xu et al. fabricated the hierarchical $NaTi_2(PO_4)_3$/C mesoporous microflowers via solvothermal and annealing treatment (Figure 7.23b) [87]. The unique mesoporous microflowers exhibit good structure stability with high charge transfer kinetics and low polarization. The hierarchical $NaTi_2(PO_4)_3$/C mesoporous microflowers exhibit an initial capacity

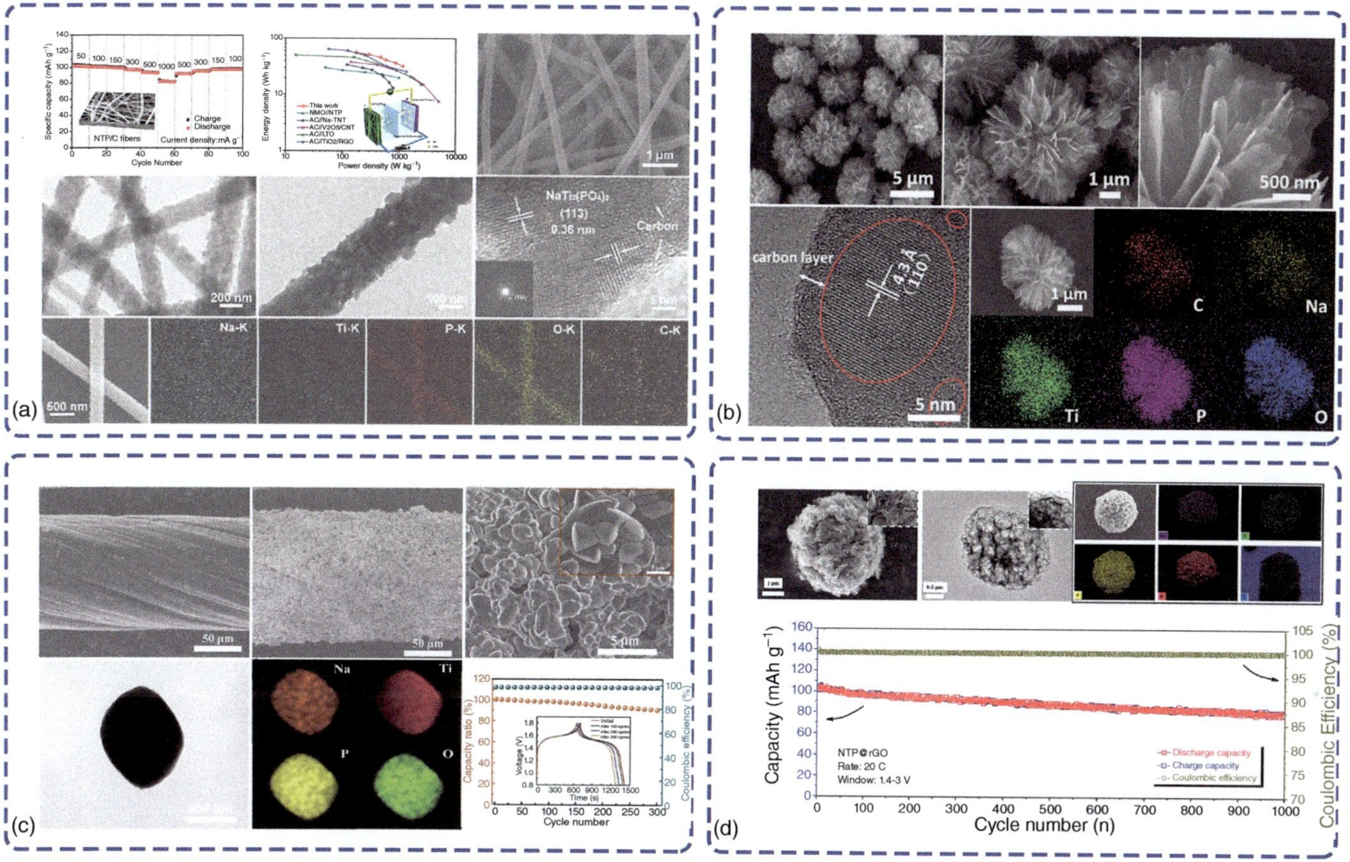

Figure 7.23 (a) Morphology and electrochemical performance of $NaTi_2(PO_4)_3$/C nanofibers composite materials. Source: Wei et al. [85a]/with permission from American Chemical Society. (b) Morphology and EDS mappings of the hierarchical $NaTi_2(PO_4)_3$/C mesoporous microflowers. Source: Xu et al. [87]/with permission from Elsevier. (c) Morphology and electrochemical performance of binder-free $NaTi_2(PO_4)_3$; Source: Zhang et al. [88]/with permission from Elsevier. (d) Morphology and electrochemical performance of 3D graphene decorated $NaTi_2(PO_4)_3$. Source: Fang et al. [85b]/with permission from Wiley-VCH.

of 125 mAh g^{-1} for SIBs, and present excellent cycle stability after 10,000 cycles at 20 C (capacity retention: 77.3%). Even at a high current rate of 100 C, it can also deliver a reversible capacity of 95 mAh g^{-1}. Zhang et al. fabricated a novel binder-free $NaTi_2(PO_4)_3$-based electrodes by solvothermal treatment (Figure 7.23c) [88]. The electrode material is used to assemble a high-voltage coaxial-fiber aqueous rechargeable SIBs, and it presents a high specific capacity (37.84 mAh cm^{-3}) and energy density (57.66 mWh cm^{-3}) owing to the advantages of the unique coaxial architecture and the synergy of novel electrode materials. Fang group synthesized a 3D graphene-decorated $NaTi_2(PO_4)_3$ via spray-drying method (Figure 7.23d) [85b]. The graphene-coated nanosized particles present a 3D network to provide high electronic conductivity and accommodate the structural changes for sodium storage. This electrode material delivers a high reversible capacity of 130 mAh g^{-1}, and it also can deliver a capacity of 38 mAh g^{-1} even at an ultra-high current density of 200 C. Furthermore, it can exhibit excellent cycle stability (capacity retention: 77%) with a long cycling test (1000 times).

7.3.4.2.2 Surface Modifications Yan group synthesized $NaTi_2(PO_4)_3$@polypyrrole (PPy) via sol–gel method and self-assembly procedure (Figure 7.24a) [85c]. The PPy coated on $NaTi_2(PO_4)_3$ particles not only increases its electronic conductivity but also provides strong physical protection to avoid effective components falling off the current conductor. The $NaTi_2(PO_4)_3$@PPy composite presents a specific capacity of 95.9 mAh g^{-1} (20 C) after 400 cycles. Wang group fabricated $NaTi_2(PO_4)_3$ embedded mesoporous carbon nanoparticles via sol–gel method (Figure 7.24b) [89]. The electrode material exhibits a low working voltage for half cells and presents a high capacity of 208 mAh g^{-1}. Besides, the anode exhibits an excellent cycle stability over 10 000 cycles at 4 A g^{-1} (capacity retention: 68%). Yang et al. in situ synthesized $NaTi_2(PO_4)_3$/C nanocomposites (20–40 nm) via a facile solvothermal method [34b]. The nanoparticles can improve the availability of active material and present an excellent rate performance (specific capacity: 66 mAh g^{-1} at 50 C).

7.3.4.2.3 Doping Wei group synthesized F-doped $Na_{1-2x}Ti_2(PO_4)_{3-x}F_x$ nanoparticles/3D carbon by sol–gel method (Figure 7.25a) [90]. In this work, fluorine atoms are successfully doped into the $NaTi_2(PO_4)_3$ crystal and the $NaTi_2(PO_4)_3$ structure is stabilized with improved electronic/ionic conductivity. Furthermore, Na^+ diffusion kinetics is also improved after fluorine doping significantly, resulting in excellent rate performance. Even at 30 C, it still can deliver a reversible capacity of 62.5 mAh g^{-1}. Furthermore, it presents a high capacity retention of 70% after 1000 cycles at 10 C. Xu et al. successfully introduced Gd^{3+} in mesoporous $NaTi_2(PO_4)_3$ nanocrystals to substitute Ti atoms to tailor the texture and conductivity simultaneously (Figure 7.25b) [91]. The optimized porous structure can enhance intrinsic electron conductivity, and the Gd^{3+}-doped mesoporous $NaTi_2(PO_4)_3$ nanocrystals show better Na ion mobility. The mesoporous $NaTi_2(PO_4)_3$ nanocrystals with 5% Gd^{3+} present a high initial Coulombic efficiency of 97% and capacity retention over 75% at 5 C after 2000 cycles. In full cells, a stable capacity is maintained after 30 000 cycles without obvious degradation at 5 A g^{-1}. Voronina group studied the effect of

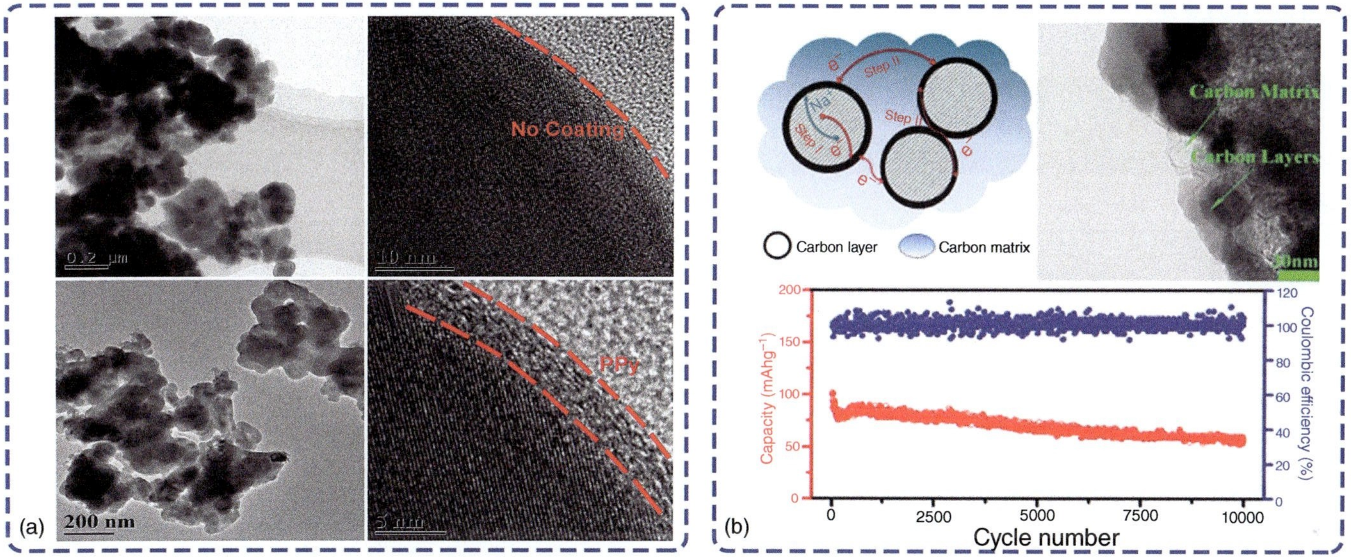

Figure 7.24 (a) Comparison of morphology of $NaTi_2(PO_4)_3$ and $NaTi_2(PO_4)_3$@PPy. Source: Yan et al. [85c]/with permission from Elsevier. (b) Morphology and electrochemical performance of $NaTi_2(PO_4)_3$ embedded mesoporous carbon nanoparticles. Source: Wang et al. [89]/with permission from American Chemical Society.

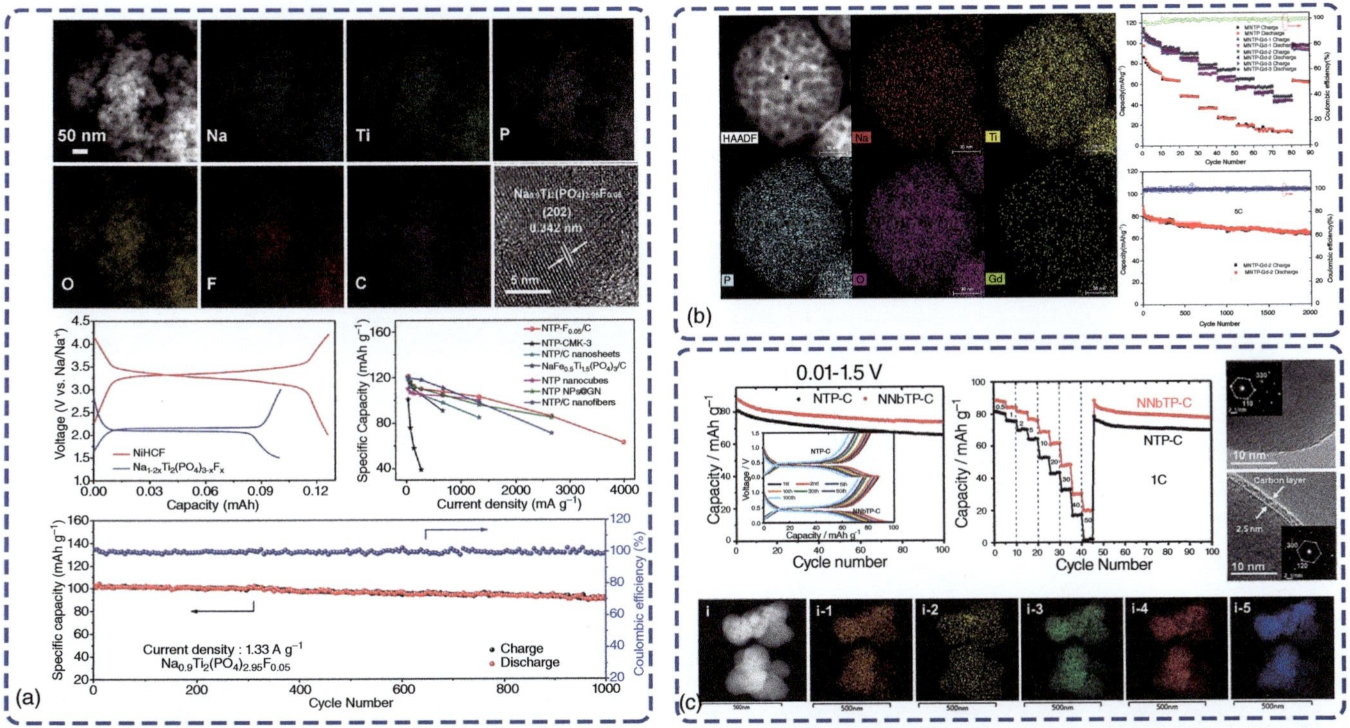

Figure 7.25 Morphology and electrochemical performance of (a) NTP-F_x/C composite. Source: Wei et al. [90]/with permission from American Chemical Society. (b) Gd^{3+} doped mesoporous $NaTi_2(PO_4)_3$ nanocrystals. Source: Xu et al. [91]/with permission from Elsevier. (c) Nb^{5+} doped in the $NaTi_2(PO_4)_3$ nanocrystals. Source: Voronina et al. [92]/with permission from Royal Society of Chemistry.

Nb^{5+} substitution in the $NaTi_2(PO_4)_3$ nanocrystals (Figure 7.25c) [92]. Nb^{5+} substitution can reduce about one-half of the band gap energy to 1.4 eV, and the electrical conductivity of $NaNb_{0.05}Ti_{1.95}(PO_4)_3$ can be enhanced by the carbonization of pitch carbon on the surface. For electrode materials, two redox reactions are related to the $Ti^{4+/3+}$ and $Ti^{3+/2+}$ couple. In a symmetric cell, it presents a high specific capacity of 105 mAh g^{-1} after 100 cycles (capacity retention: 83%).

7.3.5 $TiNb_2O_7$

7.3.5.1 Structure and Properties of $TiNb_2O_7$

$TiNb_2O_7$ belongs to monoclinic system with a space group of $C2/m$ derived from perovskite structure [93]. Ti and Nb ions are distributed in the form of octahedra, which are linked by corner or edge, as shown in Figure 7.26. These octahedra arrange into layer structures (A–B–A) along *b*-axis direction [94]. The wide 2D space in the crystal structure of $TiNb_2O_7$ ensures high diffusion rate of sodium ions. $TiNb_2O_7$ presents a theoretical capacity of 387.6 mAh g^{-1} owing to a reaction with five electrons transfer (Ti^{4+}/Ti^{3+}, Nb^{5+}/Nb^{4+}, and Nb^{4+}/Nb^{3+}) during the cycle process [34h]. Notably, the crystal structure of $TiNb_2O_7$ shows slight volume change for sodium-ion intercalation reaction, suggesting the excellent electrochemical performance of anode materials for SIBs [94, 95].

7.3.5.2 Modification Strategies of $TiNb_2O_7$

Despite $TiNb_2O_7$ has great potential for sodium storage, it also suffers from the drawbacks of poor transmission capability of electrons and ions [4c]. So far, there are a few reports about $TiNb_2O_7$ as anode materials for sodium storage, and some strategies to improve its electric conductivity have been explored to enhance the sodium storage performance, such as composite with carbon materials [34h, 96]. Li et al. synthesized layered $TiNb_2O_7$/graphene composite via a freeze drying and calcination process (Figure 7.27a) [34h]. The $TiNb_2O_7$ nanoparticles are embedded into the interlayer of graphene, and the graphene sheets can improve its electric conductivity. The electrode material exhibits a reversible capacity of 340.2 mAh g^{-1}, and a stable capacity of 200 mAh g^{-1} can also be obtained at 200 mA g^{-1}. Huang group reported a method to improve electrochemical performance of $TiNb_2O_7$ for SIBs anode by ball-milled method (Figure 7.27b) [4c]. The ball-milled $TiNb_2O_7$ provides a specific capacity of 180 mAh g^{-1} at 15 mA g^{-1}, and it presents a high capacity retention

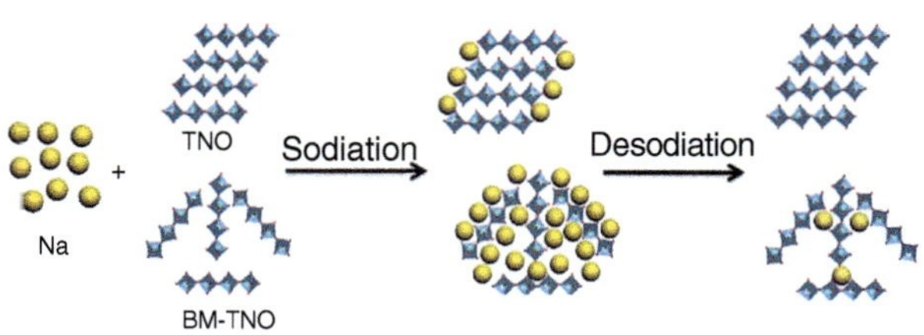

Figure 7.26 Structure diagram of $TiNb_2O_7$. Source: Huang et al. [4c]. Reproduced with permission, 2017, American Chemical Society.

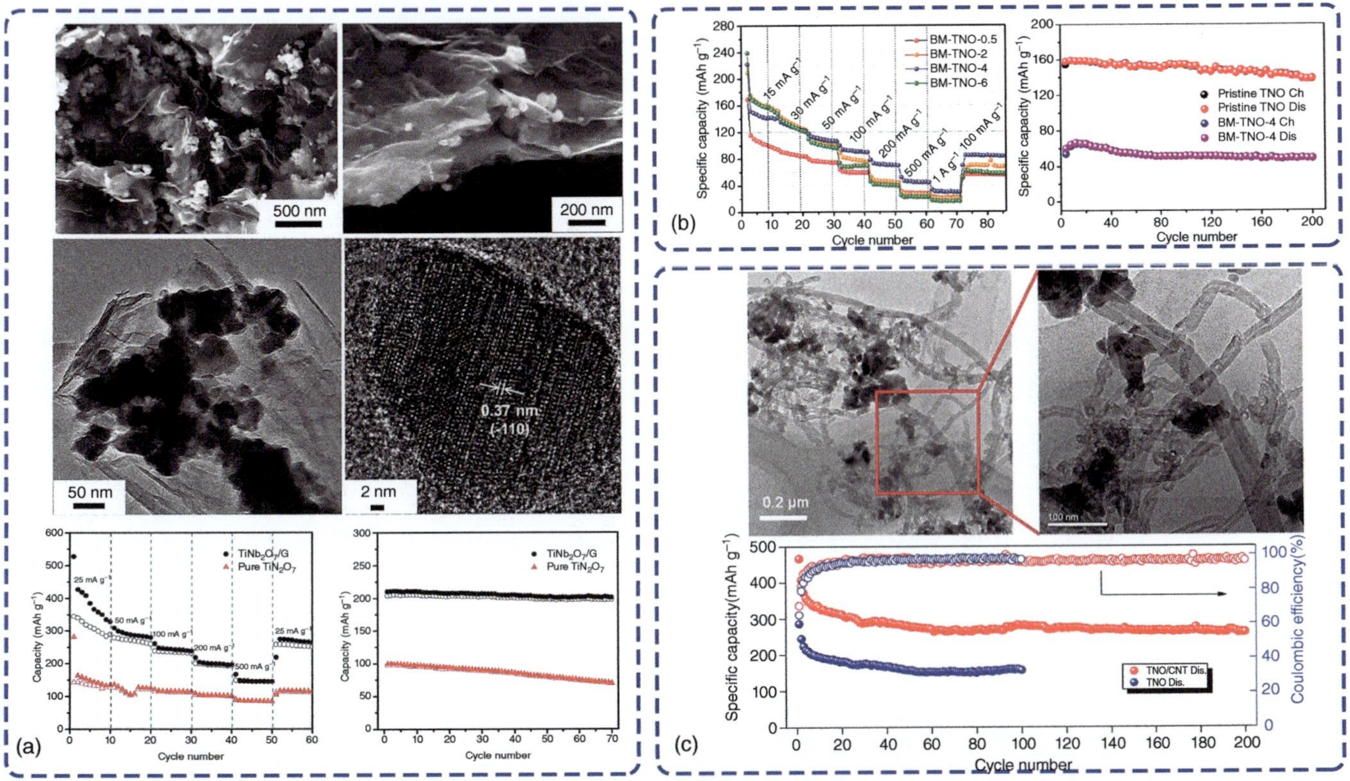

Figure 7.27 (a) Morphology and electrochemical performance of $TiNb_2O_7$/graphene composite. Source: Li et al. [34h]/with permission from Royal Society of Chemistry. (b) Electrochemical performance of ball-milled $TiNb_2O_7$. Source: Huang et al. [4c]. Reproduced with permission, 2017, American Chemical Society. (c) Morphology and electrochemical performance of $TiNb_2O_7$/carbon nanotubes composite. Source: Shang et al. [96]/with permission from Springer.

of 95% after 500 cycles at 500 mA g^{-1}. Shang et al. fabricated $TiNb_2O_7$/carbon nanotubes composite by ultrasonic dispersion and solvothermal method (Figure 7.27c) [96]. The electrode material presents a high specific capacity of 261.1 mAh g^{-1} after 200 cycles. After 1000 cycles, it still can retain a specific capacity of 110 mAh g^{-1} at 500 mA g^{-1}.

7.4 Concluding Remarks

In summary, the understanding of intercalation-type anode materials operating in the Na system is still in its infancy, and challenges remain in large-scale and high-power applications. Among the above-mentioned types of intercalation-type materials used as anode of SIBs, carbon-based materials are the cornerstone for the commercialization of SIBs. In addition, the presence of Ti^{3+}/Ti^{4+} redox pairs in the titanium-based materials makes them possible to avoid dangerous sodium plating in SIBs with redox potentials typically between 0.5 and 1.0 V, ensuring high safety in large-scale operation. And the low lattice strain of titanium-based compounds greatly facilitates the long-term cycling for SIBs.

Intercalation-type anode materials, however, suffer from sluggish Na^+ diffusion kinetics, and rate capability, which inhibit their further commercial application. Various improvement strategies are discussed in this chapter: the design of micro-nano structures can shorten the ion and electron transport distances; followed by doping strategies to introduce defects or improve the electronic conductivity; and surface modification provides strong physical protection against active ingredient shedding. Therefore, these effective strategies can significantly improve the performance of sodium storage of intercalation-type anode materials.

Although great progress has been made in the application of intercalation-type anode in SIBs, there are still many challenges to overcome. The improvement of intercalation-type anode materials and future development focus can be expanded from the following directions: (i) control the microstructure of the material at the molecular level that is more suitable for the rapid (de)intercalation of Na^+; (ii) investigation of the structure-property relationship of the electrodes by combining theoretical calculations and advanced in-situ characterization techniques; (iii) improving the electronic and ionic conductivity of the host material; and (iv) optimization of other components of the battery (e.g., electrolyte and separator) to improve its compatibility and stability.

References

1 (a) Shen, L., Shi, S., Roy, S. et al. (2021). *Adv. Funct. Mater.*, 31: 2006066. (b) Jiang, J., Nie, G., Nie, P. et al. (2020). *Nano Micro Lett.* 12: 183.

2 Zhai, H.F., Xia, B.Y., and Park, H.S. (2019). *J. Mater. Chem. A* 7: 22163.

3 Lou, S.F., Zhao, Y., Wang, J.J. et al. (2019). *Small* 15: 1904740.

4 (a) Chen, C.J., Wen, Y.W., Hu, X.L. et al. (2015). *Nat. Commun.* 6: 6929. (b) Kim, K.T., Yu, C.Y., Yoon, C.S. et al. (2015). *Nano Energy* 12: 725. (c) Huang, Y.Y., Li, X., Luo, J.H. et al. (2017). *ACS Appl. Mater. Interfaces* 9: 8696. (d) Li, M., Xiao, X.Z., Fan, X.L. et al. (2017). *J. Alloys Compd.* 712: 365.

5 (a) Hou, H.S., Qiu, X.Q., Wei, W.F. et al. (2017). *Adv. Energy Mater.* 7. (b) Zhang, S.W., Zhang, J., Wu, S.D. et al. (2017). *Acta Chim. Sin.* 75: 163. (c) Ou, J.K., Yang, L., Zhang, Z., and Xi, X.H. (2017). *Micropor. Mesopor. Mater.* 237: 23. (d) Qu, Y.H., Deng, Y.M., Li, Q. et al. (2017). *J. Mater. Sci.* 52: 2356. (e) Shi, X.D., Chen, Y.X., Lai, Y.Q. et al. (2017). *Carbon* 123: 250.

6 Yue, L.C., Zhao, H.T., Wu, Z.G. et al. (2020). *J. Mater. Chem. A* 8: 11493.

7 Saurel, D., Orayech, B., Xiao, B. et al. (2018). *Adv. Energy Mater.* 8: 1703268.

8 Jache, B. and Adelhelm, P. (2014). *Angew. Chem. Int. Ed.* 53: 10169.

9 Wen, Y., He, K., Zhu, Y. et al. (2014). *Nat. Commun.* 5: 4033.

10 Wang, P.Z., Zhu, X.S., Wang, Q.Q. et al. (2017). *J. Mater. Chem. A* 5: 5761.

11 Yamamoto, H., Muratsubaki, S., Kubota, K. et al. (2018). *J. Mater. Chem. A* 6: 16844.

12 Zhang, Y., Li, L., Xiang, Y.E. et al. (2020). *ACS Appl. Mater. Interfaces* 12: 30431.

13 Govindaraj, D., Rajan, M., Munusamy, M.A. et al. (2018). *Mater. Chem. Phys.* 220: 470.

14 Prabakar, S.J.R., Jeong, J., and Pyo, M. (2015). *Electrochim. Acta* 161: 23.

15 Meyer, F.W., Vogel, N., Diele, K. et al. (2016). *PLoS One* 11: e0149598.

16 (a) Katsuyama, Y., Nakayasu, Y., Kobayashi, H. et al. (2020). *ChemSusChem* https://doi.org/10.1002/cssc.202001837. (b) Pozio, A., Aurora, A., and Prosini, P.P. (2020). *J. Power Sources* 449. (c) Alvin, S., Yoon, D., Chandra, C. et al. (2019). *J. Power Sources* 430: 157.

17 (a) Cheng, H., Garcia-Araez, N., and Hector, A.L. (2020). *ACS Appl. Energy Mater.* 3: 4286. (b) Beda, A., Villevieille, C., Taberna, P.L. et al. (2020). *J. Mater. Chem. A* 8: 5558. (c) Shirvanimoghaddam, K., Ghasali, E., Pakseresht, A. et al. (2019). *J. Alloys Compd.* 775: 601.

18 Talekar, S., Patti, A.F., Vijayraghavan, R., and Arora, A. (2018). *ACS Sustain. Chem. Eng.* 6: 16363.

19 Jayaraman, S., Jain, A., Ulaganathan, M. et al. (2017). *Chem. Eng. J.* 316: 506.

20 Dou, X.W., Hasa, I., Hekmatfar, M. et al. (2017). *ChemSusChem* 10: 2668.

21 Lv, W.M., Wen, F.S., Xiang, J.Y. et al. (2015). *Electrochim. Acta* 176: 533.

22 Lin, X.Y., Liu, Y.Z., Tan, H., and Zhang, B. (2020). *Carbon* 157: 316.

23 Kamiyama, A., Kubota, K., Nakano, T. et al. (2020). *ACS Appl. Energy Mater.* 3: 135.

24 Hasegawa, G., Kanamori, K., Kannari, N. et al. (2015). *ChemElectroChem* 2: 1917.

25 Li, N., Yang, Q., Wei, Y. et al. (2020). *J. Mater. Chem. A* 8: 20486.

26 Li, Z.F., Chen, Y.C., Jian, Z.L. et al. (2018). *Chem. Mater.* 30: 4536.

27 (a) Jian, Z.L., Bommier, C., Luo, L.L. et al. (2017). *Chem. Mater.* 29: 2314. (b) Yao, X.H., Ke, Y.J., Ren, W.H. et al. (2019). *Adv. Energy Mater.* 9: 1900094.

28 Luo, W., Jian, Z.L., Xing, Z.Y. et al. (2015). *ACS Central Sci.* 1: 516.

29 Miao, Y.L., Zong, J., and Liu, X.J. (2017). *Mater. Lett.* 188: 355.

30 Cao, B., Liu, H., Xu, B. et al. (2016). *J. Mater. Chem. A* 4: 6472.

31 (a) Qiu, B.C., Xing, M.Y., and Zhang, J.L. (2014). *J. Am. Chem. Soc.* 136: 5852. (b) Ding, J., Hu, W.B., Paek, E., and Mitlin, D. (2018). *Chem. Rev.* 118: 6457. (c) Deng, D., Kim, M.G., Lee, J.Y., and Cho, J. (2009). *Energy Environ. Sci.* 2: 818. (d) Zhao, L., Hu, Y.S., Li, H. et al. (2011). *Adv. Mater.* 23: 1385.

32 (a) Jung, H.G., Myung, S.T., Yoon, C.S. et al. (2011). *Energy Environ. Sci.* 4: 1345. (b) Park, K.S., Benayad, A., Kang, D.J., and Doo, S.G. (2008). *J. Am. Chem. Soc.* 130: 14930. (c) Zhu, G.N., Liu, H.J., Zhuang, J.H. et al. (2011). *Energy Environ. Sci.* 4: 4016.

33 Wang, C., Wang, S.A., He, Y.B. et al. (2015). *Chem. Mater.* 27: 5647.

34 (a) Roh, H.K., Kim, H.K., Kim, M.S. et al. (2016). *Nano Res.* 9: 1844. (b) Yang, J., Wang, H., Hu, P.F. et al. (2015). *Small* 11: 3744. (c) Feng, W.L., Maca, R.R., and Etacheri, V. (2020). *ACS Appl. Mater. Interfaces* 12: 4443. (d) Cha, G., Mohajernia, S., Nguyen, N.T. et al. (2020). *Adv. Energy Mater.* 10: 1903448. (e) Wang, P.Y., Lu, X.X., Boyjoo, Y. et al. (2020). *J. Power Sources* 451: 227756. (f) Li, Z., Young, D., Xiang, K. et al. (2013). *Adv. Energy Mater.* 3: 290. (g) Wu, X.Y., Cao, Y.L., Ai, X.P. et al. (2013). *Electrochem. Commun.* 31: 145. (h) Li, S., Cao, X., Schmidt, C.N. et al. (2016). *J. Mater. Chem. A* 4: 4242.

35 Pan, T., Song, G., Cong, S. et al. (2020). *J. Phys. Chem. C* 124: 20530.

36 Mendoza, R., Rodriguez-Gonzalez, V., Oliva, A.I. et al. (2020). *Mater. Chem. Phys.* 255: 123602.

37 Wang, F., Yang, M., Zhang, Y. et al. (2020). *Chem. Eng. J.* 402: 126303.

38 (a) Khan, S.U.M., Al-Shahry, M., and Ingler, W.B. (2002). *Science* 297: 2243. (b) Zhou, W.J., Yin, Z.Y., Du, Y.P. et al. (2013). *Small* 9: 140. (c) Fujishima, A., Zhang, X.T., and Tryk, D.A. (2008). *Surf. Sci. Rep.* 63: 515.

39 (a) Gao, F., Wang, Y., Shi, D. et al. (2008). *J. Am. Chem. Soc.* 130: 10720. (b) van de Lagemaat, J., Park, N.G., and Frank, A.J. (2000). *J. Phys. Chem. B* 104: 2044.

40 (a) Seh, Z.W., Li, W.Y., Cha, J.J. et al. (2013). *Nat. Commun.* 4: 1331. (b) Wang, D.H., Choi, D.W., Li, J. et al. (2009). *ACS Nano* 3: 907. (c) Bavykin, D.V., Friedrich, J.M., and Walsh, F.C. (2006). *Adv. Mater.* 18: 2807. (d) Hu, Y.S., Kienle, L., Guo, Y.G., and Maier, J. (2006). *Adv. Mater.* 18: 1421.

41 (a) Xu, Y., Lotfabad, E.M., Wang, H.L. et al. (2013). *Chem. Commun.* 49: 8973. (b) Wang, N.N., Bai, Z.C., Qian, Y.T., and Yang, J. (2016). *Adv. Mater.* 28: 4126. (c) Zhang, Y., Foster, C.W., Banks, C.E. et al. (2016). *Adv. Mater.* 28: 9391. (d) Yang, Y.C., Ji, X.B., Jing, M.J. et al. (2015). *J. Mater. Chem. A* 3: 5648. (e) Wang, X.Y., Fan, L., Gong, D.C. et al. (2016). *Adv. Funct. Mater.* 26: 1104. (f) Tahir, M.N., Oschmann, B., Buchholz, D. et al. (2016). *Adv. Energy Mater.* 6: 1501489. (g) Longoni, G., Cabrera, R.L.P., Polizzi, S. et al. (2017). *Nano Lett.* 17: 992. (h) Kim, H., Kim, J.C., Bianchini, M. et al. (2018). *Adv. Energy Mater.* 8: 1702384.

42 Xiong, H., Slater, M.D., Balasubramanian, M. et al. (2011). *J. Phys. Chem. Lett.* 2: 2560.

43 Bi, Z., Paranthaman, M.P., Menchhofer, P.A. et al. (2013). *J. Power Sources* 222: 461.

44 Bella, F., Muñoz-García, A.B., Meligrana, G. et al. (2017). *Nano Res.* 10: 2891.

45 Ni, Q., Dong, R., Bai, Y. et al. (2020). *Energy Storage Mater.* 25: 903.

46 Hwang, J.-Y., Myung, S.-T., Lee, J.-H. et al. (2015). *Nano Energy* 16: 218.
47 Yang, X., Wang, C., Yang, Y. et al. (2015). *J. Mater. Chem. A* 3: 8800.
48 Harada, Y., Hoshina, K., Inagaki, H., and Takami, N. (2013). *Electrochim. Acta* 112: 310.
49 Dawson, J.A. and Robertson, J. (2016). *J. Phys. Chem. C* 120: 22910.
50 Huang, J.P., Yuan, D.D., Zhang, H.Z. et al. (2013). *RSC Adv.* 3: 12593.
51 Wu, L.M., Bresser, D., Buchholz, D., and Passerini, S. (2015). *J. Electrochem. Soc.* 162: A3052.
52 Zhang, Y., Ding, Z.Y., Foster, C.W. et al. (2017). *Adv. Funct. Mater.* 27: 1700856.
53 Krtil, P. and Fattakhova, D. (2001). *J. Electrochem. Soc.* 148: A1045.
54 Baudrin, E., Cassaignon, S., Koesch, M. et al. (2007). *Electrochem. Commun.* 9: 337.
55 Usui, H., Yoshioka, S., Wasada, K. et al. (2015). *ACS Appl. Mater. Interfaces* 7: 6567.
56 Feckl, J.M., Fominykh, K., Doblinger, M. et al. (2012). *Angew. Chem. Int. Ed.* 51: 7459.
57 Zhao, L., Pan, H.-L., Hu, Y.-S. et al. (2012). *Chinese Phys. B* 21: 028201.
58 Sun, Y., Zhao, L., Pan, H. et al. (2013). *Nat. Commun.* 4: 1870.
59 Roh, H.K., Lee, G.W., Haghighat-Shishavan, S. et al. (2020). *Chem. Eng. J.* 385: 123984.
60 Yang, L.Y., Li, H.Z., Cheng, L.Z. et al. (2017). *J. Alloys Compd.* 726: 837.
61 Zhou, Q., Liu, L., Tan, J.L. et al. (2015). *J. Power Sources* 283: 243.
62 Tian, Y., Wu, Z.L., Xu, G.B. et al. (2017). *RSC Adv.* 7: 3293.
63 Gao, L., Wang, L.C., Dai, S.R. et al. (2017). *J. Power Sources* 344: 223.
64 Kitta, M. and Kohyama, M. (2018). *Electrochemistry* 86: 194.
65 Wu, Z.L., Xu, G.B., Wei, X.L., and Yang, L.W. (2016). *Electrochim. Acta* 207: 275.
66 Ge, Y.Q., Jiang, H., Fu, K. et al. (2014). *J. Power Sources* 272: 860.
67 Yun, B.N., Du, H.L., Hwang, J.Y. et al. (2017). *J. Mater. Chem. A* 5: 2802.
68 Liu, J., Tang, K., Song, K. et al. (2013). *Phys. Chem. Chem. Phys.* 15: 20813.
69 Liu, Y., Liu, J., Hou, M. et al. (2017). *J. Mater. Chem. A* 5: 10902.
70 (a) Senguttuvan, P., Rousse, G., Seznec, V. et al. (2011). *Chem. Mater.* 23: 4109. (b) Pan, H.L., Lu, X., Yu, X.Q. et al. (2013). *Adv. Energy Mater.* 3: 1186. (c) Rudola, A., Saravanan, K., Mason, C.W., and Balaya, P. (2013). *J. Mater. Chem. A* 1: 2653. (d) Fu, S.D., Ni, J.F., Xu, Y. et al. (2016). *Nano Lett.* 16: 4544. (e) Li, H.S., Peng, L.L., Zhu, Y. et al. (2016). *Energy Environ. Sci.* 9: 3399.
71 Yan, Z., Liu, L., Shu, H. et al. (2015). *J. Power Sources* 274: 8.
72 Chen, S., Pang, Y., Liang, J., and Ding, S. (2018). *J. Mater. Chem. A* 6: 13164.
73 (a) Wu, M., Zhang, G., Chen, N. et al. (2020). *Energy Storage Mater.* 24: 272. (b) Zhang, F. and Qi, L. (2016). *Adv. Sci.* 3: 1600049.
74 Kong, D.Z., Wang, Y., Huang, S.Z. et al. (2019). *J. Mater. Chem. A* 7: 12751.
75 Li, Z.H., Shen, W., Wang, C. et al. (2016). *J. Mater. Chem. A* 4: 17111.
76 Xie, F., Zhang, L., Su, D. et al. (2017). *Adv. Mater.* 29: 1700989.
77 Wang, N., Xu, X., Liao, T. et al. (2018). *Adv. Mater.* 30: 1804157.
78 Liu, D.-S., Jin, F., Huang, A. et al. (2019). *Chem. Eur. J.* 25: 14881.
79 Chen, J., Zhou, X., Mei, C. et al. (2017). *Electrochim. Acta* 224: 446.

80 Xia, J., Zhao, H., Pang, W.K. et al. (2018). *Chem. Sci.* 9: 3421.

81 (a) Patoux, S., Rousse, G., Leriche, J.B., and Masquelier, C. (2003). *Chem. Mater.* 15: 2084. (b) Li, N., Wang, Y.P., Rao, R.C. et al. (2017). *Appl. Surf. Sci.* 399: 624. (c) Kabbour, H., Coillot, D., Colmont, M. et al. (2011). *J. Am. Chem. Soc.* 133: 11900. d Li, X.N., Zhu, X.B., Liang, J.W. et al. (2014). *J. Electrochem. Soc.* 161: A1181.

82 Delmas, C., Cherkaoui, F., Nadiri, A., and Hagenmuller, P. (1987). *Mater. Res. Bull.* 22: 631.

83 Pang, G., Nie, P., Yuan, C.Z. et al. (2014). *J. Mater. Chem. A* 2: 20659.

84 Ye, S.C., Li, Z.H., Song, T.B. et al. (2017). *RSC Adv.* 7: 56743.

85 (a) Wei, P., Liu, Y.X., Wang, Z.H. et al. (2018). *ACS Appl. Mater. Interfaces* 10: 27039. (b) Fang, Y.J., Xiao, L.F., Qian, J.F. et al. (2016). *Adv. Energy Mater.* 6: 1670112. (c) Yan, H.Y., Xue, X.X., Chen, W.X. et al. (2020). *Appl. Surf. Sci.* 530: 147295.

86 (a) Wei, T.Y., Yang, G.Z., and Wang, C.X. (2017). *ACS Appl. Mater. Interfaces* 9: 31861. (b) Ding, Y.L., Kopold, P., Hahn, K. et al. (2016). *Adv. Mater.* 28: 7774. (c) Fan, X.Y., Han, J.X., Ding, Y.L. et al. (2019). *Adv. Energy Mater.* 9: 1900673. d) Ding, Y.L., Wu, C., Kopold, P. et al. (2015). *Small* 11: 6026.

87 Xu, C., Xu, Y.N., Tang, C.J. et al. (2016). *Nano Energy* 28: 224.

88 Zhang, Q.C., Man, P., He, B. et al. (2020). *Nano Energy* 67: 104212.

89 Wang, D.X., Liu, Q., Chen, C.J. et al. (2016). *ACS Appl. Mater. Interfaces* 8: 2238.

90 Wei, P., Liu, Y.X., Su, Y.R. et al. (2019). *ACS Appl. Mater. Interfaces* 11: 3116.

91 Xu, G.B., Chen, Z., Liu, X. et al. (2019). *Electrochim. Acta* 309: 177.

92 Voronina, N., Jo, J.H., Choi, J.U. et al. (2019). *J. Mater. Chem. A* 7: 5748.

93 Cava, J., Murphy, D.W., and Zahurak, S.M. (1983). *J. Electrochem. Soc.* 130: 2345.

94 Lu, X., Jian, Z.L., Fang, Z. et al. (2011). *Energy Environ. Sci.* 4: 2638.

95 Guo, B.K., Yu, X.Q., Sun, X.G. et al. (2014). *Energy Environ. Sci.* 7: 2220.

96 Shang, B., Peng, Q.M., Jiao, X. et al. (2019). *Ionics* 25: 1679.

8

Phosphorus/Phosphide Anodes for Sodium–Ion Batteries on Alloy and Conversion Reactions

8.1 Introduction

Compared with insertion-type anodes, some materials are capable of storing more Na^+ through alloy or conversion reactions and thus deliver two or three times higher theoretical capacity, showing more competitive in high-energy storage. For example, phosphorus material processes the highest theoretical capacity (2596 $mAh\,g^{-1}$) for Na-storage among all kinds of anode materials based on the alloy reaction mechanism, making it an ideal candidate for practical application of high-energy sodium–ion batteries (SIBs). In addition, the metal phosphide materials display high theoretical capacities based on the multiple electrons transfer behaviors via alloy reaction, which are much higher than that of insertion-type carbonaceous anodes for SIBs.

Owing to the high theoretical capacities, anode materials for SIBs based on alloy and conversion reaction have gained growing research interests and achieved significant progress. Nevertheless, further improvements on long-term cycling stability of these alloy- and conversion-type anode materials are still required for successful commercialization. The undesired cyclability can be ascribed to the following three main reasons. Firstly, the huge volume expansion during sodiation/desodiation process due to the larger ionic radius of Na^+ than that of Li^+ leads to severe problems in suppressing the mechanical stress generated within the whole electrode. Notably, the great volume expansion effect results in the pulverization of active particles, which has two main negative impacts on cycling performance. On the one hand, it causes the exfoliation of active particles from the current collector and thereby the loss of electrical contact. On the other hand, it makes the fresh surface of active materials exposed to the liquid electrolyte all the time, which is accompanied by continuous formation of solid electrolyte interphase (SEI) film on the solid–liquid interface. The unstable SEI film dramatically blocks the charge transfer on the solid/liquid interface and results in the quick capacity loss during cycling. Secondly, the low Coulombic efficiency induced by the unstable SEI film presents another significant challenge for anode materials based on alloy reaction and conversion reaction for SIBs. Thirdly, the sluggish redox reaction kinetics due to the inferior electronic conductivity of some alloy- and conversion-type anode

Sodium-Ion Batteries: Energy Storage Materials and Technologies, First Edition. Yan Yu.

materials usually lead to unsatisfied rate performance and low power density, thus dramatically restricting their further applications for SIBs.

To date, numerous efforts have been devoted to solving these intrinsic issues faced by alloy/conversion-type anode materials for SIBs and overwhelming progress has been achieved. The proposed strategies toward enhancing the sodium-storage performance of these materials for SIBs mainly focus on the structural design of nanostructures and nanoarchitectures of electrode materials, compositing conductive compound host/matrix (e.g. carbon nanotubes, graphene), and developing innovative electrode configuration, all of which could effectively boost the electrode reaction kinetics and the cycling stability.

8.2 Phosphorus Anodes

8.2.1 Phosphorus Allotropes

Phosphorus is a high reserve-abundant element from pnictogens (symbol P, atomic number 15, period 3) in the period table of elements. Phosphorus prefers to lose the outer electrons leading to high reactivity, therefore the straightforward access of free P element from the natural Earth is significantly hard. With few exceptions, it exists in the maximally oxidized state in minerals containing phosphorus as inorganic phosphate rocks, $Ca_3(PO_4)_2$, for example.

Phosphorus can take several forms that exhibit strikingly diverse properties, which fall into four allotropes: white, red, black, and violet, as shown in Figure 8.1. In the point of structural dimension view, phosphorus can be classified into: low

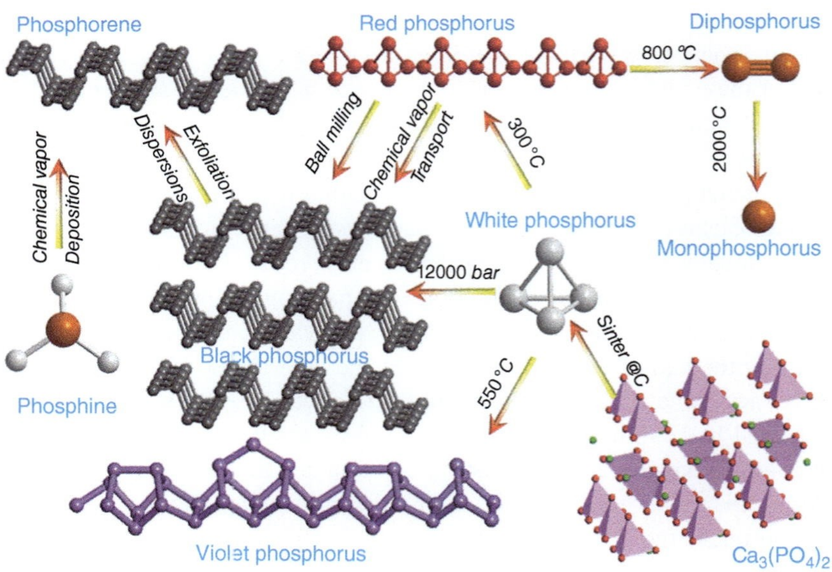

Figure 8.1 Phosphorus allotropes with transformation reactions from different phosphorus structures. Source: Pang et al. [1]. Reproduced with permission, 2017, Wiley-VCH.

dimensional molecular structures (0D, white phosphorus, abbreviated as WP), the polymeric structures (1D, red phosphorus, abbreviated as RD), the layered (2D, black phosphorus, abbreviated as BP), and tubular structures (2D and 3D, crystalline forms of RD and violet phosphorus, abbreviated as VP).

From the perspective of chemical literatures and applications, WP is the most common and important form, which was firstly discovered by Henning Brand in 1669. WP crystal structure is composed of tetrahedral P_4 molecules bonded with P—P single bond. This structure could be well maintained in liquid state or gaseous state until the temperature reaches 800 °C, when it starts thermolysis into diphosphorus, P_2. So far, three different modifications (α-, β-, and γ-P_4) of WP have been reported, due to the different orientations of the P_4 tetrahedron.

WP is the least stable, the most reactive, and the most volatile among all phosphorus allotropes. The P—P bonds in P_4 tetrahedron with very weak bonding energy result in a low oxidation barrier [2]. WP gradually transforms into RP, especially when there is light and heat. WP samples always contain small amount of RP and therefore appear yellow so as to be called yellow phosphorus. WP is highly incendiary and pyrophoric, which self-ignites spontaneously upon exposure to air, and thus liquid sealing is required for its storage. WP is insoluble in water but can be dissolved in various organic solvents. In addition, WP is the most toxic modification. Great care must be taken for its handling. Considering the battery safety and the viewpoint of practical application, WP is actually not suitable as electrode materials. However, WP can be regarded as the raw material for the production of all the other phosphorus polytypes.

RP is a more stable amorphous phosphorus allotrope, originally synthesized in 1848 by heating WP in closed vessels [3]. It exhibits a derivative of P_4 tetrahedra, in which one P—P bond in a P_4 tetrahedron dissociates and forms an additional bond with P atoms in neighboring P_4 tetrahedron, constructing a chain-like polymeric structure. Today a variety of RPs (type I–V) are known, all of which show a reddish color. RP can be obtained by heating WP to 250 °C or when WP exposed to sunlight. Under this treatment, RP tends to be amorphous and starts to crystalline upon further heating. 1D RP nanowires were first prepared by Ruck et al. in 2005 by heating RP in an evacuated ampoule at around 590 °C for several days [3]. Ruck's fibrous RP, designated as type IV, displays a tube-like structure with an arrangement of double phosphorus chain formed by the interlinked P_8 and P_9 cages. With comparison of WP, RP exhibits several advantages as SIB electrode material, such as higher chemical stability, less reactivity, and lower volatility, due to the strong covalent bonds in its crystal structure.

VP, is an allotropic modification of phosphorus, obtained as a dark red powder though a longtime phase transformation progress of RD. In 1865, Hittorf first discovered this red/purple form of phosphorus, therefore known as Hittorf's phosphorus (type V) [4]. VP exhibits a complex 3D tubular structure. Zigzag chains of phosphorus atoms formed tubes packed parallel to one another to form layers. Every two layers are connected by sharing phosphorus atoms and formed into a perpendicular orientation.

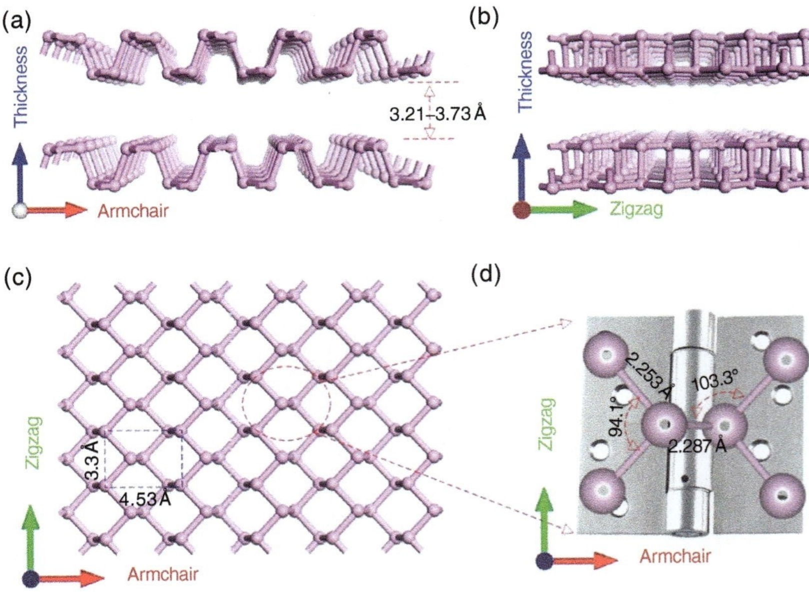

Figure 8.2 Structure of BP: side views from (a) the zigzag and (b) the armchair directions, (c) top view, and (d) zoomed-in local atomic structure. Source: Kou et al. [5]. Reproduced with permission, 2015, American Chemical Society.

Among all the phosphorus allotropes, BP is recognized as the most thermodynamically stable modification under ambient conditions. The typical 2D layered structure of BP is displayed in Figure 8.2. Bulk BP exhibits a stacked multilayer structure along with the armchair direction, which are interconnected by weak van der Waals forces. At the same time, P atoms strongly bonded to each other through covalent P—P bonds consist of a parallel zig-zag chain in the same plane. Each P atom possesses five valence electrons in its 3p orbitals that saturated by forming covalent bonds, yielding sp^3 hybridization of the P atom. Two-fifths are used to form two covalent bonds with neighboring P atoms in the same layer, while one-fifth generates another covalent bond with P atom above or below. As a result, there is a lone pair of electrons in the P structure, leading to the higher reactivity of BP surface than that of other 2D layered structure materials, such as graphite/graphene and hexagonal boron nitride (*h*-BN).

Generally, BP possesses an orthorhombic crystal structure with space group of *Cmca* (lattice constants $a = 4.47$ Å, $b = 3.34$ Å) [6]. The unit cell can be viewed as bilayers of parallel P atomic rows with different P—P distance: the short P—P bond distance connecting the nearest neighboring atoms is 2.25 Å, whereas the long P—P bond length holding the upper and lower P atoms is 2.28 Å. The corresponding angle is 94.1° and 103.3°, respectively (Figure 5.2d). BP is p-type semiconductor with an intrinsically direct band gap of ~0.34 eV, high electronic conductivity of $\sim 10^2$ S m^{-1}, and suitable density of ~2.7 g cm^{-3} [7].

Due to the difficult synthetic process, a few works have been devoted to the studies of BP since the discovery of graphene renewed the research interests in few-layer or monolayer BP. In 2014, few-layered BP (<10 layers) was (also termed "phosphorene") successfully obtained through a sticky-tape technique [8]. Phosphorene exhibits a high carrier mobility and its band gap is layer-dependent. The energy gap of monolayer phosphorene is ~2.0 eV, with a direct band gap shape. There is slight change in the lattice constants from bulk to monolayer (phosphorene: $a = 4.58$ Å, $b = 3.32$ Å). Due to the intrinsic puckered bilayer structure, 2D phosphorene demonstrated various novel anisotropic features, such as integer quantum Hall effect, superior mechanical flexibility, and negative Poisson's ratio, etc. Therefore, phosphorene has shown potential for many applications, including thermoelectric, transistor, solar cell, sensing, photovoltaic devices, and photodetectors.

Simple, low-cost, and reliable fabrication of 2D phosphorene with uniform size and thickness is essential for its future application in high-performance devices in different fields. Driven by its unique structural and physiochemical properties and promising applications of phosphorene, significant achievements in phosphorene fabrication have been realized in the past several years. In the following section, the merits and weaknesses of current synthetic methods for phosphorene and phosphorene-based materials are reviewed. In addition, some possible new approaches for preparing high-quality phosphorene are proposed. Generally, the currently available preparation strategies for fabrication of phosphorene can be classified into two categories. (i) Top–down approach, such as liquid exfoliation and mechanical cleavage. During a typical synthesizing process, mono- or few-layer BP nanosheets are usually obtained by using chemical intercalation or mechanical force to break the weak van der Waals interaction between the stacked layers in the bulk. (ii) Bottom–up method, aiming to the direct synthesis of 2D layered structure relying on chemical reactions, mainly includes wet-chemistry method and chemical vapor deposition (CVD).

8.2.2 Na-Storage Mechanism for Phosphorus-Based Materials

Phosphorus-based materials have received significant research concerns for SIB applications due to their abundant earth reserves and a relatively safe working potential of ~0.4 V vs. Na^+/Na. Among phosphorus allotropes, WP is the most reactive with air and toxic and thus not suitable for serving as the anode material for SIBs. RP exhibits relatively structural stability, and both the amorphous and crystalline RPs have shown electrochemical active with sodium ions in SIBs. In addition, due to the unique physicochemical characteristics, BP has triggered intensive exploration for energy storage devices (such as secondary batteries and supercapacitors).

8.2.2.1 Na-Storage Mechanism for Red Phosphorus

Ever since the discovery of the capability for Na-storage, studies in P-based materials have raised the great interest of research in high-performance SIBs through innovative strategies and optimized modifications. So far, however, limited works have been focused on the in-depth elucidation of the sodiation mechanism. Kim and coworkers took the first attempt on the sodium storage mechanism of P-based anode materials [9]. The final sodiation product of Na_3P phase was detected via ex situ X-ray diffraction (XRD) measurements, indicating the reversible three-electron transfer mechanism ($3Na^+ + 3e^- + P \leftrightarrow Na_3P$) of the obtained RP/C anode. That means, unlike Si, each P takes a reversible alloying reaction with three Na to form Na_3P product, resulting in the theoretical specific capacity as high as 2596 mAh g^{-1}. This value is regarded as the highest one among all the available SIB anode materials so far. Furthermore, the equilibrium redox potential for the insertion of Na^+ in RP/C anode was found to be around 0.45 V, which is ~0.4 V lower than that of Li^+ reduction for lithium-ion batteries (LIBs). It is considered to be an ideal sodiation potential for SIB anodes, offering the capability of preventing the generation of Na-metal dendrite and improving the total energy density of a full sodium–ion cell, giving much potential of phosphorus-based material for large-scale energy storage.

This Na-storage mechanism was further confirmed by Yabuuchi et al. in 2013 by conducting synchrotron XRD (SXRD) [10]. The amorphous P structure was transformed into crystalline Na_3P phase upon fully discharged, and converted back reversibly into amorphous P material after the successive dealloying process in SIBs. Qian et al. reported for the first time an intermediate compound Na_xP during the Na^+ interaction reaction and a stepwise desodiation process. As evidenced by the cyclic voltammogram (CV) results, a series of anodic peaks located between 0.5 and 1.5 V could correspond to Na_xP intermediate products, demonstrating a stepwise Na^+ extraction from the final sodiation product of Na_3P to Na_xP phases during the charging process [11]. In addition, Morrris's group conducted a first-principles study and nuclear magnetic resonance (NMR) calculations to predict structural evolutions during cycling of the phosphorus anode for both LIBs and SIBs [12]. Of particular interest, a various new stable and metastable structures were suggested via the ab initio random structure searching along with the atomic species swapping, which is displayed in Figure 8.3. Nevertheless, the detailed intermediate products during the sodiation/desodiation process remain unclear because most of the sodiation products are amorphous.

8.2.2.2 Na-Storage Mechanism for Black Phosphorus

Compared with RP material, it seems much easier for the exploitation of the electrochemical Na-storage mechanism of BP because of its crystalline structure. Many experimental studies and theoretical calculations have been explored to elucidate the sodiation mechanism of BP in order to further understand the chemical and electrochemical reactions during the sodiation/desodiation process and propose novel strategies to mitigate the performance deterioration.

Sun et al. [13] demonstrated a two-step sodiation mechanism when a phosphorene–graphene composite served as the anode for SIBs. Na^+ ions preferred to first intercalate into the phosphorene layers along the x-axis direction because the adequate space between the phosphorene channels (3.08 Å) is wide enough for the accommodation of Na^+ (2.04 Å). The characteristic diffraction peaks of phosphorene in the XRD pattern (16.9° and 34.2°, respectively) were found to be shifted toward lower 2θ degree with Na^+ intercalation, indicating the enlarged layer distance of the sodiated phosphorene layers. Both the characteristic diffraction

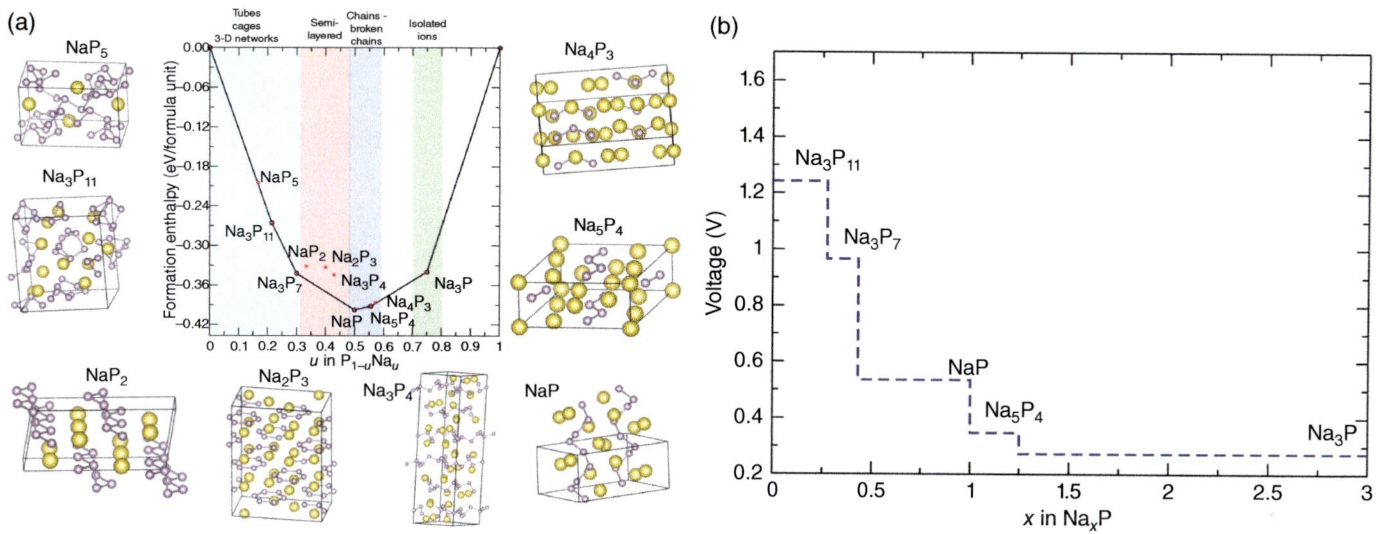

Figure 8.3 (a) Formation enthalpy per atom versus the fractional sodium concentration in the Na−P compound. (b) Average voltages relative to sodium calculated for the structures found on the convex hull. Source: Mayo et al. [12]. Licensed under CC BY 4.0.

reflections of phosphorene were always visible in the XRD pattern upon sodiation, demonstrating that the layered structure of phosphorene could be well maintained until the potential decreased to 0.54 V. With further Na^+ insertion, the reflection signals of phosphorene disappeared. Here, the authors speculated the generation of the intermediate Na_xP phase, although the corresponding diffraction peaks of Na_xP phase were hard to identify. When the potential reached 0.1 V (vs. Na^+/Na), the Na_3P phase was detected as the new diffraction peaks located at 36° and 37° were shown up, which can be well indexed to the (110) and (103) reflections of Na_3P phase.

According to the theoretical calculation of the formation energy of Na_xP during the sodiation process, Na ions tent to first intercalate into the same layer until the coverage per BP layer reaches 25% and subsequently enters into different BP layers along x-axis until forming the $Na_{0.25}P$ phase (Figure 8.4). During this intercalation process, layered structure of BP could be maintained but with sliding occurring between the layers, because the Na atoms tended to bind with neighboring four P atoms without P—P bond broken. When the Na concentration exceeds $Na_{0.25}P$, intercalation reaction starts to change into alloying reaction, resulting in the P—P bond breakage that causes the huge volume expansion of BP. Moreover, further sodiation into the layer structured Na_xP phases results in the preferable staggered P—P bonds cleavage, whereas the planar P—P bonds are essentially unaffected,

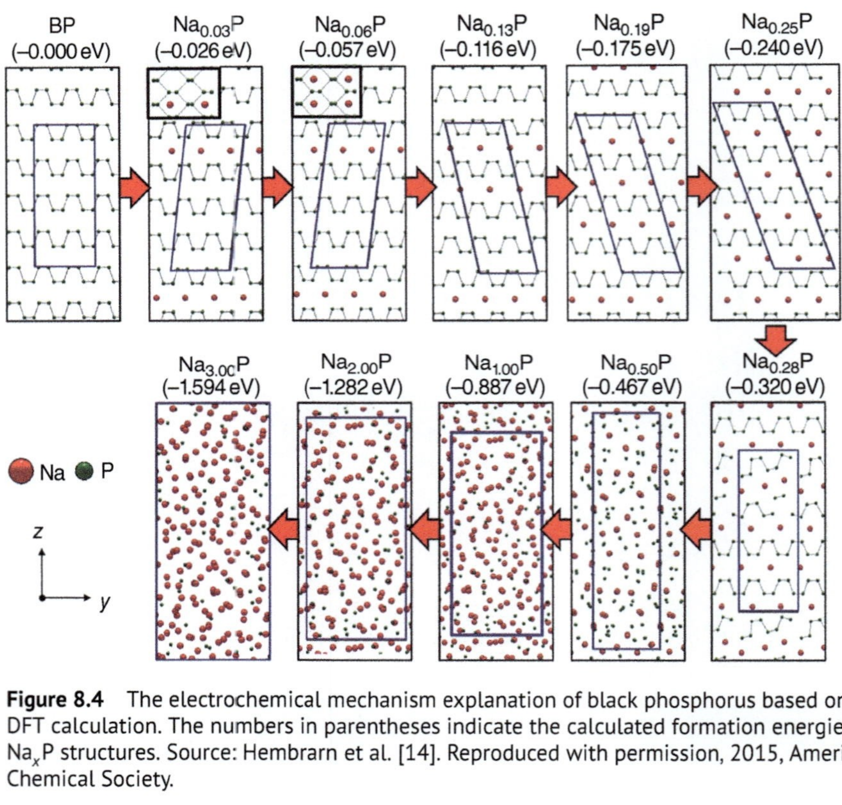

Figure 8.4 The electrochemical mechanism explanation of black phosphorus based on the DFT calculation. The numbers in parentheses indicate the calculated formation energies of Na_xP structures. Source: Hembrarn et al. [14]. Reproduced with permission, 2015, American Chemical Society.

generating the P_2 dumbbells. It indicated that the break of the P—P bonds in the Na_xP phase is much easier than that in the pristine BP, supporting the hypothesis of the intermediate production of the sodiated amorphous Na_xP structure.

Based on these results, it is commonly accepted that sodiation/desodiation mechanism of BP for Na-storage involves two steps: (i) one-dimensional Na^+ intercalation into the interlayer spaces until the formation of $Na_{0.25}P$ phase; (ii) a reversible alloying reaction with further sodiation to transform into a layered composition of Na_xP and a final sodiation product of Na_3P. Therefore, the whole sodiation/desodiation reaction mechanism of BP anode for SIBs can be expressed as: $P + Na \leftrightarrow Na_xP \leftrightarrow Na_3P$.

8.2.3 Phosphorus-Based Materials for Na-Ion Batteries

8.2.3.1 Red Phosphorus for Na-Ion Batteries

RP possesses a chain-like structure, well known for polymers, consisting of covalently bonded P_4 tetrahedra. Compared with other allotropes, RP is more commercially available with ease, relatively stable in air and nontoxic, showing great promising for its practical SIBs implementations. Furthermore, the high covalent characteristics of Na—P bonds in Na_3P leads to the anomalously smaller Na molar volume than that of other Na-metal-alloy compounds. Therefore, it is reasonable that amorphous RP has the highest theoretical volumetric capacity than that of any other alloying-type SIB anode materials. Nevertheless, microscale RP particles face two major obstacles of low electronic conductivity ($\sim 10^{-14}\,S\,cm^{-1}$) and large volume expansion (~440%) upon full sodiation, resulting in sluggish redox kinetics that severely limit its practical application [9]. Another severe issue of electrolyte decomposing caused by the highly reactive Na_3P surface results in fast capacity fading and poor cyclability of RP anodes. To address above-mentioned issues, several effective strategies have been proposed, including hybridizing RP with a conductive matrix, fabricating nanoarchitectures, as well as developing suitable electrolyte additives and electrode binders.

8.2.3.1.1 Combining RP with Conductive Matrix Combining RP particles with highly conductive matrix has been demonstrated as an effective approach not only to fasten the electrons transfer rate but also to alleviate the volume expansion resulted from repeated Na^+ insertion/extraction. Various conductive carbonaceous materials, such as carbon black, carbon nanotubes (CNTs), carbon nanofibers, and graphene, etc. have been proposed to modify the electrochemical performance of RP anode.

The pioneering study of RP as SIB anode was first carried out by Lee and co-workers [9]. An RP/C composite was prepared by ball milling with the commercial phosphorus as precursor, in which nanoscale amorphous RP particles were uniformly distributed in the carbon matrix of super P. The obtained RP/C material delivered a high capacity of $1890\,mAh\,g^{-1}$ with considerable cycling stability (<7% capacity decay after 30 cycles) when applied with poly acrylic acid (PAA) binder. In addition, the RP/C sample delivered a specific capacity as high as $1540\,mAh\,g^{-1}$ at $2860\,mA\,g^{-1}$, demonstrating an outstanding rate capability. Ex situ

XRD measurements indicated that a reversible phase transition from amorphous RP to Na_3P phase was performed during the sodiation process.

The choice of carbon materials is one of the key factors for the enhancement of Na-storage performance for RP materials. CNTs, showing high electronically conductivity and outstanding mechanical properties, have been widely used as an effective conductive network to modify RP particles for high Na-storage performance. RP/multiwalled CNTs composite (RP/MWCNTs) obtained via a simply hand grinding exhibited an improved reversible capacity of 1283 mAh g^{-1} at the 10th cycle, which was much higher than that of the P/C composite (90 mAh g^{-1} at the third cycle) [15]. Zhu et al. reported an RP coupled with single-walled CNTs composite (RP-SWCNTs) fabricated via a vaporization–condensation method (Figure 8.5a) [16]. The RP particles were uniformly mixed with the tangled single-walled CNT (SWCNT) bundles with intimate attachment, enabling high ionic/electronic conductivities of the whole electrode, a stable solid–liquid interface for effective charge transfer, as well as robust skeleton strength to alleviate the mechanical stress induced by the Na^+ insertion/extraction. Benefiting from those, excellent rate performance (reversible capacity of ~300 mAh g^{-1} at 2 A g^{-1}) with stable cycling performance (capacity retention of 80% after 2000 cycles) of the RP-SWCNT composites has been achieved.

Pei et al. first reported and selected graphene material for the modification of RP anodes for the enhancement in sodium-storage properties [21]. Nanosized RP particles were encapsulated into graphene scrolls to form an RP/rGO composite. The graphene matrix provides an interconnected conductive network of the RP/rGO composite. It could also accommodate the volume expansion and particle aggregation during the charge–discharge processes. The obtained RP/rGO composite (RP content: 50.8%) demonstrated a high reversible capacity of 2355 mAh g^{-1} and exhibited ultra-stable cyclability capacity retention of 92% at the 150th cycle at 0.25 A g^{-1}).

It follows that a relatively strong force beyond physical one between RP particles and carbon matrixes can be expected to improve the cycling performance. Song et al. reported a chemically bonded RP/rGO composite via ball milling under an argon atmosphere (Figure 8.5b) [17]. It was demonstrated that the nanoscale RP particles intimately attached to the well-exfoliated graphene nanoflakes through P—O—C bonds. This structure offers several advantages for Na-storage performance: (i) enhancing the electronic conductivity on the RP surface and improving the effective electrical contact between the active particles; (ii) stabilizing the overall electrode structure against repeated current attacks; and (iii) favoring for the formation of stable SEI layer on the solid–liquid interface. As a result, a high initial capacity of 2077 mAh g^{-1} with ~99% initial Coulombic efficiency was achieved. The prepared RP/rGO hybrid maintained 95% of its initial capacity after 150 cycles at 260 mA g^{-1}. Heteroatoms doping strategy was also proposed to obtain advanced high-performance RP-based electrodes. A flexible RP/N-doped graphene (donated as RP/N-rGO), fabricated by Zhang et al. [22], showed good rate performance (809 mAh g^{-1} at 1.5 A g^{-1}) with stable cycling behavior (85% capacity retention after 350 cycles).

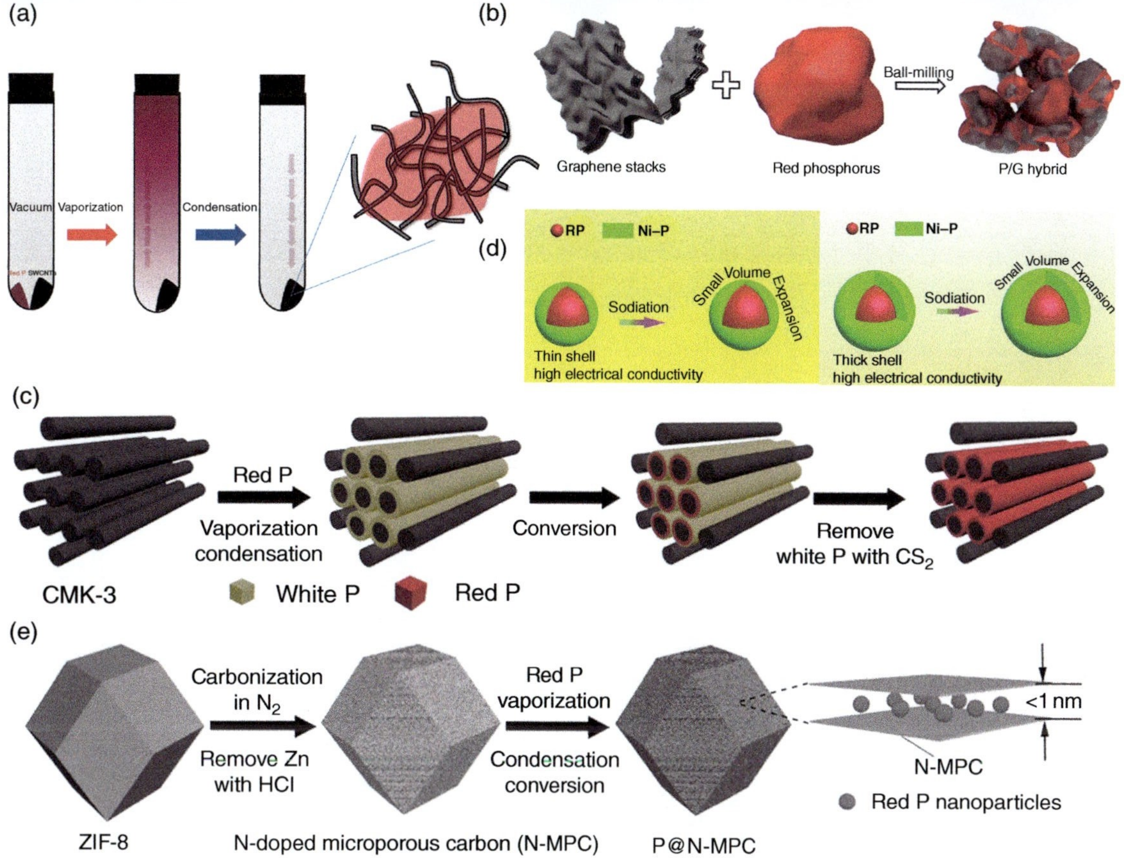

Figure 8.5 Schematic illustration of the coupling RP particles with various conductive matrix. (a) Synthesis process for RP/SWCNT composite. Source: Zhu et al. [16]. Reproduced with permission, 2015, American Chemical Society. (b) RP compositing with graphene stacks. Source: Song et al. [17]. Reproduced with permission, 2014, American Chemical Society. (c) P@CMK-3 material. Source: Li et al. [18]. Reproduced with permission, 2016, American Chemical Society. (d) Electrodeless deposition of Ni on an RP nanoparticle and the evolution of the RP@Ni–P core@shell nanostructure through chemical dealloying. Source: Liu et al. [19]. Reproduced with permission, 2017, Royal Society of Chemistry. (e) P@N-MPC material. Source: Li et al. [20]. Reproduced with permission, 2017, Wiley-VCH.

A great variety of carbonaceous materials, such as mesoporous carbon [18], carbon nanofibers [23], porous carbon [24], metal-organic frameworks (MOFs)-derived microporous carbon [20], were used as the conductive skeletons to modulate the issue of volume expansion for RP electrodes. Highly ordered mesoporous carbon (CMK-3) exhibits uniform diameter-size pores, large pore volume, and high electronic conductivities and thus has been widely used as the conductive matrix for electrode modification [25]. A CMK-3 confined amorphous RP nanocomposite (donated as P@CMK-3) was reported by Yu and co-workers (Figure 8.5c), which was synthesized by vaporization–condensation conversion method [18]. When applied as SIB anode material, an initial specific capacity as high as 2594 mAh g^{-1} (calculated by the RP mass) was achieved, and a reversible capacity of 1020 mAh g^{-1} was retained after 210 cycles when cycling at 5 C. Compared with the serious cycling decay of RP particle anode or RP@C anodes, the excellent sodium-storage performance of P@CMK-3 composites can be ascribed to the advantages of the designed nanostructure, including: (i) CMK-3 matrix constructs highly conductive network and alleviates the volume change of RP; (ii) nanoscale particle size of RP greatly reduces the diffusion distance of Na^+; (iii) RP nanoparticles intimately attached to CMK-3, beneficial for suppressing the exfoliation and agglomeration of RP particles; and (iv) the unique porous structure of CMK-3 provides open ionic migration channels and easy interpenetration of electrolyte.

Additionally, the vaporization–condensation conversion approach was also employed to realize the confinement of amorphous RP into the ZIF-8-derived N-doped microporous carbon (P@N-MPC) [20]. The highly porous structure offers easy pathways for electrolyte permeation that assures the sufficient contact between the electrolyte and the RP particles, so as to enhance the electrochemical activity (Figure 8.5e). N-doped carbon matrix constructs conductive skeleton for fast electron transfer. P@N-MPC material displayed superior sodium storage performance, e.g. a specific capacity of 600 mAh g^{-1} at 150 mA g^{-1} with superior cycling stability (0.02% capacity loss per cycle over 1000 cycles at 1000 mA g^{-1}).

Despite various carbonaceous materials being proposed to improve the ionic/electronic conductivities and mitigate the volume expansion for RP material, incorporation of carbon usually causes low initial Coulombic efficiency, high specific surface area, low tap density, and low operation potential (<0.1 V vs. Na^+/Na). The high carbon content in the RP/carbon hybrids reduces the specific capacity and volumetric capacity. In addition, it is currently difficult to realize the uniform coverage of carbon layer on the RP surface and accurate thickness control.

Recently, combination of RP with noncarbonaceous materials had shown to be promising for enhancing the sodium-storage performance. Mechanical mixing of RP and appropriate amount of Fe powder by using high-energy ball milling has been demonstrated a better rate capability than that of pure RP electrode material [26]. Additionally, a RP/Sb/Cu composite was also reported to achieve a competitive Na-storage performance by combining RP with Sb nanoparticles and Cu nanowires [27]. Benefitted from the enhanced electronic conductivity and mechanical strength, the RP/Sb/Cu composite exhibited prolonged life span and precedent rate performance (900 mAh g^{-1} retained at 2000 mA g^{-1}).

Apart from metals, metal oxides and metal phosphides have also been reported to be used to modify the electrochemical performance of RP material. A core–shell RP@Ni–P nanostructure as anode material for SIBs has demonstrated high Na-storage performance [19]. The core–shell RP@Ni–P nanostructure was realized through a two-step method: core–shell RP@Ni nanocomposite was firstly obtained via a simple electroless deposition approach, which was then converted into RP@Ni–P by using chemical dealloying. Especially, the thickness of the Ni–P layer could be regulated by tuning the dealloying time. The schematic of the formation of the core–shell RP@Ni–P nanostructure is shown in Figure 8.5d. Such structural design endows a couple of advantages for increasing the Na-storage performance of RP material: (i) the outer Ni–P shells with high mechanical strength were one of the key factors for achieving long-term cycling stability, which effectively inhibited the RP pulverization and endowed the electrode structural integrity during charge–discharge process; (ii) the Ni–P layer that in situ phase-transformed from Ni_2P tightly coated to the RP surface, which also facilitated improvement of the structural integrity of the electrode material; (iii) the high conductivity of the outer Ni–P layer provided fast electrons/ions transportation. As a result, the core–shell RP@Ni–P nanocomposite presented a superior high reversible capacity (~1250 mAh g^{-1} after 200 cycles at 260 mA g^{-1}), precedent rate capability (~490 mAh g^{-1} at 5.2 A g^{-1}), and excellent high-rate long-life cycling performance (~400 mAh g^{-1} during the 2000th cycle at 5 A g^{-1}).

8.2.3.1.2 Modulating Nanostructures Manufacturing the particle size from microscale into nanoscale has been widely considered as another effective strategy for improving the Na-storage properties of RP, because the nanosized particles are better at buffering the mechanical stress from cracking and improving the charge transfer kinetics by shortening the diffusion distance. Although ball-milling method is most commonly utilized for the fabrication of RP-based nanocomposites due to the low cost and feasibility, the large particle size and uneven size distribution consequently lead to poor rate capability and short cycling life span of RP-based anode materials. In view of this, Liu et al. fabricated a RP nanodot/rGO nanocomposite via a vaporization–condensation approach [28]. The tiny RP nanodots were observed to be tens to several hundred nanometers. The authors suggested that the ultrasmall nanosized RP particles could facilitate the ionic conductivity by reducing the transport length in solid phase, whereas the rGO nanosheets with high electronic conductivity enhanced the electrons transfer and effectively mitigated the volume expansion of RP particles. Consequently, the RP nanodot/rGO nanocomposite as the anode for SIBs received significantly enhanced rate capability (1165 mAh g^{-1}@159 mA g^{-1} and 510 mAh g^{-1}@31879 mA g^{-1}, respectively) with more prolonged cyclability (~914 mAh g^{-1} during the 300th cycle at 1593.9 mA g^{-1}) than that of the conventional RP anode materials. Based on the similar structural design concept, Gao et al. proposed a vapor-redistribution strategy to prepare an integrated C@RP/graphene porous aerogel, in which the RP nanoparticles with a particle size range of 10–20 nm were uniformly distributed in the C@graphene aerogel matrix [29]. The 3D porous C@RP/graphene aerogel anode material exhibited an excellent Na-storage performance (1867 mAh g^{-1} during the 100th cycle at 0.1 C; 1096 mAh g^{-1} after 200 cycles at 1 C).

Concerning the low RP content and the toxic WP remnants generated by the vaporization–condensation method, liquid-phase solutions have been proposed for solving these challenges. Hollow RP nanospheres (HPNs) materials were successfully prepared by Zhou et al. using a solvothermal method [30]. The formation of HPNs was based on a gas-bubble-directed mechanism, during which the NaN_3 precursor reacted with PCl_5 and generated large quantity of N_2 gas. The average diameter of the HPNs was highly depended on the NaN_3 concentration. The optimized HPNs sample with average diameter of 300 nm and the shell thickness of around 40 nm showed the best sodium-storage performance, including a high specific capacity of ~1365 mAh g^{-1} at 0.52 A g^{-1} and good cyclability over 600 cycles at 2.6 A g^{-1}. In order to further enhance the electronic conductivity of RP material, Liu et al. explored a boiling strategy to fabricate porous RP nanoparticles anchored on the rGO nanosheets (denoted as NPRP@rGO). A concentrated RP@rGO solution was boiled at 200 °C in a vacuum drying chamber. Along with the gas bubble growing, the RP particles self-assembled on the interface between the liquid and the gas bubble and finally turn to be porous nanostructure after the gas-bubbles separation. Unsurprisingly, the NPRP@rGO nanocomposite obtained a high specific capacity of 1250 mAh g^{-1} during the 150th cycle at 1 $C_{composite}$ (1 C = 256 mA ${g_{RP}}^{-1}$; 1 C = 173 mA ${g_{composite}}^{-1}$), superior rate performance of 657 mAh g^{-1} at 10 $C_{composite}$, and ultralong life cycle of 776 mAh g^{-1} at 20 C_{RP} over 1500 cycles [31].

8.2.3.2 Black Phosphorus and Phosphorene for Na-Ion Batteries

BP, as the most thermodynamically stable phosphorous allotrope, has attracted great attention for many applications. BP exhibited higher electronic conductivity (~66–300 S m^{-1}) than that of RP (10^{-12} S m^{-1}) and much more suitable operation voltage for sodiation (~0.45 V vs. Na^+/Na). Furthermore, the large interlayer distance of BP (~3.08 Å) is adequate for both Li^+ and Na^+ insertion/extraction in the channels. However, the huge volume variation (~300–500%) and the sluggish reaction kinetics are the two key obstacles for the practical applications of BP anode materials. Currently, the proposed fabrication strategies mainly include high-energy milling, high-pressure conversion from RP, and CVD synthesis, which are critically dependent on the strict experimental conditions. Therefore, developing high cost-effective synthetic methods for high-quality BP material on a large scale is also of great importance.

Similar to RP, the combination with carbonaceous material is the most commonly used strategy for the modification of the Na-storage performance of BP material. For example, BP/Ketjen-black/MWCNTs composite obtained by ball milling demonstrated an exciting initial capacity of ~2011 mAh g^{-1} and stable cycling behavior (1700 mAh g^{-1} during the 100th cycle at 1300 mA g^{-1}) [32]. In addition, Haghighat-Shishavan et al. demonstrated an enhanced electrochemical performance of a BP/MWCNTs composite when applying a binary polymeric binder (sodium carboxyl methyl cellulose–poly acrylic acid, NaCMC–PAA) [33]. A high initial capacity of 2073 mAh g^{-1} with capacity retention of 75.3% after 200 cycles at 0.2 C along with a high rate performance of 609 mAh g^{-1} at 2 C was achieved.

2D graphene material was also reported to be used for the enhancement of the charge transfer kinetics and accommodation for volume change during repeated sodium-ion intercalation/deintercalation process for BP materials. Li and co-workers used a co-exfoliation/electrophoretic deposition approach for the preparation of BP/graphene composite. Remarkably, without using any binders or additives, the free-standing BP/graphene anode for SIBs exhibited superior rate performance (2365, 1894, and 1456 mAh g^{-1} were obtained at 0.1, 0.2, and 0.5 A g^{-1}, respectively) and excellent long-term cycling stability. In addition, a room-temperature pressurization strategy has been proposed to prepare BP/graphene hybrids directly from an RP-graphene precursor [34]. As a self-supporting and additive-free SIB anode material, the obtained BP/graphene composite delivered a high reversible capacity of ~1250 mAh g^{-1} at 1 A g^{-1} with excellent cyclability of 640 mAh g^{-1} during the 600th cycle at 40 A g^{-1}. It is noteworthy that it is an availably low-cost and low-energy consumption synthetic approach for scalable production of BP-based electrode materials toward practical large-scale sodium-storage applications. Apart from graphene, 2D Mxene material also demonstrate their promising potential for energy storage because of their high electrical conductivity and mechanical strength. Meng et al. reported a BP-Mxene composite of which the BP quantum dots successfully anchored on thin Ti_3C_2 nanosheets via a self-assembling process through sonication and stirring [35]. Benefitted from the combination of Ti_3C_2 nanosheets, the BP quantum dots/Ti_3C_2 composite showed superior long-life cyclability with almost no capacity loss over 1000 cycles. In another study, a BP-cubic boron nitride (c-BN)-rGO composite was proposed by Ding et al. [36]. The hard c-BN decreased the particle size of BP but also acted as the buffer shelter to suppress the mechanical stress and accommodate the volume variation. After compositing with graphene to further enhance the electronic conductivity, the BP-cBN-rGO hybrid anode endowed excellent cyclability with 90% capacity retained after 100 cycles at 50 mA g^{-1}.

Furthermore, surface modification is also a valid approach for improving the reaction kinetics and cycling stability of BP. Dou and co-workers reported a composite material of BP nanolayers encapsulated by poly(3,4-ethylenedioxythiophene) (PEDOT), which exhibited enhanced redox kinetics and surface wettability [37]. In addition, 4-nitrobenzenediazonium (4-NBD) was utilized to realize the surface functionalization of BP, which was then further chemically bonded to reduced graphene oxides [38]. The authors suggested that the strong interaction between the surface modified BP and rGO nanosheets endowed the high Na-storage properties of BP-rGO hybrid anodes, displaying a reversible capacity of 1472 mAh g^{-1} during the 50th cycle at 0.1 A g^{-1} and maintaining ~650 mAh g^{-1} after 200 cycles at 1 A g^{-1}. The results indicate that the buildup of strong chemical bonds between the BP and host matrix is effective strategy to obtain high-performance BP-based SIB anode material. Inspired by this thought, Song et al. reported a novel $Na_2Ti_3O_7$/C-BP composite anode material and demonstrated that the P—O—Ti bonds and/or P—C bonds could be established between the BP and the $Na_2Ti_3O_7$ nanoparticles through high-energy ball-milling process [39]. Benefiting from that, the $Na_2Ti_3O_7$/C-BP composite showed a reversible capacity of ~183 mAh g^{-1} at 0.1 A g^{-1} over 100 cycles. Such low specific capacity can be ascribed to the low P content (~20%) in the hybrid material. Xia's group fabricated a stable sulfur-doped TiO_2/BP (denoted as S-TiO_2/BP) nanocomposite as the SIBs anode material and achieved a specific capacity of ~290 mAh g^{-1} at 0.5 A g^{-1} over 600 cycles [40].

Phosphorene, as a new fashion of 2D material, has shown great promising for energy-storage applications due to its unique structural properties such as large specific surface area and interlayer distance, as well as fast electron/ion Na^+ diffusivity. Kulish et al. demonstrated that the theoretical Na^+ diffusion barrier in monolayer BP is only 0.04 eV [41]. Sun et al. [13] used a facile sonication process to fabricate a phosphorene/rGO hybrid material, which demonstrated a high capacity of 2440 mAh g^{-1} at 50 mA g^{-1} and retained ~83% of this value at the 100th cycle. In this work, a two-step sodiation mechanism was investigated by in situ transmission electron microscopy (TEM) and ex situ XRD measurements. At the first stage of sodiation, Na^+ intercalated into the interlayers of phosphorene along the *x*-axis, and an alloying process took place that resulted in an anisotropic volume expansion in the directions of *y*- and *z*-axis and eventually formed Na_3P phase (Figure 8.6). Furthermore, the graphene layers in the phosphorene/rGO composite were found to act as mechanical skeletons to buffer the huge expansion of phosphene layers and

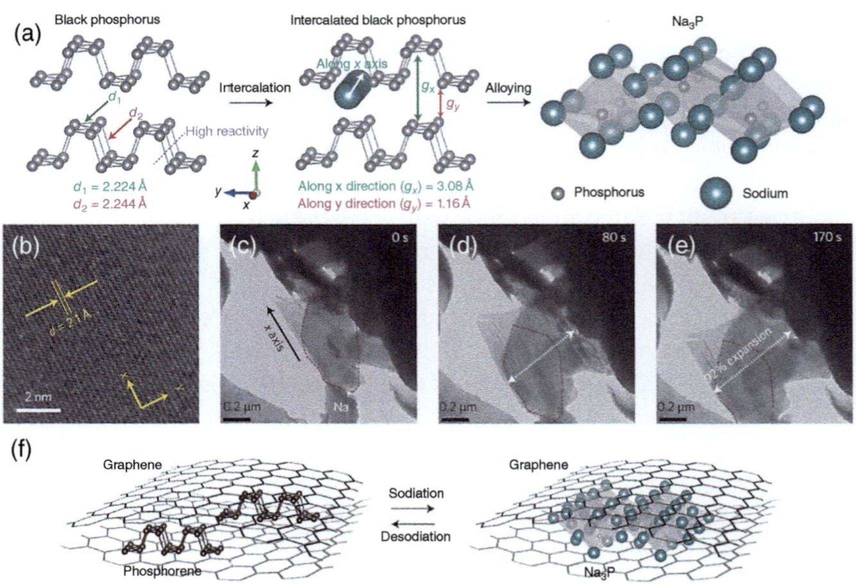

Figure 8.6 (a) Schematics of BP before sodiation, with the first step of sodium-ion intercalation, and the second step of alloy reaction to form Na_3P. (b) High-resolution and bright-field TEM image of black phosphorus before sodiation. (c–e) Time-lapse TEM images of sodiation in black phosphorus. Sodium ions transport along the *x*-axis channel, causing a volume expansion along the *y*-axis direction. (f) Structural evolution of the sandwiched phosphorene-graphene structure during sodiation. Source: Sun et al. [13]/with permission from Springer Nature.

provide electrical highway for fast charge transfer, leading to ultra-stable cyclability and superior rate performance.

Inspired by the pioneering work of phosphorene for SIBs, Huang et al. proposed a intercalation–exfoliation approach to synthesize few-layer phosphorene from bulk BP [42]. Tetraalkylammonium cations were chosen to intercalate the BP and a work voltage range of −5 to −15 V was applied for the exploitation. The results demonstrated that the number of phosphorene sheets can be tuned by manipulating the applied potential. When served as the anode material for SIBs, the exploited phosphorene delivered a discharged capacity as high as 1190 mAh g^{-1} at 100 mA g^{-1} after 50 cycles. So far, the lack of scalable and cost-effective methods for large-scale production of phosphorene limits its practical applications for SIBs. Therefore, innovative strategies, such as mechanical cleavage and liquid-phase exfoliation, and continuous efforts for the development of phosphorene are greatly encouraged.

8.3 Metal Phosphide Anodes

8.3.1 Na-Storage Mechanism for Metal Phosphides

Generally, the electrochemical reaction mechanisms of metal phosphides as anode material for SIBs are based on the conversion reactions. Comparing with the insertion-type reaction, more electrons are involved in the conversion and/or alloying reaction process, leading to much higher specific capacity.

Based on whether the metal (M) in metal phosphides (MP_x) reacts with Na through the alloying reaction, MP_x can be classified into two types: Na-active MP_x and Na-inactive MP_x.

For Na-active MP_x, both M and P possess the capability to take an alloy reaction with Na to transform into Na_yM and Na_3P in the initial discharge process. During the subsequent process, M nanocrystalline and P nanoparticles are eventually formed from the Na_yM and Na_3P phases through reversible alloying reactions. The sodiation/desodiation mechanism can be expressed as:

$$MP_x + (3x+y)Na^+ + (3x+y)e^- \rightarrow Na_yM + xNa_3P \text{ (initial discharge process)} \tag{8.1}$$

$$Na_yM \leftrightarrow yNa^+ + ye^- + M \text{ (subsequent cycles)} \tag{8.2}$$

$$Na_3P \leftrightarrow 3Na^+ + 3e^- + P \text{ (subsequent cycles)} \tag{8.3}$$

Typical samples of Na-active MP_x include tin phosphide (Sn_4P_3) and GeP_5 [43]. Additionally, for some kind of Na-active MP_x, such as GeP_3 [44], GeP, and Ge_4P_4 [45], the MP_x phase can be reformed after the full oxidation process through a reversible phase conversion process between Na_yM and Na_3P (Eq. (8.4)).

$$MP_x + (3x+y)Na^+ + (3x+y)e^- \leftrightarrow Na_yM + xNa_3P \tag{8.4}$$

In respect of Na-inactive MP_x, based on whether the original MP_x phase can be reversibly reformed, the electrochemical reactions are also categorized into two types. During the reduction process with Na^+ sodiation, the Na inactive (i.e. not alloy with Na) metal phosphides (MP_x) converted into nanosized metal particles (M) and Na phosphides (Na_3P) with the broken of the M—P bonds. This reaction process can be written as Eq. (8.5). The generated M nanocrystals are uniformly distributed in the Na_3P matrix, which serves not only as highly conductive media to accelerate the electrons transport but also as a buffer matrix to mitigate the volume expansion of P. The generated Na_3P phase further dealloys into Na and P nanocrystals, which recombined together to reform Na_3P phase during the subsequent cycles. Representatives for this kind of Na-inactive MP_x mainly include CoP [46] and FeP [47]. It should be noted that the MP_x cannot be reversibly reformed at the end of desodiation process.

$$MP_x + 3xNa^+ + 3xe^- \rightarrow M + xNa_3P \text{ (initial discharge process)} \quad (8.5)$$

For another case, M nanocrystals react with Na_3P nanoparticles to reversibly generate MP_x phase after the charging process (Eq. (8.6)). NiP_3 [48], CoP_3 [49], CuP_2 [50], and Cu_3P [51] are the typical samples of this case of Na-inactive MP_x materials.

$$M_xP_y + 3yNa^+ + 3ye^- \leftrightarrow xM + yNa_3P \quad (8.6)$$

8.3.2 Metal Phosphides for Na-Ion Batteries

8.3.2.1 Tin Phosphide Materials

Sn_4P_3 has received significant research concerns as a SIB anode material due to its high theoretical gravimetric capacity (1132 mAh g^{-1}), high theoretical volumetric specific capacity (6650 mAh cm^{-3}), as well as the high electronic conductivity (30.7 S cm^{-1}). Furthermore, Sn_4P_3 can alloy with Na^+ to form $Na_{15}Sn_4$, presenting a lower redox potential (~0.3 V vs. Na^+/Na) than that of phosphorous (~0.4 V vs. Na^+/Na).

Nevertheless, similar to RP and metal Sn anodes, Sn_4P_3 also suffers from poor cycling performance because of the huge volume expansion during the sodiation/desodiation processes. Sodium-ion storage behavior of Sn_4P_3 was first examined by Kim et al. [52]. The ball-milled Sn_4P_3 material delivered a Na^+-storage capacity of 718 mAh g^{-1} and a good cyclability with almost 100% capacity retention during the 100 cycles when applied an 1-fluoroethylene carbonate (FEC)-assisted electrolyte. Using the same synthetic method, Qian et al. [53] reported an Sn_4P_3/C nanocomposite, which demonstrated an enhanced electrochemical performance far beyond those of P/C and Sn/C composites (a high specific capacity of 850 mAh g^{-1} at 50 mA g^{-1} with 86% capacity retention during the 150th cycle at 100 mA g^{-1}). The elevated sodium-storage properties could be ascribed to the synergistic Na-storage effects between the discharge products of P and metallic Sn. After the full discharge process, the uniform Sn nanocrystals are confined in the amorphous P matrix that could enable high conductivity of the host and compensate the electrochemical activities of P phase, meanwhile the discharged products of Na_3P and P could be used as buffering components to inhibit the Sn nanoparticles aggregation and further alleviate the volume variation during the subsequent cycling.

Core–shell nanostructure construction has been proven to be a useful modification method to address the rapid capacity fading, which is ascribed to the massive volume variation during the Na^+ insertion/extraction process. Li et al. [43a] fabricated a core–shell $Sn_{4+x}P_3$@(Sn–P) nanocomposite via a facile low-speed ball-milling approach. The $Sn_{4+x}P_3$ core component with an average size of 30 nm was coated with an amorphous Sn–P shell component, forming a unique core–shell $Sn_{4+x}P_3$@(Sn–P) nanostructure. When $Sn_{4+x}P_3$@(Sn–P) nanocomposite tested as a SIB anode using a carboxymethyl cellulose (CMC) binder, a high specific capacity of 890 mAh g^{-1} at 0.1 A g^{-1} was achieved. Moreover, FEC additive facilitates the generation of a stable SEI layer that can improve the cycling stability significantly. Unfortunately, it will cause partial capacity loss due to the increased polarization that is ascribed to the high interface resistance from the thick NaF-component SEI film. Benefitting from the usage of 5 vol% FEC agent, the $Sn_{4+x}P_3$@(Sn–P) nanocomposite exhibited a stable cycling behavior (465 mAh g^{-1} at the 100th cycle, corresponding to a capacity retention of 92.6% of the capacity of the second cycle).

Furthermore, yolk–shell nanostructures have also been widely utilized to modify the huge volume expansion of the alloy-type anodes in the Na^+ insertion/extraction processes. A yolk–shell nanosphere Sn_4P_3@C composite was developed via a top-down phosphorization strategy by Yu's group [43b], as illustrated in Figure 8.7. The synthesis process mainly includes two separate steps: manufacturing nanosphere-shaped Sn@C precursors and the further transformation of yolk–shell nanospheres Sn_4P_3@C composite through a facile solvothermal phosphorization. As a result, numerous nanocrystals constituted Sn_4P_3 nanoparticles were totally encapsulated in uniformly thin carbon shells. The defined yolk–shell structure with internal void space provides obvious merits for buffering the volume change and keeping the structural integrity of the electrodes during sodiation/desodiation. Therefore, the yolk–shell Sn_4P_3@C nanospheres anodes showed an initial reversible capacity of 790 mAh g^{-1} and considerable cycling stability (360 mAh g^{-1} over 400 cycles at 1500 mA g^{-1}), as well as superior rate performance.

To enhance the electrical conductivity of the Sn-based phosphates, combination with conductive carbonaceous material has been wildly demonstrated as an effective approach. A nanostructured Sn_4P_3 and BP embedded into a graphene matrix was explored via a mechanochemical transformation strategy from SnP_3. The obtained nanocomposite exhibited a stable high-rate cyclability with a retained capacity as high as 550 mAh g^{-1} during the 1000th cycle at 1 A g^{-1} and excellent rate performance (585 and 315 mAh g^{-1} obtained at 2 and 10 A g^{-1}, respectively). Similarly, a Sn_4P_3/graphene composite, prepared via a low-temperature liquid-based synthesis route [54], also presented enhanced sodium-storage performance. The monodispersed Sn_4P_3 nanoparticles were completely encapsulated by graphene nanosheets, forming a 3D mesoporous architecture. The obtained Sn_4P_3/rGO hybrid delivered a stable cyclability with 362 mAh g^{-1} during the 1500th cycle at 1 A g^{-1}. The precedent Na-storage properties of the Sn_4P_3/rGO nanohybrids can

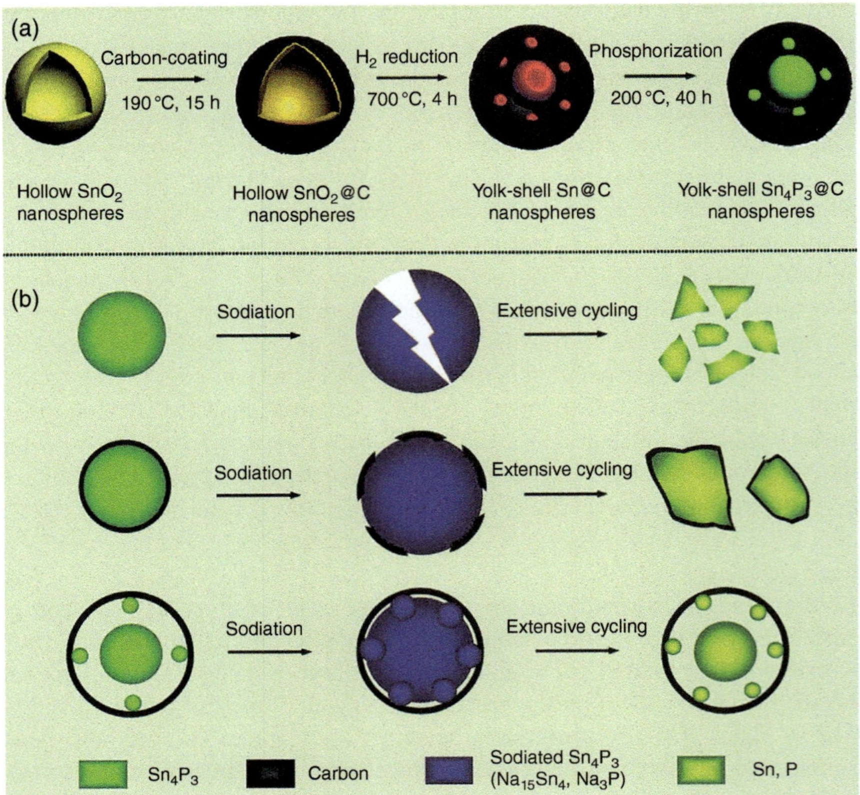

Figure 8.7 (a) Schematic illustration of the fabrication of Sn_4P_3@C yolk-shell nanosphere; (b) Schematic illustration of the sodiation process in Sn_4P_3 nanoparticles, core-shell Sn_4P_3@C nanoparticles, and the current yolk-shell Sn_4P_3@C nanoparticles. Source: Liu et al. [43b]. Reproduced with permission, 2015, Royal Society of Chemistry.

be attributed to: (i) graphene nanosheets construct highly conductive networks for e^-/Na^+ on the surface of the electrochemical active Sn_4P_3 nanoparticles; and (ii) both the graphene nanosheets and the 3D mesoporous architecture offer an elastic matrix and buffering space for alleviating the volume expansion of Sn_4P_3 during the Na^+ insertion/extraction process. All above-mentioned results indicated that the Sn_4P_3 can be one of the potential candidates for SIB anodes with high performance.

Phosphorus-rich tin phosphide SnP_3, exhibiting higher theoretical gravimetric capacity (1616 mAh g^{-1}) and volumetric capacity (6890 mAh g^{-1}) than those of Sn_4P_3, has also shown promising for SIB applications. Wang's group reported an SnP_3/C composite and proposed a novel reversible conversion reaction with the self-healing mechanism [55]. In the first sodiation stage, the SnP_3 crystalline firstly occurs conversion reaction with the breaking of Sn—P bonds to decompose to metallic Sn nanoparticles (Eq. (8.7)), which are then uniformly distributed in the nanocrystalline/amorphous Na_3P matrix. Upon further Na^+ inversion, the Sn nanoparticles subsequently evolve into $Na_{15}Sn_4$ by alloying with Na^+ (Eq. (8.8)). During the desodiation process, $Na_{15}Sn_4$ and Na_3P undergo reversible dealloying reactions to convert back to SnP_3, respectively.

$$SnP_3 + 9Na^+ + 9e^- \leftrightarrow Sn + 3Na_3P \tag{8.7}$$

$$4Sn + 15Na^+ + 15e^- \leftrightarrow Na_{15}Sn_4 \tag{8.8}$$

As a result, such SnP_3/C composite archived a specific capacity as high as 810 mAh g^{-1} at 150 mA g^{-1} and retained ~400 mAh g^{-1} at a high rate density of 2560 mA g^{-1}.

8.3.2.2 Cobalt Phosphide Materials

Cobalt phosphides have also exhibited electrochemical activity when served as anode material for SIBs. CoP is the most common compound used for SIBs. The electrochemical reaction mechanism of CoP can be expressed as:

$$CoP + 3Na^+ + 3e^- \rightarrow Co + Na_3P \text{ (initial discharge process)} \tag{8.9}$$

$$Na_3P \leftrightarrow 3Na^+ + 3e^- + P \text{ (subsequent cycles)} \tag{8.10}$$

Upon discharging, CoP firstly converts into Co and Na_3P through interaction of Na^+. Although Co is inactive for alloying with Na^+, it can serve as conductive substrate using its metallic feature to buffer volume change and improve the electronic conductivity. In the subsequent charge/discharge cycling process, P reversibly transformed into Na_3P with possible intermediate phases (Na_2P, NaP, and NaP_7). Li et al. synthesized CoP material through ball-milling commercial RP and metal Co. [46a] A high initial capacity of 770 mAh g^{-1} was obtained, indicating CoP to be one of the potential SIB anode candidates. Unfortunately, it displayed a capacity retention of only 70% after 25 cycles.

Similar to RP, the CoP material prepared through ball-milling approach usually suffers from fast capacity fading and unsatisfied rate capability because of the large particle size and uneven size distribution. To solve these problems, effective strategies toward nanostructure construction of CoP/carbon composite materials have been proposed, such as one-dimensional CoP/C nanocomposites [56] and CoP@C-rGO core–shell nanostructures [46b]. Ge et al. proposed core–shell CoP@C polyhedrons compositing with graphene nanosheets anchored on nickel foam (NF) (denoted as CoP@C-rGO/NF) as a self-supporting SIB anode through a low-temperature solid-phase phosphorization route [46b]. Polyhedron ZIF-67 precursor was transformed into core–shell nanostructured CoP@C with the shape well maintained. The MOFs-derived carbon shell and rGO nanosheets not only ensure facile pathways for easy electrons transfer but also provide adequate space to modify the mechanical stress resulting from the huge volume expansion during charging/discharging. Therefore, the core–shell nanostructured CoP@C-rGO/NF showed remarkable rate performance and superior cyclability (retaining 473 mAh g^{-1} over 100 cycles at 0.1 A g^{-1}).

8.3.2.3 Iron Phosphide Materials

Iron phosphides (FeP, FeP_2, and FeP_4), due to the reserve abundance, low cost, and environmentally friendly characteristics of element Fe, have also been investigated as promising SIB anode materials. Especially, among various iron phosphides, FeP is the most commonly studied, which possesses a high theoretical capacity of $473\,mAh\,g^{-1}$. FeP exhibits similar Na-storage mechanism to that of cobalt phosphides.

Upon the insertion of Na^+ during the first discharge process, the FeP is transformed into Fe, which can be served as the conductive medium to boost the electronic conductivity and accommodate the volumetric variation of P. Upon the subsequent charging/discharging process, P takes reversible reaction with Na to convert into Na_3P.

Li et al. investigated Na-storage behavior of FeP and FeP/graphite composites [24, 47a]. When CMC/PAA binder with 5 vol% FEC additive was applied, the obtained FeP showed an initial specific capacity close to the theoretical value. In addition, FeP nanorod arrays grew on carbon cloth (denoted as FeP NAs/CC) as electrode were achieved by two-step synthesizing process. Fe_2O_3 nanorod arrays firstly grew on the carbon cloth via a facile hydrothermal treatment and then converted into FeP nanorod arrays through phosphidation. As SIB anode material, the FeP NAs/CC demonstrated a capacity as high as $829\,mAh\,g^{-1}$ at $0.1\,A\,g^{-1}$, long-term cyclability with 99.8% capacity retention after 100 cycles along with superior rate capability. Furthermore, Han et al. proposed a self-sacrificing template approach to manufacture FeP with amorphous carbon anchored on CNT (donated as CNT@FeP@C) nanocomposite [57]. Attributed from the mesoporous carbon framework, the prepared CNT@FeP@C nanocomposite displayed much improved electrochemical performance with a high specific capacity of $415\,mAh\,g^{-1}$ during the 100th cycle at $100\,mA\,g^{-1}$ and $295\,mAh\,g^{-1}$ during the 500th cycle at $500\,mA\,g^{-1}$. CoP-coated FeP@C microcubes encapsulated by graphene nanosheets were exploited to modify the sodium-storage properties of FeP (denoted as CoP@FeP@rGO) [58]. FeP@C microcubes with an average size of 400 nm, derived by Prussian blue through phosphidation process, consisted of abundant FeP nanoparticles and lots of pores on the surface. CoP-shell coating not only contributed the sodium-storage capacity but also alleviated the volume change during the cycling process. Compared with the CoP@C–Fe and C–Fe samples, the CoP@FeP@rGO microcubes exhibited improved cyclability (specific capacity of $456\,mAh\,g^{-1}$ at the 200th cycle) and much better rate performance ($341\,mAh\,g^{-1}$ at $2\,A\,g^{-1}$), demonstrating that the 3D graphene skeleton and double coating of carbon and CoP play key roles on the improvement of electrochemical performance, which could construct high conductive pathway for fast ion/electron transport, but also effectively keep the structural ingredient during the repeated Na^+ sodiation/desodiation process.

P-rich iron phosphides have also been regarded as potential SIB anode materials because of their low cost and high theoretical capacity resulting from the high P content. Zhang et al. fabricated P-rich FeP_2 and FeP_4 compounds via simple ball-milling method and explored their Na-storage properties for the first time. Contrary to the storage behavior in LIBs, FeP_2 was nonreactive with Na when served as a SIB anode, which was possibly ascribed to the low sodiation potential nearing 0 V vs. Na/Na^+. Remarkably, FeP_4 material exhibited excellent Na-storage performance, delivering an initial capacity as high as 1137 mAh g^{-1} with the initial Coulombic efficiency of 84%, achieving excellent cyclability with 1000 mAh g^{-1} after 30 cycles, and good rate performance. Unfortunately, the sodiation/desodiation mechanism of FeP_4 is still unclear [59].

8.3.2.4 Nickel Phosphide Materials

Nickel phosphides (Ni_2P, NiP_3, and $Ni_{12}P_5$), exhibiting high electronic conductivity of around 10^{-6} S m^{-1} and low cost, demonstrate promising SIB applications. Nevertheless, the insufficient cycling life inhibits their further application. Constructing various novel nanostructures has been devoted to endow nickel phosphides with enhanced sodium-storage performance, such as Ni_2P nanocrystals, nanosheets, nanorods, nanoarrays, and NiP_3 films. One should note that high Ni content in nickel phosphides is favorable for achieving long-term cyclability, but low P content leads to low value of theoretical capacity.

In 2014, Fullenwarth et al. first fabricated two types of NiP_3 materials via high temperature calcination and ball milling, respectively, using metal nickel and RP as the raw materials [48]. The obtained NiP_3 compound through high-temperature treatment, as the anode for SIBs, delivered high reversible capacity of 1079 mAh g^{-1} over 15 cycles with a relatively low working potential of ~0.2 V, demonstrating great potential of NiP_3 for sodium storage field. In addition, the sodium insertion/extraction reaction mechanism was also investigated ($NiP_3 \rightarrow Ni + Na_3P$, and $Na_3P \leftrightarrow Na + P$). During the initial discharge process, NiP_3 decomposes into Na_3P and metal Ni nanoparticles that uniformly distributed in the conductive matrix. No possible intermediate phases, such as NaP and Na_3P_{11}, have been identified so far. During the subsequent cycles, the Na_3P reversibly transforms into amorphous P by dealloying reaction process.

Yu's group proposed an assembly and self-templating approach to synthesize a 3D yolk–shell-like Ni_2P nanocomposite, in which Ni_2P nanoparticles were encapsulated in 3D graphene skeleton (denoted as $Ni_2P\subset pGN$) [60]. Benefitted from the conductive graphene network with abundant void space, $Ni_2P\subset pGN$ demonstrated good Na-storage behavior when serving as a SIB anode, delivering an initial capacity of 214 mAh g^{-1} and stable cyclability with 89% capacity retention after 100 cycles at 200 mA g^{-1}. In addition, a sandwich-like Ni_2P nanoarchitecture was developed by Dong et al. as a SIB anode material, in which the Ni_2P nanoarrays hybrids with N-doped rGO nanosheets (denoted as $Ni_2P/NG/Ni_2P$) [61]. The novel nanostructured composite delivered a capacity retention of 57% over 300 cycles at 500 mA g^{-1}. Li et al. reported a microporous Ni_2P@C–N polyhedron-shaped nanocomposite embedded in highly rough CNTs as flexible SIB electrode material. Benefitted from the host structural design and ultrasmall Ni_2P nanoparticles with size of ~5 nm, Ni_2P@C–N polyhedrons exhibited excellent long-life cyclability with 0.04% capacity loss per cycle after 1000 cycles at 1 A g^{-1} and superior rate performance (retaining 197 mAh g^{-1} at 2 A g^{-1}) [62].

It is acknowledged that building strong chemical bonding between the transition metal phosphides and the conductive matrix is helpful for enhancing the electrochemical performance. Kim and co-workers developed nanostructured NiP_3 chemical bonded to functionalized CNTs via ball-milling method [63]. The results suggested that the P—C and P—O—C bonds interconnecting NiP_3 nanoparticles and the CNTs functionalized surface are important factors for the improvement of NiP_3 anode for sodium-storage performance. Inspiringly, the obtained NiP_3/CNTs nanocomposite demonstrated a high initial reversible capacity of 853 mAh g^{-1} at 200 mA g^{-1} and maintained 80% of this value over 120 cycles, as well as superior cyclability of 364 mAh g^{-1} during the 200th cycle at 1600 mA g^{-1}.

8.3.2.5 Copper Phosphide Materials

Copper phosphides are another type of phosphides widely regarded as promising SIB anode materials. Cu_3P and CuP_2, possessing hexagonal and monoclinic structure, respectively, are the most commonly studied recently. Ball-milling method was exploited to prepare P-rich CuP_2 SIB anode material by Kim and Manthiram [64] The strong P—O—P chemical bonds between CuP_2 nanoparticles and the carbon matrix enable the long-term cycling stability (an initial capacity of 450 mAh g^{-1} with 95.5% capacity retained after 100 cycles). Zhao et al. reported a CuP_2/carbon black nanocomposite through an optimized two-step ball-milling method, which obtained a high initial capacity (>500 mAh g^{-1}) and decent rate performance [50]. Based on the reports, the sodiation/desodiation process of CuP_2 electrode proceeds via a reversible conversion reaction mechanism, which can be concluded as follows:

$$CuP_2 + 6Na^+ + 6e^- \leftrightarrow Cu + Na_3P \tag{8.11}$$

Additionally, solution-phase method followed by phosphorization treatment have been proposed to design novel nanostructure for CuP_2 material endowing high-performance for SIBs. Yu and co-workers reported a $Cu_3(PO_4)_2{\cdot}3H_2O$ derived CuP_2/carbon sheets nanocomposite, where the CuP_2 nanoparticles were completely encapsulated by cross-linked hollow carbon sheets (denoted as CuP_2@CHCS). Benefitted from the carbon shielding and the hollow structure providing the buffer for alleviating the huge volumetric change during charging/discharging along with the significant improvement in electrical conductivity, the as-obtained CuP_2@CHCS achieved a reversible capacity as high as 451 mAh g^{-1} and retained 91% of this capacity during the 200 cycles at 80 mA g^{-1} [65].

Cu_3P was also developed as anode material for SIB applications. As shown in Eq. (8.12), Cu_3P undergoes the similar Na^+-storage mechanism to that of CuP_2 through reversible conversion reactions during the sodiation/desodiation process.

$$Cu_3P + 3Na^+ - 3e^- \leftrightarrow 3Cu + Na_3P \tag{8.12}$$

Fan et al. proposed a Cu_3P nanowire array (denoted as CPNW) electrode through the direct growth of $Cu(OH)_2$ nanowires on the copper current collector followed by in situ phosphidation treatment [66]. The nanowire structure greatly reduced the diffusion distance for both e^- and Na^+, which significantly enhanced the charge transfer kinetics. In addition, the array-structured electrode construction offered abundant open void for electrolyte penetration and continuous electrolyte supply. As a result, CPNW electrode demonstrated superior cycling stability with only 0.12% capacity loss per cycle over 260 cycles at 1 A g^{-1}.

8.4 Concluding Remarks

In general, due to the advantages of higher theoretical capacities coming from the alloy/conversion reactions with multiple electrons transfer behaviors, the abundant natural reserves and relatively appropriate working potential (~0.4 V vs. Na^+/Na), phosphorus/phosphide anodes for SIBs have gained growing research interests.

For phosphorus, especially RP and BP possessing high electrochemical active with sodium ions, they show similar alloy- or conversion-type Na-storage mechanism (i.e., $3Na^+ + 3e^- + P \leftrightarrow Na_3P$ for RP and $P + Na \leftrightarrow Na_xP \leftrightarrow Na_3P$ for BP), which provides large Na-storage capacity. However, because of the multiple Na^+ ions involved in the reaction, phosphorus anodes suffer from the huge volume expansion problem, which leads to electrode fragmentation and pulverization and thus undesired cycling performance. In addition, the short slabs of low Coulombic efficiency and low electronic conductivity also exist in the phosphorus anodes. Therefore, to solve these issues, several effective strategies have been proposed, for example, combining with conductive matrix to enhance the electronic conductivity of electrode and provide soft matrix for volume expansion; modulating nanostructures to create extra volume for volume expansion; developing suitable electrolyte additives and electrode binders to optimize battery configuration for enhancing Coulombic efficiency.

Compared with phosphorus anodes, the Na-storage mechanisms of metal phosphides anodes are also based on the alloy- or conversion-type reaction, but are more complex, which is mainly caused by the different Na^+ ions reaction activity of metal elements. In addition, some metal phosphides anodes possess a few unique advantages over phosphorus, for example, Sn_4P_3 possesses higher electronic conductivity and lower redox potential; the reserve abundance, low-cost and environmentally-friendly characteristics of Fe increase competitiveness of FeP. However, metal phosphides also suffer from poor cycling performance caused by huge volume expansion. Since the fundamental reason for the volume expansion of metal phosphides anodes is the conversion between P and Na_3P, the strategies for modified phosphorus anodes are also suitable for metal phosphides anodes.

Although great progress has been achieved for the phosphorus and metal phosphides anodes in SIBs based on their inherent advantages and ingenious modification, there are still some challenges to overcome. And we make bold predictions about the research direction in this field, as follows: (i) developing more advanced and novel method for fabricating phosphorus/phosphides materials, achieving the control of molecules and even atoms to meet the high demand of SIBs; (ii) using more advanced *in-situ* testing method and theoretical calculations to investigate the Na-storage behavior in phosphorus/phosphides anodes at molecular and even atomic levels, guiding the modified direction; and (iii) matching phosphorus/phosphides anodes with other components of the battery, such as cathode materials, electrolyte, separator and so on, to improving the compatibility.

References

1 Pang, J., Bachmatiuk, A., Yin, Y. et al. (2018). *Adv. Energy Mater.* 8: 1702093.
2 Mal, P., Breiner, B., Rissanen, K., and Nitschke, J.R. (2009). *Science* 324: 1697.
3 Roth, W.L., Dewitt, T.W., and Smith, A.J. (1947). *J. Am. Chem. Soc.* 69: 2881.
4 Wang, Y., Qian, B., Li, H. et al. (2015). *Mater. Lett.* 141: 35.
5 Kou, L., Chen, C., and Smith, S.C. (2015). *J. Phys. Chem. Lett.* 6: 2794.
6 Qiao, J., Kong, X., Hu, Z.-X. et al. (2014). *Nat. Commun.* 5: 4475.
7 Sun, J., Zheng, G., Lee, H.-W. et al. (2014). *Nano Lett.* 14: 4573.
8 Li, L., Yu, Y., Ye, G.J. et al. (2014). *Nat. Nanotechnol.* 9: 372.
9 Kim, Y., Park, Y., Choi, A. et al. (2013). *Adv. Mater.* 25: 3045.
10 Yabuuchi, N., Matsuura, Y., Ishikawa, T. et al. (2014). *ChemElectroChem* 1: 580.
11 Qian, J., Wu, X., Cao, Y. et al. (2013). *Angew. Chem. Int. Ed.* 52: 4633.
12 Mayo, M., Griffith, K.J., Pickard, C.J., and Morris, A.J. (2016). *Chem. Mater.* 28: 2011.
13 Sun, J., Lee, H.-W., Pasta, M. et al. (2015). *Nat. Nanotechnol.* 10: 980.
14 Hembrarn, K., Jung, H., Yeo, B.C. et al. (2015). *J. Phys. Chem. C* 119: 15041.
15 Li, W.-J., Chou, S.-L., Wang, J.-Z. et al. (2013). *Nano Lett.* 13: 5480.
16 Zhu, Y., Wen, Y., Fan, X. et al. (2015). *ACS Nano* 9: 3254.
17 Song, J., Yu, Z., Gordin, M.L. et al. (2014). *Nano Lett.* 14: 6329.
18 Li, W., Yang, Z., Li, M. et al. (2016). *Nano Lett.* 16: 1546.
19 Liu, S., Feng, J., Bian, X. et al. (2017). *Energy Environ. Sci.* 10: 1222.
20 Li, W., Hu, S., Luo, X. et al. (2017). *Adv. Mater.* 29: 1605820.
21 Pei, L., Zhao, Q., Chen, C. et al. (2015). *ChemElectroChem* 2: 1652.
22 Zhang, C., Wang, X., Liang, Q. et al. (2016). *Nano Lett.* 16: 2054.
23 Liu, Y., Zhang, N., Liu, X. et al. (2017). *Energy Storage Mater.* 9: 170.
24 Yang, Q.-R., Li, W.-J., Chou, S.-L. et al. (2015). *RSC Adv.* 5: 80536.
25 Yang, C.-P., Xin, S., Yin, Y.-X. et al. (2013). *Angew. Chem. Int. Ed.* 52: 8363.
26 Chin, L.-C., Yi, Y.-H., Chang, W.-C., and Tuan, H.-Y. (2018). *Electrochim. Acta* 266: 178.
27 Walter, M., Erni, R., and Kovalenko, M.V. (2015). *Sci. Rep.* 5: 8418.
28 Liu, Y., Zhang, A., Shen, C. et al. (2017). *ACS Nano* 11: 5530.
29 Gao, H., Zhou, T., Zheng, Y. et al. (2016). *Adv. Energy Mater.* 6: 1601037.
30 Zhou, J., Liu, X., Cai, W. et al. (2017). *Adv. Mater.* 29: 1700214.
31 Liu, S., Xu, H., Bian, X. et al. (2018). *ACS Nano* 12: 7380.
32 Xu, G.L., Chen, Z., Zhong, G.M. et al. (2016). *Nano Lett.* 16: 3955.
33 Haghighat-Shishavan, S., Nazarian-Samani, M., Nazarian-Samani, M. et al. (2018). *J. Mater. Chem. A* 6: 10121.
34 Liu, Y., Liu, Q., Zhang, A. et al. (2018). *ACS Nano* 12: 8323.
35 Meng, R., Huang, J., Feng, Y. et al. (2018). *Adv. Energy Mater.* 8: 1801514.
36 Ding, X., Huang, Y., Li, G. et al. (2017). *Electrochim. Acta* 235: 150.
37 Zhang, Y., Sun, W., Luo, Z.-Z. et al. (2017). *Nano Energy* 40: 576.
38 Liu, H., Tao, L., Zhang, Y. et al. (2017). *ACS Appl. Mater. Interfaces* 9: 36849.
39 Song, T., Chen, H., Xu, Q. et al. (2018). *ACS Appl. Mater. Interfaces* 10: 37163.
40 Song, T., Chen, H., Li, Z. et al. (2019). *Adv. Funct. Mater.* 29: 1900535.

41 Kulish, V.V., Malyi, O.I., Persson, C., and Wu, P. (2015). *Phys. Chem. Chem. Phys.* 17: 13921.

42 Huang, Z., Hou, H., Zhang, Y. et al. (2017). *Adv. Mater.* 29: 1702372.

43 (a) Li, W., Chou, S.-L., Wang, J.-Z. et al. (2014). *Adv. Mater.* 26: 4037. (b) Liu, J., Kopold, P., Wu, C. et al. (2015). *Energy Environ. Sci.* 8: 3531.

44 Nam, K.-H., Jeon, K.-J., and Park, C.-M. (2019). *Energy Storage Mater.* 17: 78.

45 Lu, Y., Zhou, P., Lei, K. et al. (2017). *Adv. Energy Mater.* 7: 1601973.

46 (a) Li, W.-J., Yang, Q.-R., Chou, S.-L. et al. (2015). *J. Power Sources* 294: 627. (b) Ge, X., Li, Z., and Yin, L. (2017). *Nano Energy* 32: 117.

47 (a) Li, W.-J., Chou, S.-L., Wang, J.-Z. et al. (2015). *Chem. Commun.* 51: 3682. (b) Wang, Y., Wu, C., Wu, Z. et al. (2018). *Chem. Commun.* 54: 9341.

48 Fullenwarth, J., Darwiche, A., Soares, A. et al. (2014). *J. Mater. Chem. A* 2: 2050.

49 Zhao, W., Ma, X., Wang, G. et al. (2018). *Appl. Surf. Sci.* 445: 167.

50 Zhao, F., Han, N., Huang, W. et al. (2015). *J. Mater. Chem. A* 3: 21754.

51 Kong, M., Song, H., and Zhou, J. (2018). *Adv. Energy Mater.* 8: 1801489.

52 Kim, Y., Kim, Y., Choi, A. et al. (2014). *Adv. Mater.* 26: 4139.

53 Qian, J., Xiong, Y., Cao, Y. et al. (2014). *Nano Lett.* 14: 1865.

54 Li, Q., Li, Z., Zhang, Z. et al. (2016). *Adv. Energy Mater.* 6: 1600376.

55 Fan, X., Mao, J., Zhu, Y. et al. (2015). *Adv. Energy Mater.* 5: 1500174.

56 Guo, G., Guo, Y., Tan, H. et al. (2016). *J. Mater. Chem. A* 4: 10893.

57 Han, F., Tan, C.Y.J., and Gao, Z. (2016). *ChemElectroChem* 3: 1054.

58 Cai, R., Zhang, B., Shi, J. et al. (2017). *ACS Sustainable Chem. Eng.* 5: 7690.

59 Zhang, W., Dahbi, M., Amagasa, S. et al. (2016). *Electrochem. Commun.* 69: 11.

60 Wu, C., Kopold, P., van Aken, P.A. et al. (2017). *Adv. Mater.* 29: 1604015.

61 Dong, C., Guo, L., He, Y. et al. (2018). *Energy Storage Mater.* 15: 234.

62 Li, H., Hao, S., Tian, Z. et al. (2019). *Electrochim. Acta* 321: 134624.

63 Ihsan-Ul-Haq, M., Huang, H., Cui, J. et al. (2018). *J. Mater. Chem. A* 6: 20184.

64 Kim, S.-O. and Manthiram, A. (2016). *Chem. Commun.* 52: 4337.

65 Chen, S., Wu, F., Shen, L. et al. (2018). *ACS Nano* 12: 7018.

66 Fan, M., Chen, Y., Xie, Y. et al. (2016). *Adv. Funct. Mater.* 26: 5019.

9

Metal Oxides/Chalcogenides/Alloys for Sodium-Ion Batteries on Alloy and Conversion Reactions

9.1 Introduction

In Chapter 8, the modification strategies, electrochemical performances, and reaction mechanisms of phosphorus and phosphide anodes based on alloy and conversion reactions for sodium-ion batteries (SIBs) are introduced detailly. Benefiting from the high theoretical capacity of these reactions, anode materials of metal alloys and metal oxides/chalcogenides with similar electrochemical behaviors have also attracted more and more research interest. For example, metal chalcogenides with special intrinsic properties, such as weaker M—X bond than M—O bond, are beneficial to the reaction kinetics in conversion reaction and smaller volume expansion in alloy reaction, showing promising sodium storage properties.

Despite these advantages, it is difficult to escape the same shortcoming as phosphorus and phosphide, that is, the unsatisfactory cyclability. The strategies of optimizing sodium storage performance of the above-mentioned alloy/conversion-type anode materials have been explored for a long time. The main ones include nanostructuring, morphology control, lattice engineering, construction of hybrid composite, etc. Besides, engineering bimetallic systems (bimetal alloys or bimetal chalcogenides) and utilizing compatible electrolyte (additives) systems with suitable binders also turn to be valid approaches to optimize their Na storage performance.

9.2 Metal Oxides

9.2.1 Conversion-type Oxides

Transition metal oxides (TMOs) with a conversion mechanism exhibit much higher capacities. And TMOs including iron oxide (Fe_3O_4 and Fe_2O_3), cobalt oxide (CoO and Co_3O_4), manganese oxide (MnO, Mn_3O_4, and MnO_2), molybdenum oxides (MoO_3), and vanadium oxide (V_2O_5 and VO_2), by far, have already been widely applied in the application of lithium-ion batteries (LIBs). These well-known LIB anodes have also been explored for analogous sodium storage, and the performances are attractive. Moreover, the relatively lower potential renders these conversion-type

Sodium-Ion Batteries: Energy Storage Materials and Technologies, First Edition. Yan Yu.

oxides as suitable anode materials for SIBs despite their relatively lower reversible capacities compared with those in LIBs. However, most TMOs suffer from large volume variation during Na^+ ion insertion/extraction and inherent low electronic conductivity, leading to unsatisfactory performance. Thus, fabrication of nanostructured composite is regarded effective in enhancing the electrochemical properties.

As anode materials for SIBs, iron oxides are attractive and promising because of their low cost, environmental benignity as well as the abundance in the earth crust. However, the degradation of the crystal structure and the decomposition of electrolyte occur during the sodiation–desodiation processes, resulting in capacity fading. Therefore, shortening ion transport pathways and mitigating the volume variation by designing various morphologies or compositing with other materials were adopted to enhance the electrochemical performance of iron oxides. Modafferi et al. reported α-Fe_2O_3@reduced graphene oxide (rGO) nanocomposites for SIBs [1]. Higher rGO content exhibited better electrochemical performance and various doping agents led to different effects. Ti doping is conducive to the enhancement of rate capability, and Mn doping improves the electrode stability. Amorphous Fe_2O_3 confined in 3D nitrogen-doped carbon nanofiber (CNF) networks also exhibited superior sodium properties [2]. A reversible capacity of 408 mAh g^{-1} was delivered after 350 cycles at 100 mA g^{-1}, and 183 mAh g^{-1} was remained at 3 A g^{-1}. The relief of electrode internal stress and more active sites for accommodating Na^+ were achieved by both synergistic effects of amorphous structure of Fe_2O_3 and 3D interconnected high-nitrogen-doping (10 at%) carbon network, thus ameliorating the volume changes and the electronic and ionic diffusion. Fe_3O_4 is also a very common anode material in SIBs. Ming et al. reported porous carbon–Fe_3O_4/$Na_2FeP_2O_7$ configuration for SIBs [3], and the full cell exhibited a superior capacity of 93 mAh g^{-1} and capacity retention of 93.3% after 100 cycles. The average work potential was about 2.28 V and the corresponding energy density was 203 Wh kg^{-1}. However, the development of iron oxides still needs to overcome some key issues. Firstly, the large voltage hysteresis should be alleviated and the in-depth electrochemical reaction mechanism needs to be clearly clarified. In addition, the low overall energy density during the full cell application resulted from the wide operation voltage range needs to be enhanced.

Due to the outstanding redox properties of cobalt oxides, typically Co_3O_4 and CoO are widely used in SIBs. Co_3O_4 is able to undergo a reversible conversion reaction as following: $Co_3O_4 + 8Na^+ + 8e^- \rightarrow 4Na_2O + 3Co$. Wu et al. have reported a novel 3D porous hierarchical Co_3O_4 structure as anode for SIBs [4]. The specular structure was composed of mesoporous Co_3O_4 nanosheet arrays grown on rGO coating on surface-treated Ni foam, which offered outstanding binding and electrical contact with the current collector. The impressive capacity of 757 mAh g^{-1} was retained at 100 mA g^{-1} after 400 cycles for SIBs. The results showed that the Li^+ ion intercalation/deintercalation processes were dominated by an ultrafast supercapacitor-like storage behavior caused by the 3D porous ion transport network and the graphene-mediated electronic transport pathway. In another work, rambutan-like hybrid hollow spheres of carbon confined Co_3O_4 were reported by

Xu et al. [5]. The rationally designed architecture renders high structural robustness and rapid electrode kinetics. When tested in SIBs, a high reversible specific capacity of 712 mAh g^{-1} at 0.1 A g^{-1} and 223 mAh g^{-1} even at a 5 A g^{-1} were retained. Meanwhile, satisfactory capacity retention of 74.5% was achieved after cycling 500 times. Recently, nano-frames of CoO_x encapsulated in carbonized electrospuned fibers by electrospinning ZIF-67 nanoparticles with polyacrylonitrile (PAN) were reported to exhibit outstanding sodium storage capacity [6]. With the help of in situ Raman and ex situ microscopic characterization, the reversible conversion of CoO_x during the sodiation–desodiation process was unveiled. After extreme long cycling over 6000 times, the sodium cell could achieve a high reversible capacity even at 20 A g^{-1}, arising from the unique CoO_x framework and the confinement effect of carbon fibers which alleviated the volume variation during cycling. A Ni-doped Co/CoO/NC hybrid anode for SIBs has been synthesized by Kaneti et al. [7]. They used bimetallic Ni–Co-ZIF as the starting precursor and the resulting product exhibited high specific surface area with micropores. Various factors including minimized aggregation and pulverization by introducing carbon matrix, faster insertion/de-insertion and shortened diffusion distance of Na^+ by designing micro- and mesopores, and creating extra electrochemically active sites through Ni doping, synergistically enhanced the Ni-doped Co/CoO/NC hybrid, which released high discharge capacity of 218 mAh g^{-1} at 500 mA g^{-1} and 218.7 mAh g^{-1} at 500 mA g^{-1} after cycling 100 times. In general, although cobalt oxides possess merits of high reversibility and specific capacities, the large voltage hysteresis and the wide operational potential range still hinder its commercial applications. The underlying mechanisms should be further explored for designing new modified cobalt oxides anode.

In spite of low redox potential, relatively low polarization, high theoretical capacity, and environmental friendliness, manganese oxides including MnO, Mn_3O_4, and MnO_2 still face the similar issues like other conversion-type oxides. The large volume fluctuation, low electronic conductivity together with easy dissolution of Mn^{2+} in the electrolyte negatively affect the electrochemical properties, especially cyclability. Therefore, the research progress on manganese oxide mainly focuses on how to improve the mechanical strength and electronic conductivity. He et al. have fabricated uniformly MnO nanoparticles with sizes of 4 nm on nitrogen-doped carbon (NDC) nanotubes [8]. The introduction of NDC materials (served as conductive scaffolds) facilitated the acceleration of electron transport and Na^+ ion adsorb ability in the charging process, thus leading to an improvement of electronic conductivity and inhibition of volume variation. The obtained product exhibited 709 and 292 mAh g^{-1} at 0.1 and 5 A g^{-1}, respectively. And after 3000 cycles at 5 A g^{-1}, the capacity retention could maintain 91% corresponding to a very slow capacity fading, illustrating that the structure changes could be controlled via this unique configuration. Despite many studies on the MnO anodes, there are not many reports on the Mn_3O_4 and MnO_2 anodes. Li et al., for the first time, reported a unique feather-like MnO_2 nanostructure in situ grown on a carbon paper through a simple hydrothermal method [9]. Thanks to a large pore size and the high surface area, the anode delivered approximately 300 mAh g^{-1} after cycling

400 times at $0.1\,A\,g^{-1}$. Another nanosized flower-like γ-MnO_2 was reported by Si et al. [10]. In the EC-based electrolytes, the capacity decays rapidly due to the inevitable dissolution of Mn^{2+}. However, the dissolution of Mn^{2+} was effectively alleviated when applied with ionic liquid electrolyte such as 1-butyl-3-methyl imidazolium bis[(trifluoromethyl)sulfonyl] imide (BMIMTFSI), and the cell exhibited a capacity retention of 76% after 100 cycles while only 56% for an EC/PC electrolyte tested under the same condition.

Except for the above-mentioned oxides, molybdenum oxide (MoO_3) with its attractive structure, low cost, nontoxicity, and abundance was regarded as a promising anode candidate in SIB field. Layered orthorhombic α-MoO_3 is stacked by bilayer sheets of [MoO_6] octahedron connected through wan der Waals forces. And similar to the reaction mechanism in LIBs, both the spacings from layer-to-layer [MoO_6] octahedron and the interlayer of [MoO_6] octahedron are able to accommodate Na^+. Although it has been used in LIBs for many years, the first report associated with its application in sodium storage was released in 2003 by Hariharan et al. [11]. The rocking chair full cell composed of $Na_3V_2(PO_4)_3$ cathode and MoO_3 anode revealed a potential of about 1.4 V with an initial discharge capacity of $164\,mAh\,g^{-1}$. Xia et al. have in situ characterized the sodiation–desodiation behavior of layer-structured α-MoO_3 nanobelt in the aspects of phase transformation and reaction mechanisms [12]. By means of in situ transmission electron microscopy (TEM) and electron diffraction pattern (EDP), α-MoO_3 nanobelts were found to exhibit a multi-step phase transformation process during the first cycling. In the initial sodiation, the amorphous Na_xMoO_3 phase was firstly formed and transformed into an intermediate phase of crystalline $NaMoO_2$, and the resulting product was a composite composed of dispersed crystalline Mo nanograins with Na_2O matrix. However, in the following initial desodiation, the obtained Mo nanograins were firstly converted to $NaMoO_2$ before the formation of amorphous Na_2MoO_3, showing an irreversible phase transformation with a large capacity loss after initial cycle. As the cycle continues, Mo nanograins and amorphous Na_2MoO_3 transformed reversibly. Thus, achieving a reversible transformation with a high Coulombic efficiency is the key to enhance electrochemical performance of α-MoO_3. Very recently, MoO_3 with a nanotube architecture fabricated by Jiang et al. exhibited satisfactory electrochemical performance [13]. By creating oxygen deficiency and surface phosphorylation, the resulting MoO_3 arrays sustained $265\,mAh\,g^{-1}$ at a rate of $2\,A\,g^{-1}$ over 1500 cycles with capacity retention of about 90%, which outperformed those in other literatures. Furthermore, a sodium-ion full cell was assembled to demonstrate the compatibility in practice. Paired with $Na_{0.67}(Ni_{0.23}Mg_{0.1}Mn_{0.67})O_2$ cathode, the assembled $MoO_{3-x}//Na_{0.67}(Ni_{0.23}Mg_{0.1}Mn_{0.67})O_2$ battery delivered an average cell voltage of 2.7 V with a specific energy of $128\,W\,h\,kg^{-1}$, as well as a power density of $1740\,W\,kg^{-1}$ at $68\,W\,h\,kg^{-1}$, verifying its potential in the application of SIBs.

Due to the layered structure, high energy density, and easy preparation, V_2O_5 is also studied as an anode for SIBs. V_2O_5 aerogel as anode material was attempted by Moretti et al. [14]. Combined with the carbon-coated $Na_3V_2(PO_4)_3$ cathode, the full cell presented an average voltage of 2.5 V and a maximum specific capacity of $113\,mAh\,g^{-1}{}_{anode}$ after 200 cycles at $100\,mA\,g^{-1}$. The widely utilized and feasible modification way is to composite with carbon materials such as rGO [15]. The introduction of rGO could prevent the aggregation of V_2O_5 nanowires and enhance the whole electronic conductivity of the composite anode. Thus, the as-prepared V_2O_5/rGO composite presented 0.62 mAh cm^{-2} after 100 cycles at 0.4 mA cm^{-2}. Another work reported VO_2 microspheres encapsulated into crumpled rGO as another promising anode for SIBs [16]. The unique bilayer structure VO_2 with large lattice spacing endowed it with good electrical conductivity and high capacity. Employing crumpled rGO to encapsulate VO_2 nanomaterials could extend its electrochemical potential window to the region of 0.01–3.0 V, which led to a higher capacity. A high reversible capacity of 383 and 214 mAh g^{-1} at 0.1 and 4 A g^{-1} could be obtained, respectively. Moreover, the composite is able to undergo 2000 cycles at 4 A g^{-1} with a capacity fade of 0.013% per cycle. Besides, a novel multi-shelled V_2O_3/C composite was synthesized to boost the electrochemical performance [17]. Adopting a simple solvothermal process and a calcination treatment, the composite, as expected, exhibited an outstanding specific capacity of 173 mAh g^{-1} at 1000 mA g^{-1} after 2000 cycles, ascribing to multiple positive effects induced by the high electrical conductivity, rich available redox sites, fast charge carrier transfer rate, and good structural stability.

In general, conversion-type oxides as a large anode family contribute satisfactory sodium storage properties to the whole anode field. Due to various reaction mechanisms, their phase stabilities and sodiation–desodiation processes differ from each other. Thus, designing new materials or adopting novel modification methods to solve the intrinsic drawbacks including low electronic conductivity and severe volume fluctuation still need to be further explored.

9.2.2 Conversion-alloy-type Oxides

Oxides based on alloy reactions have aroused much attention due to their high capacities. Taking SnO_2 as an example, it could react with Na^+ to generate Sn and Na_2O, and the resulting Sn could further react with Na^+ via alloying reaction. The complete process delivers a high capacity of 667 mAh g^{-1} theoretically and a low reaction potential when been applied in SIBs. Although the agglomeration and volume expansion are suppressed to some extent during the first alloy stage since Sn and Na_2O dispersing in each other, SnO_2 still bears severe volume change due to the expansion of Sn. Additionally, the low conductivity of SnO_2 anode also results in inferior rate capability and cycling life. Conventional and effective methods including reducing ionic/electronic diffusion, compositing with conductive materials, and constructing novel heterostructures have afforded it much-improved performance. For example, 3D carbon foam-supported Mo-doped SnO_2 nanoflake array was designed [18]. The carbon foam provided a 3D conductive network and also served as a buffer skeleton to suppress the volume variation, resulting in improved rate performance and cycling stability. Moreover, doping of Mo could accelerate the kinetics of sodium-ion transfer, and interlaced SnO_2 nanoflake arrays are conducive to the promotion of conversion reactions. Mo–SnO_2@C foam with 38.41 wt% SnO_2 and 3.7 wt% Mo content delivered an initial capacity of 1017.1 mAh g^{-1} at

0.1 A g^{-1}. In addition, Liang et al. employed 1D tubular TiO_2 as a structural-stable substrate to enhance the stability of SnO_2 [19]. The reaction kinetics in the process of sodiation–desodiation was accelerated by forming a layer of SnO_2 particles on the surface of the TiO_2 nanotubes, leading to excellent sodium storage capacity. The SnO_2@TiO_2 anode released 316 mAh g^{-1} after 50 cycles at 50 mA g^{-1} and the initial Coulombic efficiency (ICE) of 49.7%, which were much superior to the pristine SnO_2 (200 mAh g^{-1} and 29%) and TiO_2 nanotube (179 mAh g^{-1} and 32%).

Similarly, the use of Sb_2O_3 exhibits better performance on alleviating volume expansion compared with metallic Sb, and the capacity could be thus improved. Sb_2O_3 first converts into Sb nano-precipitates in an amorphous Na_2O matrix during sodiation, and the stress generated during the following alloy process from Sb to Na_3Sb could be alleviated to some extent. Fei et al. have adopted a simple solvothermal method to obtain a flexible Sb_2O_3 composite anode [20]. With employing carbon cloth as substrate, the uniformly distributed Sb_2O_3 nanospheres exhibited higher electronic conductivity due to the conductive pathway provided by carbon cloth for fast electron transfer. Based on alloying and conversion reactions, the Sb_2O_3/carbon cloth anode demonstrated a high initial discharge capacity of 1248 mAh g^{-1}, and 900 mAh g^{-1} was retained at 0.05 A g^{-1} after 100 cycles, resulting from the good conductivity of carbon cloth and strong chemical bonds between Sb_2O_3 and carbon cloth. Another 3D Ni-supported Sb_2O_3 anode has also been reported to deliver a high gravimetric and volumetric capacity [21]. Large pore interconnection in the 3D scaffold could be achieved via the colloidal template, which was beneficial to accommodate a large volume expansion and held a high active material loading. At a loading of 1.1 g cm^{-3}, a specific capacity and volumetric capacity of about 445 mAh g^{-1} and 488 mAh cm^{-3} were delivered, respectively. Excellent cycling performance with capacity retention of 89% after 200 cycles at 200 mA g^{-1} was also achieved.

Furthermore, since Bi_2O_3 possesses the advantages of large abundance, environmental benignity, stable chemical properties, and a high theoretical capacity of 690 mAh g^{-1}, its usage as anode for SIBs has also been explored. However, its inferior electrical conductivity and huge volume variation during the electrochemical conversion would cause rapid capacity fading. Compositing with carbon materials, the electrochemical performance of Bi_2O_3 is able to be enhanced distinctly [22]. With decorating Bi_2O_3 nanoparticles on CNFs by electrospinning method, this self-supporting composite demonstrated high rate capability of 504 mAh g^{-1} after 100 cycles at 25 mA g^{-1} and 393 mAh g^{-1} at 625 mA g^{-1}. Besides, amorphous carbon, and rGO could also be used as modifiers to boost the electrochemical performance [23].

9.3 Metal Chalcogenides

9.3.1 Metal Sulfides

Recently, as a big family of energy storage materials, metal sulfides (MS_x) have also gained a lot of attention for SIBs. Except for the merits of low working voltage, long life cycle, and high specific capacities, they exhibit higher electrochemical reversibility during redox reactions compared with the corresponding metal oxides. The reactions between metal sulfides and Na^+ ions are mainly conversion and/or alloying reactions, similar to those with Li^+ ions. In this segment, various MS_x will be thoroughly discussed.

9.3.1.1 SnS/SnS_2

Tin sulfides with unique layered structures and easy preparation process have been investigated in SIBs. Orthorhombic SnS and hexagonal SnS_2 as typical semi-conductive materials are very attractive. SnS exhibits respective high theoretical capacity of 1022 mAh g^{-1}, but it still experiences severe volume variation during repeated cycling processes and relatively poor electrical conductivity, leading to capacity fading. Because of the inherent larger radius of Na^+ (0.102 nm) than Li^+ (0.076 nm), the reaction of SnS with Na is more complicated since it bears a much larger volume change than the reaction in LIBs. Large endeavors have been tried to improve the electrochemical properties including design of a stable structure, morphological control, doping with heterogeneous atoms, compositing with conductive materials to form hybrid materials, and so on. Yan et al. have designed self-standing SnS/C nanofibers membrane anode to exhibit high sodium storage performance [24]. It exhibited a high specific discharge capacity of 324 mAh g^{-1} after 200 cycles with a capacity retention of 84.8%. By introducing interconnected nanosized building blocks, a micro/nanostructured SnS/few-layer graphene (SnS/FLG) composite fabricated by a facile scalable plasma milling method was reported [25]. The nanosized building blocks of the SnS/FLG composite combined with SnS nanoparticles and FLG matrix ensured excellent electron/ion conductivity. Moreover, the cycle-induced stress could be effectively relieved thanks to the FLG matrix and high-density boundaries, preventing SnS from pulverization after repeated cycling. Therefore, the SnS/FLG delivered a high gravimetric capacity of 494.2 mAh g^{-1} after 200 cycles, and high gravimetric/volumetric capacities of 388.1 mAh g^{-1}/768.5 mAh cm^{-3} after 500 cycles even at 1 A g^{-1}. A unique structure of N, S co-doped rGO and SnS sandwiched together by selective vulcanization of $ZnSn(OH)_6$ and subsequent in situ decomposition was reported (Figure 9.1a) [26]. The electronic conductivity could be enhanced by introducing rGO substrate and chemical bonds between SnS and rGO, and the co-doping of N and S rendered abundant sites for ions adsorption, inducing a strong pseudocapacitive effect and favoring fast electrochemical kinetics. The SnS/rGO/SnS electrode delivered high charge capacities of 573.2, 533.2, 482.0, 419.5, 359.2, 304.3, and 265.8 mAh g^{-1} at 0.05, 0.1, 0.2, 0.5, 1, 2, and 3 A g^{-1}, respectively, much superior to the pure SnS.

Adopting a conversion and alloying process toward sodium storage, SnS_2 stands out with a high theoretical capacity of 1136 mAh g^{-1}. However, the drawbacks especially poor electronic conductivity and enormous volume variation of 324% during sodiation–desodiation still hinder its application. The introduction of S-doped graphene (SG) and K doping to obtain K-doped-SnS_2@SG could greatly improve the electrochemical performance of SnS_2 [30]. S-doped graphene could act as a 2D conductive matrix to boost electrical transportation. And the K^+ ions with

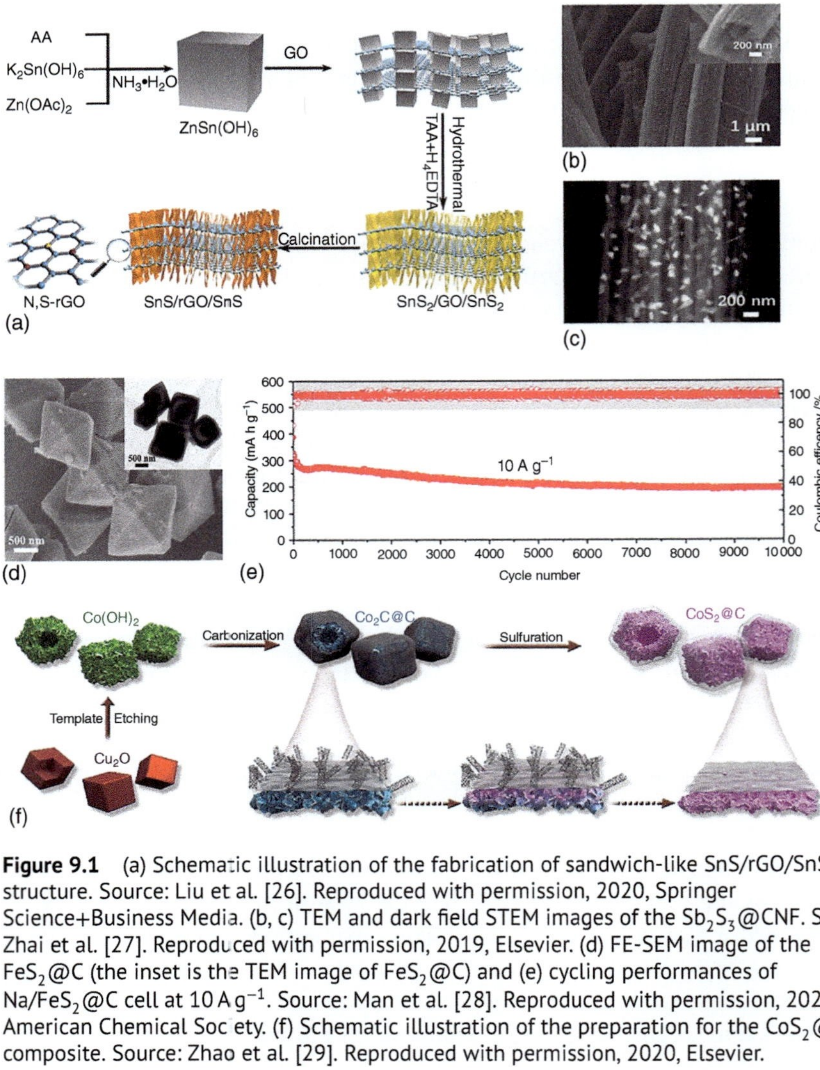

Figure 9.1 (a) Schematic illustration of the fabrication of sandwich-like SnS/rGO/SnS structure. Source: Liu et al. [26]. Reproduced with permission, 2020, Springer Science+Business Media. (b, c) TEM and dark field STEM images of the Sb_2S_3@CNF. Source: Zhai et al. [27]. Reproduced with permission, 2019, Elsevier. (d) FE-SEM image of the FeS_2@C (the inset is the TEM image of FeS_2@C) and (e) cycling performances of Na/FeS_2@C cell at 10 A g^{-1}. Source: Man et al. [28]. Reproduced with permission, 2020, American Chemical Society. (f) Schematic illustration of the preparation for the CoS_2@C composite. Source: Zhao et al. [29]. Reproduced with permission, 2020, Elsevier.

positive charge can be electrostatically adsorbed on the negatively charged graphene oxide sheets, and the resulting repulsive force also prevented graphene from stacking and maintained a dispersion of a few-layered structure. Thus, due to the synergistic effect of the dispersed S-doped graphene and uniformly distributed SnS_2 quantum dots, the K-doped–SnS_2@SG exhibited superior electrochemical performances of 743 mAh g^{-1} at 45 mA g^{-1} and 236 mAh g^{-1} even at 27 A g^{-1}, respectively. Wang and coworkers found that hierarchical SnS_2 microspheres with the created S vacancy could boost the charge transfer kinetics for SIBs [31]. The theoretical calculations have suggested that the presence of S vacancies could decrease the band gap and favor the sodium storage performance. On the one hand, nanosheet units could effectively reduce the diffusion pathway of sodium ions. On the other hand, the hierarchical SnS_2 microspheres possessed a higher conductivity, which favored the transport of electrons. Therefore, the hierarchical SnS_2 microspheres with S vacancies delivered long-term cycle life with a large capacity of 486.2 mAh g^{-1} after 1000 cycles.

9.3.1.2 Sb_2S_3/Bi_2S_3

Since Sb_2S_3 possesses the merits of a high theoretical capacity of 946 mAh g^{-1} with low toxicity, it has been widely accepted as a promising anode for SIBs. Sb_2S_3 has an orthorhombic and anisotropic-layered structure, where the 1D structures of $(Sb_4S_6)_n$ connecting by weak van der Waals force align parallelly to the [011] direction [32]. Zhai and coworkers have applied a one-pot electrospinning process to obtain Sb_2S_3 nanocrystals hosted in multichannel porous N-doped CNF composite [27]. Sb_2S_3 nanoparticles with 10–50 nm in size were uniformly embedded in CNF (Figure 9.1b,c), and the multichannel pores of CNF acted as a buffer to restrict the structural variation of Sb_2S_3 during cycling. Sb_2S_3@CNF exhibited a specific capacity of 178 mAh g^{-1} at 5 A g^{-1} with excellent cycle life (87% capacity retention after 1000 cycles at 1 A g^{-1}). By adopting a novel vaporization–condensation method, Chang et al. have investigated Sb_2S_3/active carbon nanostructured composite as an efficient anode for SIBs [33]. The confinement of Sb_2S_3 within the nanopores of active carbon accommodated the large volume of Sb_2S_3 variation and enabled the fast electron/Na-ion transfer, resulting in excellent sodium storage performance. The obtained Sb_2S_3 with nanosized pores exhibited a capacity of 799.5 mAh g^{-1} at 1162 mA g^{-1}, and 476.5 mAh g^{-1} was retained after 1000 cycles. Zhao et al. employed another ultrasound sonochemical method to fabricate amorphous Sb_2S_3–graphene composites for SIBs [34]. Because of the opening frame of ion diffusion channels and higher reversibility in thermodynamics, the amorphous composite displayed superior electrochemical properties in comparison with the crystalline counterpart for SIBs. Consequently, the amorphous Sb_2S_3–graphene composite delivered the first discharge capacity of 1867.1 mAh g^{-1} and a high reversible capacity of over 880 mAh g^{-1} after 50 cycles.

Bismuth (Bi), belonging to the same group as Sb, is an attractive anode for SIBs as well. Layered orthorhombic bismuth sulfides (Bi_2S_3) with a direct band gap of 1.3 eV is a semiconductor and isostructural to Sb_2S_3. Likewise, Bi_2S_3 often undergoes unsatisfactory cycling property resulting from poor conductivity together with the structural destruction induced by severe volume variation, leading to severe compromise on its application in advanced SIBs. Designing novel morphology to endow Bi_2S_3 with fast mass transfer and high conductivity is one of the effective methods for elevating sodium storage capacity. A hierarchically structured Bi_2S_3/graphene aerogel composite with a flower-like morphology was proposed by Zhang et al. [35]. Adopting a one-step hydrothermal reaction and the following freeze-drying process, the Bi_2S_3/graphene (23 wt%) aerogel composite exhibited a reversible specific capacity of 587 mAh g^{-1} at 100 mA g^{-1} and maintained 397 mAh g^{-1} after 50 cycles. A reversible specific capacity of 336 mAh g^{-1} was achieved even at 2 A g^{-1}. Moreover, 348 mAh g^{-1} after 120 cycles at 1 A g^{-1} was still maintained. SEM observation showed that the microsized flower-like Bi_2S_3 was composed of 1D nanorods with

700 nm in length uniformly anchored between the graphene layers, and the detailed microstructure observed from TEM illustrated the distribution of Bi_2S_3 nanorods (300 nm in width) within wrinkled graphene layers. Large amounts of pores together with a continuous graphene framework would facilitate the rapid transportation of electrons and sodium ions, and at the same time, the excellent elasticity of graphene aerogel was conducive to alleviate volume changes during sodiation–desodiation. Moreover, oxygen bridges by Bi—O—C bonds between Bi_2S_3 and the graphene aerogel further provided pathways for electron diffusion. Another attractive solution is constructing novel heterostructures which would alter the chemical properties of these functional materials. Inspired by the built-in electric field effect, Luo et al. designed heterostructured Bi_2S_3–Bi_2O_3 nanosheets via an easy water bath approach accompanying by a controllable thioacetamide-directed surfactant-assisted reaction process [36]. When applied in sodium storage, the Bi_2S_3–Bi_2O_3 heterostructure displayed a promising performance with a reversible discharge capacity of 630 mAh g^{-1} at 100 mA g^{-1} with relatively good stability.

9.3.1.3 MoS_2/WS_2

As for the group VIB metal, MoS_2 and WS_2 are the ones that have been most explored. Based on the conversion reaction mechanism, sodium ion could insert in MoS_2 and WS_2 to generate Na_2S and metallic Mo or W. Similar to graphite, MoS_2 and WS_2 demonstrate layered structures connecting by weak van der Waals forces, where S layers covalently bond with Mo(W) layers. This layered structure is able to promote a rapid diffusion of charge carriers and cushion the volume changes during redox process. Nevertheless, the inferior electrical conductivity together with the problem of restacking need to be avoided to obtain satisfactory electrochemical performance. Thus, several methodologies such as morphology optimization, size reduction, lattice expansion, as well as carbon materials incorporation have been utilized. Zhan et al. for the first time reported all-in-one MoS_2 nanosheets tailored by porous nitrogen-doped graphene (N-RGO) as an anode for SIBs [37]. The obtained material combined all the favorable qualities including vertical alignment of ultrathin layered MoS_2 with vacancy defects and expanded spacing. Furthermore, porous N-RGO would restrict the growth of all-in-one MoS_2 by a strong coupling effect, leading to the enhancement of cycling reversibility of conversion reaction. Therefore, the MoS_2@N-RGO displayed an excellent reversible capacity of 365, 219, and 149 mAh g^{-1} at 0.2, 10, and 20 A g^{-1}, respectively. And capacity of 250 mAh g^{-1} was retained after 300 cycles at 1 A g^{-1}. Moreover, when coupled with $Na_3V_2(PO_4)_3$ cathode, the full cell displayed a high reversible capacity of 265 mAh g^{-1} at 0.1 A g^{-1} with a remarkable energy density of 141 Wh kg^{-1}. Liu et al. used bacterial cellulose (BC) as a carbon source to form highly conductive CNF to synthesize nanoflower-like MoS_2/CNFs [38]. 3D porous architecture of CNFs provided sufficient room to adapt the volume change of MoS_2, while the conductive network facilitated electron and ion transfer. In this design, the formation of few-layered ultrathin MoS_2 nanosheets crosslinked with CNFs due to the confined growth, leading to a high active surface area of 35.1 m^2 g^{-1}. The alternation of CNFs content would cause various morphology evolution of MoS_2/CNFs. When the content of CNFs was 10.24 wt%, the optimized MoS_2/CNFs possessed an enlarged lattice spacing of 0.71 nm, facilitating ion and electron diffusion. Thus, it delivered a capacity of 509 mAh g^{-1} at 0.1 A g^{-1} and maintained a high discharge capacity of 368 mAh g^{-1} after 400 cycles at 6.25 A g^{-1} as anodes for SIBs.

Aiming at exploring the sodium storage performance of 3D graphene–WS_2 hybrid structure, Mohammadi et al. adopted the first-principle calculations to calculate the structural properties of Na adsorption and diffusion on 3D graphene–WS_2 as well as the electronic structure and charge transfer properties of Na adsorbed on 3D graphene–WS_2 systems [39]. In contrast to graphene and graphene nanoribbon, larger binding energy was delivered in the case of Na atom intercalating into the interlayers of the 3D GW hybrid structure, which indicated that this composite structure exhibiting better cycling stability. The estimated Na diffusion barriers between 0.05 and 0.17 eV showed that the Na atoms would readily diffuse on the 3D GW system during battery operation. Moreover, satisfactory electrical conductivity was exhibited before and after Na adsorption, which is conducive for anode performance. Additionally, Xu et al. have reported a novel additive-free anode with direct construction of vertically aligned WS_2 columnar platelets on an Al foil with low-tortuosity pores (VA/LT-WS_2) via an innovative sputtering strategy [40]. This unique structure provided high efficiency of ion diffusion. An additional carbon layer with 6 nm in thickness also covered the composite film electrode to enhance the electron transportation pathway; meanwhile, it also relieved the cracking of solid electrolyte interface (SEI) film resulting from the volume expansion. At 100 mA g^{-1}, the 2.8-μm composite film anode exhibited a high capacity of 1339 mAh cm^{-3} volumetrically. When paired with $Na_2V_3(PO_4)_3$/C cathode, the full cell demonstrated a reversible volumetric capacity of 554 mAh cm^{-3} at 50 mA g^{-1}, verifying its potential for further battery development.

9.3.1.4 FeS_x/CoS_x/NiS_x

As two widely investigated iron sulfides, pyrite FeS_2 and FeS, have drawn a lot of attention in the energy storage field for their high theoretical capacity, natural abundance, cost-effectiveness, and environmentally friendliness. In cubic FeS_2 structure, Fe takes up the octahedral sites of FeS_6 and sulfur is in the form of S_2^{2-}. Although intercalation ($FeS_2 + xNa^+ + xe^- \rightarrow Na_xFeS_2$ $(x < 2)$) and conversion reactions ($Na_xFeS_2 + (4-x)Na^+ + (4-x)e^- \rightarrow Fe + 2Na_2S$) could both happen during the discharging process in theory, experiments characterization revealed the intercalation reaction is irreversible, implying that FeS_2 cannot be regenerated during the charging process. Thus, the sodium storage mechanism of FeS_2 could be illustrated based on the conversion reaction as following: $Na_xFeS_2 + (4-x)Na^+ + (4-x)e^- \leftrightarrow Fe + 2Na_2S$, and the reaction delivers a high theoretical specific capacity of 894 mAh g^{-1}. Similarly, FeS exhibits the theoretical capacity of 609 mAh g^{-1}. However, it should be noted that severe volume change owing to their conversion reactions causes inferior cyclic performance, which hinders their applications. Extensive efforts during recent years have been dedicated to improve the performance. FeS_2@carbon (FeS_2@C) yolk–shell nanostructures fabricated via a facile carbonation–sulfidation strategy were reported by Man et al. [28] for SIBs. The yolk–shell structure was able to improve the reaction kinetics of the high-capacity FeS_2 and meanwhile provide enough space to tolerate large volume changes (Figure 9.1d). As a result, the FeS_2@C nanocomposite with the unique structure delivered a reversible capacity of 616 mAh g^{-1} after 100 cycles at 0.1 A g^{-1} and retained 220 mAh g^{-1} even after 10 000 cycles at 10 A g^{-1} (Figure 9.1e). Similar morphology of a unique dual carbon-decorated FeS_2-based hybrid (denoted as GF/FeS_2@C) was also prepared by Jing et al. [41]. Furthermore, by using Prussian blue microcubes as the self-sacrificial templates, yolk–shell FeS_2@C microbox structure was preserved, which was considered to bear repeating volume change and guarantee the mechanical integrity. And the introduction of N, S co-doped graphene framework could homogeneously cover the yolk–shell FeS_2@C microboxes, offering not only abundant defects to enrich the number of electrochemically active sites but also fast electron pathways for sodium diffusion. Benefiting from the synergistic effect, the resulting anode could deliver a reversible capacity of 200 mAh g^{-1} at 10 A g^{-1}, and a satisfactory cycle life with a capacity of 257 mAh g^{-1} after 1000 cycles at 0.5 A g^{-1}. Haridas et al. prepared a freestanding hybrid electrode incorporating FeS and sulfurized polyacrylonitrile (SPAN) for the first time [42]. They used electrochemically active SPAN fiber network to replace the inactive carbon matrix, and the sulfur atoms were covalently bonded to the carbon atoms in the cyclized and dehydrogenated PAN backbone. On the one hand, the high capacity of both the active materials, FeS and SPAN, would lead to high energy density. And conductive FeS nanoparticles (10^2–10^3 S cm^{-1}) could also compensate the inherently low conductivity of SPAN. On the other hand, the SPAN matrix was able to keep the structural stability and enhance the Na^+ transport by reducing the diffusion pathways. This configuration at 200 mA g^{-1} displayed a capacity of 782.8 mAh g^{-1}, and 327.5 mAh g^{-1} was retained after cycling 500 times at 5 A g^{-1}. High loading FeS_2 nanoparticles anchoring on biomass-derived carbon tube as anode for SIBs was prepared by Chen et al. [43]. Natural herb juncus was selected as the carbon source and mesh-like FeS_2/carbon tube/FeS_2 composites (FCF) were obtained via simple heat treatment of juncus, ferric nitrate, and sulfur. The biomass materials exhibited merits of renewability, low cost, and complicated structure. With the introduction of a carbon tube network of only 5.3 wt%, the composite demonstrated a high capacity of 542.2 mAh g^{-1} and good stability (capacity loss was less than 0.005% per cycle over 1000 cycles) and the excellent rate performance of 426.2 mAh g^{-1} at 2 A g^{-1}.

Cobalt sulfides, such as Co_9S_8, CoS, CoS_2, and Co_3S_4, have been studied as the anodes for SIBs. Among them, CoS_2 possessing the maximum S ratio demonstrates a theoretical capacity of 872 mAh g^{-1}. CoS_2 embedded in N-doped porous carbon spheres interlinking by rGO was synthesized to improve the poor cycling stability and rate capability [44]. The combination of CoS_2 nanoparticles, graphitic-structured N-doped porous carbon spheres, and rGO nanosheets could form interconnected multidimensional nanostructure to offer enough and rapid transfer pathway for electrons and ions, leading to high electric conductivity and fast reaction kinetics. In addition, the unique structure and morphology facilitated volume variation and maintained structural stability. As a consequence of the synergistic effect, the resulting composite delivered outstanding electrochemical properties in the voltage window of 0.4–2.9 V using 1.0 M $NaCF_3SO_3$ in diglyme as electrolyte. A reversible capacity of 523.9 mAh g^{-1} at 100 mA g^{-1} after 150 cycles and 423.3 mAh g^{-1} at 2 A g^{-1} was presented. Moreover, it also delivered an extraordinary long-term cycling stability, i.e. 465.9 mAh g^{-1} at 500 mA g^{-1} after 600 cycles and 412.2 mAh g^{-1} at 1 A g^{-1} after 400 cycles. Zhao et al. have prepared the hollow CoS_2@C microcubes as an anode for SIBs [29]. In this design, Cu_2O cubes served as the template and carbon coating was achieved via the pyrolysis of acetylene gas and the following sulfidation using thiourea as sulfur source (Figure 9.1f). This unique cubic hollow morphology was beneficial to buffer the volume variations accompanied by sodiation–desodiation processes. At the same time, complete carbon shells uniformly covering the CoS_2 cubes could enhance electrical conductivity by promoting charge transfer. Finally, CoS_2@C exhibited superior rate capacities with 367.6 mAh g^{-1} at 5 A g^{-1} after 200 cycles and 260 mAh g^{-1} at 15 A g^{-1}.

Co_9S_8 with a theoretical capacity of 544 mAh g^{-1} consisting of two types of polyhedrons including CoS_6 octahedra and CoS_4 tetrahedra, where each CoS_6 octahedron shares its six corners with 24 CoS_4 tetrahedra. Lin et al. combined hydrothermal synthesis and chemical vapor deposition to obtain the Co_9S_8 nanorod encapsulated by NDC shell (r-Co_9S_8@NC) as an anode for SIBs [45]. The NDC shell could form a chamber-confined effect and its sodiophilic interface well restrained the Co_9S_8 redox in the shell accompanied with improved kinetics. On the one hand, both the Co_9S_8/electrolyte side reaction and large volume expansion could be alleviated. On the other hand, sodium storage/diffusion ability was enhanced benefiting from the advantages of this structure. Consequently, high capacities of 675 mAh g^{-1} at 50 mA g^{-1} and 342 mAh g^{-1} at 10 A g^{-1} were delivered. After 1500 cycles at 5 A g^{-1}, it also exhibited a capacity of 317 mAh g^{-1}. An elaborate composite with Co_9S_8 nanoclusters embedded in honeycomb-like sulfur-doped carbon foam (Co_9S_8@S–CF) was prepared [46]. They adopted a facile sulfur-assisting calcination strategy, which tactfully induced the co-occurrence of in situ pore forming, sulfidation, sulfur doping, and carbonization. The abundant voids provided by sulfur-doped carbon foam (S–CF) could lead to 3D ionic/electronic pathways, resulting in high sodium-ion accessibility and ultrafast sodium-ion/electron transportation toward Co_9S_8 nanoclusters. A remarkable capacity of 373 mAh g^{-1} over 1000 cycles at 0.25 C with superior capacity retention of 80% and a rate capability of 180 mAh g^{-1} at 50 C (20 A g^{-1}) were delivered when it served as anode for SIBs.

Co_3S_4 has a high theoretical capacity of 702 mAh g^{-1}. Dong et al. employed a simple solvothermal and subsequent annealing approach to synthesize CoS_2/Co_4S_3@N-doped carbon (CoS_2/Co_4S_3@NC) microspheres [47]. In a half-cell configuration, they exhibited 650 and 217 mAh g^{-1} at 0.3 and 6 A g^{-1}, respectively, and ultra-long cycling performance of 256 mAh g^{-1} after 1500 cycles at 4 A g^{-1}, together with 239 mAh g^{-1} after 3000 cycles at 6 A g^{-1}. As assembled as a full cell with $FeFe(CN)_6$ cathode, it released a reversible capacity of 520 mAh g^{-1} at 500 mA g^{-1}. The extraordinary electrochemical performance resulted from the superior structural stability of the composite as well as the pseudocapacitive behavior, which accelerated the diffusion of ion and electron to further improve the high-rate capability. When S-doped graphene was used to wrap SnS_2/Co_3S_4 hollow nanocubes, the composite delivered distinguished specific capacities of 845.7 mAh g^{-1} at 0.5 A g^{-1} after cycling 100 times. And it should be noted that 392.9 mAh g^{-1} can be obtained at 10 A g^{-1} during an ultrafast sodium storage process, and the charging time is less than three minutes [48]. CoS with a theoretical capacity of 589 mAh g^{-1} endures poor intrinsic electrochemical properties as well. CoS/MXene composite with a heterolayered architecture was obtained by Zhang et al. [49] CoS nanoparticles with 6 nm in size were uniformly anchored on the Ti_3C_2 MXene flakes, leading to excellent electrochemical performance, i.e. a high reversible capacity of 267 mAh g^{-1} after 1700 cycles at 2 A g^{-1} with an extremely low capacity decay rate of only ~0.0072% per cycle, and an outstanding rate performance with 272 mAh g^{-1} at 5 A g^{-1}. The introduction of MXene flakes was able to offer a stable conductive network and the open 2D structure can expose more active sites for fast electrochemical reactions. Moreover, MXene could also prevent CoS nanoparticles from agglomeration and reduce the size of CoS nanoparticles. The small CoS nanoparticles would result in the reduction of the sodium-ion diffusion path.

Nickel-based sulfides such as NiS_2, NiS, and Ni_3S_2 are promising anodes as their merits of high theoretical specific capacity, low cost, naturally abundant, and environmentally friendly. Ni_3S_2 has a theoretical capacity of 446 mAh g^{-1}. Shuang et al. adopted sulfuration of a Ni-based metal–organic framework (Ni-MOF) and the following coating with polypyrrole (PPy) to synthesize Ni_3S_2 composite anode [50]. Ni_3S_2 nanoparticles were uniformly embedded in NDC nanosheets and the particle aggregation of Ni_3S_2 could be prevented by the NDC shell. The transportation pathway of the electrons and ions could also be shortened and the volume expansion was alleviated by the NDC shell. As expected, high discharge capacities of 432.8 mAh g^{-1} at 0.2 A g^{-1} and 371.6 mAh g^{-1} at 6.4 A g^{-1} were achieved by the composite. NiS_2 exhibits a theoretical capacity of 873 mAh g^{-1}. Zhao et al. reported the fabrication of porous CNFs with the embedment of ultrasmall NiS_2 nanoparticles [51]. The combination of NiS_2 nanoparticles and conductive CNF has a positive effect for reducing the volume expansion and offered rapid transport channel for electron and ion. Thus, the composite exhibited capacity of 500 mAh g^{-1} at 0.1 A g^{-1} and maintained 200 mAh g^{-1} at 2.0 A g^{-1}. NiS has a theoretical capacity of 590 mAh g^{-1}. Fan et al. constructed complex NiS multi-shelled hollow structures for high-performance SIB anode [52]. This delicate structure consisting of multi-shell nitrogen-doped carbon nanotube (NCNT) scaffold with NiS nanoparticles encapsulation facilitated the structural stability and rapid electronic/ionic channels. Consequently, significant improvement on cycle life and rate capability was delivered, i.e. 531.5 mAh g^{-1} at 0.05 mA g^{-1} with 89.3% capacity retention over 500 cycles.

9.3.1.5 Other Monometal Sulfides Including CuS_x/VS_x/TiS_2

CuS exhibits promising sodium storage performance owing to its good thermal stability and large specific capacity. However, the practical use of CuS is hindered by serious pulverization of electrode materials upon cycling. Zhao et al.

fabricated 1D CuS nanowires wrapped in NDC (CuS NWs@NC) for SIBs [53]. Combining the virtue of a coaxial 1D structure and a conductive carbon layer, the fast conductive highway was constructed to reduce the ion/electron transport length and accommodate the volume change as well as to enlarge the contact interface between electrolyte and electrode. Therefore, CuS NWs@NC anode with fast sodium-ion diffusion of 2.43×10^{-10} to 3.90×10^{-13} $cm^2\ s^{-1}$ exhibited an ultra-long cycling stability with a reversible capacity of 216.7 $mAh\ g^{-1}$ at 20 $A\ g^{-1}$ after 10 000 cycles, and the full cell also demonstrated 220 $mAh\ g^{-1}$ at 0.2 $A\ g^{-1}$ after 200 cycles.

Vanadium sulfides (for example, VS_4, VS_2, and V_5S_8) enjoy good electrochemical activities. Li et al. synthesized three novel 3D VS_4 nano-architectures including nanorods (nanorod-VS_4), nanocones, and nanobelts self-assembled VS_4 microspheres [54]. Among them, nanorod-VS_4 exhibited the best reversible capacity of 225 $mAh\ g^{-1}$ at 0.5 $A\ g^{-1}$ after 200 cycles and even achieved 203 and 168 $mAh\ g^{-1}$ at 1.0 and 2.0 $A\ g^{-1}$, respectively. The analysis revealed that these superior electrochemical performances were originated from the self-assembled structure of radial nanorods and the preferred orientation growth along (110) planes. Additionally, through a facile solvothermal method, VS_2 nanosheets were hierarchically assembled into flower-like structures [55]. Kinetic studies indicated a pseudocapacitance dominant mechanism. Moreover, ex situ Raman, high-resolution transmission electron microscopy (HRTEM), and selected area electron diffraction (SAED) characterizations verified the high reversibility of the electrochemical process. Tested within 3.0–0.3 V, the flower-like VS_2 exhibited superior electrochemical performance. At 0.1 $A\ g^{-1}$, a reversible capacity of around 600 $mAh\ g^{-1}$ was delivered. And 83% and 87% of its initial capacities were retained after 700 cycles at 2 and 5 $A\ g^{-1}$, respectively. Moreover, a discharge capacity of 277 $mAh\ g^{-1}$ was released at 20 $A\ g^{-1}$, verifying the excellent performance at high rates. Yang et al. have synthesized monoclinic-structured V_5S_8 nanosheets combined with graphite [56]. Inside the V_5S_8 nanosheets, VS_2 monolayer stacks layer-by-layer with the additional V atoms locating 1/4 of the available crystallographic sites. This unique morphology and structure were beneficial to the fast transport of electronic/ionic charge carriers and promoted fast Na^+ ion diffusion for reversible charge–discharge cycles. Therefore, a high reversible discharge capacity of 682 $mAh\ g^{-1}$ at 0.1 $A\ g^{-1}$ was delivered and 496 $mAh\ g^{-1}$ was retained at 1.0 $A\ g^{-1}$ after 500 cycles.

TiS_2, which features a layered structure, could accommodate guest atoms in layers to form intercalation compounds. To in-depth investigate the electrochemical behaviors of TiS_2 in SIBs, Lin et al. adopted computational modeling and a multi-modal synchrotron approach: a combination of operando X-ray absorption spectroscopy (XAS) and ex situ X-ray powder diffraction (XPD) [57]. They showed that during sodiation of TiS_2, both Ti and S were the electrochemical redox-active sites where three phases including TiS_2, $Na_{0.55}TiS_2$, and $NaTiS_2$ formed unlike that occurring in lithiation. Moreover, the decay in Coulombic efficiency resulted from the incomplete recovery of the coordination number of Ti after cycling.

9.3.1.6 Bimetallic Sulfides

Exhibiting higher electrical conductivity and richer redox reactions, bimetallic sulfides have a significant enhancement of the electrochemical performances as compared to the monometal sulfides.

M–Co–S (M = Ni, Cu). Ternary nickel cobalt sulfides are demonstrated with excellent electrochemical performance resulting from their diversity in crystal structures and morphologies. $NiCo_2S_4$, as the most widely investigated electrode, has aroused a lot of attention because of its high conductivity and multiple redox reactions. Approaches to improve the cycling and rate performance include constructing 3D electrodes with grown nanoarrays, designing hollow or novel morphologies, compositing with conductive materials, and so on. For example, coating rGO onto $NiCo_2S_4$ nanoparticles was designed (Figure 9.2a) [58]. The obtained 3D micro/mesoporous network increased intrinsic electrical conductivity which promoted the access of sodium ions and electrons and provided sufficient reaction active sites together with enhanced structural stability. The NCS@rGO electrode delivered 621 and 532 mAh g^{-1} at 0.1 and 0.5 A g^{-1}, respectively, and a high capacity retention of 95% was obtained at 0.1 A g^{-1} after 100 cycles. They revealed that the conversion products of NiS_2 and CoS_2 from the phase transformation of $NiCo_2S_4$ were successfully confined in the rGO coating. Moreover, the pseudocapacitive-dominated behavior contributed to the measured superior rate capability (Figure 9.2b). Similarly, $CuCo_2S_4$ demonstrated a relatively low resistivity (10^{-4} Ω). After compositing with conductive materials, the $CuCo_2S_4$/rGO also exhibited a high reversible capacity of 433 mAh g^{-1} at 100 mA g^{-1} after 50 cycles, and excellent rate capability with a specific capacity of 336 mAh g^{-1} at 1000 mA g^{-1} [60].

M–Fe–S (M = Co and Ni). Except for the wide application of electrocatalysts, ternary cobalt iron sulfides Co–Fe–S have been rarely utilized as anode for SIBs. Various cobalt iron sulfides including different Co/Fe ratios have been investigated by doping Co into FeS_2 nanospheres [61]. The samples were denoted as FeS_2, Fe0.9, Fe0.7, Fe0.5, and Fe0.0 based on Fe concentrations of 1.0, 0.9, 0.7, 0.5, and 0.0, respectively. The results showed that the Fe0.5 had the highest reversible capacity and best rate capability. After 5000 cycles, the Fe0.5 sample delivered a discharge capacity of 220 mAh g^{-1}, and even at 20 A g^{-1}, the Fe0.5 could still deliver a discharge capacity of 172 mAh g^{-1}. Moreover, coupled with $Na_3V_2(PO_4)_3$ cathode, an initial capacity of 340 mAh g^{-1} at 200 mA g^{-1} with an average output voltage of 1.6 V was released. Zhang et al. have synthesized heterostructured Ni/Fe-bimetallic sulfide encapsulating in a matrix of interconnecting N-doped porous cubic carbon and carbon nanotubes (CNTs) (Ni–Fe–S–CNT) via a convenient co-precipitation and subsequent sulfurization technique of the corresponding Prussian blue analog nanocage [62]. The 3D hierarchically high-conductivity 3D networks provided fast $Li^+/Na^+/K^+$ insertion/extraction, maintained the electrical contact, and guaranteed structural integrity. As anode for SIBs, a capacity of 431 mAh g^{-1} at 0.1 A g^{-1} was delivered.

M–Mo(W)–S (M = Fe, Co, Ni and Cu). Bimetallic phase $M_xMo_6S_{8-y}$ (M = Cu, Ni, Co, Fe, Ag, Cr) with 3D frameworks (rhombohedral or triclinic) has been developed

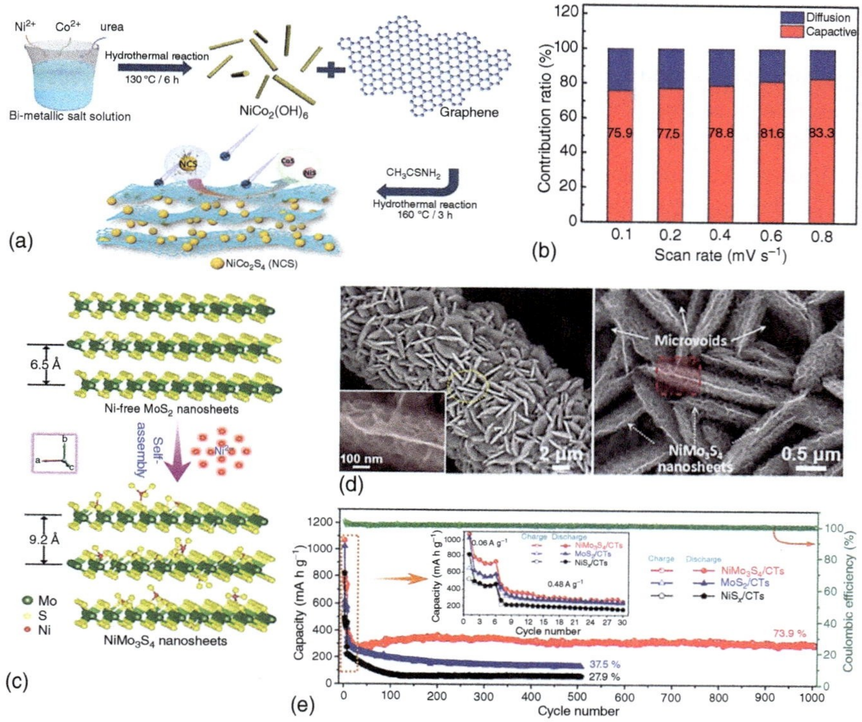

Figure 9.2 (a) Illustration of the fabrication process of the NCS/rGO composites, and (b) bar chart showing the contribution ratio of capacitive- and diffusion-controlled capacities at different scan speeds. Source: Sun et al. [58]. Reproduced with permission, 2020, Elsevier. (c) Schematic illustrations of the $NiMo_3S_4$ laminar structure. (d) Low- and high-magnification SEM images of the $NiMo_3S_4$/CTs sample. (e) Long-term cycling performances of $NiMo_3S_4$/CTs, MoS_2/CTs, and NiS_x/CTs electrodes at current densities of 0.06 and 0.48 $A\,g^{-1}$. Source: Kong et al. [59]. Reproduced with permission, 2018, Elsevier.

with high electronic conductivity (10^2–10^3 $S\,cm^{-1}$) and high Na^+ diffusion coefficient. Among them, Cu_2MoS_4 with two allotropes including P phase (*Pm*42 space group) and I phase (*Im*42 space group) exhibits tetragonal crystal structure and large interlayer spacings. Chen et al. incorporated glucose and rGO with Cu_2MoS_4 [63]. The combination of ex situ X-ray photoelectron spectroscopy and in situ XAS revealed that dominant intercalation mechanism was achieved using ether-based electrolyte with narrowed voltage window, which enabled fast Na^+ storage performances. After compositing with rGO, the anode exhibited 205.7 $mAh\,g^{-1}$ after 2000 cycles. And the full cell (paired with stable NVP-based cathode) demonstrated a capacity retention of 75.5% after 500 cycles with a high CE of about 100%. A novel $NiMo_3S_4$/carbon textiles electrode was prepared by forming 3D hierarchical defect-rich $NiMo_3S_4$ nanosheet arrays tightly anchored on flexible carbon textiles (Figure 9.2c,d) [59]. By the means of incorporating Ni ions into the MoS_2 lattice structure, the lattice spacing of the (002) crystal plane was enlarged and plenty of electrochemical active defects were generated.

Moreover, large surface area, good electron transport, and high ion diffusion rates were also conducive to the improvement of SIBs. The high-conductive carbon textile could serve as a solid mechanical backbone to support $NiMo_3S_4$ nanosheets and prevent aggregation as well. Therefore, a high specific capacity of ~480 mA h g^{-1} at 0.12 A g^{-1} with high Coulombic efficiency of 100% was delivered as anode for SIBs. As shown in Figure 9.2e, after a long-term cycling of 1000 times, reversible capacity of ~302 mA h g^{-1} with 73.9% was retained at 0.48 A g^{-1}, showing its promising application in SIBs.

M–Sb–S (M = Cu, Co, and Bi). Typically, $CuSbS_2$ offers a theoretical value of 751 mAh g^{-1}, which could also be considered as a promising energy storage material. Marino et al. explored the sodiation mechanism of $CuSbS_2$, which contained multiple reactions [64]. The conversion reaction happened first with the sequential formation of intermediate phases of $CuSb_{(1-x)}S_{(2-y)}$, Na_xS alloy and Sb. And in the following step, Sb alloyed with Na^+ ions. During the desodiation process, a Sb- and S-deficient ternary phase $CuSb_{(1-x)}S_{(2-y)}$ was formed. Wen et al. reported facile hydrothermal method and self-polymerization of dopamine to modify the $BiSbS_3$ nanorods with NDC coating [65]. Traditional and facile carbon coating would lead to fast ion/electron transfer and buffer volume expansion, thus the resulting $BiSbS_3$@NC electrode delivered high sodium storage capacity of 771.5 mAh g^{-1} in the second cycle and excellent rate performance of 518.4 mAh g^{-1} at 1000 mA g^{-1}.

M–Sn–S (M = Cu). Cu_2SnS_3 has interlayer spaces and tunnels which benefit fast Na^+ diffusion, and the good conducting ability also render it with ideal candidates for energy storage materials. Noted, its supercell crystal exhibits interlayer distance of 2.280 Å and tunnel size of 3.921 × 5.587 × 4.210 Å^3 in structure, which are larger than the diameter of sodium ion (radius 0.95 Å), and thus nanostructured Cu_2SnS_3 materials for SIBs anode have been reported including different morphologies. For example, single-crystalline Cu_2SnS_3 nanosheets were synthesized by Shi et al. [66]. The face-centered cubic-structured Cu_2SnS_3 nanosheets exhibited size of 50–70 nm in thickness and oriented parallel to the (220) crystal plane to form 2D structure. The resulting anode demonstrated initial sodiation capacity of 586 and 178 mAh g^{-1} after 50 cycles. Additionally, benefiting from the flower-like morphology, the Cu_2SnS_3 delivered high initial reversible capacity of 447.7 and 200.6 mAh g^{-1} after 50 cycles at 100 mA g^{-1} [67].

Although various structures and configurations of bimetallic sulfides have been widely explored to date, controllable size and shape via simple and facile methods are still hard to be achieved. Additionally, various chemical combinations in different bimetallic sulfides are still complicated to be understood. Thus, elaborate calculation and modeling could be carried out to further explore the inner mechanism during sodiation–desodiation process.

9.3.2 Metal Selenides

Although metal selenides exhibit lower theoretical gravimetric capacities, comparing with metal sulfides due to the heavier atomic weight of Se than S, the striking merits of metal selenides also attract researchers' attention. Firstly, as the density of Se is about 2.5 times higher than S, a comparable theoretical volumetric capacity density of metal selenide to sulfide would be achieved. Additionally, Se exhibits higher conductivity than S and metal–Se bonds are weaker than metal–S bonds, leading to high activity for sodium storage.

9.3.2.1 SnSe/$SnSe_2$

Similar to the tin sulfide counterparts, the study of tin selenides mainly focuses on SnSe and $SnSe_2$, whose theoretical capacities are 780 and 756 mAh g^{-1}, respectively, based on conversion and alloying reactions. Although their volume expansion is less than metallic Sn, they still suffer from the fading cycling performance due to volume variation. Therefore, plenty of efforts have been carried out to improve the electrochemical properties. Zhao et al. have fabricated SnSe quantum dots sheathed in nitrogen-doped CNF using a modified method to avoid irregular shapes and micron-sized agglomerations produced by conventional methods such as simple ball milling [68]. The obtained SnSe exhibited a large discharge capacity of 268 mAh g^{-1} after 750 cycles at 2 A g^{-1}. Further exploration on full cell performance coupled with $Na_3V_2(PO_4)_3$ cathode showed an output voltage of above 2.0 V, and ultra-long cycle life of 1500 cycles with a high reversible discharge capacity of about 100 mAh g^{-1} at 1 A g^{-1} together with the superior rate capability. Additionally, $SnSe_2$/rGO sandwich structure was proved to exhibit improved electrochemical properties [69]. With only introducing 7.3% rGO, the formed abundant Sn—O—C bonds were conductive to accelerate the charge transfer and enhance the mechanical strength to accommodate volume variation. Thus, the $SnSe_2$/rGO composite delivered improved ICE of 73.7% and high capacity of 402.0 mAh g^{-1} after 150 cycles at 0.1 A g^{-1} with a retention of 86.2%.

9.3.2.2 Sb_2Se_3/Bi_2Se_3

Kim et al. have studied the hetero-interfacial interactions and SEI formation between Sb_2Se_3 and functionalized CNTs [70]. In contrast to the irregular SEI formed on the commercial Sb_2Se_3 electrode (~71.8 nm in thickness), ex situ cryo-TEM illustrated a uniform SEI with ~35.7 nm in thickness was formed on the Sb_2Se_3/CNT composite (Figure 9.3a,b). And EIS analysis verified that Sb_2Se_3/CNT composite has much reduced R_{ct}, more than four orders of magnitude lower than commercial Sb_2Se_3 (Figure 9.3c). Benefiting from the reason mentioned above, Sb_2Se_3/CNT composite electrode demonstrated a reversible capacity of 428 mAh g^{-1} at 10 000 mA g^{-1} with a capacity retention of over 62% after 200 cycles. The as-prepared full cell would also lead to a high energy density of ~175 Wh kg^{-1} at 0.5 C and capacity retention of 84.5% at 20 C after 120 cycles. Li et al. have synthesized graphene-loaded Bi_2Se_3, exhibiting superior electrochemical performance as anode for SIBs [72]. Through a combined conversion–alloying reaction, a high gravimetric specific capacity of 346 mAh g^{-1} and volumetric specific capacity of 716 mAh cm^{-3} at 0.1 A g^{-1} were displayed. Even at 10 A g^{-1}, the composites still displayed 183 mAh g^{-1} gravimetrically and 379 mAh cm^{-3} volumetrically with ultrahigh stability up to 1000 cycles.

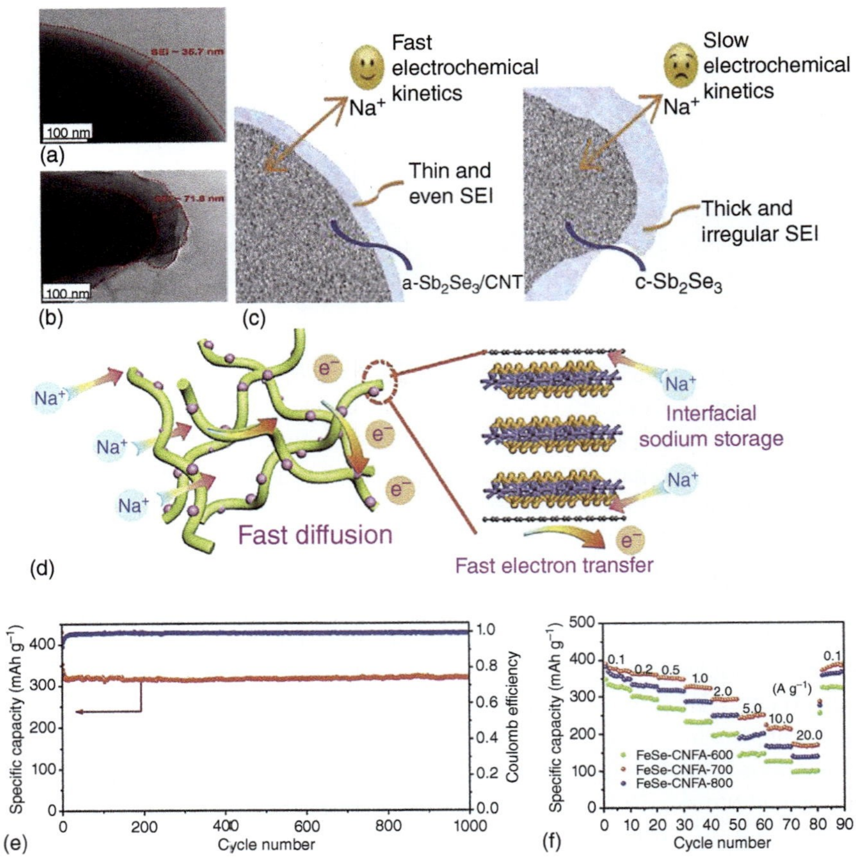

Figure 9.3 Ex situ cryo-TEM images of (a) the a-Sb_2Se_3/CNT composite and (b) c-Sb_2Se_3 electrodes after first cycle, revealing different thicknesses and morphologies of the formed SEI layers. (c) Schematic illustration of different SEI layers formed on the a-Sb_2Se_3/CNT composite and c-Sb_2Se_3 electrodes and their correlation with electrochemical kinetics of Na ion. Source: Ihsan-Ul-Haq et al. [70]. Reproduced with permission, 2020, Elsevier. (d) Schematic illustration of FeSe nanoparticles within 3D carbon nanofiber aerogels and FeSe with carbon layer. (e) Long-term cycle stability of FeSe-CNFA-700 at a 2000 mA g^{-1} over 1000 cycles. (f) Rate performances of FeSe-CNFA-T electrodes. Source: Lv et al. [71]. Reproduced with permission, 2018, Elsevier.

9.3.2.3 $MoSe_2$/WSe_2

$MoSe_2$, as typical layered transition metal selenide, exhibits a large interlayer distance of 0.646 nm for Na^+ (de)insertion. Wu et al. have prepared a composite of $MoSe_2$ with nitrogen/phosphorus (N/P) co-doped carbon (NPC) and rGO by a simple polymerization reaction followed by selenization [73]. The obtained $MoSe_2$@NPC/rGO demonstrated ~340 mAh g^{-1} after 500 cycles at 0.5 A g^{-1}, and a high reversible capacity of ~100 mAh g^{-1} at 50 A g^{-1}. The introduction of NPC/rGO sheets intercalated within the composites functioned as channels for fast electron transfer. Pseudocapacitance-dominated Na^+ storage mechanism at high rates was obtained via first-principles calculations and quantitative kinetics analysis.

WSe_2 has also been investigated as an anode candidate for SIBs. However, the difficulties in the synthesis and other related issues still hinder its exploration. Zhang et al., for the first time, adopted a simple solid-state reaction and a following high-energy ball milling method to obtain carbon-coated WSe_2 nanomaterials [74]. Crystalline WSe_2 nanoparticles with a weight ratio of 75.57% are uniformly dispersed on a carbon matrix, which was proved to be effective in alleviating the huge structural change of WSe_2. Consequently, a capacity of 270 mAh g^{-1} at 200 mA g^{-1} was obtained.

9.3.2.4 $FeSe_x/CoSe_2/NiSe_2$

Iron selenides mainly including $FeSe_2$, FeSe, and Fe_7Se_8 have been widely investigated ascribing to their advantages such as environmentally friendly, earth-abundant, and high theoretical capacity. FeSe could accommodate 2 Na^+ to form Na_2Se during the discharge process and is recovered in the following charge process. FeSe/CNF aerogel was reported by Lv et al. [71]. The FeSe exhibited ultra-small particles which uniformly embedded in interconnect 3D CNF (Figure 9.3d), and thus the obtained composite demonstrated large surface area as well as robust structural stability benefiting from a highly conductive network. The superior electrochemical performance of capacity as high as ~313 mAh g^{-1} at 2000 mA g^{-1} after 1000 cycles and ultrahigh rate capability up to 20 000 mA g^{-1} (Figure 9.3e,f) also verified that the intercalation property of Na^+ into the composite was enhanced through the strong interaction between FeSe nanocrystals and the carbon layer.

Compared with FeSe, more works are concentrating on the electrochemical performance of $FeSe_2$. A four-electron reaction occurs as for $FeSe_2$ during cycling with conversion reaction formula of $FeSe_2 + 4Na^+ + 4e^- \leftrightarrow Fe + 2Na_2Se$. Like other conversion-type anodes, $FeSe_2$ also experiences a large pulverization during repeated cycles leading to relatively poor cycling performance and rate capability. Incorporation of carbonaceous material has proved to be effective to solve issues caused by the volume changes. Taking advantage of the Kirkendall effect, Lan et al. have prepared hollow carbon-coated $FeSe_2$ nanospheres [75]. It revealed a high capacity of 474.1, 364.5, and 316.5 mAh g^{-1} at 0.05, 2.0, and 5.0 A g^{-1}, respectively, which was superior to that of bare $FeSe_2$ nanoparticles. Prussian blue microcubes were also utilized by Fan et al. to form 3D hierarchical hollow nanocubes constructing by 1D $FeSe_2$@C core–shell nanorods through a thermally induced selenization process [76]. The resulting $FeSe_2$@C composite exhibited good rate capability and stable cycling stability of 212 mAh g^{-1} after 3000 cycles at 10 A g^{-1} benefiting from the unique morphology and the coating layer.

In addition to FeSe and $FeSe_2$, a novel Fe_7Se_8 has also been firstly reported recently by Wan et al. [77]. Combined with various characterization methods including ex situ X-ray diffraction (XRD), SAED, and CV analysis, it revealed that an activation process involving structural evolution of Fe_7Se_8@NC electrode occurred during the initial 200 cycles, which would exert influence on the sodium storage mechanisms. As a result, in the following 1000 cycles, an average capacity of 339 mAh g^{-1} at 1000 mA g^{-1} was delivered with a capacity retention of 99.5%.

Reversible discharge capacities of 367 and 251 mAh g^{-1} were delivered at 500 and 2000 mA g^{-1}, respectively. Very recently, Zhang et al. have reported a hybrid material of ultra-small Fe_7Se_8 nanoparticles/N-doped CNFs (Fe_7Se_8/N-CNFs)) through a facile electrospinning method [78]. A large surface/volume ratio of electrodes and full utilization of active materials could be achieved due to the ultrasmall diameter (12 nm) of the material. And the CNFs also improved the electronic conductivity of the whole electrode, while the unique open structure of the hybrid ensured rapid electron/ion transfer to contribute to the fast kinetics. Consequently, the multiple synergistic effects led to a much better rate performance with 286.3 mAh g^{-1} at 20 A g^{-1}, superior to 101 mAh g^{-1} at 10 A g^{-1} of core–shell structural Fe_7Se_8@NC, and 150 mAh g^{-1} at 5 A g^{-1} of peapod-like Fe_7Se_8@C nanorods.

Till now, the sodium storage properties of several cobalt selenides, exampled as $CoSe_2$ and CoSe, have been studied. Zhang and coworkers have investigated the electrochemical mechanism of the $CoSe_2$ nanorods [79]. They adopted a simple solvothermal method to obtain the urchin-like $CoSe_2$ nanorod assemblings. It was found that insertion reaction firstly occurred at discharged 1.56 V, and the other two plateaus around 0.98 and 0.65 V appeared subsequently. In the following charging process, the reaction between Co and Na_2Se happened to form Na_xCoSe_2 and then to $CoSe_2$. A high discharge capacity of 410 mAh g^{-1} at 1 A g^{-1} after 1800 cycles was thus achieved, corresponding to a capacity retention of 98.6%. Full cell with the configuration of $Na_3V_2(PO_4)_3/CoSe_2$ displayed a reversible capacity of 380 mAh g^{-1}, illustrating its potential for application. Two effective strategies to improve the electrochemical performance of $CoSe_2$ have been widely recognized, i.e. the increase in the conductivity and the tuning of the nanostructure. Fang et al. have selected Co–Co Prussian blue analog microcubes as the starting material to obtain a high-performance anode for SIBs [80]. With adopting a facile two-step sequential ion exchange method, hierarchically nanosheet-assembled Cu-doped $CoSe_2$ microboxes were successfully fabricated. Benefiting from both the unique structural and compositional merits, superior sodium storage properties of the Cu-doped $CoSe_2$ microboxes were delivered, i.e. a high reversible capacity of 492 mAh g^{-1}, superior rate capability of 185 mAh g^{-1} at 3 A g^{-1}, and long cycle life of 94% capacity retention over 500 cycles.

Compared to $CoSe_2$, CoSe exhibits a relatively smaller volume change, which endows it with better stability upon repeated cycles. Zhang et al. chose ZIF-67 as sacrificial templates to fabricate nitrogen-doped, yolk–shell-structured CoSe/C mesoporous dodecahedra as anode for SIBs [81]. The cobalt species in the metal-organic framework (MOF) reacted with Se powder to achieve the in situ formation of CoSe, and the organic species were simultaneously converted into NDC. Therefore, the CoSe nanoparticles with 15 nm in size coherently confined in the mesoporous carbon framework exhibiting the specific capacities of 597.2 and 361.9 mAh g^{-1} at 0.2 and 16 A g^{-1}, respectively. A 2D composite of ultrafine nano-CoSe embedded in NDC matrix was designed to attain excellent sodium storage [82]. The composite exhibited a synergistic positive effect on the enhancement of reaction kinetics, alleviation of volume change, and the inhibition of unstable SEI. Thus, a high initial specific capacity of 530 mAh g^{-1} at 0.2 A g^{-1}

was delivered. Additionally, $Co_{0.85}Se$ embedded in carbon matrix ($Co_{0.85}Se$@CSs) has been reported [83]. The $Co_{0.85}Se$@CSs exhibited stable cycling property at $4000\,mA\,g^{-1}$ between 0.5 and 2.8 V. Co_9Se_8 has been investigated as well by Wang and coworkers [84]. During discharging process, Co_9Se_8 could convert to Co nanoparticles dispersing in Na_2Se matrix and was recovered during the following charge process. Benefiting from the surface-controlled pseudocapacitive behavior, Co_9Se_8/rGO showed $295\,mAh\,g^{-1}$ at $5000\,mA\,g^{-1}$.

Nickel selenides have been widely used in many fields including solar cells, supercapacitors, and batteries. Among them, two typical compounds including $NiSe_2$ and NiSe have been utilized as anode materials in batteries. The main strategies to obtain satisfactory performance of nickel selenides are concentrating on constructing unique nanostructure and compositing with conductive materials. In situ XRD has been applied in studying the reaction mechanism of $NiSe_2$ during sodiation–desodiation process [85]. When discharged to 1.1 V, Na^+ ion inserted into $NiSe_2$ host material and intermediate Na_xNiSe_2 was then formed. As the discharge went on, the formation of Na_2Se and Ni^0 metal could be observed. Thus, just like FeS_2 and MoS_2, the conversion reaction of Na_xNiSe_2 with Na^+ ion as well as the formation of Na_2Se and Ni metal occurred during the discharge process. During charging back to 1.9 V, Na_2Se fade away and finally disappeared and the low crystallinity phase of Na_xNiSe_2 could be observed, illustrating the reversible conversion reaction between Na_2Se and Ni^0 metal. After fully charged to 3.0 V, the main $NiSe_2$ peaks regained, suggesting a completely reversible structure evolution. Zhu et al. utilized a hydrothermal method to prepare $NiSe_2$ nanooctahedra [86]. The ICE of $NiSe_2$ was very impressive (over 90%), and outstanding long-term cyclic stability of $313\,mAh\,g^{-1}$ after 4000 cycles at $5\,A\,g^{-1}$ together with excellent high-rate capability of $175\,mAh\,g^{-1}$ at $20\,A\,g^{-1}$ were displayed. As for NiSe anodes, a few literatures have been reported. Zhang et al. designed core–shell NiSe/C composites via a facile two-step method [87]. By adopting an in situ polymerization process to form a thin polydopamine layer on NiSe, the core–shell NiSe/C composites were fabricated in the following calcination process. This core–shell NiSe/C composite exhibited a capacity of $339\,mAh\,g^{-1}$ after five cycles and maintained $280\,mAh\,g^{-1}$ even after 50 cycles, but quickly faded to $62\,mAh\,g^{-1}$ after 50 cycles. Thus, the electrochemical performance needs to be further improved.

9.3.2.5 Other Monometal Selenides

Cu_2Se with a superionic feature known for thermoelectrics and optoelectronics is also regarded as a promising candidate for SIBs. Cu_2Se intercalation compound fabricated using a green growth approach was reported by Xiao et al. [88]. By the means of inserting quaternary ammonium salt CTAB on Cu foil at room temperature, the interlayer space of Cu_2Se can increase the buffer space, stabilize the polyselenide shuttle, and prevent the fast growth of Cu nanoparticles. When served as anode for SIBs, the Cu_2Se–CTAB nanosheet arrays with expanded interlayer space delivered capability of $426.0\,mAh\,g^{-1}$ at $0.1\,A\,g^{-1}$ and $238.1\,mAh\,g^{-1}$ at $30\,A\,g^{-1}$ and a high capacity retention of $\approx 90\%$ at $20\,A\,g^{-1}$ after 6500 cycles. Moreover, compositing Cu_2Se with ZnSe to form heterojunction structure could also improve the electrochemical performance [89]. The type-I heterojunction between Cu_2Se and ZnSe was conducive to the acceleration of electron transport, and additional CNFs were able to maintain the long cycling life. Consequently, the Cu_2Se-ZnSe-CNFs showed a 310 mAh g^{-1} after 100 cycles at 0.1 A g^{-1} with a rate capacity of 237 mAh g^{-1} at 2 A g^{-1}.

Moreover, ZnSe with a theoretical capacity of 557 mAh g^{-1} has the merits of environment-friendly, less toxic, and low cost. Zhou et al. employed a one-step hydrothermal strategy to form ZnSe nanoparticles on a uniform 3D interconnected multi-walled carbon nanotubes (MWCNTs) conductive network [90]. The introduction of MWCNTs was favorable to the enhancement of electrical conductivity, ionic diffusivity, and structural stability. The optimal ZnSe/MWCNTs delivered 398.8 mAh g^{-1} at 0.1 A g^{-1} after 50 cycles and 279.4 mAh g^{-1} at 10 A g^{-1}. After 300 cycles at 4 A g^{-1}, the capacity retention can still be 93%. This superior rate performance was achieved by the pseudocapacitance in redox processes. Moreover, it should be noted that at low temperature of −10 °C, it also showed good long-term cycling stability of 246.7 mAh g^{-1} at 1 A g^{-1} after 600 cycles and 180.5 mAh g^{-1} at 5 A g^{-1}.

9.3.2.6 Bimetallic Selenides

It has been widely known that the intrinsic electronic conductivity could be improved via the introduction of binary metals or the heterogeneous species. And the rich redox sites provided by multi-metal centers are favorable to increase adsorption capabilities. Therefore, introducing multiple metals in monometallic selenides has been regarded as an effective method to boost the performance.

M–Fe–Se (M = Co and Ni). Single-phase Fe_2CoSe_4 nanoparticles were encapsulated in carbon matrix by Ali et al. (Figure 9.4a) [91] with a simple hydrothermal reaction followed by gas selenization. Compared with the Fe-Co sulfide counterpart, Fe_2CoSe_4 nanospheres possess superior electrical conductivity. And due to the synergistic effect by combining higher conductivity, intimate wrapping of carbon matrix, and unique porous structure, the Fe_2CoSe_4 anode demonstrated excellent rate capability, which delivered 816.3 mAh g^{-1} at 0.5 A g^{-1} and 400.2 mAh g^{-1} at 32 A g^{-1}, and very long cycle life, i.e. 350 mAh g^{-1} even after 5000 cycles at 4 A g^{-1} (Figure 9.4b,c). An adequately high energy capacity of 615 mAh g^{-1} after 100 cycles at 1 A g^{-1} could also be achieved. They also reported Fe_2NiSe_4 nanostructures as an anode for SIBs and concluded that the iron incorporation improved the conductivity and the performance of the sample [93].

M–Co–Se (M = Ni). Designing novel morphology and introducing conductive materials could be employed to improve the poor cycle stability and large volumetric expansion during the cyclic process of layered Ni–Co diselenides. Hou et al. have encapsulated $Ni_{1.8}Co_{1.2}Se_4$ nanoparticles inside NDC nanoboxes using the bimetallic MOF as raw materials [94]. The 3D dual continuous NDC conductive networks facilitated the promotion of highly efficient, and ultrafast transport of electrons and sodium ions, and volume changes could be buffered during the redox processes. Moreover, the inner voids could also alleviate the

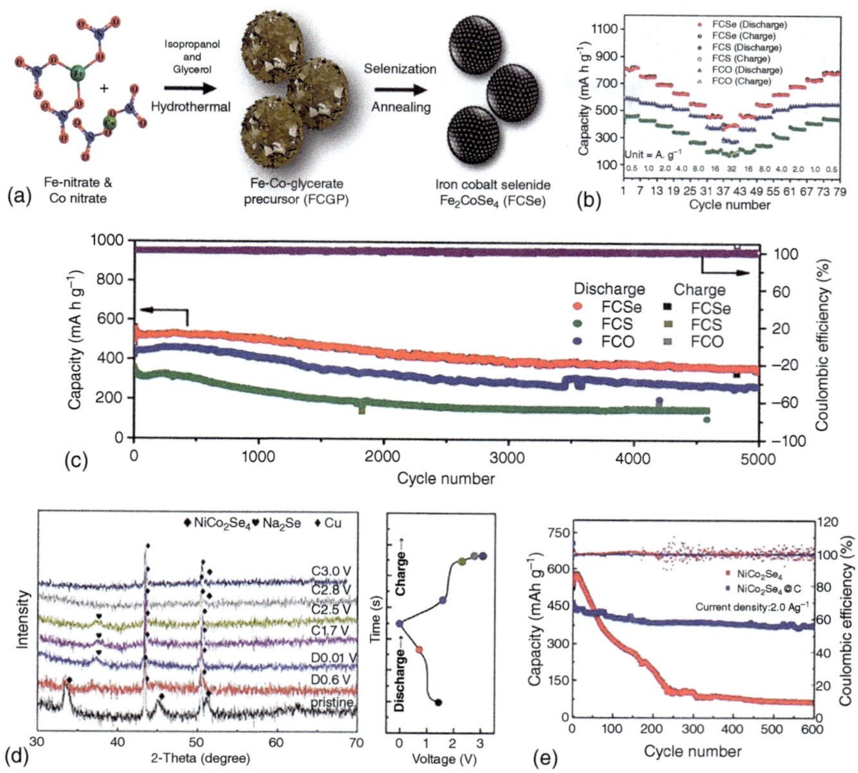

Figure 9.4 (a) Schematic illustration of the formation process of Fe_2CoSe_4. (b) Charge–discharge profiles at various current densities and (c) comparison of cycling performances of FCSe, FCS, and FCO at $4\,A\,g^{-1}$. Source: Ali et al. [91]. Reproduced with permission, 2018, Wiley-VCH. (d) Ex situ XRD of $NiCo_2Se_4$@C at various voltages and corresponding voltage profile during the first cycle. (e) Cycle performance of $NiCo_2Se_4$@C at $2.0\,A\,g^{-1}$. Source: Qui et al. [92]. Reproduced with permission, 2020, Elsevier.

expansion of $Ni_{1.8}Co_{1.2}Se_4$ nanoparticles. As a result, the $Ni_{1.8}Co_{1.2}Se_4$@C anode delivered excellent rate performance of 211 and $153\,mAh\,g^{-1}$ at 30 and $50\,A\,g^{-1}$, respectively. Moreover, a full cell with $Na_3V_2(PO_4)_2O_2F$ cathode exhibited an energy density of $227\,Wh\,kg^{-1}$ and capacity retention of above 97.6% even after 60 cycles at $0.4\,A\,g^{-1}$ between 1.2 and 4.3 V at room temperature. By an easy hydrothermal method, hollow $NiCoSe_2$ microspheres@NDC was achieved [95]. After cycling at $100\,mA\,g^{-1}$, it exhibited a reversible capacity of $464.7\,mAh\,g^{-1}$ over 200 times. Moreover, reversible capacities of 425.3, 420.8, 403.2, 394.7, 378.7, 367.8, and $337.5\,mAh\,g^{-1}$ were achieved at 100, 200, 500, 1000, 2000, 3000, and $5000\,mA\,g^{-1}$, respectively, indicating its outstanding rate performance. $NiCo_2Se_4$ has been also reported as a new anode for SIBs [92]. Investigations showed that the formed Na_2Se, Ni, and Co nanoparticles dispersed in Na_2Se matrix after fully discharge. And during the recharge process, $NiCo_2Se_4$ phase can be regenerated with a small amount of $CoSe_2$ and NiSe phases left, which would lead to the irreversible capacity loss during the initial cycle (Figure 9.4d). When hybrid with conductive carbon, the $NiCo_2Se_4$@C composite delivered a reversible capacity of 603.2 mAh g^{-1} corresponding to initial CE of 85.79% at 0.5 A g^{-1}. Reversible capacity of 377.5 mAh g^{-1} can be obtained after 600 cycles at 2 A g^{-1} (Figure 9.4e).

9.3.3 Metal Tellurides

Although metal tellurides have relatively lower theoretical gravimetric capacities compared to the metal sulfides and selenides, they exhibit higher theoretical volumetric capacity and electrical conductivity. However, the study on metal telluride anodes for SIBs is still scarcely explored. Recently, a few literatures are focusing on the improvement of their electrochemical performance including morphology design and hybrid electrode formation, and so on.

Recently, with the help of advanced characterization methods, some in-depth explorations on reaction mechanisms were performed. In situ TEM was carried out to investigate the rhombohedral Sb_2Te_3/C anode, which was prepared via a high-energy ball milling method [96]. It revealed that Na_3Sb and Na_2Te were formed at the end of discharge, and carbon served as a glue to bond the Sb_2Te_3 nanoparticles together suppressing the pulverization during cycling. However, cracks grew at the boundary between the main and attached particle for pure Sb_2Te_3 particle, leading to a severe capacity fade. On the contrary, the Sb_2Te_3/C showed a high capacity of 360 mAh g^{-1} and 93% capacity retention after 400 cycles at 1000 mA g^{-1}. Moreover, it delivered 437 mAh g^{-1} at 20 mA g^{-1} and retained 314 mAh g^{-1} at 2000 mA g^{-1}. In addition, cryogenic TEM was adopted to capture the structural change of Sb_2Te_3/CNT [97]. A uniform and thin SEI layer of ~19.1 nm was observed on the Sb_2Te_3/CNT composite while irregular and thicker SEI of ~67.3 nm for the pure Sb_2Te_3. Thus, it illustrated that uniform and thin SEI layer induced by the functionalized CNTs facilitated the amelioration of high-rate capability and cyclic performance. Therefore, the Sb_2Te_3/CNT composite anode delivered excellent reversible gravimetric and volumetric capacities of 422 mAh g^{-1} and 1232 mAh cm^{-3}, respectively, at 100 mA g^{-1} with ~97.5% capacity retention after 300 cycles. Paired with $Na_3V_2(PO_4)_2F_3$ cathode, full cell demonstrated remarkable energy density of ~229 Wh kg^{-1} at 0.5 C and excellent cyclic stability of over 71% and 66% capacity retention after 200 cycles at 5 and 10 C, respectively. Moreover, Sun et al. synthesized layered Bi_2Te_3 nanoplates/graphene (Bi_2Te_3/G) via one-pot solvothermal method [98]. Bi_2Te_3 with quintuple layer sandwich structure of a large interlayer spacing of 1.02 nm was anchored on graphene matrix. And it exhibited high reversible gravimetric and volumetric capacity as high as 416 mAh g^{-1} and 648.9 mAh cm^{-3} at 0.1 A g^{-1}. A high gravimetric capacity of 203 mAh g^{-1} was still achieved even at 5.0 A g^{-1} by Bi_2Te_3/G.

Other tellurides such as GeTe and CoTe have also been reported. Sung et al. prepared a GeTe compound with a layered structure by a simple solid-state method [99]. Employing amorphous carbon as a modifier, the GeTe/C anode demonstrated a high reversible first volumetric capacity of 662 mAh cm^{-3} with a stable capacity retention of 98.5% after 100 cycles, and capacity of 704 and 630 mAh cm^{-3} at 1 and 3 C, respectively. By examining via various ex situ analytical methods, the reaction mechanisms of GeTe/C and GeTe control sample were investigated as follows. During sodiation, both anodes could convert into NaGe and Na_2Te phases. However, in the desodiation process, only the converted NaGe and Na_2Te phases in GeTe/C were able to completely recover to their original state, which was attributed to their higher reversible capacity. Semi-conductive CoTe has been used in a variety of electronic devices and photovoltaic cells. CoTe with high electrical conductivity is expected to be a potential anode of SIBs. Ding et al. have studied the electrochemical properties of CoTe anode by a simple solvothermal route [100]. After hybridizing with rGO to further improve electrical conductivity and mechanical strength, CoTe/rGO exhibited a specific capacity of 306 mAh g^{-1} at 50 mA g^{-1} after 100 cycles and a capacity of 200 mAh g^{-1} even after 200 cycles at 0.1 A g^{-1}.

9.4 Metal Alloys

In the past decade, Li alloys have been well researched for LIBs due to their high capacity. In contrast, much less research has been concerning with the application in SIBs. After the exploration on sodium alloys carried out by Chevrier and Ceder in 2011 [101], attention has been paid on the sodium storage of diverse alloys. In this section, alloy-type electrodes (for example Sn, Sb, Bi, and their intermetallic compounds) are discussed. Different from the insertion-type anodes of carbon-based and Ti-based anodes which show relatively small volume change upon cycling, alloying reaction with sodium occurs in terms of alloy-type materials, delivering much higher capacities in contrast to the above-discussed insertion-type materials, the severe volumetric fluctuation upon the repetitive alloying–dealloying reaction caused by large sodium ions restricts the practical application of these alloy-type anodes. The reason results from the peeling off of electrode from the current collector due to loss of electrical contact induced by electrode fracture or pulverization. To date, strategies such as constructing novel morphologies and compositing conductive carbon matrix have been developed to tackle this problem. Additionally, engineering intermetallic compounds and selecting optimized binders or electrolytes (additives) are verified to enhance the electrochemical performance as well.

9.4.1 Tin (Sn)

Sn shows its potential as anode for SIBs since it has merits of low reaction potential (<0.6 V), high theoretical specific capacity of 847 mAh g^{-1} [101], inexpensive as well as environmentally benignity. Huang and coworkers studied the electrochemical evolution of Sn accompanied with volumetric variation during sodiation using in situ TEM [102]. They found that two-step sodiation process occurs initially with the formation of Na-poor, amorphous Na_xSn (a-Na_xSn, $x \sim 0.5$) alloy phase (56% expansion) via a two-phase reaction, and followed by the generation of several Na-rich amorphous phases, and ultimately the crystalline c-$Na_{15}Sn_4$ phase is formed (420% expansion) via a single-phase reaction (Figure 9.5a). Actually, the electrochemical insertion process of sodium into Sn has not been fully elucidated and it has been suggested to be more complicated than that for LIBs. The structure and composition of the intermediate Na_xSn phases have been further investigated by using various techniques and density functional theory (DFT) calculations to better understand the sodium storage mechanism of Sn [8, 110]. Recently, Stratford and coworkers adopted both in and ex situ characterizations and the theoretical calculation to elucidate some intermediates during reaction which has not been identified [8]. Based on their experimental results, a successive phase transition (β-Sn $\rightarrow NaSn_2 \rightarrow$ amorphous $Na_{1.2}Sn \rightarrow Na_{5-x}Sn_2 \rightarrow Na_{15+x}Sn_4$) has been proposed. It should be noted that the insertion/exaction of 3.75 Na atoms accompanied with volume variation plays leading role on the limitation of Sn application as anodes. The severe structural change leads to poor cycling performance due to both pulverization and loss of electrical contact.

Wang et al. [103] has carried out in situ synchrotron hard X-ray nanotomography to illustrate the structural/chemical evolution of Sn anode during cycling (Figure 9.5b, left), which shows that small particle exhibits robustness without cracks, fracture, or pulverization after insertion/extraction of Na ions. And two critical sizes including 0.5 μm for low complexity (negligible structural degradation) and 1.6 μm for high complexity (Figure 9.5b, right) are able to offer clear information in order to clarify the failure mechanisms. Obviously, designing materials to nanosize has been regarded as one of the most effective methods to accommodate volume variation upon cycling. In this approach, electrodeposited Sn thin film anodes consisted of fine grains demonstrated outstanding cycle stability with delivering 607.51 mAh g^{-1} after cycling 40 times, much superior to that composed of coarse and isolated particles showing obvious capacity fading [110c]. Furthermore, Sn nanofiber electrode synthesized by Nam et al. [110e] displayed reversible capacity of 776.3 mAh g^{-1} over cycling 100 times corresponding to capacity retention of 95.1%, ascribing to the fact that the synergistic effect of porous properties as well as anisotropic expansion would effectively alleviate the volume variation.

Constructing Sn/C nanocomposites by introducing conductive carbon host/substrate is shown to be another effective approach which can physically buffer the volume strain, prevent the nanoparticles from migrating and agglomerating, and create the electric conducting pathways. To this end, various nanostructured Sn/C nanocomposites with different morphologies like nanospheres, nanofibers, nanosheets, and nanopillars (Figure 9.5c–i) have been fabricated and displayed outstanding cycling stability with satisfactory capacity and superior rate capabilities [104–109]. For example, Sn nanodots encapsulated in porous nitrogen-doped CNFs (Figure 9.5d) could release a specific capacity of 483 mAh g^{-1} over 1300 cycles at 2000 mA g^{-1} [105]. Another core–sheath-structured Sn@CNT nanopillars aligned on carbon paper (Figure 9.5h,i) are able to achieve specific capacities of 445 and 377 mAh cm^{-2} after 100 cycles at 250 and 500 mA cm^{-2}, respectively. And good rate performance with capacities of 341 and 299 mAh cm^{-2} at 750 and 1000 mA cm^{-2} is also delivered. Such superior electrochemical performance results from the dispersing effect of carbon paper and the buffering effect provided by the carbon sheath [109].

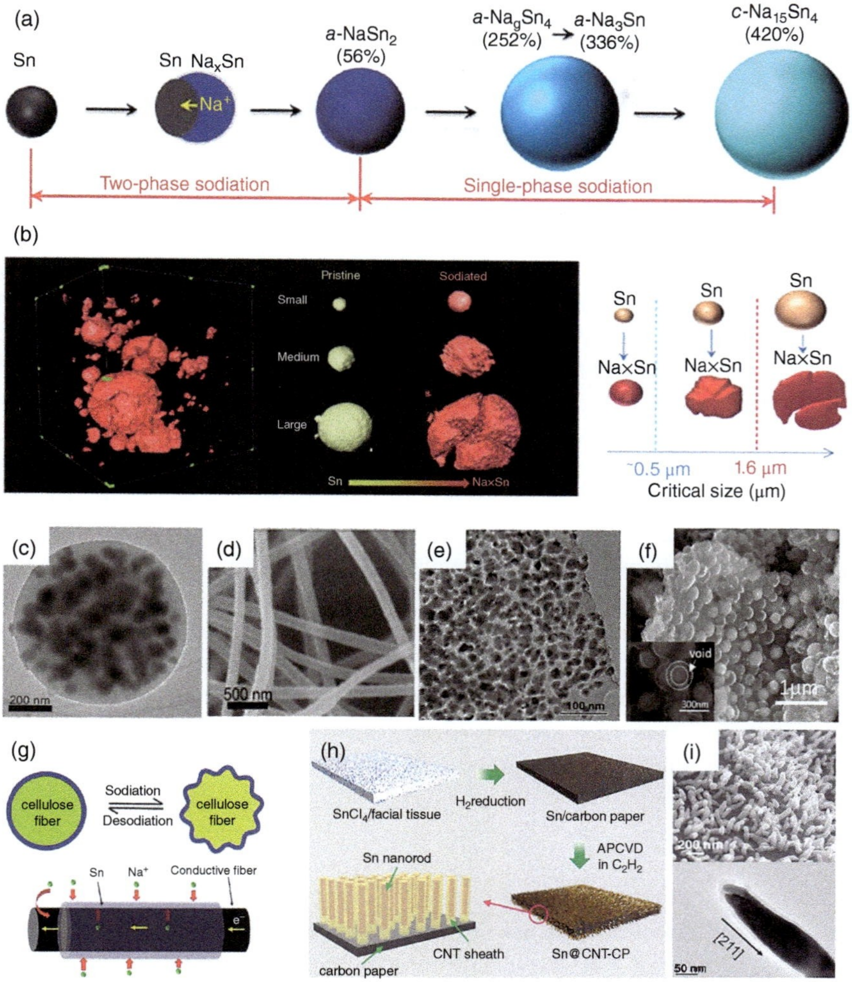

Figure 9.5 (a) Structural evolution of Sn during sodiation. Source: Wang et al. [102]. Reproduced with permission, 2012, American Chemical Society. (b) 3D images of the sodiated anode and three particles with different sizes and fracture, as well as the schematic illustration of two critical sizes for Sn fracture in SIB. Source: Wang et al. [103]. Reproduced with permission, 2015, Springer Nature. (c) TEM image of Sn@C sphere. Source: Liu et al. [104]. Reproduced with permission, 2014, Wiley-VCH. (d) SEM image of Sn NDs@PNC nanofibers. Source: Liu et al. [105]. Reproduced with permission, 2015, Wiley-VCH. (e) TEM image of F-G/Sn@C composites. Source: Luo et al. [106]. Reproduced with permission, 2016, Elsevier. (f) TEM image of yolk–shell Sn@C eggette-like nanostructure. Source: Li et al. [107]. Reproduced with permission, 2016, American Chemical Society. (g) Unique buffering effect for releasing volume stress of soft wood fiber substrates. Source: Zhu et al. [108]. Reproduced with permission, 2013, American Chemical Society. (h) Illustration to show the preparation of Sn@CNT-CP. (i) SEM (up) and TEM (down) images of the Sn@CNT-CP. Source: Xie et al. [109]. Reproduced with permission, 2015, Elsevier.

Furthermore, it has been demonstrated that binders in the electrode and electrolyte (additives) also play critical roles in battery performance [111]. For instance, Komaba et al. [111a] found that an improvement on reversibility of Sn anodes was achieved by adopting polyacrylate (PAA)) as a binder instead of the mainstream binder of polyvinylidene fluoride (PVDF), although it still suffers from capacity decay which results from the volume change and electrolyte decomposition. They also found that the introduction of small amount of fluoroethylene carbonate (FEC) into the propylene carbonate-based electrolyte could further improve the performance of the Sn electrode with PAA binder. Zhang et al. attempted to investigate the performances of microsized Sn particles directly as anode for SIBs by applying different electrolytes [111c]. In their research, glyme-based electrolyte (1 M $NaPF_6$ in diethylene glycol dimethyl ether [diglyme, DGME]) was selected to compare with carbonate-based electrolyte system (1 M $NaPF_6$ in PC electrolytes with 5% FEC). In sharp contrast to that cycled in PC with 5% FEC, the Sn anode in DGME exhibited much smoother surface, indicating the structural integrity. Furthermore, thin SEI film formed on surface of particles could work as protective layer to prevent the generation of insulating domains, giving rise to an improvement in stability. Once coupled with $Na_3V_2(PO_4)_3$/C cathode, the corresponding microsized Sn//1 M $NaPF_6$ in DGME// $Na_3V_2(PO_4)_3$/C full battery delivered initial charge–discharge capacities of 102/83 mA h g^{-1} at an average voltage of 3 V. After 50 cycles, capacity retention of 90% was maintained. Taking both cathode and anode into account, the gravimetric energy density and volumetric energy density are 200 Wh kg^{-1} and 703 Wh L^{-1}, respectively, 19% and 61% higher than the hard carbon//1 M $NaPF_6$ in EC/DMC// $Na_3V_2(PO_4)_3$/C configuration.

9.4.2 Antimony (Sb)

Similar to Sn, Sb electrode has been explored as an alloying-type anode for SIBs as well. It has theoretical specific capacity of 660 mAh g^{-1} with forming full sodiation product of Na_3Sb at low potential of about 0.5 V. As for the reaction mechanism, Sb would alloy with Na to form amorphous intermediates (Na_xSb with $x \sim 1.5$) initially in the discharge process. After fully transformation of Sb into amorphous phase, cubic-hexagonal Na_3Sb is formed and later stabilized in the existence of hexagonal Na_3Sb [112]. In the most recent study by Allan et al., alloying mechanism of Sb was further studied by utilizing operando pair distribution function analysis and ex situ Na magic-angle spinning solid-state nuclear magnetic resonance (NMR) spectroscopy. Two previously unknown amorphous of a-$Na_{3-x}Sb$ ($x \approx 0.4$–0.5) with similar structure to crystalline Na_3Sb and highly amorphous a-$Na_{1.7}Sb$ intermediate phases were identified [113].

Sb also meets the challenge of huge volume change (~390%) during sodiation–desodiation process, which could cause capacity fading. Constructing anodes with nanometer size has been recognized as common and facile method to solve the issue. Recently, various nanostructures such as porous hollow microspheres (PHMSs) [114], nanorod arrays [115], cypress leaf-like structures [116], porous structures [24, 117], and nanotubes [118] were reported to optimize the performance of Sb anode while inhibiting volume change. For instance, Hou et al. reported Sb PHMSs as advanced anode for SIBs via employing Zn microspheres as templates [114]. The Sb PHMSs delivered 617 mAh g^{-1} at 100 mA g^{-1} after 100 cycles corresponding to capacity retention of 97.2%, and 312.9 mAh g^{-1} at 3200 mA g^{-1}. The authors ascribed this excellent performance to the novel structural design including hollow structure with high porosity, thus leading to alleviation of volume change and rapid Na^+ transportation upon cycling. Later, Liu et al. proposed a top-down process to obtain nanoporous Sb [117a]. By the means of adopting chemical dealloying of Al–Sb alloy ribbon precursors to fabricate Sb particles with different size, the coral-like nanoporous Sb electrode exhibited 573.8 mAh g^{-1} after 200 cycles at 100 mA g^{-1} as well as excellent rate property up to 3300 mA g^{-1}. The innovative electrode of 3D porous structure assures high sodium-ion accessibility, strong structural integrity, and fast electrode transport, thus resulting in outstanding cycling stability and rate performance.

It is noted that innovative exploration of Sb-based anodes also concentrates on designing various Sb–carbon composites. As far, various Sb–carbon composites with novel structures, for example Sb@C yolk–shell spheres [119], peapod-like Sb@N–C hybrid [120], freestanding porous Sb–C framework films [121], Sb/C spheres [122], Sb nanoparticles@porous carbon [123], Sb@C coaxial nanotubes [124], Sb@(N, S–C) hybrid [125], Sb/C fibers [126], and Sb/graphene nanosheets [127], have been reported. These Sb–carbon composites exhibited expressive performance. Some typical examples are shown in Figure 9.6. Sb@C nanosphere anode with biomimetic yolk–shell structure has been prepared [119]. As shown in Figure 9.6a, a layer of conductive carbon with protection effect covered around the Sb hollow yolk to form the yolk–shell microstructure. The yolk–shell Sb@C exhibited large amount of void space which could accommodate the volume change; moreover, the carbon shell could also induce the formation of stable SEI film. As expected, as anode for SIBs, these Sb@C yolk–shell nanospheres exhibited good rate capability, i.e. 600, 548, 463, 408, 329, and 279 mAh g^{-1} at 50, 100, 200, 500, 1000, and 2000 mA g^{-1}, respectively (Figure 9.6b). In addition, long cycle performance with about 280 mAh g^{-1} at 1000 mA g^{-1} over 200 cycles was also obtained, indicating the favorable effect of yolk–shell structure on stabilizing active Sb. Doping heteroatoms such as N, S, F, and P into carbon frameworks have been intensively explored to enhance sodium storage capacity of the carbonaceous materials. The heteroatoms-doped strategy of carbon was introduced into Sb to obtain peapod-like Sb@N–C hybrid [120] (Figure 9.6c,d) and Sb@(N, S–C) hybrid [125] (Figure 9.6e,f). Benefiting from the synergistic merits of sufficient N-doping, 1D conductive carbon coating as well as extensive inner void space, the peapod-like Sb@N–C hybrid displayed excellent long-term cycle performance (345.6 mAh g^{-1} after 3000 cycles at 2 A g^{-1}) (Figure 9.6d). Novel Sb/C hybrid encapsulating the Sb nanorods into highly conductive N and S co-doped carbon (Sb@(N, S–C)) frameworks has performed capacity of 374.3 mAh g^{-1} at 10 A g^{-1} (Figure 9.6f).

The effect of binders and electrolyte additive on performance of Sb was studied as well. Lu and coworkers examined and compared sodium storage process of

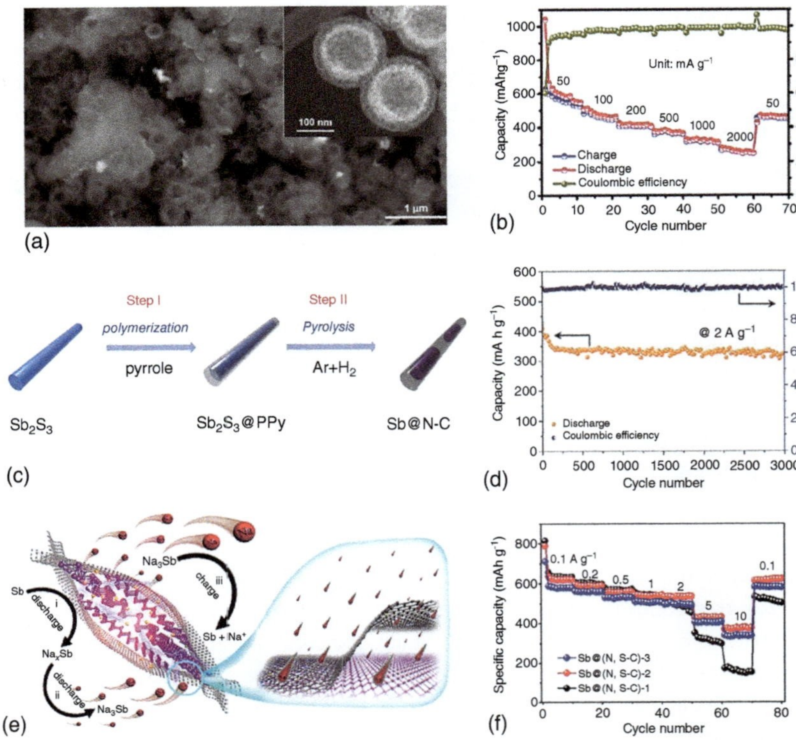

Figure 9.6 (a) Low-magnification SEM images of hollow Sb@C yolk–shell nanospheres with dark-field TEM image in inset. (b) Rate capability at various rates. Source: Liu et al. [119]. Reproduced with permission, 2017, American Chemical Society. (c) Schematic illustration of preparing peapod-like Sb@N–C hybrid. (d) Prolonged cycling stability of Sb@N–C at 2 A g^{-1}. Source: Luo et al. [120]. Reproduced with permission, 2018, Wiley-VCH. (e) The reaction mechanism during the charge–discharge process of Sb@(N, S–C). (f) Rate capabilities at various current densities. Source: Cui et al. [125]. Reproduced with permission, 2018, American Chemical Society.

SiC–Sb–C electrode in different electrolytes [128]. They found that the presence of FEC could lead to the generation of SEI film with the thin, chemically/mechanically stable, and structurally compact properties, leading to the greatly enhanced cycling performance. Another polymer with crosslinked chitosan network has been reported by Gao et al. exhibited its advantages as binder to effectively alleviate volume variation in contrast to PVDF or carboxymethyl cellulose (CMC) [129].

9.4.3 Bismuth (Bi)

Bi has attracted considerable attention as anode for SIBs because of its unique layered structure with large crystal lattice along *c*-axis ($d(003) = 0.395$ nm) for ion insertion, low melting point, low thermal conductivity, non-toxicity, and moderate theoretical gravimetric capacity of 385 mAh g^{-1} by forming Na_3Bi. However, works reporting its sodiation mechanism are somehow ambiguous. Ellis et al.

proposed that the sodiation process of Bi undergoes the formation of NaBi and Na_3Bi [130]. Similar mechanism was also reported by Sottmann et al. based on the quasi-simultaneous in operando synchrotron XRD/XAS and DFT calculations [131]. It was found that NaBi was formed initially, and subsequently the mixture of cubic or hexagonal Na_3Bi were generated. The authors proposed that sodiation would go on with following two obvious structural mechanisms according to crystallite size of Bi. While Su et al. adopted DFT calculations to verify the rapid Na^+ transportation and volume accommodation arising from substantial sites of Bi, demonstrating its intercalation mechanism other than alloy reaction [132]. Similar result was also reported by Liu et al. with employing ex situ XRD measurements on arrayed bismuth nanorod bundle [133]. However, in a more recent study by Gao et al. [134] the operando XRD method verified that Bi anodes with different morphologies, size, and sources exhibited the same sodium storage mechanism based on alloying/dealloying process ($Bi \rightarrow NaBi \rightarrow Na_3Bi$), corresponding to two-step process with delivering two voltage plateaus. Moreover, both Na_3Bi and NaBi phases exhibited metastable property, which tends to decompose by ex situ conditions with the equation of $Na_3Bi \rightarrow NaBi \rightarrow Bi$. Therefore, the phenomenon that only Bi phase could be detected under ex situ XRD conditions was able to be made clear, which also explains the inaccurate mechanism that reported previously.

Based on previous reports on Bi anode, it has been widely known that the sodiation of Bi proceeds with sequential formation of NaBi and Na_3Bi. Huge volume change of 250% is still the main obstacle to hinder its application due to the slow kinetics and capacity fading resulted from the lost contact with current collector. Targeted at this, Bi with various nanostructures is reported to promote the electrochemical performance. Liu et al. [133] initially put forward the preparation of arrayed Bi nanorod bundles via chemical dealloying method. Finally, the materials used as anode could demonstrate 301.9 mAh g^{-1} after cycling 150 times at 50 mA g^{-1} with clear potential profiles. The outstanding electrochemical properties could be attributable to the following reasons: on the one hand, severe volume variation is able to be alleviated due to the design of arrayed Bi nanorod bundles with large interval spacing; on the other hand, the novel structure also facilitates the diffusion of sodium ions. Bi nanoflakes on Ni foam (Bi/Ni) with 3D interconnected structure have been reported by Wang et al. via a facile replacement reaction [135]. The obtained material used as anode is conductive to bear the large volume expansion and the diffusion of Na^+, which exhibited 302.4 mAh g^{-1} after 100 cycles at 200 mA g^{-1}.

In addition, Bi nanoparticles composited with conductive carbon matrices have also been prepared, like Bi@graphene nanocomposites [132], Bi@C microspheres [136], Bi–NS@C composite [137], Bi@graphite [138], Bi@N-C [52], yolk–shell Bi@Void@C nanospheres [139], Bi nanosheets/carbon fiber cloth [140], and Bi@C composite nanosheets [141]. These rational designed Bi/carbon composites have exhibited good performance because of the improved conductivity and structural stability by introducing carbon matrices. Particularly, nanostructure equipped with substantial void spaces for buffering the volume variation seems to be effective in order to obtain ultrafast cycling life. For example, Xue et al. [52] have fabricated Bi nanorods encapsulated in NDC nanotubes (Bi@N–C) (Figure 9.7a). Assembled

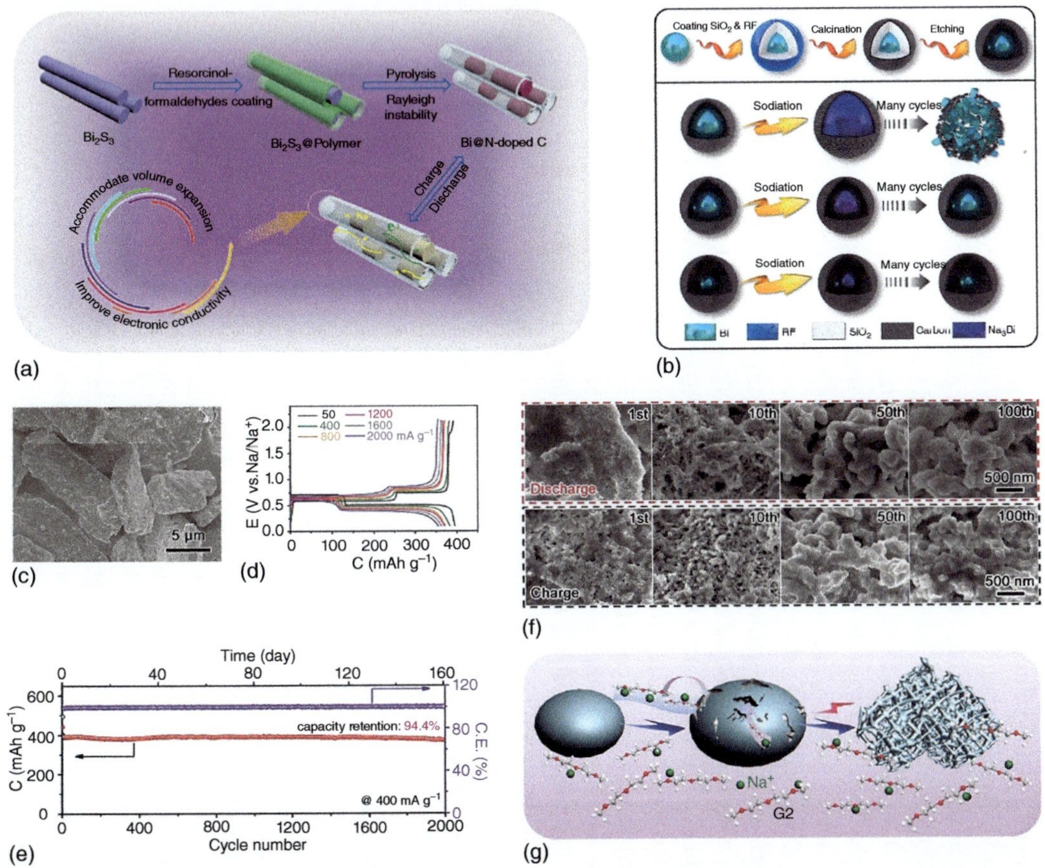

Figure 9.7 (a) Fabrication illustration of Bi@N−C and the corresponding operation mechanism. Source: Xue et al. [52]. Reproduced with permission, 2019, American Chemical Society. (b) Schematic illustration of Bi@Void@C and the Bi@Void@C with various void sizes during cycling. Source: Yang et al. [139]. Reproduced with permission, 2019, American Chemical Society. (c) SEM image of the bulk Bi; (d) the discharge/charge curves at various current densities and (e) cycling performance at 400 mA g^{-1} in the $NaPF_6$-G2 electrolyte; (f) SEM images of the discharged and charged bulk Bi anode free of carbon additives for selected cycles; (g) schematic illustration of morphological evolution of the Bi anode during cycles. Source: Wang et al. [142]. Reproduced with permission, 2017, Wiley-VCH.

in a half cell, the as-prepared Bi@N–C anode exhibited capacity of 410 mAh g^{-1} at 50 mA g^{-1} and 368 mAh g^{-1} at 2 A g^{-1}. Moreover, it could go through 1000 repeated sodiation–desodiation at 1 A g^{-1}, verifying its long life span. While tested in full cell with $Na_3V_2(PO_4)_3$/C cathode, a satisfactory power density of 1190 W kg^{-1} together with energy density of 119 Wh kg^{-1} based on total weight was released. Finally, the authors summarized the reason leading to excellent sodium storage performance including unique novel structure and stable SEI layer, which facilitates both accommodation of volume variation and improvement of electronic transportation. Another yolk–shell Bi@C nanospheres has been reported by Yang et al. [139] to exhibit large amounts of void volume (Figure 9.7b), which could be regulated to adapt the volume variation. In contrast to structural crack induced by limited void space and lower volumetric capacity resulted from excess void space, the optimized Bi@Void@C-2 sample with appropriate void demonstrated superior electrochemical properties, i.e. 198 mAh g^{-1} at 20 A g^{-1} even after 10 000 cycles and 170 mAh g^{-1} at ultrahigh current density of 100 A g^{-1}. This result indicates that appropriate void space plays important role in accommodating the volume variation and maintain high volumetric energy density.

Interestingly, commercial Bi anode for SIBs delivering comparable electrochemical performance once glyme-based electrolytes was used has been reported by Wang et al. [142]. Adopting glyme-based electrolytes ($NaPF_6$ in diglyme), the commercial bulk Bi of less than 20 μm in size (Figure 9.7c) demonstrated stable cycling performance up to 2 A g^{-1} and ultra-long cycle span of 2000 cycles with 400 mAh g^{-1} at 400 mA g^{-1} (Figure 9.7d,e). In addition, Bi anode equipped with a variety of glyme-based electrolytes and binders has been investigated as well. Regardless of the involved Na salts or binders, performance adopting glyme-based electrolytes exhibited little difference, verifying the synergistic effect of Bi anode and glyme-based electrolytes on enhancing the electrochemical performance. Of note, two obvious plateaus located at 0.67/0.77 V and 0.46/0.64 V could also confirm the highly reversible alloying/dealloying mechanism, corresponding to the reactions of Bi $\leftrightarrow$ NaBi and NaBi $\leftrightarrow$ Na_3Bi. In addition, in contrast to carbonate-based electrolytes, the bulk Bi anode was found to gradually develop into a porous integrity upon cycling, which results from the facile Na^+ transport and structural stability influenced by glyme-based electrolytes (Figure 9.7f,g). Based on their comprehensive investigations, the excellent performance of bulk Bi arises from the unique porous integrity evolved in the glyme-based electrolytes, stable SEI, and good electrode wettability of glymes. Again, this study further confirms the importance of electrolytes options in order to obtain anodes with high performance.

9.4.4 Intermetallic Compounds

Except for single-element alloy-based materials, engineering binary intermetallic compounds has proved to be resultful in improving electrochemical performance of alloy-type metal anodes. Once compositing with appropriate additives, the intermetallic compounds would display additional favorable property in contrast to individual component. As for M–Sn/Sb/Bi, M could be either an electrochemically active component (Sn, Sb, Bi, Zn, etc.) or inactive component (Ni, Cu, Fe, Co, Mo, Mn, etc.). Compared with electrochemically inert metals, electrochemically active metals can deliver additional capacity to the whole electrode. During the sodiation process, a displacement reaction occurs to extrude one of the elements, and subsequently forming Na_xM alloy while followed by alloying reaction with the incoming Na^+. Specifically, Sn/Sb/Bi and active M metals react with Na at different potentials thus can serve as mutual buffers to moderate the volume changes. Owing to the synergistic effects and the combined advantages of these two elements, the intermetallic compounds possess advantages of excellent specific capacity and prolonged cycling span, which outperforms the single-metal anode.

Various binary compounds containing active M metal like Sn–Sb nanostructures [143], Bi–Sb nanostructures [144], Sn–Sb/C [72, 145], Bi–Sb/C [146], and Zn–Sb [147] have been investigated as SIB anodes. However, they still suffer from capacity fading due to the formation of single-phase element which tends to peeling off from the current collector, leading to poor electronic conductivity. Compositing active materials with carbon materials is an effective method to solve this issue. For instance, Li et al. [145a] fabricated SnSb–core/carbon–shell nanocables anchored on graphene sheets. The outer carbon nanocable layer played great influence on accommodating the volume expansion, and the involved conductive graphene would isolate the agglomeration of SnSb. When adopting electrolyte of 1 M $NaClO_4$ in PC + 5% FEC, better electrochemical performance of 360 mAh g^{-1} upon 100 cycles was demonstrated in comparison to 1 M $NaClO_4$ in PC, 1 M $NaClO_4$ in PC/FEC (1 : 1 v/v), and 1 M $NaPF_6$ in PC. Bi–Sb alloy/CNFs with various Bi/Sb ratios were prepared by Guo et al. [146b] and the bonding nature of Bi–Sb was studied. Benefiting from the unique lattice softening mechanism and conductive CNFs, the $BiSb_3/C$ nanofibers anode exhibited prolonged cycling stability and sodium storage of 233.2 mAh g^{-1} at 2 A g^{-1} after 2500 cycles.

Intermetallic compounds with electrochemically inactive metals including M–Sn (M = Ni, Cu, Co, Fe, Mn) [84, 148], M–Sb (M = Ni, Cu, Co, Fe, Mo) [149], and Ni–Bi [95] have also been studied. Inert M metal could be used as buffer for alleviating the volume changes of active components during sodiation–desodiation and improve electrical conductivity. For instance, Yu and coworkers [148a] reported Ni_3Sn_2 microcages with large porosity via template-free solvothermal method. Finally, the anode composed of tiny Ni_3Sn_2 nanoparticles exhibited about 270 mAh g^{-1} at 1 C for 300 cycles, and satisfactory rate performance even at 10 C. The reason for the excellent performance is owing to the following aspect: firstly, both unique hollow-core structural design and the introduction of Ni matrix would ease the mechanical stress induced by volume fluctuations. Secondly, the formation of uniformly dispersed Ni arising from Ni_3Sn_2 provides excellent electronic conductivity for the whole anode. $Sn_{0.9}Cu_{0.1}$ nanoparticles with 100 nm in size reported by Mullins and coworkers [148b] has been prepared by surfactant-assisted wet chemistry route. As anode for SIBs, over 420 mAh g^{-1} at 0.2 C rate after 100 cycles was retained. With a high capacity retention of 97%, it has more promising application than the microparticles (13%) and nanoparticles (49%) counterparts. The good performance arises from the copper addition which could effectively lower the charge transfer resistance in the interfaces and suppress aggregation among nanoparticles. In principle, examples discussed-above provide ideal candidates to replace pure elemental electrodes. However, inactive M metal is unable to offer sodium storage stability, leading to decreased capacity density of such Sn/Sb/Bi-based intermetallic compound. Beside the above-discussed binary intermetallic compounds, some ternary alloys, e.g. $Sn_{10}Bi_{10}Sb_{80}$ film [150], $Fe_{1.0}$–SnSb ribbon [151], and $Sn_{50}Ge_{25}Sb_{25}$ film [152], were fabricated as energy storage materials and exhibited excellent cycling performance for SIBs.

This section gives the latest research advances of Sn/Sb/Bi-based alloy-type anodes for SIBs. Strategies to solve the inherent disadvantages of these anodes with high capacity but unstable structure have been proposed. Most of them are concentrating on the buffering of volume fluctuation, inhibiting aggregation as well as accelerating the ionic/electronic diffusion kinetics. And the common methods could be summarized as following: constructing unique nanostructures with large porosity, compositing with carbonaceous materials to form anode with stable structural robustness, as well as choosing appropriate binder and electrolyte for different anode categories.

References

1 Modafferi, V., Triolo, C., Fiore, M. et al. (2020). *Nanomater.* 10: 1588.
2 Shi, L., Li, Y., Zeng, F. et al. (2018). *Chem. Eng. J.* 356: 107.
3 Ming, J., Ming, H., Yang, W. et al. (2015). *RSC Adv.* 5: 8793.
4 Wu, K., Geng, B., Zhang, C. et al. (2020). *J. Alloys Compd.* 820: 153296.
5 Xu, M., Xia, Q., Yue, J. et al. (2019). *Adv. Funct. Mater.* 29: 1807377.
6 Yang, C., Li, Y., Zhang, B. et al. (2020). *Nanoscale* 12: 8922.
7 Kaneti, Y.V., Zhang, J., He, Y.-B. et al. (2017). *J. Mater. Chem. A* 5: 15356.
8 Stratford, J.M., Mayo, M., Allan, P.K. et al. (2017). *J. Am. Chem. Soc.* 139: 7273.
9 Li, H., Liu, A., Zhao, S. et al. (2018). *ChemElectroChem* 5: 3266.
10 Si, H., Li, L., Hao, W. et al. (2019). *ACS Appl. Energy Mater.* 2: 5050.
11 Hariharan, S., Saravanan, K., and Balaya, P. (2013). *Electrochem. Commun.* 31: 5.
12 Xia, W., Xu, F., Zhu, C. et al. (2016). *Nano Energy* 27: 447.
13 Jiang, Y., Sun, M., Ni, J., and Li, L. (2019). *ACS Appl. Mater. Interfaces* 11: 37761.
14 Moretti, A., Secchiaroli, M., Buchholz, D. et al. (2015). *J. Electrochem. Soc.* 162: A2723.
15 Wang, Z., Gao, L., Chen, S. et al. (2021). *J. Solid State Electrochem.* 25: 721.
16 Yan, B., Li, X., Bai, Z. et al. (2017). *J. Mater. Chem. A* 5: 4850.
17 Li, Y., Zhang, S., Wang, S. et al. (2019). *J. Mater. Chem. A* 7: 19234.
18 Wang, M., Wang, X., Yao, Z. et al. (2020). *J. Colloid Interface Sci.* 560: 169.
19 Liang, J., Zhang, L., XiLi, D., and Kang, J. (2020). *Electrochim. Acta* 341: 136030.
20 Fei, J., Cui, Y., Li, J. et al. (2017). *Chem. Commun.* 53: 13165.

21 Kim, S., Qu, S., Zhang, R., and Braun, P.V. (2019). *Small* 15: 1900258.
22 Demir, E., Soytas, S.H., and Demir-Cakan, R. (2019). *Solid State Ionics* 342: 115066.
23 (a) Fang, W., Fan, L., Zhang, Y. et al. (2017). *Ceram. Int.* 43: 8819. (b) Nithya, C. (2015). *ChemPlusChem* 80: 1000.
24 Yan, H., Yang, M., Liu, L. et al. (2020). *J. Alloys Compd.* 843: 155899.
25 Cheng, D., Yang, L., Hu, R. et al. (2021). *Energy Environ. Mater.* 4: 229.
26 Liu, G., Sun, D., Li, X. et al. (2020). *J. Mater. Sci.* 55: 14477.
27 Zhai, H., Jiang, H., Qian, Y. et al. (2020). *Mater. Chem. Phys.* 240: 122139.
28 Man, Z., Li, P., Zhou, D. et al. (2020). *Nano Lett.* 20: 3769.
29 Zhao, Z., Li, S., Li, C. et al. (2020). *Appl. Surf. Sci.* 519: 146268.
30 Liang, S., Cheng, Y., Zhu, J. et al. (2020). *Small Methods*: 4, 2000218.
31 Wang, J., Huang, J., Huang, S. et al. (2020). *ACS Sustain. Chem. Eng.* 8: 9519.
32 Liu, Y., Yang, C., Zhang, Q., and Liu, M. (2019). *Energy Storage Mater.* 22: 66.
33 Chang, G., Yin, X., Shi, S. et al. (2020). *J. Electrochem. Soc.* 167: 140531.
34 Zhao, W., Li, M., Qi, Y. et al. (2021). *J. Colloid Interface Sci.* 586: 404.
35 Zhang, Y., Fan, L., Wang, P. et al. (2017). *Nanoscale* 9: 17694.
36 Luo, W., Li, F., Li, Q. et al. (2018). *ACS Appl. Mater. Interfaces* 10: 7201.
37 Zhan, W., Zhu, M., Lan, J. et al. (2020). *ACS Appl. Mater. Interfaces* 12: 51488.
38 Li, Z., Zheng, Y., Liu, Q. et al. (2020). *J. Mater. Chem. A* 8: 19113.
39 Mohammadi, M. (2020). *J. Phys. Chem. C* 124: 3536.
40 Xu, S., Gao, X., Hua, Y. et al. (2020). *Energy Storage Mater.* 26: 534.
41 Jing, P., Wang, Q., Wang, B. et al. (2020). *Carbon* 159: 366.
42 Haridas, A.K., Heo, J., Li, X. et al. (2020). *Chem. Eng. J.* 385: 123453.
43 Chen, K., Li, G., Wang, Y. et al. (2020). *Green Energy Environ.* 5: 50.
44 He, X., Bi, L., Li, Y. et al. (2020). *Electrochim. Acta* 332: 135453.
45 Lin, X., Cui, X., Yan, H. et al. (2019). *ACS Sustain. Chem. Eng.* 7: 15729.
46 Wang, Y., Wang, Y., Wang, Y.-x. et al. (2019). *ACS Appl. Mater. Interfaces* 11: 19218.
47 Dong, C., Guo, L., Li, H. et al. (2020). *Energy Storage Mater.* 25: 679.
48 Wu, Y.Q., Yang, H.X., Yang, Y. et al. (2019). *Small* 15: 1903873.
49 Zhang, Y., Zhan, R., Xu, Q. et al. (2019). *Chem. Eng. J.* 357: 220.
50 Shuang, W., Huang, H., Kong, L. et al. (2019). *Nano Energy* 62: 154.
51 Zhao, W., Ci, S., Hu, X. et al. (2019). *Nanoscale* 11: 4688.
52 Xue, P., Wang, N., Fang, Z. et al. (2019). *Nano Lett.* 19: 1998.
53 Zhao, D., Yin, M., Feng, C. et al. (2020). *ACS Sustain. Chem. Eng.* 8: 11317.
54 Li, W., Huang, J., Cao, L. et al. (2018). *Electrochim. Acta* 274: 334.
55 Yu, D., Pang, Q., Gao, Y. et al. (2018). *Energy Storage Mater.* 11: 1.
56 Yang, C., Ou, X., Xiong, X. et al. (2017). *Energy Environ. Sci.* 10: 107.
57 Zhao, L., Rong, X., Niu, Y. et al. (2020). *Small* 16: 2004925.
58 Sun, Z., Zhao, C., Cao, X. et al. (2020). *Electrochim. Acta* 338: 135900.
59 Kong, D., Wang, Y., Von Lim, Y. et al. (2018). *Nano Energy* 49: 460.
60 Gong, Y., Zhao, J., Wang, H., and Xu, J. (2018). *Electrochim. Acta* 292: 895.
61 Zhang, K., Park, M., Zhou, L. et al. (2016). *Angew. Chem. Int. Ed.* 55: 12822.
62 Zhang, S., Wang, G., Wang, B. et al. (2020). *Adv. Funct. Mater.* 30: 2001592.

63 Chen, J., Mohrhusen, L., Ali, G. et al. (2019). *Adv. Funct. Mater.* 29: 1807753.
64 Marino, C., Block, T., Pöttgen, R., and Villevieille, C. (2017). *J. Power Sources* 342: 616.
65 Wen, S., Zhao, J., Chen, J. et al. (2019). *Dalton Trans.* 48: 10448.
66 Shi, L., Wang, W., Wu, C. et al. (2017). *J. Alloys Compd.* 699: 517.
67 Fu, L., Bi, Z., Wei, B. et al. (2018). *Nanomater.* 8: 475.
68 Zhao, W., Ma, X., Li, Y. et al. (2020). *Appl. Surf. Sci.* 504: 144455.
69 Wang, T., Yang, K., Shi, J. et al. (2020). *J. Energy Chem.* 46: 71.
70 Ihsan-Ul-Haq, M., Huang, H., Wu, J. et al. (2021). *Carbon* 171: 119.
71 Lv, C., Liu, H., Li, D. et al. (2019). *Carbon* 143: 106.
72 Youn, D.H., Park, H., Loeffler, K.E. et al. (2018). *ChemElectroChem* 5: 391.
73 Wu, J., Yu, J., Liu, J. et al. (2020). *J. Power Sources* 476: 228660.
74 Zhang, Z., Yang, X., and Fu, Y. (2016). *RSC Adv.* 6: 12726.
75 Lan, Y., Zhou, J., Xu, K. et al. (2018). *Chem. Commun.* 54: 5704.
76 Fan, H., Yu, H., Zhang, Y. et al. (2018). *Energy Storage Mater.* 10: 48.
77 Wan, M., Zeng, R., Chen, K. et al. (2018). *Energy Storage Mater.* 10: 114.
78 Zhang, D.M., Jia, J.H., Yang, C.C., and Jiang, Q. (2020). *Energy Storage Mater.* 24: 439.
79 Zhang, K., Park, M., Zhou, L. et al. (2016). *Adv. Funct. Mater.* 26: 6728.
80 Fang, Y., Yu, X.Y., and Lou, X.W. (2018). *Adv. Mater.* 30: 1706668.
81 Zhang, Y., Pan, A., Ding, L. et al. (2017). *ACS Appl. Mater. Interfaces* 9: 3624.
82 Li, X., Zhang, W., Feng, Y. et al. (2019). *Electrochim. Acta* 294: 173.
83 Ali, Z., Tang, T., Huang, X. et al. (2018). *Energy Storage Mater.* 13: 19.
84 Zhang, R., Wang, Z., Ma, W. et al. (2017). *RSC Adv.* 7: 29458.
85 Ou, X., Li, J., Zheng, F. et al. (2017). *J. Power Sources* 343: 483.
86 Zhu, S., Li, Q., Wei, Q. et al. (2017). *ACS Appl. Mater. Interfaces* 9: 311.
87 Zhang, Z., Shi, X., and Yang, X. (2016). *Electrochim. Acta* 208: 238.
88 Xiao, Y., Zhao, X., Wang, X. et al. (2020). *Adv. Energy Mater.* 10: 2000666.
89 Gao, Q., Li, P., Ding, S. et al. (2020). *Ionics* 26: 5525.
90 Zhou, Y., Sun, X., Fan, A. et al. (2021). *Appl. Surf. Sci.* 538: 148194.
91 Ali, Z., Asif, M., Huang, X. et al. (2018). *Adv. Mater.* 30: 1802745.
92 Qiu, L.-C., Wang, Q.-C., Yue, X.-Y. et al. (2020). *Electrochem. Commun.* 112: 106684.
93 Ali, Z., Asif, M., Zhang, T. et al. (2019). *Small* 15: 1901995.
94 Hou, B.H., Wang, Y.Y., Liu, D.S. et al. (2018). *Adv. Funct. Mater.* 28: 1805444.
95 Gao, H., Song, L., Niu, J. et al. (2019). *J. Mater. Chem. A* 7: 13602.
96 Yang, Z., Sun, J., Ni, Y. et al. (2017). *Energy Storage Mater.* 9: 214.
97 Ihsan-Ul-Haq, M., Huang, H., Wu, J. et al. (2020). *Nano Energy* 71: 104613.
98 Sun, D., Zhang, G., Li, D. et al. (2019). *Sustain. Energy Fuels* 3: 3163.
99 Sung, G.-K., Nam, K.-H., Choi, J.-H., and Park, C.-M. (2020). *Electrochim. Acta* 331: 135393.
100 Ding, Y., Wang, W., Bi, M. et al. (2019). *Electrochim. Acta* 313: 331.
101 Chevrier, V.L. and Ceder, G. (2011). *J. Electrochem. Soc.* 158: A1011.
102 Wang, J.W., Liu, X.H., Mao, S.X., and Huang, J.Y. (2012). *Nano Lett.* 12: 5897.

103 Wang, J., Eng, C., Chen-Wiegart, Y.-c.K., and Wang, J. (2015). *Nat. Commun.* 6: 1.

104 Liu, Y., Zhang, N., Jiao, L. et al. (2015). *Adv. Funct. Mater.* 25: 214.

105 Liu, Y., Zhang, N., Jiao, L., and Chen, J. (2015). *Adv. Mater.* 27: 6702.

106 Luo, B., Qiu, T., Ye, D. et al. (2016). *Nano Energy* 22: 232.

107 Li, S., Wang, Z., Liu, J. et al. (2016). *ACS Appl. Mater. Interfaces* 8: 19438.

108 Zhu, H., Jia, Z., Chen, Y. et al. (2013). *Nano Lett.* 13: 3093.

109 Xie, X., Kretschmer, K., Zhang, J. et al. (2015). *Nano Energy* 13: 208.

110 (a) Ellis, L.D., Hatchard, T.D., and Obrovac, M.N. (2012). *J. Electrochem. Soc.* 159: A1801. (b) Baggetto, L., Ganesh, P., Meisner, R.P. et al. (2013). *J. Power Sources* 234: 48. (c) Nam, D.-H., Hong, K.-S., Lim, S.-J. et al. (2014). *J. Phys. Chem. C* 118: 20086. (d) Baggetto, L., Bridges, C.A., Jumas, J.-C. et al. (2014). *J. Mater. Chem. A* 2: 18959. (e) Nam, D.-H., Kim, T.-H., Hong, K.-S., and Kwon, H.-S. (2014). *ACS Nano* 8: 11824.

111 (a) Komaba, S., Matsuura, Y., Ishikawa, T. et al. (2012). *Electrochem. Commun.* 21: 65. (b) Dai, K., Zhao, H., Wang, Z. et al. (2014). *J. Power Sources* 263: 276. (c) Zhang, B., Rousse, G., Foix, D. et al. (2016). *Adv. Mater.* 28: 9824.

112 Darwiche, A., Marino, C., Sougrati, M.T. et al. (2012). *J. Am. Chem. Soc.* 134: 20805.

113 Allan, P.K., Griffin, J.M., Darwiche, A. et al. (2016). *J. Am. Chem. Soc.* 138: 2352.

114 Hou, H., Jing, M., Yang, Y. et al. (2015). *J. Mater. Chem. A* 3: 2971.

115 Liang, L., Xu, Y., Wang, C. et al. (2015). *Energy Environ. Sci.* 8: 2954.

116 Hou, H., Jing, M., Zhang, Y. et al. (2015). *J. Mater. Chem. A* 3: 17549.

117 (a) Liu, S., Feng, J., Bian, X. et al. (2017). *Energy Environ. Sci.* 10: 1222. (b) Fan, X.-Y., Han, J., Jiang, Y. et al. (2018). *ACS Appl. Energy Mater.* 1: 3598. (c) Shi, L. and Wang, W. (2020). *J. Alloys Compd.* 846: 156369.

118 Wang, N., Bai, Z., Qian, Y., and Yang, J. (2016). *Adv. Mater.* 28: 4126.

119 Liu, J., Yu, L., Wu, C. et al. (2017). *Nano Lett.* 17: 2034.

120 Luo, W., Li, F., Gaumet, J.J. et al. (2018). *Adv. Energy Mater.* 8: 1703237.

121 Kong, B., Zu, L., Peng, C. et al. (2016). *J. Am. Chem. Soc.* 138: 16533.

122 (a) Qiu, S., Wu, X., Xiao, L. et al. (2016). *ACS Appl. Mater. Interfaces* 8: 1337. (b) Liu, X., Tian, Y., Cao, X. et al. (2018). *ACS Appl. Energy Mater.* 1: 6381. (c) Zhang, N., Liu, Y., Lu, Y. et al. (2015). *Nano Res.* 8: 3384. (d) Ko, Y.N. and Kang, Y.C. (2014). *Chem. Commun.* 50: 12322.

123 Cheng, N., Zhao, J., Fan, L. et al. (2019). *Chem. Commun.* 55: 12511.

124 Liu, Z., Yu, X.-Y., Lou, X.W.D., and Paik, U. (2016). *Energy Environ. Sci.* 9: 2314.

125 Cui, C., Xu, J., Zhang, Y. et al. (2018). *Nano Lett.* 19: 538.

126 (a) Zhu, Y., Han, X., Xu, Y. et al. (2013). *ACS Nano* 7: 6378. (b) Zhu, M., Kong, X., Yang, H. et al. (2018). *Appl. Surf. Sci.* 428: 448.

127 (a) Liu, X., Gao, M., Yang, H. et al. (2017). *Nano Res.* 10: 4360. (b) Yi, Y., Shim, H.-W., Seo, S.-D. et al. (2016). *Mater. Res. Bull.* 76: 338.

128 Lu, H., Wu, L., Xiao, L. et al. (2016). *Electrochim. Acta* 190: 402.

129 Gao, H., Zhou, W., Jang, J.H., and Goodenough, J.B. (2016). *Adv. Energy Mater.* 6: 1502130.

130 Ellis, L., Wilkes, B., Hatchard, T., and Obrovac, M. (2014). *J. Electrochem. Soc.* 161: A416.

131 Sottmann, J., Herrmann, M., Vajeeston, P. et al. (2016). *Chem. Mater.* 28: 2750.

132 Su, D., Dou, S., and Wang, G. (2015). *Nano Energy* 12: 88.

133 Liu, S., Feng, J., Bian, X. et al. (2016). *J. Mater. Chem. A* 4: 10098.

134 Gao, H., Ma, W., Yang, W. et al. (2018). *J. Power Sources* 379: 1.

135 Wang, L., Wang, C., Li, F. et al. (2018). *Chem. Commun.* 54: 38.

136 Yang, F., Yu, F., Zhang, Z. et al. (2016). *Chem. Eur. J.* 22: 2333.

137 Qiu, J., Li, S., Su, X. et al. (2017). *Chem. Eng. J.* 320: 300.

138 Chen, J., Fan, X., Ji, X. et al. (2018). *Energy Environ. Sci.* 11: 1218.

139 Yang, H., Chen, L.-W., He, F. et al. (2019). *Nano Lett.* 20: 758.

140 Liu, S., Luo, Z., Guo, J. et al. (2017). *Electrochem. Commun.* 81: 10.

141 Xiong, P., Bai, P., Li, A. et al. (2019). *Adv. Mater.* 31: 1904771.

142 Wang, C., Wang, L., Li, F. et al. (2017). *Adv. Mater.* 29: 1702212.

143 (a) Darwiche, A., Sougrati, M.T., Fraisse, B. et al. (2013). *Electrochem. Commun.* 32: 18. (b) He, M., Walter, M., Kravchyk, K.V. et al. (2015). *Nanoscale* 7: 455. (c) Walter, M., Doswald, S., and Kovalenko, M.V. (2016). *J. Mater. Chem. A* 4: 7053. (d) Choi, J.-H., Ha, C.-W., Choi, H.-Y. et al. (2018). *J. Power Sources* 386: 34. (e) Xie, H., Tan, X., Luber, E.J. et al. (2018). *ACS Energy Lett.* 3: 1670. (f) Ma, J. and Prieto, A.L. (2019). *Chem. Commun.* 55: 6938.

144 (a) Gao, H., Niu, J., Zhang, C. et al. (2018). *ACS Nano* 12: 3568. (b) Zhang, W., Yan, W., Jiang, H. et al. (2019). *ACS Sustain. Chem. Eng.* 8: 335.

145 (a) Li, L., Seng, K.H., Li, D. et al. (2014). *Nano Res.* 7: 1466. (b) Jia, H., Dirican, M., Chen, C. et al. (2018). *ACS Appl. Mater. Interfaces* 10: 9696. (c) Ji, L., Gu, M., Shao, Y. et al. (2014). *Adv. Mater.* 26: 2901. (d) Yi, Z., Han, Q., Geng, D. et al. (2017). *J. Power Sources* 342: 861. (e) Kim, I.T., Kim, S.-O., and Manthiram, A. (2014). *J. Power Sources* 269: 848. (f) Qin, J., Wang, T., Liu, D. et al. (2018). *Adv. Mater.* 30: 1704670.

146 (a) Zhao, Y. and Manthiram, A. (2015). *Chem. Mater.* 27: 3096. (b) Guo, S., Li, H., Lu, Y. et al. (2020). *Energy Storage Mater.* 27: 270.

147 (a) Nie, A., Gan, L.-y., Cheng, Y. et al. (2016). *Adv. Funct. Mater.* 26: 543. (b) Fan, L., Liu, Y., Tamirat, A.G. et al. (2017). *New J. Chem.* 41: 13060. (c) Liao, S., Sun, Y., Wang, J. et al. (2016). *Electrochim. Acta* 211: 11. (d) Jackson, E., Green, S., and Prieto, A. (2015). *ACS Appl. Mater. Interfaces* 7: 7447.

148 (a) Liu, J., Wen, Y., van Aken, P.A. et al. (2014). *Nano Lett.* 14: 6387. (b) Lin, Y.-M., Abel, P.R., Gupta, A. et al. (2013). *ACS Appl. Mater. Interfaces* 5: 8273. (c) Edison, E., Satish, R., Ling, W.C. et al. (2017). *J. Power Sources* 343: 296. (d) Kim, I.T., Allcorn, E., and Manthiram, A. (2015). *J. Power Sources* 281: 11. (e) Yui, Y., Ono, Y., Hayashi, M. et al. (2015). *J. Electrochem. Soc.* 162: A3098. (f) Huang, B., Yang, J., Li, Y. et al. (2018). *Mater. Lett.* 210: 321. (g) Vogt, L.O. and Villevieille, C. (2016). *J. Mater. Chem. A* 4: 19116. (h) Zhu, J. and Deng, D. (2015). *J. Phys. Chem. C* 119: 21323. (i) Yamamoto, T., Nohira, T., Hagiwara, R. et al. (2016). *Electrochim. Acta* 211: 234.

149 (a) Liu, J., Yang, Z., Wang, J. et al. (2015). *Nano Energy* 16: 389. (b) Baggetto, L., Allcorn, E., Unocic, R.R. et al. (2013). *J. Mater. Chem. A* 1: 11163. (c) Wang,

L., Wang, C., Zhang, N. et al. (2017). *ACS Energy Lett.* 2: 256. (d) Nam, D.-H., Hong, K.-S., Lim, S.-J., and Kwon, H.-S. (2014). *J. Power Sources* 247: 423. (e) Zheng, W., Yu, X., Guo, Z. et al. (2019). *Mater. Res. Express* 6: 056410. (f) Han, J., Zhu, K., Liu, P. et al. (2019). *J. Mater. Chem. A* 7: 25268.

150 Xie, H., Kalisvaart, W.P., Olsen, B.C. et al. (2017). *J. Mater. Chem. A* 5: 9661.

151 Edison, E., Sreejith, S., Ren, H. et al. (2019). *J. Mater. Chem. A* 7: 14145.

152 Farbod, B., Cui, K., Kalisvaart, W.P. et al. (2014). *ACS Nano* 8: 4415.

10

Effective Strategies to Restrain Dendrite Growth of Na Metal Anodes

10.1 Introduction

Nowadays, rechargeable lithium ion batteries (LIBs) have been widely used in electric vehicles and portable electronic devices since the first commercialization in 1991, due to its magnificent advantages like high energy density, no memory effect, and little self-discharge [1]. However, the limited abundance of lithium sources (0.0065%) is not conducive to the rapid increasing need of super-large-scale energy storage devices in the future [2]. As the same group element (Group IA), sodium element shares similar physical and chemical properties with lithium in many aspects. What is more, sodium sources are rich (2.8%) and worldwide availability. Therefore, sodium-based batteries are considered as ideal candidates for super-large-scale and inexpensive energy storage devices and attracted considerable attention [3]. Meanwhile, the existing theories of LIBs can contribute to the development of sodium-based batteries.

Among the many candidate anode materials, Na metal gains most attention because of its high theoretical capacity (1166 mAh g^{-1}) and low electrode potential. There are many new types of rechargeable Na–metal-based batteries like sodium–sulfur (Na–S) batteries, sodium–oxygen (Na–O_2) batteries, and ZEBRA-type sodium–metal halide batteries [1b]. However, Na metal batteries still suffer from many serious issues. Like lithium metal anode, the major issue of sodium metal anode is uneven plating/stripping behavior which will cause wild dendrite growth, dead Na, and even cell interior short-circuit [4]. Another problem is the continuous side reaction between Na and electrolyte without the protection by stable solid electrolyte interphase (SEI) [5]. The irreversible reaction will consume too much of the electrolyte and Na, resulting in low Coulombic efficiency (CE).

To solve the above-mentioned issues, various strategies have been carried out (Figure 10.1). Typically, it is useful to optimize electrolytes or introduce electrolyte additives for stabilizing reaction interface and SEI film. Electrode architectures are also practicable ways that both reduce the current density and serve as effective nucleation templates. Besides, alloy designs to tune the bulk properties of the metal demonstrate some appreciable improvements [3]. This chapter will focus on these three classical strategies and discuss in depth. Hope this chapter can help to comprehend the protection strategies of Na metal anode and offer some guidance.

Sodium-Ion Batteries: Energy Storage Materials and Technologies, First Edition. Yan Yu.

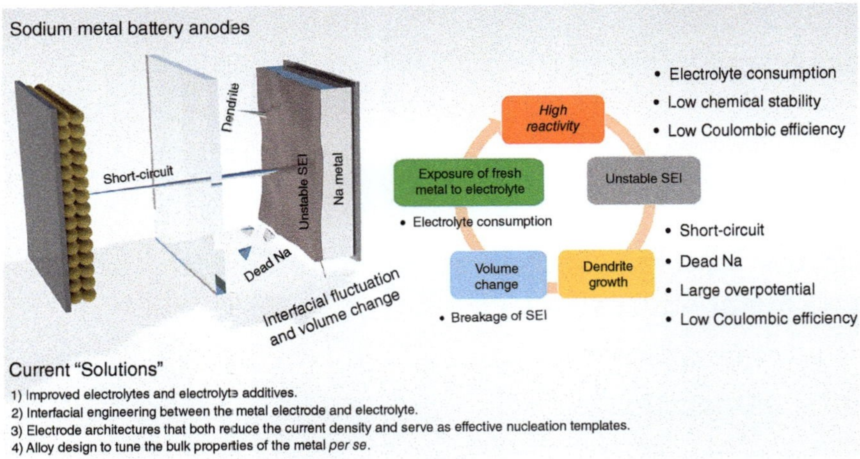

Figure 10.1 The summary of the aspect involved in this chapter. Source: Lee et al. [3]. Reproduced with permission, 2019, American Chemical Society.

10.2 Liquid Electrolyte Optimization for Na Metal Anodes

Electrolytes serve as an important bridge to connect anode and cathode, providing a transport pathway for Na ions. Thus, it is imperative to optimize the electrolyte for enhancing the performance of Na metal anodes. For liquid electrolyte, the reaction with the highly active Na has some severe problems (e.g., uncontrolled dendrite growth), which will cause superfluous consumption of both electrolyte and Na metal. To improve it, plenty of researches focusing on solvents, salts, and additives in electrolyte have been carried out [2, 6].

10.2.1 Traditional Electrolyte

Carbonate solvents are commonly applied to the sodium-ion battery (SIB) electrolyte due to wide electrochemical stability and high ionic conductivity. Iermakova's group assembled Na||Na symmetrical cells with the electrolytes of ethylene carbonate (EC)–dimethyl carbonate (DMC) (1 : 1,v/v) and EC–DMC–propylene carbonate (PC) (4.5 : 4.5 : 1,v/v/v) [5]. They found a large overpotential occurred at 0.1 mA cm^{-2} for Na||Na cells while comparing to Li||Li cells, which was due to an enhanced interfacial resistance but no ionic conductivity. However, the typical SEI in these carbonate electrolytes is consisted of organic products like HCOONa, RCH_2OCO_2Na, $R(OCO_2Na)_2$, and $ROCO_2Na$, which cannot prevent penetration of the electrolyte nor passivate Na metal anode. To stabilize the SEI on Na metal surface, fluoroethylene carbonate (FEC) is selected as the additive, which has already been reported in lithium metal batteries [7]. For example, Choi's group added 1 wt% FEC into the electrolyte consisting of 1 M sodium bis(fluorosulfonyl)imide

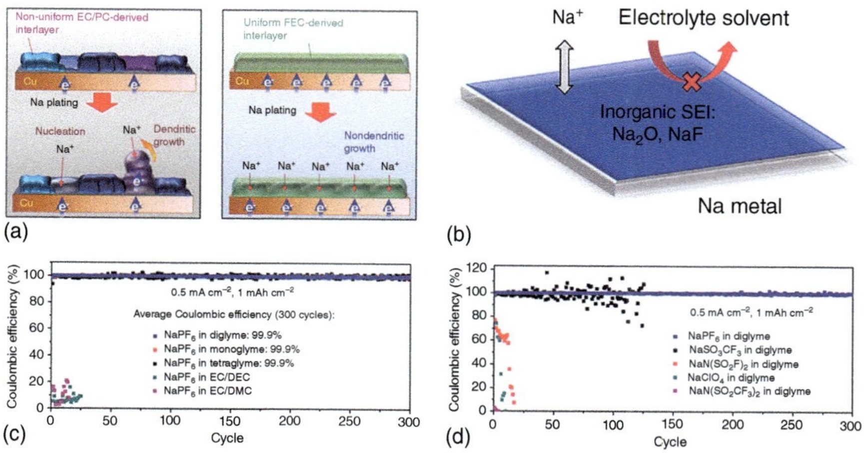

Figure 10.2 (a) Schematic drawing of Na plating process in 1 M NaFSI-EC/PC and 1 M NaFSI-FEC. Source: Lee et al. [8]. Reproduced with permission, 2018, American Chemical Society. (b) Schematic illustration of the inorganic SEI in Na anode. Coulombic efficiencies of sodium metal anodes cycled in (c) various electrolyte solvents with 1 M $NaPF_6$ and (d) 1 M of various electrolyte salts in diglyme. Source: Seh et al. [9]. Reproduced with permission, 2015, American Chemical Society.

(NaFSI) in PC : EC (1 : 1), and an ion-permeable and mechanically strong SEI comprised of Na_2CO_3, NaF, and sodium alkylcarbonates was demonstrated, which could restrain the growth of Na dendrite, as shown in Figure 10.2a [8]. Moreover, Tarascon and coworkers confirmed that the additive of FEC could minimize the irreversible capacity of Na-half cells [10]. Kong and his coworkers chose sodium-difluoro(oxalato)borate (NaDFOB)-based electrolytes to replace common $NaPF_6$-based electrolytes and achieved preferable rate performance and longer cycle life in Prussian blue||Na cells [11].

On the other hand, ether-based solvents with the higher lowest unoccupied orbital (LUMO) energy than that of carbonate solvents have wider electrochemical stability window, making them stand a more reductive environment without decomposition [12]. Thus, ether-based electrolytes are more befitting for SIBs [1b, 4, 13]. Cui and his coworker reported a comprehensive study on the electrolyte of $NaPF_6$ with mono-, di-, and tetraglyme solvents and found glyme-typed solvents decomposed after Na salts decomposition [9]. As shown in Figure 10.2b, an even and compact inorganic SEI was formed in $NaPF_6$–glyme-based electrolyte. Such an inorganic SEI made of NaF and Na_2O not only had good ionic conductivity to facilitate uniform Na deposition without Na dendrite but also efficiently prevented the permeation of electrolytes, resulting in a high CE (Figure 10.2c,d). Besides, Li's group investigated the additive of sodium polysulfide (Na_2S_6) in $NaPF_6$–diglyme electrolyte and indicated Na_2S_6 could serve as a pre-passivation agent which can enhance long-term anode stability [14]. Meanwhile, they also found that the use of Na_2S_6-$NaNO_3$ as the co-additive had disadvantageous impact on Na anode, which was different in the lithium anode system.

Self-healing electrostatic shield mechanism is also a fantastic method to suppress the growth of Na dendrite. Chen and his coworkers proposed a Li–Na hybrid battery, where the Li^+ could form an electrostatic shield layer during Na plating (Figure 10.3a) [15]. The reduction potential of Li-ions (−3.04 V vs. SHE) is lower than that of Na-ions (−2.71 V vs. SHE), so Li-ions will not be reduced during the process of plating. Instead, Li-ions will assemble on the tips of the plating regions to build a positively charged electrostatic shield due to the accumulated negatively charged electric field. When Na-ions approach, they will diffuse and plate into adjacent regions because of repulsive force from the electrostatic shield. As a result, the Na deposition showed cubic morphology without growth of dendrites (Figure 10.3b). In addition, Wang's group used a bifunctional electrolyte additive, potassium bis(trifluoromethylsulfonyl)imide (KTFSI) in tetraethylene glycol dimethyl ether (TEGDME) containing 1 M sodium trifluoromethylsulfonate

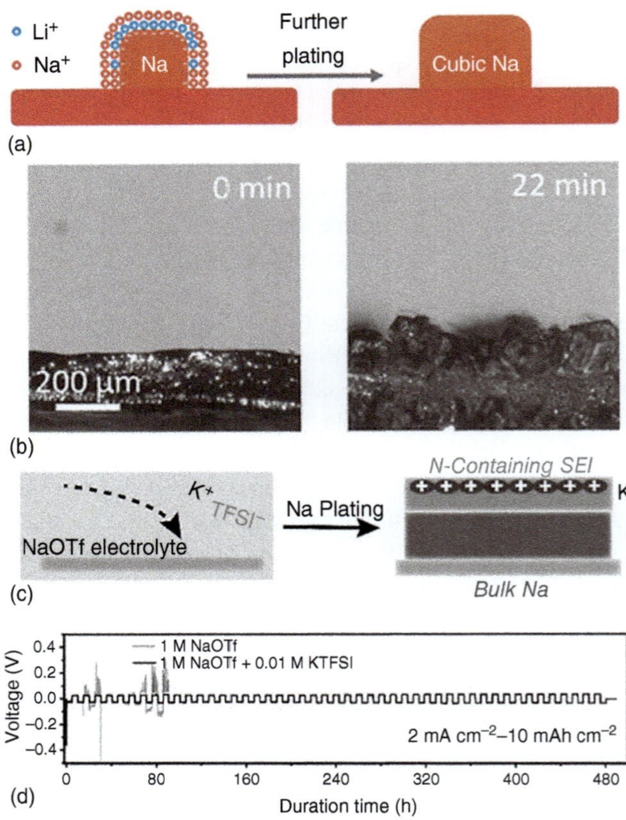

Figure 10.3 (a) Schematic of Na plating patterns with Li^+ electrostatic shield layer. (b) *In situ* optical observation of Na depositing in hybrid electrolyte. Source: Zhang et al. [15]. Reproduced with permission, 2018, Wiley-VCH. (c) Schematic of the stabilization effects of KTFSI additive for Na deposition. (d) Charge-discharge voltage profiles of Na||Na symmetric cells with and without 0.01 M of KTFSI added in the 1 M NaOTf in TEGDME electrolyte. Source: Shi et al. [16]. Reproduced with permission, 2018, Wiley-VCH.

(NaOTf) [16]. As exhibited in Figure 10.3c, K-ions (−2.93 V vs. SHE) could form an electrostatic shield layer while $TFSI^-$ would contribute to form a desirable SEI, rendering Na anode with a long cycle life even at a high capacity of 10 mAh cm^{-2} (Figure 10.3d).

10.2.2 High-concentration Electrolyte

The properties of electrolyte are affected by the concentration of salt, which will directly impact the performance of batteries. High salt concentration electrolytes possess some preferable advantages like the enhanced thermal stability, reductive/oxidative stability, and ion transference number [12]. Meanwhile, the increase in initial metal ion concentration will prolong the Sand's time and therefore postpone dendrite formation [17]. Besides, a large number of Na-ions in the electrolyte will coordinate with more free solvent molecules to form a unique solvation structure, which can mitigate Na anode corrosion induced by the organic solvents [18]. Zhang's group reported a high concentrated electrolyte consisted of 4 M NaFSI in 1,2-dimethoxyethane (DME) [19]. The Na||Cu cells with this electrolyte exhibited an excellent performance with high efficiency (>99%). Afterwards, many researches on NaFSI–DME-based high-concentration electrolyte were reported [20]. Moreover, Choi's group further increased the concentration of NaFSI to 5 M in DME solvent, where they found it could also mitigate the corrosion of Al current collectors (Figure 10.4a) [21]. Besides, they noted that high concentration of salt was only effective in DME solvents and ineffective in carbonated-based electrolyte (Figure 10.4b,c).

High-concentration electrolyte, however, also brings some inevitable defects, such as high viscosity, high salt cost, and poor wettability. To overcome those problems, a localized high-concentration electrolyte was proposed by Zhang and his coworkers [22]. They used hydrofluoroether as an "inert" diluent to remarkably reduce the salt concentration in whole electrolyte. The bis(2,2,2-trifluoroethyl)ether diluent would not break the solvation structure of NaFSI–DME and still maintain a high Na-ion transference number, as displayed in Figure 10.4d. With this special electrolyte, large amounts of nodule-like Na were deposited on the Cu electrode without the growth of needle-like Na (Figure 10.4e). In addition, Na||$Na_3V_2(PO_4)_3$ batteries can reach high CE (>99%), fast charging (20 C), and stable cycling (90.8% retention after 40 000 cycles) (Figure 10.4f). This work offers a novel strategy to customize the electrolyte for developing high-performance SIBs.

Except for the conventional organic electrolyte, there are also many significant efforts in designing novel electrolytes to improve battery performance. For example, Kim's group reported dendrite-free Na anode for sodium rechargeable battery, engineered by employing nonflammable and highly Na^+-conductive $NaAlCl_4{\cdot}2SO_2$ inorganic electrolyte [23]. They dissolved 5.3 M $NaAlCl_4$ in SO_2-solvated molten complex, demonstrating a high ionic conductivity of 0.1 S cm^{-1} and a high cation concentration. Compared to the electrolyte of 1 M $NaPF_6$ in EC : PC, the high concentrated $NaAlCl_4{\cdot}2SO_2$ electrolyte could remain an initial impedance for long storage time in Na||Na symmetric cells. The X-ray photoelectron spectroscopy (XPS) analysis revealed that a dense and even inorganic layer of NaCl was formed

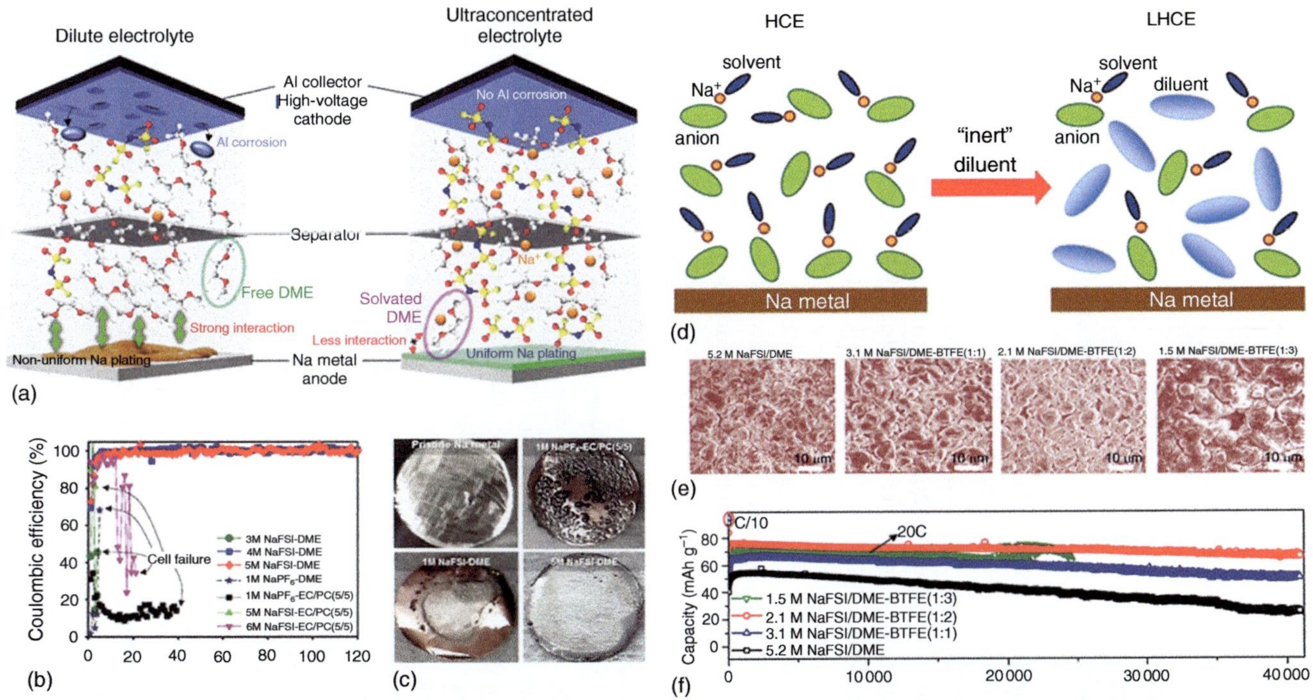

Figure 10.4 (a) Schematic illustration of the effect of concentrated electrolyte in protecting the Al current collector and guiding Na ions deposition. (b) Coulombic efficiency for Na plating/stripping in Na/stainless steel (SS) cells. (c) Photographs of a pristine Na anode and Cu substrates after the initial Na plating with different concentration of electrolytes. Source: Lee et al. [21]. Reproduced with permission, 2017, American Chemical Society. (d) Schematic illustration of dilution from a high-concentration electrolyte (HCE) to a localized high-concentration electrolyte (LHCE). (e) SEM images of Na deposited on Cu electrodes with a capacity of 2 mAh cm^{-2} in different concentration of electrolytes. (f) Cycling performance of Na||NVP batteries. Source: Zheng et al. [22]. Reproduced with permission, 2018, American Chemical Society.

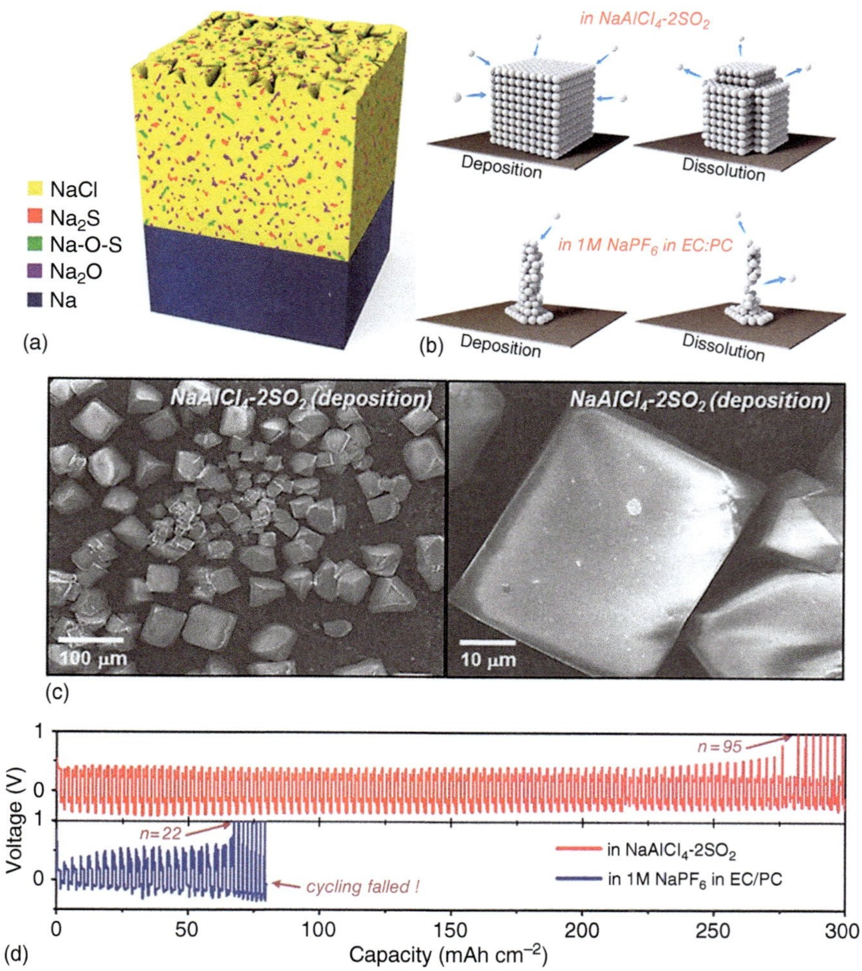

Figure 10.5 (a) Schematic for the surface layer formed on the Na anode after contacting $NaAlCl_4{\cdot}2SO_2$. (b) Schematic illustration of deposition and dissolution of Na metal (upper) in $NaAlCl_4{\cdot}2SO_2$ inorganic electrolyte and (lower) in 1 M $NaPF_6$–EC : PC organic electrolyte. (c) SEM results of the Na deposit in $NaAlCl_4{\cdot}2SO_2$ inorganic electrolyte. (d) Cycling performance of Na||Na symmetric cells in a $NaAlCl_4{\cdot}2SO_2$ inorganic electrolyte and 1 M $NaPF_6$–EC : PC organic electrolyte. Source: Song et al. [23]. Reproduced with permission, 2015, American Chemical Society.

on the surface of anode, which could passivate Na metal from electrolyte corrosion and offer a high Na ion diffusion rate (Figure 10.5a). The scheme of Na deposition/dissolution is illustrated in Figure 10.5b and a polygonal morphology of plated Na is observed in Figure 10.5c, which is desirable as such Na deposition cannot pierce the separator [1a]. Applied this inorganic electrolyte, Na||Na symmetric cells showed stable voltage profiles up to 95 cycles, while the cells in $NaPF_6$–EC: PC failed after 22 cycles (Figure 10.5d). In addition, another novel inorganic electrolyte based on liquid ammonia was reported by Gómez and his coworkers, which could be shorted as $NaY{\cdot}xNH_3$ (Y: I^-, BF_4^-, BH_4^-) [24]. The optimized electrolyte could achieve ultrahigh sodium concentration of 7 M and high conductivity of 0.1 S cm^{-1} with low density, low viscosity, and low cost [3]. Using $NaBF_4$–$2.5NH_3$, the plating-stripping test of Na on Cu foil showed a low-voltage hysteresis of 4.5 mV for 100 cycles.

10.2.3 Ionic Liquids

Ionic liquids (ILs) are kinds of room-temperature molten salts consisted of organic cations and organic/inorganic anions existing in a liquid form under 100 °C. The characters of ILs include high thermal stability, low vapor pressure, low flammability, wide electrochemical widow, and improved safety, so it is an intense topic for ILs to apply as SIB electrolyte (Figure 10.6a) [25]. Hosokawa and his coworkers investigated the stability of $[C_2C_1im][FSA]$- and $[C_2C_1im][TFSA]$-based ILs against Na metal [27]. After storing Na metal in the Na[FSI]–$[C_2C_1im][FSI]$ electrolyte for four weeks, the Na surface maintained smooth with metallic luster while the surface became rough and dark after immersion in $TFSI^-$-based counterpart, which was due to the different property of SEI. Besides, Archer's group reported SEI films generated directly on Na metal through electropolymerization of functional IL membranes, as shown in Figure 10.6b [26]. They found that the ionic membranes can prevent parasitic reactions of Na with electrolyte but not compromise ion transport in SEI. In addition, the ionic membranes can reduce the electric field on Na surface, causing the even deposition of Na metal, high CE, and stable long-term cycling.

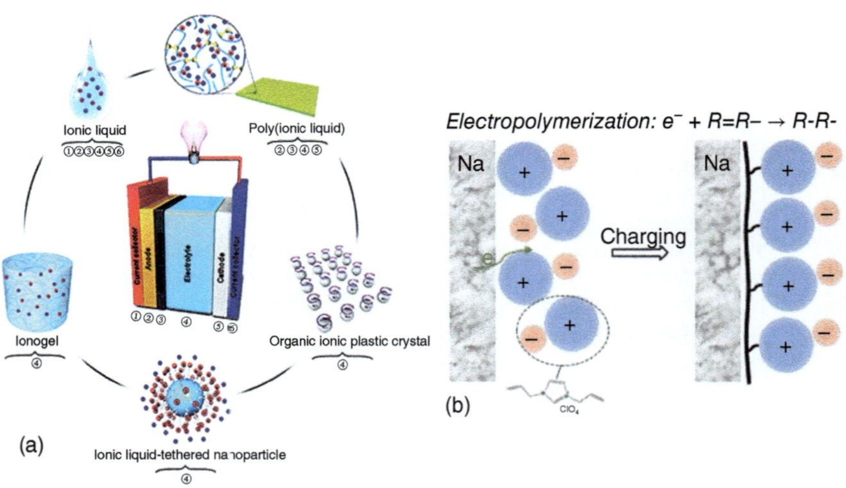

Figure 10.6 (a) Schematic of the structures of ionic liquids, IL-derived materials, and the battery components for which they can be applied in SIBs. Source: Yang et al. [25e]. Reproduced with permission, 2018, The Royal Society of Chemistry. (b) Schematic of the polymeric ionic liquid film forms on the electrode. Source: Wei et al. [26]. Reproduced with permission, 2017, Wiley-VCH.

10.3 Construction of Novel Current Collectors for Na Metal Anodes

Since the uneven distribution of sodium ion on electrode surface would directly lead to sodium dendrites growth, it is urgently need to explore new strategies to generate uniform sodium ion flux [1a]. The development of novel current collectors to inhibit the sodium dendrite growth by dissipating the local current density of the electrode surface area is accepted as an efficient strategy [19, 28]. These current collectors should have high chemical stability without a reaction with sodium, high mechanical stability to recover the initial shape and excellent electrical conductivity with fast electronic transmission [6a, 29]. Therefore, current collectors with a unique nanostructure for the stabilization of sodium metal anodes have been designed and explored.

10.3.1 Metallic Current Collectors

Frequently used Cu current collector tends to form sodium dendrites due to its rough surface and uneven electron distribution during the plating/stripping cycles [28, 30]. It is thus necessary to design or modify the traditional current collectors with a unique nanostructure to dissipate the local current density of sodium anodes and prevent the growth of sodium dendrites. Wang and his colleagues developed a porous Cu composite matrix by utilizing commercial Cu foam current collector for Na metal anode [31]. As shown schematically in Figure 10.7a, the oxygen-(or sulfur-) treated Cu matrix with excellent mechanical properties and the abundant porous structure was used as the current collector to stabilize the metal sodium anode, which promoted to form homogeneous impregnation and distribution of Na metal by the chemical interaction, restraining the volume change and the growth of sodium dendrite. The composite sodium metal significantly improved the stability of cyclic behavior, showing a superior rate performance at a high rate of 5 C, which is much better than the bare sodium metal electrode in carbonate ester-based electrolyte without additive. Chen's group reported the usage of 3D Cu nanowires as Na current collector via a facile hydrothermal treatment of the planar Cu foil (Figure 10.7b), which greatly promotes the uniform ion flux distribution and reduces the uneven charges distribution, resulting in stable Na deposition, long life with a high efficiency, and low overpotential [32].

Al foil can also serve as the current collector for Na anode due to the lower cost and weight than Cu foil, and no alloy reaction with sodium [3, 33]. However, similar to Cu foil, the planar Al foil also tends to form sodium dendrites because of its low surface areas and nucleation sites. Luo's group proposed an interconnected porous Al current collector as the plating substrate to suppress the growth of Na dendrites [33], which greatly improved the cycling stability due to its enough Na nucleation surface area (Figure 10.8a). Detailedly, the sodium current collector of 3D porous Al nanostructure showed long-term cycling stability with a small voltage hysteresis over 1000 hours at 0.5 mA cm^{-2} (Figure 10.8b) and high average CEs of above 99.9% (Figure 10.8c).

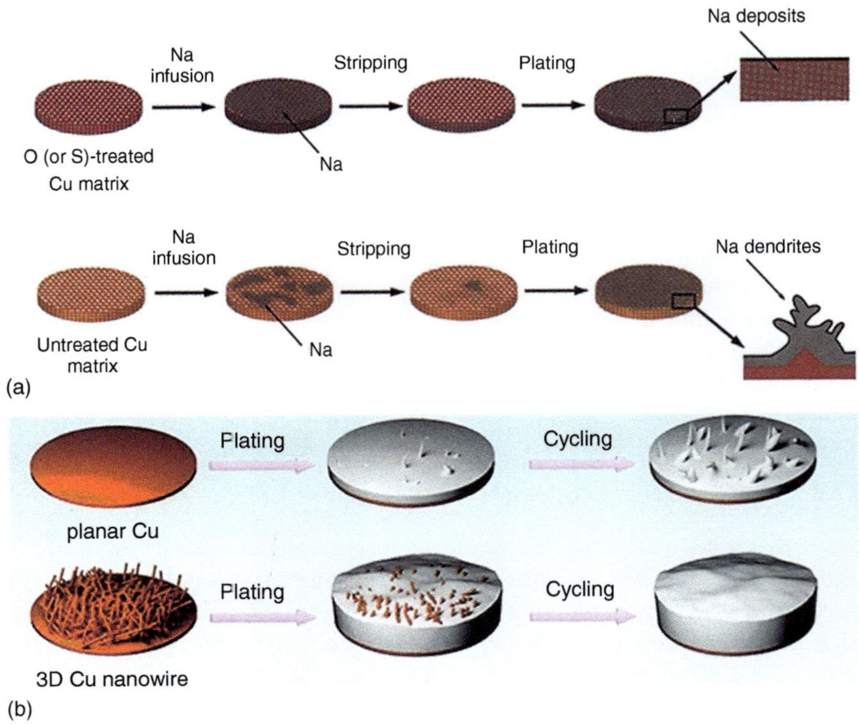

Figure 10.7 (a) Schematic illustrations of oxygen-(or sulfur-) treated Cu matrix for Na stripping/plating. Source: Wang et al. [31]. Reproduced with permission, 2018, Wiley-VCH. (b) Na-plating models of planar Cu foil and 3D Cu nanowires. Source: Lu et al. [32]. Reproduced with permission, 2017, The Royal Society of Chemistry.

10.3.2 Carbon-Based Current Collectors

Carbonaceous materials have been extensively used in various rechargeable battery systems because of the benefits of high electrical conductivities, excellent chemical, and mechanical stability [34]. Recently, carbonaceous materials have been developed to replace the Cu/Al-based current collectors for sodium metal anodes [35]. Chu and coworkers proposed an oxygen-doped carbon nanotube framework (a-CNT) by acidifying, which induced the homogeneous Na nucleation and deposition due to the sodiophilic peculiarity of oxygenic functional groups (Figure 10.9a), and suppressed dendrite growth [36]. In addition, the a-CNT/Na anode revealed excellent electrochemical performances with 99.8% CE after 1000 cycles at 3 mA cm^{-2}. Peng's group also presented an oxygen-functionalized 3D carbon nanotube network by mediating sodiophilic interphase for ultrahigh capacity and cycle life over 3000 cycles in Na metal batteries [37]. A robust and sodiophilic interphase was formed on oxygen-functionalized 3D carbon nanotube network electrode, which promoted the uniform Na deposition and the formation of dendrite-free morphology during recurring Na plating/stripping processes (Figure 10.9b). In brief, because of the synergistic effect arising from these optimizing features, the

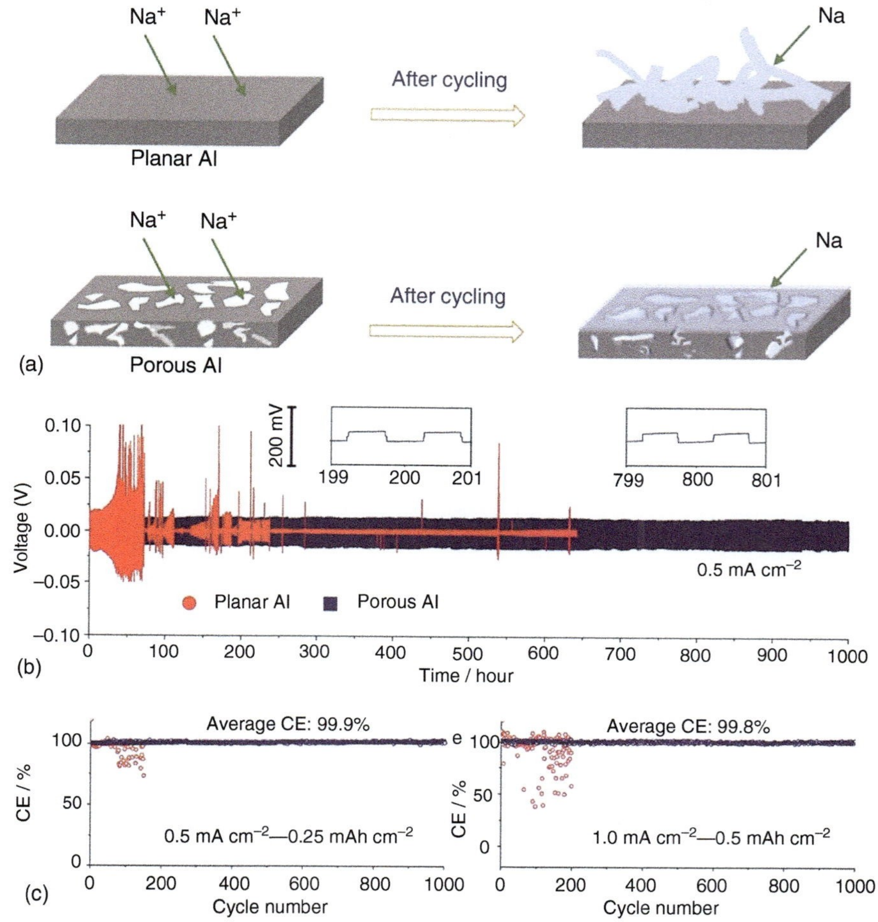

Figure 10.8 (a) Schematic of Na deposition on two different Al foils. (b) The cycling performance of symmetric cells, and (c) the Coulombic efficiency at different current densities. Source: Liu et al. [33]. Reproduced with permission, 2017, American Chemical Society.

carbon-based current collector electrodes with sodiophilic peculiarity realized long-term stable cyclic performance without dendrite formation.

10.3.3 3D Scaffolds/Na Metal

To reduce the local current density of the electrode surface area for uniform Na nucleation and suppress dendrite growth, 3D carbon-based scaffolds and metal Na composites were used as effective host materials, which had great significance for the safety of Na metal anode [38]. Luo et al. proposed a Na carbonized wood composite electrode by a melt infusion strategy (Figure 10.10a) [39]. Benefiting from conductive, porous framework and good mechanical stability, the carbonized wood was proved to be an ideal host for Na metal anode, which displayed stable cycling

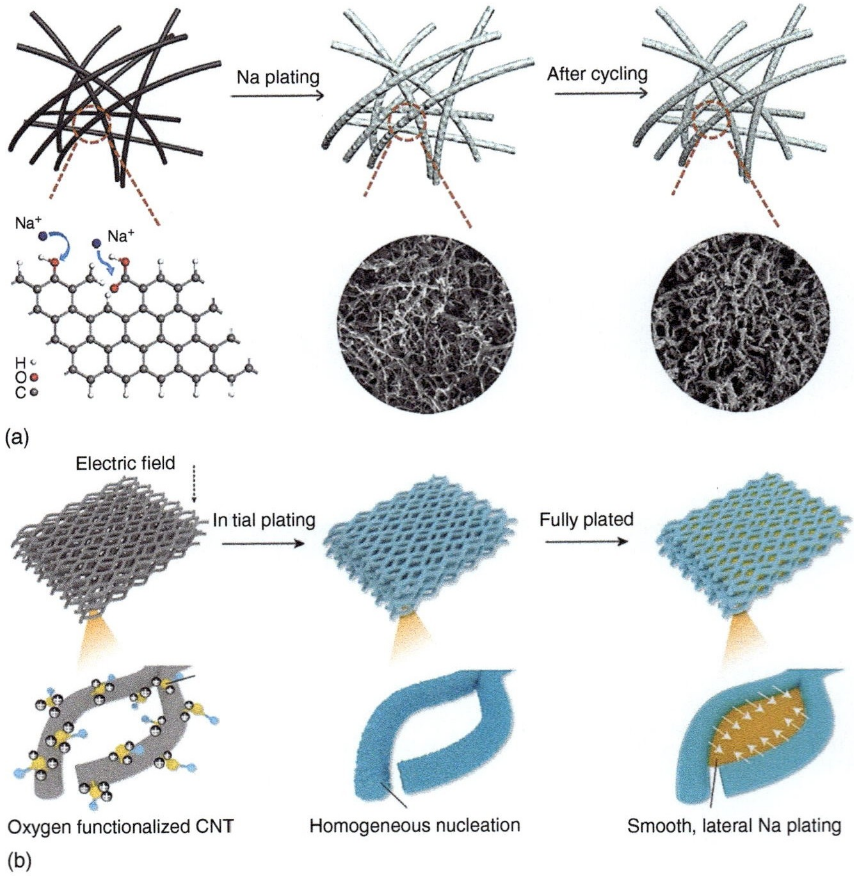

Figure 10.9 (a) Schematic illustrations of the homogeneous Na deposition on the oxygen-doped carbon nanotube framework. Source: Chu et al. [36]. Reproduced with permission, 2019, Elsevier. (b) Schematic illustrations of uniform Na nucleation on oxygen-functionalized 3D carbon nanotube network. Source: Ye et al. [37]. Reproduced with permission, 2019, Wiley-VCH.

performance and flat plating/stripping profiles. Moreover, large-scale commercial 3D flexible carbon felt as the stable host to pre-store sodium to form a Na/C composite by a melt infusion strategy was demonstrated by Chi and his colleagues (Figure 10.10b–e) [40]. The anode significantly inhibited Na dendrite growth and alleviated large volume change, demonstrating a good suitability in full cells.

Additionally, some 3D metallic scaffolds have been recently utilized to steady Na anode and significantly enhanced cycling stability by suppressing dendritic Na growth due to the advantages of excellent mechanical features, rich surface functional groups, and good electronic conductivity [41]. Cao's group designed Ti_3C_2Tx-coated carbon cloth and Na metal composites (Na–$Ti_3C_2T_x$–CC) as good host for Na metal anodes (Figure 10.11a–d) [42]. This unique electrode was constructed by a tough skeleton of CC and functional $Ti_3C_2T_x$ with sodiophilicity and fast electron channels, which facilitated uniform infusion of molten Na, resulting

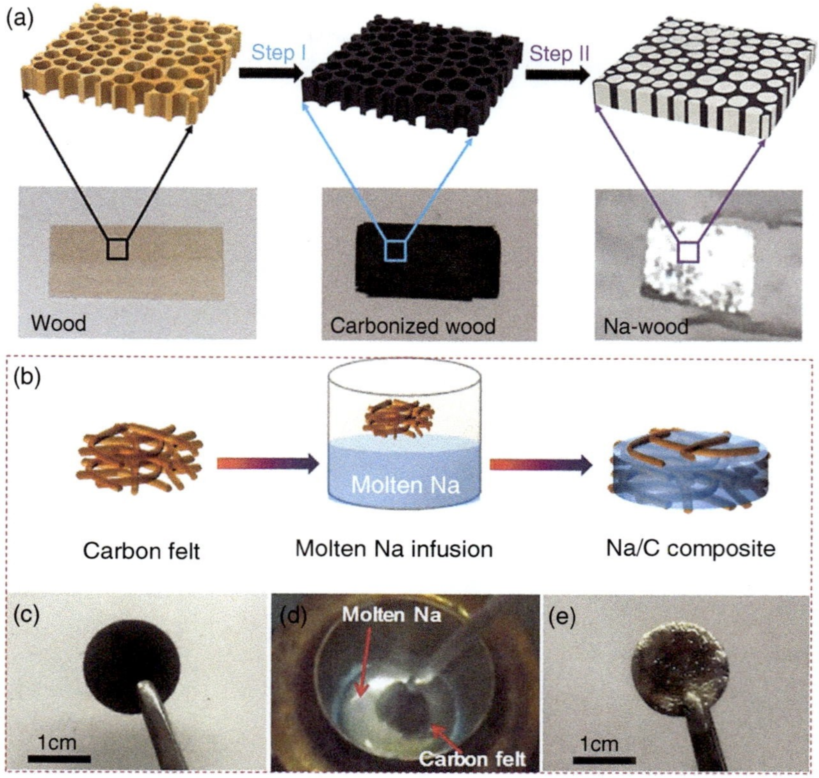

Figure 10.10 (a) Schematics of the construction of Na-carbonized wood composite electrodes. Source: Luo et al. [39]. Reproduced with permission, 2017, American Chemical Society. (b) Schematic of Na/C anode fabrication process and the corresponding images of (c) carbon felt, (d) molten Na infusion, and (e) the Na/C composite. Source: Chi et al. [40]. Reproduced with permission, 2018, Wiley-VCH.

in excellent cycling performance and high stripping/plating capacity in ether and carbonate-based electrolytes. Zhang et al. reported a SnO_2–carbon fiber (SnO_2–CF) and Na composites for Na metal batteries (Figure 10.11e), which decreased local current density and effectively repressed the growth of dendrites [43].

10.4 Alloy-Based Na Metal Anodes

10.4.1 Alkali-metal Alloys

It is reported that no dendrites were observed in commercial sodium–sulfur batteries using high temperature molten alkali metal as anode [44]. However, the operating temperature of liquid sodium exceeds 300°, which imposes strict requirements on other components of the batteries [45]. Since the eutectic point of alkali metal alloys is lower than room temperature, liquid alloys such as Li–Na, K–Na, Na–Rb, and Cs–Na can be realized as anodes at room temperature. Because

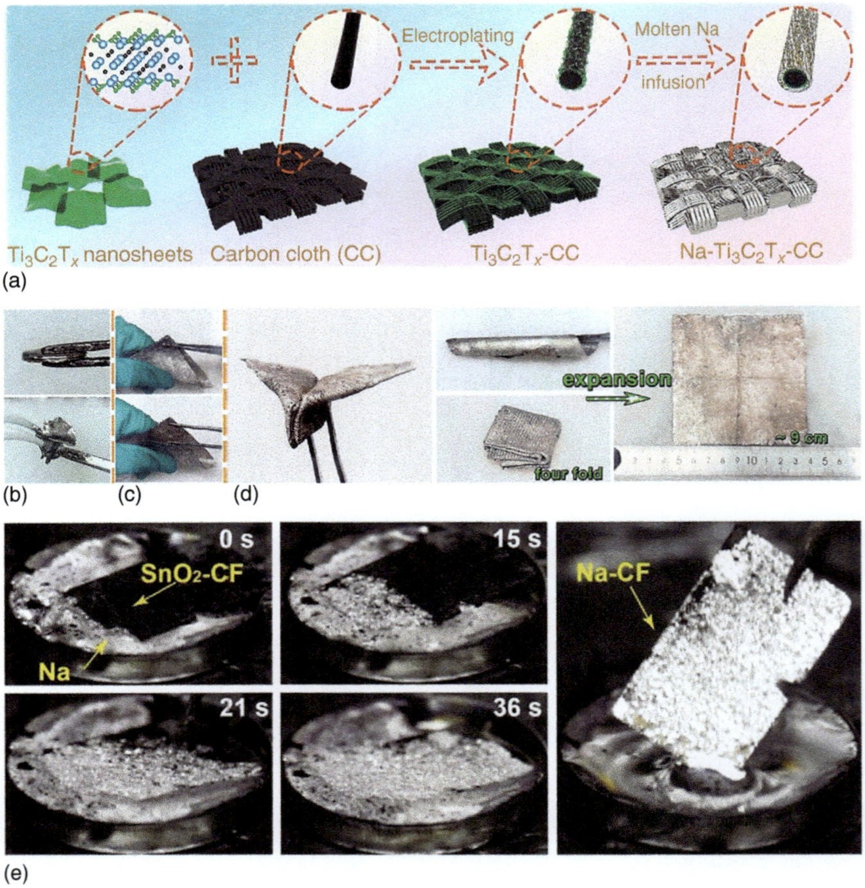

Figure 10.11 (a) Schematics of the fabrication for Na–$Ti_3C_2T_x$–CC metal anodes, and optical images of (b) bare Na foil, (c) Na–$Ti_3C_2T_x$–CC stickiness, and (d) foldable and bendable Na–$Ti_3C_2T_x$–CC metal anodes. Source: Fang et al. [42]. Reproduced with permission, 2020, American Chemical Society. (e) Optical image of the SnO_2–CF framework and molten Na. Source: Zhang et al. [43]. Reproduced with permission, 2018, American Chemical Society.

Na^+ is an alkali ion with the highest redox potential, theoretically it can form alloys with any other alkali ions, resulting in electrostatic shielding [46]. With the increase in ion radius, the electrostatic shielding effect is weakened, so the lithium ion is the most favorable. Jiang and coworkers reported that the Li–Na alloy anode combined with the electrolyte additive greatly improved the cycle stability of bimetallic Li–Na alloy—O_2 battery [46]. The strong and soft passivation film formed by the reaction of the alloy with the additive effectively inhibited the dendrite and alleviated the bulk expansion of the alloy anode, thus ensuring a stable cycle of the battery. On the sodium electrode, Na^+ is easy to deposit on the tip to form dendrite, while on the Li–Na alloy electrode, because the Li^+ is adsorbed at the tip, the Na^+ is excluded from the tip area to avoid the formation of dendrite (Figure 10.12a). They point out that the Na/Li ratio is an important factor in inhibiting dendrite growth and

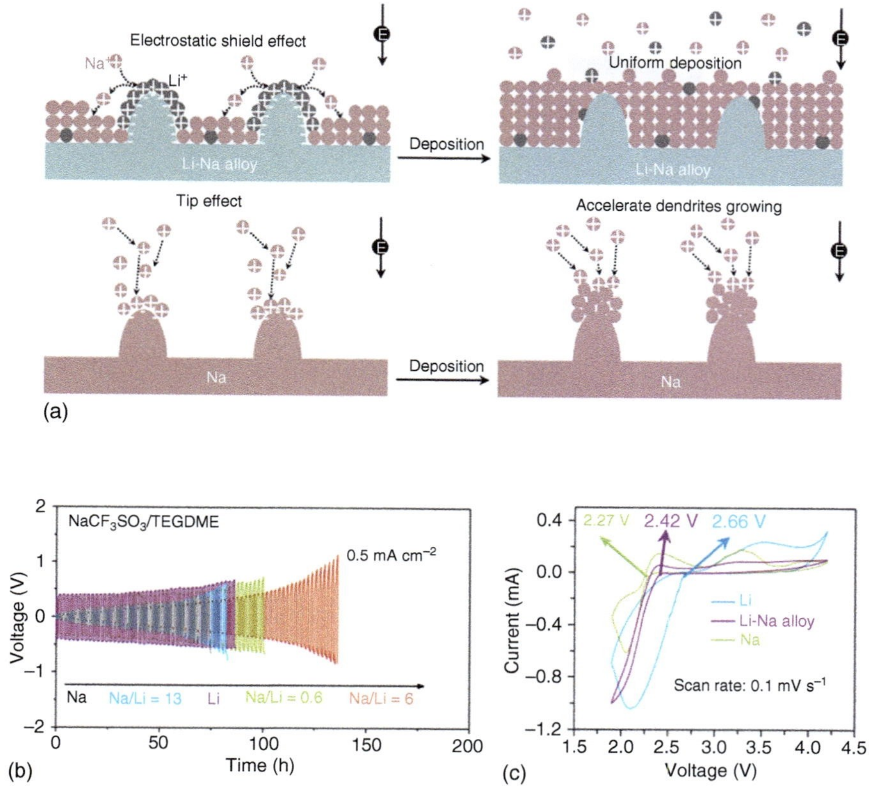

Figure 10.12 (a) Illustration of dendrite and crack suppression for Li-Na alloy anode. (b) Voltage profiles for symmetric batteries in 0.5 M $NaCF_3SO_3$/TEGDME. (c) CV curves for metal–O_2 batteries at a scan rate of 0.1 mV s^{-1} between 1.9 and 4.2 V. Source: Ma et al. [46]. Reproduced with permission, 2019, Springer.

improving battery cycle stability. A small amount of dendrites can still be observed when the alloy with ultra-low lithium. If the Na/Li ratio reaches 6, the alloy anode shows the best electrochemical stability. When the symmetrical cells with 0.5 M $NaCF_3SO_3$/TEGDME electrolyte were tested at 1 mA cm^{-2}, Li-Na alloy (Na/Li = 6) symmetrical cells stable cycle over 130 hours, while the other cells run less than 130 hours, which indicates that alloy anode can stabilize sodium electrode to some extent (Figure 10.12b). The electrochemical properties of metal–O_2 batteries were tested using commercial carbon nanotubes as cathode and lithium sodium alloy, lithium, and sodium as anode. As displayed in Figure 10.12c, the cyclic voltammetry (CV) graph shows that the reduction potential in Li–Na alloy battery is 2.42 V, which is between Li–O_2 battery (2.66 V) and Na–O_2 battery (2.27 V), implying that Li and Na participate in redox reaction during charge and discharge processes.

In recent years, there is a great progress on the research of liquid Na–K alloy anode. Liquid Na–K alloy is immiscible with the most of hydrocarbon electrolytes, so it can form a stable liquid–liquid anode–electrolyte interface. However, the high surface tension of sodium–potassium alloy causes the poor wettability of the liquid electrolyte surface. Goodenough and coworkers solved this problem

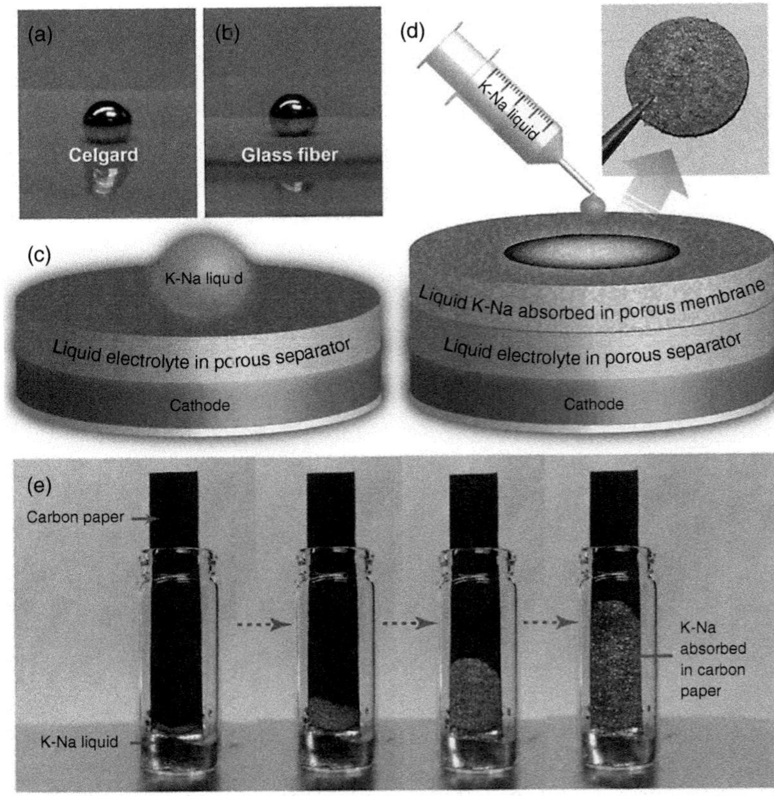

Figure 10.13 Photos of a liquid K–Na dropped on (a) Celgard and (b) glass fiber membranes containing an organic liquid electrolyte at room temperature. Schematics of a cell with a liquid K–Na drop (c) not absorbed and (d) absorbed by a porous membrane. (e) Photos of liquid K–Na alloy being absorbed in a carbon paper strip at 420 °C. Source: Xue et al. [47]. Reproduced with permission, 2016, Wiley-VCH.

by making the Na–K alloy into sheets [47]. They found that liquid Na–K alloy could be absorbed into carbon paper at 420 °C easily (Figure 10.13), and then well recycled in the solvent of TEGDME. This flexible porous membrane containing liquid Na–K alloy can be used as a dendrite-free anode in contact with organic electrolyte. Yu's group has developed a simple method to fabricate sodium and potassium electrode at room temperature [48]. By immersing the pretreated graphite intercalation compound (GIC) in liquid Na–K alloy (removing oxide layer at high temperature), because the diffusion rate of K is much faster than that of Na, the network of KC_8 is formed at first, which induced the uniform diffusion of Na–K alloy. Compared with solid anode, Na–K–GIC–carbon (Na–K–GC) coincidence electrode, which combines the fast electron transport characteristics of GIC and liquid metal anode with self-repair ability, has superior effect on inhibiting dendrite growth and improving electrochemical stability. Through the above discussion, we can clearly understand that Na–K liquid alloy can not only realize dendrite growth inhibition but also have stable electrochemical performance in the full cells. However, there is a question about when liquid Na–K alloy is used as sodium anode or potassium anode. Goodenough's work shows that the alkali metal peeled from liquid Na–K anode depended on the kind of cathode [49]. If the main structure of the cathode selectively accepts sodium ion, it is used as the anode of the sodium battery, and if the potassium ion is preferred, it is the anode of the potassium rechargeable battery. For example, when the cathode material is $Na_{2/3}Ni_{1/3}Mn_{2/3}O_2$ and NASICON-structured $Na_3V_2(PO_4)_3$, because these materials are more receptive to Na^+, Na–K anode acts as a sodium anode, and when together with a cathode body favoring K^+, such as $Na_2MnFe(CN)_6$, $MnFe(CN)_6$, and $K_2MnFe(CN)_6$, it acts as a potassium anode. This provides a guide for the later research of Na–K alloy negative electrode.

In Yu's another work, they dipped the carbon paper into the liquid Na–K alloy at room temperature, and then liquid alloy was soaked into the carbon paper through a pressure differential created by vacuuming [50]. Compared with high temperature treatment, this process is more safe and effective, and can avoid the side reaction at high temperature. The mechanism of the Na–K alloy anode was explored through comprehensive characterization and theoretical calculations. When sodium rhodizonate dibasic (SR) was chose as cathode material, Na–K alloy||SR battery can serve as both Na-ion battery and K-ion battery, and all of them showed excellent cycle performance. The theoretical calculation (Figure 10.14a,b) demonstrated that the selectivity of organic cathode to different carriers is negligible, and the deposition–dissolution reaction of different alkali metal ions in Na–K alloy anode was dominated by the SEI film formed by the corresponding electrolyte. As displayed in Figure 10.14c,d, compared with pure Na or K, Na–K alloy and Li metal cell displayed a stable symmetrical voltage curve, which indicates that liquid Na–K metal has self-recovery property and can effectively eliminate dendrite growth. The CE of Na-ion battery and K-ion battery was 100% (Figure 10.14e,f), which implies that the reversibility of SR/Na–K redox reaction is very excellent. When commercial $Ni_{1/3}Mn_{1/3}Co_{1/3}O_3$ (NMC) was used as cathode material and Na–K alloy as anode to design mixed ion battery, the characteristics of high energy density and dendritic inhibition of the battery are also realized [51]. In Li^+/K^+ mixed electrolyte (3 : 1 Li/K-TFSI or 1 : saturated Li/K–ClO_4 salts in EC/DEC), K^+ in the anode of Na–K alloy was redoxed, and Li^+ was embedded and removed in NMC, which achieved stable cycle and outstanding rate performance.

When it comes to other alkali/Na alloys, where they have the same advantages as Na/K, Liu's group studied the liquid phase sodium metal anode for the liquid Na/BASE (β''-Al_2O_3 solid electrolyte)/S system, showing a better cycle life than pure solid sodium at 95–150 °C [52]. They also investigated the wettability of liquid Na–Cs, Na–K, and Na–Rb to the BASE. The contact angles of Na–Cs and Na–Rb to BASE are 62.4° and 88.4° at 100 °C, respectively (Figure 10.15a). However, for practical applications, wettability needs to be improved at lower temperatures.

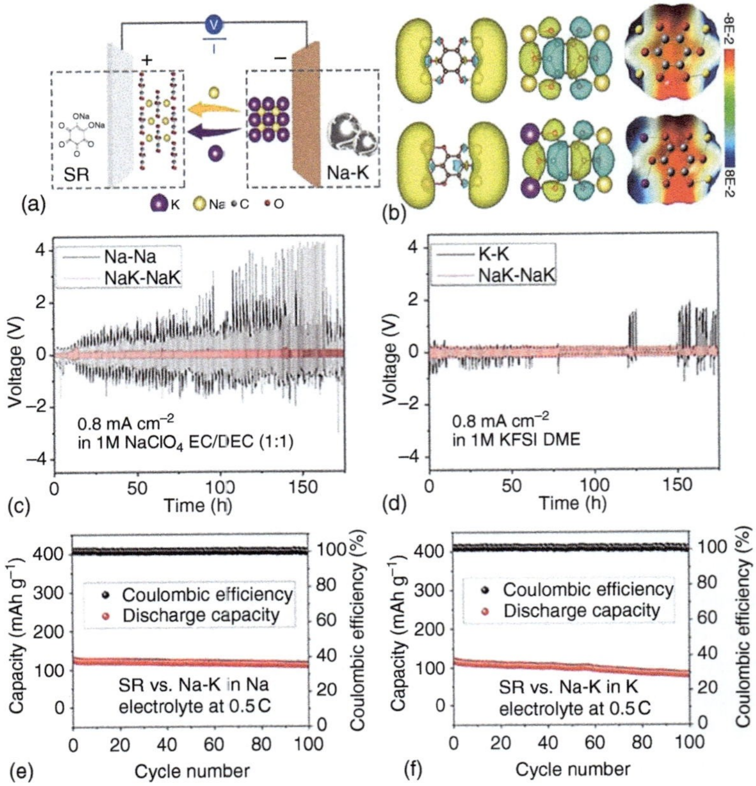

Figure 10.14 Schematic and DFT calculation. (a) Schematic of SR insertion mechanism of Na^+ and K^+, and (b) DFT calculations of cathode products, including: LUMO states, HOMO states, and the ESP distribution with the isosurface scale bar in unit of 26.568 kcal mol^{-1}, for Na- or K-intercalated SR: RNa_4 and RNa_2K_2, respectively. Battery stability measurement. Na–K anode stability in (c) Na electrolyte and (d) K electrolyte. Full cell cycling performances of SR versus Na–K in (e) Na electrolyte and (f) K electrolyte after stabilization. Source: Ding et al. [50]. Reproduced with permission, 2019, Wiley-VCH.

10.4.2 Other Metals/Na Alloys

In addition to the alloy formed by Na and alkali metals, anodes composed of other metals and Na have also been investigated. Wu and coworkers reported a novel nucleation buffer layer containing core–shell carbon–antimony nanoparticles (C@Sb NPs) (Figure 10.15b), which allows uniform deposition of sodium metals [53]. These C@Sb NPs increase the initial nucleation sites of Na by forming Sb–Na alloys, and the carbon layer keeps these nuclei stable. The asymmetric cells (Na||Cu and Na||C@Sb@ Cu) can reach a stable CE with an average value of 99.72% for nearly 738 cycles at 1 mA cm^{-2} with a capacity of 4 mAh cm^{-2} (Figure 10.15c). Mao et al. reported a simple method for depositing sodium by introducing a sodiophilic Cu@Au layer on a copper substrate as a collector [54]. By sputtering ultra-thin Au layer on Cu surface, and then alloying with Na in the incipient cycle, sodiophilic Au/Na alloy layer was introduced on the surface of Cu (Figure 10.16a). The low

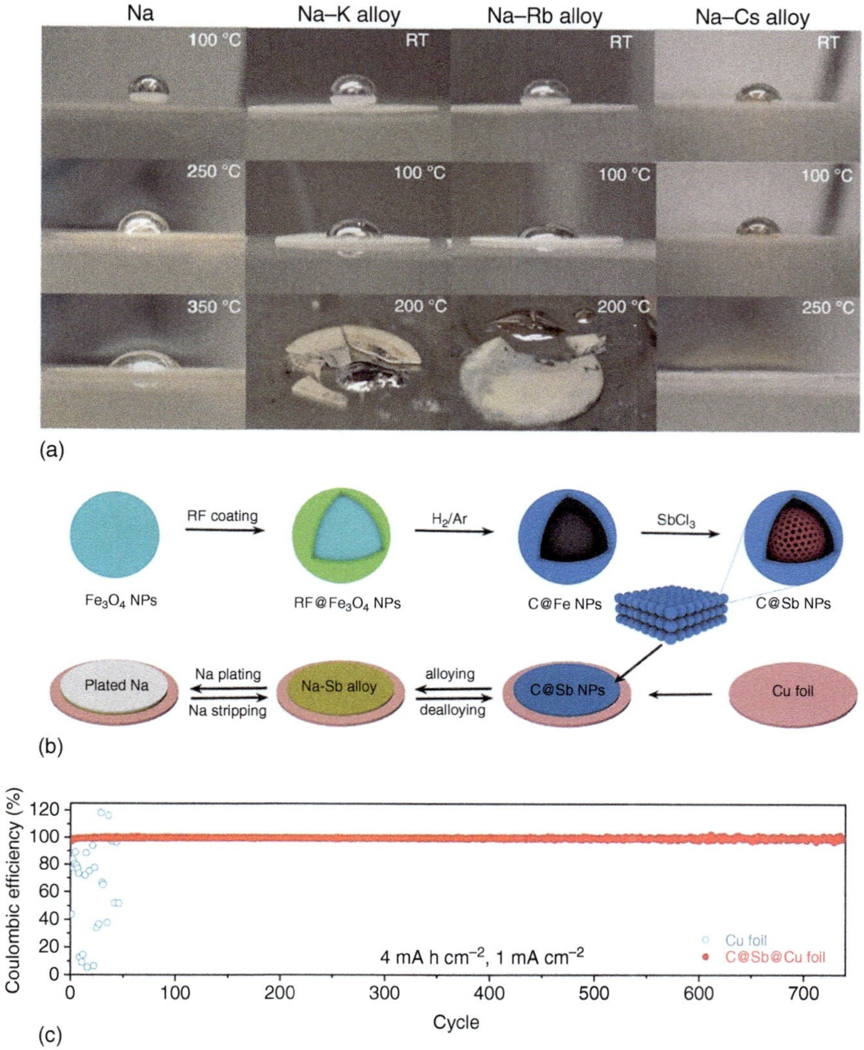

Figure 10.15 (a) Wetting behavior of Na and Na alloys on BASE surface. Source: Lu et al. [52]. Reproduced with permission, 2014, Springer Nature. (b) Schematic for the synthesis of C@Sb NPs and the sodium plating/stripping on C@Sb@ Cu foil. (c) CE of sodium plating/stripping for the cells using bare Cu and C@Sb@Cu foils. Source: Wang et al. [53]. Reproduced with permission, 2020, American Chemical Society.

binding energy of Cu and Na causes the high nucleation barrier of Na on Cu, which makes the deposition of Na uneven. However, the binding energy of Au/Na alloy layer with Na is high, which greatly increases the number of nucleation sites for Na deposition, which finally enables the deposition of uniform and dense Na layer. The results show that the sodiophilic substrate can significantly improve the deposition of Na anode and can remarkably reduce the nucleation overpotential of Na. Through the SEM images (Figure 10.16b) of Na deposition on Cu and

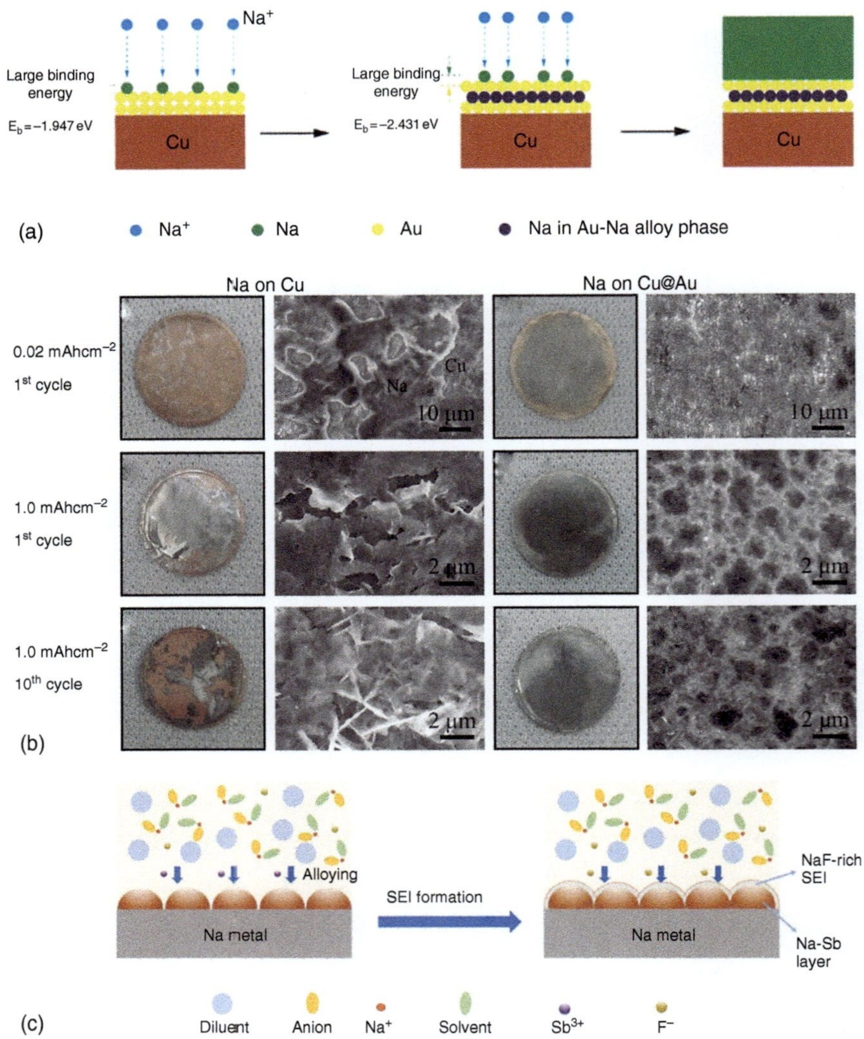

Figure 10.16 (a) Schematic diagram of Na plating on Cu@Au substrate. (b) Optical photographs of Na plating on Cu and Cu@Au. Source: Tang et al. [54]. Reproduced with permission, 2018, Elsevier. (c) Schematic diagram of SEI formation mechanism of Na–Sb layer. Source: Fang et al. [55]. Reproduced with permission, 2021, Springer.

Cu@Au substrates, it is clear that Na deposition is very uneven on the Cu surface. With repeated plating/stripping, obvious dendrite appears on the surface of Cu. In contrast, the Na deposition on the surface of Cu@Au is very uniform, which indicates that Na nucleates compactly. In addition to improving the collection fluid of the alloy to inhibit the growth of sodium dendrite, the method of forming the alloy by adding the corresponding additives and using the ion exchange reaction in the solution has also proved to be a very effective method for uniform deposition of sodium. Yu's group inhibits the growth of dendrite by adding SbF_3 to form hard

Na–Sb alloy on the surface of Na metal (Figure 10.16c) [55]. This method is simple and effective and provides a valuable reference for alloying inhibition of dendrite.

10.5 Conclusions

The development of high efficiency and stability of Na-based anode is one of the final goals of high energy and low-cost energy storage system. In summary, this chapter summarized various strategies to restrain Na dendrite growth and improve long life of Na metal anodes.

For liquid electrolytes, appropriate electrolyte solvents and salts are crucial to the formation of SEI, which is the key for Na metal anode. Among various salts and solvents, $NaPF_6$ and glymes are the best choices for electrolytes, which can facilitate the formation of an even inorganic SEI layer consisting of Na_2O and NaF that can restrain dendrite growth. In addition, increasing the concentration of electrolyte salts could also make prominent influences to the stabilization of SEI layer and suppression of dendrites. Furthermore, electrolyte additives exhibit an immediate way to adjust the components of SEI because they decompose preferentially on the anode. The development of electrolyte additives, especially those that are compatible with both Na anodes and high energy density cathodes, presents great opportunities. However, the side reactions of additives between electrolytes and other cell components as well as increased costs are the inevitable problems that have to explore. Moreover, to further solve the current problems, some novel electrolyte systems are worthy of further study.

Construction of novel current collectors and hosts for sodium metal anode is also an important engineering strategy. In brief, 3D porous structure current collectors with large surface area (up to several orders of magnitude compared to the planar current collector) can promote to dissipate the localized current density, leading to the suppression of Na dendritic growth. Furthermore, introducing sodiophilic compounds or surface functional groups on 3D carbon scaffolds with high electronic conductivity and high-surface area, can increase the sodium ion deposition sites to decrease the effective current density and result in an even Na nucleation. Moreover, the different kinds of alloy-based sodium anodes can also bring excellent performance improvement for Na metal anodes. It is a priority to develop high-speed charging/discharging with high current density and safety for Na metal batteries to ensure its practical applications.

References

1 (a) Zhao, Y., Adair, K.R., and Sun, X. (2018). *Energy Environ. Sci.* 11: 2673.
(b) Sun, B., Xiong, P., Maitra, U. et al. (2019). *Adv. Mater.* 32: 1903891.

2 Fan, L. and Li, X. (2018). *Nano Energy* 53: 630.

3 Lee, B., Paek, E., Mitlin, D., and Lee, S.W. (2019). *Chem. Rev.* 119: 5416.

4 Zhang, J., Wang, D.-W., Lv, W. et al. (2018). *Adv. Energy Mater.* 8: 1801361.

5 Iermakova, D.I., Dugas, R., Palacín, M.R., and Ponrouch, A. (2015). *J. Electrochem. Soc.* 162: A7060.
6 (a) Ma, L., Cui, J., Yao, S. et al. (2020). *Energy Storage Mater.* 27: 522. (b) Wang, Y.-X., Zhang, B., Lai, W. et al. (2017). *Adv. Energy Mater.* 7: 1602829.
7 Zhang, X.-Q., Cheng, X.-B., Chen, X. et al. (2017). *Adv. Funct. Mater.* 27: 1605989.
8 Lee, Y., Lee, J., Lee, J. et al. (2018). *ACS Appl. Mater. Interfaces* 10: 15270.
9 Seh, Z.W., Sun, J., Sun, Y., and Cui, Y. (2015). *ACS Central Sci.* 1: 449.
10 Dugas, R., Ponrouch, A., Gachot, G. et al. (2016). *J. Electrochem. Soc.* 163: A2333.
11 Gao, L., Chen, J., Liu, Y. et al. (2018). *J. Mater. Chem. A* 6: 12012.
12 Zheng, X., Bommier, C., Luo, W. et al. (2019). *Energy Storage Mater.* 16: 6.
13 (a) Su, D., Kretschmer, K., and Wang, G. (2016). *Adv. Energy Mater.* 6: 1501785. (b) Wang, C., Wang, L., Li, F. et al. (2017). *Adv. Mater.* 29: 1702212.
14 Wang, H., Wang, C., Matios, E., and Li, W. (2018). *Angew. Chem. Int. Ed.* 57: 7734.
15 Zhang, Q., Lu, Y., Miao, L. et al. (2018). *Angew. Chem. Int. Ed.* 57: 14796.
16 Shi, Q., Zhong, Y., Wu, M. et al. (2018). *Angew. Chem. Int. Ed.* 57: 9069.
17 Brissot, C., Rosso, M., Chazalviel, J.-N., and Lascaud, S. (1999). *J. Power Sources* 81-82: 925.
18 Chen, C.-Y., Kiko, T., Hosokawa, T. et al. (2016). *J. Power Sources* 332: 51.
19 Cao, R., Mishra, K., Li, X. et al. (2016). *Nano Energy* 30: 825.
20 (a) Basile, A., Makhlooghiazad, F., Yunis, R. et al. (2017). *ChemElectroChem* 4: 986. (b) Schafzahl, L., Hanzu, I., Wilkening, M., and Freunberger, S.A. (2017). *ChemSusChem* 10: 401.
21 Lee, J., Lee, Y., Lee, J. et al. (2017). *ACS Appl. Mater. Interfaces* 9: 3723.
22 Zheng, J., Chen, S., Zhao, W. et al. (2018). *ACS Energy Lett.* 3: 315.
23 Song, J., Jeong, G., Lee, A.-J. et al. (2015). *ACS Appl. Mater.* 7: 27206.
24 Ruiz-Martínez, D., Kovacs, A., and Gómez, R. (2017). *Energy Environ. Sci.* 10: 1936.
25 (a) Fukunaga, A., Nohira, T., Kozawa, Y. et al. (2012). *J. Power Sources* 209: 52. (b) Matsumoto, K., Hosokawa, T., Nohira, T. et al. (2014). *J. Power Sources* 265: 36. (c) Monti, D., Jónsson, E., Palacín, M.R., and Johansson, P. (2014). *J. Power Sources* 245: 630. (d) Forsyth, M., Girard, G.M.A., Basile, A. et al. (2016). *Electrochim. Acta* 220: 609. (e) Yang, Q., Zhang, Z., Sun, X.-G. et al. (2018). *Chem. Soc. Rev.* 47: 2020.
26 Wei, S., Choudhury, S., Xu, J. et al. (2017). *Adv. Mater.* 29: 1605512.
27 Hosokawa, T., Matsumoto, K., Nohira, T. et al. (2016). *J. Phys. Chem. C* 120: 9628.
28 Hou, Z., Wang, W., Yu, Y. et al. (2020). *Energy Storage Mater.* 24: 588.
29 He, Y., Xu, H., Shi, J. et al. (2019). *Energy Storage Mater.* 23: 418.
30 (a) Xu, Y., Menon, A.S., Harks, P.P.R.M.L. et al. (2018). *Energy Storage Mater.* 12: 69. (b) Choi, H., Hwang, D.Y., Park, J. et al. (2018). *J. Mater. Chem. A* 6: 19672.
31 Wang, C., Wang, H., Matios, E. et al. (2018). *Adv. Funct. Mater.* 28: 1802282.
32 Lu, Y., Zhang, Q., Han, M., and Chen, J. (2017). *Chem. Commun.* 53: 12910.

33 Liu, S., Tang, S., Zhang, X. et al. (2017). *Nano Lett.* 17: 5862.
34 (a) Sun, Z., Jin, S., Jin, H. et al. (2018). *Adv. Mater.* 30: 1800884. (b) Zheng, Z., Zeng, X., Ye, H. et al. (2018). *ACS Appl. Mater. Interfaces* 10: 30417.
35 Sun, B., Li, P., Zhang, J. et al. (2018). *Adv. Mater.* 30: 1801334.
36 Chu, C., Wang, N., Li, L. et al. (2019). *Energy Storage Mater.* 23: 137.
37 Ye, L., Liao, M., Zhao, T. et al. (2019). *Angew. Chem. Int. Ed.* 58: 17054.
38 (a) Go, W., Kim, M.-H., Park, J. et al. (2018). *Nano Lett.* 19: 1504. (b) Zhao, Y., Yang, X., Kuo, L.-Y. et al. (2018). *Small* 14: 1703717.
39 Luo, W., Zhang, Y., Xu, S. et al. (2017). *Nano Lett.* 17: 3792.
40 Chi, S.-S., Qi, X.-G., Hu, Y.-S., and Fan, L.-Z. (2018). *Adv. Energy Mater.* 8: 1702764.
41 (a) Fang, Y., Zhang, Y., Zhu, K. et al. (2019). *ACS Nano* 13: 14319. (b) Luo, J., Wang, C., Wang, H. et al. (2019). *Adv. Funct. Mater.* 29: 1805946.
42 Fang, Y., Lian, R., Li, H. et al. (2020). *ACS Nano* 14: 8744.
43 Zhang, Y., Wang, C., Pastel, G. et al. (2018). *Adv. Energy Mater.* 8: 1800635.
44 Kim, H., Boysen, D.A., Newhouse, J.M. et al. (2013). *Chem. Rev.* 113: 2075.
45 Skilton, R.A., Bourne, R.A., Amara, Z. et al. (2014). *Nat. Chem.* 7: 1.
46 Ma, J.-l., Meng, F.-l., Yu, Y. et al. (2019). *Nat. Chem.* 11: 64.
47 Xue, L., Gao, H., Zhou, W. et al. (2016). *Adv. Mater.* 28: 9608.
48 Zhang, L., Peng, S., Ding, Y. et al. (2019). *Energy Environ. Sci.* 12: 1989.
49 Xue, L., Gao, H., Li, Y., and Goodenough, J.B. (2018). *J. Am. Chem. Soc.* 140: 3292.
50 Ding, Y., Guo, X., Qian, Y. et al. (2019). *Adv. Mater.* 31: 1806956.
51 Guo, X., Ding, Y., Gao, H. et al. (2020). *Adv. Mater.* 32: 2000316.
52 Lu, X., Li, G., Kim, J.Y. et al. (2014). *Nat. Commun.* 5: 5.
53 Wang, G., Zhang, Y., Guo, B. et al. (2020). *Nano Lett.* 20: 4464.
54 Tang, S., Qiu, Z., Wang, X.-Y. et al. (2018). *Nano Energy* 48: 101.
55 Fang, W., Jiang, R., Zheng, H. et al. (2020). *Rare Metals* 433.

11

Organic Liquid Electrolytes for Sodium-Ion Batteries

11.1 Introduction

Electrolyte of sodium-ion batteries (SIBs) is Na^+ conductor but electron insulator, which only allows Na^+ ions transfer between cathodes and anodes. As one of the most important components, the electrolyte plays a great role on the performance of batteries, such as cycling stability, rate capability, temperature suitability, safety, etc. The chemical and electrochemical stability of electrolytes influence the electrochemical window, output voltage and cycling lifetime of SIBs. The ionic conductivity of electrolyte and compatibility to electrode materials determine the migration rates of Na^+ ions between anode and cathode, which has a very large impact on the rate capability and power density of the batteries. More importantly, the electrolyte is the main cause of safety issues, since it generally involves flammable organic solvents. Thus, developing high-performance and safe electrolytes is one of the most popular topics for SIBs. The electrolyte physicochemical properties are determined by the compositions or its fine structure, and thus most of the research studies aim to optimize the microstructures of the electrolyte components to achieve better performance. In this chapter, organic liquid electrolytes are summarized and discussed, including Na salts, solvents, and functional additives.

11.2 Electrolyte Properties

Organic liquid electrolytes are currently the most commonly used electrolytes for SIBs and mainly contain sodium salts as solute and organic solvents to dissolve sodium salts. In addition, small amount of functional additives, which helps to improve the performance, is possibly included. The evaluation on liquid electrolyte is mainly based on the following properties:

(1) *Liquid range.* Electrolytes should be liquid at the operating temperature range, with low melting points and high boiling points. Also, the viscosity should be low even at low temperature to guarantee fast ion motion.
(2) *Conductivity.* A high ionic conductivity of the electrolyte allows Na ions to migrate fast, which is critical for the power density of the battery. On the other

Sodium-Ion Batteries: Energy Storage Materials and Technologies, First Edition. Yan Yu.

hand, electron must be not conducted in the electrolyte between the cathode and anode.

(3) *Electrochemical and chemical stability*. In principle, electrolytes should be inert to all the other component of SIBs, including active (electrodes) and inactive (separator, binder, current collector, etc.) parts. However, in practice, electrolytes are favorable to forming an interfacial layer on surfaces of both cathode and anode. The electrochemical stability of electrolytes depends on whether protective layer is stable or not.

In addition, low cost and low toxicity are also necessary for large-scale applications of electrolytes. These properties are mainly determined by the electrolyte compositions, including sodium salts, solvents, and functional additives. Here, different compositions in organic liquid electrolytes and how they affect the performance are going to be discussed below.

11.3 Sodium Salts

As a main component of the electrolyte, sodium salt has great effects on the ionic conductivity and electrochemical stability. Several key points should be taken into consideration in the selection of sodium salt: (i) high solubility is supposed to acquire adequate charge carriers [1]; (ii) the salt should have appropriate oxidative and reductive stability [2]; (iii) the compatibility between the electrolyte and other components in the battery is largely determined by the chemical inertness of the anions [3]; (iv) the thermal stability of the salts influences the battery safety [4]; (v) the salt should be low-cost and environmentally friendly.

Na^+ has larger ionic radius than Li^+, so that sodium salts are soluble in solvents with low dielectric constant and more sodium salts are available than analogue lithium salts in the electrolyte [1]. Usually electronegative peripheral ligands coupled with stable central atoms constitute the anion group of sodium salts, which is beneficial to the delocalized negative charges for Na^+ transfer [5]. The advantages and disadvantages of electrolytes largely depend on the anions. Hence the regulation of anion group is essential in developing new sodium salts. The structure and properties of some sodium salts are shown in Table 11.1. $NaClO_4$ has been successfully applied because of its fast ion transport, low cost, and good compatibility with electrode [12]. However, several intrinsic drawbacks, including high toxicity, explosion hazard, and high-water content, hinder the practical application of $NaClO_4$. Similar to $LiPF_6$, $NaPF_6$ is commonly used in SIBs and demonstrates high conductivity in PC-based electrolytes [6]. However, it also inherits the shortcomings of $LiPF_6$, such as low decomposition temperature and toxicity along with HF release. Moreover, the solubility of $NaPF_6$ in many single solvents is low. For instance, the solubility of $NaPF_6$ in ethylene carbonate (EC), propylene carbonate (PC), diethyl carbonate (DEC), and dimethyl carbonate (DMC) is only 1.4, 1.0, 0.8, and 0.6 M, respectively. Therefore, it is necessary to use EC-containing multicomponent solvent for preparing $NaPF_6$-based electrolytes. Anions with large

Table 11.1 Structure, physical, and chemical characters of various sodium salts.

Salt	Molecular structure of anion	Molecular weight	Al-corrosion	Decomposition temperature, T_m (°C)	Conductivity, $\sigma^{a)}$ (m S cm^{-1})	Toxicity	Reference
$NaClO_4$	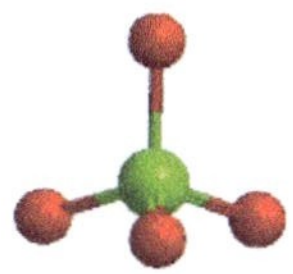	122.4	No	480	6.4	Highly toxic	[6]
$NaPF_6$	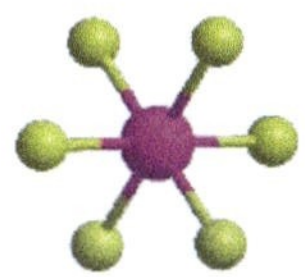	167.9	No	300	7.98	Low toxicity	[6]
$NaBF_4$	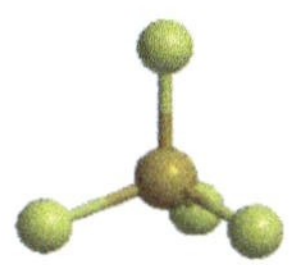	109.8	No	384	–	Highly toxic	[1]
NaFSI	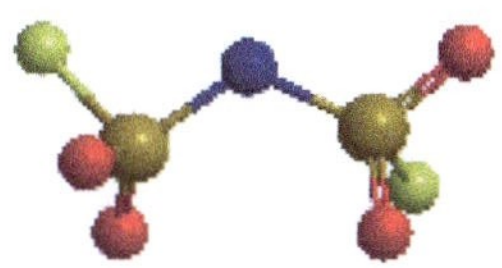	203.3	High	118	–	Nontoxic	[1]

(*continued*)

Table 11.1 (Continued)

Salt	Molecular structure of anion	Molecular weight	Al-corrosion	Decomposition temperature, T_m (°C)	Conductivity, $\sigma^{a)}$ (m S cm^{-1})	Toxicity	Reference
NaTFSI	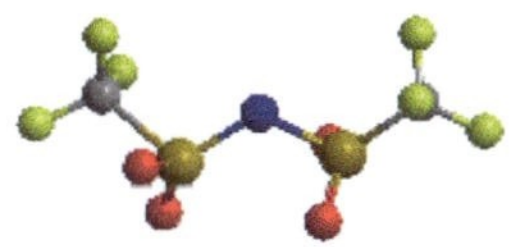	303.1	High	257	6.2	Nontoxic	[6]
NaOTf	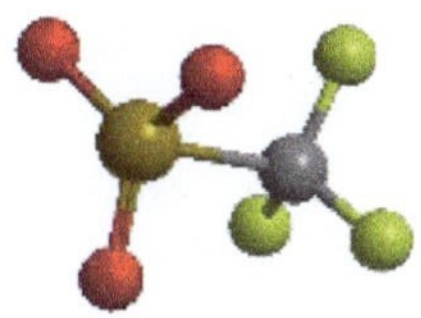	172.1	High	248	3 (1 mol NaOTf dissolved in 1 kg TEGDME)	Nontoxic	[7b]
NaDFOB	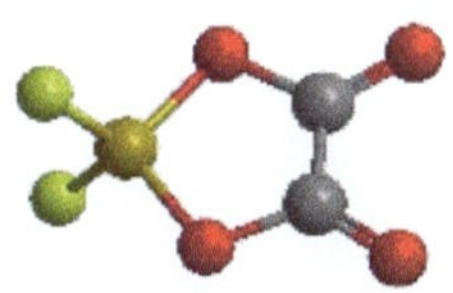	159.8	No	–	≈7	Nontoxic	[8]
NaTDI	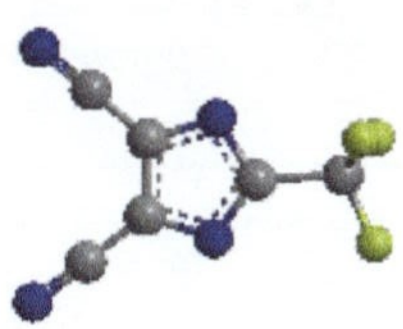	208.1	No	–	4.47	Nontoxic	[9]

NaPDI	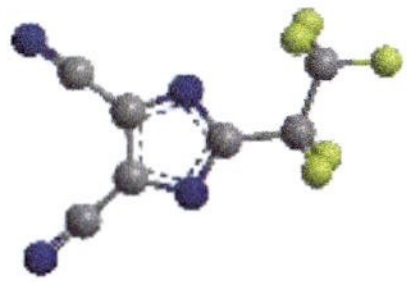	258.1	No	–	4.65	Nontoxic	[9]
NaBOB	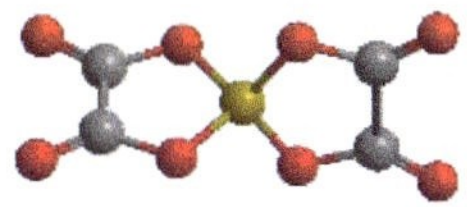	209.8	No	345	0.256 (0.025 M)	Nontoxic	[10]
NaBSB	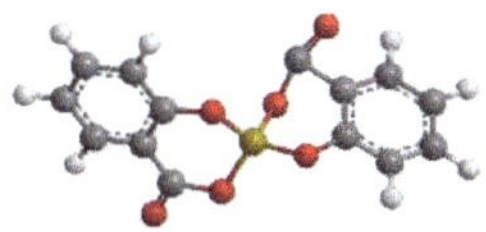	306.0	No	353	0.239 (0.025 M)	Nontoxic	[10]
NaBDSB	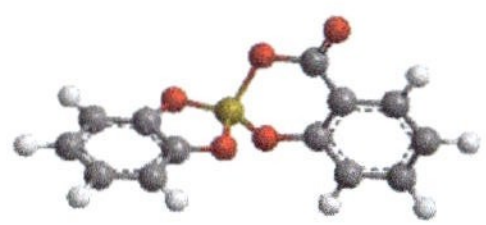	278.0	No	304	0.071 (0.025 M)	Nontoxic	[10]
$NaBPh_4$	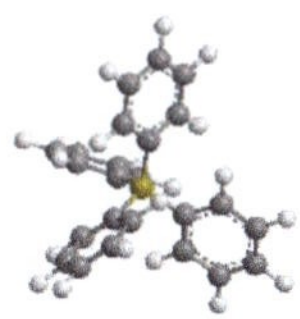	342	No	–	6.3 (0.5 M in DME)	Nontoxic	[11]

a) The ionic conductivity of 1M sodium salt was measured in pure PC solvent at 25 °C.

ionic radius and good chemical stability, such as bis(fluorosulfonyl) imide (FSI^-), bis(trifluoromethylsulfonyl) imide ($TFSI^-$) [6], tetrafluoroborate (BF_4^-) [13], and trifluoromethane sulfonate (OTf^-) [7], have large ionic radius and can realize high ionic conductivity of the electrolyte. The main defect to NaFSI and NaTFSI is that Al corrosion occurs significantly in carbonate-based electrolytes [14]. Goktas et al. investigated the electrolyte properties of $NaPF_6$, $NaClO_4$, NaOTf, NaFSI, and NaTFSI salts in diglyme-based solvent [15]. The performances of Na|graphite cells and Na|Na symmetric cells demonstrate that NaOTf and $NaPF_6$ are superior to $NaClO_4$ and NaFSI (Figure 11.1), as the latter two salts bring more parasitic reactions and increase the cell resistance. NaTFSI exhibits the worst cycling stability and Coulombic efficiency (Figure 11.1), since the thickness of the interfacial layer on the electrode increased a lot due to the excessive side reactions during cycling. In this diglyme-based electrolyte, the performance of these salts can be ranked as: NaOTf $\geq$ $NaPF_6$ > $NaClO_4$ > NaFSI $\gg$ NaTFSI.

The high electronegativity and inductive effects enable halogen atoms, e.g. F, to be used in the formula of sodium salts despite their high toxicity. To acquire higher conductivity and better safety, a series of sodium salts containing F atom have been developed. Sodium difluoro(oxalato) borate (NaDFOB) as a novel salt

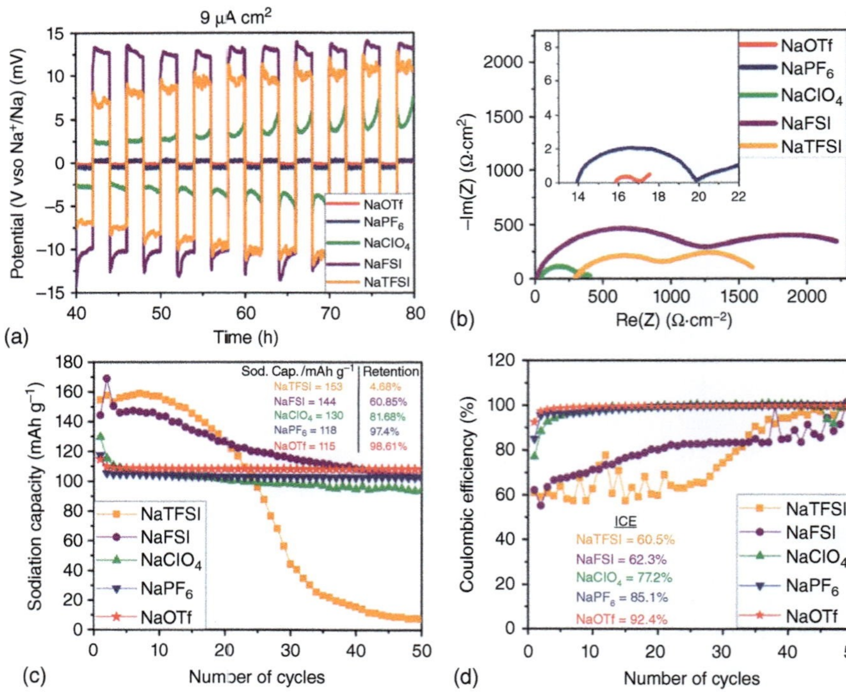

Figure 11.1 (a) Plating/stripping performance of Na|Na symmetric cells at 9 μA cm^{-2} and (b) the impedance after cycling in diglyme-based electrolytes with different salts. (c) Cycle stability and (d) Coulombic efficiency of graphite|Na cells in diglyme-based electrolyte with different salts. Source: Goktas et al. [15]. Reproduced with permission, 2019 American Chemical Society.

is synthesized and exhibits satisfactory conductivity in the electrolyte [16]. On the other hand, NaDFOB possesses better compatibility with various solvents and exhibits improved electrochemical performance compared with that of $NaPF_6$ and $NaClO_4$. Salts containing F and heterocyclic ring structures possess similar advantages. For example, both sodium 4,5-dicyano-2-(trifluoromethyl)imidazolate (NaTDI) and sodium 4,5-dicyano-2-(pentafluoroethyl)imidazolate (NaPDI) [17] are thermally stable even under 300 °C and can construct a stable interphase protection layer on Al foil, preventing the corrosion. These salts with F atoms and heterocyclic ring structures also show a wide electrochemically stable window (ESW) to over 4.5 V vs. Na^+/Na to obtain salts with better stability and less toxicity, organic-ligand doping method has been adopted to fabricate halogen-free sodium salts (Figure 11.2) [10]. Several new salts, such as sodium bis(oxalate)borate (NaBOB), sodium (salicylate-benzenediol)borate (NaBDSB), sodium bis[salicylato (2-)]-borate (NaBSB), and sodium tetraphenylborate ($NaBPh_4$), are synthesized and exhibit high ionic conductivity [10]. From the cyclic voltammetry (CV) analysis, a preferential reduction peak located at around 1.2 V is observed for the NaBOB electrolyte in the first cathodic sweep [18] and the significantly reduced peak current in the later cycles proves the effective passivation. In addition, from the first CV curve, the increased current at the voltage of above 3 V is attributed to the decomposition of some reductive products, while the oxidative decomposition delays to approximately 4 V and is greatly suppressed during the latter two cycles. For the $NaPF_6$ electrolyte, the reduction peak occurs at around 1 V in the first scan, then shifts to lower voltages in subsequent scans [18]. One defect of this kind of sodium salts is their low conductivity because of low solubility.

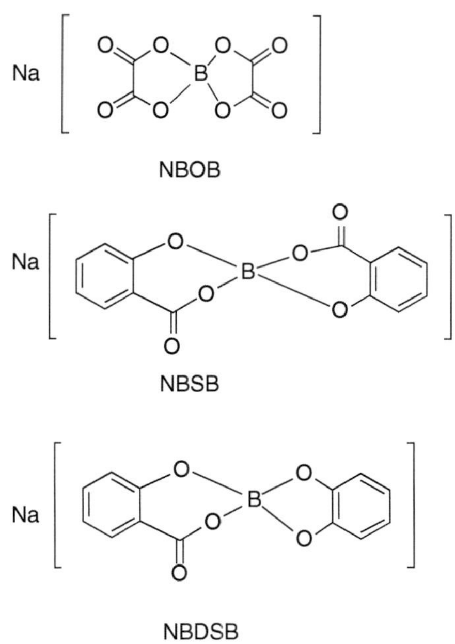

Figure 11.2 The structures of NaBOB, NaBSB and NaBDSB molecules. Source: Ge et al. [10]. Reproduced with permission, 2014, Elsevier.

Dual-salt electrolytes are very promising for their potential to form stable interphase on both anode and cathode, which has been widely investigated in Li metal batteries (LMBs). According to Jiao's research, the capacity retention of Li | $LiNi_{1/3}Co_{1/3}Mn_{1/3}O_2$ batteries could reach 90.5% after 300 cycles in the dual-salt electrolyte containing 2 M LiTFSI and 2 M LiDFOB [19]. Xiang et al. found that the dual-salt electrolyte containing LiTFSI and LiBOB significantly strengthens the stability of solid electrode interface (SEI) layer on Li metal anode [20]. LiBOB improves the stability of SEI and suppresses the corrosion of Al foil, while LiTFSI increases the ionic conductivity, which synergistically help to form a robust and conductive SEI. Based on LiTFSI/LiBOB electrolyte, the capacity retention of the Li|$LiNi_{0.8}Co_{0.15}Al_{0.05}O_2$ cell reaches 80%, while the cell with only $LiPF_6$ electrolyte shows the retention of 15%. The formation of polycarbonates in SEI is obtained with further adding 0.05 M $LiPF_6$ in the dual-salt electrolyte. Other SEI component and isolated Li are cohered together and stuck on Li metal anode with polycarbonates [21]. The life span of Li | $LiNi_{0.4}Co_{0.2}Mn_{0.4}O_2$ cells could reach 450 cycles at 1.75 mA cm^{-2} with this reinforced SEI. Successful application of dual-salt electrolytes under high and conventional concentrations in LMBs has been confirmed. Some sodium salts with similar structures have also been synthesized, but the feasibility of dual-salt electrolytes for sodium metal batteries (SMBs) is still rarely discussed. It is foreseeable that the application of dual-salt electrolyte in SIBs is also very promising.

11.4 Solvents

The solvents in SIBs are basically developed from the successful experiences in lithium-ion batteries [22]. In practical applications, there will be some differences in the selection criteria of solvents due to the different purposes, but basically obey the following principles:

(1) The organic solvent should be inert to the electrode. It does not react with the cathode and anode during the cycling of the battery [23].
(2) The solvent should have the ability to dissolve enough salt to ensure high conductivity of the electrolyte.
(3) The solvent needs to remain liquid within a wide temperature range [24].
(4) The components of solvents need to be environmentally friendly [25].

11.4.1 Carbonate Ester-Based Electrolytes

There are mainly two types of cyclic carbonate solvents, which are often used in commercial electrolyte of 1 M (mol L^{-1}) $NaClO_4$-EC/PC. The linear carbonates, such as ethyl methyl carbonate (EMC), DEC, and DMC, are often mixed with cyclic carbonate solvents to get high ionic conductivity and wide liquid range temperature [23]. Their physical and chemical characteristics are summarized in Table 11.2.

DEC, DMC, and EMC are three commonly used linear carbonates. The viscosity and melting point of cyclic carbonates are higher than those of linear carbonates,

Table 11.2 Physicochemical characteristics of the carbonate solvents for SIBs.

Solvents	Dielectric constant at 25 °C	Viscosity at 25 °C (cP)	Heat of vaporization (kJ mol^{-1})	HOMO value (eV)	LUMO value (eV)
EC	89.78	1.90 (40 °C)	51.68	−0.2585	−0.0177
PC	64.92	2.53	45.73	−0.2547	−0.0149
DMC	3.107	0.59	36.53–38.56	−0.2488	−0.0091
EMC	2.958	0.65	34.63	−0.2457	−0.0062
DEC	2.805	0.75	41.96–44.76	−0.2426	−0.0036

thus the linear carbonates are usually combined with cyclic carbonates in order to obtain better performance of electrolytes.

The melting point of DMC is 4.6 °C, with a boiling point of 90 °C, and a flash point of 18 °C. It is slightly toxic and can form an azeotrope with water or alcohol. The molecular structure of DMC contains some functional parts, so it is reactive and can be applied as carbonylation and methylation reagents, gasoline additives, and raw materials for polycarbonate synthesis.

DEC has a similar structure to DMC. It is a liquid at room temperature and has a very low melting point of −74.3 °C. Its toxicity is stronger than that of DMC. Both DMC and DEC have low viscosity (0.58 and 0.75 mPa·s) with a dielectric constant of 3.107 and 2.805, respectively.

EMC is another linear carbonate solvent, which has similar physical and chemical characteristic to DMC and DEC. However, it has poor thermal stability and easily undergoes a transesterification reaction under alkaline conditions to DMC and DEC.

11.4.2 Carboxylate Ester-Based Electrolytes

PC and EC are the two important organic solvents in the electrolyte, both of which belong to the category of cyclic carbonates. PC is a colorless and transparent liquid with aromatic odor at ambient temperature and pressure, and has a freezing point of −49.27 °C. In addition, PC also exhibits high chemical and electrochemical stabilities. However, PC has slight hygroscopicity, which will have an adverse effect on the control of moisture in the electrolyte. EC is a colorless crystal at room temperature and its structure is analogous to PC. The dielectric constant of EC is much higher than that of PC, even higher than that of water. Furthermore, EC has high thermal stability, and only decomposition occurs when the temperature reaches 200 °C. However, it is easily decomposed under alkaline conditions.

EC is one of the most prominent solvents that attributes to its excellent properties. For instance, EC has the highest dielectric constant [26], which enables EC to dissolve enough salts. Because of the role of EC, a sturdy protective layer can be formed on the surfaces of various electrodes [6, 27], which has been verified in SIBs [28]. PC is also used in SIBs, which remains liquid within wide temperature range and has lower dielectric constant (64.92) [6]. Furthermore, it has good compatibility with hard carbon (HC), which is a common anode in SIBs. On the contrary, for the traditional graphite anode in PC-based electrolytes, the structure is always damaged due to the coprecipitation of PC solvent [27b, 29]. The cosolvent systems are thus desirable, and the synergy enables practical enhancements in conductivity, viscosity, electrochemical stability, and even the safety of SIBs [6].

As far as the present situation is concerned, various studies have been carried out on the modification of electrolytes. In 2012, Ponrouch tested the performances of 1 M NaTFSI, $NaClO_4$, and $NaPF_6$ with different solvents. In Figure 11.3a [6], it revealed similar ionic conductivity and viscosity of various salts in PC. Whereas for the $NaClO_4$ salt in different solvents (Figure 11.3b), more obvious changes of viscosity and ionic conductivity can be perceived. Compared with single solvent electrolytes, the ionic conductivity in binary mixed solvent is significantly higher. Figure 11.3c shows the thermal stability and electrochemical window of different electrolytes, which are affected by both Na salts and solvents. It is found that most Na salts and organic solvent are not stable against Na metal, which should be reduced to form a SEI layer, and their cycle performances are closely related to the stability of SEI layer. Figure 11.3d,e shows the cycle performance of HC | Na cells in $NaClO_4$– based electrolytes. The capacity remains 185 mAh g^{-1} after 70 cycles in EC/PC, which is better than that in other solvents, indicating more stable SEI layer in EC/PC electrolyte.

In 2014, Komaba and coworkers further studied the effect of $NaClO_4$ on HC|Na cells in different solvents. As shown in Figure 11.4 [30], HC | Na batteries in PC

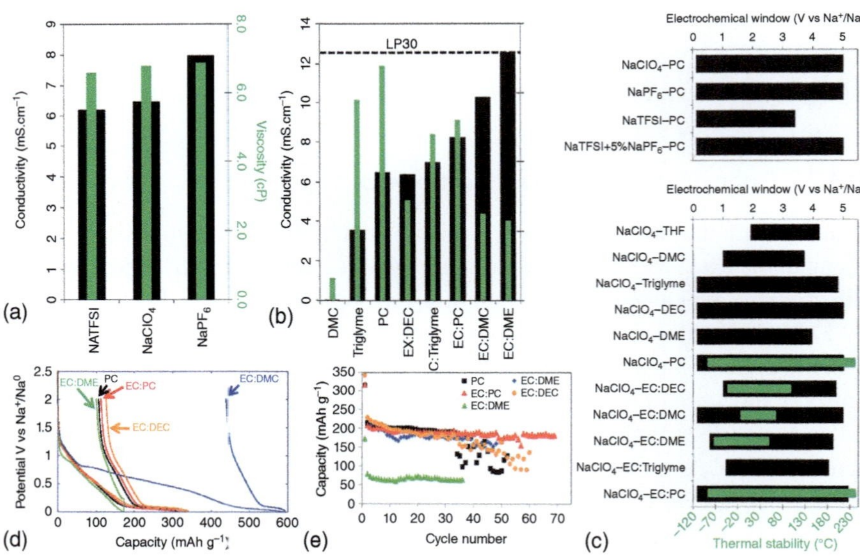

Figure 11.3 (a) Conductivity and viscosity of different salts in PC, and (b) 1 mol L^{-1} $NaClO_4$ in different solvents. (c) Cutoff voltage and thermal stability of different electrolytes. (d) Capacity–voltage curves of HC | Na batteries in various solvents with 1 M $NaClO_4$ during the initial cycle, and (e) their cycling performances. Source: Ponrouch et al. [6]. Reproduced with permission, 2012, Royal Society of Chemistry.

or EC-DEC exhibited superior electrochemical performances compared with other electrolytes. Furthermore, they investigated different salts with pure PC solvents. The cycling performance of electrolytes using $NaPF_6$ and NaTFSI is more superior to $NaClO_4$ [6, 30]. These results indicate that different salts need to be used with different solvents.

The most important cyclic carboxylic ester solvent is γ-butyrolactone (BL), and it can remain liquid in a wide temperature range. However, BL is easily decomposed in water, and it has high toxicity with unsatisfactory cycle efficiency. Linear carboxylic esters mainly include methyl formate (MF), methyl acetate (MA), methyl butyrate (MB), ethyl propionate, etc. The freezing point of these esters is lower than that of carbonates, and their viscosity is lower, which can significantly improve the low temperature performance of the electrolyte. In 2018, Tarascon and coworkers reported a complete survey that aimed to assess the $Na_3V_2(PO_4)_2F_3$|HC performance in the cyclic and linear carbonates (Figure 11.5) [31]. They discussed pathways for decomposition reactions and proposed shuttle mechanism, namely the dissolved species stemming from the reduction of linear carbonates on carbon electrode shuttle to the positive side to be oxidized, hence causing a huge amount of Na ions lost in the full cell. These findings provide fundamental clues to rationally design new electrolyte formulation, which will lead to a stable rather than a soluble SEI. This research is successively pursued with guidance from computational calculations and a novel electrolyte formulation relied on the synergy between additives, enabling high interfacial stability even at high temperature (55 °C).

The compatibility of the cathode and the electrolyte is also worthy of consideration. Vidal-Abarca et al. demonstrated that Na|Na_2FePO_4F battery showed the highest capacity retention in $NaPF_6$-EC/DEC (Figure 11.6) [32]. In addition, they developed a better electrolyte of 1 M $NaClO_4$-EMS (ethyl methane sulfonate) +2% 1-fluoroethylene carbonate (FEC) with excellent electrochemical performance [34]. This electrolyte is able to maintain a stable cycling when the voltage is up to 5.6 V. Bhide et al. also investigated various salts in EC/DMC for Na | $Na_{0.7}CoO_2$ cells [33]. The better cycling ability in $NaPF_6$-EC/DMC electrolyte is believed to be due to stable SEI.

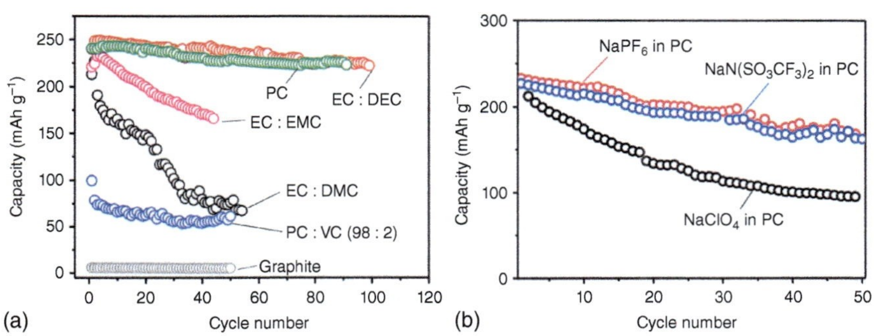

Figure 11.4 Performances of HC | Na batteries in (a) various solvents with 1 mol L^{-1} $NaClO_4$ and (b) different salts in PC. Source: Dahbi et al. [30]. Reproduced with permission, 2014, Royal Society of Chemistry.

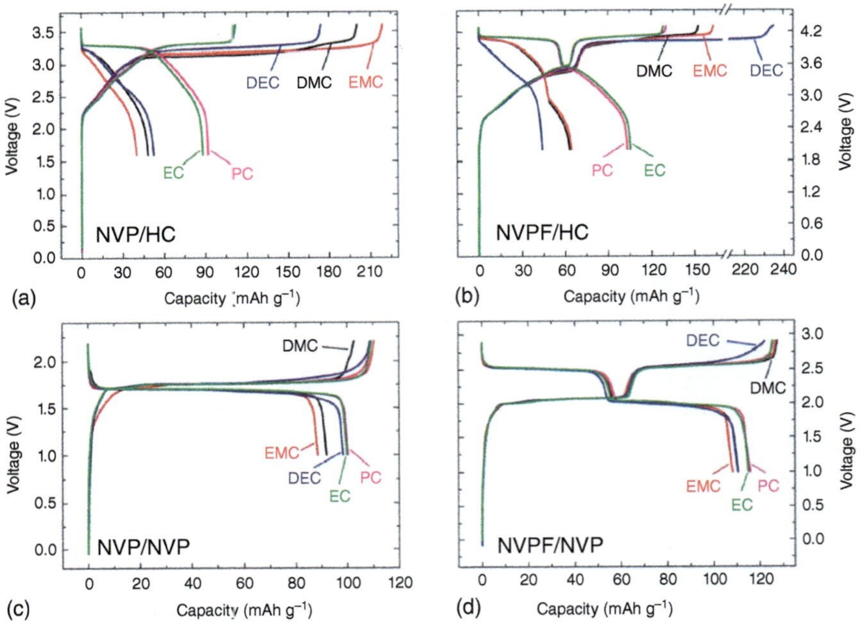

Figure 11.5 The first charge–discharge curves of different solvents in various batteries at 0.1 C. Source: Yan et al. [31]. Reproduced with permission, 2018, Electrochemical Society.

Nowadays, the safety of SIBs has aroused widespread attention. In 2016, Cao and coworkers came up with a safer SIB using trimethyl phosphate (TMP) [35]. The performance of this kind of electrolyte was evaluated by combustion experiments, cyclic voltammetry, ionic conductivity, and so on (Figure 11.7). These results indicate that TMP-based electrolytes containing FEC are nonflammable and have a wide electrochemical window. The $NaNi_{0.35}Mn_{0.35}Fe_{0.3}O_2$ | Na cell in TMP + 10 vol% FEC electrolytes demonstrated to be nonflammable, with considerable capacity and cyclability, which ensures the safety of SIBs for practical applications.

In 2020, Ma and coworkers concentrated on electrolyte optimization. The basic physicochemical properties of electrolytes in PC or EC, including ionic conductivity, wettability, viscosity, and thermochemical stability, were systematically studied by using $NaPF_6$ as the salt and EMC, DMC, and DEC as mixed solvents. The electrolyte properties were tested in $NaNi_{1/3}Fe_{1/3}Mn_{1/3}O_2$ (NFM) | HC cells. It is found that the electrolytes using PC performed more superior to EC-based electrolytes in terms of capacity retention and safety. The PC/EMC electrolyte, combined with a variety of additives, further facilitates the formation of a dense interfacial layer, thereby reducing the side reactions and inhibiting chemical dissolution of transition metal ions. Therefore, the 0.8 M $NaPF_6$-PC/EMC mixed with additives (such as FEC) can not only make the reversibly cycle of NFM|HC cell more than 2500 times, but also show excellent safety [36].

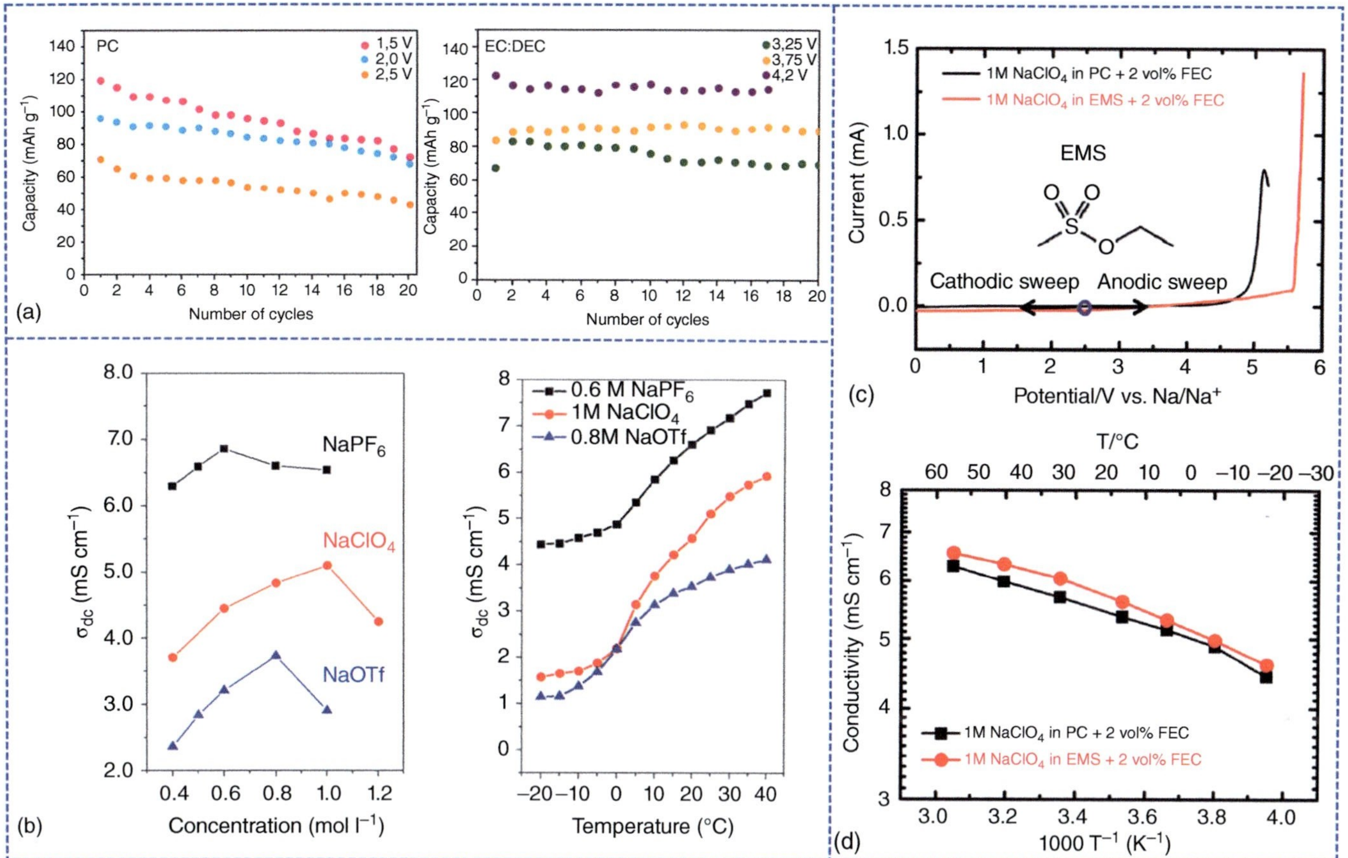

Figure 11.6 (a) Cycling performance at 0.05 C in PC (left) and EC/DEC (right) at various cutoff voltages. Source: Vidal-Abarca et al. [32]. Reproduced with permission, 2012, Elsevier. (b) Ionic conductivity of electrolytes containing various salts in EC/DMC (left) and ionic conductivity of different electrolytes (right). Source: Bhide et al. [33]. Reproduced with permission, 2013, Royal Society of Chemistry. (c) LSV of a sodium cell using two kinds of electrolytes and (d) conductivity of different electrolytes at different temperatures. Source: Oh et al. [34]. Reproduced with permission, 2014, American Chemical Society.

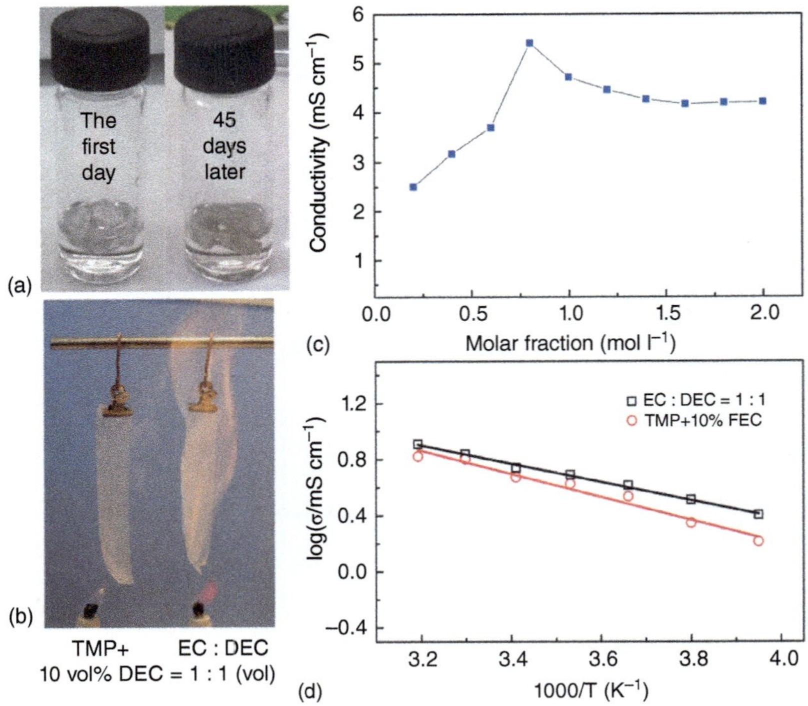

Figure 11.7 (a) The stability of the TMP + 10% FEC at 25 °C. (b) Combustion experiment of the TMP and carbonate electrolyte. (c) The ionic conductivities of the TMP-based electrolytes with different concentrations at 25 °C. (d) The ionic conductivities of 0.8 M $NaPF_6$ in TMP and 1 M $NaPF_6$ in EC/DEC (1 : 1) at different temperatures. Source: Zeng et al. [35]. Licensed under CC BY 4.0.

11.4.3 Ether-Based Electrolytes

In LIBs, the ether electrolyte has a poor passivation effect on the anode and is also unstable up to 4 V vs. Li/Li^+, so it is considered to be poor solvent for electrolyte [23]. However, in recent years, due to their higher reduction potential, ether electrolyte has attracted researchers' attention again. Compared with ester-based electrolytes, the SEI layer derived from ether-based electrolyte is thinner and the initial Coulombic efficiency (ICE) is higher [37].

In many anodes with ether-based electrolytes, the storage performance of Na has been improved [26a, 38]. Cohn et al. proved that highly crystalline few-layered graphene can be used in ether-based electrolyte system, in which the diglyme solvent shell promotes quick ion intercalation (Figure 11.8a,b) [38a]. Xu's group found a scheme to promote the cycling performance of HC anode (Figure 11.8c) [38b]. Specifically, the HC anode was stable with a "foreign SEI," which was predesigned from the ester-based electrolyte (Figure 11.8d), which achieved unprecedented performance [39]. Yang's team started a research on the difference of SEI layer on reduced graphene oxide (rGO) between ester and ether-based electrolytes. The results show that the reversible capacity and cycle life of SIBs after

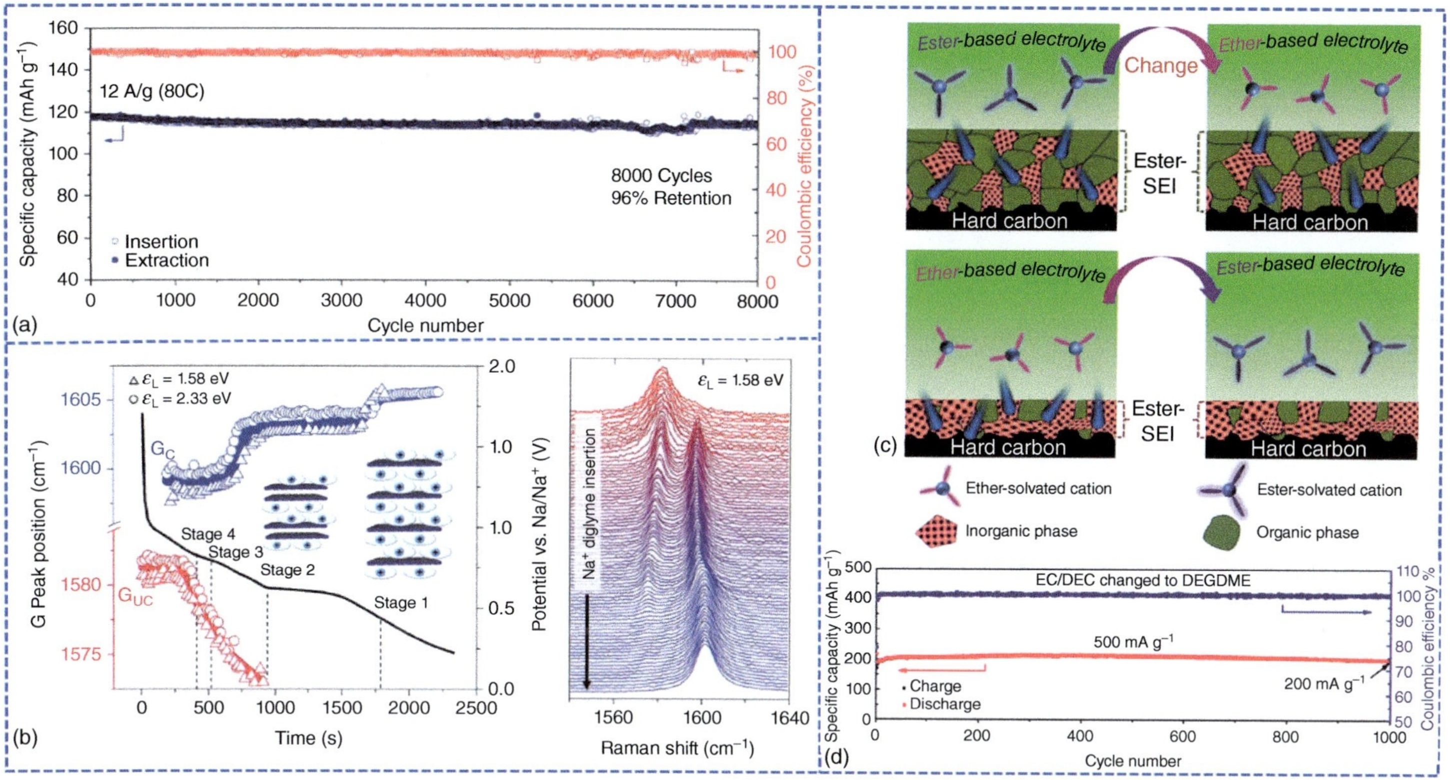

Figure 11.8 (a) Cycling life span in ether electrolyte, and (b) in situ Raman measurement of the anode during discharge. Source: Cohn et al. [38a]. Reproduced with permission, 2016, American Chemical Society. (c) Schematic of Na^+ transfer to the SEI layers in various solvents, and (d) cycling stability of the HC with a preformed SEI in ester electrolyte. Source: Bai et al. [38b]. Reproduced with permission, 2018, Elsevier.

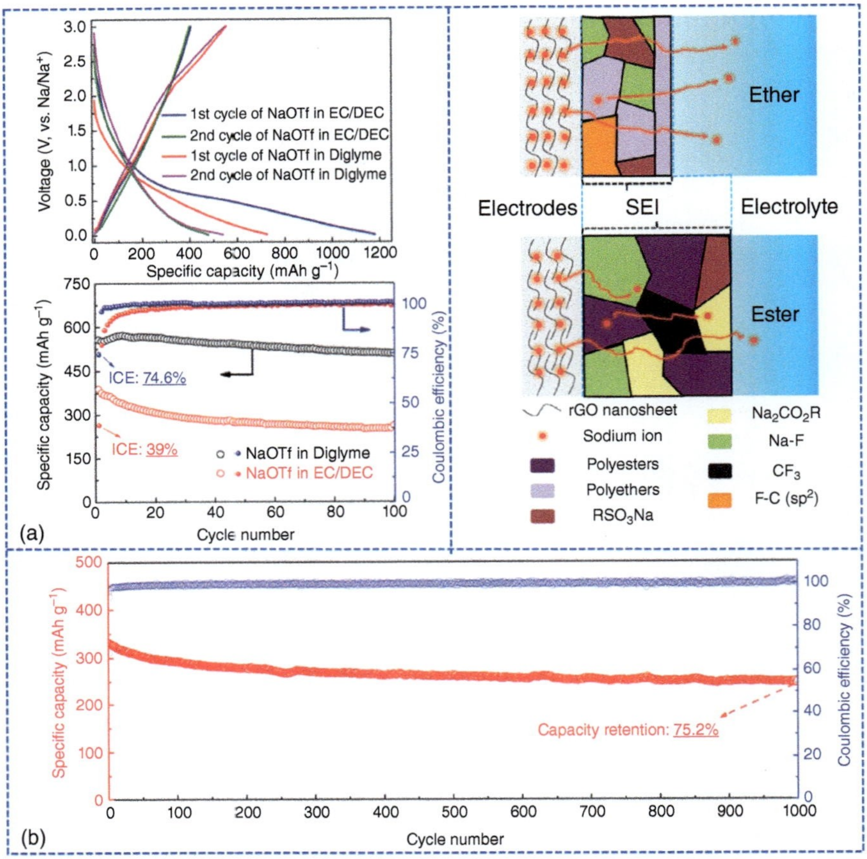

Figure 11.9 (a) Voltage–capacity curves and cycling life span of rGO in different electrolytes at 100 mA g^{-1}. (b) Cycling life span of rGO in different electrolytes. (c) Schematic diagram of the various components of SEI in different electrolytes. Source: Zhang et al. [40]. Reproduced with permission, 2017, Royal Society of Chemistry.

cycling in a electrolyte of 1 M sodium trifluoromethanesulfonate (NaOTF)-diglyme are better than those of the batteries cycling in 1 M NaOTF-EC/DEC electrolyte (Figure 11.9a,b) [40]. The poor electrochemical properties in ester-based electrolyte are mainly due to the excessive side reactions, which finally lead to the irregular SEI film (Figure 11.9c). Moreover, they studied the progress of the interface layer between the ether solvents and TiO_2 anode [41], and found that when TiO_2 is coupled with ether solvents, superior capacity retention is achieved (Figure 11.10a). In situ physical characterization shows that the sodiation process of TiO_2 in ether electrolytes is more complete than that in carbonate electrolytes (Figure 11.10b). In addition, replacing the EC/DEC with ether electrolyte can obtain a lower potential barrier and interface charge transfer resistance, thereby significantly improving the rate performance of TiO_2 (Figure 11.10c). Since the ether-based electrolyte forms a denser SEI than the EC/DEC-based electrolyte, its exterior is composed of a denser inorganic layer, which facilitates fast ion transport [40, 41]. At the same time, it was

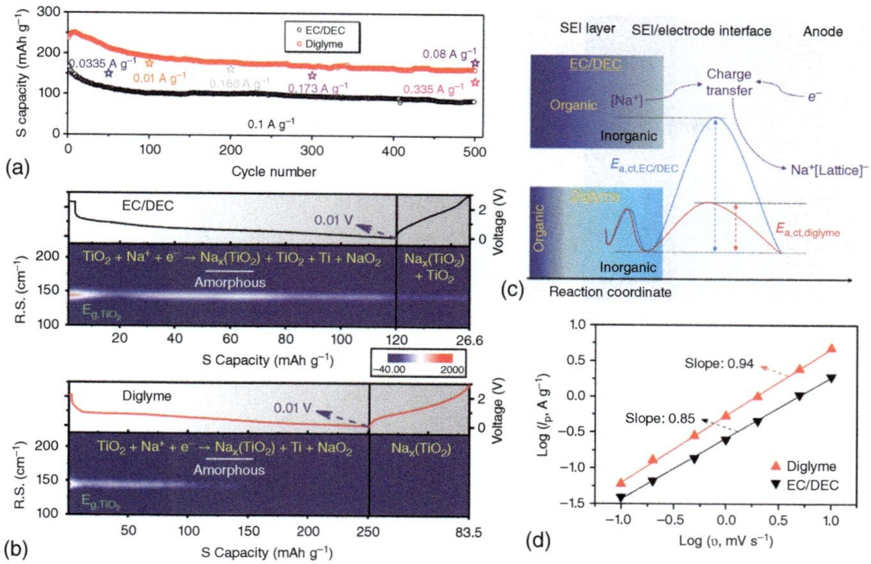

Figure 11.10 (a) Cycling performance and (b) in situ Raman spectroscopy of the TiO_2 anode during the initial cycle in different electrolytes. (c) Comparison of SEI composition and energy barriers in various electrolytes. (d) Dynamic analysis of the TiO_2 anode in different electrolytes. Source: Li et al. [41]. Licensed under CC BY 4.0.

also found that faster kinetics further increases the pseudo-capacitance behavior of Na^+ intercalation (Figure 11.10c,d), which greatly promotes the electrochemical stability of TiO_2.

In 2019, Santhanagopalan and coworkers demonstrated the influence of electrolyte formulations on the lithium titanate (LTO) anode. They used electrolytes with $NaPF_6$ salt dissolved in three different solvents (diglyme, PC, and EC/DMC) to study the electrochemical performance of surface-modified LTO samples (Figure 11.11) [42]. LTO-diglyme-based electrode has unprecedented rate ability (up to 300 C), cycle stability (1000 cycles), and wide operating temperature range (−10 to 55 °C), which is the best performance of LTO-based anodes in SIBs so far. Through X-ray photoelectron spectroscopy (XPS), a detailed surface chemical analysis of the original electrode and cycled electrode in each electrolyte was carried out to understand the evolution and formation of the SEI. The results reveal that a thin SEI is formed in the DiG-based electrolyte and has excellent rate and cycle performance. On the contrary, ester-based electrolytes form a thick SEI, resulting in poor cycling stability.

In 2020, Jiao reported a high-stability sodium metal anode based on ether electrolyte. The cell delivered stable cycling over 400 cycles, and the Coulombic efficiency was up to 99.93% (Figure 11.12) [43]. The results show that the hybrid structure containing B-O species and NaF in the ultrathin SEI is the main reason for its excellent electrochemical performance. In addition, Na | $Na_3V_2(PO_4)_3$ cells successfully achieved stable long-cycle performance, promoting the practical application of sodium metal batteries.

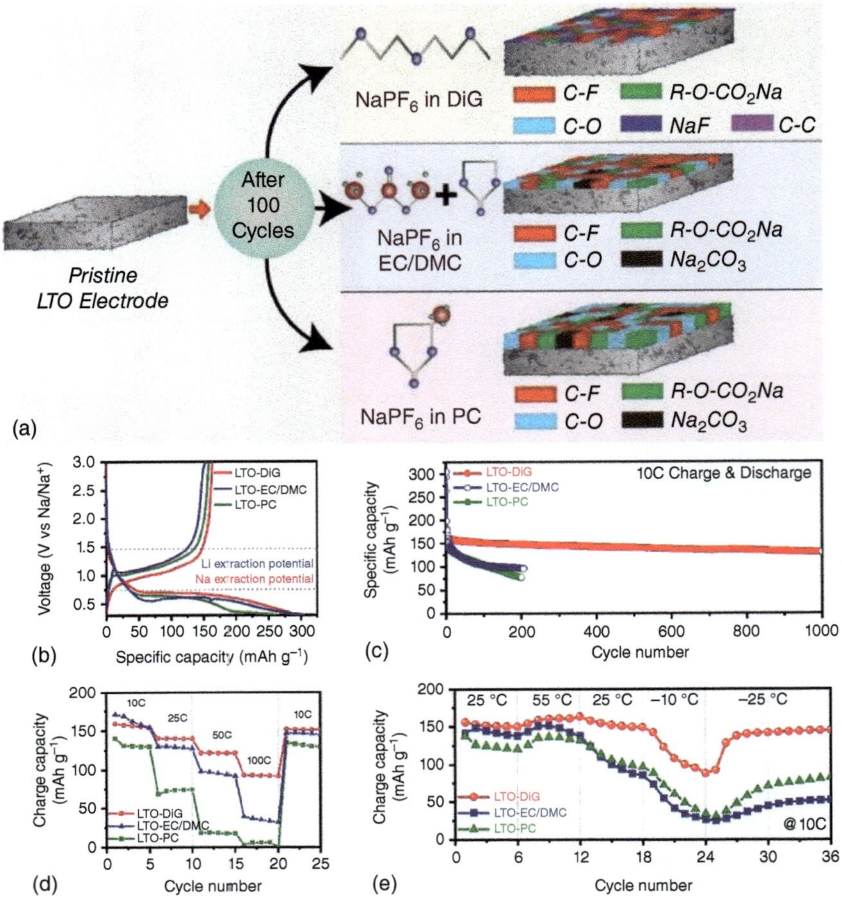

Figure 11.11 (a) Schematic of interfacial engineering of LTO in three electrolytes, (b) initial curve at 10 C, (c) long cycling performance, (d) rate performance, and (e) temperature-dependent cycling performance. Source: Reproduced with permission from Gangaja et al. [42]. Reproduced with permission, 2019, Royal Society of Chemistry.

In 2020, Hong and coworkers studied the deposition of Na^+ in carbon anodes in pure ether electrolyte and the kinetic characteristics of ion diffusion on the electrode surface. Although the porous nanocarbon layer has a small spacing, the specific surface area is high. It thus displays a high initial Coulombic efficiency of 91.1% and an excellent sodium ion storage rate in the ether-based electrolyte (Figure 11.13) [44]. This is because the energy barrier in ether-based electrolytes is significantly reduced, which is less than one-third of that in ester electrolytes. Through the DFT calculation, it can be found that the Gibbs free energy of the solvation and desolvation of Na^+ is relatively low, so the use of ether electrolyte can stabilize the electrode/electrolyte interface.

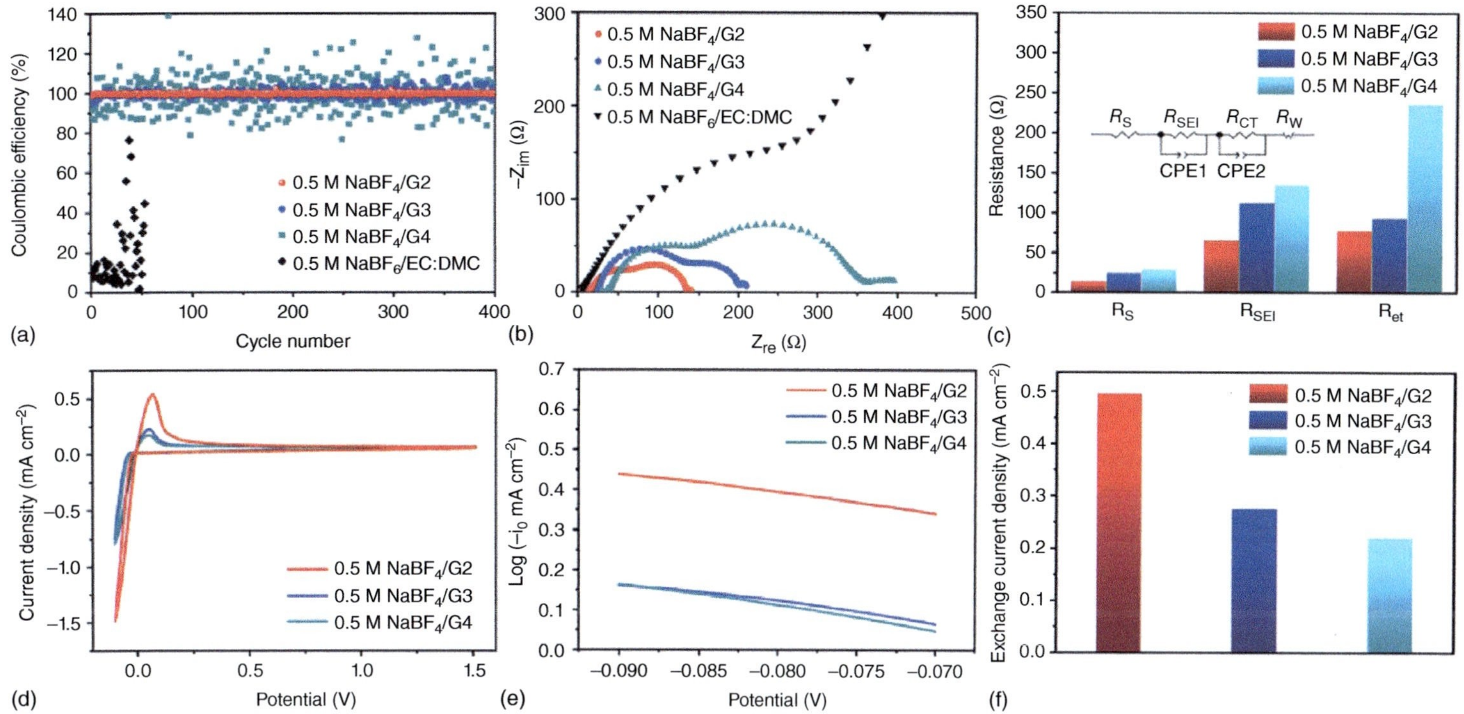

Figure 11.12 Electrochemical performance of sodium metal anode in different electrolytes. (a) Coulombic efficiency at 0.5 mA cm^{-2}. (b) Nyquist plots of Na | Cu batteries. (c) The corresponding resistance values. (d) CV curves, (e) Tafel curves, and (f) the exchange current density. Source: Wang et al. [43]. Reproduced with permission, 2020, Wiley-VCH Verlag.

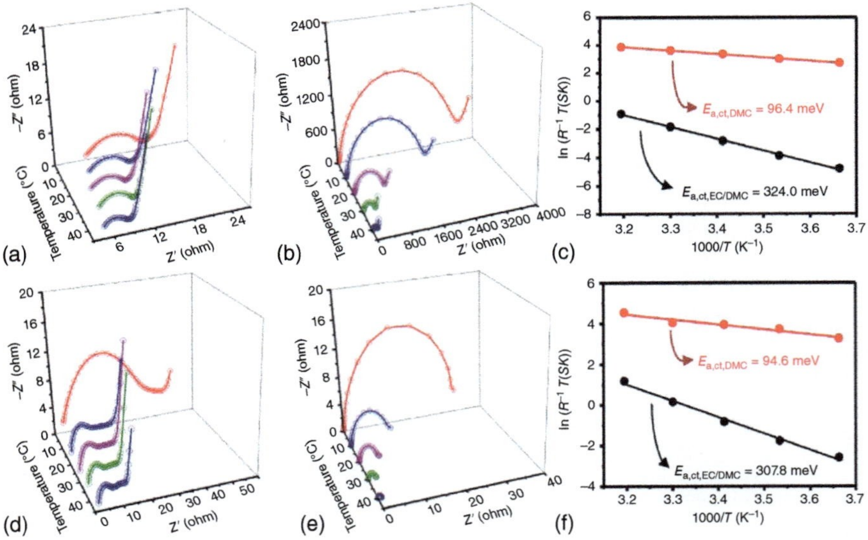

Figure 11.13 Nyquist plots of the NVP at different temperatures in (a) ether and (b) ester-based solvents. (c) Arrhenius plot of activation energies of NVP with two electrolytes. Nyquist plots of PNC at different temperatures in (d) ether and (e) ester-based electrolyte. (f) Arrhenius plot of activation energies for PNC in two solvents. Source: Zhen et al. [44]. Reproduced with permission, 2020, Elsevier BV.

11.5 Functional Additives

Additives are considered to be the easiest, and most effective way to regulate targeted properties of the electrolyte. The choice of the additives may vary from system to system depended on their prospective applications. The additives herein are sorted on the basis of their core functions, such as electrode passivation, safety enhancer, mechanical property improver, etc. An upper limit of 10 vol% or wt% is used in this section to define additives.

11.5.1 Basic Characteristics of Additives

As described above, the features for additives vary according to their intended functions. For example, the HOMO/LUMO energies, chemical hardness (η), bonding energy, and electron affinity (E_A) are valuable parameters to help track/screen film-forming additives effectively. Theoretically, the chemical potential of positive electrode (μ_C) is higher than the HOMO energy of an ideal electrolyte, and the chemical potential of negative electrode (μ_A) should be lower than the LUMO energy of an ideal electrolyte to minimize parasitic reactions between electrodes and electrolytes (Figure 11.14a). Therefore, Na salts and solvents are supposed to be stable kinetically at the polarized electrode to obtain a maximum ESW of the electrolyte [45]. Practically, according to Peled's research, μ_A of most anode materials is higher

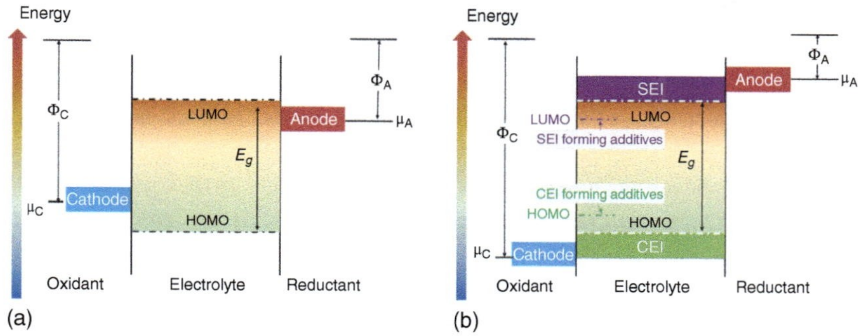

Figure 11.14 Schematic illustration of the chemical energy for the (a) ideal electrolyte and (b) practical electrolyte. Source: Eshetu et al. [45]. Reproduced with permission, 2018, John Wiley and Sons.

than the LUMO energy of electrolyte, resulting in reductive decomposition of electrolyte and the formation of SEI layer on anode [46]. The high-voltage cathodes are usually strong oxidizers that can cause electrolyte decomposition and CEI formation, which is similar to the reduction behavior of the anode [47].

The target to design electrolyte additives is establishing stable electrode/electrolyte interphase on cathode and anode. The E_g of the additives is expected to be in the range of the HOMO–LUMO gap of electrolyte (Figure 11.14b). The LUMO energy of the additive should be lower than that of polymers/solvents and salts, allowing the SEI-building additives to be reduced preferentially to tailor the property of the SEI layer. For the CEI-forming additives, the HOMO energy should be higher than that of the electrolyte. Nevertheless, the orbital energy illustrated above does not guarantee a stable interphase.

Other additives are often designed according to the specific purpose. For instance, to improve the safety of SIBs, flame-retardant additives composed of fluorine, phosphorus, or a combination of them that can trap free radicals generated in the process of electrolyte combustion, are developed [48]. In addition, it is necessary to consider the following basic criterions when screening suitable additives:

(1) The additive should be facile to synthesize and environmentally friendly.
(2) The cost of additives should be low.
(3) The electrolyte additives should have the desired effect, then it is supposed to have minimal passive influence on the whole performance of electrolyte.
(4) It is significantly important for additives to possess the positive synergistic effect.

11.5.2 Additives for Na-Ion Batteries

Electrolyte/electrode interphase (EEI) is regarded as a decisive factor for the overall battery performance. Thus, tremendous efforts have been devoted to developing additives that can construct an effective interphase. The EEI layers with several nanometers thickness allow Na^+ ions to migrate facilely but prevent the electrons transporting. Herein, the EEI-forming additives and associated effects are discussed.

11.5.2.1 SEI-Forming Additives for Anodes

Additives are employed to protect original maternal electrolytes from decomposition via a sacrificial reduction. The reduction potential of the additives should be higher than that of the electrolyte components, thus they can be firstly reduced to form the SEI layer [49]. Several filming additives studied so far are shown in Table 11.3.

In addition to protecting the electrolyte composition from further reduction, the SEI formed must meet a number of requirements, including: (i) during the process of

Table 11.3 Several SEI-forming additives and their effects on SIB anodes.

Additive	Content	Electrolyte	Anode material	Improvement of capacity retention (%)	Reference
FEC	2 vol%	1 M $NaClO_4$-PC	HC	+46	[50]
FEC	0.5%	1 M $NaPF_6$-PC	HC	+43	[51a]
FEC	2%	1 M $NaClO_4$-EC/PC	HC	+10	[52]
FEC	2%	1 M $LiPF_6$-EC/PC	HC	+1.2	[52]
$RbPF_6$	0.05 M	0.8 M $NaPF_6$-PC/ EC + 2 wt% FEC	HC	+15	[53]
$CsPF_6$	0.05 M	0.8 M NaPF6-PC/ EC + 2 wt% FEC	HC	+17	[53]
EMImFSI	10%	1 M $NaPF_6$-EC/PC	HC	+2.8	[52]
FEC	2 vol%	1 M $NaClO_4$-PC	N-doped hollow carbon nanospheres	+60	[51b]
FEC	5 wt%	1 M $NaClO_4$-PC	AlSb	+67	[54]
FEC	5 wt%	1 M $NaClO_4$-PC	Ge	+94	[55]
FEC	5%	1 M $NaPF_6$-EC/DEC	Sb/C	+94	[56]
FEC	5 wt%	1 M $NaClO_4$-EC/DEC	SiC-Sb-C	+50	[57b]
FEC	5 wt%	1 M $NaPF_6$-EC/DMC	Sn-S-C	+78	[57a]
FEC	5%	1 M $NaClO_4$-EC/DEC	Sn_4P_3	+97	[58]
FEC	5 vol%	1 M $NaClO_4$-PC/EC	Mo_3Sb_7/AB	+37	[57c]
FEC	2 vol%	1 M $NaClO_4$-EC/PC	FeSb-TiC-AB	+52	[57e]
FEC	2 vol%	1 M $NaClO_4$-PC/EC	Cu_6Sn_5-TiC-C	+60	[57d]
FEC	10%	1 M $NaPF_6$-EC/DEC	Amorphous phosphorus/ CB	+11	[59c]
FEC	5%	1 M $NaClO_4$-EC/PC	Sb_2Te_3-TiC	+63	[60]
VC	1 vol%	1 M $NaPF_6$-EC/DEC	Black phosphorus/AB	+27	[59a]
VC	2 wt%	1 M $NaClO_4$-PC	MoO_2	+67	[61]

(de-)sodiation, an appropriate path is provided for the transfer of sodium ions; (ii) it also works as a barrier for other species at the same time, such as solvent molecules; and (iii) it is required to exhibit good chemical stability in all battery environments [62]. According to the above requirements, FEC is the most widely studied and effective film-forming additive, which can improve the life span and property of SIBs.

Designing and developing novel anode materials is of great significance to replace LIBs with SIBs. The anodes currently used in SIBs mainly include carbonaceous, alloying, conversion ones, etc.

Graphite is a well-known anode material for LIBs, but it exhibits a poor insert ability for sodium storage [63]. Then, HC receives great attention. However, the low capacity and sodium insertion issues limit the application of HC [50]. FEC is applied to form a robust SEI layer on HC to prevent parasitic reactions and rapid volume attenuation, prolonging the life span [51]. Theoretical and experimental results have jointly demonstrated that the insertion of Na^+ ions into HC is controlled by the SEI formation. $RbPF_6$ and $CsPF_6$ are able to form a interphase rich in P— and C—F categories on HC, which renders better capacity retention and lower impedance [53]. Ionic liquids (ILs) are also employed as additives or electrolytes in SIBs. However, the application of ILs as additives or cosolvents is confined by their high cost and high viscosity. The effect of 1-ethyl-3-methylimidazolium bis(fluoromethanesulfonyl)imide (EMImFSI) as additive (2 wt%) or cosolvent (10 wt%) was explored in EC/PC electrolyte with various Na salts [52]. The experimental results were compared with the electrolytes with FEC as additive (Figure 11.15). The CV results indicate that FSI^- anions, mainly coming from IL, introduce new cathodic peaks related to the reduction of FSI^- anions. The cycling performance is greatly determined by the EEI induced by the degradation of anions. It is evidenced that EMImFSI introduces a larger polarization in the discharge/charge curves, and improved Coulombic efficiency and capacity retention are only observed in $NaPF_6$-based electrolyte. EMImFSI performs worse in $NaClO_4$ electrolyte, and the beneficial effect in the first cycle is eliminated after 25 cycles in NaFSI electrolytes. As a contrast, 2 wt% FEC affords better cycling stability and Coulombic efficiency with lower price.

For intermetallic alloys, the large volume fluctuation during (de-)sodiation causes severe problems. The widely studied Ge [55], Sb [54], and Sn [64] anodes and their composites call for robust SEI layer with high mechanical stability and ionic conductivity. The SEI layer rich in NaF and Na_2CO_3 alleviates the continuous electrode degradation and prevents further electrolyte degradation [55, 64]. However, it is still unable to cope with the large volume changes [64, 65].

Composites are thus developed, in which the active material is distributed in conductive substrate evenly, having a buffering effect to accommodate the volume change of composites. The property requirements for the additives remain similar. The employment of FEC can also render a thin SEI layer with low impedance, which avoids particle agglomeration [57]. In addition, improved rate performance can be acquired under the protection of the high ionic-conductive SEI layer induced by FEC [57d, 64].

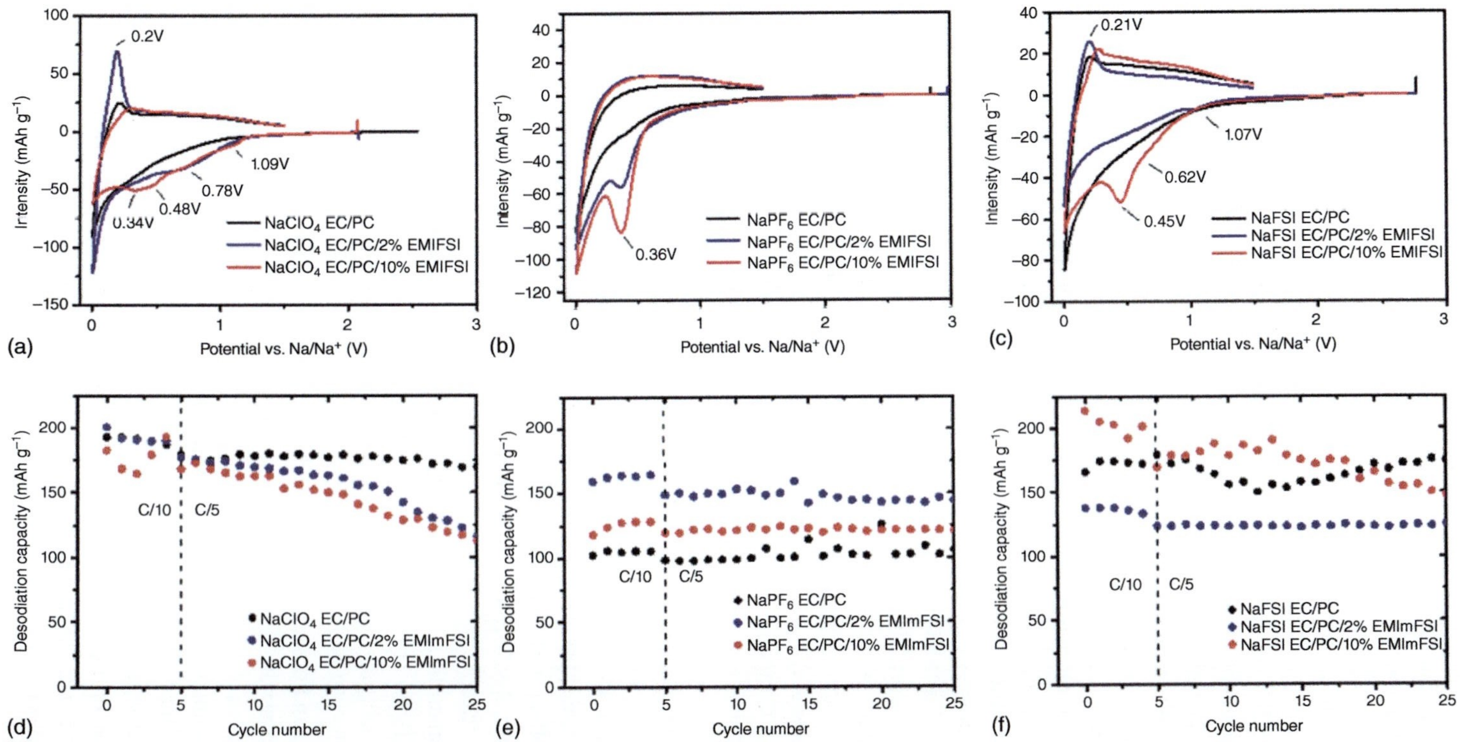

Figure 11.15 (a-c) The cyclic voltammetry measurements and (d-f) cycling stabilities of Na|HC half cells in EMImFSI/EC/PC-based electrolytes. Source: Benchakar et al. [52]. Reproduced with permission, 2020, Elsevier.

The use of phosphorus-containing components would result in ionic Na_3P, which will help to alleviate volume change and maintain high capacity [59]. This is a very promising idea to construct high-quality SEI. Introducing FEC will alter the main components of SEI formed in phosphorus-containing electrolyte by preferential reduction [59b, c].

Apart from FEC, some fluorinated [50] and non-fluorinated [57c] carbonates are also investigated extensively as additives in SIBs. Vinyl carbonate (VC) also performs well in certain systems [59a]. Time-of-flight secondary ion mass spectrometry (TOF-SIMS) reveals that the VC contains high organic compounds content and commonly observed inorganic species, including NaF, Na_2CO_3, and Na_2O in SEI layer. The $CHCHOCO_2^-$ free radical originated from VC decomposition can undergo further coordination with Na^+ ions [59a]. The cycling stability of MoO_2 anode in SIB can be greatly improved by adding VC (Figure 11.16) [61]. After 1000 cycles, the capacity of the MoO_2 electrode is retained by 96.5% with 2 wt% VC, while without VC, the capacity retention is only 20.1%. Due to the serious electrolyte decomposition, in the absent of VC, the polarization of MoO_2 | Na cells is increased significantly, which impairs the cycling performance. The SEI layer formed by sacrificial reduction of VC also demonstrates the properties of inhibiting the decomposition of electrolyte and improving the stability of MoO_2.

11.5.2.2 CEI-Forming Additives for Cathodes

Over the past few years, researchers have put considerable effort into developing new types of electrolytes for room temperature SIB cathodes [66]. Similar to the additives for anodes, the effect of cathode additives is also quite related to the characteristics of cathode materials. Some important results are shown in Table 11.4.

Na_xMO_2 (M = transition metal) is one of the most studied cathode materials due to its high energy density [73]. Komaba et al. pointed out that FEC can also form a passivation layer and inhibit the oxidation of electrolyte [50, 74]. Generally, FEC is used as an additive to achieve smaller polarization and overall improvement in electrochemical performance, including Coulombic efficiency, discharge capacity, and capacity retention [59b, 67, 74].

Another category of cathode material is polyanionic compound [$NaM(XO_n)$] (M refers to transition metal atom and X = B, S, P, or Si) [73]. Many polyanionic materials have been reported, especially for the iron- and vanadium-based compounds. For carbon-coated $Na_xFe_y(P_2O_7)_2$ cathode, the CEI layer formed by FEC is beneficial to the inhibition of transition metal dissolution, thus improving the cycling stability [68]. Adding FEC into DEC-based electrolyte can effectively enhance the electrochemical performance of Na | $Na_4Fe_3(PO_4)_2(P_2O_7)$ cells [69].

With the addition of FEC, $Na_3V_2(PO_4)_3$ (NVP), and $Na_3V_2(PO_4)_2F_3$ (NVPF) cathodes exhibit improved capacity retention, Coulombic efficiency, and power density [74, 75]. VC also performs well in EC/PC electrolyte with Na_2MnSiO_4 cathode [70]. The CEI layer formed by VC possesses low impedance and can inhibit the dissolution of transition metal. In 10 M $NaClO_4$ aqueous solution, 2 wt% VC can form a robust CEI layer on cathode, thus preventing parasitic reactions and dissolution of transition metal [71].

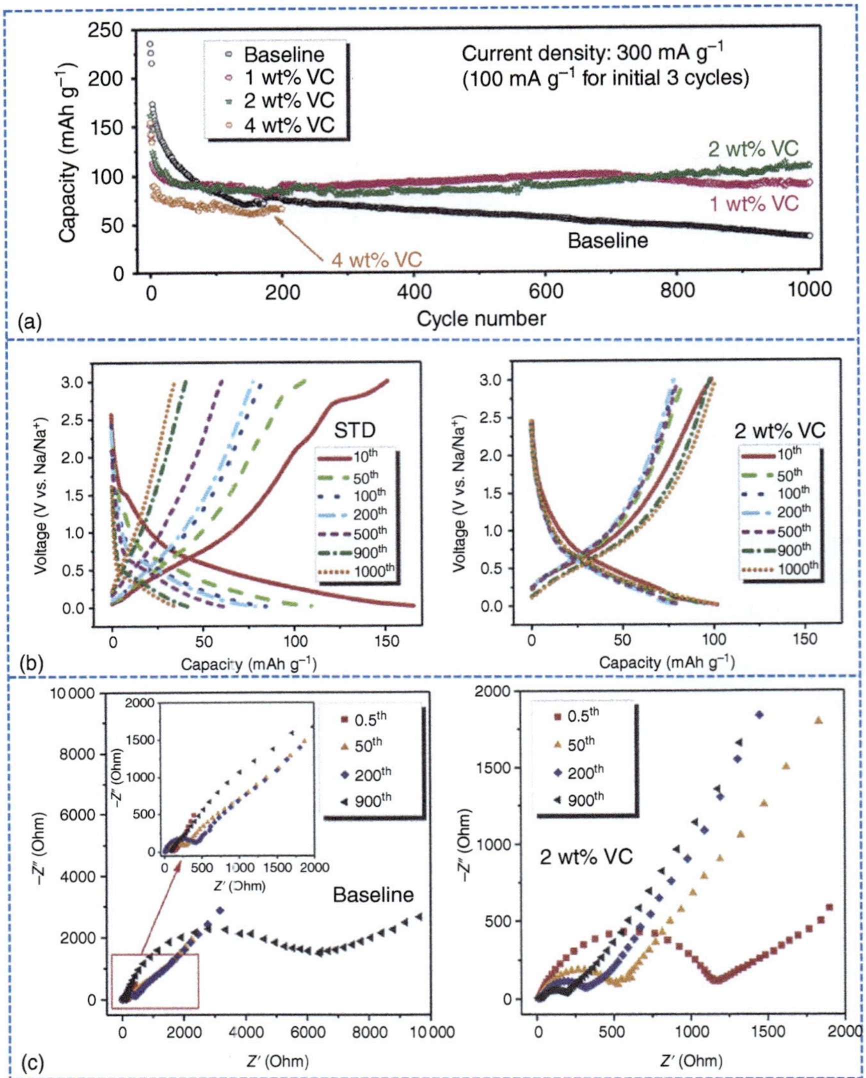

Figure 11.16 Comparison of (a) cyclic stability, (b) selected charge/discharge curves, and (c) EIS results of the MoO_2|Na half-cells with VC. Source: Zhang et al. [61]. Reproduced with permission, 2020, Elsevier.

Rapid Na^+ transport can be acquired in sodium-based Prussian blues (PB) [76]. The optimized electrochemical performance of $Na_{0.75}Fe_{2.08}(CN)_6\cdot 3.4H_2O$ cathode is obtained with 2 wt% FEC [77]. This is attributed to the conductive and mechanically stable CEI layer formed by FEC, which improves the Coulombic efficiency and cycle stability [78].

It is also a promising way to expand the rapid development of sodium-ion technology by adding appropriate additives to form a stable interface both on cathode and

Table 11.4 Several CEI-forming additives and their effects on SIB cathodes.

Additive	Content	Electrolyte	Cathode material	Improvement of capacity retention (%)	Reference
FEC	10%	1 M $NaClO_4$-PC	$NaNi_{1/2}Mn_{1/2}O_2$	+37	[50]
FEC	2 vol%	1 M $NaClO_4$-PC	$NaNi_{0.5}Mn_{0.5}O_2$	+24	[67a]
FEC	2 wt%	1 M $NaClO_4$-EC/PC	$Na_{0.44}MnO_2$	+8	[67c]
FEC	2 wt%	1 M $NaPF_6$-EC/DEC	$Na_{0.44}MnO_2$	+3	[67c]
FEC	5 vol%	1 M $NaClO_4$-EC/PC	$Na_{3.32}Fe_{2.34}(P_2O_7)_2$	+7	[68b]
FEC	5 wt%	0.5 M $NaClO_4$-EC/PC/DEC	$Na_4Fe_3(PO_4)(P_2O_7)$	+4	[69]
VC	5 vol%	1 M $NaPF_6$-EC/PC	Na_2MnSiO_4	+61	[70]
VC	2%	10 M $NaClO_4$-H_2O	$Na_3V_2O_{2x}(PO_4)_2F_{3-2x}$	+25	[71]
VC	5%	1 M $NaPF_6$-EC/PC	$Na_3V_2(PO_4)_2F_3$	−3	[72]
CMC	0.5 wt%	10 M $NaClO_4$-H_2O	$Na_3V_2O_{2x}(PO_4)_2F_{3-2x}$	−4	[71]
Agarose	0.5 wt%	10 M $NaClO_4$-H_2O	$Na_3V_2O_{2x}(PO_4)_2F_{3-2x}$	+3	[71]
NaDFOB	0.5%	1 M $NaPF_6$-EC/PC	$Na_3V_2(PO_4)_2F_3$	+10	[72]
PS	3%	1 M $NaPF_6$-EC/PC	$Na_3V_2(PO_4)_2F_3$	+2	[72]
SN	1%	1 M $NaPF_6$-EC/PC	$Na_3V_2(PO_4)_2F_3$	−1	[72]

on anode. At 55 °C, the effects of four additives, including 1,3-propane sultone (PS), succinonitrile (SN), NaDFOB, and VC, have been rationalized (Figure 11.17) [72]. The dosage is required to be strictly controlled. For example, an appropriate amount of VC can produce a stable SEI on the anode, but due to its oxidative decomposition, a large amount of sediment will be generated on the cathode, thus the effect of excessive VC may be counterproductive. Similarly, when the amount of NaODFB > 3%, it causes a sharp increase in battery impedance, which is due to the growth of thick deposits on the cathode. The effect of PS is determined to be an enrichment of sulfates in SEI layer. The combination of SN and VC contributes to a robust CEI formed on cathode.

11.5.3 Additives for Na Metal

Sodium metal with lower potential and high capacity is considered to be a very promising anode. However, sodium metal has encountered similar obstacles in practical applications as lithium metal. Although lithium and sodium share many similar physical and chemical properties and both batteries work in a similar way, their electrochemical behavior in metal cells is quite different [79]. For instance, it is more difficult for sodium to nucleate and grow than the lithium and its reaction kinetics is slower. Sodium-metal-based batteries demonstrate lower working potential due to its higher overpotential than lithium-metal-based batteries [80]. However, pursuing

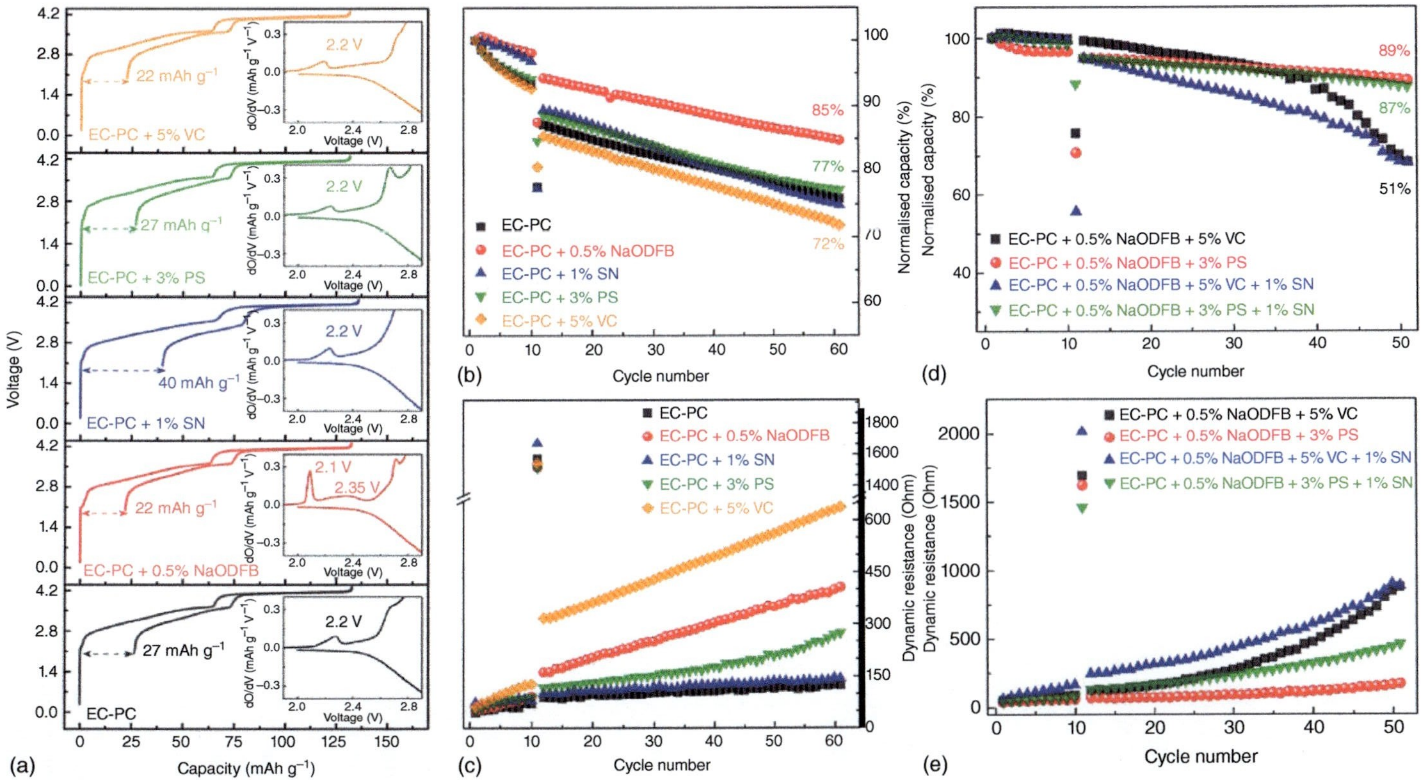

Figure 11.17 (a) The voltage–capacity and *dQ/dV* curves of NVPF|C cells. (b–e) The cycling stability and dynamic resistance evolution of NVPF|C cells with different additives. (f) The representative charge–discharge curves of NVPF|C cells. (g) The frontier orbital energies of the solvents, anions, and additives. Source: Yan et al. [72]. Reproduced with permission, 2019, John Wiley and Sons.

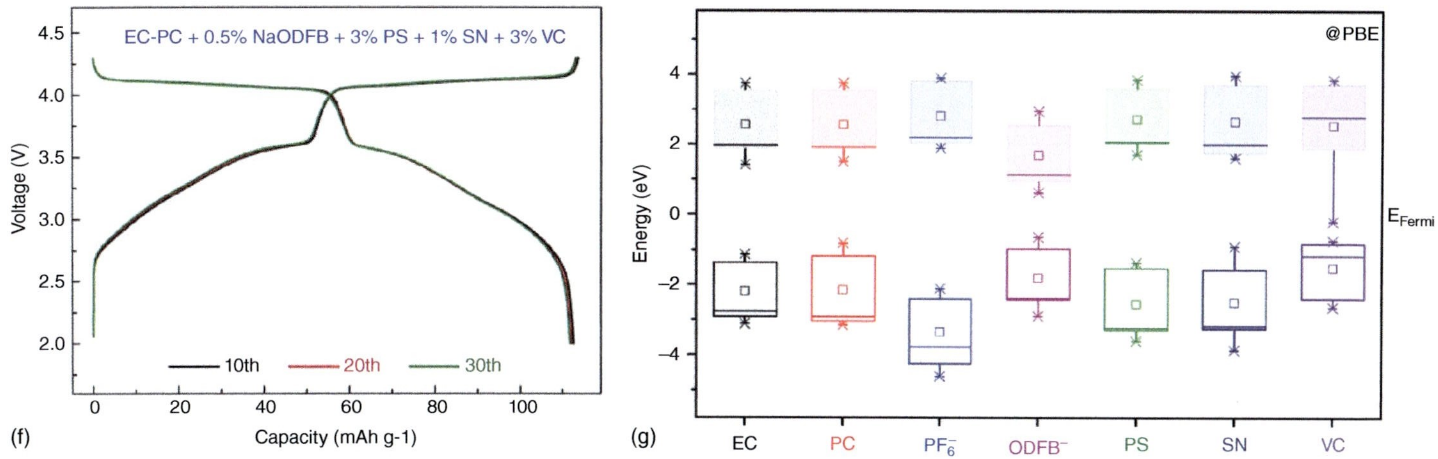

Figure 11.17 *(continued)*

the application of lithium-metal-based batteries will certainly enlighten the research of sodium-metal-based batteries. In addition, sodium metal also faces some unique barriers, including high reactivity between Na metal and electrolyte and low melting point.

Na metal is widely used in Na–O_2, Na–S, and Na-insertion-type cathode cells. Dendrite formation and parasitic reactions are needed to be addressed. Applying additives to form a SEI layer with high surface coverage, high mechanical stability, and ionic conductivity is considered to be the most facial and economic method.

Inspired by the self-healing electrostatic shield mechanism working in lithium metal, Wang et al. developed potassium bis(trifluoromethylsulfonyl)imide (KTFSI) additive for 1 M $NaOSO_2CF_3$-triethylene glycol dimethyl ether (TEGDME) electrolyte to passivate Na metal anode [81]. KTFSI demonstrates superior performance due to the following two aspects: (i) stable SEI layer with highly conductive, electrochemically stable Na_3N and oxynitrides compounds formed by the decomposition of $TFSI^-$ anion, and (ii) K^+ preferentially adsorbed on the Na deposition process, generating electrostatic shielding effect and inhibiting the formation and growth of dendrites.

Inspired by the effect of Li_2S_n–$LiNO_3$ in Li–S cells, sodium polysulfide (Na_2S_6) and $NaNO_3$ were used as additives in ether-based electrolyte. It is found that Na_2S_6 alone is beneficial to the electrochemical performance of Na metal anode, but Na_2S_6–$NaNO_3$ is harmful to sodium metal unexpectedly [82]. XPS analysis of Na anode showed that when Na_2S_6 was used alone, the SEI was mainly composed of inorganic components, including Na_2O, Na_2S_2, and Na_2S, which enhance the mechanically strength of SEI layer and inhibit dendrite growth. Once $NaNO_3$ added, the SEI layer was dominated by RCH_2ONa and Na_2S, which is insufficient to inhibit dendrite growth.

Constructing a Na-alloy layer is another efficient method to passivate the Na interface. Zheng et al. added 50 mM $SnCl_2$ to the carbonate electrolyte to form Na–Sn alloy layer and SEI layer is rich in NaCl spontaneously (Figure 11.18a,b) [83]. Under the protection of this dual-layer interphase, rapid Na^+ ion transfer and suppressed parasitic reaction are achieved. As a result, the non-dendritic morphology and reduced polarization voltage are observed. Xu and coworkers dissolved 0.01 M SbF_3 in DMC to pretreat the Na metal foil to acquire sodium-alloy-fluoride (SAF-Na) anode [84]. The formation of the Na–Sb alloy layer and the chemical/electrochemical complementary SEI provides high interface strength and ionic conductivity. The anode is particularly compatible with different cathodes and exhibits improved cycling performance. (Figure 11.18c–f) [84]. Application of SbF_3 in high-concentration electrolyte (HCE) also boosts the cycling stability of Na|Na symmetric cells and Na|$Na_3V_2(PO_4)_3$ cells significantly [85].

In addition to the additives above mentioned, FEC also performs well in sodium metal. When FEC is added to 1 M NaFSI–DME, a highly stable and dense SEI layer is formed by the sacrifice of FEC during long electrochemical cycles (Figure 11.19) [86].

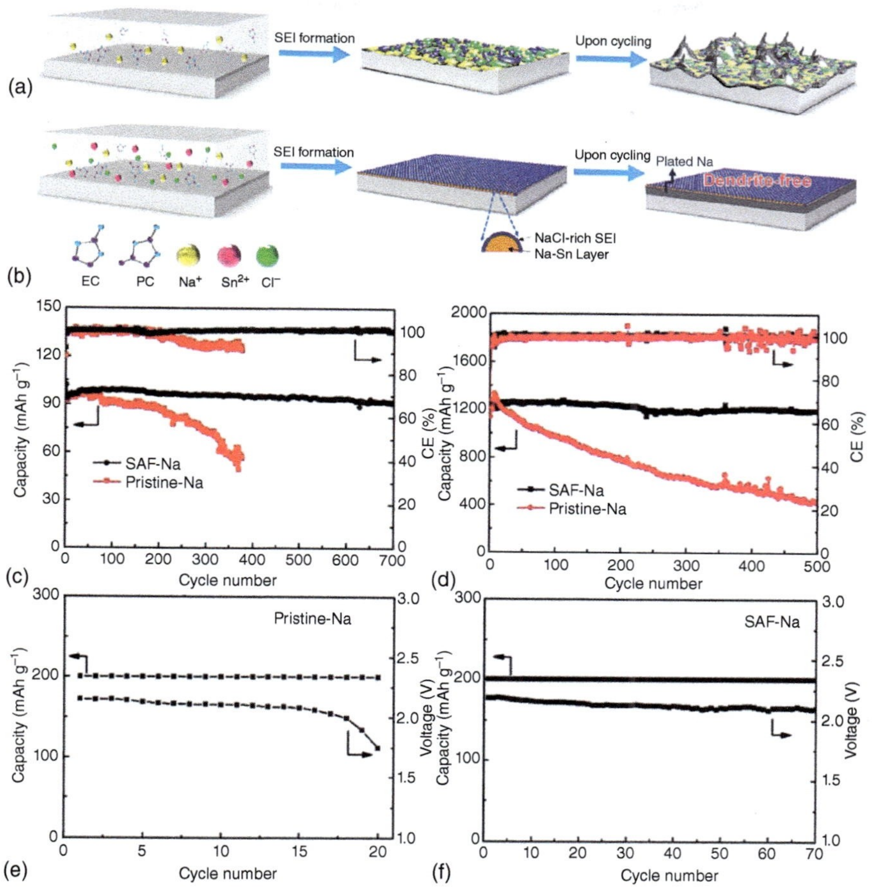

Figure 11.18 Graphical illustration of the SEI layer formed in (a) conventional carbonate electrolyte and (b) electrolyte with $SnCl_2$ additive. Source: Zheng et al. [83]. Reproduced with permission, 2019, American Chemical Society. The cycle stability of (c) Na-NVP in $NaPF_6$-diglyme, (d) Na-S@pPAN in $NaClO_4$-EC/PC + 2% FEC, and (e, f) $Na-O_2$ cells in $NaPF_6$-diglyme. The SAF-Na is the abbreviation of sodium-alloy-fluoride anode. Source: Xu et al. [84]. Reproduced with permission, 2019, Wiley-VCH.

11.5.4 Safety Inspired Additives

An in-depth evaluation for the safety-related concerns is required by the widespread application of Na-based rechargeable batteries. Generally, incorporating additives into the electrolyte as standalone or reactive environments is an effective method to eliminate the safety hazards. Some functionalities are required for these additives, such as thermal stability improvers, overcharge protectors, flame retardancy, etc.

For the overcharge protection additives, ESW is a key characteristic that should be emphasized. Overcharge inhibitors, including redox shuttles and shutdown-type additives, are designed and investigated extensively [87]. Redox shuttles refer to

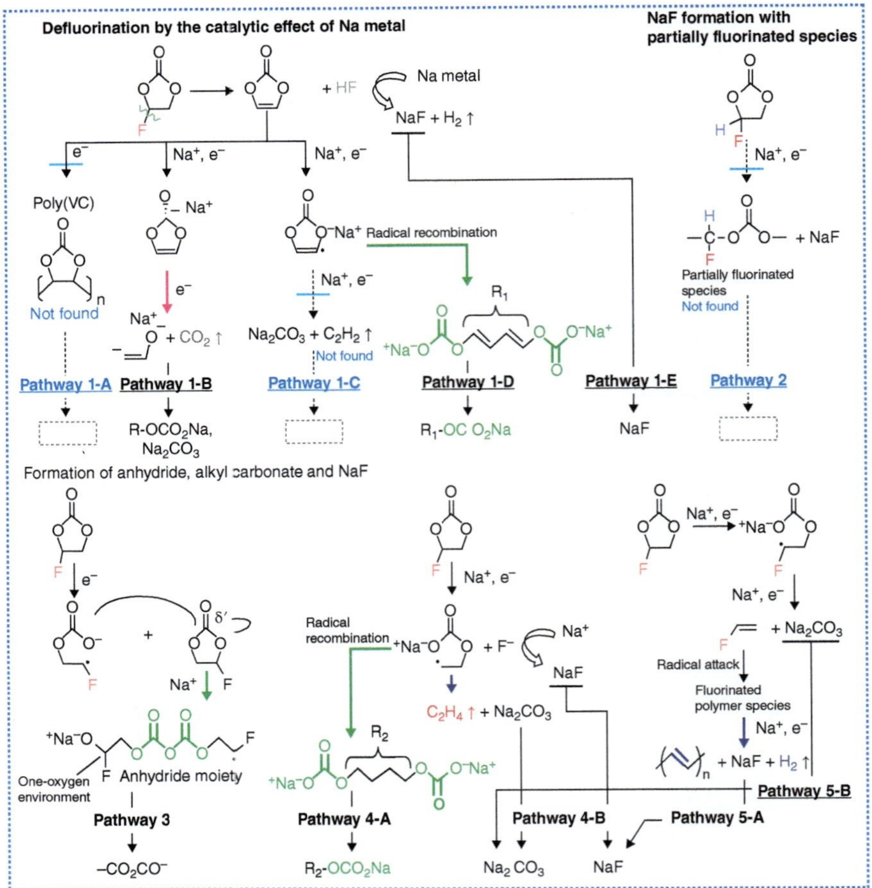

Figure 11.19 Possible formation mechanisms of the SEI layer derived by FEC. Source: Lee et al. [86]. Reproduced with permission, 2018, American Chemical Society.

certain kind of electroactive molecules that can achieve reversible overcharge protection as electrolyte additives. Redox shuttles transform into corresponding radical cations after being oxidized at cathode surface under overcharge conditions. The resulting radicals diffuse to the anode surface and then be reduced. Voltage of the cells can be maintained at the reverse redox potential of the shuttles with the radicals shuttle between cathode and anode continuously. Therefore, potential damages of all the cell components will be restrained. Sodium anodes and other components in the SIBs are different from their LIB counterparts as they are more reactive. Thus, state-of-the-art redox shuttles employed in LIBs maybe not function as well in SIBs. Trisaminocyclopropenium perchlorate (TAC·ClO_4), a functional organic salt, was investigated by Ji et al. as a redox shuttle for overcharge protection in SIBs [88]. TAC·ClO_4 shows unique characteristics in both redox states, including fast diffusion and high chemical/electrochemical stability due to its smallest aromatic ring structure. 0.1 M TAC·ClO_4 additive enables NVP cathode to be overcharged even up to 10 C or 400% SOC (Figure 11.20a).

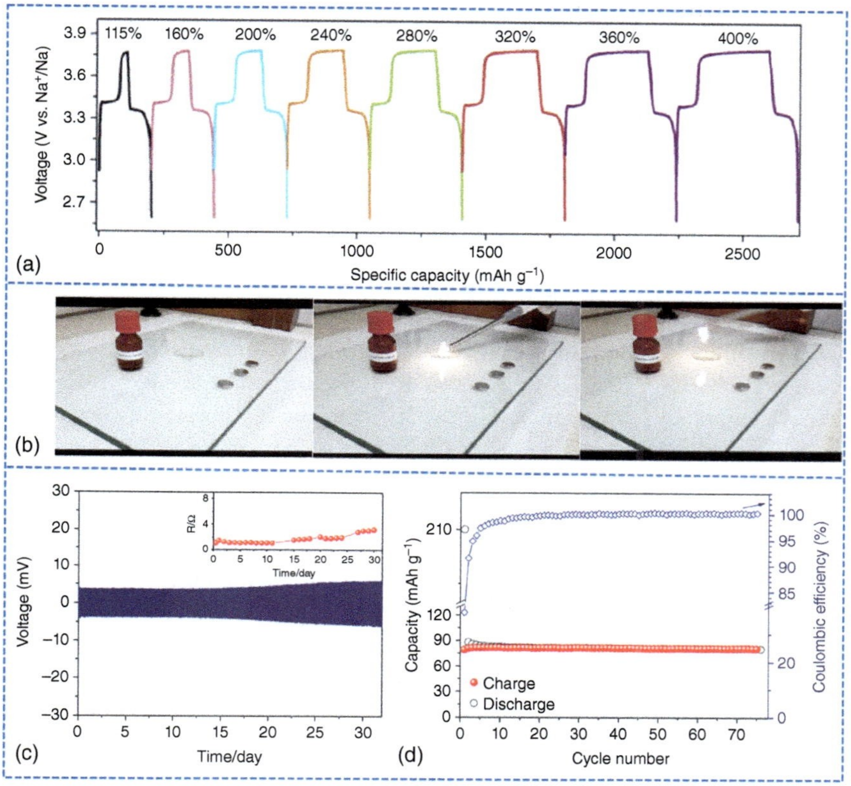

Figure 11.20 (a) The galvanostatic charge–discharge profiles for NVP cathode in electrolyte with 0.1 M TAC additive overcharged to different capacities. Source: Ji et al. [88]. Reproduced with permission, 2020, Elsevier. (b) Flammability tests of the 0.5 M NaBOB-TMP electrolyte and PC solvent. Source: Yamada and Yamada [89]/ Springer Nature. (c) Ignitability tests of the $NaCF_3SO_3$–TEGDME electrolyte exposure to flame for 30 s. (d) Na|Na symmetric cell performance and interphase impedance evolution during time, and (e) cycling stability of a Na | MCMB cell with $NaCF_3SO_3$–TEGDME electrolyte. Source: Di Lecce et al. [7b]. Reproduced with permission, 2019, Springer Berlin Heidelberg.

Shutdown-type additives polymerize to form an insulating layer on the electrode at a characteristic potential, which can avoid further degradation reaction. The application of shutdown-type additives is the most cost-efficient and facile method applied so far. For example, the electrochemical polymerization of biphenyl (BP) at 4.3 V (vs. Na/Na^+) prevents $Na_{0.44}MnO_2$/Na battery from voltage runaway by depleting the excess charge even under 800% overcharge capacity [90]. Through generating a conductive or isolated polymer, the overcharge current flow can be bypassed or interrupted. The effects of the optimal addition of BP beyond its intended effect were further investigated and the addition of BP is negligible for the capacity attenuation.

Since Wang et al. reported the application of trimethyl phosphate (TMP) [91], various flame retarding (FR) additives have been developed for LIBs. However, less effort is devoted to exploring FRs additives for SIBs. The FRs usually remove the reactive radicals formed during thermal decomposition of electrolytes [92]. The initiation and duration of the combustion reaction require cascading chain propagation reaction [93]. Therefore, FRs should mitigate/prevent the catalytic reaction by chemical action or physical blocking. The effect of four widely used FRs, including TMP, dimethyl methylphosphonate (DMMP), tri(2,2,2-trifluoroethyl)phosphite (TFEP), and methyl nonafluorobutylether (MFE), were evaluated by Feng et al. [94] Among these FRs, MFE exhibits the best electrochemical stability toward Na metal. Stable cycling performance is also acquired in carbon nanotube | PB battery. Nevertheless, the ionic conductivity of some measured FRs is relatively low (5×10^{-4} S cm^{-1}), which maybe not enough for practical application. The same group also studied the flame-retardant property of ethoxy (pentafluoro)cyclo-triphosphazene (EFPN). The flammability of the 1 M $NaPF_6$-EC/DEC electrolyte can be quenched by introducing 5% EFPN [95]. The low-cost, fluoride-free single-solvent TMP electrolyte is completely nonflammable. Furthermore, the TMP electrolyte exhibits high ionic conductivity and good electrochemical performance in SIBs with Prussian white cathode and HC anode [18]. The EFPN-based electrolyte has been confirmed to be chemically stable with Na metal anode and exhibits improved cyclability for acetylene black | $Na_{0.44}MnO_2$ battery. Di Lecce et al. reported that the TEGDME–$NaCF_3SO_3$ electrolyte was totally nonflammable. This may be due to the relatively low vapor pressure of the glyme solvents, which further reduces as the chain length raises. Thus it is beneficial to improve the electrolyte safety (Figure 11.20b–d) [7b]. The impact of FEC on the thermal stability of 1 M $NaPF_6$-EC/DEC electrolyte was evaluated in the progress of Na plating/stripping [96]. Unpredictably, after adding FEC, the heat flow was found to be higher.

11.6 Novel Concentration Electrolyte Systems

11.6.1 High-Concentration Electrolytes

HCE means that as the concentration of the salt increases, the interaction between the cation and anion in the solvent is enhanced, and the content of free solvent molecules is greatly reduced. Among nonaqueous solvents, when the salt concentration increases to 3.5 M, the free solvent molecules basically disappear, forming a new type of electrolyte with a special structure, which is called a high-concentration electrolyte. As shown in Figure 11.21a, generally, in a dilute electrolyte, most salt ions will form a solvated structure with solvent molecules, only a part of anions is directly coordinated with cations to form solvent-separated ion pairs (SSIP), and the rest is mostly free solvent. In this case, the components in SEI mainly come from the decomposition of solvents. When the salt concentration is exceeded, salt anions begin to participate in solvation, forming contact ion pairs (CIPs) and anion–cation aggregates (AGG) due to insufficient solvent, as shown in Figure 11.21b. Thus, the main component of SEI is inorganic. They are completely different from the traditional electrolyte (1 M), and demonstrate theadvantages: (i) high ion migration number; (ii) less corrosion of aluminum current collectors; (iii) effectively inhibited dendrites; (iv) broader electrochemical window; and (v) improved the flame-retardant

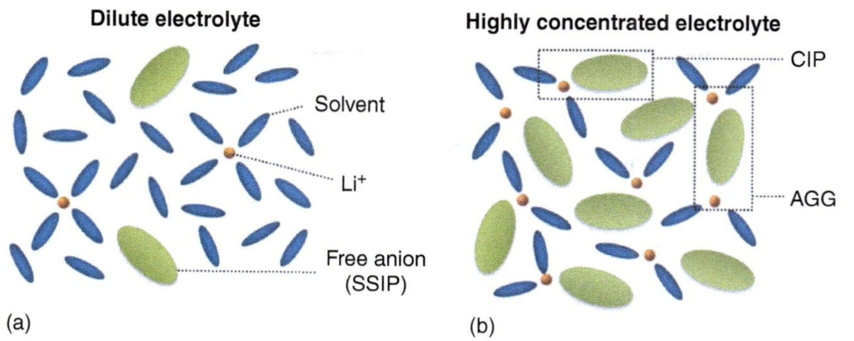

Figure 11.21 Schematic diagram of the structure in (a) dilute electrolyte and (b) high concentration electrolyte. Source: Yamada and Yamada [89].

performance and safety. SIBs are composed of highly active cathode materials and organic electrolytes and are prone to "thermal runaway" under heating conditions, which may cause dangerous accidents. The strong solvation in the HCE reduces free solvent molecules and inhibits flammable and volatile solvents, thereby improving flame retardancy.

In 2015, Chen and coworkers reported an electrolyte of 4 M NaTfO–TEGDME for Na | 9,10-anthraquinone (AQ) batteries [97]. The capacity is 214 mAh g^{-1}, and it maintains 190 mAh g^{-1} after 50 cycles at 0.2 C. In 2016, Zhang and coworkers developed a highly concentrated electrolyte for Na cells [98]. In 4M-NaFSI/DME electrolyte, the plating/stripping of Na is stable without dendrite growth, and the Coulombic efficiency is as high as 99%. XRD and XPS characterizations show that the main component of SEI is NaF. Chang and coworkers announced a highly concentrated electrolyte of 3.0 M NaFSI-EC/PC in Na | HC batteries [12b]. In 2017, Yamada and coworkers reported a electrolyte composed of 50% NaFSI-SN (1 : 1 by molar ratio) [99]. The ionic conductivity of this electrolyte is 0.31 m S cm^{-1}. When used in Na|HC cells, the battery exhibits a capacity of 200 mAh g^{-1} after 100 cycles, which is higher than that in 1.0 M $NaPF_6$-EC/DEC. In 2018, Wang et al. found a new safe electrolyte of 3.3 M NaFSI-TMP (Figure 11.22) [100]. This electrolyte shows a SET of 0 s g^{-1} and is stable under the boiling point (197 °C). Furthermore, the electrolyte loses only 11% of its mass after heating to 150 °C. In the HCE of 3.3 M NaFSI-TMP, the ICE of the HC|Na battery is 75%, which is very close to the ICE (79%) of the 1.0 M $NaPF_6$-EC/DEC electrolyte. The TMP-based electrolyte has also been used in SIBs [101]. In 2018, aqueous/nonaqueous electrolyte was applied by blending the equivalent weight of 7 M NaTfO with H_2O and 8 M NaTfO in PC. It stays stable from −1.25 to 1.55 V with high ionic conductivity at room temperature. The existence of PC promotes the stability of the electrolyte.

11.6.2 Local High-Concentration Electrolytes

The local high-concentration electrolyte (LHCE) refers to the addition of a diluent that is miscible with the solvent but not miscible with the salt in the HCE. After

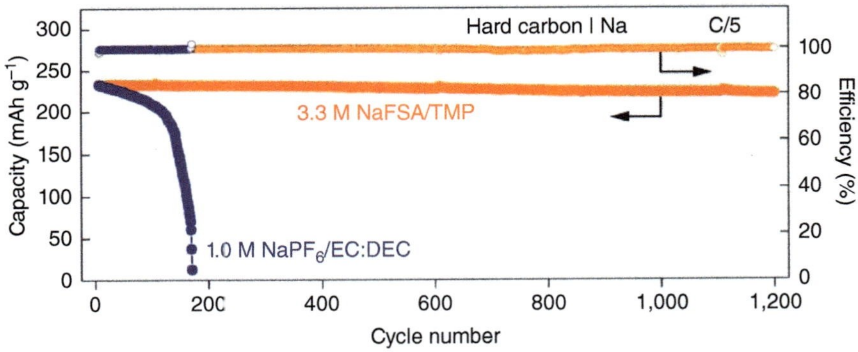

Figure 11.22 Cycling stability of HC|Na batteries in different electrolytes. Source: Wang et al. [100].

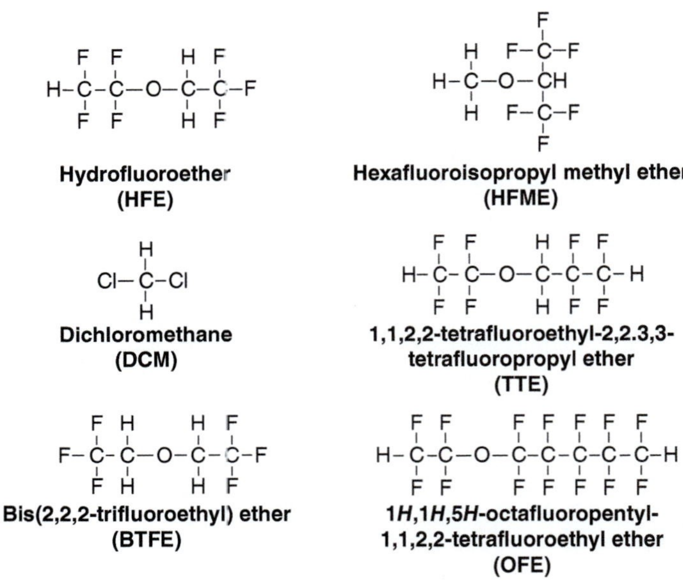

Figure 11.23 The molecular structures of the common diluents. Source: Doi et al. [102]. Licensed under CC BY 4.0.

the diluent is added, it does not affect the special coordination structure of the salt solvent in the original HCE, thus forming the LHCE. This not only guarantees the characteristics of HCE, but also has its own unique characteristics. Its viscosity is much lower than that of HCE, and it has excellent wettability. The diluent used for the LHCE needs to meet the following characteristics: (i) low dielectric constant and low complexing capacity, compatible with the solvent, and does not affect the special coordination structure of the salt-solvent in the original HCE; (ii) low viscosity and high wettability; (iii) low cost; (iv) flame retardant; and (v) conducive to the formation of good SEI. Some diluents are listed in Figure 11.23 [102].

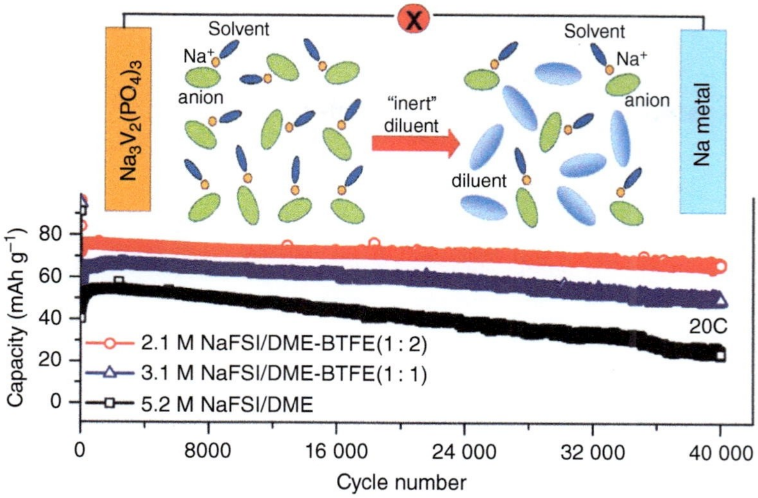

Figure 11.24 (a) Schematic of the solvation structure in HCE and LHCE. (b) Cycling stability of Na | $Na_3V_2(PO_4)_3$ batteries at 20 C in various concentrated electrolytes. Source: Zheng et al. [104]. Reproduced with permission, 2018, American Chemical Society.

In 2018, Xu and coworkers added solvents to "dilute" the HCE [103]. The diluents are electrochemically inert and have little influence on the solvation structure of HCEs. Nevertheless, they promoted the interface reaction kinetics, thus enhancing the electrochemical performance of the battery. As shown in Figure 11.24, the LHCE even showed better electrochemical performance than the original concentrated electrolyte [104].

In 2020, Yu et al. reported an electrolyte based on TMP and 1,1,2,2-tetrafluoroethyl-2,2,3,3-tetrafluoropropyl ether (F-EPE) with FEC as an additive, and EC-DEC-FEC electrolyte was used as the blank electrolyte (Figure 11.25) [105]. The results of flammability tests show better safety of electrolyte with TMP and F-EPE. Meanwhile, the prepared electrolytes also increase the capacity of the NFM cathode (129.9 mAh g^{-1}) and improve the cycle stability of the NFM | HC pouch battery (70.8% is retained after 500 cycles). They studied the dissolution behavior of NFM cathode in various electrolytes, as well as the XPS spectra of NFM cathodes after cycling in the prepared electrolyte to explore the principle of capacity attenuation. The upshots prove that this safe electrolyte has great development prospects in SIBs.

HCE has the advantages of low combustibility, good thermal stability, and good electrochemical performance. It is an ideal choice for the development of safe advanced electrolytes. It is also urgent to use noncombustible flame retardants to replace traditional organic solvents to develop fire-extinguishing electrolytes. The introduction of the intercalation solvent as a diluent can overcome the disadvantages of high viscosity and poor wettability and promote the interface reaction kinetics and interface stability of the electrode. However, its high cost still limits its practical application.

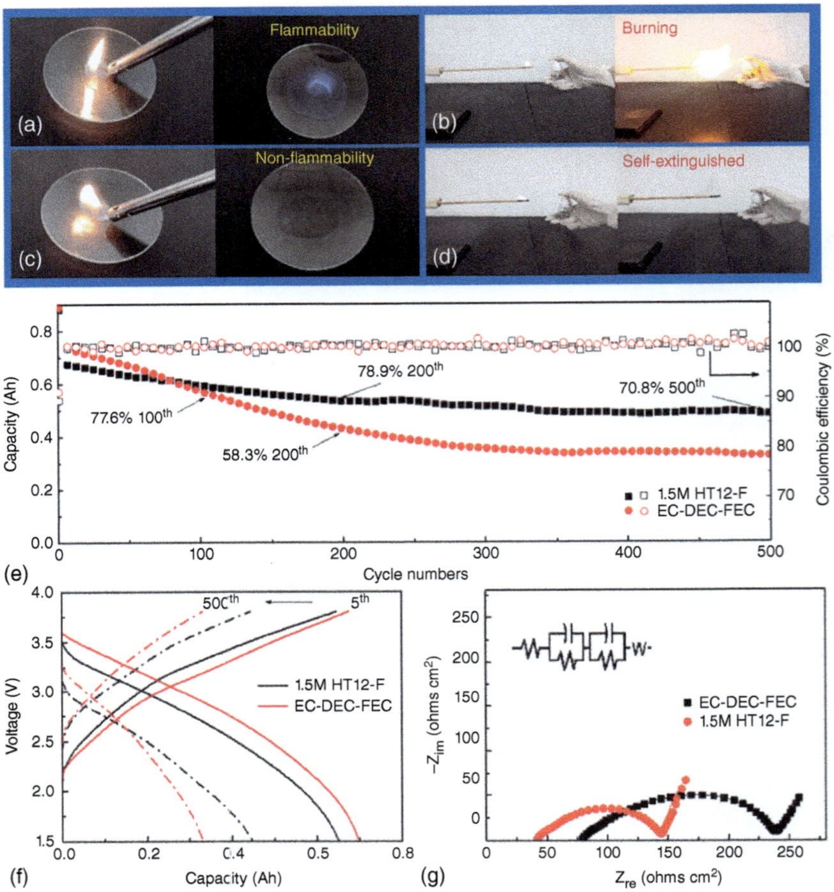

Figure 11.25 Photos of combustion experiment. (a, b) EC/DEC/FEC, (c, d) the HT12 and HT12-F. (e) Cycling life span and CE of HC|NFM batteries at 1C. (f) Charge–discharge curves in different cycles, and (g) Nyquist plots of NFM | HC battery at 3.8 V after 500 cycles. Source: Yu et al. [105]/Elsevier.

11.6.3 Low-Concentration Electrolytes

Adjusting the electrolyte concentration is the key to the functional design of energy storage equipment. In recent years, HCE is used in various batteries [106]. On the other hand, since low ion conductivity may cause concentration polarization, the research on reducing the salt concentration to form an ultra-dilute electrolyte has not been carried out. Since the Stokes radius and desolvation energy of Na^+ are smaller than that of Li^+, it is possible to use ultra-dilute electrolyte to gain sufficient kinetic capability [107]. What is more, reducing the salt content is beneficial for practical application, with lower price.

In 2020, Hu and coworkers reported a low-concentration electrolyte (0.3 M) in SIBs for the first time (Figure 11.26) [108]. Due to the chemical properties of the diluted electrolyte, the low-concentration electrolyte not only significantly

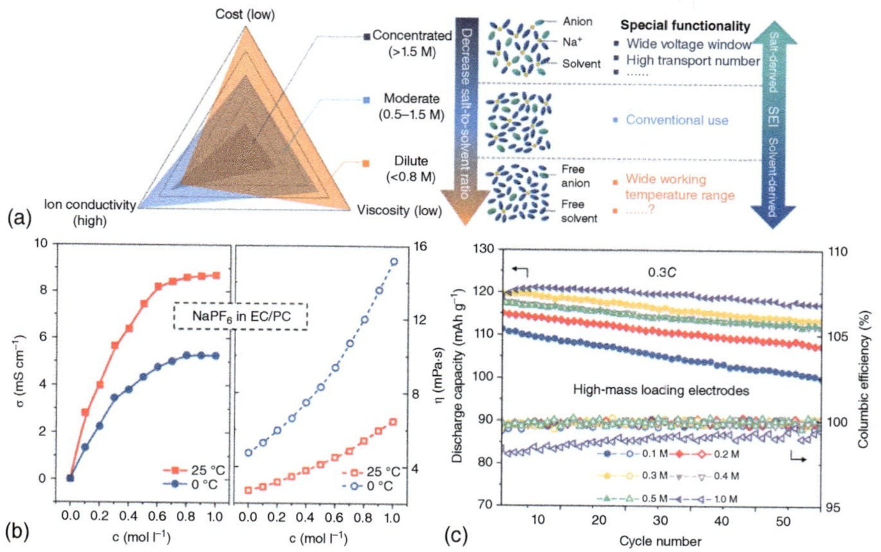

Figure 11.26 (a) Epitome of the electrolytes with different concentration and the related properties. (b) Viscosity and ionic conductivity at different concentration at 25 and 0 °C. (c) Cycling performance of cells with different concentrations electrolytes. Source: Li et al. [108]. Reproduced with permission, 2020, American Chemical Society.

reduces the cost of salt, but also expands the operating temperature range of the safety-enhanced SIB (−30 to 55 °C). Under low- and high-temperature conditions, low viscosity and corrosion risk (less hydrofluoric acid attack) will help improve the wettability of the separator and the CE of the battery. In addition, the formed stable organic-rich SEI/CEI has superior kinetic characteristics, which enables SIBs to work sustainably at extreme temperatures. The new dilute electrolyte is expected to be used in other electrolyte systems. Effective additives can further improve the capacity and interface regulation performance of the ultralow-concentration electrolyte.

11.7 Concluding Remarks

Organic liquid electrolytes with Na salts dissolving in the organic solvents are currently the most promising electrolytes for commercial SIBs. For organic liquid electrolytes, liquid range, ionic conductivity, electrochemical/chemical stability are the most important parameters that affect the performance and operating temperature range, which are determined by the organic solvent, Na salt and concentration.

In addition to organic solvent and Na salts, small amount functional additives can improve the performance of organic liquid electrolytes. Functional additives can work as SEI-forming additives for anodes, CEI-forming additives for cathodes and safety inspired additives, depending on their properties. SEI-forming additives should be reduced at higher potential than the main component of organic liquid electrolytes that can protect organic liquid electrolytes from decomposing. On the other side, CEI-forming additives should be oxidized at lower potential than the main component of organic liquid electrolytes. Safety inspired additives include overcharge protection additives to introduce shuttle effect at high voltages, shutdown-type additives to form an insulating layer on the electrode at a characteristic potential to avoid further degradation reaction, and flame retarding (FRs) additives to remove the reactive radicals for combustion reaction.

Furthermore, high concentration and low concentration liquid electrolytes with special properties are also developed. Although the electrochemical performance is enhanced with high concentration of Na salts, the cost and viscosity are much higher than those of regular electrolytes. To solve these problems, localized high-concentration electrolytes are developed with addition of a diluent that is miscible with the solvent but not miscible with the Na salt. Low concentration electrolytes with lower price are beneficial for practical application. Smaller desolvation energy of Na+ make it possible to use ultra-dilute electrolyte to gain sufficient kinetic capability and acceptable electrochemical performance. Furthermore, low concentration electrolytes with low viscosity broaden the operating temperature range (e.g., at low temperatures of at least −20°).

References

1 Chen, S., Ishii, J., Horiuchi, S. et al. (2017). *Phys. Chem. Chem. Phys.* 19: 17366.

2 Zhang, J., Yao, X., Misra, R.K. et al. (2020). *J. Mater. Sci. Technol.* 44: 237.

3 Lin, Z., Xia, Q., Wang, W. et al. (2019). *InfoMat* 1: 376.

4 Huang, Y., Zhao, L., Li, L. et al. (2019). *Adv. Mater.* 31: e1808393.

5 Jonsson, E. and Johansson, P. (2012). *Phys. Chem. Chem. Phys.* 14: 10774.

6 Ponrouch, A., Marchante, E., Courty, M. et al. (2012). *Energy Environ. Sci.* 5: 8572.

7 (a) Hashmi, S.A., Bhat, M.Y., Singh, M.K. et al. (2016). *J. Solid State Electrochem.* 20: 2817. (b) Di Lecce, D., Minnetti, L., Polidoro, D. et al. (2019). *Ionics* 25: 3129.

8 Chen, J., Huang, Z., Wang, C. et al. (2015). *Chem. Commun.* 51: 9809.

9 Plewa-Marczewska, A., Trzeciak, T., Bitner, A. et al. (2014). *Chem. Mater.* 26: 4908.

10 Ge, C., Wang, L., Xue, L. et al. (2014). *J. Power Sources* 248: 77.

11 Morikawa, Y., Yamada, Y., Doi, K. et al. (2020). *Electrochemistry* 88: 151.

12 (a) Ao, H., Chen, C., Hou, Z. et al. (2020). *J. Mater. Chem. A* 8: 14190. (b) Patra, J., Huang, H.-T., Xue, W. et al. (2019). *Energy Storage Mater.* 16: 146.

13 Wu, F., Zhu, N., Bai, Y. et al. (2016). *ACS Appl. Mater. Interfaces* 8: 21381.

14 Eshetu, G.G., Grugeon, S., Kim, H. et al. (2016). *ChemSusChem* 9: 462.

15 Goktas, M., Bolli, C., Buchheim, J. et al. (2019). *ACS Appl. Mater. Interfaces* 11: 32844.

16 Gao, L., Chen, J., Liu, Y. et al. (2018). *J. Mater. Chem. A* 6: 12012.

17 Bitner-Michalska, A., Krzton-Maziopa, A., Zukowska, G. et al. (2016). *Electrochim. Acta* 222: 108.
18 Mogensen, R., Colbin, S., Menon, A.S. et al. (2020). *ACS Appl. Energy Mater.* 3: 4974.
19 Jiao, S., Ren, X., Cao, R. et al. (2018). *Nat. Energy* 3: 739.
20 Xiang, H., Shi, P., Bhattacharya, P. et al. (2016). *J. Power Sources* 318: 170.
21 Zheng, J., Engelhard, M.H., Mei, D. et al. (2017). *Nat. Energy* 2: 17012.
22 Ponrouch, A., Monti, D., Boschin, A. et al. (2015). *J. Mater. Chem. A* 3: 22.
23 Xu, K. (2014). *Chem. Rev.* 114: 11503.
24 Goodenough, J.B. and Kim, Y. (2010). *Chem. Mater.* 22: 587.
25 Chayambuka, K., Mulder, G., Danilov, D.L., and Notten, P.H.L. (2018). *Adv. Energy Mater.* 8: 1800079.
26 (a) Zhang, B., Rousse, G., Foix, D. et al. (2016). *Adv. Mater.* 28: 9824. (b) Jiang, X., Liu, X., Zeng, Z. et al. (2018). *iScience* 10: 114.
27 (a) Shakourian-Fard, M., Kamath, G., Smith, K. et al. (2015). *J. Phys. Chem. C* 119: 22747. (b) Xing, L., Zheng, X., Schroeder, M. et al. (2018). *Acc. Chem. Res.* 51: 282.
28 Xu, K. (2004). *Chem. Rev.* 104: 4303.
29 Tasaki, K., Goldberg, A., and Winter, M. (2011). *Electrochim. Acta* 56: 10424.
30 Dahbi, M., Yabuuchi, N., Kubota, K. et al. (2014). *Phys. Chem. Chem. Phys.* 16: 15007.
31 Yan, G., Alves-Dalla-Corte, D., Yin, W. et al. (2018). *J. Electrochem. Soc.* 165: A1222.
32 Vidal-Abarca, C., Lavela, P., Tirado, J.L. et al. (2012). *J. Power Sources* 197: 314.
33 Bhide, A., Hofmann, J., Katharina Dürr, A. et al. (2014). *Phys. Chem. Chem. Phys.* 16: 1987.
34 Oh, S.-M., Myung, S.-T., Yoon, C.S. et al. (2014). *Nano Lett.* 14: 1620.
35 Zeng, Z., Jiang, X., Li, R. et al. (2016). *Adv. Sci.* 3: 1600066.
36 Che, H., Yang, X., Yu, Y. et al. (2020). *Green Energy Environ.* https://doi.org/10.1016/j.gee.2020.04.007.
37 Lee, M., Hong, J., Lopez, J. et al. (2017). *Nat. Energy* 2: 861.
38 (a) Cohn, A.P., Share, K., Carter, R. et al. (2016). *Nano Lett.* 16: 543. (b) Bai, P., He, Y., Xiong, P. et al. (2018). *Energy Storage Mater.* 13: 274. (c) Liu, X., Qin, B., Zhang, H. et al. (2019). *ACS Appl. Energy Mater.* 2: 2786. (d) Hu, Z., Zhu, Z., Cheng, F. et al. (2015). *Energy Environ. Sci.* 8: 1309. (e) Firouzi, A., Qiao, R., Motallebi, S. et al. (2018). *Nat. Commun.* 9: 861.
39 Kajita, T., Noji, T., Imai, Y. et al. (2018). *J. Electrochem. Soc.* 165: A3582.
40 Zhang, J., Wang, D.-W., Lv, W. et al. (2017). *Energy Environ. Sci.* 10: 370.
41 Li, K., Zhang, J., Lin, D. et al. (2019). *Nat. Commun.* 10: 725.
42 Gangaja, B., Nair, S., and Santhanagopalan, D. (2019). *Sustain. Energy Fuels* 3: 2490.
43 Wang, S., Chen, Y., Jie, Y. et al. (2020). *Small Methods*: 4.
44 Zhen, Y., Sa, R., Zhou, K. et al. (2020). *Nano Energy*: 74.
45 Eshetu, G.G., Martinez-Ibañez, M., Sánchez-Diez, E. et al. (2018). *Chem. Asian J.* 13: 2770.

46 Peled, E. (1979). *J. Electrochem. Soc.* 126: 2047.
47 Gauthier, M., Carney, T.J., Grimaud, A. et al. (2015). *J. Phys. Chem. Lett.* 6: 4653.
48 (a) Eshetu, G.G., Grugeon, S., Laruelle, S. et al. (2013). *Phys. Chem. Chem. Phys.* 15: 9145; (b) Harris, S.J., Timmons, A., and Pitz, W.J. (2009). *J. Power Sources* 193: 855.
49 Kumar, H., Detsi, E., Abraham, D.P., and Shenoy, V.B. (2016). *Chem. Mater.* 28: 8930.
50 Komaba, S., Ishikawa, T., Yabuuchi, N. et al. (2011). *ACS Appl. Mater. Interfaces* 3: 4165.
51 (a) Dahbi, M., Nakano, T., Yabuuchi, N. et al. (2016). *Chemelectrochem* 3: 1856. (b) Wen, Y., Wang, B., Luo, B., and Wang, L. (2016). *Eur. J. Inorg. Chem.* https://doi.org/10.1002/ejic.2015011722051.
52 Benchakar, M., Naéjus, R., Damas, C., and Santos-Peña, J. (2020). *Electrochim. Acta*: 330.
53 Che, H., Liu, J., Wang, H. et al. (2017). *Electrochem. Commun.* 83: 20.
54 Baggetto, L., Marszewski, M., Gorka, J. et al. (2013). *J. Power Sources* 243: 699.
55 Baggetto, L., Keum, J.K., Browning, J.F., and Veith, G.M. (2013). *Electrochem. Commun.* 34: 41.
56 Qian, J., Chen, Y., Wu, L. et al. (2012). *Chem. Commun.* 48: 7070.
57 (a) Pervez, S.A., Kim, D., Lee, S.M. et al. (2016). *J. Power Sources* 315: 218; (b) Lu, H., Wu, L., Xiao, L. et al. (2016). *Electrochim. Acta* 190: 402. (c) Hur, J. and Kim, I.T. (2015). *Bull. Kor. Chem. Soc.* 36: 1625. (d) Kim, I.T., Allcorn, E., and Manthiram, A. (2015). *J. Power Sources* 281: 11. (e) Kim, I.T., Allcorn, E., and Manthiram, A. (2014). *Phys. Chem. Chem. Phys.* 16: 12884.
58 Kim, Y., Kim, Y., Choi, A. et al. (2014). *Adv. Mater.* 26: 4139.
59 (a) Dahbi, M., Yabuuchi, N., Fukunishi, M. et al. (2016). *Chem. Mater.* 28: 1625. (b) Yabuuchi, N., Matsuura, Y., Ishikawa, T. et al. (2014). *Chemelectrochem* 1: 580; (c) Qian, J., Wu, X., Cao, Y. et al. (2013). *Angew. Chem. Int. Ed.* 52: 4633.
60 Nagulapati, V.M., Yoon, Y.H., Kim, D.S. et al. (2019). *J. Ind. Eng. Chem.* 76: 419.
61 Zhang, W., Xing, L., Chen, J. et al. (2020). *J. Alloys Compd.*: 822.
62 Mogensen, R., Brandell, D., and Younesi, R. (2016). *ACS Energy Lett.* 1: 1173.
63 Mogensen, R., Maibach, J., Brant, W.R. et al. (2017). *Electrochim. Acta* 245: 696.
64 Shimizu, M., Usui, H., and Sakaguchi, H. (2014). *J. Power Sources* 248: 378.
65 Webb, S.A., Baggetto, L., Bridges, C.A., and Veith, G.M. (2014). *J. Power Sources* 248: 1105.
66 Hwang, J.-Y., Myung, S.-T., and Sun, Y.-K. (2017). *Chem. Soc. Rev.* 46: 3529.
67 (a) Komaba, S., Yabuuchi, N., Nakayama, T. et al. (2012). *Inorg. Chem.* 51: 6211. (b) Yabuuchi, N., Kajiyama, M., Iwatate, J. et al. (2012). *Nat. Mater.* 11: 512. (c) Dall'Asta, V., Buchholz, D., Chagas, L.G. et al. (2017). *ACS Appl. Mater. Interfaces* 9: 34891.
68 (a) Shen, B., Xu, M., Niu, Y. et al. (2018). *ACS Appl. Mater. Interfaces* 10: 502. (b) Chen, M., Chen, L., Hu, Z. et al. (2017). *Adv. Mater.*: 29.
69 Lee, Y., Lee, J., Kim, H. et al. (2016). *J. Power Sources* 320: 49.
70 Law, M., Ramar, V., and Balaya, P. (2017). *J. Power Sources* 359: 277.

71 Kumar, P.R., Jung, Y.H., Moorthy, B., and Kim, D.K. (2016). *J. Electrochem. Soc.* 163: A1484.

72 Yan, G., Reeves, K., Foix, D. et al. (2019). *Adv. Energy Mater.*: 9.

73 (a) Fang, Y., Chen, Z., Xiao, L. et al. (2018). *Small*: 14. (b) Han, M.H., Gonzalo, E., Singh, G., and Rojo, T. (2015). *Energy Environ. Sci.* 8: 81.

74 Simone, V., Lecarme, L., Simonin, L., and Martinet, S. (2017). *J. Electrochem. Soc.* 164: A145.

75 (a) Serras, P., Palomares, V., Goni, A. et al. (2013). *J. Power Sources* 241: 56; (b) Zhang, H., Hasa, I., Qin, B. et al. (2017). *Chemelectrochem* 4: 1256.

76 Xu, Y., Zheng, S., Tang, H. et al. (2017). *Energy Storage Mater.* 9: 11.

77 Jose Piernas-Munoz, M., Castillo-Martinez, E., Gomez-Camer, J.L., and Rojo, T. (2016). *Electrochim. Acta* 200: 123.

78 (a) Huang, Y., Xie, M., Zhang, J. et al. (2017). *Nano Energy* 39: 273. (b) Jose Piernas-Munoz, M., Castillo-Martinez, E., Bondarchuk, O. et al. (2016). *J. Power Sources* 324: 766.

79 (a) Tian, H., Seh, Z.W., Yan, K. et al. (2017). *Adv. Energy Mater.* 7: 1602528. (b) Wang, Y., Wang, Y., Wang, Y.-X. et al. (2019). *Chem* 5: 2547.

80 Ma, L., Cui, J., Yao, S. et al. (2020). *Energy Storage Mater.* 27: 522.

81 Shi, Q., Zhong, Y., Wu, M. et al. (2018). *Angew. Chem. Int. Ed.* 57: 9069.

82 Wang, H., Wang, C., Matios, E., and Li, W. (2018). *Angew. Chem. Int. Ed.* 57: 7734.

83 Zheng, X., Fu, H., Hu, C. et al. (2019). *J. Phys. Chem. Lett.* 10: 707.

84 Xu, Z., Yang, J., Zhang, T. et al. (2019). *Adv. Funct. Mater.* 29: 1901924.

85 Fang, W., Jiang, H., Zheng, Y. et al. (2020). *J. Power Sources*: 455.

86 Lee, Y., Lee, J., Lee, J. et al. (2018). *ACS Appl. Mater. Interfaces* 10: 15270.

87 Yabuuchi, N., Kubota, K., Dahbi, M., and Komaba, S. (2014). *Chem. Rev.* 114: 11636.

88 Ji, W., Huang, H., Zhang, X. et al. (2020). *Nano Energy* 72: 104705.

89 Yamada, Y. and Yamada, A. (2015). *J. Electrochem. Soc.* 162: A2406.

90 Feng, J., Ci, L., and Xiong, S. (2015). *RSC Adv.* 5: 96649.

91 Wang, X., Yasukawa, E., and Kasuya, S. (2001). *J. Electrochem. Soc.* 148: A1058.

92 Eshetu, G.G., Elia, G.A., Armand, M. et al. (2020). *Adv. Energy Mater.* 10: 2000093.

93 Yang, C., Xin, S., Mai, L., and You, Y. (2021). *Adv. Energy Mater.* 11: 2000974.

94 Feng, J., Zhang, Z., Li, L. et al. (2015). *J. Power Sources* 284: 222.

95 Feng, J., An, Y., Ci, L., and Xiong, S. (2015). *J. Mater. Chem. A* 3: 14539.

96 Kondou, H., Kim, J., and Watanabe, H. (2017). *Electrochemistry* 85: 647.

97 Guo, C., Zhang, K., Zhao, Q. et al. (2015). *Chem. Commun.* 51: 10244.

98 Cao, R., Mishra, K., Li, X. et al. (2016). *Nano Energy* 30: 825.

99 Takada, K., Yamada, Y., Watanabe, E. et al. (2017). *ACS Appl. Mater. Interfaces* 9: 33802.

100 Wang, J., Fu, G., Li, W. et al. (2018). *Aquac. Fisheries* 3: 22.

101 Jiang, X., Liu, X., Zeng, Z. et al. (2018). *Adv. Energy Mater.* 8: 1802176.

102 Doi, T., Shimizu, Y., Hashinokuchi, M., and Inaba, M. (2017). *J. Electrochem. Soc.* 164: A6412.

103 Chen, S., Zheng, J., Yu, L. et al. (2018). *Joule* 2: 1548.
104 Zheng, J., Chen, S., Zhao, W. et al. (2018). *ACS Energy Lett.* 3: 315.
105 Yu, Y., Che, H., Yang, X. et al. (2020). *Electrochem. Commun.* 110: 106635.
106 (a) Borodin, O., Self, J., Persson, K.A. et al. (2020). *Joule* 4: 69. (b) Cao, X., Xu, Y., Zhang, L. et al. (2019). *ACS Energy Lett.* 4: 2529.
107 (a) Hu, Y.-S. and Lu, Y. (2019). *ACS Energy Lett.* 4: 2689. (b) Li, Y., Lu, Y., Adelhelm, P. et al. (2019). *Chem. Soc. Rev.* 48: 4655.
108 Li, Y., Yang, Y., Lu, Y. et al. (2020). *ACS Energy Lett.* 5: 1156.

12

Ionic Liquid Electrolytes for Sodium-Ion Batteries

12.1 Introduction

Ionic liquids (ILs) are liquids with only anions and cations at or near room temperature. As an ionic compound, the factor that causes the low melting point of ILs is the asymmetry of some substituents in their structure prevented ions from being regularly accumulated into crystals. ILs usually consist of organic cations and organic/inorganic anions. ILs are generally classified according to their cations or anions. The positive charge position of the cations is N, P-based, and the main categories are imidazole, pyrrolidine, pyridine, morpholine, piperidine, quaternary ammonium, quaternary phosphonium, guanidine, etc. Among them, imidazole-type ILs generally have a low melting point. The anions include halogen ion, tetrafluoroborate ion, and hexafluorophosphate ion [1].

Organic electrolytes in combination with Na salts have been commonly used in Na secondary batteries and have achieved good performance. However, the main hidden dangers of secondary batteries are related to the volatility and flammability of organic electrolytes. ILs not only have the safety advantages, but also show excellent cycling and multiplication properties. IL electrolyte can form a more stable solid electrode interface (SEI) film, which makes it easier for Na ion diffusion, as well as a stable layer on the Al collector, to prevent the Al collector from being corroded by the electrolyte [2]. In addition, ILs can operate at high temperatures, which can promote the diffusion rate of Na in the electrode and electrolyte.

ILs offer many advantages over organic electrolytes commonly used in sodium metal batteries. For example: (i) the liquid has a wide application temperature range and can exist stably under both high and low temperatures; (ii) low vapor pressure, and chemically stable; (iii) high recycling rate, green and environmental; (iv) high conductivity; and (v) high adaptability, as well as adjustable acidity and alkalinity. Figure 12.1 shows the specific energy-power diagrams of various energy storage devices (ESDs). It can be clearly seen that the Na secondary battery with IL additive electrolyte has high specific energy (equivalent to the specific energy of LIB) and power (exceeding the actual specific energy of the supercapacitor) at the optimum temperature, which is difficult to achieve for general energy equipment [3].

Sodium-Ion Batteries: Energy Storage Materials and Technologies, First Edition. Yan Yu.

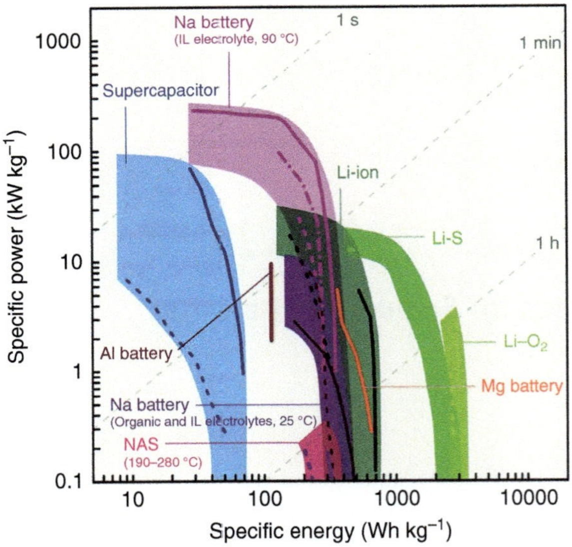

Figure 12.1 Specific energy-power diagram of SIBs with IL electrolytes in comparison with various energy storage devices (Li-ion, Li–O_2, Li–S, Na-ion, Na–S, Al-ion and Mg-ion batteries, and supercapacitors). Source: Matsumoto et al. [1]. Reproduced with permission, 2019, Royal Society of Chemistry.

12.2 The Cationic Species in Ionic Liquids

Common cations in ILs include quaternary ammonium salt ions, quaternary phosphonium salt ions, imidazolium salt ions, etc. (Table 12.1), and anions include halogen ions, tetrafluoroborate ions, and hexafluorophosphate ions. Among the ILs currently studied, the cations are mainly imidazole cations.

Benchakar et al. [4] explored the use of a small amount of 1-ethyl-3-methylimidazolium bis-(fluoromethylsulfonyl) imide (EMImFSI) IL in electrolyte. The experimental results show that the EMImFSI additive has far-reaching impact on all aspects of electrolytes with $NaClO_4$, $NaPF_6$, and NaFSI. The EMImFSI additive helps to promote the ionic conductivity of the electrolyte but increases the viscosity. For electrolytes without EMImFSI ILs, more salt dissociation occurs, along with a higher extension of sodium plating/exfoliation. The introduction of EMImFSI leads to the appearance of thicker SEI, which can affect the Na plating. Electrochemical cycling of mixed hard carbon electrodes shows that FSI-anions can appear new peaks in cyclic voltammetry or new steps in constant current discharge process. The addition of EMImFSI to the $NaPF_6$ system helps improving the Coulombic efficiency and capacity retention (50%) (Figure 12.2a,b).

Electrochemical properties of different sodium salts added to IL electrolytes of 1-butyl-1-methylpyrrolidinium bis (trifluoromethanesulfonyl) imide (BMPTFSI) were reported in SIBs with spherical P_2-$Na_{0.6}Co_{0.1}Mn_{0.9}O_2$ (NCO) as cathode. Experimental results showed that BMPTFSI with sodium bis(fluorosulfonyl)imide (NaFSI) was able to exhibit the best electrochemical performance with significantly better cycling stability (90% half-cell capacity retention) compared to the conventional sodium metal cell electrolyte (1M $NaClO_4$). The cyclic voltammogram (CV) results showed that the BMPTFSI-based IL electrolyte is stable

Table 12.1 List of abbreviations and molecular structures of typical IL cations.

Abbreviation	Molecular structure
$[Emim]^+$	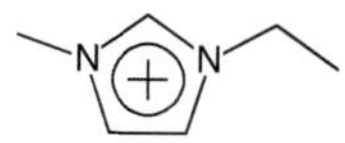
$[C_n mim]^+$	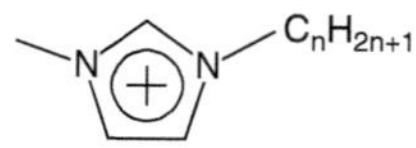
$C_nC_1pyrr^+$	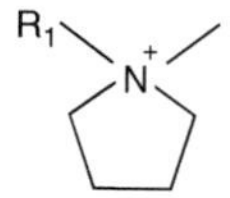
$[Pyr_{13}]^+$	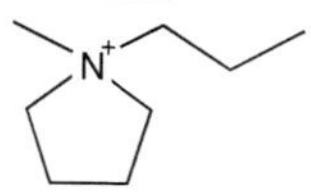
$[Pyr_{14}]^+$	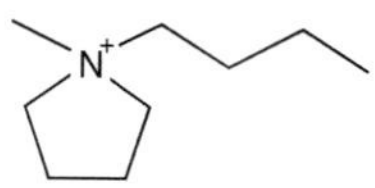
$N_{nnnn}{}^+$	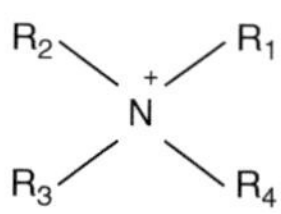
$[P_{111i4}]^+$	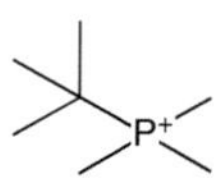
$[P_{1i444}]^+$	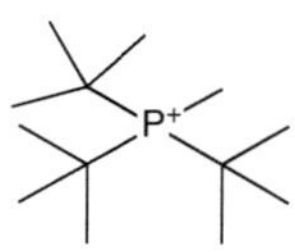
$[DEME]^+$	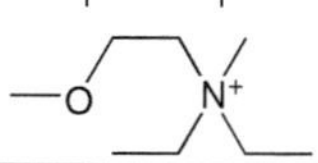

up to 4.8 V vs. Na/Na^+, especially using NaFSI and NaTFSI as conducting salts (Figure 12.2c,d) [5].

12.3 The Anionic Species in Ionic Liquids

According to the different anions, ILs can be divided into two camps: halide salts and non-halide salts (Table 12.2):

(1) The halogenated salt ILs were studied earlier, and solid halogenated salts can be mixed with $AlCl_3$ to obtain ILs. Halide ILs have many advantages but are extremely sensitive to water that need to be handled under inert atmosphere.

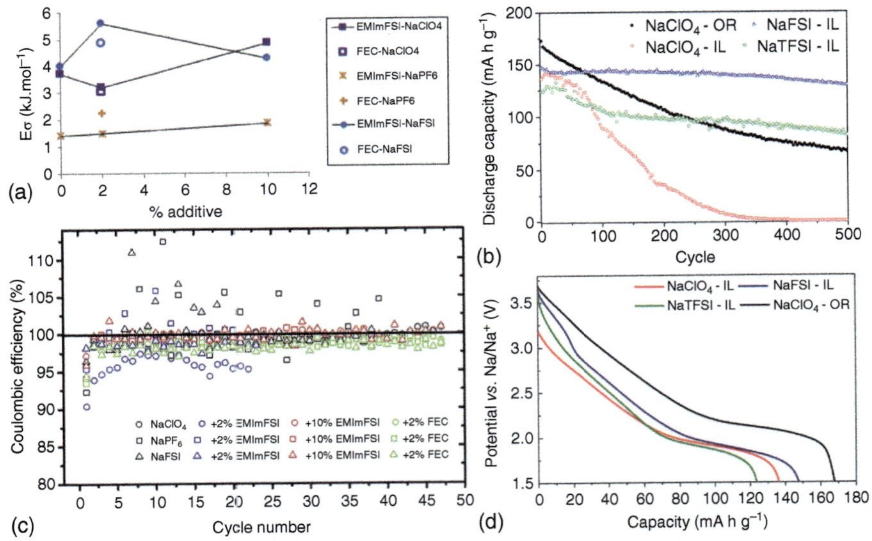

Figure 12.2 Effect of additives content on (a) the activation energy of the conductivity, and (b) the Coulombic efficiency of cells with EMImFSI/EC/PC or FEC/EC/PC ternary mixtures. Source: Benchakar et al. [4]. Reproduced with permission, 2020, Elsevier. (c) Cycling performance and (d) discharge profile of NCO in ionic liquid-based electrolytes at 50 mA g^{-1} in the voltage range of 1.5–3.8 V vs. Na/Na^+. Source: Do et al. [5]. Reproduced with permission, 2019, American Chemical Society.

(2) Non-halogenated salt ILs have a fixed composition, and most of them are stable to water and air, mainly including: CF_3COO^-, $C_3F_7COO^-$, $CF_3SO_3^-$, $C_4F_9SO_3^-$, $(C_4F_9SO_2)N^-$, $(C_2F_5SO_2)N^-$, $CB_{11}H_{12}^-$, $MeSO_4^-$, $C_8H_{17}SO_4^-$, etc.

Hwang et al. [6] prepared $Na_3V_2(PO_4)_3$ cathode, which displayed superior electrochemical properties in Na[bis(fluorosulfonyl)-amide]-[1-ethyl-3-methylimidazolium] [bis (fluorosulfonyl) amide] IL electrolyte with good heat resistance, high voltage stability, and other properties. In 1C (116 mA g^{-1}) cycling test, the capacity retention was 99% at 90 °C. It also reached 89.2% after 5000 cycles at a high rate of 20 C under 184 °C (Figure 12.3a,b).

Zhang et al. [7] successfully synthesized carbon and aluminum oxide co-coated $Na_3V_2(PO_4)_2F_3$ (NVPF) cathode material (Figure 12.3c,d). The cells with NVPF/C cathode and [EMIM]TF_2N ILs electrolyte exhibited high reversible specific capacity (117.5 mA h g^{-1} at 1 C), good rate performance (94 mA h g^{-1} at 10 C), and excellent cycling stability (90% capacity retention after 1000 cycles at 1 C). Experiments have also shown that doping of C with elements such as N, S, and F induces the formation of an effective conducting network of electrons and Na ions, leading to excellent electrochemical properties of NVPF/C-ILs.

Basile et al. [8] reported the practical use of pyrrolidine ILs containing dicyandiamide anions as electrolyte additives in sodium metal batteries, which is inexpensive. A comparison was made with a reference organic electrolyte consisting of propylene carbonate-fluorocarbonate. The experimental results

Table 12.2 List of abbreviations and molecular structures of typical IL anions.

Abbreviation	Molecular structure
$[BF_4]^-$	
$[FSA]^-$	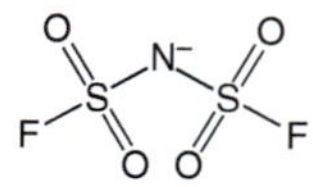
$[TFSA]^-$	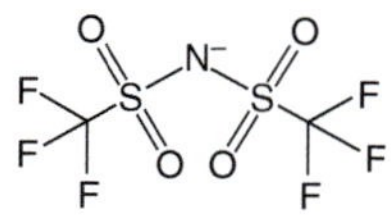
$[N(CN)_2]^-$	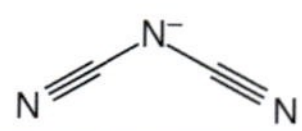

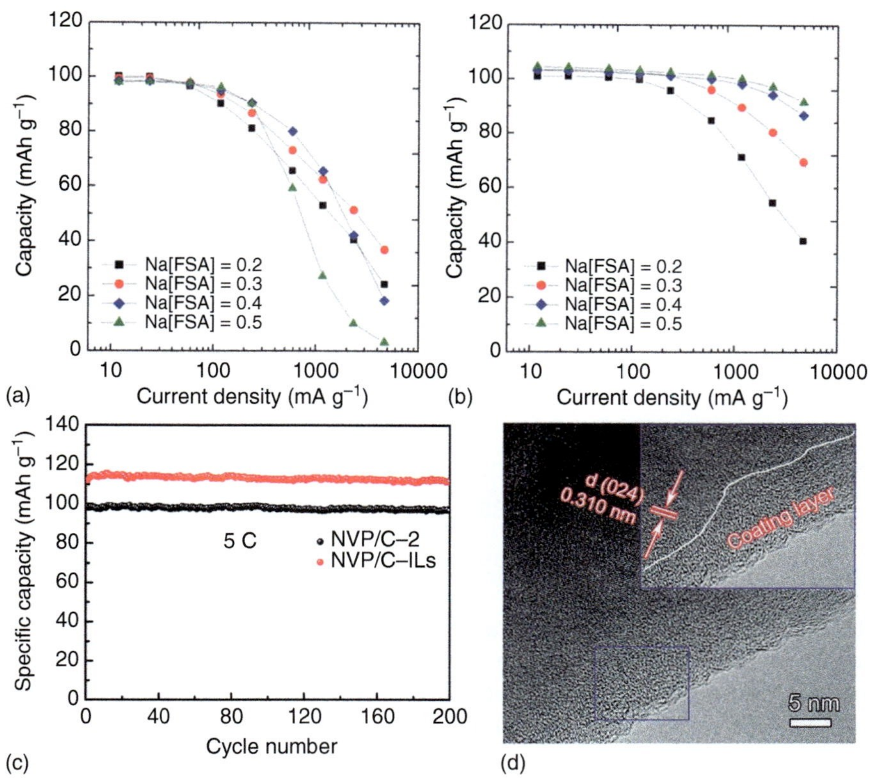

Figure 12.3 Rate performance of $Na/Na_3V_2(PO_4)_3$ cells in Na[FSA]-[C2C1im] [FSA] electrolytes with different Na[FSA] molar fractions (0.2–0.5 M) at (a) 148 °C and (b) 184 °C (charge current density: 0.1 C, and cutoff voltage: 2.4–3.8 V). Source: Hwang et al. [6]. Reproduced with permission, 2018, Wiley-VCH. (c) Cycling stability and (d) corresponding TEM images of carbon-coated $Na_3V_2(PO_4)_2F_3$ under an electrolyte containing ionic liquids. Source: Zhang et al. [7]/Royal Society of Chemistry.

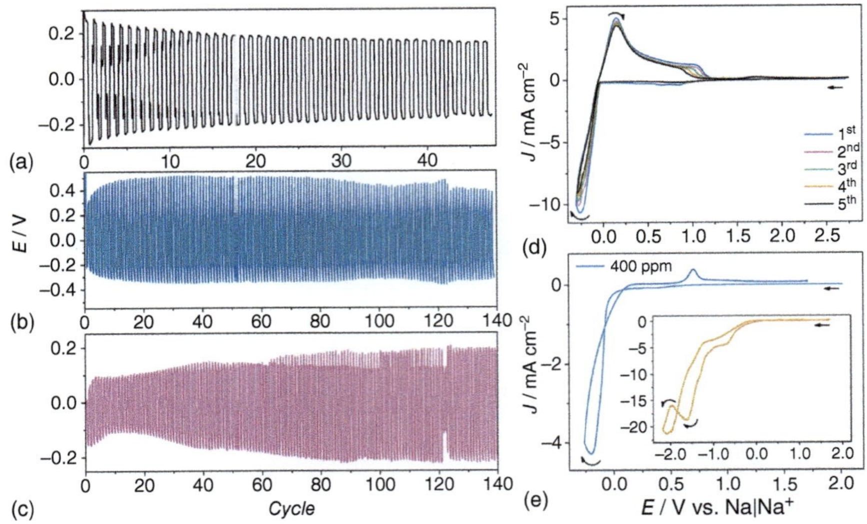

Figure 12.4 Cycling of a Na||Na symmetrical cell in PC/FEC electrolyte containing 0.5 M $Na[NTf_2]$ (a), Na[dca] in C_4mpyr[dca] with different content: 400 ppm (b) and 90 ppm (c). Cyclic voltammetric of the reduction of Na^+ to Na0 onto Ni electrodes from 0.5 M Na[dca] in C_4mpyr[dca] at room temperature with a scan rate of 20 mV s^{-1} under a moisture content of 90 ppm (d) and 400 ppm (e). Inset: second scan with wider potential window. Source: Basile et al. [8]. Reproduced with permission, 2016, Elsevier.

indicate that the presence of trace amounts of water in the electrolyte with ILs affects the Na|Na symmetric cell significantly. When the water content of an *N*-butyl-*N*-methyl-pyrrolidinium dicyanamide IL electrolyte including Na[dca] salt is maintained at 90 ppm, the enhanced cycling performance for the sodium electrode is achieved. Repetitive stripping/plating cycling studies show that a Na|Na symmetrical cell can cycle effectively for >100 times at a current density of 0.01 mA cm^{-2} without short circuits. By maintaining a low but not zero water content, sodium metal electrodes will not undergo formation of a resistive passivating layer as that generated in the electrolyte comprising 400 ppm water content (Figure 12.4).

12.4 Electrolyte Properties

12.4.1 Physicochemical Properties

The melting point of ILs is an important physical property that determines the application temperature range of ILs. The composition structure of ILs, such as elements and functional groups, has a great influence on their melting points. Factors affecting the melting point of ILs are: (i) the structure of the cation; (ii) the coordination of the anion; (iii) the symmetry of the anion; (iv) the H-bond interaction; (v) the planarity of the cation; and (vi) the symmetry effect of the molecular structure. Research indicates that the smaller the volume of the organic cation, the more concentrated the charge, the better the molecular symmetry, and the higher melting point of the IL are demonstrated. The larger the proportion and space occupied by anions, the lower the melting point of the IL is acquired, e.g. showing the order of $Cl^- > [PF_6]^- > [NO_2]^- > [NO_3]^- > [AlCl_4]^- > [BF_4]^- > [CF_3SO_3]^- > [CF_3CO_2]^-$. For quaternary ammonium salt ILs, the effect of cations on the melting point is greater than that of anions [9].

Density is one of the physical features of ILs that can be easily adjusted. There are billions of anions and cations for ILs. The density of the IL mainly depends on the types of anions and cations. Therefore, the anion and cation species of the IL need to be mastered to regulate the required density.

High viscosity is a very typical physical characteristic of ILs, which is usually higher than that of the conventional organic solvents. At the same time, the viscosity of ILs is affected by the temperature. As the temperature rises, the interaction between anions and cations in ILs gradually decreases, resulting in reduced viscosity. Studies have shown that as the organic content increases, the viscosity of the IL mixture gradually decreases, and the decrease is mainly due to the stronger hydrogen bonds between the anions and cations in ILs [10]. Seddon et al. [11] have suggested the following formula to express the relationship between the viscosity of IL + organic system and its composition:

$$\eta = \theta \exp(-x/b)$$

In this formula, η and θ represent the viscosity of the mixture and the IL, respectively, x is the mole fraction of organic matter in the mixture, and b is the parameter to be determined in the experiment.

12.4.2 Electrochemical Properties

"Electrochemical window" mainly refers to the range of potential changes from the reduction limit to the oxidation limit of the electrolyte. Theoretical studies show that the main limited factors are the highest occupied molecular orbital and the lowest unoccupied molecular orbital of the molecules in the electrolyte, but they are often kinetically dominated in practical situations, as clearly demonstrated by the importance of the SEI [12]. In the case of Na secondary cells, an electrochemical window from the Na^+/Na redox potential to approximately 5 V is ideal for high energy density operation, and many ILs containing BF^-, $TFSA^-$, and FSA^- meet this requirement (or provide a window close to it). Therefore, ILs have a wide range of applications in SIBs.

Figure 12.5 shows the electrochemical windows for some common ILs. Studies combining molecular dynamics (MD) simulations and density functional theory (DFT) calculations indicate that the stability of ionic and counter-ion combinations in ILs should be discussed in a specific theoretical model. Therefore, when considering the electrochemical stability of ILs, it is valuable to fully consider the chemical stability of multiple ionic combinations [13]. TFSA is proved to have a higher reduction potential than $C_3C_1pyrr^+$. Thermodynamic theoretical calculations show that the alkyl chain length of the 1-alkyl-3-methylimidazole cation is not directly related to the electrochemical window, and ILs containing BF_4^- and PF_6^- anions

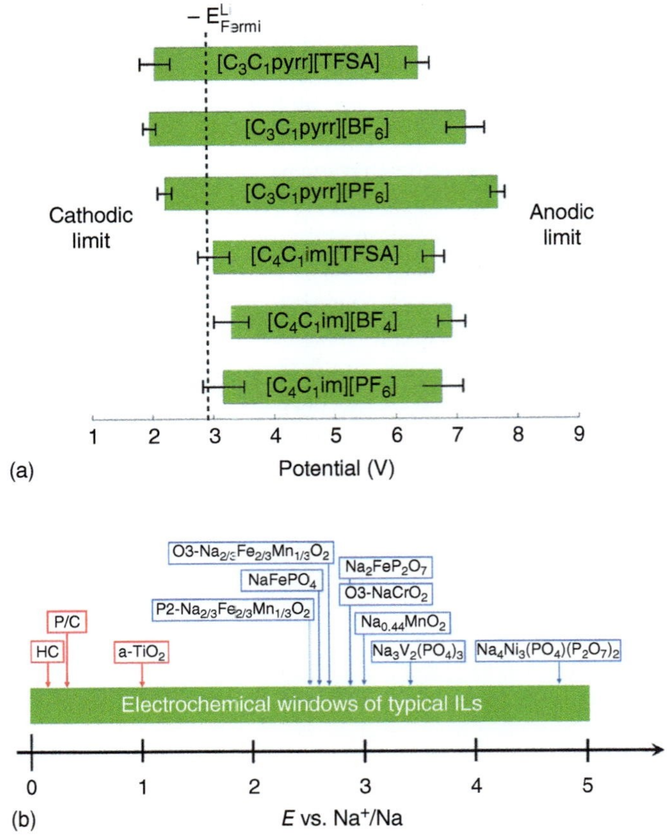

Figure 12.5 (a) Electrochemical windows for different classes of ILs. Source: Ong et al. [13]. Reproduced with permission, 2011, American Chemical Society. (b) Electrochemical window of a Na secondary cell with different IL additives in the electrolyte. Source: Matsumoto et al. [1]. Reproduced with permission, 2019, Royal Society of Chemistry.

have better electrochemical stability than those containing TfO or TFSA anions [1, 14].

Cyclic voltammetry and linear scanning voltammetry are also commonly used methods to characterize the electrochemical stability of ILs. Two-electrode cells (mainly coin cells) and three-electrode cells are commonly used to test performances in Li^+ and Na^+ ILs. Regardless of the model of the anions, the addition of Li or Na salts to the IL will generally extend its reduction limit. Although aromatic cations, including alkylimidazolium and alkylpyridinium, are electrochemically unstable compared to non-aromatic or linear alkylammonium cations and are difficult to reduce in pure IL, the application of FSA as a counter anion can improve electrochemical stability [15]. For FSA-based ILs, the differences of electrochemical stability based on the cationic structure are minimal. The reduction of FSA^- and $TFSA^-$ generates radical anions, and the radical anion reactivity of FSA^- is abnormally low and cannot form a stable surface film. When contacted and reacted with different cations, such as Na^+ and Li^+, it still failed to form a stable SEI film, which verified the above conclusion [16].

Cells with IL in the electrolyte have a low probability of suffering corrosive effects on the cathode at high potentials. Unlike the mechanism of instantaneous formation of fluorinated SEI film on the cathode by PF_6-based electrolytes, electrolytes containing only $TFSA^-$ and FSA^- ILs take longer time to form SEI film and do not protect the cathode well [17]. However, in IL electrolyte, the formed SEI film is able to minimize the corrosion of Al current collector by the electrolyte after long cycles. The same behavior was also observed in the Na battery system. In the FSA^- and $TFSA^-$ IL electrolytes, the aluminum collector for the cathode is not corroded [18]. At high potentials, ILs form a stable film on Al, which prevents Al from being corroded and enhances the battery performance.

12.4.3 Thermal Properties

Thermogravimetric (TG) and differential scanning calorimeter are commonly used analytical methods to study the thermal stability properties of IL electrolytes [19]. In general, IL is more thermally stable than organic solvents. When using TG for testing, all aspects of the test must be strictly controlled to ensure that the measured thermal decomposition temperature is obtained under the same conditions. In general, the operating temperature range of organic solvents is limited, but ILs can extend the maximum operating temperature of the battery and enhance the high temperature performance of the battery.

The thermal decomposition temperature of IL in the SIB system was investigated. The TFSA-based IL reduced the decomposition temperature to 330–360 °C (about 400 °C for pure salts) [20]. It can be found that the addition of $NaPF_6$ to the IL reduces the decomposition temperature (weight loss starts at about 100 °C). The IL system consisting of $NaBF_4$-$[C_2C_1im]$ $[BF_4]$ has non-combustibility even with direct flame contact and reaches temperatures up to 360 °C, regardless of the concentration of $NaBF_4$ (0.10–0.75 mol dm^{-3}) [20a]. It is thermally stable at high temperatures. Further studies on the thermal stability of FSA-based salts are needed because of their low solubility in organic solvents. Even if the melting point of IL is higher than the theoretical decomposition temperature [21], slow melting with slow decomposition rate is sometimes observed. TG studies of IL systems have shown that the thermal decomposition temperature of electrolytes containing FSA salts increases continuously with increasing volume size of alkali metal cations [22].

Figure 12.6 summarizes the theoretical temperature range of IL electrolytes and the actual operating temperature range for various practical applications. The suitable temperature range of IL electrolytes in commercial SIBs is also summarized and compared. Except for some studies on FSA-based IL which show that it can operate at low temperatures, most studies on Na secondary cells have been performed at room temperature. In fact, some studies at room temperature have shown that electrochemical testing above the IL melting point may improve the performance of electrode materials. There are also studies in the literatures that have attempted to

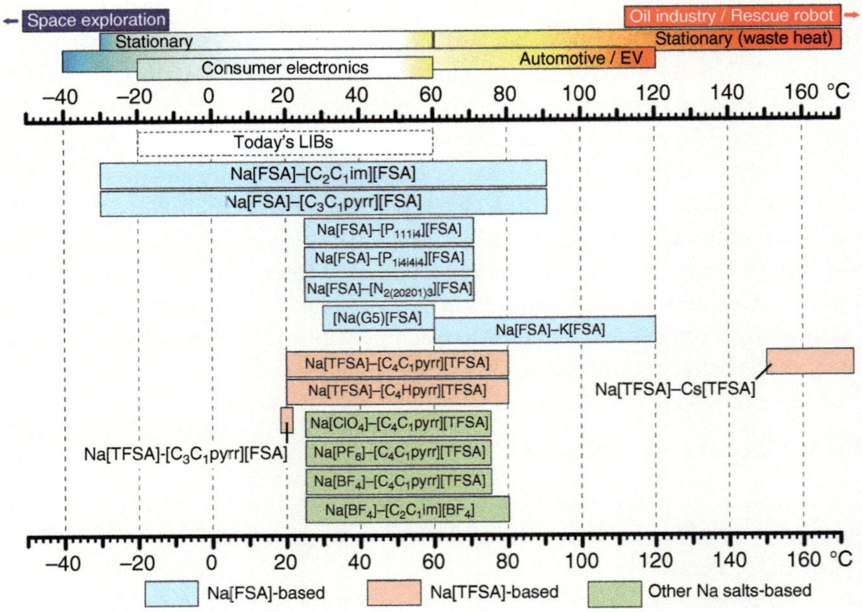

Figure 12.6 The actual temperature range of the IL applied to sodium secondary batteries. Source: Matsumoto et al. [1]. Reproduced with permission, 2019, Royal Society of Chemistry.

use IL for neutralization temperature operation (up to 150 °C) in waste heat and high temperature environments so as to improve the mobility of Na ions and the electrode reaction. For example, when the temperature was changed from 25 to 85 °C, the ionic conductivity was increased from 1.9 to 17.3 mS cm^{-1}. The following sections will discuss in detail the performance of the IL at the appropriate temperature [6, 23].

12.5 Stability of Ionic Liquids

12.5.1 Thermal and Electrochemical Stability

The design of battery systems and related applications using mixtures of IL, H_2O, and organic solvents have been extensively investigated. The use of IL to achieve the temperature regulation characteristics of battery systems has several advantages. Firstly, ILs have high tunability. Both their cationic and anionic properties can be easily controlled as needed. Secondly, ILs are highly polar and insoluble in many organic solvents, making them easier to separate from non-aqueous systems. The addition of ILs to the electrolyte can enhance the temperature limit and voltage range of the electrochemistry in the cell.

Figure 12.7 displays the effect of temperature on the charge/discharge properties of some SIB cathode materials. $Na_3V_2(PO_4)_3$ and $NaCrO_2$ were paired with 20 mol% Na[FSA]-[C_2C_1im][FSA] and Na[FSA]-[C_3C_1pyrr][FSA] IL electrolytes, respectively, and cycled at a wide temperature range of −30 to 90 °C [6, 25]. $Na_3V_2(PO_4)_3$

cells are stable over the wide temperature range (Figure 12.7a,b). As the test temperature decreases, it leads to an increasing resistance and polarization of the cell, and then the reversible capacity is decreased. This is because that the ionic conductivity is decreased and charge transfer resistance is increased, which hinder the diffusion of Na ions in the electrolyte. Nevertheless, the IL electrolyte can still support the full operating temperature range of the Na secondary batteries.

Previous study showed that mallet-$NaFePO_4$ is an electrochemically inert material. It is constructed by FeO_6-FeO_6 and PO_4-FeO_6 (horn-shared) units, which hinder the effective diffusion of Na^+ [26]. The application of IL electrolytes allows to increase the operating temperature, thus enhancing the diffusion of Na^+ in this structural framework, which effectively makes the mallet-$NaFePO_4$ electrochemically active. The sodiation/desodiation of mallet-$NaFePO_4$ was achieved at an elevated operating temperature [24, 27]. The charging and discharging of mallet-$NaFePO_4$ in the IL electrolyte showed that the reversible capacity increased sharply from only 23 mA h g^{-1} at 25 °C (Figure 12.7c) to 107 mA h g^{-1} at 90 °C (Figure 12.7d) [24].

12.5.2 Electrochemical Properties

Related studies confirmed that high-temperature operation using ILs can significantly improve the electrochemical performance. Figure 12.8 shows the electrochemical performance of the Na[FSA]-[C_2C_1im][FSA] electrolyte under different electrode mass loading and operating temperatures [28]. The synergistic effect of $Na_3V_2(PO_4)_3$ and IL electrolyte at 25 °C can lead to better performance of the cell. However, when the mass loading of the electrodes is high, the rate performance at room temperature deteriorates sharply due to the increased geometric current density. For example, electrode loadings of 3.0 and 6.0 mg cm^{-2} show capacity retention of 90% and 63% at 5 C, respectively (Figure 12.8a). However, capacity retention is almost the same when applying the same geometric current density (68% at 10 C for 3.0 mg cm^{-2} and 63% at 5 C for 6 mg cm^{-2}). The rate performance is maintained even when the mass load of $Na_3V_2(PO_4)_3$ is 6 mg cm^{-2} at 90 °C (Figure 12.8a). Therefore, the rate performance mainly depends on the geometric current density of the active material of the pole piece. As shown in Figure 12.8b, at 500, 1000, 1600, and 2000 C, the capacity retention is 69%, 58%, 36%, and 19%, respectively. At room temperature of about 25 C, the organic solvent electrolyte never reaches this level of rate performance. The EIS measurements show that the interfacial resistance (characteristic frequency 60–700 Hz) represented by the semicircle is significantly lower (Figure 12.8c) [6].

Most studies reported that IL electrolytes have excellent wide temperature performance, maintaining good cycling stability in both high and low temperature environments [28a, 29]. It is well known that IL electrolyte can create a specific SEI film on the surface of Na metal. Figure 12.9 shows the cycling performance of the selected electrode material. Cycle testing of Na/$NaFePO_4$ cells reveals better cycling performance using Na[TFSA]-[C4C1pyrr] [TFSA] IL electrolyte than that

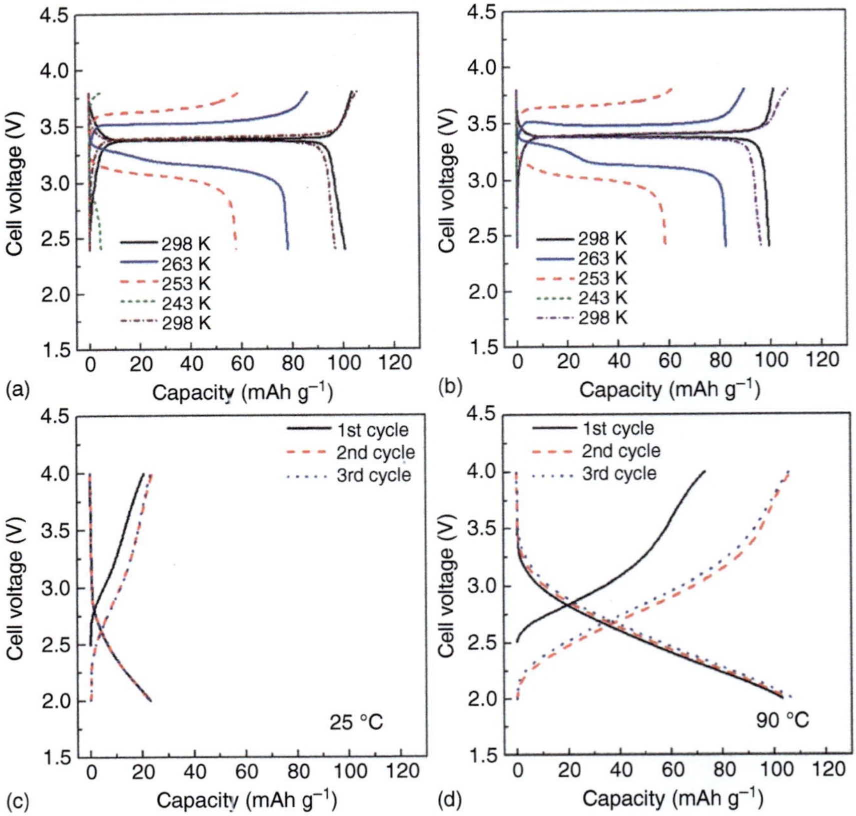

Figure 12.7 Charge/discharge curves of Na/20 mol% Na [FSA]-[C_2C_1im] [FSA]/carbon coated $Na_3V_2(PO_4)_3$ cells at (a) −30 to 25 °C and (b) −20 to 90 °C. Source: Reproduced with permission Hwang et al. [6]. Copyright 2018, Wiley-VCH. Charge/discharge curves of Na/30 mol% Na [FSA]-[C_2C_1im] [FSA]-$NaFePO_4$ cells at (c) 25 °C and (d) 90 °C. Source: Hwang et al. [24]. Reproduced with permission, 2018, Elsevier.

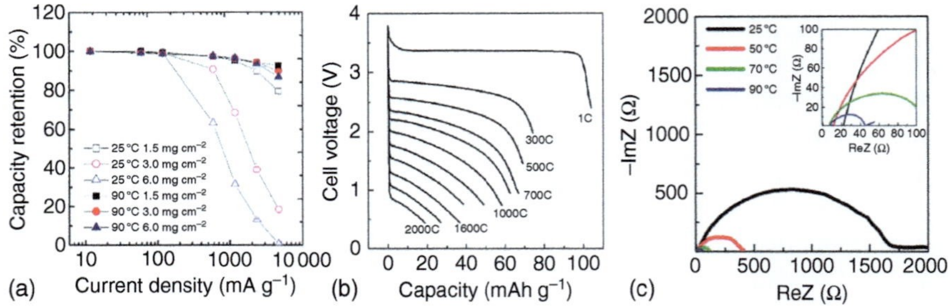

Figure 12.8 Electrochemical properties of IL added to carbon coated $Na_3V_2(PO_4)_3$: (a) rate performance of Na/40 mol% Na [FSA]-[C_2C_1im] [FSA]/carbon coated $Na_3V_2(PO_4)_3$ cells at 25 and 90 °C with different mass loading. Source: Hwang et al. [28a]. Reproduced with permission, 2018, Wiley-VCH. (b) Rate performance of Na/50 mol% Na [FSA]-[C_2C_1im] [FSA]/carbon coated $Na_3V_2(PO_4)_3$ cell at 90 °C, and (c) EIS of Na/50 mol% Na [FSA]-[C_2C_1im] [FSA]/carbon coated $Na_3V_2(PO_4)_3$ cell. Source: Hwang et al. [6]. Reproduced with permission, 2018, Wiley-VCH.

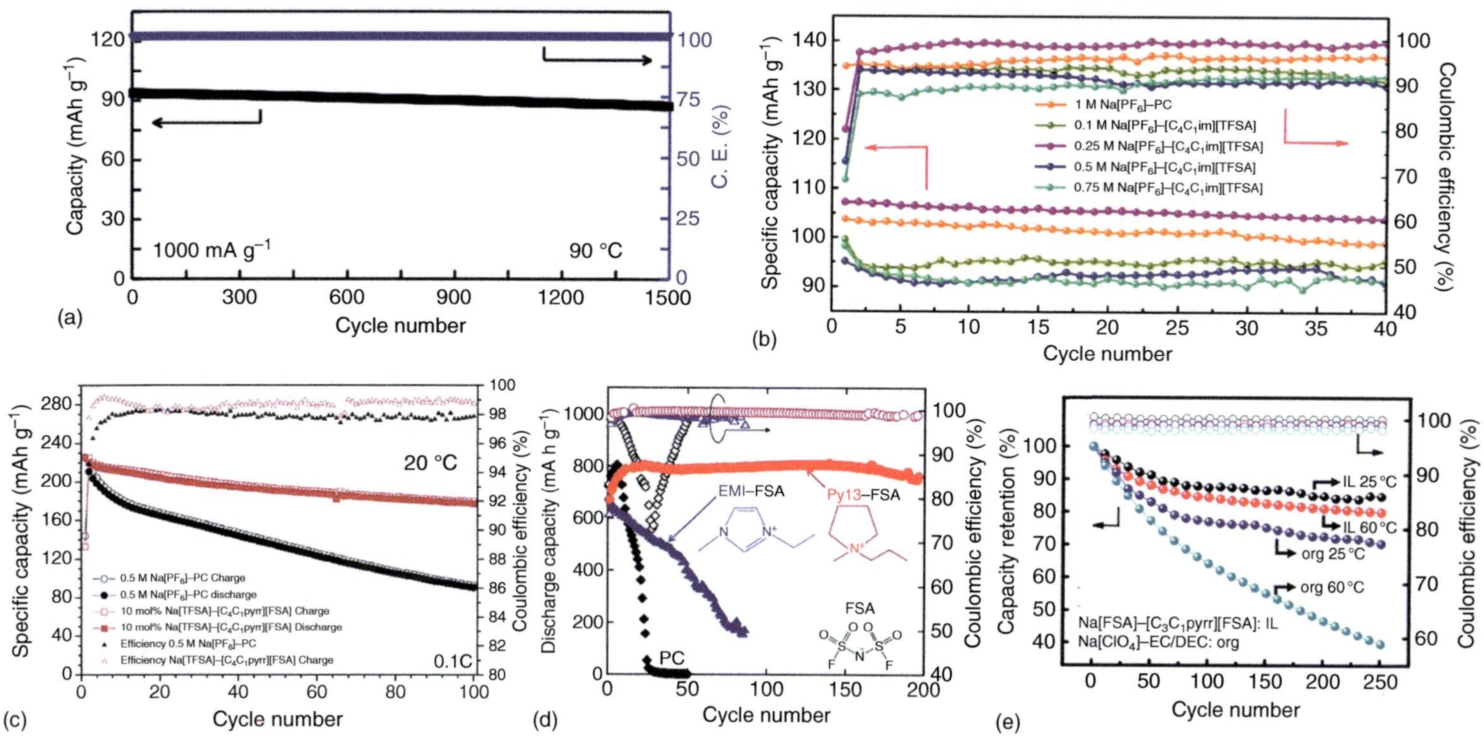

Figure 12.9 (a) Cycling performance of Na/30 mol% Na [FSA]-[C_2C_1im] [FSA]/$Na_2FeP_2O_7$ cells at 90 °C. Source: Chen et al. [29c]. Reproduced with permission, 2016, Elsevier. (b) Cycling performance of Na/1 mol dm^{-3} Na [PF_6]-PC/$Na_3V_2(PO_4)_3$ and different proportion of sodium (0.1, 0.25 and 0.5) and 0.75 mol dm^{-3} Na [PF_6]-[C_2C_1im] [TFSA]/$Na_3V_2(PO_4)_3$ cell systems. Source: Wu et al. [30]. Reproduced with permission, 2018, Elsevier. (c) Cycling performance in the Na/0.5 mol dm^{-3} Na [PF_6]-PC/$Na_{0.45}Ni_{0.22}Co_{0.11}Mn_{0.66}O_2$ and Na/10 mol% Na [TFSA]-[C_4C_1im] [TFSA]/$Na_{0.45}Ni_{0.22}Co_{0.11}Mn_{0.66}O_2$ cell systems. Source: Chagas et al. [29b]. Reproduced with permission, 2014, Elsevier. (d) Cycling performance and CE of 1 mol dm^{-3} Na [FSA]-[C_3C_1pyrr] [FSA], Na [FSA]-[C_2C_1im] [FSA], and Na [FSA]-PC at 30 °C. Source: Usui et al. [31]. Reproduced with permission, 2017, American Chemical Society. (e) Capacity and CE of 3D CF electrodes in 1 mol dm^{-3} Na [ClO_4]-EC/DEC (1 : 1, V) and 1 mol dm^{-3} Na [FSA]-[C_3C_1pyrr] [FSA] at 100 mA g^{-1} under 25 and 60 °C. Source: Luo et al. [32]. Reproduced with permission, 2018, Elsevier.

of Na[TFSA]-[C4C1pyrr] [TFSA] electrolyte [29c]. The cycling stability of the Na [FSA]-[C_2C_1im] [FSA] electrolyte in Na/$Na_2FeP_2O_7$ full cells was tested at 90 °C (Figure 12.9a), which shows a capacity retention of 93% and an average Coulombic efficiency of 99.9% after 1500 cycles. Another experiment using the same battery system also shows excellent cycling capabilities. After 300 cycles at 1 C (117 mA g^{-1}) under 25 and 90 °C, the capacity retention was 99.9%. The excellent cycling performance of this battery system is attributed to the stability of this cathode material on the one hand, and also depends on the stability of this IL electrolyte above 90 °C on the surface of Na metal and electrode sheet (Figure 12.9b) [30].

The study of $Na_3V_2(PO_4)_3$ in Na [PF_6]-[C_4C_1im] [TFSA] IL electrolyte showed that the salt concentration of the electrolyte was optimized, which could simultaneously improve the capacity and cycling performance. As shown in Figure 12.9b, the specific discharge capacity of the sodium metal cell with 0.25 mol dm^{-3} Na [PF_6]-[C_4C_1im] [TFSA] ILs was 107.2 mA h g^{-1} and the capacity retention was 97% after 40 cycles. It is worth noting that sufficient Na^+ can facilitate the (de-)alkalinization process, but excess sodium ions reduce the ionic conductivity of the electrolyte, which in turn leads to poor electrochemical performance of the battery (Figure 12.9c–e) [29b, 31, 32].

12.5.3 Electrolyte/Electrode Interfaces

The SEI film formed by the reductive decomposition of the commercial carbonate electrolyte on the surface of anodes has poor stability and cannot effectively mitigate the phenomenon of swelling of the electrode active material, which in turn leads to a rapid decrease in capacity [33]. The success of conventional HC anodes depends on the introduction of complex electrolyte formulations and SEI-forming additives. Therefore, the exploration of electrolytes that are compatible in all aspects of performance is essential for the development of anode.

The electrochemical performance and safety of the battery heavily depend on the properties of the SEI film (Figure 12.10a) [34], which is sensitive to ambient temperature, air humidity, and the viscosity of the binder. A detailed study of the physical–mechanical properties and chemical reaction characteristics of the SEI layer help us to understand the degradation mechanism of the cell in more detail. However, progress on this aspect is slow due to the complex structure of the SEI layer, and limited characterization tools. Furthermore, the solubility of SEI membranes in conventional electrolytes used in Na secondary cells exacerbates the self-discharge properties.

Recently, the electrochemical properties of IL electrolytes interacting with metal Na have been further investigated. Typically, IL electrolytes in SIBs produce up to a large interfacial impedance, which may be caused by the surface film on the Na metal [35]. Studies of ILs systems using Na/Na symmetric cells have shown that cell resistance is inversely proportional to ambient temperature but increases with the increase of Na salt content in the electrolyte. Eshetu et al. have extensively studied the structure and properties of the SEI layer formed by the decomposition of electrolytes on the HC surface, and it has been reported that the anion of the

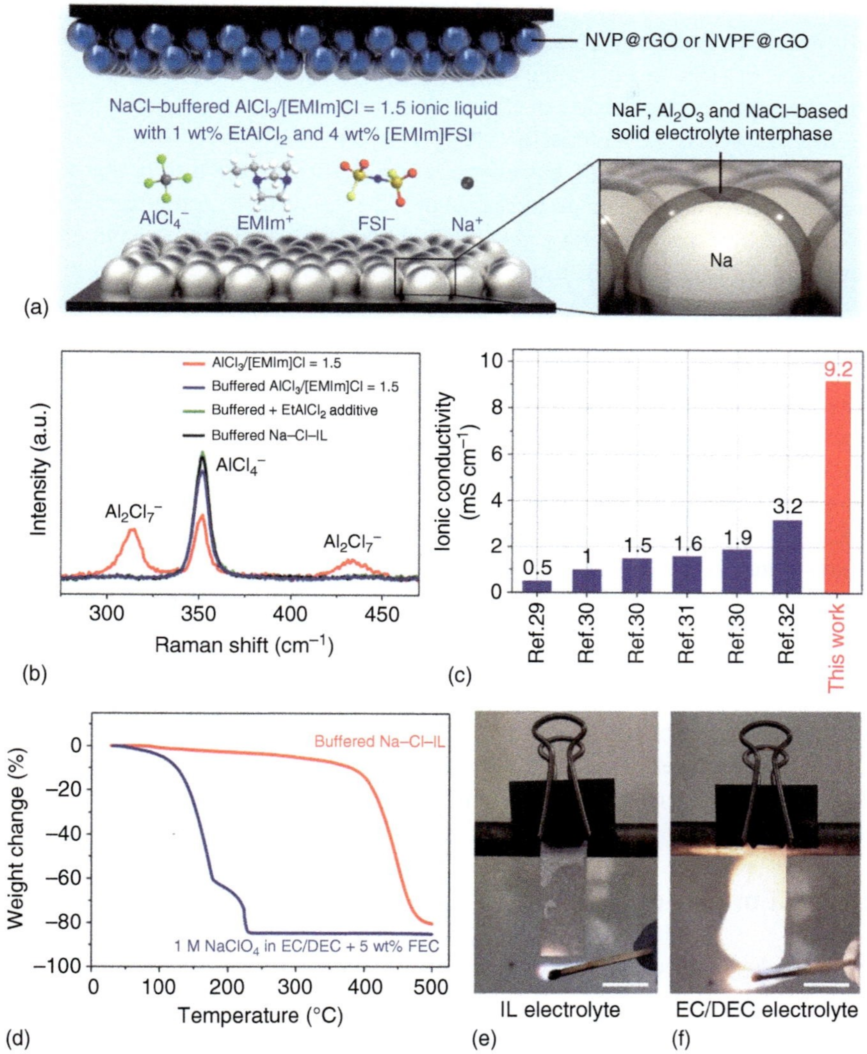

Figure 12.10 (a) Schematic diagram of cell structure and electrolyte composition of IL electrolyte. (b) Raman spectra of IL paired with AlCl3/[EMIm]Cl = 1.5 for different types of IL additives. (c) Ionic conductivity of Na–Cl–IL and different IL-based electrolytes at room temperature. (d–f) Thermal stability (d) and flame retardancy tests using Na–Cl–IL (e) and 1.0 M $NaClO_4$ in EC: DEC (1 : 1 V) with 5 wt% FEC electrolyte (f). Source: Sun et al. [34a]/Springer Nature.

electrolyte salt dominates the SEI film characteristics. This group also studied the effect of reductive decomposition of ILs on the thermal stability of SEI film. For conventional carbonate solvents, the initial temperature of the first exothermic peak (related to the SEI membrane cleavage) is mainly in the following order: $Na[ClO_4] < Na[PF_6] < Na[TFSA] \approx Na[TFSA] < Na[FSA]$ (Figure 12.10b) [28a, 34a]. In the conventional carbonate solvent EC/DEC, there is a more severe corrosion of the aluminum collector. However, the presence of Na[FSA] salt in different solvents slows down the corrosion of Al collectors, and the order of stability is EC/DEC < EC/DEC + 5% $NaPF_6$ < [C_4C_1pyrr] [FSA]. The experimental results show that the ILs electrolyte can effectively maintain the structural stability of the SEI layer as well as the Al collector.

The SEI layer formed in IL electrolyte is significantly different from that formed in organic electrolytes. The performance of SEI layer formed on the surface of HC modulated by 1 mol dm^{-3} Na [FSA]-[C_3C_1pyrr] [FSA] IL electrolyte and conventional organic electrolyte at room temperature were experimentally investigated. The results showed that the charge transfer resistance of the SEI films formed in the HC/Na symmetric cells containing IL electrolyte was lower than that in the conventional organic electrolyte. SEM and XPS measurements confirmed the formation of dense SEI films enriched with Na_2CO_3 and $NaCO_2R$ inorganics in the organic electrolyte. These inorganics are considered to be resistive components in the SEI layer, which affect the cycling performance of the cell. Similar inorganics were also observed in the three-dimensional carbon framework using 1 mol L^{-1} $NaClO_4$+EC/DEC (1 : 1, v/v), while polyolefins ($(CH_2)_n$) and SQCQO from FSA anions were found in the SEI layer formed in the IL (Figure 12.10c) [34a, 35b].

Dai and co-workers [34a] reported a chloroaluminate IL electrolyte consisting of two important new additives, ethylaluminum dichloride and 1-ethyl-3-methylimidazolium bis(fluorosulfonyl)imide. The fabricated cell achieves a high level of energy and power density of approximately 420 Wh kg^{-1} and 1766 W kg^{-1}, respectively, with a Coulombic efficiency of 99.9% and a voltage of approximately 4 V. The experimental results show that the sodium metal battery has high energy, high power density, long cycle life, and high safety (Figure 12.10e,f).

12.6 Concluding Remarks

ILs offer their own distinct advantages over conventional organic carbonate electrolytes. Their low volatility and wide temperature properties have a very high commercial value. To further explore the need for a suitable electrochemical window, fast ion conduction, and wide temperature range, further research is desirable to develop ideal ILs. Therefore, the ionic and dissolved structures, sodium salt morphology and concentration, cell structure, and operating procedures must be further optimized in order to obtain the best performance of SIBs.

Novel ILs with superior functionality of anions are important targets for future electrolyte research. For instance, IL gel polymer electrolytes prepared by combining IL with inorganic ceramics are very promising, which not only have suitable mechanical strength and excellent conductivity but also can be used to develop high safety and high specific energy all-solid Na secondary batteries. In addition, eutectic solutions (DES) and other complex salt-based electrolytes may be inexpensive alternatives, although there is still a gap in performance with ILs. ILs and their analogs will be used more and more widely in the SIB field.

References

1 Matsumoto, K., Hwang, J., Kaushik, S. et al. (2019). *Energy Environ. Sci.* 12: 3247.

2 (a) Molina Piper, D., Evans, T., Xu, S. et al. (2016). *Adv. Mater.* 28: 188. (b) Elia, G.A., Ulissi, U., Jeong, S. et al. (2016). *Energy Environ. Sci.* 9: 3210.

3 Yang, Q., Zhang, Z., Sun, X.-G. et al. (2018). *Chem. Soc. Rev.* 47: 2020.

4 Benchakar, M., Naéjus, R., Damas, C., and Santos-Peña, J. (2020). *Electrochim. Acta*: 330.

5 Do, M.P., Bucher, N., Nagasubramanian, A. et al. (2019). *ACS Appl. Mater. Interfaces* 11: 23972.

6 Hwang, J., Matsumoto, K., and Hagiwara, R. (2018). *Adv. Sustain. Syst.* 2: 1870033.

7 Zhang, L.-L., Zhou, Y.-X., Li, T. et al. (2018). *Dalton Trans.* 47: 4259.

8 Basile, A., Yoon, H., MacFarlane, D.R. et al. (2016). *Electrochem. Commun.* 71: 48.

9 (a) Welton, T. (1999). *Chem. Rev.* 99: 2071. (b) Wilkes, J.S. (2002). *Green Chem.* 4: 73.

10 (a) Ferdousi, S.A., Hilder, M., Basile, A. et al. (2019). *ChemSusChem* 12: 1700. (b) Basile, A., Ferdousi, S.A., Makhlooghiazad, F. et al. (2018). *J. Power Sources* 379: 344.

11 Seddon, K.R., Annegret, S., and María-José, T. (2000). *Pure Appl. Chem.* 72: 2275.

12 (a) Peled, E. and Menkin, S. (2017). *J. Electrochem. Soc.* 164: A1703. (b) Bommier, C. and Ji, X. (2018). *Small* 14: 1703576.

13 Ong, S.P., Andreussi, O., Wu, Y. et al. (2011). *Chem. Mater.* 23: 2979.

14 Kazemiabnavi, S., Zhang, Z., Thornton, K., and Banerjee, S. (2016). *J. Phys. Chem. B* 120: 5691.

15 (a) Fukuda, Y., Tanaka, R., and Ishikawa, M. (2007). *Electrochemistry* 75: 589. (b) Hosokawa, T., Matsumoto, K., Nohira, T. et al. (2016). *J. Phys. Chem. C* 120: 9628. (c) Borgel, V., Markevich, E., Aurbach, D. et al. (2009). *J. Power Sources* 189: 331.

16 (a) Iwahashi, T., Miwa, Y., Zhou, W. et al. (2016). *Electrochem. Commun.* 72: 54. (b) Yamagata, M., Nishigaki, N., Nishishita, S. et al. (2013). *Electrochim. Acta* 110: 181.

17 (a) Dahbi, M., Ghamouss, F., Tran-Van, F. et al. (2011). *J. Power Sources* 196: 9743. (b) Han, H.-B., Zhou, S.-S., Zhang, D.-J. et al. (2011). *J. Power Sources* 196: 3623. (c) Krause, L.J., Lamanna, W., Summerfield, J. et al. (1997). *J. Power Sources* 68: 320.

18 (a) Monti, D., Ponrouch, A., Palacín, M.R., and Johansson, P. (2016). *J. Power Sources* 324: 712. (b) Nohira, T., Ishibashi, T., and Hagiwara, R. (2012). *J. Power Sources* 205: 506.

19 (a) Maton, C., De Vos, N., and Stevens, C.V. (2013). *Chem. Soc. Rev.* 42: 5963. (b) Xue, Z., Qin, L., Jiang, J. et al. (2018). *Phys. Chem. Chem. Phys.* 20: 8382.

20 (a) Wongittharom, N., Wang, C.-H., Wang, Y.-C. et al. (2014). *ACS Appl. Mater. Interfaces* 6: 17564. (b) Wang, C.-H., Yeh, Y.-W., Wongittharom, N. et al. (2015). *J. Power Sources* 274: 1016.

21 Kubota, K., Nohira, T., and Hagiwara, R. (2010). *J. Chem. Eng. Data* 55: 3142.

22 Matsumoto, K., Taniki, R., Nohira, T., and Hagiwara, R. (2015). *J. Electrochem. Soc.* 162: A1409.

23 (a) Lin, X., Salari, M., Arava, L.M.R. et al. (2016). *Chem. Soc. Rev.* 45: 5848. (b) Sakaebe, H., Matsumoto, H., and Tatsumi, K. (2007). *Electrochim. Acta* 53: 1048. (c) Guidotti, R.A., Reinhardt, F.W., and Odinek, J. (2004). *J. Power Sources* 136: 257.

24 Hwang, J., Matsumoto, K., Orikasa, Y. et al. (2018). *J. Power Sources* 377: 80.

25 Ding, C., Nohira, T., Fukunaga, A., and Hagiwara, R. (2015). *Electrochemistry* 83: 91.

26 Avdeev, M., Mohamed, Z., Ling, C.D. et al. (2013). *Inorg. Chem.* 52: 8685.

27 Hwang, J., Matsumoto, K., Nohira, T., and Hagiwara, R. (2017). *Electrochemistry* 85: 675.

28 (a) Hwang, J., Matsumoto, K., and Hagiwara, R. (2019). *ACS Appl. Energy Mater.* 2: 2818. (b) Wongittharom, N., Lee, T.-C., Wang, C.-H. et al. (2014). *J. Mater. Chem. A* 2: 5655.

29 (a) Zuo, L., Ao, X., and Guo, Y. (1628). *J. Chromatogr. A* 2020: 461446. (b) Chagas, L.G., Buchholz, D., Wu, L. et al. (2014). *J. Power Sources* 247: 377. (c) Chen, C.-Y., Kiko, T., Hosokawa, T. et al. (2016). *J. Power Sources* 332: 51.

30 Wu, F., Zhu, N., Bai, Y. et al. (2018). *Nano Energy* 51: 524.

31 Usui, H., Domi, Y., Fujiwara, K. et al. (2017). *ACS Energy Lett.* 2: 1139.

32 Luo, X.-F., Helal, A.S., Hsieh, C.-T. et al. (2018). *Nano Energy* 49: 515.

33 Zhang, J., Wang, D.-W., Lv, W. et al. (2017). *Energy Environ. Sci.* 10: 370.

34 (a) Sun, H., Zhu, G., Xu, X. et al. (2019). *Nat. Commun.* 10: 3302. (b) Pinson, M.B. and Bazant, M.Z. (2012). *J. Electrochem. Soc.* 160: A243.

35 (a) Wang, C.-H., Yang, C.-H., and Chang, J.-K. (2016). *Chem. Commun.* 52: 10890. (b) Matsumoto, K., Hosokawa, T., Nohira, T. et al. (2014). *J. Power Sources* 265: 36.

13

Solid-State and Gel Electrolytes for Sodium-Ion Batteries

13.1 Introduction

Liquid electrolyte used in traditional secondary battery systems has good ion transport capacity and excellent electrochemical performance. However, liquid electrolyte is not perfect with several disadvantages. First, the thermal stability and electrochemical stability of liquid electrolyte are not good enough, especially at elevated temperatures or against Na metal anode or high-potential cathode. Second, liquid electrolyte suffers from concentration polarization, due to low Na^+ transport number. Third, liquid electrolyte poses safety risks due to its flammability. Lastly, batteries with liquid electrolyte are easy to leak and gas, which also leads to safety problems.

Because of these disadvantages in liquid electrolytes, the use of solid-state electrolytes (SSEs) to replace liquid electrolytes and separators to fabricate all solid-state batteries (ASSBs) has attracted more and more attention from researchers. In theory, SSEs are safer and more stable, including thermal stability and electrochemical stability. Also, SSEs can simplify the battery assembly process (Figure 13.1), since they can act as both ion conductor and separator. Therefore, the evolution of solid-state batteries may make breakthroughs in electrochemical storage systems such as mobile/portable electronic products and grid storage. However, there are several drawbacks of SSEs comparing to liquid electrolytes: the most important one is poor interfacial contact between solid particles (i.e. huge interfacial resistance). Even the interfacial contact at the beginning is acceptable, the volume changed during cycling would cause contact loss and resistance increase. Therefore, good cycling performance for ASSBs requires certain pressure to improve and maintain the physical contact. According to different material, SSEs for sodium-ion batteries (SIBs) can be divided into inorganic, organic, and composite solid electrolytes [1].

13.2 Electrolyte Characteristics

13.2.1 Energy Density

Besides safety, higher energy density is another important advantage for ASSBs. Qiao et al. conducted the energy density and technical feasibility of lithium and sodium

Sodium-Ion Batteries: Energy Storage Materials and Technologies, First Edition. Yan Yu.

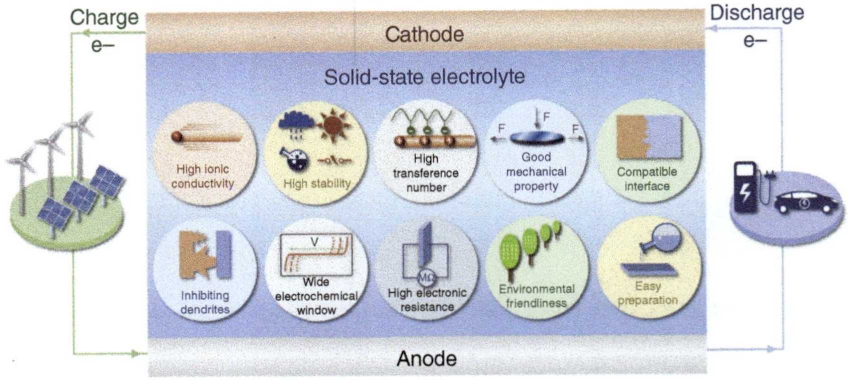

Figure 13.1 The structure of solid-state sodium batteries and the required properties of SSE. Source: Lu et al. [1c]. Reproduced with permission, 2021, Elsevier.

batteries based on liquid electrolyte (LEs) and SSEs, and the results are summarized and shown in Figure 13.2 [2].

(1) The potential difference between the sodium anode and the lithium anode is only 500 mV, so the energy density of sodium-based battery is not much lower than that of lithium-based batteries, especially for those high-capacity, high-potential cathodes.
(2) SSE will slightly increase the energy density (about 10%). And depending on the density of different solid electrolytes (1.2 g cm^{-3} for polymer electrolytes,

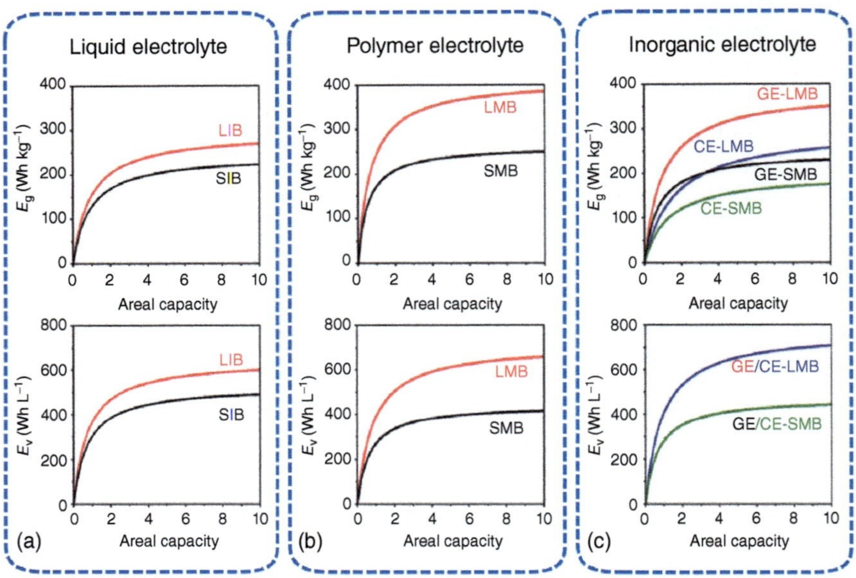

Figure 13.2 Estimated gravimetric (E_g) and volumetric (E_v) energy densities of lithium and sodium batteries with various kinds of electrolytes: (a) liquid electrolyte, (b) polymer electrolyte, and (c) inorganic electrolyte. Source: Qiao et al. [2]. Licensed under CC BY 4.0.

2.4 g cm^{-3} for glass electrolytes, and 5.15 g cm^{-3} for ceramic electrolytes), ASSBs with polymer electrolytes show the highest mass energy density.

(3) Replacing hard carbon anode with Na anode only increases a little bit energy density, which is different from replacing graphite anode with Li anode. For example, Li|SPE|$LiFePO_4$ battery has an energy density of 350 Wh kg^{-1}, while graphite|LE|$LiFePO_4$ battery has only 250 Wh kg^{-1}, showing a decrease of over 30% over the former. On the other hand, the energy density of Na|SPE|$NaFePO_4$ and HC|LE|$NaFePO_4$ batteries is similar (around 200 Wh kg^{-1}).

It should be noted that here only electrode films, current collectors, and electrolytes/separators are considered to calculate energy density, so it is still higher than that of commercial batteries. Considering that ASSB single cells can stack for several layers and seal in one pack, while batteries with liquid electrolyte cannot, the energy density of ASSBs can be even higher than that of liquid-electrolyte-based batteries. In addition, ASSBs make anode-free batteries possible, which would further increase the energy density.

13.2.2 Ionic Conductivity

The requirements for electrolytes are high ionic conductivity and low electron conductivity. The solid electrolyte is composed of a main chain structure of mobile ions and immobile framework. In general, an increase in the valence of adjacent mobile ions causes a stronger Coulomb interaction, which raises the activation energy barrier. When ions with the same charge are moving, the interaction with neighboring ions is more sensitive to the influence of ion mobility, especially when two ions are close. Because the Coulomb repulsive force has a tendency of mutual principle during the movement, there is a reduction in the number of ion migration. Therefore, the open crystal framework is essential for ion conduction. In addition, the relative size of vacancies in the host crystal also affects ion migration. If the intermediate vacancy for the mobile ion is too small, it may be hindered by a large electrostatic trap, and if it is too large, the framework may be not stable. Moreover, the ionic conductivity is also related to the temperature. That is, the natural logarithm of conductivity is proportional to the reciprocal of absolute temperature, and the slope represents the diffusion activation energy of the solid electrolyte.

For polymer solid electrolyte, it embodies certain ionic conductive properties under the action of external electric field. The matrix materials of traditional polymer electrolytes include polyethylene oxide (PEO), polyacrylonitrile (PAN), polyvinyl alcohol (PVA), polyvinylpyrrolidone (PVP), poly(methyl methacrylate (PMMA), etc. Polymers with electronegative groups can form coordination with ions, accompanied by local and long-range thermal motion of the molecular chain. The directional transport occurs under the action of an external electric field. Selecting a polymer matrix with a higher dielectric constant (ε) can reduce the force between the anions and cations of the salt, improve the dissociation of the salt, and theoretically result in a higher ionic conductivity. In SIBs, the transfer of Na^+ in PEO occurs mainly in the amorphous region. With the creep of the PEO segment,

Na^+ and O on the EO monomer are continuously complexed and decomposed, thus achieving the migration of Na^+. However, PEO has high ionic conductivity only in the amorphous state. Due to the presence of crystalline EO segments, the creep of molecules is restricted, which leads to difficulties in ion migration and low ionic conductivity. Therefore, the subsequent improvement of PEO properties is mainly reflected in the reduction of its crystallinity. Commonly used methods include co-mixing, cross-linking, addition of inorganic nanoparticles, and compounding with ceramic electrolytes.

The concentration of mobile ions and microstructure also has important effects on ionic conductivity [3]. When the concentration exceeds the critical point, the host lattice is deformed, and vacancies are reduced, resulting in a decrease of the ionic conductivity. Moreover, the resistance generated by the microstructure grain boundaries must be mitigated. It is well known that the conductivity of the grain boundaries is lower than the conductivity of the bulk. Thus, it is necessary to adjust the grain size, structure, and the synthesis method to reduce the grain boundary resistance.

Structure and composition of SSE not only have a significant impact on properties such as ionic conductivity, but also seriously affect battery performance. There is not yet a clear understanding of SSEs in SIBs, but various electrolyte systems have been investigated, which are mainly derived from lithium-ion batteries (LIBs).

13.2.3 Chemical Stability

The chemical stability of the electrolyte itself and the interfaces is also a vital aspect affecting the cycle life and safety of the battery. Chemical stability is the property of a material to keep its properties intact during preparation and storage. For materials with poor chemical stability, it is detrimental to battery performance and can easily lead to safety issues. Therefore, the chemical stability of the material and interface dictates the production conditions and configuration.

Liquid electrolytes have been widely used until now due to their high interface contact with solid electrodes. However, liquid electrolytes have safety concerns, especially after a series of high-profile fire hazards caused by thermal runaway. SSEs thus came into the public eye. The use of SSEs will have a number of the advantages in addition to fundamental addressing the safety of batteries [4]. For example, it can circumvent the well-known polysulfide shuttle effect in lithium-sulfur batteries [5]. So far, some significant progress has been made in the field of SSEs including polymer, oxide, and sulfide systems. If some of the existing problems of these SSEs can be solved, it would be possible to achieve battery systems with high energy density [6].

In solid-state battery systems, interface issues present pressing challenges that need to be addressed. Achieving the ideal fixed solid–solid contact during operation (e.g. the poor garnet SSEs-Na interface [7]) is very difficult, and severe volume changes are sometimes occurred during this process. Koerver et al. constructed a high-energy solid-state battery, which showed that at high potentials, the sulfide electrolyte decomposes and forms an adverse interfacial phase, causing an

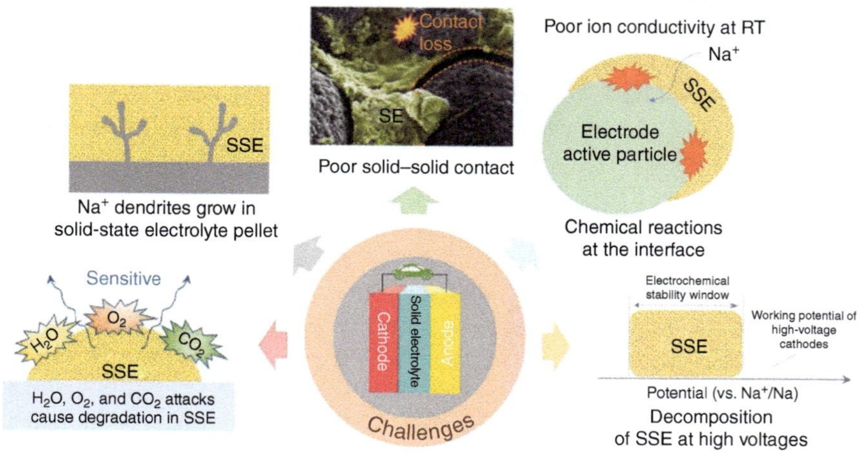

Figure 13.3 Remaining challenges of solid-state electrolytes: (i) unsatisfied ion conductivity at room temperature, (ii) potential Na dendrite growth inside the SSE, (iii) poor solid–solid contact during operation, (iv) high air sensitivity of SSE, (v) unfavorable interactions at the solid–solid interface, and (vi) decomposition of SSE at high voltages. Source: Zhao et al. [8]. Reproduced with permission, 2020, Royal Society of Chemistry.

irreversible capacity loss in the first cycle. The results also displayed that the chemical–mechanical shrinkage of the active material during desodiation results in poor solid–solid contact, hence leading to high interfacial resistance and significant capacity loss (see SEM image in Figure 13.3). Decomposition of SSEs also usually occurs at low voltages, and a stable electrode/electrolyte interface is desired to form to avoid chemical side reactions. The SEI problem will become even more challenging for future high-energy SMBs using aggressive high-voltage cathodes and sodium metal anodes. Unfortunately, in order to perform safer operations at more extreme potentials, higher chemical stability (larger electrochemical stability window) usually accompanies lower ionic conductivity. Another serious challenge for solid SMBs is the harmful Na dendrites that even produced in dense SSEs samples [9]. In addition, air sensitivity, mechanical stability, electronic conductivity, and thermal stability should be considered [10]. For example, in many cases, the presence of H_2O, O_2, and CO_2 in the air may lead to irreversible degradation during manufacturing or even operation. Two points should be mentioned here: (i) as shown in the detailed morphological analysis of storage [11], looking for electrodes with a high ionic contribution (e.g. NASION structure) can minimize the number of critical electrode/electrolyte contacts; (ii) the use of synergistic solid/liquid composite electrolytes [11b, 12] can provide an elegant solution to the above dilemma [8].

In summary, the stability of the solid electrolyte and the electrolyte/electrode interface has a significant impact on the design and manufacture of the actual battery. The environmentally sensitive nature of the SSEs determines the conditions of synthesis and transport [13]. Therefore, from phenomenon to a comprehensive understanding of the fundamental mechanisms in solid electrolytes is essential.

13.2.4 Mechanical Stability

Mechanical stability refers to the ability of the material to remain unchanged under certain stress. Some of these stresses arise from the unavoidable volume changes of the electrodes. Close contact between the electrode and the SSEs allows for better mechanical stability of the cell. However, dendrite growth in SIBs often causes cracking or chalking of SSEs during cycling. These findings suggest that mechanical stability issues cannot be ignored in SIBs [14].

Highly branched grafted polymers and hyperbranched polymers have poor mechanical properties. Blending with polymers of high mechanical strength can improve electrolyte properties. For inorganic solid electrolytes, the bonding energy between the cation and anion affects the elastic modulus. The elastic modulus of $75Na_2S–25P_2S_5$ is lower than that of $75Li_2S-25P_2S_5$ [15]. In addition, sulfide glasses can be pressurized and sintered at RT, while oxide glasses require high-temperature sintering due to their high bond energy. As a result, all-solid-state batteries with higher energy density can be obtained by reducing the modulus of elasticity, increasing the denseness of the SSEs, and adapting the volume changes in electrode.

13.2.5 Thermal Stability

Thermal stability also has a significant impact on the practical application of ASSBs, which should be not easy to decompose or react at high temperatures. However, recently there have been reports of thermal runaway, heat generation, and gas release in ASSBs. Thus, for the commercialization, the thermal stability of SSEs in ASSB is needed to further improve.

The required characteristics for the SSEs as mentioned above are summarized in Figure 13.4 [17].

13.3 Polymer Electrolytes

13.3.1 Solid Polymer Electrolytes (SPEs)

The discovery of ion conducting polymers can be traced back to 40 years ago, when Wright and Armand reported that alkali metal salts (i.e. Li, Na, K, Rb, and Cs) could combine with PEO to form composite materials, which demonstrate ion conduction, and can be used in electrochemical devices. Polymer electrolytes (PEs) have both solid ionic conductivity and mechanical flexibility, and their good interface contact with solid electrodes makes them an ideal substitute for liquid electrolytes in electrochemical cells. Solid polymer electrolytes (SPEs) have relatively low cost and lightweight, which are beneficial to increase energy density. In addition, they own good safety and workability and are advantageous for forming, patterning, and integration with electrodes. Also, they have a high degree of flexibility in battery design and are easy to manufacture ultrathin membrane. Finally, SPEs display good resistance to suppress the electrode volume change during the charging and

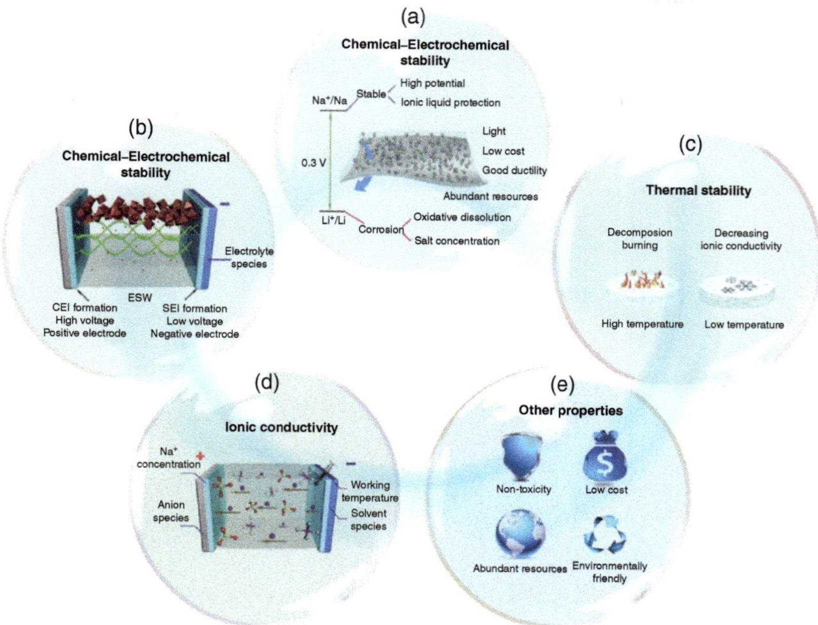

Figure 13.4 Schematic summary of the fundamental properties of SSEs in SIBs. Source: Huang et al. [16]. Reproduced with permission, 2019, WILEY-VCH Verlag GmbH & Co. KGaA, Weinheim.

discharging processes. However, the conductivity of SPEs at room temperature is only 10^{-5}–10^{-7} S cm^{-1}, which cannot meet the requirements of battery operation. Therefore, SPE-based batteries usually work at elevated temperatures of 60–90 °C.

PEO is the most widely used polymer for SPE preparation. In addition to PEO-based electrolytes, several other polymer electrolytes have been studied such as PVP, polyvinyl chloride (PVC), PVA, PAN, and polycarbonate (Table 13.1).

13.3.1.1 PEO-Based Electrolyte

PEO is an industrial crystalline polymer, whose molecular weight can vary within a wide range, e.g. the higher degree of polymerized PEO having relative molecular mass of 1×10^5–1×10^6. PEOs are white flowable powders with a molecular structure of $(CH_2CH_2O)_n$, but these resins have a low concentration of active end groups. Due to the presence of C—O—C bonds, they are usually pliable and can form associates with electron acceptors or certain inorganic electrolytes. In addition, PEOs are water-soluble polymers due to the formation of hydrogen bonds. The abovementioned structural characteristics make PEOs versatile.

PEOs can form complexes with many low-molecular organic compounds, polymers, and some inorganic electrolytes. Among them, inorganic compounds mainly include amines fluoride, sodium fluoride, bromine, iodine, potassium, mercury halides, ammonium thiocyanate, potassium thiocyanate, etc. Because of good water solubility, low toxicity, easy processing and molding characteristics, PEO can be

Table 13.1 Polymer molecular formula and structure for SPEs.

Polymer matrix	Molecular formula	Structure
PEO	$(CH_2CH_2O)_n$	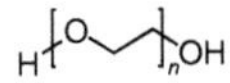
PVP	$(C_6H_9NO)_n$	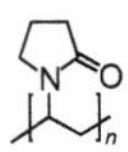
PAN	$(CH_2CH(CN))_n$	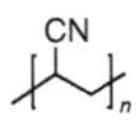
PMMA	$(CH_2C(CH_3)(CO_2CH_3))_n$	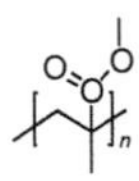
PVDF	$(CF_2CH_2)_n$	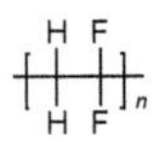
PVDF-HFP	$(CF_2CH_2)_n$-$(CF_2CF(CF_3))_m$	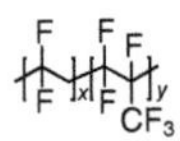
PVA	$(CH_2CHOH)_n$	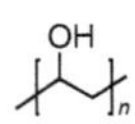
PVC	$(CH_2CHCl)_n$	

used as water-soluble membrane, textile slurries, thickeners, flocculants, lubricants, dispersants, aqueous-phase reducer, cosmetic additives, anti-static agents, etc.

PEO-based SPEs have also been researched and applied in Na secondary batteries. In 1995, Chandra and coworker [18] used solution casting technology to prepare PEO/$NaPF_6$ membrane with different Na^+/EO ratios. The highest ionic conductivity of PEO/$NaPF_6$ electrolytes is 5×10^{-6} S cm^{-1} with Na^+/EO ratio of 0.065 at room temperature. Other groups also studied the performance of electrolytes with different salts added to PEO. The types of salts include $NaClO_3$ [19], $NaLaF_4$ [20], $NaClO_4$ [19], NaFSI [21], NaFNFSI [22], and NaTFSI [21a].

Chandrasekaran et al. [19] investigated the conductive polymer electrolyte of PEO/$NaClO_4$ for Na/SPE/MnO_2 batteries with PEG as plasticizers in 2001. The PEO/$NaClO_4$ SPE has a high activation energy of 0.539 eV for sodium ion migration at room temperature, and its ion conductivity is only about 10^{-8} S cm^{-1}. Adding PEG (10%) to PEO/$NaClO_4$ (1 : 2) increases the ion conductivity to 3.40×10^{-6} S cm^{-1} and decreases the activation energy of sodium ion migration to 0.417 eV. The energy density of the Na/SPE/MnO_2 battery is close to 350 Wh kg^{-1}. Moreover, adding sodium salt of $NaLaF_4$ to PEO-based SPE also demonstrates similar benefits (Figure 13.5a).

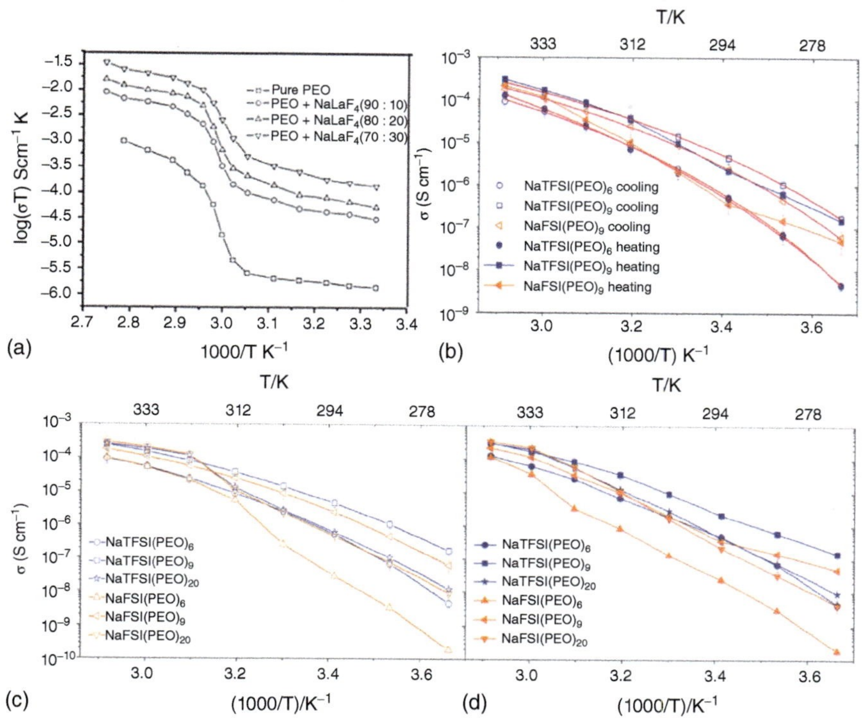

Figure 13.5 Arrhenius curves of ionic conductivities for (a) PEO + $NaLaF_4$ (0%, 10%, 20%, 30%) SPE, (b) PEO + NaTFSI (cooling and heating), and (c, d) PEO + NaFSI (cooling and heating). Source: Boschin and Johansson [21a]. Reproduced with permission, 2015, Elsevier.

Boschin et al. [21a] studied the effects of species and molar ratios of different anions (e.g. NaFSI and NaTFSI) on SPE performance. The NaTFSI-$(PEO)_n$ ($n = 9$) exhibits a higher ionic conductivity (4.5×10^{-5} S cm^{-1}) at 20 °C. It is because that the internal flexibility and large size of $TFSI^-$ inhibit crystallization, while bis(fluorosulfonyl) imide (FSI^-) crystallizes more easily at room temperature. In addition, there is a stronger interaction between FSI^- and Na^+, which reduces the ionic conductivity of NaFSI-based SPE (Figure 13.5b–d).

In particular, Armand and coworkers [23] proposed that the fluorinated anion can act as a plasticizer for PEO-based electrolytes, thereby reducing the crystallinity of the polymer and achieving higher ionic conductivity at room temperature. This effect can provide mechanical elasticity and plasticize the polymer electrolyte.

Hu and coworkers [24] researched the workability of PEO-$NaPF_6$ SPEs in SSBs, which exhibit comparatively high ionic conductivity of 6.3×10^{-4} S cm^{-1} at 80 °C (Figure 13.6a) and a large Na^+ transfer number (t_{Na+}) of 0.58 (Figure 13.6b,c). Also, PEO-$NaPF_6$ SPEs have good thermal stability in SSBs up to around 200 °C, which enables it stable enough in SSBs. As displayed in Figure 13.7, the PEO-$NaPF_6$ SPEs have excellent capacity retention at 80 °C in Na|$Na_3V_2(PO_4)_3$@C battery (85.8% after 200 cycles at 2 C) and Na|$NaNi_{2/9}Cu_{1/9}Fe_{1/3}Mn_{1/3}O_2$ battery (77% after 100 cycles

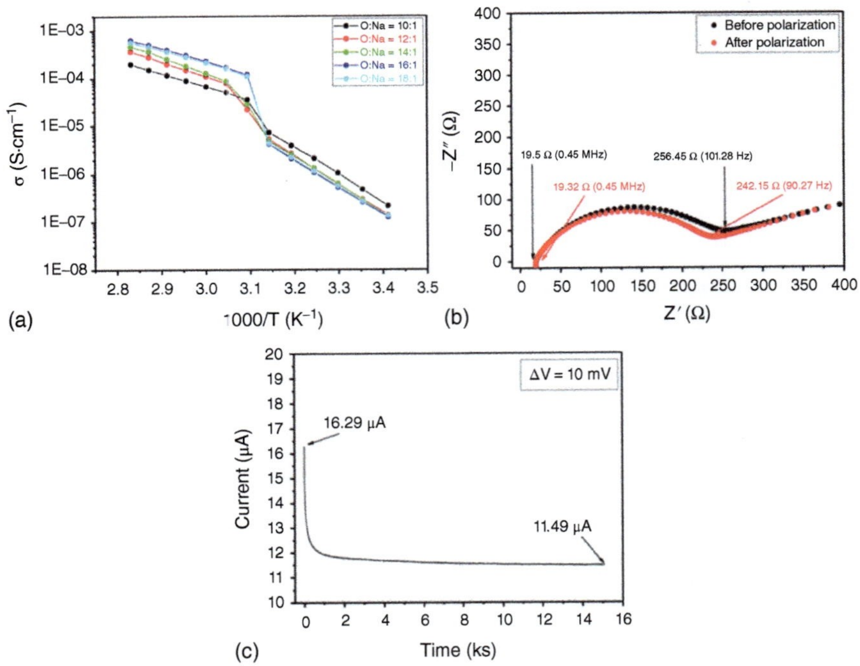

Figure 13.6 (a) The relationship between ionic conductivity and temperature of polymer electrolytes with different EO: Na molar ratios (10, 12, 14, 16, 18). (b) EIS and (c) DC polarization curves of Na|SPE|Na symmetric battery at 80 °C. Source: Reproduced with permission from Zhang et al. [24]. Copyright 2020, IOP Publishing.

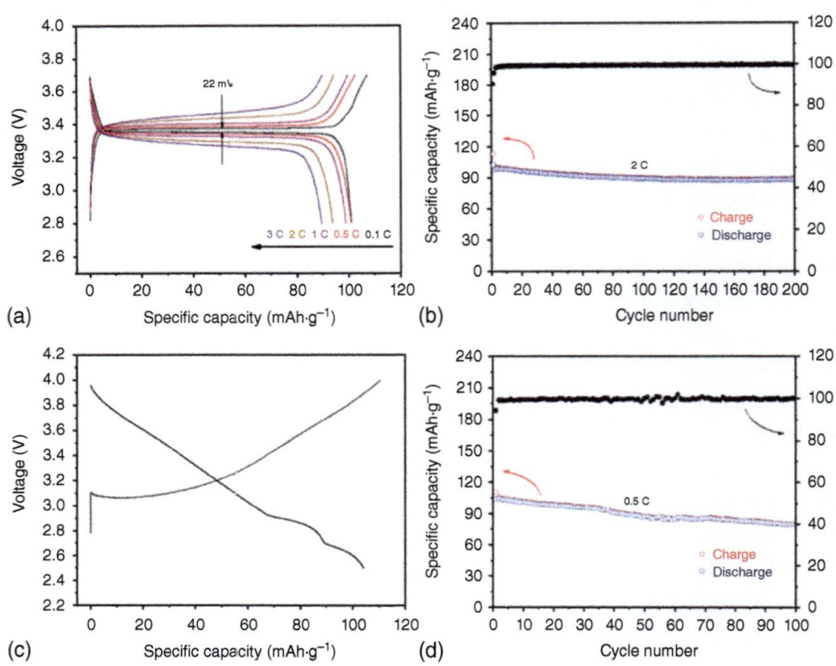

Figure 13.7 (a) Charge and discharge curves (0.1–3 C) and (b) cycle performance (2 C) of $Na_3V_2(PO_4)_3$|SPE|Na@C battery. (c) Charge and discharge curve (0.5 C) and (d) cycle performance (0.5 C) of $NaNi_{2/9}Cu_{1/9}Fe_{1/3}Mn_{1/3}O_2$|SPEs|Na battery. Source: Zhang et al. [24]. Reproduced with permission, 2020, ResearchGate GmbH. Licensed under CC BY 4.0.

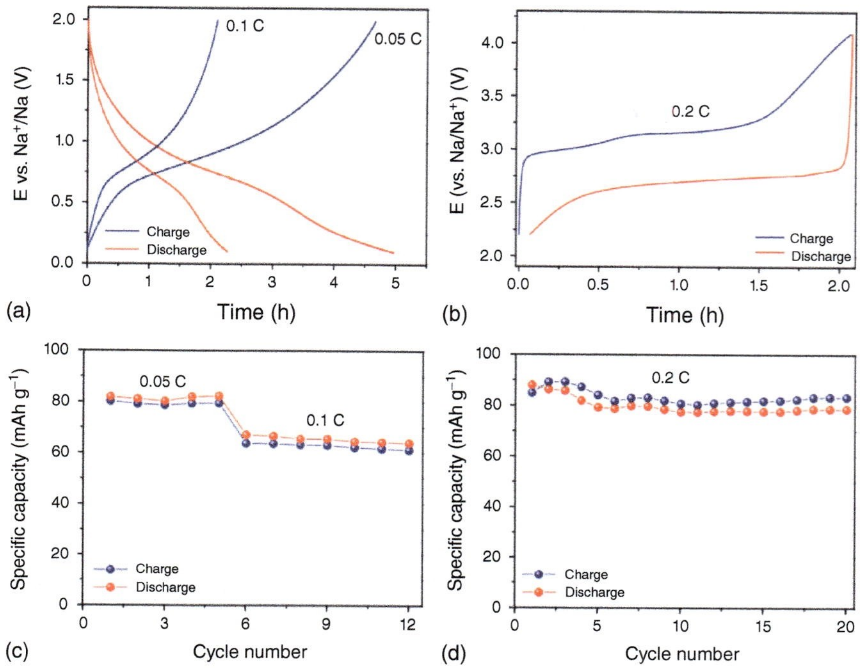

Figure 13.8 The electrochemical performance of the battery assembled by PEO-Na-CMC (60 °C). (a) Time-charge and discharge curves of Na|SPE|TiO_2 battery (0.05 C at the 5th cycle and 0.1 C at the 10th cycle), (b) Na|SPE|$NaFePO_4$ battery (0.2 C at the 10th cycle), and cycle performance at (c) 0.05 and 0.1 C and (d) 0.2 C [25]. Source: Colò et al. [25]. Reproduced with permission. Copyright 2015, Elsevier.

at 0.5 C). All these characteristics make PEO-$NaPF_6$ SPEs promising substitutes for SSBs applications.

Adding cellulose to the mixed polymer electrolyte is an effective method to improve the mechanical properties of all-solid-state SIBs. Gerbaldi and coworkers [25] prepared a PEO-based SPE with carboxymethyl cellulose Na (PEO : $NaClO_4$: carboxymethyl cellulose [CMC] = 82 : 9 : 9). Meanwhile, CMC could act as an electrode binder and improve the interface between the electrolyte and electrode. The PEO-CMC electrolyte showed lower charge transfer impedance than that of the PEO electrolyte, which meant better ion diffusion and compatibility between the PEO-CMC electrolyte and the electrode. The Na|SPE|TiO_2 and Na|SPE|$NaFePO_4$ half-cells assembled using PEO-CMC electrolytes had good reversibility, stable voltage platform, and cycling stability (Figure 13.8).

13.3.1.2 PVA-Based Electrolyte

PVA is an organic compound with the chemical formula of $[C_2H_4O]_n$. The appearance is a white flaky, flocculent, or powdered solid, odorless. It is soluble in water (above 95 °C), especially in dimethyl sulfide (DMSO), but insoluble in gasoline, kerosene, vegetable oil, and benzene. PVA is another Na ion-conducting polymer.

Owing to the strong solvent affinity, good film-forming properties, and wide temperature window, PVA-based gel electrolytes attract great attention at present.

Bhargav et al. [26] prepared PVA/NaBr electrolytes by solution casting technique and dried the film under vacuum. They found that the ionic conductivity of PVA polymers can be improved by introducing NaBr, and the as-prepared PVA/NaBr (weight ratio = 70 : 30) SPE has the maximum ionic conductivity of $1.362 \times 10^{-5}\,S\,cm^{-1}$ at 40 °C, which is three orders of magnitude larger than pure PVA. At the same time, the activation energy reduced from 0.478 to 0.326 eV. Thus, the PVA/NaBr electrolyte has potential applications in all solid-state SIBs.

Kumar and coworkers [27] synthesized polymer electrolyte of PVA/sodium citrate (9 : 1) membranes doped with different concentrations of nano-Pr_2O_3 by solution casting technique (Figures 13.9 and 13.10). The membrane with 3.0% nano-Pr_2O_3 content was found to be more homogeneous (Figure 13.9) and possessed more amorphous zones, which lead to more interactions between the functional groups of the polymer films and the nano-Pr_2O_3 particles, thus making the films more conductive. The pure PVA with sodium citrate shows a low ionic conductivity ($5.59 \times 10^{-10}\,S\,cm^{-1}$) and a high activation energy (0.42 eV). With adding minor nano-Pr_2O_3, the Na^- ion conduction is improved and optimized with 3% Pr_2O_3 (Figure 13.10a). The ionic conductivity of the composite electrolyte with 3% Pr_2O_3 is measured to be $7 \times 10^{-4}\,S\,cm^{-1}$, combined with an activation energy of 0.21 eV

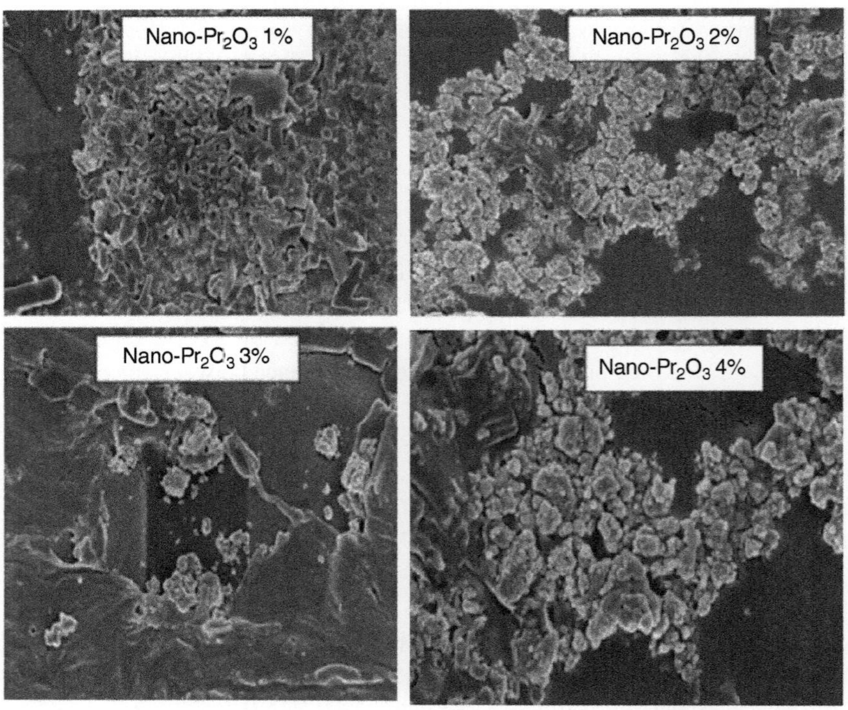

Figure 13.9 SEM images of composite polymer electrolyte (PVA: sodium citrate = 9 : 1 + 1–4% nano-Pr_2O_3). Source: Babu et al. [27]. Reproduced with permission, Copyright 2018, Hindawi.

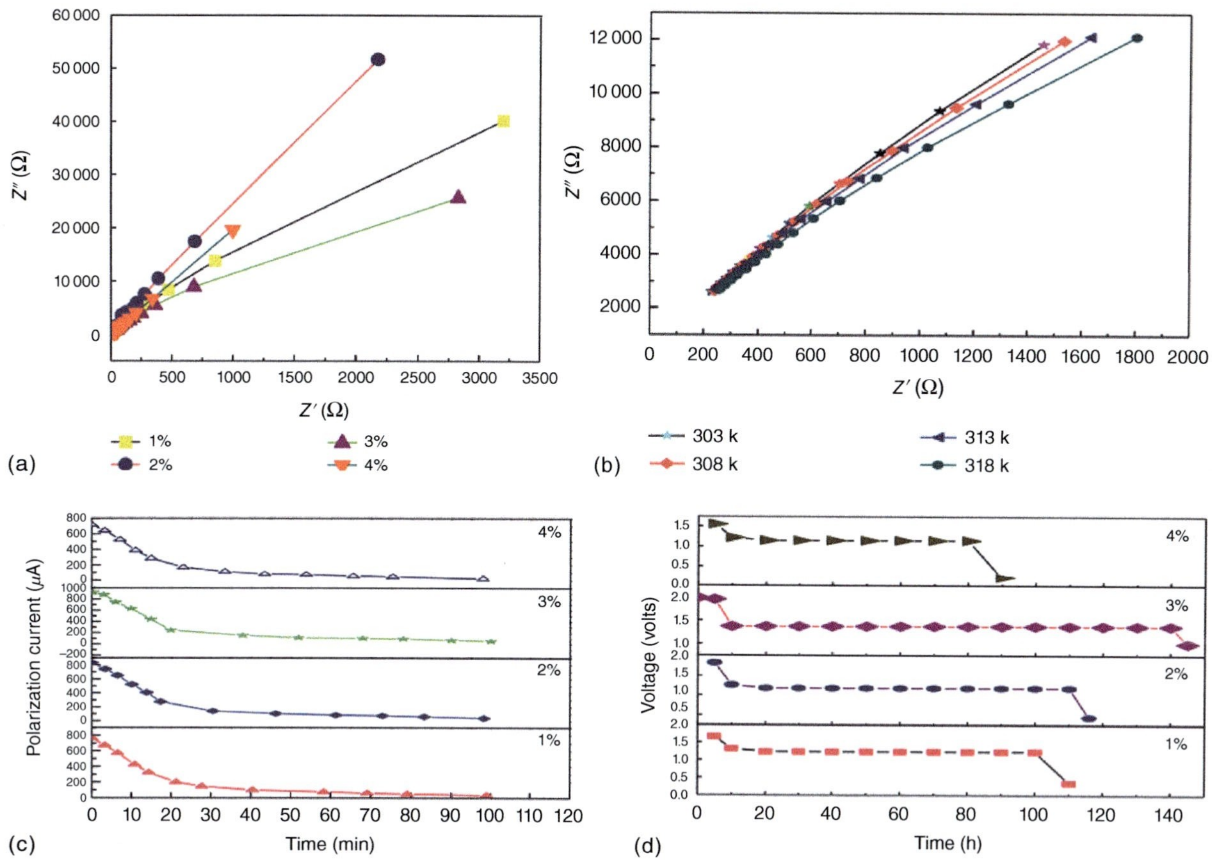

Figure 13.10 EIS of composite polymer electrolytes of (a) PVA: sodium citrate = 9 : 1 + 1–4% nano Pr_2O_3, and (b) PVA: sodium citrate = 9 : 1 + 3% nano Pr_2O_3 at different temperatures. (c) Transference number measurements for polymeric electrolyte films of compositions PVA (90%) + sodium citrate (10%) + nano-Pr_2O_3 (1–4%). **(d)** Discharge characteristics plots of the battery with the PVA: sodium citrate = 9 : 1 + 1–4% nano Pr_2O_3. Source: Babu et al. [27]. Reproduced with permission, Copyright 2018, Hindawi.

(Figure 13.10b). With increasing temperature, the conductivity becomes higher and higher. This is due to the enhanced jumping of inter-chain and intra-chain ionic motions, which further reduces the microscopic viscosity of the membrane. In addition to ionic conductivity, the transference number of the composite electrolytes with Pr_2O_3 is close to 1, indicating the composite electrolyte as a single-ion conductor (Figure 13.10c). These membranes were successfully employed as polyelectrolytes for electrochemical batteries and their discharge characteristics were evaluated (Figure 13.10d).

13.3.1.3 PAN-Based Electrolyte

Osman et al. [28] prepared PAN-based polymer electrolytes with $NaCF_3SO_3$ as Na salt and DMF as solvents. The optimized ionic conductivity is about 0.31 mS cm^{-1} when 24 wt.% $NaCF_3SO_3$ is added to PAN, combining with an activation energy of 0.28 eV. Ionic conductivity of PAN + 24 wt.% $NaCF_3SO_3$ is much higher than that of other polymer electrolytes. However, the electrochemical stability is not reported yet, which is also important for SSEs.

13.3.1.4 PVP-Based Electrolyte

Electrolyte based on polyvinylpyrrolidone (PVP) and sodium salts of NaF, $NaClO_4$, and $NaNO_3$ is another ion-conducting polymer electrolyte.

Kiran Kumar et al. [29] investigated the effect of NaF on PEO/PVP blend electrolyte. The addition of NaF increases the ionic conductivity and decreases the activation energy of PEO/PVP polymer electrolyte. The increase in the conductivity with adding more NaF is attributed to a decrease in crystallinity and an increase in amorphosity. Vahini and Muthuvinayagam [30] studied the effect of $NaNO_3$ on PVP-based SSEs and found that the 6 wt.% $NaNO_3$-doped system has the maximum ionic conductivity value (1.21×10^{-5} S cm^{-1}). The ionic conductivity of PVP + $NaNO_3$ electrolyte is positive correlated to the Na salt concentration, that is, it increases with increase of sodium salt concentration. When the weight ratio of $NaNO_3$ exceeds 6 wt.%, the decrease in ionic conductivity is due to aggregation of ions in polymer matrix.

13.3.1.5 PVDF-Based Electrolyte

Polyvinylidene fluoride (PVDF)-based polymer electrolyte was also investigated as a sodium conductor, with $NaCF_3SO_3$ as the Na salt [31]. An ionic conductivity of 2.1×10^{-6} S cm^{-1} and activation energy of 0.33 eV were achieved at 25 °C for PVDF-$NaCF_3SO_3$ (Figure 13.11a). With adding 2 wt.% SiO_2 filler, the ionic conductivity was enhanced to 0.06 mS cm^{-1} and the activation energy was reduced to 0.28 eV (Figure 13.11a). The cycling test with symmetric cell using SiO_2/PVDF composite electrolyte and Na electrode showed high stability of SiO_2/PVDF composite electrolyte against Na metal (Figure 13.11b) [31]. The battery performance was also investigated using SiO_2/PVDF composite electrolyte, Na anode, and $Na_3V_2(PO_4)_3$ (NVP) cathode. This SSB provides a specific capacity of over 100 mA h g^{-1} and a specific energy density of 126 Wh kg^{-1} at 0.5 C. The SSB maintains 70% of its initial specific capacity after 100 cycles (Figure 13.11c,d). In addition, the rate performance is good, with 50 mAh g^{-1} capacity at 5 C (Figure 13.11e,f). The results indicate that

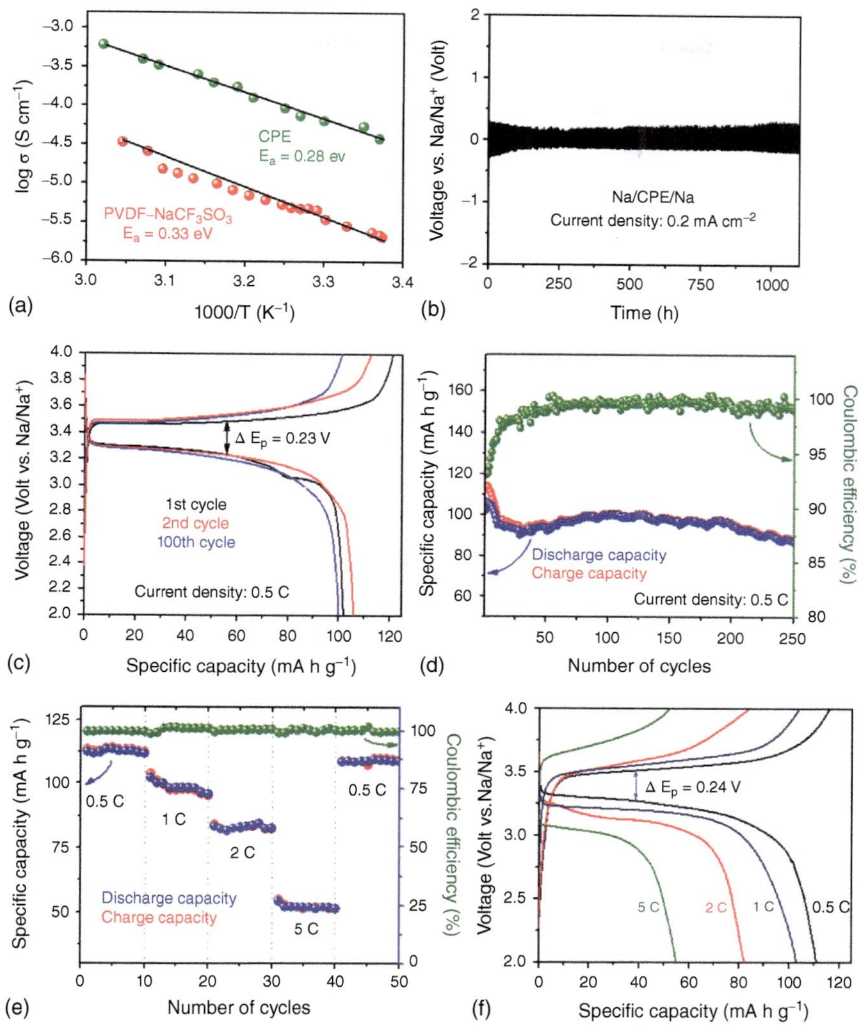

Figure 13.11 (a) Conductivity–temperature dependence of the PVDF-$NaCF_3SO_3$ and CPE. (b) Polarization curves of CPE in a symmetrical battery (Na|CPE|Na) at a current density of 0.2 mA cm^{-2}. (c) The charge and discharge curves of the battery (NVP|CPE|Na) at 0.5 C. (d) Cycle performance of NVP|CPE|Na battery (0.5 C), (e) its rate performance, and (f) charge and discharge curves at different C rates. Source: Bag et al. [31]. Reproduced with permission, Copyright 2020, Elsevier.

PVDF-based polymer electrolytes are promising with acceptable ionic conductivity and good electrochemical stability.

13.3.2 Na Polymer Single-Ion Conductors

Single-ion conductive polymer electrolyte is a newly developed one to increase the SPE ion transfer number, which usually consists of a (block copolymer) polymer.

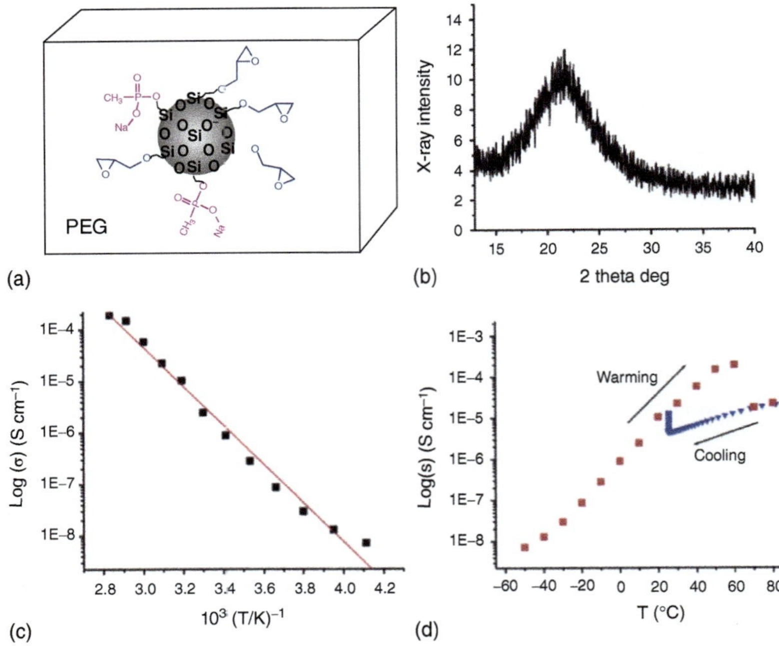

Figure 13.12 (a) Schematic representation of PEG + SiO_2 composite single-ion conductor. (b) XRD profile of the PEG + SiO_2 composite single-ion conductor. (c) Arrhenius curve of the PEG + SiO_2 composite single-ion conductor. (d) Dependence of conductivity on the thermal history of the PEG + SiO_2 composite single-ion conductor. Red squares indicate warming, while blue triangles indicate cooling. Source: Bronstein et al. [32]. Reproduced with permission, 2021, American Chemical Society.

It is characterized by the covalent linking of anions to a largely immobile polymer backbone, and the only movable ion is the cation. In this ionomer, there is no concentration gradient and high-rate charging and discharging can be applied. By immobilizing counter anions on the polymer chain, concentration polarization can be inhibited, thereby facilitating the migration of cations (sodium ions). In addition, the growth of dendrites can be restrained in single-ion conductor.

Bronstein et al. [32] developed a mono-ionic conductor composite polymer electrolyte with excellent electrochemical properties composed of PEG and SiO_2 nanoparticle, conductivity of which is 10^{-4} S cm^{-1} and cation transference number is 0.9 at room temperature (Figure 13.12). High transference number was achieved with an organic–inorganic component formed on the surface of SiO_2 (Figure 13.12a). In addition, PEG became amorphous with adding SiO_2 (Figure 13.12b), which increases the ionic conductivity to about 10^{-4} S cm^{-1} (Figure 13.12c). However, the activation energy for this single-ion conductor was measured to be 0.7 eV, limiting the application in lower temperatures (Figure 13.12c). Further research revealed that the ionic conductivity also depends on the thermal history (Figure 13.12d). During the heating process (red squares), the conductivity increases to 60 °C, then drops, and stabilizes at 80 °C. Upon cooling (blue diamonds), the conductivity

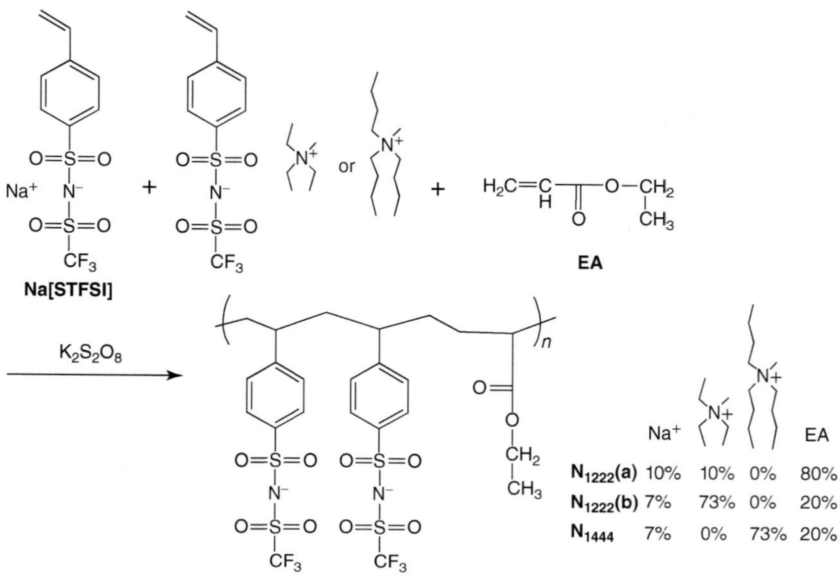

Figure 13.13 Synthesis method of ionomer and different initial molar ratio of precursor. Source: Li et al. [33]. Reproduced with permission, 2015, Royal Society of Chemistry.

decreases monotonically to room temperature and then increases back to the original value, when holding at room temperature for four days (Figure 13.12d). One possible reason for the conductivity dropping above 60 °C could be a lost contact between the SPE film and the electrode, because of differential expansion coefficient between the SPE film and gold film electrode.

Forsyth and coworkers [33, 34] prepared a neutral polymer mono-ionic conductor electrolyte consisting of trifluoromethanesulfonimide anionic groups and quaternary ammonium cations (Figure 13.13). The conductivity is 10^{-6} S cm^{-1} at 90 °C. In polymer chains, the anion is immobilized and the cation is the only movable ionic species in electrolyte, thus minimizing the effect of concentration gradients, since the ion transfer number can approach 1. The use of quaternary ammonium cations in ionic polymers leads to a reduction in the combination of sodium cations and sulfonic acid groups. Bulky ammonium cations have weaker ion–ion interactions and can act as plasticizers due to the mobility of the alkyl chains. However, these systems create additional problems due to the binding of anions. The interaction between the immobilized anion and the cation decreases the mobility of the cation, which usually results in low conductivity.

13.3.3 Adding Ceramic Additives to Polymer Electrolytes

Conventional fillers (inactive fillers), such as titanium, alumina, and silica dioxide, have been used as ceramic additives to prepare composite solid polymer electrolytes (CSPEs) for SIBs.

In 1998, Scrosati and coworkers [23, 35] investigated the effect of nano SiO_2 on PEO/NaTFSI mixtures. The CSPE based on a ratio of EO/Na = 20 and 5 wt.% SiO_2

achieves the highest ionic conductivity (10^{-5} S cm^{-1}) and sodium ion transfer number (0.51). The performance of CSPE can be further improved by functionalization of SiO_2 nanoparticles, which covalently bonded to oligomeric polyethylene glycol chains [36].

Hwang and coworkers [37] made a CSPE with nano TiO_2 (3.4 nm) and PEO/$NaClO_4$. The conductivity of CSPE (with 5 wt.% TiO_2) is 2.6×10^{-4} S cm^{-1} at 60 °C, which is twice as high as that of the pure PEO SPE (1.4×10^{-4} S cm^{-1}). The increase in ionic conductivity can be attributed to the decrease in the crystallinity of the PEO phase due to the addition of TiO_2, which increases the content of amorphous regions.

Liu et al. [38] designed a new method to prepare CSPE by a selective reaction between NaFSI, trace amount of H_2O, and inorganic filler (Al_2O_3), which forms favorable interfaces for rechargeable sodium batteries (Figure 13.14a). It was reported that due to the unstable S—F bond of FSI^-, trace amount of H_2O hydrolyzes the anion to form hydrogen fluoride, and the Al_2O_3 filler reacts spontaneously with HF to produce a stable $AlF_3 \cdot xH_2O$ product, thus forming a stable SEI layer on the sodium electrode (Figure 13.14b). At the same time, Al_2O_3 nanoparticles can reduce the crystallinity of PEO, which increases the segmented movement of PEO and improves the ionic conductivity to 1.0×10^{-3} S cm^{-1} at 80 °C. Therefore, the Na||$Na_3V_2(PO_4)_3$ battery with CSPE exhibits the capacity of 110 mA h g^{-1} and a high initial Coulombic efficiency of 93.8% at 80 °C (Figure 13.14c).

In addition to these inactive inorganic fillers, active inorganic particles were also introduced as SPE fillers. These inorganic fillers have some ionic conductivity

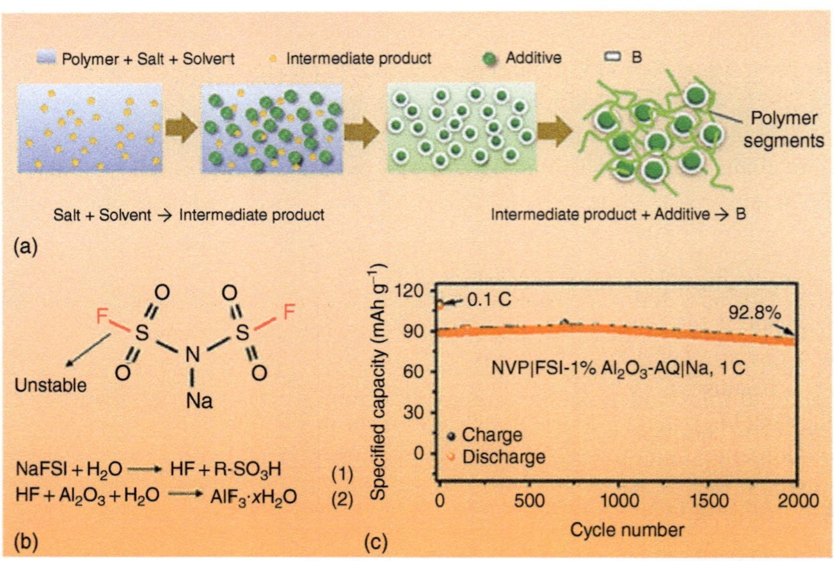

Figure 13.14 (a) Schematic diagram of the design principle of CSPE: the salt reacts with the solvent, and the intermediate product is transformed into a stable product B (e.g. $AlF_3 \cdot xH_2O$) with the help of the additive. (b) The structure diagram of NaFSI. (c) Cycle performance of Na|CSPE|Na battery (1 C, 80 °C). Source: Liu et al. [38]. Reproduced with permission, 2019, American Chemical Society.

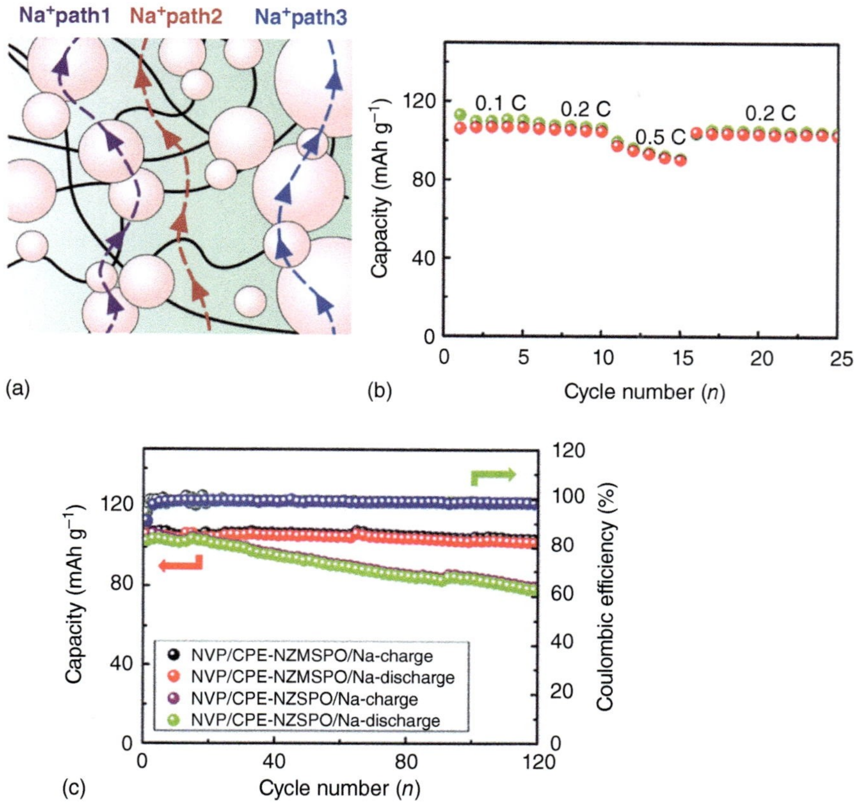

Figure 13.15 (a) Schematic diagram of the transmission path of sodium ions in the CPE. Source: Zhang et al. [39]. Reproduced with permission, 2017, Elsevier. (b) Charge and discharge capacity of the Na|CPE-NZMSPO|NVP battery at different C rates. (c) The cycle performance of the Na|CPE-NZMSPO|NVP battery and the Na|CPE-NZSPO|NVP battery (0.1 C). Source: Zhang et al. [40]. Reproduced with permission, 2016, Royal Society of Chemistry.

themselves compared with the conventional inactive inorganic fillers. Hu and coworkers [39, 40] developed two different Na superionic ion conductors (NASICON) as inorganic fillers ($Na_{3.4}Zn_{1.8}Mg_{0.2}Si_2PO_{12}$ and $Na_3Zr_2Si_2PO_{12}$) for CSPEs (Figure 13.15a). The ionic conductivity of NASION-based CSPE was improved, which is because: (i) the introduction of the filler reduces the crystallization of polymer and improves the transport of sodium ions in the amorphous region; (ii) the percolation effect can form a continuous and interconnected sodium ion transport path on the surface of the nanoparticles; and (iii) the migration of sodium ions in the NASION body contributes to improve the conductivity of the CSPE. A possible ion transport mechanism is illustrated in Figure 13.15a. As a result, a high ionic conductivity of 2.4×10^{-3} S cm^{-1} is obtained at 80 °C with 40 wt.% $Na_{3.4}Zn_{1.8}Mg_{0.2}Si_2PO_{12}$, which is much higher than that of the pure SPE (9.0×10^{-4} S cm^{-1}). The initial capacity of Na|CSPE|$Na_3V_2(PO_4)_3$ battery is 106 mA h g^{-1} with negligible capacity loss over 120 cycles and excellent rate performance (Figure 13.15b,c). The excellent performance is related to the good interface stability between the electrolyte and the electrode, as well as the high ionic conductivity of the CSPE.

Ciucci and coworkers [41] found that the CSPE composed of $Na_3Zr_2Si_2PO_{12}$, PVDF and $NaClO_4$ could not only inhibit the growth of sodium dendrites, but also effectively prevent the dissolution and migration of manganese ions from the $Na_{0.67}MnO_2$ cathode, giving the battery good cycling stability and rate performance.

13.3.4 Gel Polymer Electrolytes (GPEs)

In order to distinguish from the aforementioned electrolyte, gel polymer electrolyte (GPE) is defined as a polymer–salt complex containing liquid plasticizer and/or solvent. The matrix is similar to the polymer electrolyte, including polyethylene oxide, perfluorinated sulfonic membranes, PAN, PMMA, and PVDF.

In terms of physical state and chemical properties, GPE can be considered as an intermediate state between liquid electrolyte and solid electrolyte. Compared with liquid electrolytes, GPEs have an improved safety due to the avoidance of electrolyte leakage and have an ionic conductivity of about 10^{-3} S cm^{-1}, which is intermediate between dry polymer electrolytes (10^{-4} S cm^{-1}) and liquid electrolytes (10^{-2} S cm^{-1}) [42].

Although the safety of GPE is better than that of liquid-based electrolytes, it is worse than dry polymer electrolytes. As a result, GPE is limited in addressing the flammability and leakage issues of electrolytes. However, due to their polymer nature, gels have good flexibility and processability compared with inorganic ceramic or glass electrolytes. These characteristics and advantages make GPEs have great prospects as electrolytes in practical battery applications in the future.

13.3.4.1 PMMA-Based GPE

PMMA is one of the most extensively studied polymer matrices for battery applications due to its good affinity, amorphous nature with organic electrolytes, and high room-temperature ionic conductivity [43].

Goodenough and coworkers [44] prepared a SIB consisting of a cross-linked PMMA composite GPE, Sb anode, and $Na_3V_2(PO_4)_3$ cathode. The cross-linked PMMA GPE was prepared by in-situ polymerization with the liquid electrolytes of 1.0 M $NaClO_4$ in PC/FEC (9 : 1, v/v) (Figure 13.16). The PMMA-based GPE has a wide electrochemical window (4.8 V) because of the cross-linked PMMA having a high oxidation potential. The strong interaction between the liquid electrolyte and PMMA chains makes a high ionic conductivity of 6.2×10^{-3} S cm^{-1} at 25 °C. The activation energy of the GPE is 10.6 kJ mol^{-1}, which is much lower than that of PMMA-based electrolyte (12–27 kJ mol^{-1}) and PEO-based electrolyte (44 kJ mol^{-1}). The lower activation energy indicates that the ion mobility in the GPE is mainly due to the jumping of sodium ions rather than the movement of polymer chain segments. The Sb|GPE|$Na_3V_2(PO_4)_3$ battery has a high discharge capacity (106.8 mA h g^{-1} at 0.1 C) and good rate performance (61.1 mA h g^{-1} at 10 C). The Sb|GPE|$Na_3V_2(PO_4)_3$

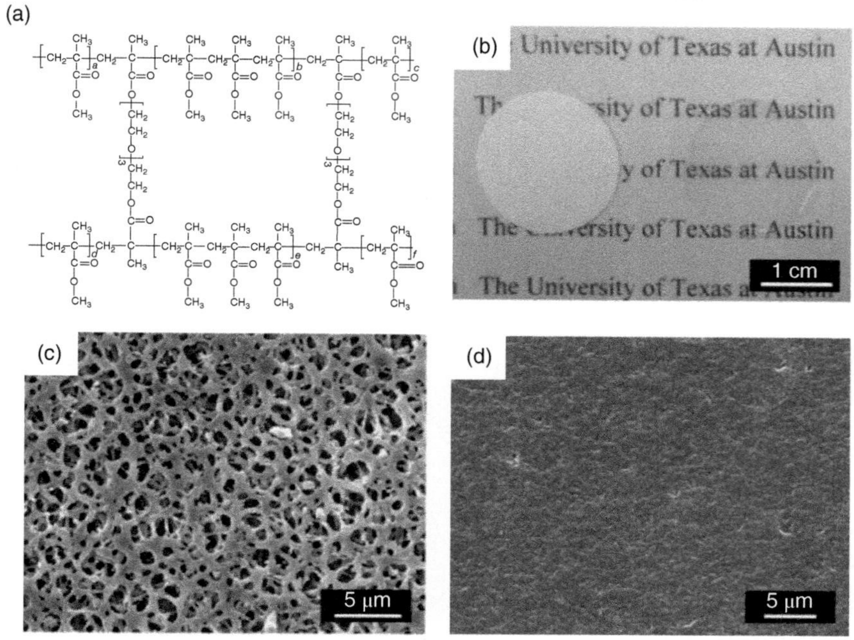

Figure 13.16 (a) Schematic diagram of PMMA structure. (b) Optical photo of cellulose membrane (left) and cellulose membrane containing cross-linked PMMA GPE (right). SEM image of (c) cellulose membrane and (d) cellulose membrane containing cross-linked PMMA GPE. Source: Gao et al. [44]. Reproduced with permission, 2016, John Wiley & Sons, Inc.

battery also has outstanding high-capacity retention compared with conventional liquid electrolyte cells.

13.3.4.2 PVDF-Based GPE

PVDF is another most extensively studied polymers for the preparation of GPEs because it has a strong polar functional group (—C—F) with a high dielectric constant ($\varepsilon = 8.4$), which facilitates the dissolution of sodium salts, giving a high concentration of charge carriers.

Park et al. [45] studied PVDF-based GPE with melamine as plasticizer and $NaCF_3SO_3$ as salt. The conductivity of the GPE is $5.1 \times 10^{-4}\,S\,cm^{-1}$ at room temperature. The as-fabricated Na|GPE|S battery can successfully operate at room temperature.

PVDF-HFP-based electrolyte is also a desirable polymer backbone because it is semicrystalline, which has good mechanical properties and special porous structure [46].

Wu and coworkers [47] fabricated a highly porous membrane of PVDF-HFP-based GPE by a simple phase inversion method using water as a nonsolvent (Figure 13.17a–c). The PVDF-HFP membrane was soaked in liquid electrolyte (1 M $NaClO_4$ in EC/DMC/DEC = 1 : 1 : 1, wt.) to obtain GPE. Its thermal stability is almost identical to that of a commercial separator (Celgard 2730), which is stable at 130 °C. The ionic conductivity of PVDF-HFP-based GPE is $0.60\,mS\,cm^{-1}$

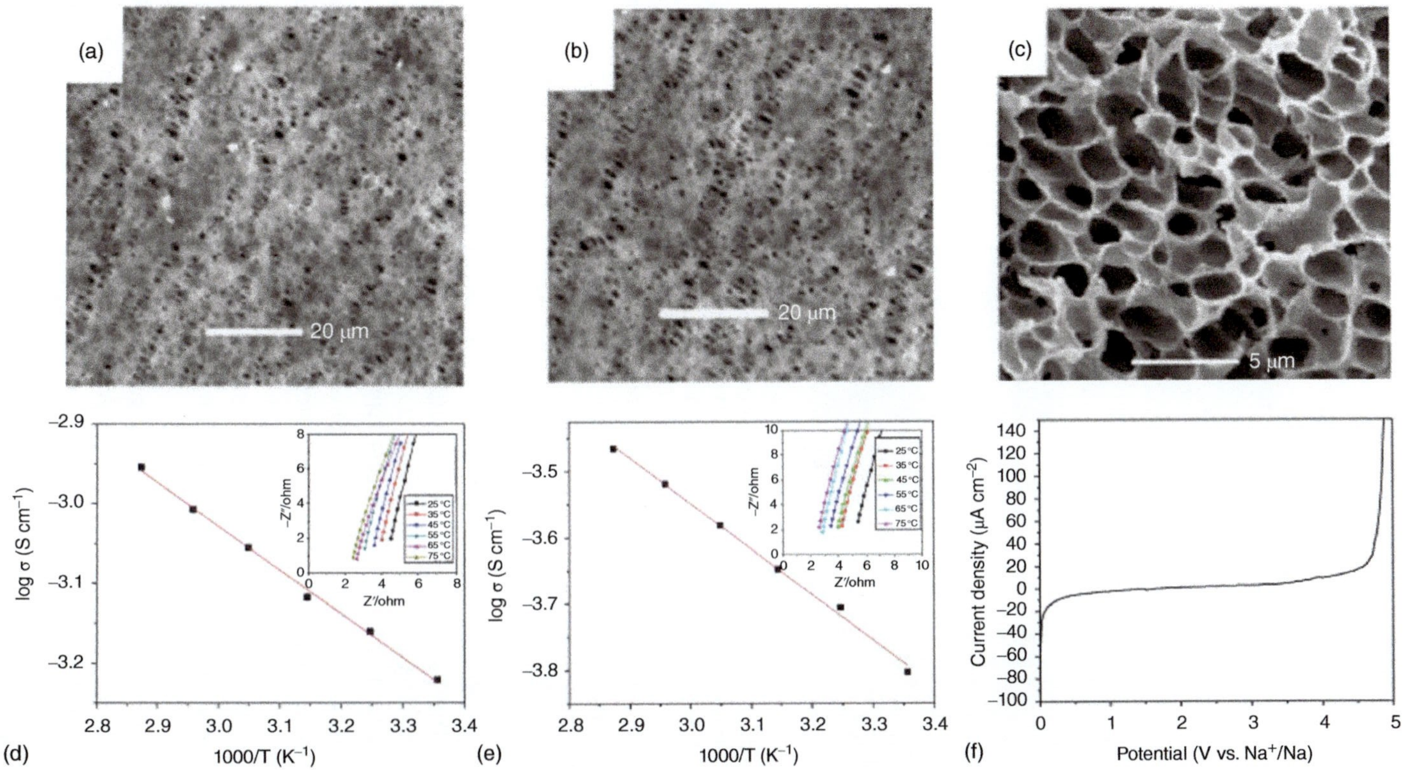

Figure 13.17 SEM images of (a) the top, (b) bottom and (c) cross section of PVDF-HFP membrane. Arrhenius curve of (d) PVDF-HFP membrane and (e) Celgard 2730 separator with 1 M $NaClO_4$ liquid electrolyte. (f) LSV curve of GPE. Source: Yang et al. [47]. Reproduced with permission, 2014, Elsevier.

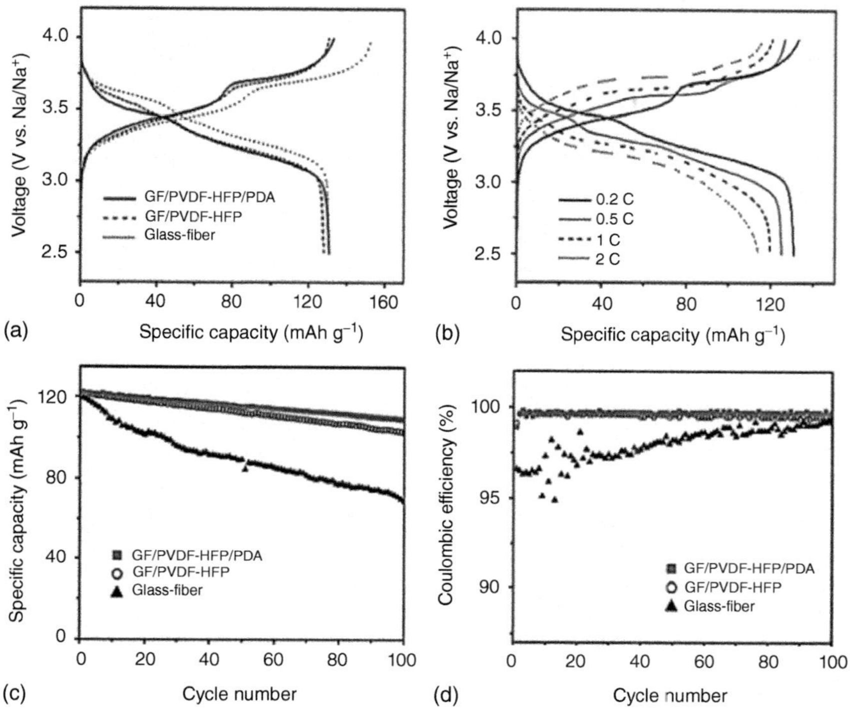

Figure 13.18 Electrochemical performance of $Na_2MnFe(CN)_6$ battery with GF, GF/PVDF-HFP and GF/PVDF-HFP/PDA separators. (a) The first charge and discharge curves (0.2 C). (b) Charge and discharge curves of the battery with CGPE at different C rates. (c) Cycle performance and (d) CE of the battery with different separators at 1 C. Source: Gao et al. [48]. Reproduced with permission, 2015, WILEY-VCH Verlag GmbH & Co. KGaA, Weinheim.

at room temperature (Figure 13.17d), which is higher than that of Celgard 2730 (0.16 mS cm^{-1}, Figure 13.17e). PVDF-HFP-based GPE also has a higher sodium ion transfer number ($t_{Na+} = 0.3$) than that of Celgard 2730 ($t_{Na+} = 0.17$), as well as a wider electrochemical window (4.6 V, Figure 13.17f). In addition, PVDF-HFP-based GPE not only can improve safety, but also exhibits excellent mechanical properties (tensile strength of 7.6 MPa).

Goodenough and coworkers [48] prepared a composite GPE consisting of polydopamine (PDA), glass fiber (GF), and PVDF-HFP with liquid electrolyte (1 M $NaClO_4$ in PC) (Figure 13.18). The composite GPE has good mechanical properties with a tensile strength (20.9 MPa) and good thermal stability (200 °C). The composite GPE also has a wide electrochemical window (4.8 V). The ionic conductivity of the PVDF-HFP/GF GPE is 4.6 mS cm^{-1} at 25 °C. When the PDA was added, the ionic conductivity of GPE was improved to 5.4 mS cm^{-1}. The increase in ionic conductivity can be attributed to the coating of polydopamine, which accelerates the transmission of sodium ions in GPE. As shown in Figure 13.18, the cycle stability and Coulombic efficiency of the $Na_2MnFe(CN)_6$||Na battery assembled with PVDF-HFP/GF/PDA GPE are significantly improved, due to its higher ionic conductivity and better stability. The specific capacity at 0.2 C is similar with different electrolytes (Figure 13.18a), while the rate performance and cycling performance were improved with PVDF-HFP/GF/PDA GPE. With PVDF-HFP/GF/PDA GPE, the capacity can be retained at 114 mA h g^{-1} at a current of 2 C (Figure 13.18b), which is higher than that of batteries with glass–fiber separators (67 mA h g^{-1}) [48]. After 100 cycles at 1 C, the capacity retention is about 89.4% for GF/PVDF-HFP/PDA GPE, while it is only 57.8% for glass fiber separators (Figure 13.18c). In addition, the Columbic efficiency is close to 100% with GF/PVDF-HFP/PDA GPE, much higher than that with glass–fiber separators (Figure 13.18d).

13.3.4.3 Nafion-Based GPE

Nafion is well known as a good proton conductor, which has been widely applied in fuel cells. Exchanging proton with Li^+ or Na^+ can make it Li^+ or Na^+ conductor. Kreuer et al. [49] developed perfluorinated sulfonic GPE membrane by membrane ion exchange in an aqueous NaOH solution with an exchange rate of close to 100%. After removing the water and drying, the GPE was obtained by polymer membranes expanding with DMSO. The ionic conductivity can reach over 1 mS cm^{-1} with adjusting [DMSO]/[Li^+] ratio to be 20.

Li and coworkers [50] used EC/PC (1 : 1, v : v) expanded Nafion to obtain a transparent homogeneous film. It shows an ion conductivity of 2.5×10^{-4} S cm^{-1} at RT and an activation energy of 0.19 eV. The battery assembled with Nafion-based GPE and $Na_{0.44}MnO_2$ cathode shows better cycle performance, comparing with conventional liquid electrolytes.

13.3.5 Adding Ceramic Filler to GPEs

The mechanical properties of GPEs are worse compared with dry polymer electrolytes. To overcome this problem, the addition of ceramic fillers to GPE is a widely explored method to improve the mechanical properties of GPE. The addition of inorganic fillers can not only improve the mechanical properties of GPE, but also improve the electrochemical properties of GPE.

Aravindan et al. [51] studied the effect of nano Sb_2O_3 as the ceramic filler in GPE with $NaCF_3SO_3$ as Na^+ salt and blending PVDF and PEMA as polymer. A maximum conductivity of 0.560 mS cm^{-1} was achieved at room temperature with adding 10 wt.% Sb_2O_3. XRD reveals the amorphous nature of the composite electrolyte, which is the main reason for high conductivity.

The Goodenough group [52] reported PVDF-HFP and PVP blended GPE with a weight ratio of 65 : 35 using 30 wt.% Sb_2O_3 nano-powders as ceramic fillers. The GPE with a conductivity of 6×10^{-3} S cm^{-1} was obtained by the immersion of polymer film in liquid electrolyte (1 M $NaClO_4$ in EC/DEC = 1 : 1). In addition, the GPE could restrain the growth of sodium dendrites, as demonstrated by the stable Na stripping/deposition (Figure 13.19a,b). The prepared GPE was successfully applied to sodium metal battery with the Prussian blue cathode material. The discharge capacity of the battery can reach 115 and 65 mA h g^{-1} at current densities of 24 and 600 mAg^{-1}, respectively.

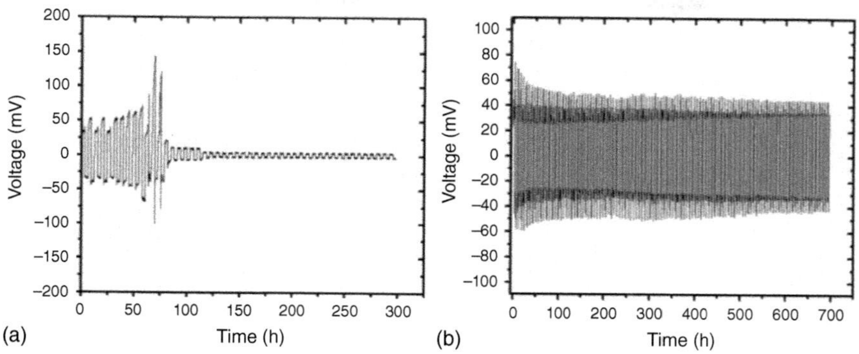

Figure 13.19 Voltage polarization of Na|GPE|Na symmetric battery (0.1 mA cm^{-2}): (a) PVDF-HFP/PVP with 1 M $NaClO_4$ in EC/DEC,1 : 1 and (b) PVDF-HFP/PVP/Sb_2O_3 with 1 M $NaClO_4$ in EC/DEC,1 : 1. Source: Ansari et al. [52]. Reproduced with permission, 2014, IOP Publishing.

13.3.6 Cross-linked GPEs

Another strategy to improve the poor mechanical properties of GPEs is to cross-link the polymers, which has been extensively explored in LIB applications [53].

Zhao and coworkers [54] reported a cross-linked GPE, which was synthesized by in situ thermal polymerization of di(2-methacryloyltrioxyethyl)phenylphosphonate (MATEPP), methyl methacrylate (MMA), and trifluoroethyl methacrylate (TFMA) (Figure 13.20a). The prepared GPEs have wide electrochemical window (4.9 V, Figure 13.20b) and high ionic conductivity (6.29×10^{-3} S cm^{-1}). A high specific capacity (110.7 mA h g^{-1}) at 1 C rate was obtained for the SIB based on cross-linked GPEs and $Na_3V_2(PO_4)_3$ cathode. Furthermore, excellent long cycling stability was observed at 5 C rate, retaining 81.8% and 69.2% of the capacity after 4500 and 10 000 cycles, respectively (Figure 13.20c). The battery with GPE also exhibited good cycle stability when cycled at high temperature of 60 °C. Compared with conventional liquid electrolytes, good cycle performance was also observed in SnS_2|GPE|Na battery and SnS_2|GPE|NVP battery. It is shown that cross-linked GPE containing phosphates holds great prospect for the production of safety battery systems.

13.3.7 Ionic Liquid-Based GPEs

In general, GPE is characterized with high ionic conductivity, but the use of liquid organic solvents as plasticizers (EC, PC, DMC, DEC, etc.) reduces its safety compared with “dry polymer electrolytes”. These liquid electrolytes are confined by a polymer matrix. But due to the presence of volatile solvents, there are problems of electrochemical instability and flammability (especially at high temperatures), which limit the broad application of the GPE. The use of nonflammable and nonvolatile ionic liquids (ILs) instead of liquid organic electrolyte is a valuable option.

Kumar and Hashmi [55] prepared IL-based gel polymer electrolytes from PVDF-HFP, 1-ethyl-3-methyltrifluoromethanesulfonic acid (EMIMTf), and sodium cyanate (NaTf) with high ionic conductivity (5×10^{-3} S cm^{-1}). Boschin and

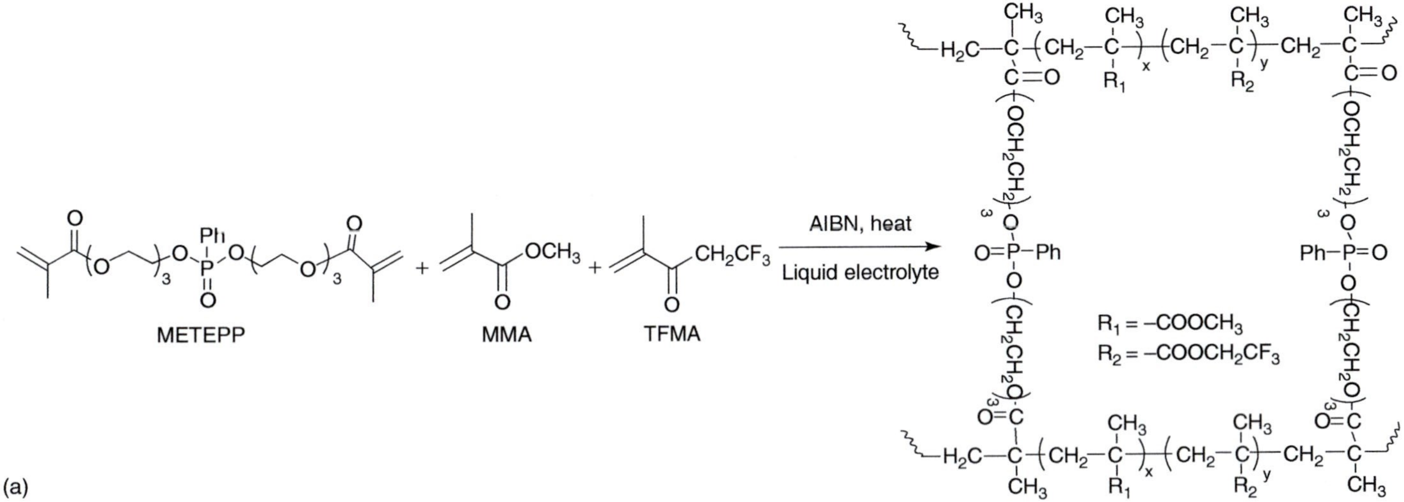

(a)

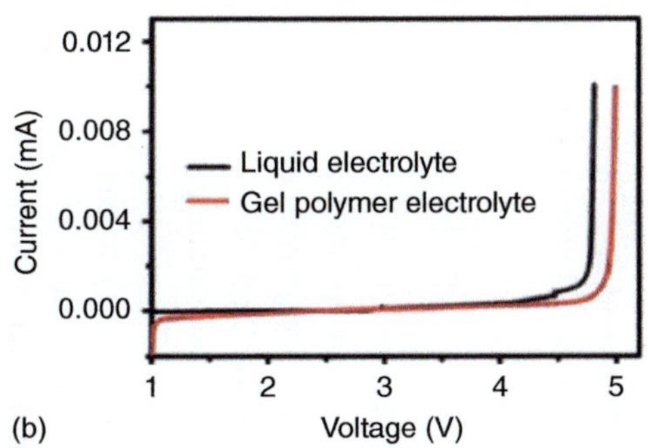

(b)

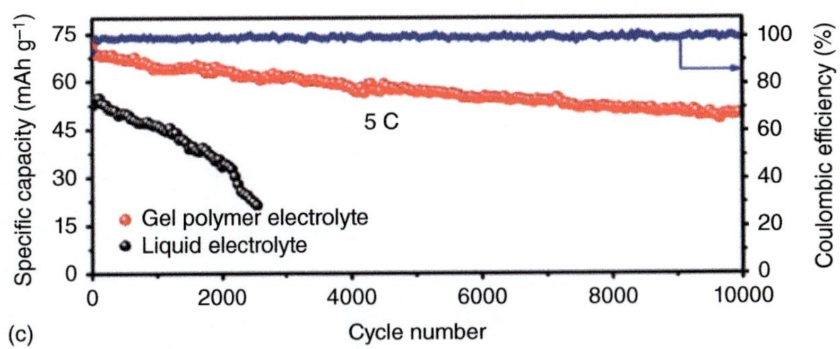

(c)

Figure 13.20 (a) Synthesis process of gel polymer electrolytes. (b) Linear sweep voltammograms of GPE and liquid electrolyte, with a scan rate of $1\ mV\ s^{-1}$. (c) Cycle performance of $Na_3V_2(PO_4)_3$/Na batteries using the GPE and liquid electrolyte. Source: Zheng et al. [54]. Reproduced with permission, 2018, Royal Society of Chemistry.

Johansson [56] reported IL-based GPE with components of Na salt (NaTFSI or NaFSI), PEO, and Pyr13TFSI with the mole ratios of EO : Na = 1 : 6, 9, 20. The mass fractions of IL were 5, 10, and 20 wt.%. This work demonstrated that the conductivity generally increases with the increase of ionic liquid content, which is attributed to the increase in sodium ion conduction pathways and the improvement of polymer chain dynamics through the plasticizing effect. However, the conductivity was lower than 10^{-5} S cm^{-1} at room temperature, which limits its practical application.

13.4 Inorganic Solid-State Electrolytes

13.4.1 Oxide-Based Solid-State Electrolytes

13.4.1.1 Beta-Alumina

Faraday explored the remarkable ionic transport properties of solid PbF_2 and Ag_2S for the first time in 1838 [57]. Nonetheless, an oxide-based SSE of beta-alumina, which has an outstanding Na^+ conductivity, got people's attention until 1960s [58]. At first, what caught the researchers' attention is its good refractory. Later, it was revealed that beta-alumina has promising ionic conductivity and electronic resistivity in 1967 [58, 59]. This discovery brought a great chance to use beta-alumina to replace liquid electrolyte in energy storage batteries.

Beta-alumina (beta-Al_2O_3), which typically operates at 300–350 °C, is principally serviced in ZEBRA batteries. What's more, it is an important component of solid electrolytes in the application of Na–S batteries (Figure 13.21) [60]. The ionic conductivity of beta-Al_2O_3 at room temperature is also high, making it possible to assemble ASSBs with beta-Al_2O_3 and operate at room temperature [61]. In addition, beta-Al_2O_3 has been proven to be stable against Na metal anodes and obtain low interfacial resistance between beta-Al_2O_3 and Na [62]. However, voids form in the Na metal/beta-Al_2O_3 interface during stripping and accumulate on cycling, making the interface unstable and interfacial resistance increase [63]. This can be restrained at high stack pressure and the interfacial stability improves [63].

In fact, beta-Al_2O_3 refers to a compound consisting of $M_2O{\cdot}xAl_2O_3$ ($M = Na^+$, K^+, Rb^+, Ag^+, etc., x = 5–11). For $Na_2O \cdot xAl_2O_3$, it is a layered compound with an overlapping structure of a spinel block $Al_{11}O_{16}$ and a conductive plane Na–O layer. Beta-Al_2O_3 has two different structures: β-Al_2O_3 ($P6_3/mmc$; a = 0.559 nm, c = 2.261 nm) and β″-Al_2O_3 ($R3m$; a = 0.560 nm, c = 3.395 nm). Their chemical composition and the order of accumulation of oxygen ions between the ion conductive layers are different. The β phase is $Na_2O{\cdot}(8\text{–}11)Al_2O_3$, and β″ phase is $Na_2O{\cdot}(5\text{–}7)Al_2O_3$. Figure 13.22 shows the structures of the β phase and β″ phase alumina [1c]. In the β-Al_2O_3 crystal structure, the spinel-based $Al_{11}O_{16}$ and the Na–O layer are overlapped. The O^{2-} accumulation at the base of the spinel is the most compact, while the Na–O accumulation is loose. In the Na–O layer, Na^+ occupies three positions called Bever–Ross (BR), anti-Bever–Ross (aBR) and mid-oxygen (mO). Compared with the β-Al_2O_3 phase, β″-Al_2O_3 phase has higher ionic conductivity of 2×10^{-3} S cm^{-1} at RT and 0.2–0.4 S cm^{-1} at 300 °C. However,

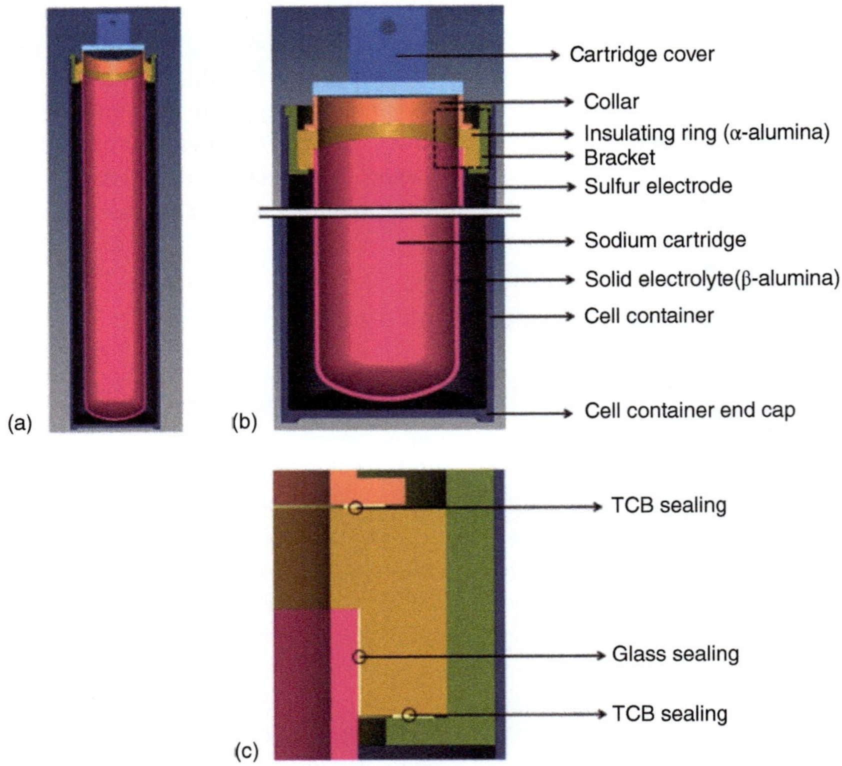

Figure 13.21 Schematic diagram of the cross-sectional structure of a representative tubular Na|S battery showing (a) external structure view, (b) the head and bottom structure view, and (c) cross-sectional structure view of the TCB and GS sections. Source: Jung et al. [60]. Reproduced with permission, 2014, Elsevier.

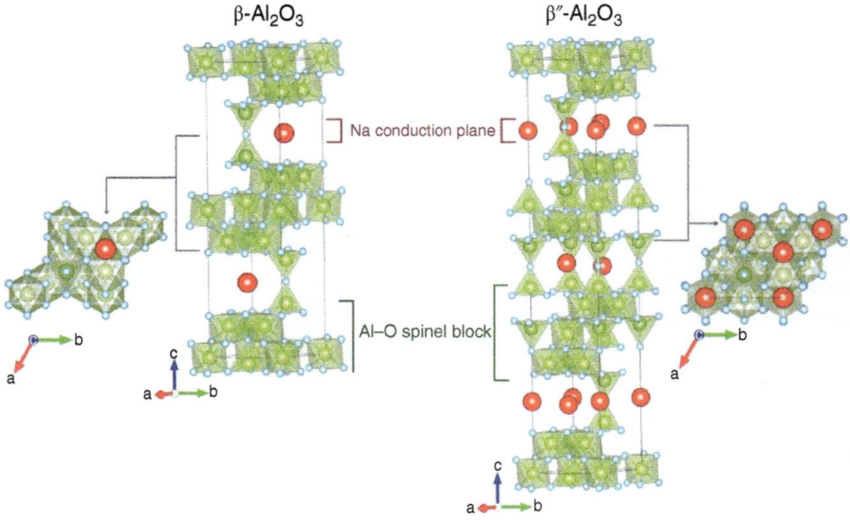

Figure 13.22 Crystal structures of β-Al_2O_3 and β″-Al_2O_3. Source: Lu et al. [1c]. Reproduced with permission, 2018, Elsevier.

it is hard to obtain high-purity β''-Al_2O_3. Chemical stability of β-Al_2O_3 was also investigated, and the results indicate β-Al_2O_3 would absorb water to get hydrated, and the amount is about one H_2O molecular per cation in the conducting layer [64].

High-temperature heat treatment of α-Al_2O_3 and Na_2CO_3 is generally used to synthesize β''-Al_2O_3, with Li_2CO_3 and/or MgO as stabilizers [65]. Specifically, the synthesis is carried out by wet ball milling, drying, and heat treatment of the precursors at ≈1200 °C for several hours, followed by re-milling and sintering at 1600 °C for 30 minutes [66]. However, this method has some drawbacks: (i) grain growth, (ii) water-sensitive $NaAlO_2$ formed at grain boundaries, and (iii) high β-Al_2O_3 percentage with decreased ionic conductivity. Furthermore, several other methods such as coprecipitation method [67], sol–gel method [68], solution combustion [69], microwave heating [70], spray freeze/freeze-drying method [71], and mechanochemical method [72] have been used to synthesize uniform β''-Al_2O_3. However, high-purity chemical precursors or complicated production processes are required to prepare high-purity β''-Al_2O_3, which is not economically friendly. The preparation of β''-Al_2O_3 using cheap and rich hydroxyalumina groups (e.g. boehmite and bayerite) as raw materials has been proposed [73]. It is reported that the method can observably lower the sintering temperature and result in less sodium loss.

Besides preparation method, cation doping such as Mg^{2+} and Li^+ can stabilize the β''-Al_2O_3 and increase ionic conductivity [74]. The addition of Mg^{2+} or Li^+ can increase the octahedral coordinating aluminum and tetrahedral coordinated aluminum $[A_{Al(IV)}/A_{Al(VI)}]$, making β''-Al_2O_3 more stable and increase the β''/β ratio. Up to 95% content of the β''-Al_2O_3 phase can be obtained by doping 2.0 wt.% MgO. At the same time, the β''-Al_2O_3 prepared by the doping method exhibits the highest bending strength (283 MPa) [75]. Furthermore, Song and coworkers revealed that TiO_2 could promote the diffusion of Al^{3+} and O^{2-}, thus improving the compactness of β''-Al_2O_3 [76]. Chen et al. stabilized the crystal orientation of β''-Al_2O_3 by TiO_2 doping and thus increased the ionic conductivity of the material [77].

13.4.1.2 NASICON

NASICON-type solid electrolyte materials were first reported by Goodenough and Hong in 1976, with the advantages of high ionic conductivity, low coefficient of expansion, and suitability for working at high temperatures [78]. The common one is $Na_{1+x}Zr_2Si_xP_{3-x}O_{12}$ ($0 \leq x \leq 3$), derived from $NaZr_2(PO_4)_3$ by replacing P with Si. $Na_3Zr_2Si_2PO_{12}$ is inexpensive and high conductive [79], which has been used as an electrolyte in low-temperature Na–S [80], Na–O_2 [81], and hard carbon-$NaFePO_4$ batteries [80b] with excellent performance.

The NASICON material system has two different structures as shown schematically in Figure 13.23, both have open 3D Na^+ transport channels [83]. One is rhombus symmetrical structure with $R\bar{3}c$ space group. The PO_4 tetrahedron and the ZrO_6 octahedron are connected at the same vertex to form a 3D skeleton. Na^+ is located in the voids of the skeleton and can be conduct tropically along the 3D channels formed by these voids. There are two crystallographically disparate sodium ion positions: Na1 and Na2. The Na1 position is between two ZrO_6 octahedra and forms

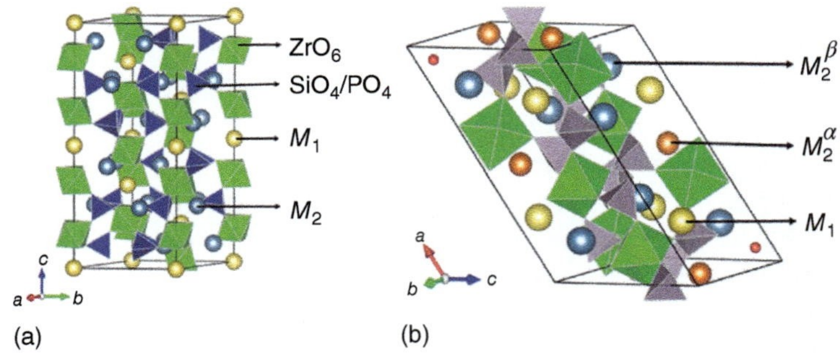

Figure 13.23 Schematic diagram of NASICON cells with different structures: (a) rhombohedral structure, and (b) monoclinic structure. Source: Samiee et al. [82]. Reproduced with permission, 2017, Elsevier.

an $O_3ZrO_3NaO_3ZrO_3$ structure along the *c*-axis, with each two $O_3ZrO_3NaO_3ZrO_3$ connected by a tetrahedron PO_4. A structural belt parallel to the *c*-axis is formed, while the position of Na2 is between the two structural belts (Figure 13.23a) [82]. When using Si to partially replace P, the crystal lattice will be slightly deformed and its symmetry will be reduced because the tetrahedron of SiO_4 is larger than PO_4. In this case, the structure would change from rhombohedral to monoclinic (Figure 13.23b), and the position of Na2 splits into two positions ($Na_{1+x}Si_xP_{3-x}O_{12}$, $1.8 < x < 2.2$). Partially replacing P by Si also introduces more Na to achieve charge equilibrium. Further increasing Si in $Na_{1+x}Si_xP_{3-x}O_{12}$ ($x > 2.2$) can change the structure back to rhombohedral.

NASICON has an interconnected polyhedral covalent framework containing a mass of interstitial sites, in which cations migrate [78a]. The strong covalent bonding framework and structural pores give the structure an increased lattice thermal conductivity and a high Debye temperature. Therefore, it is extensively believed to have high thermal and chemical stability. Generally, the formula of NASICON is $NaB_2(AO_4)_3$, where A can be P^{5+} or Si^{4+} with smaller size and B can be divalent, trivalent, tetravalent, or pentavalent metal ions with bigger size. In addition, Na^+ concentration can be adjusted to balance the charge. The valence states and ionic radii of dopant elements are vital factors when selecting B dopants. The appropriate doping has an important influence on the Coulombic interaction between the skeleton atoms and the mobile ion. As shown in Figure 13.24a, the crystal structure and Na concentration can determine the ionic conductivity of NASICON ceramics. Furthermore, the ionic conductivity is also related to composition, density, and preparation method [84].

The number of Na vacancies effects the ionic conductivity [85]. For $Na_{3+x}Sc_2Si_xP_{3-x}O_{12}$, the optimized Na vacancy is about 0.6 with the maximum ionic conductivity of 6.9×10^{-4} S cm^{-1}, when *x* equals to 0.4 ($Na_{3.4}Sc_2(SiO_4)_{0.4}(PO_4)_{2.6}$) (Figure 13.24b) [85]. The Na vacancy can also be developed with doping Zr site by transition metals or P site with Si. For example, the as-reported $Na_{3.3}La_{0.3}Zr_{1.7}Si_2PO_{12}$ with about 0.7 Na vacancies exhibits a high conductivity of 1.34×10^{-3} S cm^{-1} among

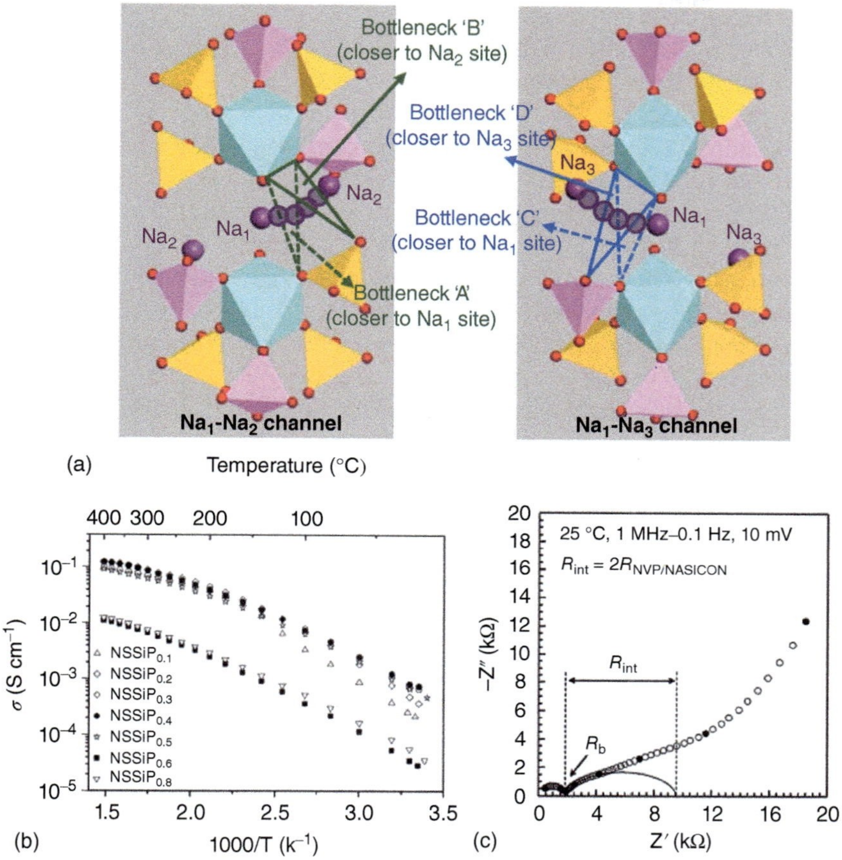

Figure 13.24 (a) Bottleneck of the Na conductivity pathway in NASICON. Source: Park et al. [84]. Reproduced with permission, 2016, American Chemical Society. (b) The conductivity of $Na_{3+x}Sc_2(SiO_4)_x(PO_4)_{3-x}$ at different temperatures. Source: Guin et al. [85]. Reproduced with permission, 2016, Elsevier. (c) EIS of NVP|NASICON|NVP solid-state batteries at RT. Source: Noguchi et al. [86]. Reproduced with permission, 2013, Elsevier.

$Na_{3+x}Zr_{2-x}La_xSi_2PO_{12}$ ($0 \le x \le 0.5$) [87]. Besides proper Na vacancies, the metal cation size is also important, which affects the lattice parameters and diffusion bottleneck. Guin and Tietz [88] deduced that ionic conductivity can reach the highest when the average size of the metal cation is about 0.72 Å, and the Na content is close to 3.3 mol (0.7 Na vacancies). In addition, the two different structures, monoclinic and rhombohedral, also affect the ionic conductivity. When converting from monoclinic to rhombohedral (around 147 °C), the activation energy of $Na_3Zr_2Si_2PO_{12}$ decreases from 0.36 to 0.21 eV, indicating that rhombohedral phase is more conductive [89]. In addition to adjusting Si/P ratio, replacing Zr^{4+} with heterovalent ions (e.g. Y^{3+}) can stabilize the more symmetrical rhombohedral phase of NASICON-type materials [90]. Up to date, the as-reported $Na_{3.4}Zr_2Si_{2.4}P_{0.6}O_{12}$

exhibits the conductivity of 5×10^{-3} S cm^{-1} at RT, which is even comparable to liquid electrolytes [91]. The high ionic conductivity of $Na_{3.4}Zr_2Si_{2.4}P_{0.6}O_{12}$ is due to its rhombohedral structure, proper cation size, and also the suitable Na vacancies (0.6).

Crystallization conditions have a critical effect on the microstructure of NASICON. $Na_{1.3}Ti_{1.7}Al_{0.3}(PO_4)_3$ ceramics with different grain sizes can be synthesized under different calcination conditions [92]. It is found that small grains can enhance the conductivity of ceramic grain boundaries [92]. Temperature and calcination time also have an effect on the microstructure of sodium aluminum germanium phosphate (NAGP) electrolytes, which in turn affects conductivity [93]. The heat treatment conditions of NASICON electrolytes are usually determined by the composition [94].

Solid-state reaction, hydrothermal, sol–gel, spray freeze, and freeze-drying methods are generally used to synthesize NASICON [78b, 95]. It is preferable to use solid-state reactions, in which a certain amount of the precursors is ball-milled, dried, pre-calcined (≈900 °C), and then sintered at higher temperatures to obtain a well-crystallized product. Sol–gel or hydrothermal methods are also adopted, which can achieve more homogeneous products compared with other synthetic methods.

In 2018, Goodenough and coworkers [96] used $Na_3Zr_2Si_2PO_{12}$ as the electrolyte to construct $Na_2MnFe(CN)_6|Na_3Zr_2Si_2PO_{12}|Na$ batteries, which not only prevented the branching phenomenon of the anode material during the cycle, but also eliminated the dissolution of the cathode material, effectively improving the performance of the battery. Although NASICON SSE would decompose to NaO_x at high potential and low potential vs. Na/Na^+, the formed interface layers are electron insulator, which can prevent further decomposing of NASICON SSE. However, Schmid et al. [80b] showed that when temperature increases to 300 °C, $Na_{1+x}Zr_2P_{3-x}Si_xO_{12}$ ($x = 2$ and 2.2) readily reacts with metallic sodium, thus making poor battery performance under high temperature conditions.

Even with high ionic conductivity and high electrochemical stability, ASSBs with NASICON-type SSE still cannot work well at room temperature, because of the poor contact between NASICON and cathode. Goodenough and coworkers added polymer electrolytes in the cathode to improve the contact and operate the ASSBs at 60 °C, while helps to make the ASSBs to work [96]. In order to improve the compatibility between the NASICON SSE and cathode, Noguchi et al. prepared all-solid-state batteries with NVP as the cathode and $Na_{3-x}V_{2-x}Zr_x(PO_4)_3$ as electrolyte (Figure 13.24c), both of which are NASICON-type materials [86]. To further reduce the interfacial resistance, the researchers prepared a single-phase ($Na_{3-x}V_{2-x}Zr_x(PO_4)_3$) all-solid-state sodium battery [97]. When $x = 0.6$, it is demonstrated a high conductivity of 1.2×10^{-5} S cm^{-1}. This cell offers a new direction for the all-solid-state battery field. Furthermore, Goodenough and coworkers incorporated a polymer film at the interface. It is demonstrated that the batteries with $Na_3Zr_2PO_4(SiO_4)_2$ as the electrolyte have a capacity of 102 mA h g^{-1} after 70 cycles at 0.2 C [98]. Recently, the addition of ionic liquids to solid electrolytes has also emerged as an effective way of solving the interface problem [99].

13.4.2 Sulfide-Based Solid-State Electrolytes

Oxide-based solid electrolytes have been widely studied, but the problems of high interfacial resistance and low ion conductivity at RT limit their applications. Sulfide-based solid electrolytes are good alternatives owing to the following reasons. Firstly, sulfide-based SSEs have higher ionic conductivity than that of oxide-based solid-state electrolytes because oxygen is more electronegative than sulfur [100]. Secondly, sulfide-based SSEs have lower production cost [101]. Thirdly, heat treatment at high temperatures is generally required to reduce the interfacial impedance for oxide-based electrolyte, while sulfide-based electrolytes only require a cold pressing to obtain a better contact interface [102].

Figure 13.25a shows the crystal structure models of different sulfides, which can be divided into two different structures, tetragonal ($P\bar{4}2_1c$) and cubic ($I\bar{4}3m$) [103]. The structures and ionic conductivities are affected by different dopants and synthesis approaches. In Figure 13.25b, it can be seen that the different dopants affect the Na site energy of the sulfides [104]. When the dopant element is Sn^{4+}, the prepared $Na_{3+x}Sn_xP_{1-x}S_4$ has the highest Na site energy, which makes it less stable. For $Na_3As_{0.38}P_{0.62}S_4$, DFT calculations indicate that the high ionic conductivity of sulfide is mainly due to the long length of the Na—S bond, causing a lower energy barrier for sodium ion migration (Figure 13.25c) [105]. When S is totally replaced by Se (Na_3PSe_4), the structure transfers from tetragonal to cubic, leading to higher ionic conductivity of 1.16 m S cm^{-1} and a lower activation energy of 0.21 eV at room temperature (Figure 13.25d) [106].

13.4.2.1 Na_3PS_4

Na_3PS_4 was first reported by Jansen and Henseler [107]. The space group of Na_3PS_4 at room temperature is tetragonal ($P\bar{4}2_1c$), and it transfers to cubic ($I\bar{4}3m$) at around 260 °C. The conductivity of tetragonal Na_3PS_4 is 4.17×10^{-6} S cm^{-1} at 50 °C, with an activation energy of about 0.42 eV. After converting to cubic phase, the activation energy is slightly reduced to 0.4 eV. However, it is difficult to prepare cubic phase at room temperature since it is not stable. Tatsumisago and coworkers successfully prepared cubic Na_3PS_4 in 2012, which is metastable glass–ceramic and delivers much higher ionic conductivity of over 10^{-4} S cm^{-1} and lower activation energy of about 0.27 eV at room temperature [108].

The modification of cubic Na_3PS_4 was focused on enhancing their ionic conductivity. Researchers have optimized the preparation processes such as grinding and heat treatment conditions [109]. The Na_2S and P_2S_5 precursors were mechanically ground in a 3 : 1 molar ratio for 1.5 hours and heat-treated at 270 °C for two hours to obtain cubic Na_3PS_4. Yubuchi et al. [110] used *N*-methylformamide (NMF) as a solvent in which Na_2S and P_2S_5 were dissolved. After the removal of the solvent, the cubic phase Na_3PS_4 was obtained. Although its room-temperature conductivity was only 2.6×10^{-6} S cm^{-1}, lower than that of the product obtained by solid-state reaction, this approach can be effective in addressing interfacial problems in solid-state SIBs by greatly increasing the contact area at the electrolyte–electrode interface.

In recent years, Meng and coworkers have improved the synthesis conditions of Na_3PS_4 [111]. The synthesis parameters including total ball milling time, rotation

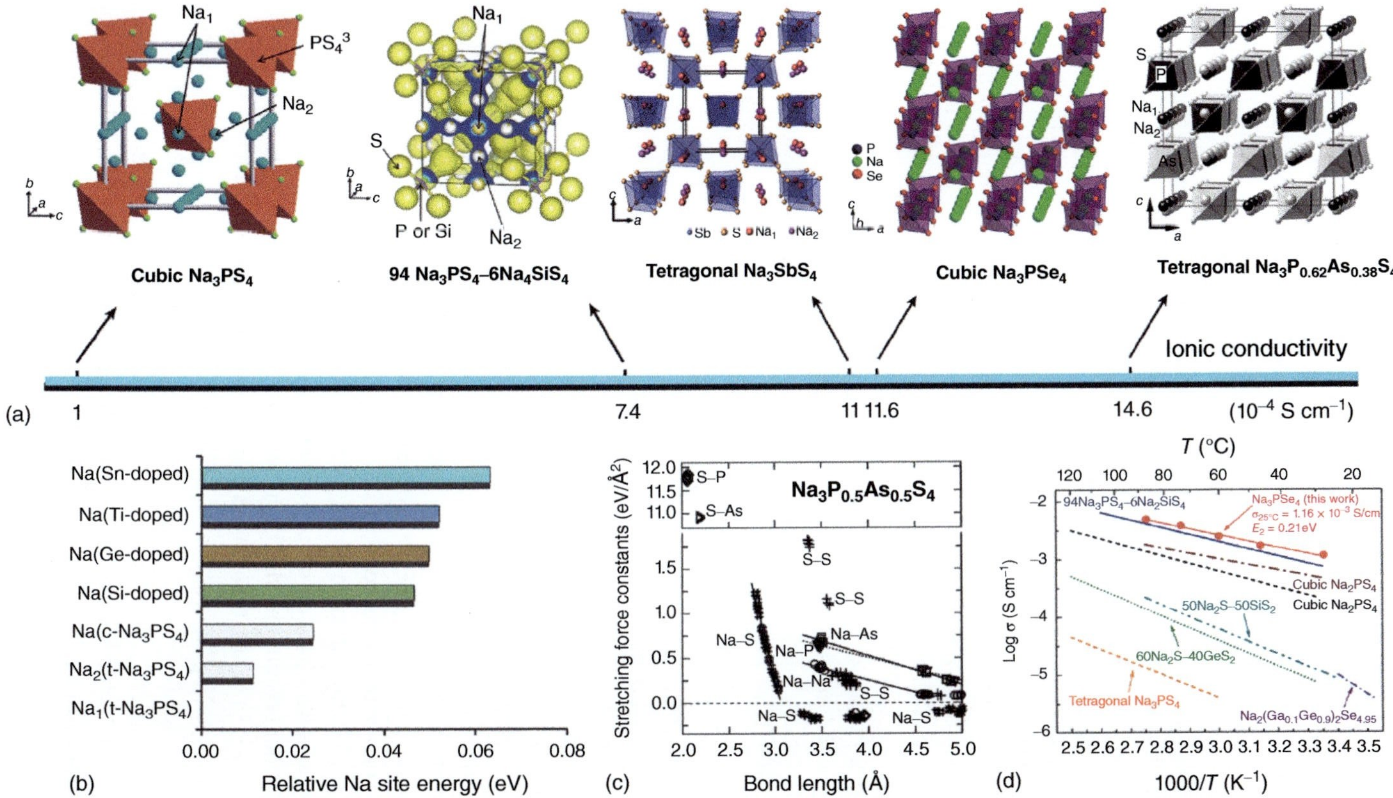

Figure 13.25 (a) Crystal structure models of the different sulfides and the corresponding ionic conductivities. Source: Tanibata et al. [103]. Reproduced with permission, 2014, Wiley-VCH. (b) Site energies of Na^+ in the sulfides obtained with different dopants. Source: Banerjee et al. [104]. Reproduced with permission, 2016, Wiley-VCH. (c) Stretching force constants of the low energy structure of $Na_3P_{0.5}As_{0.5}S_4$ calculate by DFT. Source: Zhang et al. [105]. Reproduced with permission, 2017, Wiley-VCH. (d) Arrhenius plots of different sulfide-based electrolytes. Source: Yu et al. [106]. Reproduced with permission, 2017, Wiley-VCH.

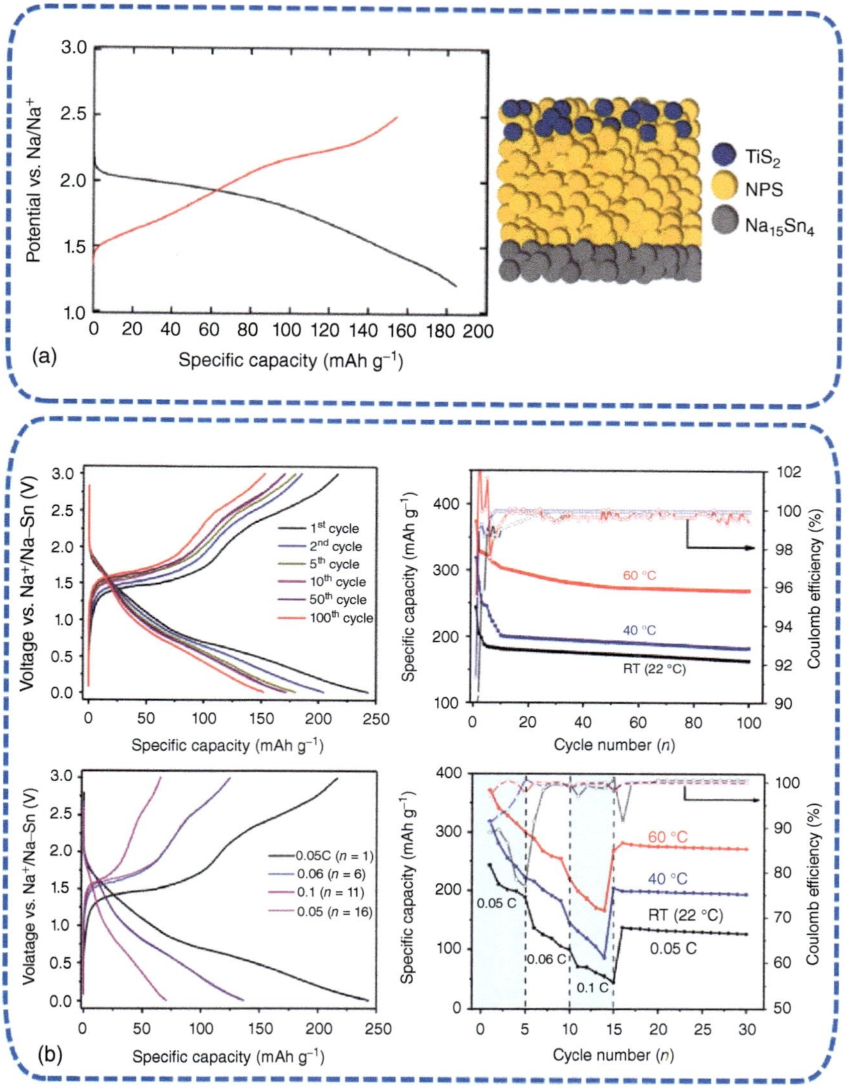

Figure 13.26 (a) Specific capacity–voltage curve at the first cycle and structural schematic of all-solid-state sodium battery using Na_3PS_4 as electrolyte. Source: Hayashi et al. [111]. Reproduced with permission, 2019, Springer Nature. (b) Charge and discharge curves and cycling stability of Na–Sn|Na_3PS_4|(MoS_2, Na_3PS_4, carbon) battery at different cycles, multiplicities, and temperatures. Source: Santhosha et al. [112]. Reproduced with permission, 2020, American Chemical Society.

speed, ball diameter, and secondary heat treatment were regulated. Figure 13.26a exhibits that the highest ionic conductivity of Na_3PS_4 can be produced using ball milling. When evaluated it as SSE, the TiS_2 battery can provide a capacity of 185 mA h g^{-1} and a better rate performance [111]. Adelhelm and coworkers developed a battery using MoS_2 nanosheets mixed with sulfide and carbon as the working electrode, Na_3PS_4 as the SE, and Na–Sn alloy as the counter electrode. Figure 13.26b indicated that the specific capacity of the battery was maintained at 163 mA h g^{-1} after 100 cycles at 0.05 C, with a capacity retention rate of 66% [112].

Besides optimizing synthesis, ion doping is another efficient way to improve the ionic conductivity, including cation and anion doping, for both cubic and tetragonal Na_3PS_4.

Studies have shown that partial substitution of P by other elements in Na_3PS_4 could improve the ionic conductivity. The effect of doping mainly depends on the radius and valence of the doped ions, with adjusting the lattice and introducing different types of defects. Equivalence ion substitution of P^{5+} by As^{5+} and Sb^{5+} would not introduce extra defect such as interstitial Na or Na vacancy but can expand the lattice. Yu et al. doped Na_3PS_4 with As to obtain $Na_3P_{0.62}As_{0.38}S_4$. Due to the weak expansion of As and its strong interaction with S, the ionic conductivity and the stability in moisture environment of the electrolyte are significantly improved (Figure 13.27a) [106]. It can be seen from the XRD pattern that $Na_3P_{0.62}As_{0.38}S_4$ still keeps the identical structure after being exposed to 15% humidity for 100 hours.

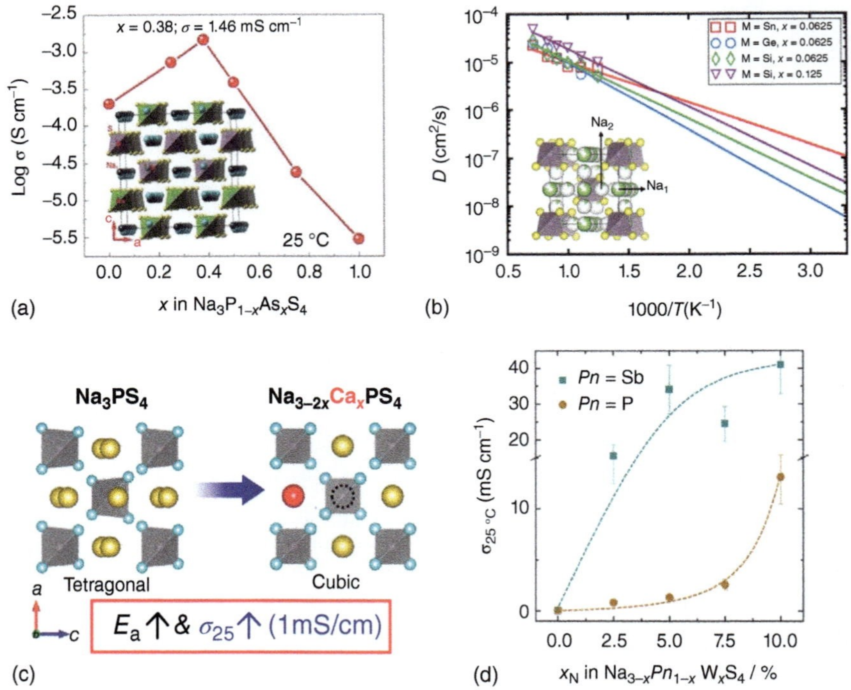

Figure 13.27 (a) A Na-ion solid-state electrolyte of $Na_3P_{0.62}As_{0.38}S_4$. Source: Yu et al. [106]. Reproduced with permission, 2017, Wiley-VCH. (b) Arrhenius plots of $Na_{3+x}M_xP_{1-x}S_4$ (M = Si, Ge, or Sn). Source: Zhu et al. [113]. Reproduced with permission, 2015, American Chemical Society. (c) A new Na^+ superionic conductor, Ca-doped cubic Na_3PS_4. Source: Moon et al. [114b]. Reproduced with permission,.2018, American Chemical Society. (d) Room-temperature conductivities of W^{6+} doped Na_3PS_4 and Na_3SbS_4. Source: Fuchs et al. [114a]. Reproduced with permission, 2019, Elsevier.

Cation doping with lower valence would reduce positive charge, thus interstitial Na is needed to balance the charge. Hayashi and coworkers proposed to replace some of the P with Si [115]. They investigated the effect of replacing P with different amounts of Si on Na_3PS_4. Based on XRD analysis, it can be concluded that no significant structural changes occurred. But when x in $Na_{3+x}Si_xP_{1-x}S_4$ exceeded 10, a new phase with a different cubic structure appeared. Further, it was revealed that the ionic conductivity reached the maximum at $x = 6$ and decreased significantly when x exceeded 10, which is related to the structure change of Na_3PS_4 [115]. Later, Ong and coworkers [113] used molecular dynamics simulation to explain why partial replacing of P by Si can enhance the ionic conductivity. The results indicate that the channel volume and correlation in Na^+ motions may play important roles in enhancing Na^+ conductivity in this structure. With Si^{4+} doping, the channel volume increases from 863 to 879 Å^3 ($Na_{3.125}Si_{0.125}P_{0.875}S_4$). Meanwhile, excess Na^+ partially occupied interstitial sites can induce Na disorder in c-Na_3PS_4, which enhances the Na^+ motions and ionic conductivity. In addition, they also studied the effect of Ge and Sn doping on the ionic conductivity. As shown in Figure 13.27b, it was found that $Na_{3.0625}Sn_{0.0625}P_{0.9375}S_4$ with cubic structure shows higher ionic conductivity and lower activation energy.

On the other hand, supervalence cation doping (e.g. replacing Na^+ with Ca^{2+} or P^{5+} with W^{6+}) would introduce Na vacancies in Na_3PS_4, which is also benefit to the ionic conductivity [114]. For example, Jung et al. have doped 0.135 Ca in t-Na_3PS_4, which generated 0.135 Na vacancy at the same time ($Na_{2.73}Ca_{0.135}PS_4$), showing the enhanced ionic conductivity of 0.94 mS cm^{-1}. In addition, after Ca^{2+} doping, tetragonal Na_3PS_4 would convert to cubic $Na_{2.73}Ca_{0.135}PS_4$. Density functional theory–based MD (DFTMD) simulations reveal the vacancy-driven Na-ion diffusion in cubic Na_3PS_4. Thus, the Na vacancy in $Na_{2.73}Ca_{0.135}PS_4$ would enhance the Na ion diffusion and ionic conductivity. However, Ca^{2+} with higher positive charge compared with Na^+ would push away neighboring Na^+ due to strong Coulombic repulsion. Thus, Na vacancies are readily formed near Ca^{2+} and easily trapped, resulting in a high migration barrier (Figure 13.27c). The comprehensive effect makes the conductivity reach the maximum with the optimized doping amount (0.135 Ca and 0.94 mS cm^{-1}). In order to reduce the adverse effects of cation doping at the Na site, Zeier et al. substituted P^{5+} by W^{6+}, which also introduce Na vacancies to Na_3PS_4 ($Na_{3-x}W_xP_{1-x}S_4$). Maximum of 0.1 W^{6+} can be doped in t-Na_3PS_4 ($Na_{2.9}W_{0.1}P_{0.9}S_4$), showing the enhanced ionic conductivity of 13 mS cm^{-1} and lower activation energy of 0.22 eV (Figure 13.27d).

Similar to lithium-ion conductors, anion substitution also has a significant implication for the diffusion of Na^+ in sodium solid electrolytes. For example, replacing S^{2-} by Se^{2-}, which has a larger ion radius and stronger polarization, can expand the unit cell parameters and ion diffusion channels, weaken the binding effect of the skeleton on Na^+ ions, reduce the migration activation energy of Na^+ ions, and thus increase the conductivity [116].

Ong et al. found that anion substitution had a more significant impact on conductivity [117]. Substitution of sulfur with elemental Se also improved the ionic conductivity of sodium sulfides (Figure 13.28a) [116]. Later, Zhang et al. confirmed

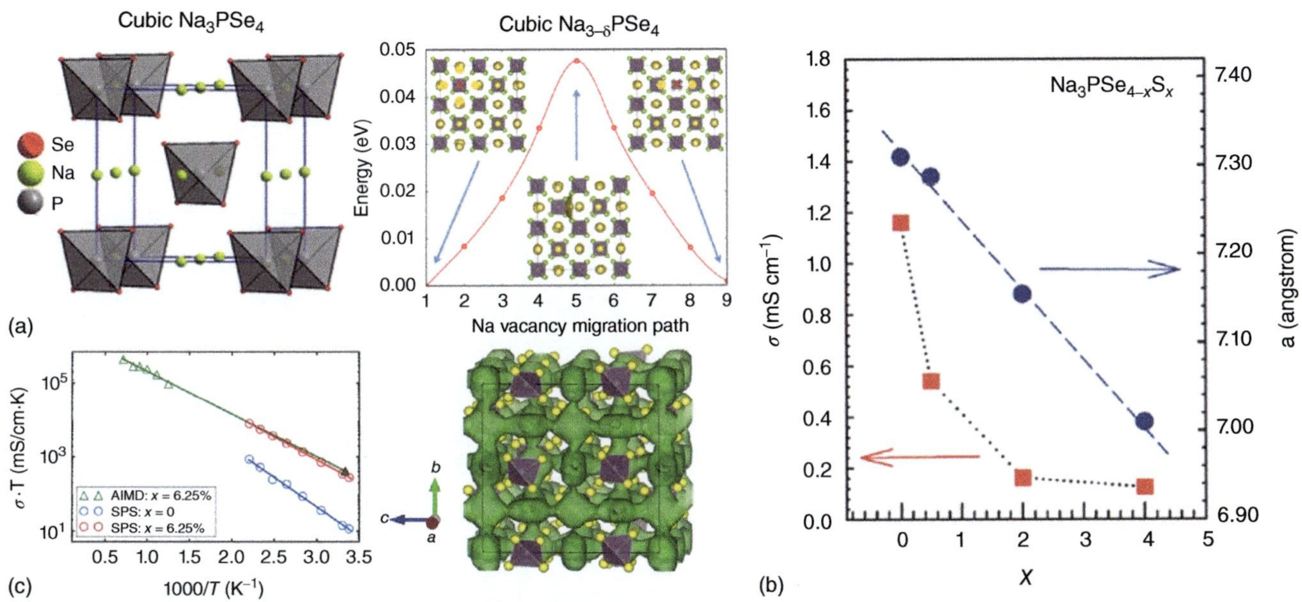

Figure 13.28 (a) The refined structure of Se-substituted sodium sulfide and calculated Na vacancy migration barrier in the cubic Na_3PSe_4. Source: Bo et al. [116]. Reproduced with permission, 2015, American Chemical Society. (b) Ionic conductivity of $Na_3PSe_{4-x}S_x$ at different x contents. Source: Zhang et al. [105]. Reproduced with permission, 2015, Wiley-VCH Verlag. (c) The left is Arrhenius plots obtained from SPS tests and AIMD tests for t-$Na_{3-x}PS_{4-x}Cl_x$ at different x contents; the right is contours of the Na^+ probability density distribution at 800 K for t-$Na_{2.9375}PS_{3.9375}Cl_{0.0625}$. Source: Chu et al. [118a]. Licensed under CC BY 4.0.

the reasons for the increased ionic conductivity after Se substitution in Na_3PS_4 (Figure 13.28b) [105]. First, the substitution of Se will cause the expansion of the crystal lattice, which may allow for smoother sodium ion transport. Second, Na^+ transfer is rapid due to the weak binding energy between Se^{2-} and Na^+.

Alternatively, halide doping has been suggested to increase ionic conductivity [118]. Chu et al. used computational and experimental methods to investigate the effect of Cl^- substitution on S^{2-} site of Na_3PS_4. The ionic conductivity at different temperatures as well as the activation energy after Cl^- doping is shown in Figure 13.28c [118a]. With Cl^- doping at S^{2-} site, ionic conductivity was improved to 1.14 mS cm^{-1} (t-$Na_{3-x}PS_{4-x}Cl_x$, $x = 0.0625$). Chu et al. also reported a Na/t-$Na_{2.9375}PS_{3.9375}Cl_{0.0625}$/$TiS_2$ battery, which provides a capacity of 80 mA h g^{-1} for 10 cycles at 0.1 C. Later, Hu and coworkers studied the function of two defects formed with Cl^- doping ($Cl_S^{\cdot}$ and V'_{Na}) [119]. Experiment results combined with DFT simulations indicate that although Na^+ ions diffuse through Na vacancies, excess Na vacancy (>2%) reduces the diffuse channel and ionic conductivity. Doping Cl^- at S^{2-} site plays more important role in enhancing ionic conductivity. First, the diffusion channel increases with Cl^- doping, thus decreasing activation energy. Second, Na^+ mobility improves with lower interaction between Cl^- and Na^+. Higher ionic conductivity of 1.96 mS cm^{-1} is achieved with optimized $Cl_S^{\cdot}$ and V'_{Na} defects ($Na_{3-\delta}PS_{3.8}Cl_{0.2}$).

13.4.2.2 Na_3SbS_4

Generally, phosphorous-containing sulfide-based solid electrolyte reacts easily with H_2O and generates H_2S gas. Therefore, chemical stability at ambient air is a key issue for practical applications. The stability of the sulfide electrolyte can be understood on the basis of the HSAB theory [120]. Accordingly, hard acids react preferentially with hard bases while soft acids are more likely to react with soft bases. Thus, Na_3PS_4 is not stable in air because phosphorus is hard acid, which prefers to react with oxygen (the hard base) rather than sulfur (the soft base). To overcome this limitation, phosphorus needs to be replaced by some soft acid, such as Sb^{5+}. Similar to Na_3PS_4, there are also two different structures for Na_3SbS_4, i.e. tetragonal phase being stable at room temperature and cubic phase being stable above 120 °C [121]. When exposed to air, Na_3SbS_4 absorbs moisture to form hydrated $Na_3SbS_4{\cdot}9H_2O$. However, this reaction is reversible, and Na_3SbS_4 can be re-obtained by vacuum drying at 150 °C to remove water [120]. This feature makes it possible to prepare Na_3SbS_4 from aqueous solution, which is scalable and low price. Na_2S, Sb_2S_3, and S were dissolved in water to form a Na_3SbS_4 solution first. $Na_3SbS_4{\cdot}9H_2O$ was then obtained by drying. Finally, Na_3SbS_4 was prepared by heating to remove H_2O. Compared with $Na_3SbS_4{\cdot}9H_2O$, Na_3SbS_4 has a higher conductivity of 1.03×10^{-3} S cm^{-1} due to its 3D tunnel network (Figure 13.29a) [120]. As shown in Figure 13.29b, the structure consists of two-dimensional planal channels and one-dimensional channels perpendicular to the plane. Here, the prepared Na_3SbS_4 is tetragonal, the one with higher activation energy and lower ionic conductivity. To improve the ionic conductivity, cubic Na_3SbS_4 was prepared by ball milling of t-Na_3SbS_4 at 250 rpm, which shows a high ionic conductivity of 2.8 mS cm^{-1} [120]. Recently, Hayashi et al. further improved the

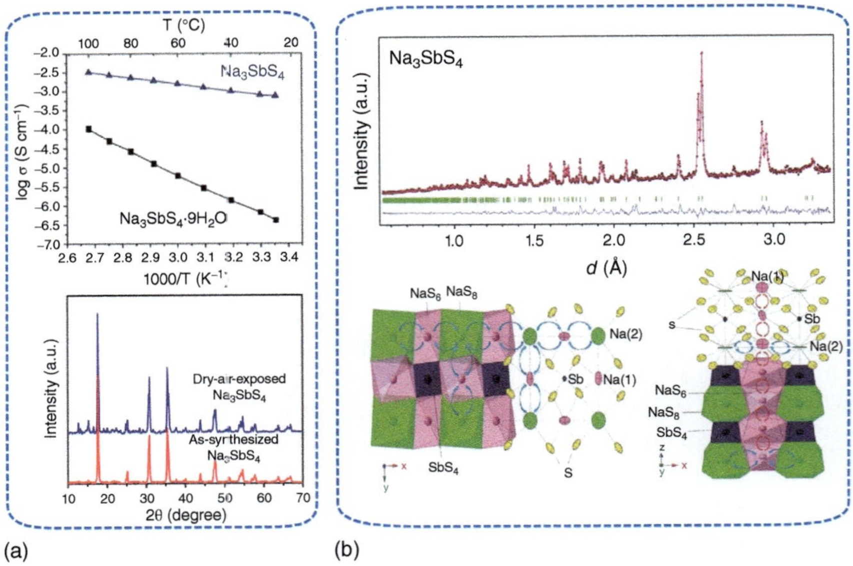

Figure 13.29 (a) Arrhenius plots and XRD patterns of Na_3SbS_4. Source: Zhang et al. [122]. Reproduced with permission, 2016, Wiley-VCH. (b) Neutron diffraction pattern and Rietveld refinement of Na_3SbS_4 at room temperature. The framework structure of Na_3SbS_4 along [001] and [010] direction. Source: Wang et al. [120]. Reproduced with permission, 2016, Wiley-VCH.

ionic conductivity of Na_3SbS_4 with W^{6+} doping. With introducing Na vacancies and the stabilized cubic structure, the ionic conductivity of $Na_{2.88}Sb_{0.88}W_{0.12}S_4$ reaches $32\,mS\,cm^{-1}$, which is higher than that of $Li_{10}GeP_2S_{12}$ and even comparable to liquid electrolytes [111].

Shortly after, Jung et al. proposed a novel processing method for Na_3SbS_4. In contrast to the conventional method of preparing composite electrodes, they manufactured a composite electrode, in which the Na_3SbS_4 was coated on the active material. To prepare solution-processable Na_3SbS_4, tests were carried out under a variety of solution and thermal treatment conditions. The conductivity of the electrolyte-coated electrode was typically lower than that of the hybrid electrode (1.1×10^{-5}–$2.6\times10^{-5}\,S\,cm^{-1}$), but the overall performance of the battery prepared by the electrolyte-coated electrode was better than that of the mixed electrode.

13.4.2.3 $Na_{10}SnP_2S_{12}$

Inspired by $Li_{10}GeP_2S_{12}$, researchers have discovered a new sulfide electrolyte $Na_{10}SnP_2S_{12}$ with high ionic conductivity for SIBs. DFT predicts that $Na_{10}GeP_2S_{12}$ would exhibit high ionic conductivity ($1.2\times10^{-2}\,S\,cm^{-1}$) [123]. On this basis, Richards et al. successfully synthesized a novel solid electrolyte, $Na_{10}SnP_2S_{12}$ (NSPS), similar to $Li_{10}GeP_2S_{12}$ (LGPS), with the assistance of calculations. Through DFT calculations, the researchers reveal that the new SIB solid electrolyte has a conductivity of $9.4\times10^{-4}\,S\,cm^{-1}$, which is close to the experimental value (Figure 13.30) [124]. However, the prepared NSPS is not pure, with a few impurities.

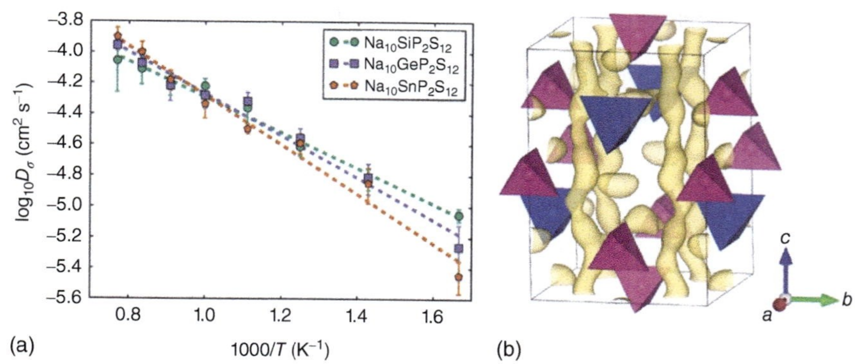

Figure 13.30 (a) Na-diffusivity of $Na_{10}SiP_2S_{12}$, $Na_{10}GeP_2S_{12}$ and $Na_{10}SnP_2S_{12}$ obtained from AIMD simulations. (b) Na^+ probability density isosurface of $Na_{10}SnP_2S_{12}$ obtained from AIMD simulations. Source: Richards et al. [124]. Licensed under CC BY 4.0.

Later, pure $Na_{11}Sn_2PS_{12}$ was successfully prepared with higher ionic conductivity over 3 mS cm^{-1} [125].

13.4.3 Complex Hydrides

Orimo and coworkers firstly reported complex hydrides as Na^+-ion solid electrolytes in 2012 [126]. It was found that the ionic conductivities of Na_3AlH_6 and $NaAlH_4$ were only 6.4×10^{-7} and 2.1×10^{-10} S cm^{-1} at room temperature, but their ionic transfer numbers nearly reach to 1. This study laid the foundation for the use of complex hydrides in solid-state sodium batteries. Later, they prepared $Na_2(BH_4)(NH_2)$ with an ionic conductivity of 3×10^{-6} S cm^{-1} at 300 K by combining $NaBH_4$ and $NaNH_2$ in a molar ratio of 1 : 1 [127]. According to the study, the reason why $Na_2(BH_4)(NH_2)$ has higher conductivity can be attributed to the Na^+ vacancy of the specific anti-perovskite-type structure.

Compared with small anions, complexed hydrides with large anions usually displayed higher ionic conductivities at above the ordered–disordered structure phase transition temperature (Figure 13.31a) [130]. Due to the special structure of cation-void-rich, $Na_2B_{12}H_{12}$ can exhibit a conductivity greater than 0.1 S cm^{-1} at 573 K. However, this phase transition temperature is so high that it poses a challenge for commercial application. For these reasons, a major modification approach for complex hydrides is to reduce the phase transition temperature. Currently, there are sorts of approaches to solve this problem. For example, the purpose of introducing C is to modify the anions of $Na_2B_{12}H_{12}$ and $Na_2B_{10}H_{10}$, which can significantly reduce their phase-transition temperatures (Figure 13.31b) [1c]. For example, C doping in $Na_2B_{12}H_{12}$ ($NaCB_{11}H_{12}$) can reduce the phase transition temperature from 529 to 380 K and C doping in $Na_2B_{10}H_{10}$ ($NaCB_9H_{10}$) can reduce the phase transition temperature from 380 to 290 K. The phase transition temperature also changes when replacing H^- with halides (X^-, X—Cl, Br, and I) [128]. It has been shown that the modification in partly mutual effect of static cation–anion, rotational dynamics, and orientation preferences can significantly change the phase transition

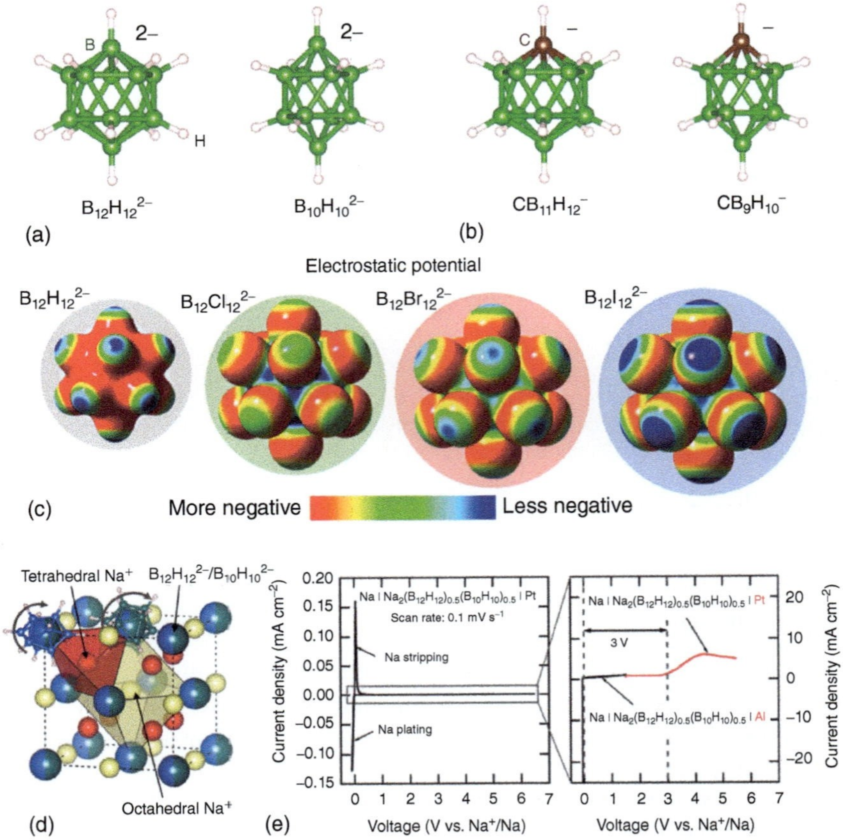

Figure 13.31 Structure of complex hydride. (a) Relative geometries of the $B_{12}H_{12}^{2-}$, and $B_{10}H_{10}^{2-}$ anions and (b) $CB_{11}H^{-}_{12}$, and $CB_9H^{-}_{10}$ anions. Source: Lu et al. [1c]. Reproduced with permission, 2018, Elsevier. (c) Electrostatic potential surface maps of the $B_{12}H_{12}^{2-}$, $B_{12}Cl_{12}^{2-}$, $B_{12}Br_{12}^{2-}$, and $B_{12}I_{12}^{2-}$ anions. Source: Hansen et al. [128]. Reproduced with permission, 2017, American Chemical Society. (d) Simplified structure of $Na_2(B_{12}H_{12})_{0.5}(B_{10}H_{10})_{0.5}$ and (e) electrochemical stability window. Source: Duchene et al. [129b]. Reproduced with permission, 2017, Royal Society of Chemistry.

temperature. When halogen atoms completely replace the H atoms, the anisotropic electron density and the size/mass of the anion will increase (Figure 13.31c), resulting in a significant increase in the phase-transition temperature (475 °C for $Na_2B_{12}Cl_{12}$, 525 °C for $Na_2B_{12}Br_{12}$, 525 and 570 °C for $Na_2B_{12}I_{12}$) [128].

Recently, some reports indicate that it is helpful to reduce the phase-transition temperature by mixing different anions [129]. For example, $Na_2(B_{12}H_{12})_{0.5}(B_{10}H_{10})_{0.5}$ is synthesized by mixing two disparate anions, which shows a high ionic conductivity of 9×10^{-4} S cm^{-1} at room temperature with the high temperature structure (Figure 13.31d) [129b]. Phase transition happens when $\Delta G = \Delta H - T\Delta S = 0$, so the phase-transition temperature is affected by both enthalpy change and entropy change during the phase transition ($T = \Delta H/\Delta S$). With mixing two different ions, the increasing of ΔS helps to reduce the phase-transition temperature. What is more, increase surface energy by reducing particle size can help to stabilize the high temperature structure. Tang et al. prepared high temperature phase with high ionic conductivity near room temperature by ball milling to reduce the particle size [131].

Except for ionic conductivity, the chemical stability of hydride is good. Although $Na_2B_{10}H_{10}$ readily absorbs water, the bound water can be removed by vacuum drying and the structure can be restored [130b]. The electrochemical stability of $Na_2(B_{12}H_{12})_{0.5}(B_{10}H_{10})_{0.5}$ is also evaluated and the electrochemical stability window is determined to between 0 and 3 V (Figure 13.31e) [129a]. Further research shows that ASSBs using $Na_2(B_{12}H_{12})_{0.5}(B_{10}H_{10})_{0.5}$ as SSE and 3 V cathode $NaCrO_2$ are able to cycle between 2 and 3.25 V vs. Na/Na^+, and 85% capacity can be retained after 250 cycles at C/5 [132]. To further broaden the electrochemical stability window, surface modification on the cathode should be necessary.

13.5 Concluding Remarks

Compared with liquid electrolytes, all solid-state batteries using solid electrolytes show higher energy density and high safety, which are promising for the next-generation energy storage devices. Polymer electrolytes with polymer matrix and Na salts are flexible and easy processing but limited in ionic conductivity, thus can only operate at elevated temperatures. Na^+ ion migration in polymer electrolytes is related to the creep of the polymer segment. Take PEO as an example, EO monomer is continuously complexed and decomposed to achieve the migration of Na^+, which normally happens at the amorphous region. Ionic conductivity can be enhanced with ceramic fillers to make organic/inorganic composite electrolytes and increase the amorphous region, or minor liquid additives to make gel electrolytes and allow Na^+ diffusing through the liquid part. In addition to ionic conductivity, thermal stability and electrochemical stability of polymer electrolytes are other limitations for real application.

Inorganic solid electrolytes are another important SSE with higher ionic conductivity, higher ion transfer number, and better thermal stability. Typical inorganic Na^+ SSEs include beta alumina, NASICON, sulfides, and complex hydrides. As a traditional sodium ion conductor material, Na-β/β''-Al_2O_3 with conductivity up to $2\times10^{-3}\,S\,cm^{-1}$ at RT is widely used in Na/S batteries. The ionic conductivity of Na-β/β''-Al_2O_3 is highly related to the β/β'' ratio and density, which are determined by the composition and synthesis procedure either. Dopant such as Mg^{2+} or high sintering temperature helps to increase β''-Al_2O_3 ratio and density, thus to achieve high ionic conductivity. The formula of NASICON fast ion conductor is $Na_{1+x}Zr_2Si_xP_{3-x}O_{12}$ $(0\leq x\leq3)$. With adjusting the structure to rhombohedral and Na^+ concentration to about 3.4 by heteroatom doping, the room-temperature conductivity can reach up to $5\times10^{-3}\,S\,cm^{-1}$. Oxides such as Na-$\beta/\beta''$-$Al_2O_3$ and NASICON are electrochemical stable against Na metal and high potential cathodes, but the compatibility with rigid cathode is bad, leading to high interface resistance and slow Na^+ migration through the interface.

Sulfide-based electrolytes with bigger S^{2-} anions always show higher ionic conductivities. There are mainly three different sulfide electrolytes, including tetragonal and cubic Na_3AS_4 (A = P, As, Sb, etc.) and $Na_{11}Sn_2PS_{12}$. Among them, cubic Na_3SbS_4 shows the highest ionic conductivity and lowest activation energy. With W doping to stabilize the cubic phase and introducing Na^+ vacancies, the ionic conductivity can reach over $40\,mS\,cm^{-1}$ at room temperature. Although the compatibility against cathodes is better, the electrochemical stability of sulfides is worse, especially against Na metal anode. Sulfides would decompose to Na_3P or Na-Sb (Sn) alloy with high electron conductivity at the Na metal anode side, which allows the decomposition to occur continuously.

In summary, although much work has been done for SSEs, multiple issues still exist to hinder the application in all solid-state batteries. For the existing sodium ion solid electrolyte material systems, ionic conductivity, interfacial resistance, and electrochemical stability are important factors limiting their application. Therefore, there are still many challenges in the basic science and key technology of solid-state batteries, which need our continuous efforts to explore.

References

1 (a) Xu, K. (2004). *Chem. Rev.* 104: 4303. (b) Xu, K. (2014). *Chem. Rev.* 114: 11503. (c) Lu, Y., Li, L., Zhang, Q. et al. (2018). *Joule* 2: 1747.

2 Qiao, L., Judez, X., Rojo, T. et al. (2020). *J. Electrochem. Soc.* 167: 070534.

3 Shin, D.O., Oh, K., Kim, K.M. et al. (2015). *Sci. Rep.* 5: 18053.

4 Manthiram, A., Yu, X., and Wang, S. (2017). *Nat. Rev. Mater.* 2: 16103.

5 (a) Han, F., Yue, J., Fan, X. et al. (2016). *Nano Lett.* 16: 4521. (b) Yamada, T., Ito, S., Omoda, R. et al. (2015). *J. Electrochem. Soc.* 162: A646.

6 Kato, Y., Hori, S., Saito, T. et al. (2016). *Nat. Energy* 1: 16030.

7 Fu, K., Gong, Y., Liu, B. et al. (2017). *Sci. Adv.* 3: e1601659.

8 Zhao, C.-D., Guo, J.-Z., Gu, Z.-Y. et al. (2020). *J. Mater. Chem. A* 8: 17454.

9 (a) Mo, F., Ruan, J., Sun, S. et al. (2019). *Adv. Energy Mater.* 9: 1970155. (b) Li, Q., Yi, T., Wang, X. et al. (2019). *Nano Energy* 63: 103895. (c) Raj, R. and Wolfenstine, J. (2017). *J. Power Sources* 343: 119.

10 Han, F., Westover, A.S., Yue, J. et al. (2019). *Nat. Energy* 4: 187.

11 (a) Usiskin, R.E. and Maier, J. (2018). *Phys. Chem. Chem. Phys.* 20: 16449. (b) Bhattacharyya, A.J. and Maier, J. (2004). *Adv. Mater.* 16: 811.

12 Pfaffenhuber, C., Göbel, M., Popovic, J., and Maier, J. (2013). *Phys. Chem. Chem. Phys.* 15: 18318.

13 Ma, J., Chen, B., Wang, L., and Cui, G. (2018). *J. Power Sources* 392: 94.

14 Koerver, R., Zhang, W., de Biasi, L. et al. (2018). *Energy Environ. Sci.* 11: 2142.

15 Nose, M., Kato, A., Sakuda, A. et al. (2015). *J. Mater. Chem. A* 3: 22061.

16 Huang, Y., Zhao, L., Li, L. et al. (2019). *Adv. Mater.* 31: e1808393.

17 Wang, Y. and Zhong, W.-H. (2015). *Chemelectrochem* 2: 22.

18 Hashmi, S.A. and Chandra, S. (1995). *Mater. Sci. Eng., B* 34: 18.

19 Chandrasekaran, R. and Selladurai, S. (2001). *J. Solid State Electrochem.* 5: 355.

20 Mohan, V.M., Raja, V., Bhargav, P.B. et al. (2007). *J. Polym. Res.* 14: 283.
21 (a) Boschin, A. and Johansson, P. (2015). *Electrochim. Acta* 175: 124. (b) Qi, X., Ma, Q., Liu, L. et al. (2016). *Chemelectrochem* 3: 1741.
22 Ma, Q., Liu, J., Qi, X. et al. (2017). *J. Mater. Chem. A* 5: 7738.
23 Gorecki, W., Jeannin, M., Belorizky, E. et al. (1995). *J. Phys.: Condens. Matter* 7: 6823.
24 Zhang, Q., Lu, Y., Yu, H. et al. (2020). *J. Electrochem. Soc.* 167: 070523.
25 Colò, F., Bella, F., Nair, J.R. et al. (2015). *Electrochim. Acta* 174: 185.
26 Bhargav, P.B., Mohan, V.M., Sharma, A.K., and Rao, V.V.R.N. (2008). *J. Appl. Polym. Sci.* 108: 510.
27 Babu, J.R., Ravindhranath, K., and Kumar, K.V. (2018). *Int. J. Polym. Sci.* 2018: 1.
28 Osman, Z., Md Isa, K.B., Ahmad, A., and Othman, L. (2010). *Ionics* 16: 431.
29 Kiran Kumar, K., Ravi, M., Pavani, Y. et al. (2011). *Phys. B* 406: 1706.
30 Vahini, M. and Muthuvinayagam, M. (2019). *J. Mater. Sci.: Mater. Electron.* 30: 5609.
31 Bag, S., Zhou, C., Reid, S. et al. (2020). *J. Power Sources* 454: 227954.
32 Bronstein, L.M., Karlinsey, R.L., Stein, B. et al. (2006). *Chem. Mater.* 18: 708.
33 Li, J., Zhu, H., Wang, X. et al. (2015). *J. Mater. Chem. A* 3: 19989.
34 Pope, C.R., Romanenko, K., MacFarlane, D.R. et al. (2015). *Electrochim. Acta* 175: 62.
35 Serra Moreno, J., Armand, M., Berman, M.B. et al. (2014). *J. Power Sources* 248: 695.
36 (a) Nugent, J.L., Moganty, S.S., and Archer, L.A. (2010). *Adv. Mater.* 22: 3677. (b) Schaefer, J.L., Moganty, S.S., Yanga, D.A., and Archer, L.A. (2011). *J. Mater. Chem.* 21: 10094.
37 Ni'mah, Y.L., Cheng, M.-Y., Cheng, J.H. et al. (2015). *J. Power Sources* 278: 375.
38 Liu, L., Qi, X., Yin, S. et al. (2019). *ACS Energy Lett.* 4: 1650.
39 Zhang, Z., Xu, K., Rong, X. et al. (2017). *J. Power Sources* 372: 270.
40 Zhang, Z., Zhang, Q., Ren, C. et al. (2016). *J. Mater. Chem. A* 4: 15823.
41 Lyu, Y.-Q., Yu, J., Wu, J. et al. (2019). *J. Power Sources* 416: 21.
42 Long, L., Wang, S., Xiao, M., and Meng, Y. (2016). *J. Mater. Chem. A* 4: 10038.
43 (a) Appetecchi, G.B., Croce, F., and Scrosati, B. (1995). *Electrochim. Acta* 40: 991. (b) Song, J.Y., Wang, Y.Y., and Wan, C.C. (1999). *J. Power Sources* 77: 183.
44 Gao, H., Zhou, W., Park, K., and Goodenough, J.B. (2016). *Adv. Energy Mater.* 6: 1600467.
45 Park, C.-W., Ahn, J.-H., Ryu, H.-S. et al. (2006). *Electrochem. Solid-State Lett.* 9: A123.
46 Abbrent, S., Plestil, J., Hlavata, D. et al. (2001). *Polymer* 42: 1407.
47 Yang, Y.Q., Chang, Z., Li, M.X. et al. (2015). *Solid State Ionics* 269: 1.
48 Gao, H., Guo, B., Song, J. et al. (2015). *Adv. Energy Mater.* 5: 1402235.
49 Kreuer, K.-D., Wohlfarth, A., de Araujo, C.C. et al. (2011). *ChemPhysChem* 12: 2558.
50 Cao, C., Liu, W., Tan, L. et al. (2013). *Chem. Commun.* 49: 11740.
51 Aravindan, V., Lakshmi, C., and Vickraman, P. (2009). *Curr. Appl. Phys.* 9: 1106.

52 Ansari, Y., Guo, B., Cho, J.H. et al. (2014). *J. Electrochem. Soc.* 161: A1655.

53 Manuel Stephan, A. and Nahm, K.S. (2006). *Polymer* 47: 5952.

54 Zheng, J., Zhao, Y., Feng, X. et al. (2018). *J. Mater. Chem. A* 6: 6559.

55 Kumar, D. and Hashmi, S.A. (2010). *Solid State Ionics* 181: 416.

56 Boschin, A. and Johansson, P. (2016). *Electrochim. Acta* 211: 1006.

57 Faraday, M. (1839). *Philos. Trans. R. Soc. London* 129: 1.

58 Yung-Fang Yu, Y. and Kummer, J.T. (1967). *J. Inorg. Nucl. Chem.* 29: 2453.

59 Baliteau, S., Sauvet, A.L., Lopez, C., and Fabry, P. (2005). *J. Eur. Ceram. Soc.* 25: 2965.

60 Jung, K., Lee, S., Park, Y.-C., and Kim, C.-S. (2014). *J. Power Sources* 250: 1.

61 (a) Bay, M.-C., Wang, M., Grissa, R. et al. (2020). *Adv. Energy Mater.* 10: 1902899. (b) Lei, D., He, Y.-B., Huang, H. et al. (2019). *Nat. Commun.* 10: 4244.

62 (a) Sudworth, J.L. (1984). *J. Power Sources* 11: 143. (b) Wenzel, S., Leichtweiss, T., Weber, D.A. et al. (2016). *ACS Appl. Mater. Interfaces* 8: 28216.

63 Spencer Jolly, D., Ning, Z., Darnbrough, J.E. et al. (2020). *ACS Appl. Mater. Interfaces* 12: 678.

64 (a) Will, F.G. (1976). *J. Electrochem. Soc.* 123: 834. (b) Flor, G., Marini, A., Massarotti, V., and Villa, M. (1981). *Solid State Ionics* 2: 195.

65 (a) Virkar, A.V., Miller, G.R., and Gordon, R.S. (1978). *J. Am. Ceram. Soc.* 61: 250. (b) Ray, A.K. and Subbarao, E.C. (1975). *Mater. Res. Bull.* 10: 583. (c) Oshima, T., Kajita, M., and Okuno, A. (2004). *Int. J. Appl. Ceram. Technol.* 1: 269.

66 Barison, S., Fasolin, S., Mortalò, C. et al. (2015). *J. Eur. Ceram. Soc.* 35: 2099.

67 (a) Hodge, J.D. (1984). *J. Am. Ceram. Soc.* 67: 183. (b) Takahashi, T. and Kuwabara, K. (1980). *J. Appl. Electrochem.* 10: 291.

68 (a) Hodge, J.D. (1983). *J. Am. Ceram. Soc.* 66: 166. (b) Zaharescu, M., Pârlog, C., Stancovschi, V. et al. (1985). *Solid State Ionics* 15: 55. (c) Jayaraman, V., Gnanasekaran, T., and Periaswami, G. (1997). *Mater. Lett.* 30: 157.

69 Mathews, T. (2000). *Mater. Sci. Eng., B* 78: 39.

70 Wang, J., Jiang, X.-P., Wei, X.-L. et al. (2010). *J. Alloys Compd.* 497: 295.

71 Pekarsky, A. and Nicholson, P.S. (1980). *Mater. Res. Bull.* 15: 1517.

72 Lin, J., Wen, Z., Wang, X. et al. (2010). *J. Solid State Electrochem.* 14: 1821.

73 Lu, X., Li, G., Kim, J.Y. et al. (2012). *J. Power Sources* 215: 288.

74 Zhu, C. and Xue, J. (2012). *J. Alloys Compd.* 517: 182.

75 Chen, G., Lu, J., Zhou, X. et al. (2016). *Ceram. Int.* 42: 16055.

76 Zhao, K., Liu, Y., Zeng, S.M. et al. (2016). *Ceram. Int.* 42: 8990.

77 Chen, G., Lu, J., Li, L. et al. (2016). *J. Alloys Compd.* 673: 295.

78 (a) Goodenough, J.B., Hong, H.Y.P., and Kafalas, J.A. (1976). *Mater. Res. Bull.* 11: 203. (b) Hong, H.Y.P. (1976). *Mater. Res. Bull.* 11: 173.

79 Wu, C., Kopold, P., Ding, Y.-L. et al. (2015). *ACS Nano* 9: 6610.

80 (a) Heetaek, P. and Byoungwoo, K. (2014, MA2014-02). *ECS Meet. Abstr.* 435. (b) Schmid, H., De Jonghe, L.C., and Cameron, C. (1982). *Solid State Ionics* 6: 57.

81 (a) Hashimoto, T. and Hayashi, K. (2015). *Electrochim. Acta* 182: 809. (b) Hayashi, K., Shima, K., and Sugiyama, F. (2013). *J. Electrochem. Soc.* 160: A1467.
82 Samiee, M., Radhakrishnan, B., Rice, Z. et al. (2017). *J. Power Sources* 347: 229.
83 Hueso, K.B., Armand, M., and Rojo, T. (2013). *Energy Environ. Sci.* 6: 734.
84 Park, H., Jung, K., Nezafati, M. et al. (2016). *ACS Appl. Mater. Interfaces* 8: 27814.
85 Guin, M., Tietz, F., and Guillon, O. (2016). *Solid State Ionics* 293: 18.
86 Noguchi, Y., Kobayashi, E., Plashnitsa, L.S. et al. (2013). *Electrochim. Acta* 101: 59.
87 Sun, F., Xiang, Y., Sun, Q. et al. (2021). *ACS Appl. Mater. Interfaces* 13: 13132.
88 Guin, M. and Tietz, F. (2015). *J. Power Sources* 273: 1056.
89 von Alpen, U., Bell, M.F., and Wichelhaus, W. (1979). *Mater. Res. Bull.* 14: 1317.
90 Jolley, A.G., Taylor, D.D., Schreiber, N.J., and Wachsman, E.D. (2015). *J. Am. Ceram. Soc.* 98: 2902.
91 Ma, Q., Tsai, C.-L., Wei, X.-K. et al. (2019). *J. Mater. Chem. A* 7: 7766.
92 Kazakevičius, E., Kežionis, A., Žukauskaitė, L. et al. (2014). *Funct. Mater. Lett.* 7: 1440002.
93 Zhu, Y.S., Li, L.L., Li, C.Y. et al. (2016). *Solid State Ionics* 289: 113.
94 Honma, T., Okamoto, M., Togashi, T. et al. (2015). *Solid State Ionics* 269: 19.
95 (a) Agrawal, D.K. (1996). *Trans. Indian Ceram. Soc.* 55: 1. (b) Perthuis, H. and Colomban, P. (1986). *Ceram. Int.* 12: 39. (c) Yue, Y. and Pang, W. (1990). *J. Chem. Soc., Chem. Commun.* **10**: 764. (d) Visco, S.J. and Kennedy, J.H. (1983). *Solid State Ionics* 9–10: 885.
96 Gao, H., Xin, S., Xue, L., and Goodenough, J.B. (2018). *Chem* 4: 833.
97 Inoishi, A., Omuta, T., Kobayashi, E. et al. (2017). *Adv. Mater. Interfaces* 4: 1600942.
98 Zhou, W., Li, Y., Xin, S., and Goodenough, J.B. (2017). *ACS Cent. Sci.* 3: 52.
99 Zhang, Z., Zhang, Q., Shi, J. et al. (2017). *Adv. Energy Mater.* 7: 1601196.
100 Bachman, J.C., Muy, S., Grimaud, A. et al. (2016). *Chem. Rev.* 116: 140.
101 Han, F., Zhu, Y., He, X. et al. (2016). *Adv. Energy Mater.* 6: 1501590.
102 Sakuda, A., Hayashi, A., and Tatsumisago, M. (2013). *Sci. Rep.* 3: 2261.
103 Tanibata, N., Noi, K., Hayashi, A. et al. (2014). *Chemelectrochem* 1: 1130.
104 Banerjee, A., Park, K.H., Heo, J.W. et al. (2016). *Angew. Chem. Int. Ed.* 55: 9634.
105 Zhang, L., Yang, K., Mi, J. et al. (2015). *Adv. Energy Mater.* 5, 1501294.
106 Yu, Z., Shang, S.-L., Seo, J.-H. et al. (2017). *Adv. Mater.* 29: 1605561.
107 Jansen, M. and Henseler, U. (1992). *J. Solid State Chem.* 99: 110.
108 Hayashi, A., Noi, K., Sakuda, A., and Tatsumisago, M. (2012). *Nat. Commun.* 3: 856.
109 Noi, K., Hayashi, A., and Tatsumisago, M. (2014). *J. Power Sources* 269: 260.
110 Yubuchi, S., Hayashi, A., and Tatsumisago, M. (2015). *Chem. Lett.* 44: 884.
111 Nguyen, H., Banerjee, A., Wang, X. et al. (2019). *J. Power Sources* 435: 126623.
112 Santhosha, A.L., Medenbach, L., Palaniselvam, T., and Adelhelm, P. (2020). *J. Phys. Chem. C* 124: 10298.
113 Zhu, Z., Chu, I.-H., Deng, Z., and Ong, S.P. (2015). *Chem. Mater.* 27: 8318.

114 (a) Fuchs, T., Culver, S.P., Till, P., and Zeier, W.G. (2020). *ACS Energy Lett.* 5: 146. (b) Moon, C.K., Lee, H.-J., Park, K.H. et al. (2018). *ACS Energy Lett.* 3: 2504.

115 Tanibata, N., Noi, K., Hayashi, A., and Tatsumisago, M. (2014). *RSC Adv.* 4: 17120.

116 Bo, S.-H., Wang, Y., Kim, J.C. et al. (2016). *Chem. Mater.* 28: 252.

117 Ong, S.P., Mo, Y., Richards, W.D. et al. (2013). *Energy Environ. Sci.* 6: 148.

118 (a) Chu, I.-H., Kompella, C.S., Nguyen, H. et al. (2016). *Sci. Rep.* 6: 33733. (b) de Klerk, N.J.J. and Wagemaker, M. (2016). *Chem. Mater.* 28: 3122.

119 Feng, X., Chien, P.-H., Zhu, Z. et al. (2019). *Adv. Funct. Mater.* 29: 1807951.

120 Wang, H., Chen, Y., Hood, Z.D. et al. (2016). *Angew. Chem. Int. Ed.* 55: 8551.

121 Zhang, Q., Zhang, C., Hood, Z.D. et al. (2020). *Chem. Mater.* 32: 2264.

122 Zhang, D., Cao, X., Xu, D. et al. (2018). *Electrochim. Acta* 259: 100.

123 Kandagal, V.S., Bharadwaj, M.D., and Waghmare, U.V. (2015). *J. Mater. Chem. A* 3: 12992.

124 Richards, W.D., Tsujimura, T., Miara, L.J. et al. (2016). *Nat. Commun.* 7: 11009.

125 (a) Duchardt, M., Ruschewitz, U., Adams, S. et al. (2018). *Angew. Chem., Int. Ed.* 57: 1351. (b) Zhang, Z., Ramos, E., Lalère, F. et al. (2018). *Energy Environ. Sci.* 11: 87.

126 Oguchi, H., Matsuo, M., Kuromoto, S. et al. (2012). *J. Appl. Phys.* 111: 036102.

127 Matsuo, M., Oguchi, H., Sato, T. et al. (2013). *J. Alloys Compd.* 580: S98.

128 Hansen, B.R.S., Paskevicius, M., Jørgensen, M., and Jensen, T.R. (2017). *Chem. Mater.* 29: 3423.

129 (a) Tang, W.S., Yoshida, K., Soloninin, A.V. et al. (2016). *ACS Energy Lett.* 1: 659. (b) Duchene, L., Kuhnel, R.S., Rentsch, D. et al. (2017). *Chem. Commun.* 53: 4195.

130 (a) Udovic, T.J., Matsuo, M., Unemoto, A. et al. (2014). *Chem. Commun.* 50: 3750. (b) Udovic, T.J., Matsuo, M., Tang, W.S. et al. (2014). *Adv. Mater.* 26: 7622.

131 Tang, W.S., Matsuo, M., Wu, H. et al. (2016). *Energy Storage Mater.* 4: 79.

132 Duchêne, L., Kühnel, R.S., Stilp, E. et al. (2017). *Energy Environ. Sci.* 10: 2609.

14

Binders for Sodium-Ion Batteries

14.1 Introduction

Binder is one of the important parts of sodium-ion batteries (SIBs) (Figure 14.1), which has played vital roles on improving the electrochemical performances of SIBs. Its main functions are listed as follows [1, 2]:

(1) To enhance the contact between the active materials, the conductive agents, and the current collector to form a stable electrode structure;
(2) To stabilize the surface of the electrodes, inhibit the volume change caused by sodium ion insertion/extraction in/from active materials, maintain the mechanical integrity of the electrodes, and improve the cycle stability of the batteries;
(3) To promote the homogeneous dispersion of active materials and conductive agents, and simplify the coating process.

Based on the above functions, binder is vital for a successful electrode, although the amount of binder in the electrode is usually only 1–10% of the total weight of electrode [3]. Moreover, the amount of binder in the electrode is also important. On the one hand, the proportion of binder should not be too low; otherwise, the bonding strength between the binder and the active material will be weak, and then the active materials are easily to peel off, leading to the capacity decay and cycle life reduction of the battery. On the other hand, the proportion of binder in the electrode should not be too high either. If adding too much binder into electrode, the chemical inertia of the electrode will be increased because the binder is not active material. Additionally, it also results in the reduction of the electrode conductivity, increase of the internal resistance, and polarization of the electrode, and the side effects of the electrochemical performances of the batteries. Therefore, the binder with strong bonding strength can reduce the amount of the binder, which can improve the electrochemical performances of SIBs [4]. In this chapter, we will introduce several popular binders in detail and some other new binders will be also mentioned briefly.

Sodium-Ion Batteries: Energy Storage Materials and Technologies, First Edition. Yan Yu.

Figure 14.1 Schematic diagram of the electrode.

14.2 Main Functions and Performance Requirements of Binders

The adhesive theory about binder is very important for the research and development of binder, which can be summarized as follows:

(1) *The adsorption theory*. The adhesion is caused by the interaction between the binder and the adherent at the interface. This adsorption is the result of the interaction between molecules and atoms.
(2) *The chemical binding theory*. The strong bonding force is due to the chemical reaction between the binder and the adhered material. However, this theory is only applicable to reactive binders [5].
(3) *The mechanical force theory*. The liquid binder first diffuses and fills the uneven parts on the surface of the adhered material and then solidifies to produce a bonding effect, so that the binder and the adhered material are combined. However, this theory cannot explain the adhesive to the smooth surface of the material.
(4) *The electrostatic adsorption theory*. The adhered material and the adherent substance contact each other, producing an electric double layer, through electrostatic attraction. However, this theory only applies to polar matter [6].
(5) *The diffusion theory*. Long-chain molecules move between the binder and the adhered material under the action of molecular thermal motion. If the binder can dissolve or swell the adhered material, the binder and the adhered material will diffuse to each other and interweave. This theory only applies to the adhered material compatible with the binder.
(6) *The polarity theory*. Adhered material needs polar binder, and nonpolar adhered material needs nonpolar binder.
(7) *The weak interface theory*. There is a weak boundary layer on the surface of the object, which decreases the bond strength, so it is necessary to remove the weak interface layer to enhance the force between the binder and the adhered material.
(8) *The coordination bond theory*. The binder and the adhered material provide electron pairs and empty orbitals at the interface of the binder and the adhered material to form coordination bonds.

It is important to notice that the adhesive theory depends on the properties of binders and adherends. Moreover, several bonding theories may exist simultaneously between the binder and the adherend. The active substance and the conductive additive such as carbon black are usually powder or particle. The electrode is usually prepared by dispersing the active substance and conducting powder evenly into the solution with dissolved binder, then coating them on the current collector, and drying subsequently. The binder attaches the active substance and conducting agent to the current collector. Since the surface of the active substance is usually uneven or porous, and the surface of the conducting agent and the current collector is not very smooth, the bonding effects among the electrode binder and the active substance, conducting agent and current collector, can be explained by the mechanism shown in Figure 14.2 [7]. Firstly, the binder wets the surface of the electrode material and diffuses and fills into bumps or holes on the surface of the electrode material (Figure 14.2a). After that, during the drying process of the electrode, the binder and the electrode material are bound together by physical or chemical forces (Figure 14.2b). Interactions between the binder and the electrode material include electrostatic adsorption, coordination, and chemical reaction (for reactive binders) through covalent bonds, hydrogen bonds, and mechanical forces, etc. (Figure 14.2c).

The main binder force is generally considered to involve four types. (i) Chemical bond force, which exists between atoms or ions, containing metal bond, ionic bond, and covalent bond. (ii) Intermolecular force, which includes van der Waals forces (dispersion, induction, orientation) and hydrogen bond forces. (iii) Interface electrostatic gravity, which is generated by formation of double electrical layer. When the electron donor (metal materials) and the electron acceptor (nonmetallic materials) contact, the electron will transfer, then the interface contact potential is generated, and the double electrical layer formed finally. (iv) Mechanical force, which can increase the bonding effect, rather than the factors that produce adhesion, from the physical and chemical point of view. To sum up, in all bonding systems, intermolecular force is considered to be universal, while the other three forces usually exist under specific conditions.

A good binder should reach the following requirements [8]. (i) It should have polar groups, such as –OH, –COOH, –OCOR, $-SO_3H$, –CN, etc. and can provide

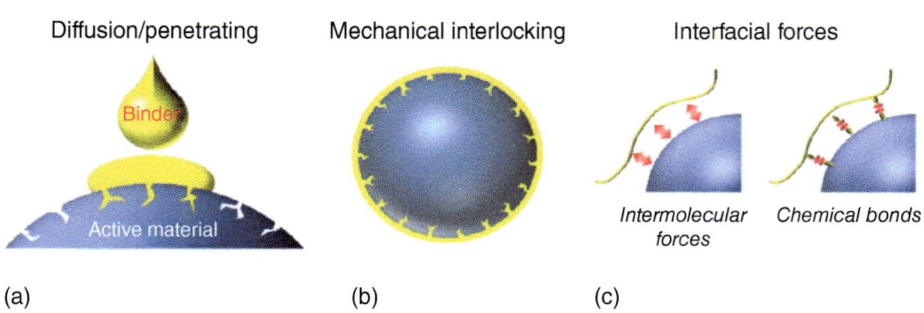

Figure 14.2 Schematic diagram of bonding mechanism: (a) diffusion and permeation process of bonding agent during the process of electrode preparation; (b) mechanical bonding process during drying; and (c) the formation process of interface layer adhesion. Source: Chen et al. [7]. Reproduced with permission, 2018, American Chemical Society.

strong adhesive force, high elastic modulus, and high tensile strength toward active materials, conductive agents, and current collectors. (ii) It should have little influence on the conduction of electrons and ions during the electrochemical reaction processes. (iii) It should be stable in the electrolyte. In addition, it should be easily wetted by electrolyte to ensure the fast transmission of sodium ions. (iv) It should have a stable electrochemical window. In the positive electrode, the binder should have a wide and stable electrochemical window, to avoid the electrochemical oxidation reaction during charge processes. While in the negative electrode, the binder should have good electrochemical reduction stability at low potential. (v) It also should be easily dissolved in a certain solvent. The active materials and conductive agents should be also dispersed uniformly in this obtained binder solution. Then after drying process, they can be firmly bonded to the current collector. (vi) The last but not the least, low price, non-toxicity, environmentally friendly production, and high safety are also very important for the choosing of binder. In a word, the electrochemical performances of SIBs, such as long cycling life, good rate capability, and high safety, can be improved by selecting the appropriate type and adding the optimal amount of binder.

Based on the solvent used to dissolve binder, the binder can be divided into oil-based binder and aqueous binder. The binders commonly used in SIBs are polyvinylidene fluoride (PVDF), sodium alginate (SA), styrene butadiene rubber (SBR), carboxymethyl cellulose (CMC), and polyacrylic acid (PAA), whose structures are shown in Figure 14.3, respectively. Among them, PVDF is oil-based binder, while PAA, CMC, SBR, and SA are aqueous binders.

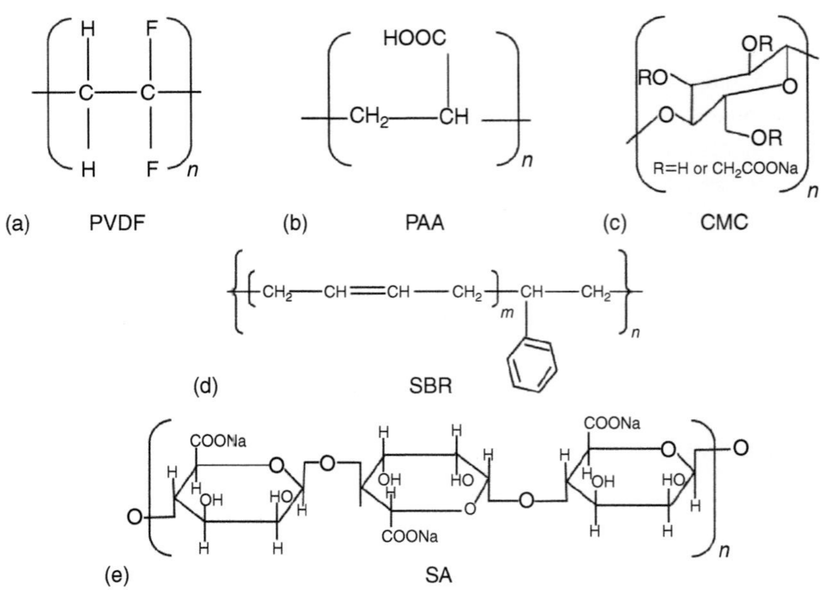

Figure 14.3 The molecular structures of the common binders used in SIBs: (a) PVDF, (b) PAA, (c) CMC, (d) SBR, and (e) SA.

14.3 Polyvinylidene Fluoride (PVDF)

14.3.1 Chemical Properties of PVDF

PVDF is translucent or white powder, which has closely arranged molecular chains and strong hydrogen bonds. It is incombustible, with oxygen index of 46%, crystallinity of 65%–78%, density of 1.77–1.80 g cm^{-3}, and thermal deformation temperature of 112–145 °C. PVDF is usually used in −40 to 150 °C. Table 14.1 has listed the main properties of various PVDF from American Solvay. PVDF mainly refers to vinylidene fluoride (VDF) homopolymer or copolymer of VDF and a small amount of fluorine-containing viny monomer, ethylene homopolymer, or copolymer of VDF, and other small amount of fluorine-containing viny monomers. Therefore, PVDF combines the features of fluorine resin and general resin. It not only has the merits of chemical corrosion resistance, high temperature resistance, radiation resistance, oxidation resistance, and weather resistance, but also performs good piezoelectric, dielectric, and thermoelectric properties. It is the second largest fluorinated plastic product in terms of output, with the annual production of more than 53 000 tons.

PVDF is a kind of nonpolar chain polymer binder, which is the most commonly used electrode binder in lithium-ion batteries (LIBs)/SIBs industry, due to its outstanding characteristics of strong antioxidant and anti-reduction ability, high thermal stability, and good dispersibility in *N*-methyl pyrrolidone (NMP). NMP has high volatilization temperature, which is widely used in the LIBs/SIBs industry to dissolve binder during the electrode preparation.

In terms of the influence of binder on electrode performances, the molecular weight of PVDF and its content and distribution in the electrode all affect the performances of the batteries. In general, the higher the molecular weight of PVDF, the better the performance of the binder. Therefore, the ratio of PVDF used in electrode is reduced, which can increase the energy density of SIBs. However, if the molecular weight is too high, the solubility of PVDF in NMP will decrease and the dispersion of PVDF in the electrode will be not easily uniform. Moreover, PVDF with high molecular weight has high crystallinity and great resistance to the migration of

Table 14.1 The main properties of various PVDF from American Solvay.

Indicators	Unit	Solef6010	Solef6020	Solef5130	Solef2121
The molecular weight			700 000	1 100 100	600 000
Melting point	°C	172	170	162	135
The melting enthalpy	J g^{-1}	60	55	44	26
Modulus	Mpa	1700–2500	1650	1300	500
The yield strength	Mpa	50–60	53–57	40–45	15–25
Elongation at yield	%	5–10	5–10	5–10	15–18
Breaking elongation	%	20–300	20–50	350–600	300–650

electrons and protons in the electrode, thus resulting in high impedance of the electrode and high overpotential during charge–discharge processes. In summary, the content of PVDF in the electrode significantly affects the electrical conductivity and mechanical properties of the electrode and thus plays important roles on the capacity, rate property, and long-term cycling performance of the batteries.

Recently, researchers began to use VDF homopolymer modification, VDF copolymerization with the second monomer and the third monomer modification, and VDF homopolymer blending with another copolymer to improve and design some PVDF copolymers as electrode binder for SIBs. In addition, the partial fluorine ethylene–hexafluoropropylene copolymer [P(VDF–HFP)], vinylidene fluoride–hexafluoropropylene–tetrafluoroethylene terpolymer [(PVDF–TFE–HFP)], partial fluorine ethylene–vinyl chloride copolymer, and VDF homopolymer–VDF copolymers have already shown more excellent mechanical properties, but the related research is rather little. Therefore, PVDF will be still the most widely used binder of the VDF-based binder for a quite longer time.

14.3.2 Application of PVDF in Na-Ion Batteries

PVDF has been the most widely used binder for LIBs and SIBs in the past two decades, due to its good electrochemical stability and strong binding ability [9–11]. However, PVDF is easily swollen by nonaqueous liquid electrolytes, to gelate, dissolve, and form viscous fluid or gel polymer electrolytes, which will lead to peeling of electrode particles and then result in volume attenuation and cycling life decay [10, 12]. In addition, the peeled off binder strongly inhibits the migration of Na^+, so the rate capability of the electrode will significantly decrease [10, 13]. Much of the research on binders has focused on enhancing the binding affinity among active materials, carbon additives, and current collectors, which is helpful to maintain a complete electronic network with a certain level of electrode uniformity, thus helping to improve the electrochemical performance of the electrodes [14].

As shown in Figure 14.4, SIBs consist of an anode and a cathode immersed in an electrolyte and separated by a separator [11]. Both electrodes contain an active material, a conducting agent, a current collector, and a binder. The active material determines the energy of the electrode. Because of the enhanced electron transport, the conductive agent can improve the power ratio of the electrode. Binders are used to bind the active material and conducting agent to the collector. Thus, electrons can flow in and out of an external circuit through an active material-conducting agent-collector chain. Although binders account for only 2–5 wt.% in electrode, the binder material is one of the most critical electrode assemblies for improving battery performances. Without a binder, the active material will lose contact with the collector fluid, resulting in a loss of capacity [11].

Zhu et al. synthesized a nitrogen–phosphorus co-doped carbon@CNT with a large number of micropores and mesopores. When PVDF is used as binder to make an anode and applied to an SIB, the as-prepared electrode shows high reversible capacity, excellent rate performance, and long-term cyclability [15]. Wu and coworkers

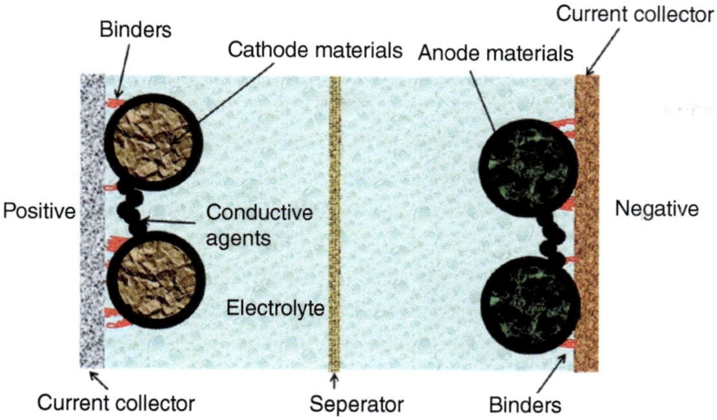

Figure 14.4 Schematic diagram of working principle of SIBs. Source: Chou et al. [11]/Royal Society of Chemistry.

used PVDF for phosphorus-based anode and showed excellent electrochemical performance. Due to the strong adhesion of PVDF, the prepared P@C-GO/MOF-5 electrode can maintain 93% of its initial capacity after 100 cycles at $2\,A\,g^{-1}$ [16]. However, PVDF is easy to wet in the nonaqueous liquid electrolyte, which will reduce the adhesion between the current collector and the active material, resulting in the increase of contact resistance, a large swelling rate in the electrolyte, and poor shape retention. On the other hand, the chemical synthesis process of PVDF is complicated and NMP is volatile and toxic. Hence, some other aqueous binders have appeared.

14.4 Polyacrylic Acid (PAA)

PVDF is widely used in SIBs due to its acceptable bonding ability and electrochemical stability. However, it is usually dissolved in NMP, which is volatile, flammable, and toxic [17]. In addition, for SIBs, it has been confirmed that PVDF will defluorinate during the process of sodiation and desodiation, damaging the integrity of electrode [18]. Moreover, considering the cost and difficulty of recycling, PVDF needs to be replaced by a better one. Hence, new aqueous binders with low cost, nontoxic, environmental friendliness, and strong bonding ability have gained extensive attention [14, 19].

PAA is an aqueous amorphous polymer, which can form stable polyacrylates with lots of metal ions, such as Na^+, Ca^{2+}, and Mg^{2+}. The molecular structure of PAA and polyacrylates contains a long carbon chain and numerous carboxyl functional groups (Figure 14.3b). In thermal performance test, PAA presents better thermal stability, smaller volume expansion, and larger thermal diffusions than those of SBR, PVDF, and CMC. Within the normal voltage window of battery, PAA is chemically inert and insoluble in the electrolyte. In addition, the oxygen-containing groups in the structure are conducive to form hydrogen bond with the surface of electrode materials, especially for silicon and carbon, which endows the active particle with strong binding force to the current collector. Thus, PAA is suitable for research and application as the binder for SIBs.

By comparing the basic properties of PAA, PVDF, and CMC, Li and coworkers came to a conclusion that PAA has strong adhesion, and it can make the active particles disperse uniformly [20]. However, the mechanism of the improved electrochemical performance of the PAA remains ambiguous and needs to be explored further. To study the influence of PAA on the performance of metal oxide anodes for SIBs, Sun et al. prepared electrodes with different binders (PAA–CMC and PVDF) and reached a similar conclusion [21]. Due to the strong adhesion of PAA, electrode using PAA–CMC as binder can effectively endure the volume expansion of oxide particles and maintain the integrity of the electrode (Figure 14.5). Therefore, the electrode with CMC–PAA binder shows more excellent cyclability and Coulombic efficiency. Patra et al. also obtained excellent mechanical and electrochemical properties by using PAA in SnO_2@CMK-8 anode for SIBs. The Na-CMC/Na-PAA binder can be effectively wetted and dispersed on the surface of active materials, binding SnO_2@CMK-8 particles tightly. This hybrid binder not only withstands volume change of electrode and maintains structural integrity during cycling but also forms a solid protective film on the surface of electrode, which is beneficial to prevent the direct contact between active particles and electrolyte. Electrochemical tests show that the R_{ct} of the electrode decreased and D_{Na^+} increased, showing excellent capacity, rate capacity, and cyclability. In addition, the use of Na-CMC/Na-PAA binder also improves the safety of SIBs by reducing the total heat generated during fast sodiation and desodiation [22]. Lee et al. used PAA as binder for natural graphite anode to study its effect on carbon materials. The peel test results show that PAA significantly improves the adhesion strength of the graphite electrode on the copper substrate. In addition, gravimetric and volumetric energy densities of graphite anode made from PAA are more than 340 mA h g^{-1} and 560 mA h cm^{-3}, respectively.

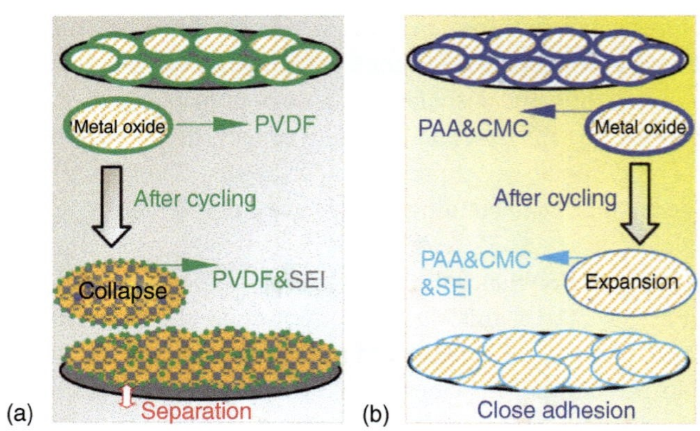

Figure 14.5 The schematic diagram shows probable behaviors of electrodes prepared with (a) PVDF and (b) PAA–CMC before and after cycling. Source: Mink et al. [21]. Reproduced with permission, 2014, Royal Society of Chemistry.

Hence, the graphite electrode with PAA exhibits excellent cyclability (90% capacity retention after 500 cycles) [23].

Fan et al. synthesized N-doped hollow carbon nanotubes through carbonizing PPA nanotubes as the anode for SIBs and compared the effect of PVDF and NA-PAA binders on the sodium storage performance. The results showed that compared with the traditional PVDF, the aqueous NA-PAA binder can enhance the initial Coulombic efficiency, cyclability, and rate capability (Figure 14.6). The anode with NA-PAA binder exhibits a high ICE of 61.2% and long-term cyclability (175.5 $mA\,h\,g^{-1}$ after 300 cycles at 0.2 $A\,g^{-1}$). As an electrochemical active binder, the capacity of electrode is closely related to its molecular weight of NA-PAA. Moderate NA-PAA can improve electrode capacity more effectively by forming smooth and stable SEI layer and reducing internal resistance. Thus, aqueous NA-PAA binder is an attractive alternative for carbonaceous anode in SIBs [24].

In addition to carbon materials, the performance of PAA in alloying anodes is also superior to PVDF. For example, Komaba and coworkers reported that Sn powder electrode exhibited improved specific capacity and Coulombic efficiency in SIBs by using PAA binder instead of PVDF (Figure 14.7), which should be due to the reduced internal resistance and more stable interface layer on the surface of electrode [25].

For high-capacity anode of black phosphorus (theoretically 2596 $mA\,h\,g^{-1}$), its application is hindered by the large volume expansion (400%), leading to the rapid attenuation of capacity. Zhang et al. reported that this issue can be addressed

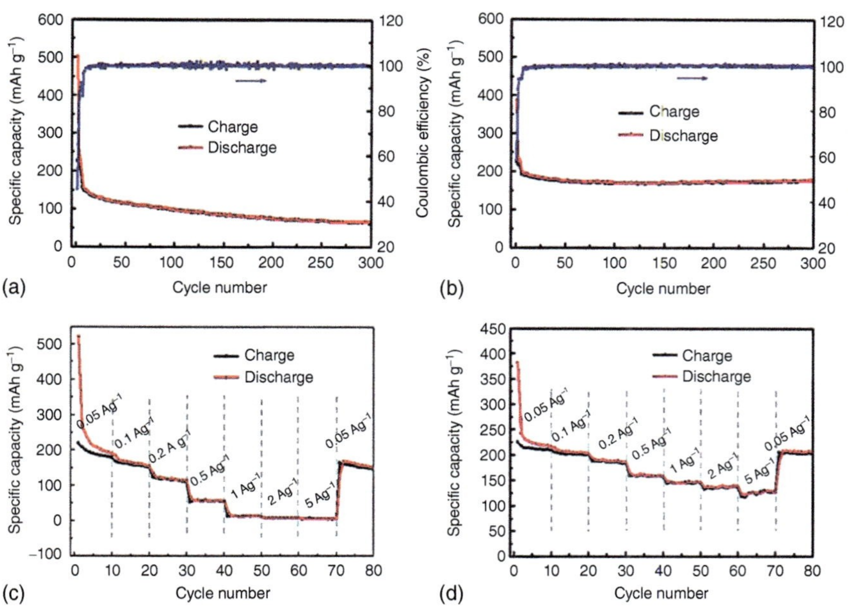

Figure 14.6 Cycle performance of N-CNTs electrodes using (a) PVDF and (b) NA-PAA as binders at 0.2 $A\,g^{-1}$. Rate capability of N-CNTs electrodes using (c) PVDF and (d) NA-PAA as binders at different current densities. Source: Fan et al. [24]. Reproduced with permission, 20154, Elsevier.

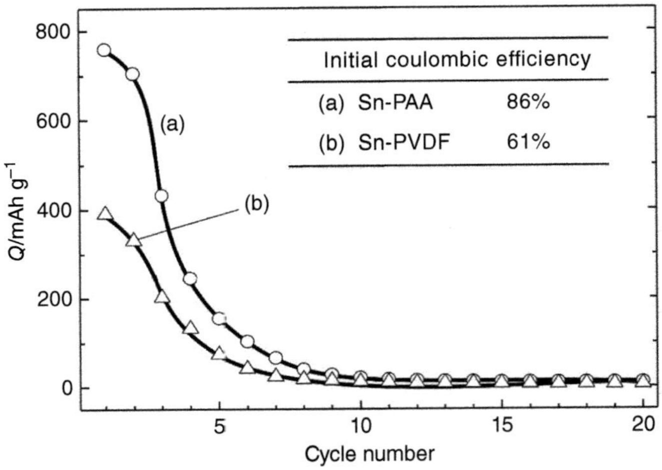

Figure 14.7 Reversible capacity variation of (a) Sn-PAA and (b) Sn-PVDF. Source: Komaba et al. [25]. Reproduced with permission, 2012, Elsevier.

by using aqueous binder. The incorporation of amylose molecules with a helical structure gives linear polyacrylic acid polymer binders extraordinary tensile and elastic properties at 400% strain. When used as binder for the negative electrode of black phosphorus for SIBs, the bonding force between the electrode and the current collector is much stronger (2.95 N) than that of PVDF binder (1.90 N). Hence, the electrode using PAA-Am as binder shows superior cyclability at $0.2\,\mathrm{A\,g^{-1}}$ and excellent rate capability from 0.2 to $6\,\mathrm{A\,g^{-1}}$, which is better than the electrode with PVDF binder. This cost-effective binder has great practical potential for SIBs [26].

14.5 Carboxymethyl Cellulose (CMC)

Different from the organic solvent used for oiliness binders, aqueous binders have the advantage of being water soluble. Due to the low cost and environmental friendliness, aqueous binders play an important role in the research and manufacture of SIBs. Among them, CMC is the most commonly used water-soluble binder material.

CMC is a kind of polyanionic polysaccharide with linear long chain, which is usually applied as adhesive, emulsifier, or stabilizer in medicine and chemical industry. It is derived from the substitution of carboxymethyl group ($-CH_2COONa$) for hydrogen atom on hydroxyl group (–OH) in cellulose structure. The molecular structure is shown in Figure 14.3c.

The number of carboxymethyl groups in the molecular chain determines the degree of substitution of CMC and partly determines the molecular weight of CMC. The aqueous solution of CMC shows certain viscoelastic properties because of its good hydrophilicity. Viscosity, degree of substitution, and molecular weight are the three important factors of CMC properties, which affect the adhesion, thickening property, and water retention of CMC solution. The performance parameters of several commercial CMC samples are listed in Table 14.2.

In the process of electrode preparation, the most important step is to make the electrode slurry and then fix the mixture on the collector. In order to obtain good electrical contact between electrode material and current collector, it is necessary to keep the uniform distribution of binder and achieve superior mechanical property after drying. When using the CMC binder, high elastic polymerized SBR is usually added into the slurry to improve the flexibility and robustness of the electrode. In addition, recent work showed that acrylic rubber and graphene oxide can also be mixed with CMC to achieve better binding effect [27, 28]. For now, there are still different opinions on CMC adhesion mechanism, mainly including covalent bond theory and hydrogen bond theory. Hochgatterer and Mazouzi et al. have observed the existence of covalent bond connection of C(O)OR in Si electrode containing CMC binder through infrared spectrum test, which is considered to be formed by the interaction between carboxymethyl group on CMC surface and hydroxyl group adsorbed on the oxide layer on Si surface [29, 30]. However, Bridel et al. showed that the chemical bond of adhesion was hydrogen bond rather than strong covalent bond by employing nuclear magnetic resonance spectrometer [31]. Similar results were presented by Munao et al., who also demonstrated that there is no covalent bond in the CMC containing electrode obtained by electrodeposition [32].

CMC is more often used in anode rather than cathode, especially for alloy and conversion anodes. Actually, it is well known that the most popular PVDF binder has an unfavorable effect on the performance of alloy anodes in LIBs. These anode materials have large volume change during charge and discharge processes, which leads to poor electrical contact between the material and the conductive matrix, serious crushing of material particles, and excessive formation of SEI film on the surface of materials, thus exacerbating the pulverization. Introduction of water-soluble binders has been proved to promote their electrochemical performance. Qian

Table 14.2 The properties of several commercial CMC samples.

Project	Daicel CMC2200	Daicel CMC1390	CP Kelco CEKOL 30000	Nippon Paper SUNROSE MAC250HC
Purity %			99.5	≥99.5
Viscosity of aqueous solution (20 g L^{-1}, 25 °C)	1500–3000	2500–5000	2500–3500	3000–5000
PH (10 g L^{-1}, 25 °C)	6.0–8.5	6–8.5	6.5–8.0	6.0–8.5
Weight loss by drying %	≤10	≤10	≤10	≤10
Chloride (as NaCl) %	≤1.0	≤1.0	≤0.5	
Degree of substitution	0.8	1.0–1.5	0.75–0.85	
Product properties	White or yellowish powder particles	White or yellowish powder particles	White to creamy powder, tasteless and neutral	White powder

et al. used CMC binder in Sb/C nanocomposites, which exhibits $610\,mA\,h\,g^{-1}$ at $2000\,mA\,g^{-1}$ and maintains 94% initial capacity after 100 cycles, showing good rate performance and cycle stability [33]. Sun et al. used water-soluble binders (SA, PAA, CMC, etc.) for high-capacity P/C materials and achieved significant improvement in reversible capacity, cycle stability, and rate performance. It has a high capacity of $1374\,mA\,h\,g^{-1}$ at $400\,mA\,g^{-1}$ and maintains $1071\,mA\,h\,g^{-1}$ after 200 cycles and has $479\,mA\,h\,g^{-1}$ at a high current density of $3200\,mA\,g^{-1}$ [34].

Compared with PVDF using the solvent of volatile and toxic NMP, CMC is water soluble, which indicates that the process is easier and safer. According to the migration control drying kinetics, the main reason for binders migrating to the upper surface of the electrode is the flow of solvent during the drying process. Due to the lower evaporation rate and the longer drying time of NMP-based slurry, the uniformity of PVDF in dry electrode is much worse than that of water-based SBR/CMC [35]. Therefore, its performance in terms of adhesion, resistance, and charge/discharge efficiency is worse than that of water-based binders.

Furthermore, researchers have carried out many comparative studies on the properties of CMC and PVDF binders in SIBs. Dahbi et al. investigated the initial irreversible capacity and cycle reversibility of hard carbon anode materials using CMC and PVDF binders [18]. The experimental results showed that CMC-containing anode inhibits the decomposition of electrolyte during the initial cycle, which has superior initial Coulombic efficiency, as compared with PVDF. Soft X-ray photoelectron spectroscopy revealed the difference between the two electrodes. It was found that hard carbon was only covered by CMC binder, which is considered to be beneficial to the stability of electrochemical performance, presenting better capacity retention and higher reversibility [36]. On the other hand, they also studied the synergistic effect of binders and electrolyte additives on the electrochemical properties of hard carbon materials. As one of the commonly used electrolyte additives, fluorinated vinyl carbonate (FEC) can significantly improve the cycling performance of PVDF-based materials, but it has adverse effects on the materials using CMC. Scanning electron microscopy (SEM) images showed that the PVDF-based electrode with FEC additive had a rough surface with small particles deriving from the decomposition of electrolyte, which is beneficial to the cycling stability of the material. For the material using CMC, although FEC was added to the electrolyte, no deposition was observed on the electrode surface, indicating that the deposition caused by electrolyte decomposition is inhibited in CMC-based electrode. This means that the disparity of battery chemistry requires to optimize binders for the different electrodes and electrolytes. Furthermore, Zhang et al. investigated the effects of different binders (PVDF, CMC, PAA) on porous $CoFe_2O_4$ nanocubes anode for SIBs [37]. Electrochemical impedance spectroscopy (EIS) in Figure 14.8 showed that the sodium ion diffusion coefficient of CMC-containing electrode was the highest, where R_{CT} and E_a were the lowest, which was conducive to the reaction of electrode/electrolyte interface. Compared with the traditional PVDF-based electrode, the water-soluble CMC and PAA binder have higher electrochemical activity, thus improving the ICE and cycle stability of the electrode. The electrode with CMC binder maintained a high specific capacity of $394\,mA\,h\,g^{-1}$ after 50 cycles, and the capacity retention was 91.4%.

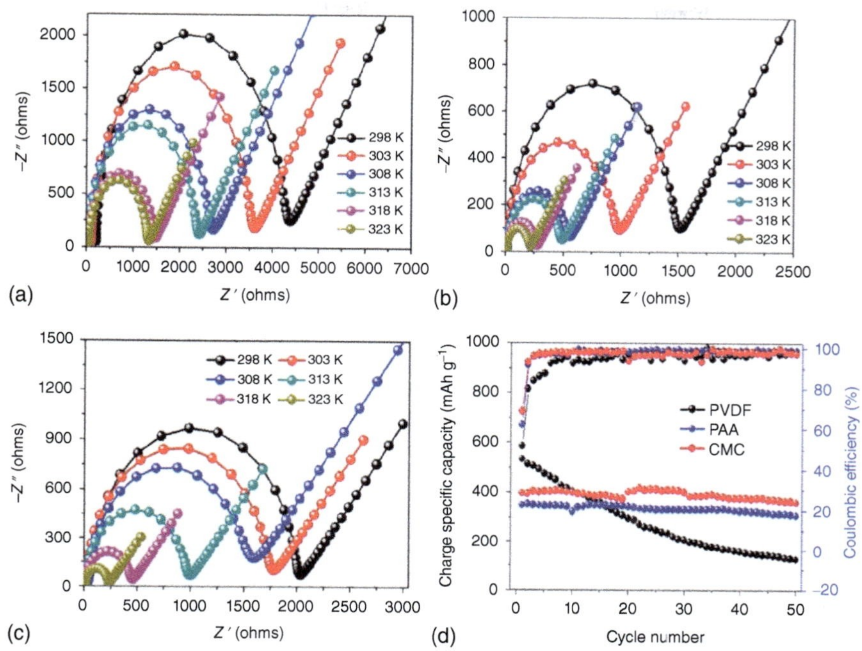

Figure 14.8 Nyquist plots of electrodes with (a) PVDF, (b) CMC, (c) PAA binders, and (d) cycling performance. Source: Zhang et al. [37]. Reproduced with permission, 2019, Elsevier.

In addition to the anode, some results have been showed that CMC binder also has positive effect on some cathode materials. $Na_3V_2(PO_4)_2F_3$ with sodium super ionic conductor (NASICON) structure is known as the most promising cathode material for SIBs because of its fast ion conduction, high working voltage, and stable structure. However, there are still many challenging problems in its rate performance and cycle performance. Recently, Zhao et al. have overcome the above shortcomings by introducing CMC binder. The conductive network composed of CMC-Super P chain is conducive to the diffusion of sodium ions in the interface and materials [38]. Moreover, the strong binding force of CMC and the stable solid permeable interface reduce the resistance and maintain the integrity of the electrode. Therefore, the reversible capacity of the material reaches $75\,mA\,h\,g^{-1}$ at a high current of 70 C, and a capacity retention of 79% after 3500 cycles at 30 C is obtained. The full battery constructed with hard carbon materials as negative electrode also exhibits an energy density of $216\,Wh\,kg^{-1}$. These advantages highlight the potential application of CMC binder in SIBs.

14.6 Styrene Butadiene Rubber (SBR)

SBR is a copolymer mainly composed of styrene and butadiene, as shown in Figure 14.3d. With the change of component ratio, the characteristics of SBR will greatly change. The mechanical properties of SBR are similar to natural rubber, but its heat resistance and aging resistance are much better. SBR is also processed

Table 14.3 The properties of several commercial SBR samples.

Project	NIPPON A&L SN-307R	Zeon BM-430B	Zeon BM-400B	JSR Corporation CB-100	BASF Binder 21-11 ap
Solid content, wt%	48	40.0 ± 0.8	40	49–50	49–51
Viscosity mPa.s		5–100	12	30–150	80–400
pH	5.0	8.0 ± 0.8	6	4.8–6.2	6.0–7.0
Glass transition temperature		0 °C	−5 °C		

in water, which belongs to environmentally friendly aqueous binder. Owing to the good elasticity of SBR, the volume changes of active materials in the process of charging and discharging can be buffered. Therefore, SBR has been widely used in negative electrode for SIBs, which is usually mixed with CMC for better electrochemical performance [39, 40]. Recently, Komaba et al. compared the effect of four kinds of SBR binders, who found there is an obvious relationship between the electrochemical performance of electrode and the degree of cross-linking of SBR. The electrode containing low degree of cross-linking modified SBR showed the best cycle and rate performance. In addition, the electrode using acrylonitrile-modified SBR exhibited a high solvent absorption rate and a prominent electrochemical performance. The performance of several commercial SBR samples are listed in Table 14.3. These samples are all the dispersion system of SBR particles in water.

14.7 Other Binders

14.7.1 Sodium Alginate (SA)

SA is a by-product of iodine and mannitol extracted from seaweed, which is nontoxic and water-soluble. It is a kind of polysaccharides consisting of β-D-mannuronic and α-L-guluronic connected by β-1,4-glycosidic bonds. The degree of polymerization and molecular weight are directly related to the viscosity of SA solution.

SA is also usually used as a binder for alloy anodes, which exhibit a great volume change during the charge/discharge. The numerous oxygen-containing functional groups in SA are conducive to the formation of stable chemical bond with the surface of alloy materials, thus alleviating the volume expansion problem, which is confirmed by Zhu in silicon anode for LIBs [41]. Similar results were reported by Zhao and coworkers, who found that red phosphorus anode containing SA showed better sodium storage capacity than that of PVDF-containing anode [42]. Furthermore, Xiao et al. recently revealed that the used binders in P/C anode determine the degree of oxidation of the active phosphorus. The anode using SA binder showed lower oxidation degree than PVDF-containing anode, thus exhibited superior electrochemical performance [34].

14.7.2 Xanthan Gum (XG)

XG is a natural nontoxic polysaccharide, which has been mass-produced and applied as a food additive, rheology modifier, and polymer stabilizer for decades. XG contains many functional groups in the monomer unit of each polymer, such as carboxyl group, hydroxyl group, and ester group, making it completely soluble in hot and cold water and giving high solution viscosity at low concentrations. These functional groups naturally exist in the polymer chain and are evenly distributed.

The concentration of carboxyl and hydroxyl groups on the XG chain is high and the distribution is uniform, suggesting more active bonding points between active materials, conductive agents, and current collectors, thereby promoting electronic and ion conduction. This feature also ensures good dispersion and electrode integrity during slurry production and electrode manufacture. XG has been used as a water-soluble binder for the negative electrodes, including carbon materials, such as mesoporous carbon microspheres [43], graphite [44], and silicon materials [45, 46], which presented excellent cycle performance, rate performance, and capacity.

14.7.3 Guar Gum (GG)

GG is a nonionic galactomannan extracted from the endosperm of the legume guar. GG and its derivatives have good water solubility and high viscosity at low mass fraction. Because of this characteristic, it has applied in many fields, such as papermaking, medicine, textile printing, dyeing, etc. XG, SA, and GG are all natural biopolymers, which are much cheaper than PVDF. In addition, a large number of –OH and –COO– functional groups can produce high adhesion between the material particles and the conductor, resulting in better electrical conductivity, which may be beneficial to cycle stability. Zhang et al. studied the effects of three water-soluble binders, i.e. XG, SA, and GG, on the P2-type cathode material of $Na_{2/3}Mn_{2/3}Ni_{1/3}O_2$ (Figure 14.9) [47]. The XG binder, which is the most viscous among them, exhibits the best electrochemical performance with the cathode. Moreover, the electrode using GG binder also shows stable cycling performance.

14.7.4 Polyimide (PI)

PI is a kind of polymer with excellent thermal stability and mechanical properties. It is widely used in gas separation, memory materials, covalent organic frameworks, hydrogen storage, and optical materials. In the field of LIBs, PI separators have been extensively studied due to their high thermal stability. As the binder, a mixture of 10 wt% co-PI (P84) and 90 wt% PVDF can significantly improve the cycle performance of $LiCoO_2$ electrode at high temperatures (60 °C) [48]. In addition, it can also be used for silicon anodes to inhibit powdering [49–53]. Thus, it has a potential application for use with alloy anode materials in SIBs. Moreover, according to the theoretical calculation results of molecular orbital method, the addition of –F or $-CF_3$ functional groups in PI can provide greater resistance to electrochemical oxidation and improved thermal stability.

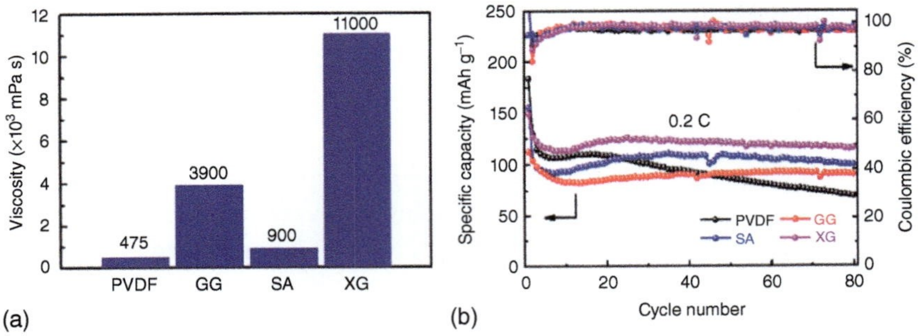

Figure 14.9 (a) The viscosity of the 1 wt% binder solutions. (b) The cycling performances of $Na_{2/3}Mn_{2/3}Ni_{1/3}O_2$ using different binders at 0.2 C. Source: Zhang et al. [47]. Reproduced with permission, 2019, Elsevier.

14.8 Concluding Remarks

In summary, binder has played important roles on improving the electrochemical performances. Binder can enhance the contact between active materials in electrode, stabilize the electrode surface, and promote the homogeneous dispersion of active materials and conductive agents, because there are four main types of binder force involved in the binder, which are chemical bond force, intermolecular force, interface electrostatic gravity, and mechanical force respectively. Therefore, a good binder usually has polar groups and a stable electrochemical window. Meanwhile it should be easily dissolved in a certain solvent and stable in the electrolyte. In addition, low price, non-toxicity, environmentally friendly production, and high safety are also good properties that a good binder should have. According to the solvent that used to dissolve binder, the binder can be classified to two main types, oil-based binder and aqueous binder. In this chapter, we have introduced the main binders used in SIBs based on these two type binders. In the oil-based binder, PVDF is discussed in detail. While, PAA, CMC, and SBR are introduced as the aqueous binders. SA, XG, GG, and PI have also been discussed as novel binder types. Deep understanding of binders' properties, mechanisms and applications is vital for improving the electrochemical performances of SIBs. Therefore, more efforts should be spent on the design and engineering of binders for the better performance of SIBs in the near future.

References

1 Shi, Y., Zhou, X., and Yu, G. (2017). *Acc. Chem. Res.* 50: 2642–2652.
2 Liu, G., Xun, S., Vukmirovic, N. et al. (2011). *Adv. Mater.* 23: 4679–4683.
3 Buqa, H., Holzapfel, M., Krumeich, F. et al. (2006). *J. Power Sources* 161: 617–622.
4 Hong, X., Jin, J., Wen, Z. et al. (2016). *J. Power Sources* 324: 455–461.
5 Pizzi, A., Mtsweni, B., and Parsons, W. (1994). *J. Appl. Polym. Sci.* 52: 1847–1856.

6 Derjaguin, B.V., Aleinikova, I.N., and Toporov, Y.P. (1994). *Prog. Surf. Sci.* 45: 119–123.
7 Chen, H., Ling, M., Hencz, L. et al. (2018). *Chem. Rev.* 118: 8936–8982.
8 Gao, Z., Zhang, Y., Song, N., and Li, X. (2017). *Mater. Res. Lett.* 5: 69–88.
9 Liu, G., Zheng, H., Battaglia, V. et al. (2019). *ECS Trans.* 6: 45–56.
10 Maleki, H., Deng, G., Kerzhner-Haller, I. et al. (2000). *J. Electrochem. Soc.* 147: 4470.
11 Chou, S.-L., Pan, Y., Wang, J.-Z. et al. (2014). *Phys. Chem. Chem. Phys.* 16: 20347–20359.
12 Fedorková, A., Oriňáková, R., Oriňák, A. et al. (2010). *J. Power Sources* 195: 3907–3912.
13 Li, J., Christensen, L., Obrovac, M.N. et al. (2008). *J. Electrochem. Soc.* 155: 234–238.
14 Wang, Y., Zheng, H., Qu, Q. et al. (2015). *Carbon* 92: 318–326.
15 Zhu, S., Dong, X., Huang, H., and Qi, M. (2020). *J. Electroanal. Chem.* 870: 114200.
16 Wu, Y., Liu, Z., Zhong, X. et al. (2018). *Small* 14: 1703472.
17 Li, J.T., Wu, Z.Y., Lu, Y.Q. et al. (2017). *Adv. Energy Mater.* 7: 30.
18 Dahbi, M., Nakano, T., Yabuuchi, N. et al. (2014). *Electrochem. Commun.* 44: 66–69.
19 Komaba, S., Okushi, K., Ozeki, T. et al. (2009). *Electrochem. Solid State Lett.* 12: 107–110.
20 Zhang, Z., Zeng, T., Lai, Y. et al. (2014). *J. Power Sources* 247: 1–8.
21 Ming, J., Ming, H., Kwak, W.-J. et al. (2014). *Chem. Commun.* 50: 13307–13310.
22 Patra, J., Rath, P.C., Li, C. et al. (2018). *ChemSusChem* 11: 3923–3931.
23 Lee, J.-H., Paik, U., Hackley, V.A., and Choi, Y.-M. (2006). *J. Power Sources* 161: 612–616.
24 Fan, Q., Zhang, W., Duan, J. et al. (2015). *Electrochim. Acta* 174: 970–977.
25 Komaba, S., Matsuura, Y., Ishikawa, T. et al. (2012). *Electrochem. Commun.* 21: 65–68.
26 Zhang, H., Lv, Z., Liang, Q. et al. (2019). *Batteries Supercaps* 3: 101–107.
27 Hitomi, S., Kubota, K., Horiba, T. et al. (2019). *ChemElectroChem* 6: 5070–5079.
28 Liu, Y., Sun, J., Du, H. et al. (2019). *Chem. Commun.* 55: 10460–10463.
29 Hochgatterer, N.S., Schweiger, M.R., Koller, S. et al. (2008). *Electrochem. Solid-State Lett.* 11: 76.
30 Mazouzi, D., Lestriez, B., Roué, L., and Guyomard, D. (2009). *Electrochem. Solid-State Lett.* 12: 215.
31 Bridel, J.S., Azaïs, T., Morcrette, M. et al. (2010). *Chem. Mater.* 22: 1229–1241.
32 Munao, D., Van Erven, J.W.M., Valvo, M. et al. (2011). *J. Power Sources* 196: 6695–6702.
33 Qian, J., Chen, Y., Wu, L. et al. (2012). *Chem. Commun.* 48: 7070–7072.
34 Xiao, W., Sun, Q., Banis, M.N. et al. (2020). *Adv. Funct. Mater.* 30: 15.
35 Jean, J.-H. and Wang, H.-R. (2001). *J. Am. Ceram. Soc.* 84: 267–272.
36 Zhang, W., Dahbi, M., and Komaba, S. (2016). *Curr. Opin. Chem. Eng.* 13: 36–44.

37 Zhang, X., Gao, X., Wu, Z. et al. (2019). *Chem. Phys.* 523: 124–129.

38 Zhao, J., Yang, X., Yao, Y. et al. (2018). *Adv. Sci.* 5: 1700768.

39 Yu, W., Qian, J., Zhang, S. et al. (2019). *ECS Trans.* 1: 79–85.

40 He, M., Yuan, L., Hu, X. et al. (2013). *Nanoscale* 5: 3298–3305.

41 Zhu, L., Du, F., Zhuang, Y. et al. (2019). *J. Electroanal. Chem.* 845: 22–30.

42 Zhao, L., Zhao, Q., Niu, Z. et al. (2016). *Chinese J. Inorg. Chem.* 32: 929–934.

43 Courtel, F.M., Niketic, S., Duguay, D. et al. (2011). *J. Power Sources* 196: 2128–2134.

44 Cuesta, N., Ramos, A., Cameán, I. et al. (2015). *Electrochim. Acta* 155: 140–147.

45 Chen, D., Yi, R., Chen, S. et al. (2014). *Solid State Ionics* 254: 65–71.

46 Jeong, Y.K., Kwon, T.-W., Lee, I. et al. (2015). *Energy Environ. Sci.* 8: 1224–1230.

47 Zhang, Y.-Y., Zhang, S.-J., Li, J.-T. et al. (2019). *Electrochim. Acta* 298: 496–504.

48 Choi, J., Ryou, M.-H., Son, B. et al. (2014). *J. Power Sources* 252: 138–143.

49 Choi, N.-S., Yew, K.H., Choi, W.-U., and Kim, S.-S. (2008). *J. Power Sources* 177: 590–594.

50 Kim, J.S., Choi, W., Cho, K.Y. et al. (2013). *J. Power Sources* 244: 521–526.

51 Qian, G., Wang, L., Shang, Y. et al. (2016). *Electrochim. Acta* 187: 113–118.

52 Uchida, S., Mihashi, M., Yamagata, M., and Ishikawa, M. (2015). *J. Power Sources* 273: 118–122.

53 Zhang, J., Lu, Q., Fang, J. et al. (2014). *ACS Appl. Mater. Interfaces* 6: 17965–17973.

15

Sodium-Ion Full Batteries

15.1 Introduction

To broaden the application of sodium-ion batteries (SIBs), it is critical to optimize the key electrode materials in Chapters 3–10, the electrolytes in Chapters 11–13 and the binders in Chapter 14, and assemble them into high-performance full batteries. Nowadays, three types of SIBs are widely studied, including nonaqueous, aqueous, and all-solid-state SIBs. The main challenge in developing high-performance aqueous SIBs lies in the relatively narrow electrochemical stable voltage window, which leads to low energy density. One common strategy to widen the window is using high-concentration electrolyte, but increasing the amount of electrolyte salts will further increase the cost, which is against the initial goal of developing low-cost and safe SIBs. On the other hand, the stability of electrode materials in aqueous electrolytes also limits the choice of cathode and anode materials. Therefore, the aqueous SIBs will have a good prospect only if the cost can be reduced, while the window can be widened to increase the energy density. The current solid-state SIBs mainly use quasi-solid and all-solid electrolytes. Among them, quasi-solid SIBs may perform well in special-purpose environments such as flexible devices. In the all-solid-state batteries, the sodium metal electrode needs to be protected, and the conductivity of the solid electrolyte and the electrode and electrolyte interface need to be optimized, to achieve high safety and high energy density. In contrary, the commercialization of nonaqueous SIBs is more feasible. The commercialization of the first-generation SIBs has started since 2015. The current commercial SIBs use hard carbon as anodes. For the cathode, three types of cathode materials have been commercially produced. The Faradion Company of the United Kingdom and the HiNa Battery of China have both developed layered oxide cathode materials with higher specific capacity which can even exceed the lithium iron phosphate in LIBs. Polyanion-based fast ion conductors and Prussian blue analog (PBA)-based cathode materials have lower energy density, but they can achieve extremely high power density, which is suitable for high-power devices. Novasis Energies of the United States and Natron Energy, a Stamford-based company, have successfully developed a SIB with PBA as the cathode material. Currently, many anode and cathode materials with excellent electrochemical properties are being developed for commercial applications.

Sodium-Ion Batteries: Energy Storage Materials and Technologies, First Edition. Yan Yu.

Typical SIBs comprise of anode and cathode materials, binders, separator membranes, and electrolytes. Cathode materials with high capacity and high crystal structure stability, as well as high-capacity alloy-type anodes in contrary to the hard carbon, are desirable. Adhesives need to be selected according to different cathode and anode material systems to ensure effective bonding and stability. Appropriate coating needs to be applied to the separator membranes to satisfy different battery systems, for more stable electrochemical performance. The solvent and additives used in the electrolyte system needs to be adjusted with consideration in the composition of solid electrode interface (SEI) in different electrolyte system and its influence on the battery performance. At present, many prototypes of SIBs have been proposed, and their progress will be summarized in this chapter.

15.2 Aqueous Sodium-Ion Full Batteries

Aqueous rechargeable sodium-ion batteries (ASIBs) are among the most promising candidates for application in the large-scale energy storage system (ESS). Compared with nonaqueous batteries, the ASIBs have the advantages of high ion conductivity, nonflammability, low manufacturing cost, therefore attract extensive research interests. However, the narrow working window (1.23 V) of water and low energy density hinder the developments of ASIBs. Therefore, people have tried to find novel higher capacity electrode materials with better cycling performance and develop new strategies to widen the stable window of aqueous electrolyte to eliminate electrolyte decomposition and other side reactions. So far, a variety of materials have been developed as cathodes or anodes. Figure 15.1 shows the insertion/extraction potential of Na ion (right) for some typical electrode materials and compares them with the reaction potential of O_2/H_2 evolution (left).

The representative cathode materials used in aqueous sodium-ion full cells include manganese-based oxides, PBAs, polyanionic compounds, carbonyl-based organic compounds, etc. (Table 15.1).

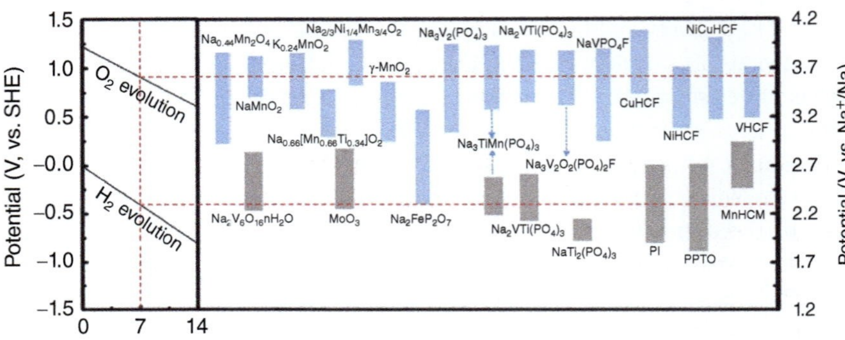

Figure 15.1 The relationship between the insertion–extraction potential of Na ion (right) in electrodes and the reaction potential of hydrogen and oxygen evolution (left). Source: Bin et al. [1]. Reproduced with permission, 2018, Wiley-VCH.

Table 15.1 The representative aqueous sodium-ion full batteries.

Type	Full batteries	Electrolyte	Voltage window (V)	Specific capacity ($mAh\,g^{-1}$)	Capacity retention (cycle numbers)	Power density ($Wh\,kg^{-1}$)
Hybrid	$AC//Na_{0.44}MnO_2$	1 M Na_2SO_4	0.4–1.8	≈ 24 (4 C)	≈100% (1000, at 4 C)	≈24.5 (5 C)
	$AC//\lambda-MnO_2$	1 M Na_2SO_4	0.6–1.6	—	≈100% (5000, 20 min per cycles)	23
	$NaTi_2(PO_4)_3$–carbon composite//AC	1 M Na_2SO_4	0.2–1.6	100 (1 C)	61% (500, at 2 C)	33.4
Insertion	$NaV_3(PO_4)_3//Na_{0.44}MnO_2$	1 M Na_2SO_4	0–1.5	≈ 115 (1 C)	84% (500, at alternate 5 C and 20 C)	—
	Nano–sized $NaTi_2(PO_4)_3$@C//$Na_{0.44}MnO_2$	1 M Na_2SO_4	0–1.6	43 (0.1 $A\,g^{-1}$)	60% (1000, at 0.2 $A\,g^{-1}$)	30 (4678 $W\,kg^{-1}$)
	$NaTi_2(PO_4)_3$@C//$Na_{0.66}[Mn_{0.66}Ti_{0.34}]O_2$	1 M Na_2SO_4	0.3–1.7	≈ 76 (2 C)	89% (300, at 2 C)	—
	$NaTi_2(PO_4)_3$//Hollow–$K_{0.27}MnO_2$	1 M Na_2SO_4	0–1.6	84.9 (0.15 $A\,g^{-1}$)	68.7 $mAh\,g^{-1}$ (100, at 0.2 $A\,g^{-1}$)	—
	$NaTi_2(PO_4)_3//Na_2NiFe(CN)_6$	1 M Na_2SO_4	0.2–1.6	100 (1 C)	88% (250, at 5 C)	42.5 (130 $W\,kg^{-1}$)
	$NaTi_2(PO_4)_3//Na_2CuFe(CN)_6$	1 M Na_2SO_4	0–1.8	104 (2 C)	97% (100, at 2 C)	48 (91 $W\,kg^{-1}$)
	Mn^{II}—N≡C—$Mn^{III/I}$//Cu^{II}—N≡C—$Fe^{III/II}$	10 M $NaClO_4$	0.65–1.35	≈ 26 (10 C)	≈100% (1000, 10 C)	27 (1 C)
	$NaTi_2(PO_4)_3//Na_3V_2(PO_4)_3$	1 M Na_2SO_4	0.5–1.6	71 (2 $A\,g^{-1}$)	50% (50, at 10 $A\,g^{-1}$)	36 (2567 $W\,kg^{-1}$)

(continued)

Table 15.1 (Continued)

Type	Full batteries	Electrolyte	Voltage window (V)	Specific capacity ($mAh\,g^{-1}$)	Capacity retention (cycle numbers)	Power density ($Wh\,kg^{-1}$)
	$NaTi_2(PO_4)_3$–C//$Na_3V_2O_{2x}(PO_4)_2F_{3-2x}$/MWCNT		1.0–1.8	40 (10 C)	≈75% (400, at 10 C)	
	$Na_3MnTi(PO_4)_3$//$Na_3MnTi(PO_4)_3$	1 M Na_2SO_4	0.4–1.8	57.9 (0.5 C)	98% (100, at 1 C)	40 (0.5 C)
	$Na_2VTi(PO_4)_3$//$Na_2VTi(PO_4)_3$	1 M Na_2SO_4	0.2–1.5	50 (1 C)	70% (1000, at 10 C)	30
	$NaTi_2(PO_{4)3}$//$Na_{2.85}K_{0.15}V_2(PO_4)_3$	$NaClO_4(H_2O)_2AN_{2.4}$	0–2.0	52 (1 $A\,g^{-1}$)	60.2% (100, at 1 $A\,g^{-1}$)	52 (1 $A\,g^{-1}$)
New	Disodium naphthalenediimide(SNDI)//$KCo_{0.5}Cu_{0.5}Fe(CN)_6$	1 M Na_2SO_4	0.5–1.5	34 (20 C charging, 10 C discharging)	88% (100, same rate)	26 (53 $W\,kg^{-1}$)
	PNTCDA//aqueous solution containing I^-/I_3^-	Nafion film	0–1.6	140 (5.5 C)	70% (50 000, at 10 $A\,g^{-1}$)	≈63.8
	PNP@CNTs//$Na_{0.44}MnO_2$	1 M Na_2SO_4	0–1.5	92 (1 C)	≈100% (200, 5 C)	25
	Alizarin nanowires//PPy	$NaClO_4$ gel	0.5–1.7	158 (1 $A\,g^{-1}$)	—	—
	TiS_2//MFCN	15 M $NaClO_4$	0.8–2.6	58 (1 C)	92% (1000, 5 C)	100 (200 $W\,kg^{-1}$)

The earliest ASIBs often employed activated carbon (AC) which has capacitive properties as the electrode, and the adsorption and desorption of Na ions occurred at its surface without redox reactions, while Na-rich compounds were mainly applied as the counter electrode that would undergo intercalation and deintercalation reactions through the charge–discharge process. This type of batteries could also be named as hybrid aqueous SIB. Whitacre et al. firstly reported the electrochemical activity of $Na_4Mn_9O_{18}$ ($Na_{0.44}MnO_2$) in aqueous SIBs in 2008, which was obtained by a simple solid-state synthesis method [2]. Then they applied it as cathode to construct an aqueous sodium-ion full battery with AC as anode in 1 M Na_2SO_4 water-based electrolyte. The cyclic voltammogram (CV) curves of the full cell showed three pairs of redox peaks, corresponding to the insertion and de-insertion of Na^+ into the orthorhombic structure of $Na_4Mn_9O_{18}$. At a C/8 discharge current density, the full cell can deliver a capacity of about 45 mAh g^{-1}. To further verify the stability of the cell, long-term cycling performance was tested as displayed in Figure 15.2a. At the rate of 4 C, the cell cycled in a potential window of 0.4~1.8 V exhibited a specific capacity nearly half of that from the low rate, and the change in voltage range compensated for the significant overpotential at the high rate. After 100 cycles or even 1000 cycles, there was no measurable loss of capacity, showing extremely superior electrochemical stability of the as-assembled device. The excellent rate performance of the cell is shown in Figure 15.2b. Correlation between the specific capacity/energy and the rates showed that nearly 20 mAh g^{-1} could still be extracted at an extremely high rate of 18 C. The appealing cycling performance and the rate capability demonstrated that $Na_4Mn_9O_{18}$ is a promising low-cost cathode material in aqueous batteries. Successively, Whitacre et al. [3] explored the properties of large format devices based on AC (6 mm thickness) as anode and λ-MnO_2 (3 mm thickness) as cathode in 1 : 1 mass ratio. Considering that the ionic conductivity of ions in aqueous electrolyte is generally at least an order of magnitude larger than that in organic-based electrolytes, so the electrodes used in aqueous cells could be much thicker than those commonly used in organic batteries. The cubic spinal λ-MnO_2 increased approximately threefold in specific energy than that obtained from $Na_4Mn_9O_{18}$, by creating $LiMn_2O_4$ firstly then followed by electrochemical or chemical delithiation. In 1 M Na_2SO_4 aqueous solution, the device can provide 25 Wh L^{-1} in five hours, corresponding to 23 Wh kg^{-1}, and operate 700 cycles without measurable capacity loss at a rate of ~six hours charge–discharge (two cycles per day) as shown in Figure 15.2c. Ultimately, a 2.4 kW h, 80 V (max voltage) battery was assembled, which consisted of two groups of 40 batteries connected in parallel. The pack was tested in a long-term 24-hours operating mode. The maximum power was about 0.5 kW, and the total energy was 1.8 kW h per day as shown in Figure 15.2d. The above results reveal that ASIB as a low-cost, stable, and environmentally friendly energy storage device is possible for stationary applications.

Likewise, in hybrid ASIBs, phosphate is also widely used as electrodes. Cao et al. [4] synthesized $NaTi_2(PO_4)_3$-carbon (NTP-C) composite by sintering to enhance electronic conductivity of $NaTi_2(PO_4)_3$ and adopted AC as cathode to fabricate a full cell in 1 M Na_2SO_4 aqueous electrolyte. The charge–discharge curves of electrodes vs. Ag/AgCl are shown in Figure 15.2e. The full cell displayed an

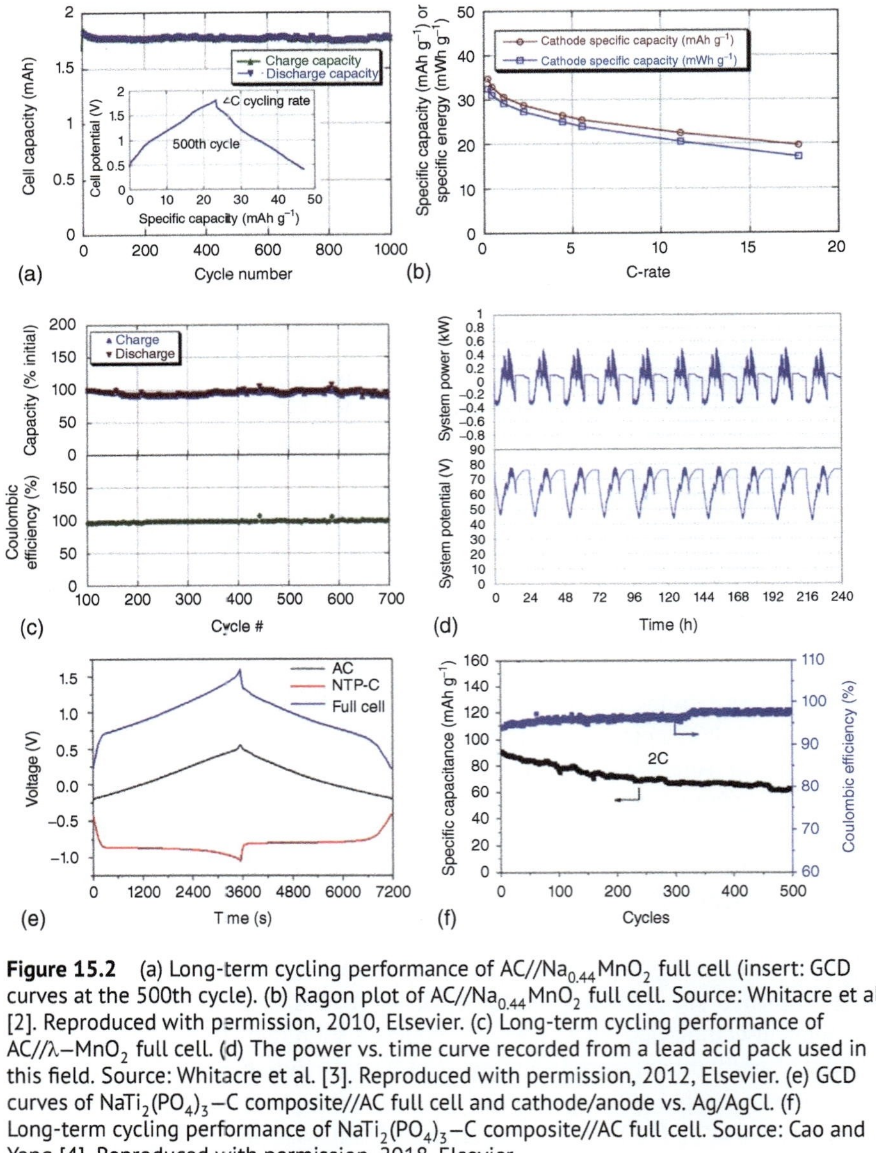

Figure 15.2 (a) Long-term cycling performance of AC//$Na_{0.44}MnO_2$ full cell (insert: GCD curves at the 500th cycle). (b) Ragon plot of AC//$Na_{0.44}MnO_2$ full cell. Source: Whitacre et al. [2]. Reproduced with permission, 2010, Elsevier. (c) Long-term cycling performance of AC//λ–MnO_2 full cell. (d) The power vs. time curve recorded from a lead acid pack used in this field. Source: Whitacre et al. [3]. Reproduced with permission, 2012, Elsevier. (e) GCD curves of $NaTi_2(PO_4)_3$–C composite//AC full cell and cathode/anode vs. Ag/AgCl. (f) Long-term cycling performance of $NaTi_2(PO_4)_3$–C composite//AC full cell. Source: Cao and Yang [4]. Reproduced with permission, 2018, Elsevier.

average operating voltage of 1.17 V within a potential window of 0.2 and 1.6 V. At the rate of 1 C, the cell can deliver a capacity of 100 mAh g^{-1}, and the value can be maintained at 89.8, 67.2, and 55.4 mAh g^{-1} when the rate increased to 2, 5, and 10 C, respectively. As shown in Figure 15.2f, the full cell exhibited a great long-term cycling performance at a rate of 2 C, maintaining a capacity of 61.9 mAh g^{-1} after 500 cycles, corresponding to a retention of 61%. The galvanostatic charge and discharge (GCD) curves after hundreds of cycles showed little difference from the initial cycle, indicating an outstanding stability of the full battery. The energy density of 33.4 Wh kg^{-1} based on the total mass of the full cell was achieved with a power density of 33.4 W kg^{-1}.

The full batteries relying on the intercalation/deintercalation reactions are much more promising in water-based SIBs thanks to the wider working potential window and relative high energy density. The mechanisms are similar to the traditional organic SIBs that store Na^+ in the electrodes by reversible insert/extract reaction. Materials including manganese-based oxides, polyanions, and PBAs, have been investigated. Among them, manganese-based oxides and the phosphates have attracted extensive attention. Ke et al. [5] prepared a nanofiber of $NaV_3(PO_4)_3$@C in high alignment by electrospinning and fabricated a full cell with $NaV_3(PO_4)_3$@C as anode paired with $Na_{0.44}MnO_2$ as cathode in 1 M Na_2SO_4 aqueous solution. Figure 15.3a shows the GCD profiles of the full cell at the rate of 1 C within a voltage window between 0.0 and 1.5 V, and a capacity of approximately 115 mAh g^{-1} was obtained. At the high rates of 5/20 C, the cell still delivered 84% of the capacity retention after 500 cycles, demonstrating an appealing rate and cycling properties of the cell, as shown in Figure 15.3b. Guo et al. [6] assembled two-dimensional (2D) belt-shaped and one-dimensional (1D) fiber-shaped ASIBs, which are highly safe and extremely flexible based on $Na_{0.44}MnO_2$ (NMO) cathode and nano-sized $NaTi_2(PO_4)_3$@C (NTPO@C) anode [8]. For the belt-shaped full cell, the GCD curves at different current densities are shown in Figure 15.3c. It could deliver a discharge capacity of 43 mAh g^{-1} at 0.1 A g^{-1} (based on cathode) within the voltage range of 0 ~1.6 V with three voltage platforms and maintained a capacity of 27 mAh g^{-1} even at a high current density of 5 A g^{-1}, corresponding to 63% of the initial capacity. Figure 15.3d shows the cycling performance of the full cell at 0.2 A g^{-1}. The capacity decreased from 42 to 26 mAh g^{-1} over 1000 cycles, corresponding to a retention of 60%. Accordingly, as calculated based on the total mass, the energy density achieved was 30 Wh kg^{-1}. Similarly, the fiber-shaped full battery delivered a specific capacity of 46 mAh g^{-1}, similar as that of belt-shaped full battery. Once the current density increased to 3 A g^{-1}, the capacity of fiber-shaped cell decreased to 12 mAh g^{-1}, and only 76% of the initial capacity can be maintained after 100 cycles at 0.2 A g^{-1}, showing slightly poorer performances than the belt-shaped full cell.

To further improve the capacity of $Na_{0.44}MnO_2$, researchers also tried to dope other elements to enhance its properties. Wang et al. [9] partially substituted Mn with Ti to synthesize a novel Na-rich tunnel-type material with a composition of $Na_{0.66}[Mn_{0.66}Ti_{0.34}]O_2$ and utilized $NaTi_2(PO_4)_3$/C (NTP/C) as anode to fabricate a full cell in 1 M Na_2SO_4 aqueous electrolyte. A discharge capacity of 76 mAh g^{-1} was obtained at 2 C (150 mA g^{-1}) within a voltage window of 0.3 and 1.7 V. Even at a high rate of 10 C, a specific capacity of 54 mAh g^{-1} could still be obtained. Long-term cycling test showed that after 300 cycles, the capacity retention was 89% at a rate of 2 C. Furthermore, Huang and coworkers [7] synthesized Binessite-type $K_{0.27}MnO_2$ with hollow structure by a template method and constructed a full cell with $NaTi_2(PO_4)_3$ as anode. In 1 M Na_2SO_4 aqueous solution, the cell delivered a capacity of 84.9 mAh g^{-1} at 0.15 A g^{-1} in a voltage range of 0–1.6 V with slope charge–discharge curves, as shown in Figure 15.3e.

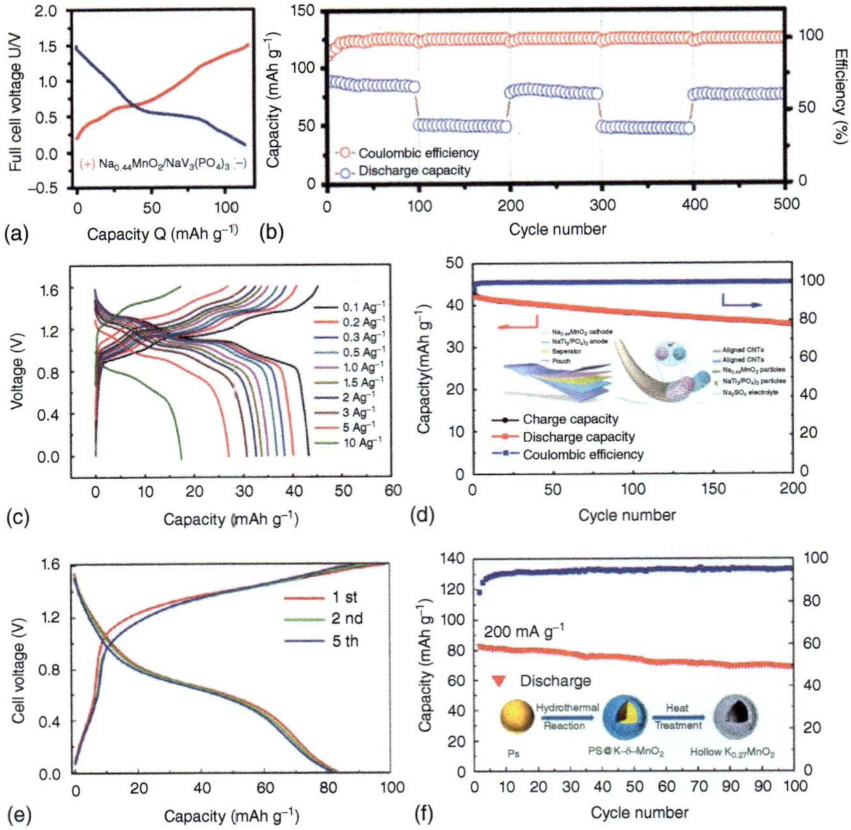

Figure 15.3 (a) GCD curves of $NaV_3(PO_4)_3$//$Na_{0.44}MnO_2$ full cell. (b) Long-term cycling performance of $NaV_3(PO_4)_3$//$Na_{0.44}MnO_2$ full cell. Source: Ke et al. [5]. Reproduced with permission, 2017, Royal Society of Chemistry. (c) GCD curves of belt-shaped nano-sized $NaTi_2(PO_4)_3$@C//$Na_{0.44}MnO_2$ full cell. (d) Long-term cycling performance of belt-shaped nano–sized $NaTi_2(PO_4)_3$@C (insert: schematic illustration of the structure of belt-shaped and fiber-shaped battery). Source: Guo et al. [6]. Reproduced with permission, 2017, Elsevier. (e) GCD curves of $NaTi_2(PO_4)_3$//hollow–$K_{0.27}MnO_2$ full cell. (f) Long-term cycling performance of $NaTi_2(PO_4)_3$//hollow–$K_{0.27}MnO_2$ full cell (insert: schematic illustration of the fabrication). Source: Liu et al. [7]. Reproduced with permission, 2016, American Chemical Society.

With the current density increased to 0.6 A g^{-1}, the cell could still achieve a discharge capacity of 56.6 mAh g^{-1}, demonstrating an outstanding rate performance. In addition, the long-term cycling performance is shown in Figure 15.3f. At a current density of 2 A g^{-1}, the cell maintained a capacity of 68.7 mAh g^{-1} after 100 cycles, corresponding to 83% capacity retention, exhibiting an excellent cycling performance.

PBAs are another promising electrode material due to their open-framework structure which can allow the Na ions or other alkali metal ions to insert/extract quickly without damaging the crystallographic lattice. The electrochemical activities of PBAs in aqueous electrolytes have been confirmed. However, most of

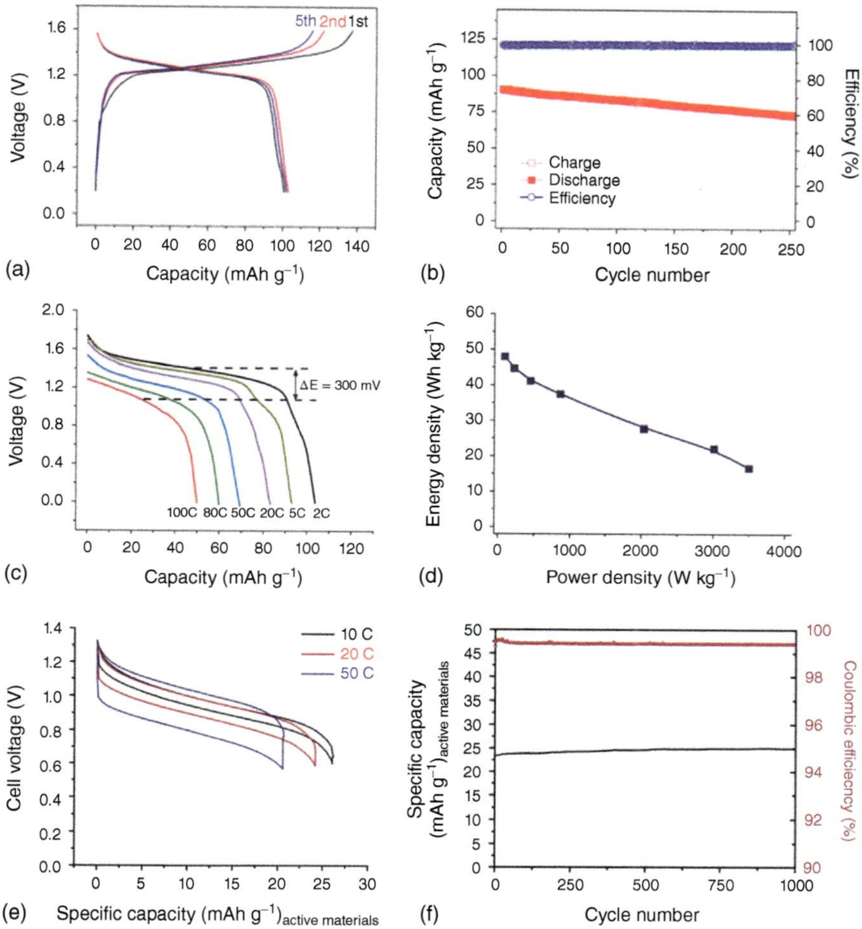

Figure 15.4 (a) GCD curves and (b) long-term cycling performance of $NaTi_2(PO_4)_3$//$Na_2NiFe(CN)_6$ full cell. Source: Wu et al. [10]. Reproduced with permission, 2013, Elsevier. (c) Discharge curves at various rates and (d) Ragon-plot of $NaTi_2(PO_4)_3$//$Na_2CuFe(CN)_6$ full cell. Source: Wu et al. [11]. Reproduced with permission, 2014, Wiley-VCH. (e) The voltage profiles at different rates and (f) long-term cycling performance of Mn^{II}−N≡C−$Mn^{III/I}$//Cu^{II}−N≡C−$Fe^{III/II}$ full cell. Source: Pasta et al. [12]. Reproduced with permission, 2013, Springer Nature.

them could not be directly considered as cathode to construct ASIBs since they nearly contain no Na. Therefore, Wu et al. successfully synthesized Na-rich hexacyanoferrate $Na_2NiFe(CN)_6$ [10] and explored its electrochemical performances in full ASIBs with $NaTi_2(PO_4)_3$ as anode. Figure 15.4a shows the GCD curves of the full cell (with a voltage platform around 1.27 V) in 1 M Na_2SO_4 water-based electrolyte at a rate of 1 C. The Coulombic efficiency was relatively low in the initial several cycles which could be ascribed to the O_2/H_2 evolution during the cycles. With the rate increased to 10 C, a discharge capacity of 86 mAh g^{-1} was maintained. And a capacity of 79 mAh g^{-1} could be delivered at 5 C after 250 cycles, as shown in Figure 15.4b. Accordingly, on the basis of the total mass weight of both electrodes, an energy density of 42.5 Wh kg^{-1} could be achieved at a power density of 130 W kg^{-1}. To realize a high-voltage aqueous system, Wu et al. successively synthesized $Na_2CuFe(CN)_6$ [11], similar to $Na_2NiFe(CN)_6$, and investigated the electrochemical performances. At a current density of 0.2 A g^{-1} (2 C based on the anode), it delivered a capacity of 104 mAh g^{-1} with a higher discharge platform of 1.4 V. Rate performance is shown in Figure 15.4c. Even at an extremely high rate of 100 C, the battery could still exhibit a discharge capacity of 50 mAh g^{-1}, with only 0.3 V voltage drop, showing an excellent rate performance among the reported materials of the ASIBs. The Ragon plot of the full cell based on the total mass of electroactive materials is shown in Figure 15.4d, and the battery could achieve an energy density of 48 Wh kg^{-1} at 91 W kg^{-1} and attain 16.6 Wh kg^{-1} at an extraordinarily high power density of 3500 W kg^{-1}. The excellent properties make it a competitive material among the conventional materials reported for ASIB technology. In addition, Cui and coworkers [12] developed a novel symmetric full ASIB based on manganese hexacyanomanganate (Mn^{II}—N≡C—$Mn^{III/II}$) as anode and Cu^{II}—N≡C—$Fe^{III/II}$ as cathode in excess of 10 M $NaClO_4$ aqueous electrolyte. The cell was operated between 0.65 and 1.35 V with little voltage hysteresis even at 10 C thanks to the impressively high ionic conductivity of the electrolyte and fast reaction rate of the electrodes (Figure 15.4e). When the current increased to 50 C, the energy efficiency could still maintain at 84.2%, showing excellent rate performance. The cycling performance of the cell at 10 C is shown in Figure 15.4f. There was no visible capacity loss after 1000 cycles, corresponding to a Coulombic efficiency of over 99.8% for each cycle. Consequently, considering the total mass weight of electrodes, a maximum energy density of 27 Wh kg^{-1} was realized at a rate of 1 C.

Besides the above-mentioned materials, another important electrode material is "polyanionic" compounds possessing the advantages of strong electrostatic interaction between the anions, structural diversity, long cycle life, and high safety. Zhang et al. [13] assembled a full battery on the basis of all NASICON structural materials, i.e. $NaTi_2(PO_4)_3$ as anode and $Na_3V_2(PO_4)_3$ as cathode with a mass ratio of 8 : 7 in 1 M Na_2SO_4 aqueous solution. The cell showed a voltage platform at around 1.3 V with some voltage hysteresis in the potential window of 0.5–1.6 V (Figure 15.5a). As the current density increasing, the capacity firstly increased from 49 to 73 mAh g^{-1} (3 A g^{-1}) and then decreased to 58 mAh g^{-1} at 10 A g^{-1}. This phenomenon can be ascribed to the weakened side reaction or the dissolution of the electrode at high voltage and high rate. In the cycling test, it showed an optimum performance at a current density of 10 A g^{-1}, as shown in Figure 15.5b. The battery was capable of maintaining a capacity retention of 50% after 50 cycles. In general, the cycling performance of the cell was pretty poor owing to the complex side reactions which needs to be further explored and addressed. Kumar et al. [14] firstly explored the feasibility of $V_2O_{2x}(PO_4)_2F_{3-2x}$ with multi-walled carbon nanotubes (MWCNTs) in aqueous system and fabricated a full cell with $NaTi_2(PO_4)_3$–C as anode. At a current density of 650 mA g^{-1} (10 C), the cell achieved a capacity of 30 mAh g^{-1} after 400 cycles in the voltage range of 1.0–1.8 V, as shown in Figure 15.5c.

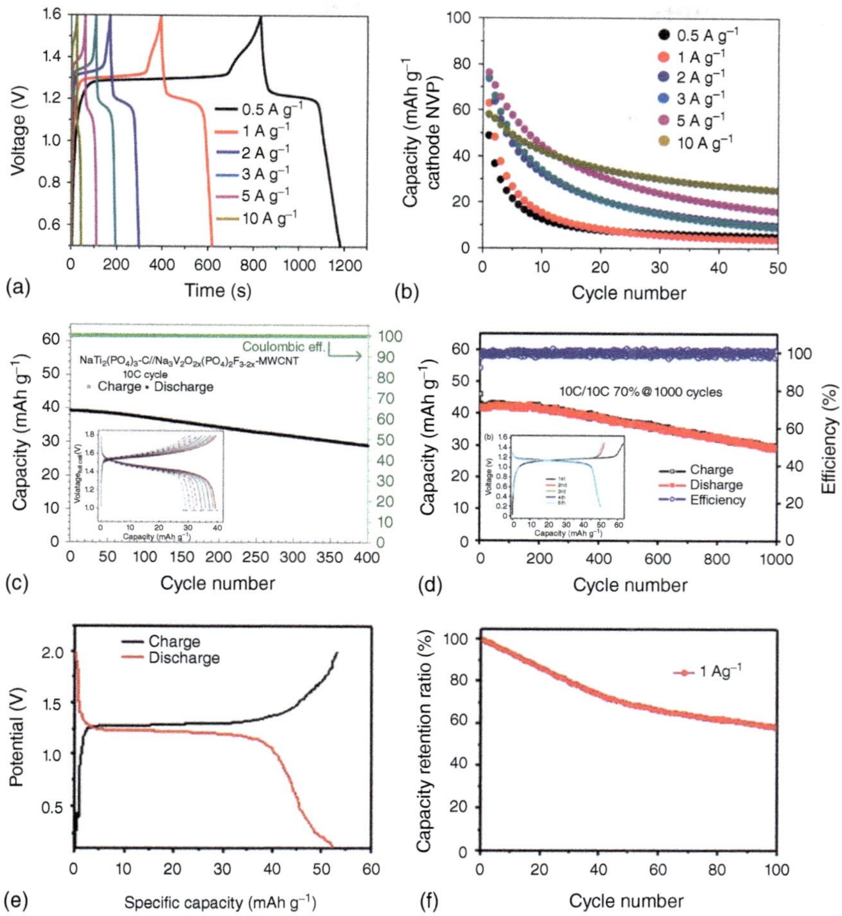

Figure 15.5 (a) GCD curves and (b) long-term cycling performance of $NaTi_2(PO_4)_3$// $Na_3V_2(PO_4)_3$ full cell. Source: Zhang et al. [13]. Reproduced with permission, 2013, Elsevier (c) Long-term cycling performance of $NaTi_2(PO_4)_3$−C//Na_3-$V_2O_{2x}(PO_4)_2F_{3-2x}$/MWCNT full cell (insert: GCD curves of full cell). Source: Kumar et al. [14]. Reproduced with permission, 2015, Royal Society of Chemistry. (d) Long-term cycling performance of $Na_2VTi(PO_4)_3$// $Na_2VTi(PO_4)_3$ full cell (insert: GCD curves of the full cell). Source: Wang et al. [15]. Reproduced with permission, 2017, Springer. (e) GCD curve and (f) long-term cycling performance of $NaTi_2(PO_{4)3}$//$Na_{2.85}K_{0.15}V_2(PO_4)_3$ full cell. Source: Shen et al. [16]. Reproduced with permission, 2020, Royal Society of Chemistry.

As the polyphosphates could compose of multiple metal ions, symmetric batteries can be constructed depending on the different redox potentials of various metal ions. Gao and Goodenough [17] reported a typical symmetric NASICON-based full battery. They synthesized $Na_3MnTi(PO_4)_3$, a material that could be employed either as negative or positive electrodes because of the Ti^{4+}/Ti^{3+} and Mn^{3+}/Mn^{2+} redox couples. A pair of redox peaks emerged at around 1.4 V in the CV curves, which was consistent with the predicted redox potential. The full SIBs in 1 M Na_2SO_4 aqueous electrolyte delivered a capacity of 57.9 mAh g^{-1} at a rate of 0.5 C within the voltage range of 0.4–1.8 V, and the calculated energy density was about 40 Wh kg^{-1} based on the mass of the total active materials. With the rate increased from 1 to 10 C, the capacity decreased from 56.5 to 46.7 mAh g^{-1}, showing an appealing rate performance. When cycled at 1 C, the cell could maintain almost 98% of the initial capacity after 100 cycles, showing impressive stability of the cell. Beyond that, Wang et al. [15] constructed a symmetric full battery with $Na_2VTi(PO_4)_3$ relying on the redox couples of Ti^{4+}/Ti^{3+} and V^{4+}/V^{3+}. As shown in Figure 15.5d, within the voltage range of 0.2–1.5 V (where no unexpected redox peaks were found in CV curves), the battery delivered a capacity of ~ 50 mAh g^{-1} with the operating voltage plateau at 1.14 V and little voltage hysteresis was observed at a rate of 1 C. The rate performance was tested at 1, 2, 5, and 10 C, and a capacity of 40.6 mAh g^{-1} could be achieved at 10 C, corresponding to 80.6% of the initial capacity. Moreover, as seen from Figure 15.5d, the cell showed excellent cycling stability, with 70% of initial capacity maintained after 1000 cycles at 10 C. Overall, the excellent electrochemical performances of the cell enabled the application of NASICON-structured $Na_2VTi(PO_4)_3$ in ASIBs for ESS. Recently, Shen et al. [16] explored a high-performance ASIB utilizing $NaTi_2(PO_4)_3$ as anode and carbon coated $Na_{2.85}K_{0.15}V_2(PO_4)_3$ as cathode with a $NaClO_4$-based acetonitrile/water hybrid solution ($NaClO_4(H_2O)_2AN_{2.4}$) as electrolyte. Thanks to the extraordinary electrolyte, the battery could be stably cycled with a relatively small internal resistance. The GCD profile of the full battery is shown in Figure 15.5e, and the cell delivered a capacity of 52 mAh g^{-1} (based on the total mass) under the current density of 1 A g^{-1} at a platform of around 1.2 V with little voltage hysteresis. Meanwhile, the cell could maintain 60.4% of the initial capacity with the current density increased from 0.5 up to 5 A g^{-1}. Figure 15.5f displays the long-term cycling performance at 1 A g^{-1}, with 60.2% of initial capacity maintained after 100 cycles, indicating the great stability of the cell even at high charge–discharge current density.

In the hope of developing a new aqueous sodium-ion full battery with higher energy density, longer cycle life and better rate performance, tremendous efforts have been made and some novel aqueous sodium battery systems with organics, ionic liquids, biomass, etc., have been explored as the electrodes. Kim et al. [18] firstly reported a new carbonyl-based organic salt, namely disodium naphthalenediimide (SNDI), a Prussian blue derivatives, and employed it as anode coupling with $KCo_{0.5}Cu_{0.5}Fe(CN)_6$ (CoCuHCF) as cathode to assemble an ASIB (Figure 15.6a). As the diffusion rate of Na ions in SNDI in desodiation process was much higher than that in the sodiation process, the full cell was cycled at a high rate of 20 C during charging and 10 C during discharging, which delivered a discharge capacity of 34 mAh g^{-1} with an average voltage platform at 1.1 V. As shown in Figure 15.6b, after 100 cycles, the cell could maintain a discharge capacity of 30 mAh g^{-1}, corresponding to a capacity retention of 88%, showing great cycling performance. Taking into account the total mass weight of both electrodes, the cell was capable of attaining an energy density of 17 and 26 Wh kg^{-1} at power density of 262 and 53 W kg^{-1}, respectively. Xia and coworkers [19] reported a novel aqueous battery system consisting of the reversible enolization in polyimide as anode and a liquid cathode that relied on the redox of I^-/I^{3-} couple. Significantly, the battery system contained no metal

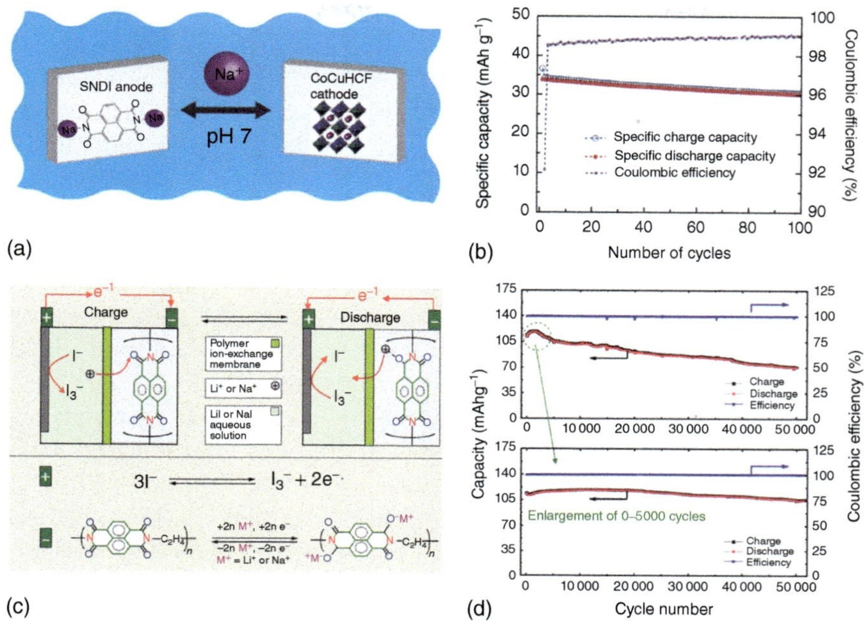

Figure 15.6 (a) Schematic illustration of the (H)SNDI//CoCuHCF full cell, and (b) its long-term cycling performance. Source: Kim et al. [18]. Reproduced with permission, 2014, Wiley-VCH. (c) Schematic illustration of cell structure and electrode reactions of PNTCDA//aqueous solution containing I^-/I^{3-}. (d) Long-term cycling performance of the full cell. Source: Dong et al. [19]. Reproduced with permission, 2016, American Association for the Advancement of Science.

elements and the ion (Na^+/Li^+) was only used as the charge carrier. A polymer membrane was employed to allow the ion exchange between the cathode and the anode. The GCD curves at various rates are shown in Figure 15.6c, and the system could deliver 140 mAh g^{-1} at 1 A g^{-1} within the voltage range of 0.0–1.6 V. When a high current density of 100 A g^{-1} was applied, the battery could still achieve a capacity of 28 mAh g^{-1}, indicative of its excellent rate performance. The long-term cyclic performance is shown in Figure 15.6d. It is surprising that the cell can maintain 70% of the initial capacity after 50 000 cycles, exhibiting outstanding stability. Besides, the rate performance was tested in a low-concentration aqueous solution (0.1 M NaI) and the mass loading on anode side was only ~1 mg cm^{-2}. As above, the novel environmentally friendly system exhibited remarkable performance, promising its potential application for future grid-scale energy storage.

Gu et al. [20] synthesized a polyimide–MWCNT composite (PNP@CNTs) from the 1,4,5,8-naphthalenet-etracarboxylic dianhydride (NTCDA) and phenylene diamine (PDA), and then successively constructed a full battery with $Na_{0.44}MnO_2$ (NMO) as cathode in 1 M Na_2SO_4 water-based electrolyte to explore the potential application of the composite in aqueous system. The full cell delivered a capacity of 92 mAh g^{-1} based on the anode mass at a rate of 1 C (100 mA g^{-1}) within the voltage window between 0.0 and 1.5 V (Figure 15.7a). Figure 15.7b shows the cycling performance

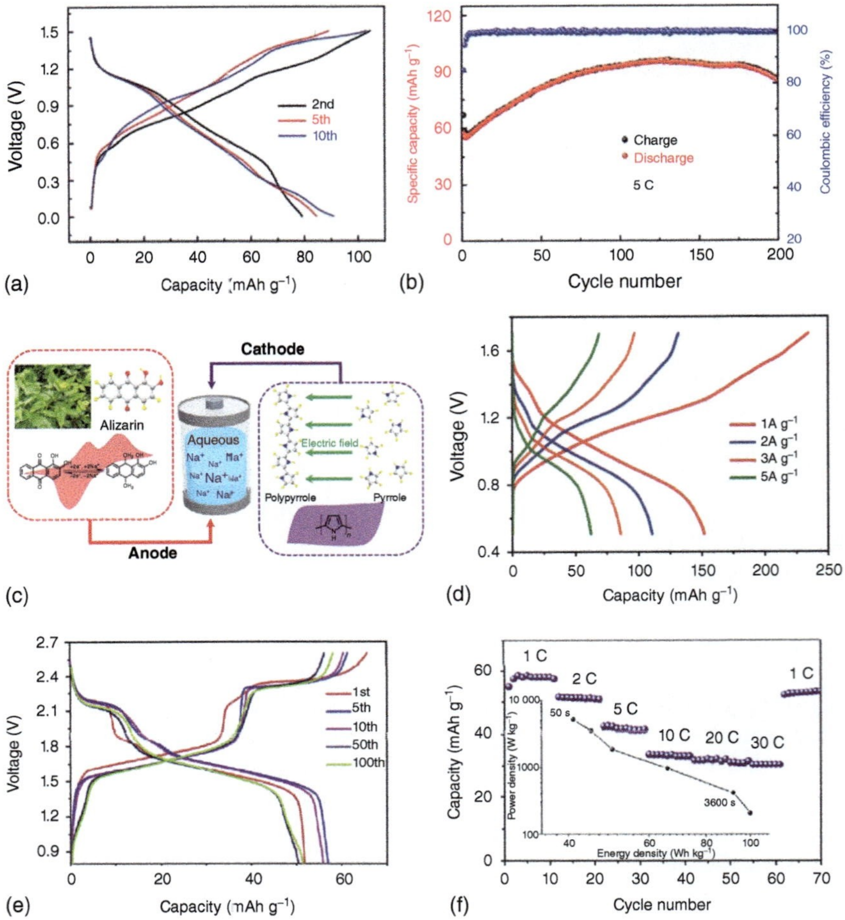

Figure 15.7 (a) GCD curves and (b) long-term cycling performance of PNP@CNTs//NMO full cell. Source: Gu et al. [20]. Reproduced with permission 2016, Royal Society of Chemistry. (c) Schematic illustration of the Alizarin nanowires//PPy full cell, and (d) its GCD curves at various rates. Source: Long et al. [21]. Reproduced with permission, 2018, Wiley-VCH. (e) GCD curves, and (f) rate-capability and Ragone plot of TiS_2//MFCN full cell. Source: Hou et al. [22]. Reproduced with permission, 2019, Elsevier.

of the cell at a rate of 5 C, and the Coulombic efficiency increased rapidly in the initial cycles due to rising capacity. After approximately 150 cycles, the capacity of the cell began to decrease, and at last the cell could obtain 25 Wh kg^{-1} of energy density over 200 cycles, demonstrating its great cycling performance. In addition to the organics and ionic liquids, biomass materials have also been investigated by researchers. Wang and coworkers [21] proposed that the biomolecule alizarin is electrochemical active and can be used as anode. As displayed in Figure 15.7c, they investigated the properties of the full cell consisting of the biomolecule alizarin as anode and polypyrrole (PPy) which is biocompatible and biodegradable as cathode. The cathode and anode were separated by a piece of filter paper which had been soaked in the $NaClO_4$ gel electrolyte for 30 minutes. The CV curves showed the good sodium storage capability of the full cell and excellent electrochemical reversibility. The cell could achieve a capacity of 146.1 mAh g^{-1} (based on the anode) at a scan rate of 20 mV s^{-1} and maintained 65.1 mA h g^{-1} when the scan rate increased to 200 mV s^{-1}. The GCD profiles are shown in Figure 15.7d, where the cell exhibited a sloping plateau with an average discharge potential at approximately 1.06 V. It can be seen that when the current density was 1 A g^{-1}, the cell could obtain an impressive capacity of 151.8 mAh g^{-1} and maintained 62.8 mAh g^{-1} at 5 A g^{-1}. Nevertheless, the relatively low-capacity retention cannot be ignored, which might be due to the dissolution of reduced anthraquinone–Na complexes in the aqueous solution. It should be noted that the stability of the electrodes in water is a universal issue which has not been addressed completely.

Besides of developing new high-performance active materials, other battery components are also worthy of attention. Qian and coworkers [22] found that depositing oxide films onto the current collectors could exhibit strong passivation effect and further broaden the stability window of the aqueous electrolyte, such as depositing TiO_2 film and Al_2O_3 film on Ti and Al current collector by atomic layer deposition (ALD). They revealed that when TiO_2 film was 5 nm and Al_2O_3 film was 3 nm, the cell in aqueous electrolyte could be cycled up to 3.5 V. Thanks to the wide operating voltage window, the TiS_2 which has low working potential was firstly used as anode in ASIBs, coupling with manganese hexacyanoferrate (MFCN) as cathode to assemble a full battery in highly concentrated aqueous solution of 15 M $NaClO_4$. The GCD profiles are shown in Figure 15.7e, and the cell delivered a capacity of 52 mAh g^{-1} (based on the total mass) at a rate of 1 C accompanied with two platforms at 1.75 and 2.45 V approximately. After 150 cycles, a discharge capacity of 52 mAh g^{-1} (corresponding to a capacity retention of 90%) could still be maintained. The rate performance and Ragon plot of energy density and power density are shown in Figure 15.7f. The cell could achieve a discharge capacity of 30 mAh g^{-1} even at a high rate of 30 C, corresponding to an extremely high specific energy density of 100 Wh kg^{-1} on the basis of total mass weight of both electrodes at a power density of 200 W kg^{-1}. The cycling performance was tested at 5 C, and after 1000 cycles, the capacity retention was about 92% and Coulombic efficiency was close to 100%. Therefore, the cell exhibited competitive electrochemical properties due to the extremely high energy density at a high rate with outstanding stability.

Aqueous SIBs are gaining more and more attention in large-scale stationary ESSs owing to the advantages of low cost, sustainability, safety, and friendliness, while there are still many technical or scientific challenges that need to be further explored to achieve better application, mainly including hydrogen and oxygen evolution in aqueous electrolyte, the dissolution of electrode in water, the reaction between electrode and the water or O_2 and proton co-intercalation, etc. Some methods have been evaluated and the directions of future development have been proposed.

First of all, optimization of electrode material is essential. The strategies to improve the properties of existing electrode materials involve: (i) material structure design; (ii) construction of artificial interface layer; and (iii) surface coating and doping. These methods could regulate voltage, increase capacity, enhance electronic and ionic conductivity, and improve structural and chemical stability during the cycles.

The second is to widen the stability window. In order to expand the electrochemical stabilization window, the following strategies can be adopted: (i) introducing additives into aqueous electrolytes; (ii) adopting highly concentrated electrolytes or gel electrolytes; (iii) constructing the artificial SEI layers; (iv) adjusting the surface characteristics of electrodes; (v) passivation oxygen evolution reaction (OER) or hydrogen evolution reaction (HER) reactivity on the current collector surface. With the expansion of the voltage window, not only the stability of the system could be improved due to the inhibition of O_2/H_2 evolution and other side reactions, but also more and more electrode materials with suitable redox potentials and high capacity can be utilized to construct a high-voltage ASIB.

The third is to improve the energy density of ASIBs with long cycle life. Exploring and optimizing electrode materials with excellent properties is the key to achieve the full battery system with energy density higher than 25 Wh kg^{-1} (voltage > 1.2 V) which can enhance the competitiveness of ASIBs in energy storage system (EES). Therefore, the full battery system with high-capacity organic materials as electrodes is a good choice to obtain high energy density. On the other hand, high voltage and good cycling stability of ASIBs can also be achieved by introducing organic additives into aqueous electrolytes to form mixed electrolyte.

In conclusion, further breakthroughs are still needed in applicable aqueous electrolyte, advanced electrode materials, devices, etc. With the improvement of comprehensive performance, as well as the optimization of electrode materials, it is believed that ASIBs will play a significant role in EES, so as to promote the application and development of clean energy and large-scale renewable energy connected to smart grids.

15.3 Nonaqueous Sodium-Ion Full Batteries

The electrolyte of nonaqueous SIBs includes salt, organic solvent, and additives. Nowadays, we pay more attention to the energy/power density, cycle stability, and the cost of the electrode materials before commercial production of SIBs. Therefore, a series of cathode and anode materials started to be widely studied as shown in Figure 15.8 [23]. As for cathode materials, polyanion compounds, transition metal oxides, Prussian blue compounds, and organics have relatively high specific capacity and the polyanion has high operating potential. Cathode materials with high operating voltage and lager discharge capacity are being pursued to achieve high-energy and high-power SIBs. Anode materials, including carbon materials, metal oxides/sulfides, alloy compounds, and phosphorus on the basis of insertion/extraction, conversion, and alloying reaction types have been studied.

Energy and power densities of aqueous-based SIBs were restricted by the potential window. Hence, nonaqueous SIBs were regarded as potential options for large-scale ESS. According to the sodium storage mechanism of different kinds of anode materials, increasing initial Coulombic efficiency and cycling stability are the key issues.

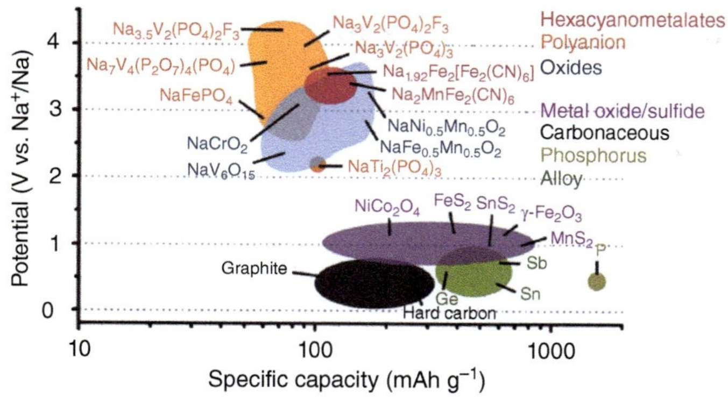

Figure 15.8 The specific capacity–potential vs. Na^+/Na graphic including major type of cathode and anode materials. Source: Ren et al. [23]. Reproduced with permission, 2017, Wiley-VCH.

The low initial Coulombic efficiency will result in active sodium loss and lead to poor cycling performance. Many works have carried out pre-sodiation on anode to eliminate initial irreversible capacity. As for cathode materials, most researches focus on increasing average working voltage and discharge capacity. There are still many problems need to be solved before SIBs become commercially available as lithium-ion batteries today. In this section, the nonaqueous SIBs are summarized in Table 15.2. Specifically, we categorize them as the battery with carbon materials and with non-carbon materials as the anode.

15.3.1 Carbon-Anode-based Sodium-Ion Full Batteries

Olivine $NaFePO_4$ with similar chemical composition with $LiFePO_4$ has been widely studied because of high theoretical capacity and operating potential. Loh and coworkers [24] used an ion exchange method to synthesize the olivine-phase $NaFePO_4$, delivering an improved electrochemical performance comparable to the organic-based exchange systems. The initial discharge capacity of the $NaFePO_4$ in half-cell was 142 mAh g^{-1} with a potential window ranging from 2 to 4 V vs. Na^+/Na. The sodium-ion full cell was assembled by combining aqueous ion-exchanged $NaFePO_4$ and hard carbon. Figure 15.9a,b show the GCD curves and the cycle performance of the full cell tested at 0.1 C, respectively. The full cell delivered a capacity of 86 mAh g^{-1} after 75 cycles at 1 C with 84% capacity retention.

A series of layered oxides, including P-type and O-type oxides, have been explored as sodium-ion full battery cathodes due to their layered structures and high capacities. Myung and coworkers [25] synthesized a carbon-coated $NaCrO_2$ (C-$NaCrO_2$) cathode material to build a HC//C-$NaCrO_2$ full cell. The half-cell performance of C-$NaCrO_2$ cathode was measured in the potential range of 2.0–3.6 V vs. Na^+/Na. The 3.4 wt% C-$NaCrO_2$ delivered a high capacity of 122 mAh g^{-1} at current density of 0.5 C and maintained a discharge capacity of 99 mAh g^{-1} even at an extremely high rate of 150 C. The cycling performance of 3.4 wt% C-$NaCrO_2$ was also impressive,

Table 15.2 The representative nonaqueous sodium-ion full batteries.

Type	Full batteries	Average working potential (V)	Specific capacity (mAh g^{-1})	Cycle performance (cycle)	Energy density (Wh kg^{-1})
Carbon-based anode	HC//$NaFePO_4$	2	140 (0.1 C)	84% (75 at 1 C)	—
	HC//$NaCrO_2$	2.9	102 (20 mA g^{-1})	89.73% (300 at 1 C)	—
	HC//P2/P3-$Na_{0.7}Li_{0.06}Mg_{0.06}Ni_{0.22}Mn_{0.67}O_2$	3.36	100.6 (0.1 C)	82.2% (50 at 0.1 C)	218
Non-carbon anode	Fe_3S_4//$Na_3V_2(PO_4)_3$@C@rGO	1.7	305 (100 mA g^{-1} based on the anode weight)	—	—
	VS_2-SNSs//$Na_3V_2(PO_4)_3$/C	1.5	232 (0.2 A g^{-1} based on the anode weight)	92% (50 at 0.2 A g^{-1})	—
	Sn//$Na_3V_2(PO_4)_3$-CNT	3.2	645 (400 mA g^{-1} based on the anode weight)	96.4% (1800 at 400 mA g^{-1})	253.4
	Sb-CNT//$Na_3V_2(PO_4)_2O_2F$	3.0	120.3 (1 C)	80% (50 at 5 C)	206.7
	$Na_2Ti_3O_7$//$VOPO_4$	2.9	114 (0.1 C)	92.4% (100 at 1 C)	220
	CNT/FeO_x—$Na_xFeFe(CN)_6$	1.75	100 (25 mA g^{-1})	81% (100 at 250 mA g^{-1})	—
	Na-$Ti_3C_2T_x$-CC//$Na_3V_2(PO_4)_3$	3.4	117 (50 mA g^{-1})	88% (300 at 500 mA g^{-1})	—
	Juglone/rGO//$Na_3V_2(PO_4)_3$/C	1.0	150 (50 mA g^{-1})	67% (40 at 50 mA g^{-1})	—

which showed a capacity retention of 90% after 300 cycle with around 100% CE. The effective carbon coating on the surface of cathode material could boost the electron transferring and result in an excellent rate capability. Sodium-ion full cells were fabricated by C-$NaCrO_2$ and hard carbon. As shown in Figure 15.9c, the initial discharge capacity of full cell reached 102 mAh g^{-1} at 20 mA g^{-1} in the voltage window of 1.8–3.4 V. The cycling performance of bare $NaCrO_2$ and C-$NaCrO_2$ cathode-based full cells was significantly varied. For bare $NaCrO_2$, after 100 cycles, the capacity retention was only 70%. In contrast, the capacity retention of C-$NaCrO_2$ cathode-based full cells was ~90% after 300 cycles (Figure 15.9d). All in all, this work showed that the O3-type layer compounds with Cr^{3+}/Cr^{4+} redox reaction can achieve excellent electrochemical performance through appropriate carbon coating.

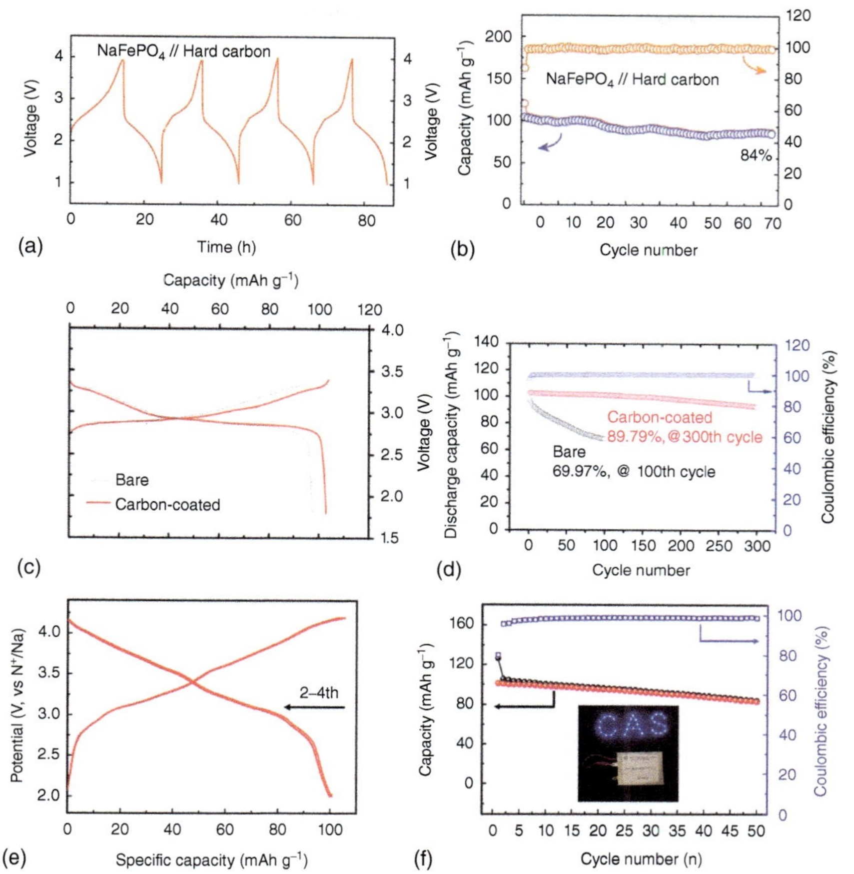

Figure 15.9 Electrochemical performance of HC//NaFePO$_4$ full cells. (a) Four cycles of charge and discharge curves at 0.1 C. (b) Cycle performance at 1 C. Source: Tang et al. [24]. Reproduced with permission, 2016, Royal Society of Chemistry. Electrochemical performances of HC//bare NaCrO$_2$ and HC//C-NaCrO$_2$ full cells. (c) Initial GCD curves. (d) Cycle tests. Source: Yu et al. [25]. Reproduced with permission, 2015, Royal Society of Chemistry. Electrochemical performance of HC//P2/P3-NLMNM pouch cell. (e) GCD profiles. (f) Cycling performance (inset: pouch cell power commercial LED). Source: Zhou et al. [26]. Reproduced with permission, 2019, Elsevier.

In Na_xTMO_2 cathode materials, the phase transition often takes place during the charge–discharge process, which results in terrible cycle performance and low initial CE. Guo et al. [26] designed a P2/P3-$Na_{0.7}Li_{0.06}Mg_{0.06}Ni_{0.22}Mn_{0.67}O_2$ (P2/P3-NLMNM) cathode material with multiphase structure for SIBs. The multiphase structure design restrained the phase transition in traditional P2-Na_xMnO_2 and improved the electrochemical performances and the discharge capacity of P2/P3-NLMNM in half-cell was ~120 mAh g^{-1} in the potential window of 2.0–4.4 V at 0.2 C. To evaluate the electrochemical properties of P2/P3-NLMNM cathode in sodium-ion full cell, the pouch cell was assembled with the pre-cycled hard carbon. The discharge capacity in the second cycle of the pouch cell reached 100.6 mAh g^{-1}

(based on cathode) with an average potential of 3.36 V (Figure 15.9e). The cycle performance of the cell is shown in Figure 15.9f, where the discharge capacity decreased from 100.6 to 82 mAh g^{-1} after 50 cycles, along with the ability to lighten up an array of light-emitting diode (LED) bulbs. It can be seen that the pre-sodiation treatment of the hard carbon anode can improve the performance of the SIB full cells.

15.3.2 Non-Carbon-Anode-based Sodium-Ion Full Batteries

In the non-carbon anode materials, the transition metal sulfide anodes can well match the cathode, showing intriguing commercialization potential. Among cathode materials, polyanion compounds with stable 3D host frameworks have attracted great attention because of long cycle life, enhanced rate capability, relatively high operating potential and high safety. A series of works on polyanion-cathode-based SIBs have been reported. As for polyanion compounds, NASICON-type $Na_3V_2(PO_4)_3$ is widely studied as cathode material for sodium-ion full cells. This NASICON-type framework provides 3D open diffusion pathways for Na^+ ion transport, and Na^+ ion diffusion coefficient of $Na_3V_2(PO_4)_3$ can achieve 10^{-11} cm^2 s^{-1}. The theoretical discharge capacity of $Na_3V_2(PO_4)_3$ rises up to 117.6 mAh g^{-1} with two Na^+ ions intercalated/extracted at 3.4 V vs. Na^+/Na. Yan and coworkers [27] synthesized $Na_3V_2(PO_4)_3$ nanocrystals coated with amorphous carbon and wrapped by rGO nanosheets (NVP@C@rGO). NVP nanocrystals shortened the lengths of Na ion diffusion, and amorphous carbon and rGO nanosheets accelerated the electron transport, as well as the macro/mesopore network enabled the full infiltration of electrolyte. The discharge capacity of NVP@C@rGO cathode reached 115 mAh g^{-1} at 20 C, and even at an ultrahigh rate of 100 C, it still maintained 86 mAh g^{-1}, corresponding to capacity retention of 73%. Meanwhile, the NVP@C@rGO cathode delivered a capacity retention of 64% after 10 000 cycles at 100 C. The full cells were constructed by combining NVP@C@rGO cathode with pre-cycled Fe_3S_4 as anode. The initial charge–discharge profiles of Fe_3S_4 at 0.1 A g^{-1} in the potential window of 0.01–3.0 V vs. Na^+/Na are displayed in Figure 15.10a. The Fe_3S_4//NVP@C@rGO full cell showed a great performance with a large discharge capacity of 305 mAh g^{-1} (based on the anode weight) at 0.1 A g^{-1} in the voltage range of 0.8–3.0 V (Figure 15.10b).

VS_2, a kind of the transition metal dichalcogenides, shows a sandwich crystal structure made up of two sulfide layers and one vanadium layer. The large open channels between the layers allow fast insertion and extraction of Na ions and the metallic property is beneficial for charge transfer. Mai and coworkers [28] synthesized a novel layer-by-layer-stacked VS_2 nanosheets structures (VS_2-SNSs), which ensured the stability of nanosheets during charge and discharge process, resulting in the long cycling stability. The conversion reaction would occur at specific potential, which can be controlled by limiting the discharge charge voltage window to improve the cycling stability. The GCD curves of the VS_2-SNSs anode at current density of 0.2 A g^{-1} in the potential window of 0.4–2.2 V are shown in Figure 15.10c. The GCD curve was slope-shaped without any obvious plateau. The

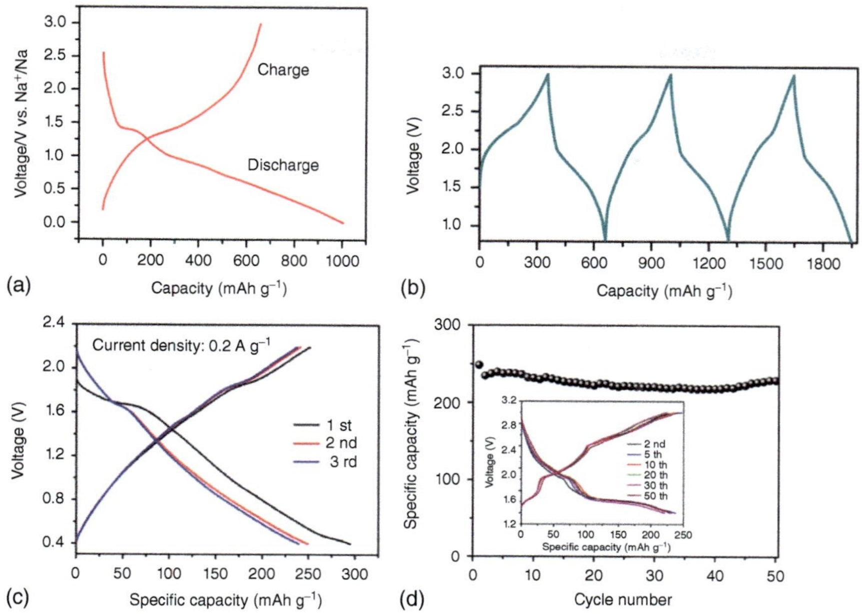

Figure 15.10 (a) The initial charge and discharge curves of the Fe_3S_4 anode. (b) GCD profiles of Fe_3S_4//NVP@C@rGO full cell. Source: Rui et al. [27]. Reproduced with permission, 2015, Wiley-VCH. (c) GCD curves of VS_2-SNSs anode. (d) Cycle performance of VS_2-SNSs//$Na_3V_2(PO_4)_3$/C full cells (inset is the GCD curves of different cycles). Source: Sun et al. [28]. Reproduced with permission, 2017, Elsevier.

VS_2-SNSs anode can deliver a capacity of as high as 252 mAh g^{-1} at 0.1 A g^{-1}, and even at an extremely high current density of 20 A g^{-1}, it still possessed a capacity of 150 mAh g^{-1}. The full cell was constructed by the usage of VS_2-SNSs as anode and $Na_3V_2(PO_4)_3$/C as cathode. In Figure 15.10d, VS_2-SNSs//$Na_3V_2(PO_4)_3$/C full cell displayed a stable platform at around 1.2 V with a discharge capacity of ~230 mAh g^{-1} at 0.2 A g^{-1} and a 92% capacity retention after 50 cycles, which indicated its excellent stability during the cycling process.

Besides of the transition metal sulfide anodes, alloy anode materials (Sn, Sb, etc.) have been widely explored by researchers because of their much higher theoretical specific capacity than that of hard carbon. For example, Sn with a large capacity of 847 mAh g^{-1} and low redox potential of ~0.2 V seems to be a promising anode for SIBs. Li and coworkers [29] constructed a high energy density sodium-ion full battery with micro-sized Sn anode and compound of $Na_3V_2(PO_4)_3$ and carbon nanotubes (NVP-CNT) as cathode in $NaPF_6$-diglyme electrolyte. The discharge capacity of Sn in half-cell reached 729 mAh g^{-1} at 50 mA g^{-1}, and even at a high current density of 3200 mA g^{-1}, it still preserved 587.8 mAh g^{-1}. The charge–discharge profiles of the Sn//NVP-CNT full cell cycle at current density of 400 mA g^{-1} (based on anode) are presented in Figure 15.11a, in which GCD curves of full cell are homologous with those of Sn and NVP-CNT in half-cells. Benefited from the excellent

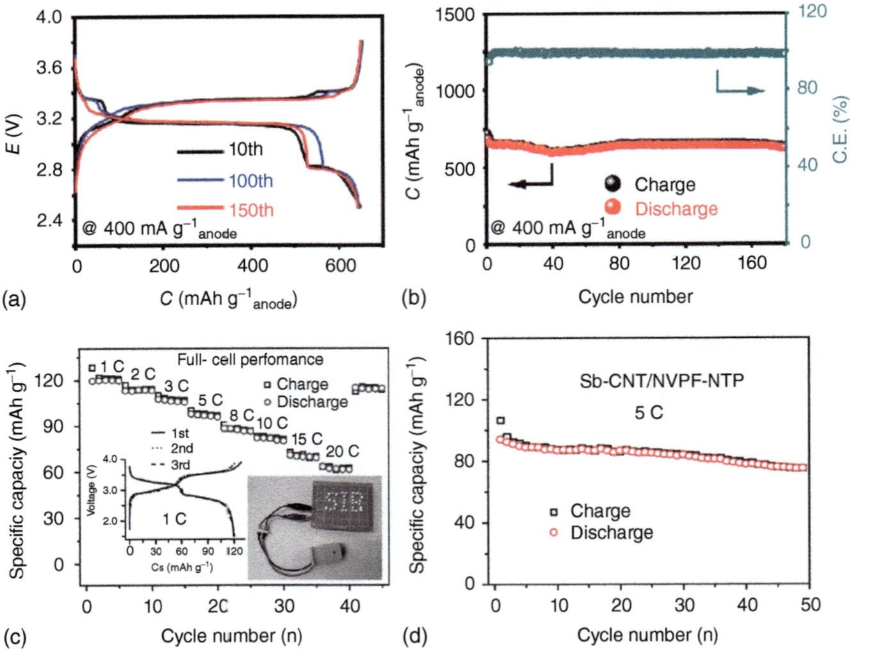

Figure 15.11 Electrochemical performance of Sn//NVP-CNT full cell. (a) GCD curves of different cycles. (b) Cycle performance. Source: Song et al. [29]. Reproduced with permission, 2019, Springer. Electrochemical performance of Sb-CNT//NVPF-NTP full cells. (c) Rate performance, GCD curves at 1 C and a digital photo of coin cell lighting on LED lights. (d) Cycle tests. Source: Guo et al. [30]. Reproduced with permission, 2017, Wiley-VCH.

rate performance in half-cells, the full cell also showed an impressive rate capability, which delivered a discharge capacity of 630.8 mAh g^{-1} at 3.6 A g^{-1}. For practical application, the excellent rate performance represents the fast charge capability of full cell. The cycling performance of full cell is presented in Figure 15.11b, which delivered a capacity of 645 mAh g^{-1} after 180 cycles with a capacity retention of 96.4%.

Many efforts have been made on the above-mentioned typical NASICON-type cathode—$Na_3V_2(PO_4)_3$, because of its crystal structure allows ultrafast sodium-ion intercalation/extraction. However, calculated from the operating potential and theoretical capacity, the energy density of the $Na_3V_2(PO_4)_3$ is only 394 Wh kg^{-1}. Thus, there is still a gap in employing $Na_3V_2(PO_4)_3$ as cathode material to build practical SIBs. In order to achieve high-energy-density full cell, cathode materials with high operating potential and large discharge capacity need to be designed. For polyanion-type cathode, introducing high electronegative anions – F^- to the anionic framework results in higher redox potential of V^{4+}/V^{5+} couple. Guo et al. [30] prepared the $Na_3V_2(PO_4)_2O_2F$ material composed of uniform and carbon-free nano-tetraprisms (NVPF-NTP) via a hydrothermal method to construct a well-performed SIB. The rate performance of NVPF-NTP cathode material was very impressive, which maintained a discharge capacity of 84.1 mAh g^{-1} at the current density of 40 C. The capacity retention of NVPF-NTP half-cell at 1 C after 1000 cycles still reached 88.5% of the initial discharge capacity of 117.6 mAh g^{-1}. The sodium ion full batteries were assembled by combining NVPF-NTP as cathode and Sb-CNT as anode. The Sb-CNT anode in half-cell delivered a large discharge capacity of 400 mAh g^{-1} at a current density of 2500 mA g^{-1}. Figure 15.11c shows the electrochemical performance of the Sb-CNT//NVPF-NTP full cells, including the GCD curves at 1 C (based on cathode) in the voltage window of 1.5–3.9 V, the rate performance, and a digital photo of the coin cell lighting on an array of LED lights. The rate performance of full cells was remarkable with a discharge capacity of 120 and 65 mAh g^{-1} at 1 and 20 C, respectively. The full cells cycled at 5 C for 50 cycles showed a capacity retention of around 80% (Figure 15.11d).

It is also a good choice to use the intercalation/deintercalation type anode for SIBs due to better stability. Yu and coworkers [31] utilized the "zero-strain" material $Na_2Ti_3O_7$ as anode and the high-performance $VOPO_4$ as cathode to construct a sodium-ion full battery. The electrochemical performance of the anode and cathode were assessed in half-cells initially. Figure 15.12a,c show their CV and GCD curves, respectively. The orange part of Figure 15.12a shows two redox couples corresponding to V^{5+}/V^{4+} in the potential window of 2.5–4.3 V. Accordingly, Figure 15.12c shows the GCD curve at a current density of 50 mA g^{-1} with an average voltage of 3.5 V. The blue area in Figure 15.12a shows the CV curve of the $Na_2Ti_3O_7$ in the potential range from 0.01 to 2.5 V with a distinctive redox peak at around 0.55 V, corresponding to the redox reaction of Ti^{4+}/Ti^{3+}. The GCD curve of $Na_2Ti_3O_7$ at 50 mA g^{-1} is shown in the blue area of Figure 15.12c, of which the capacity contribution is mostly from below 1.5 V. Figure 15.12b shows the schematic working potential of the V^{5+}/V^{4+} redox couple and Ti^{4+}/Ti^{3+} redox couple vs. Na^+/Na. From the above results, excellent electrochemical performance can be expected from the sodium ion full batteries based on $Na_2Ti_3O_7$ nanotube anode and $VOPO_4$ nanosheet cathode. Besides, pre-desodiation of the $Na_2Ti_3O_7$ anode and $VOPO_4$ cathode was carried out in advance before constructing the full cell, which reduced the polarization and irreversibility during the initial charge–discharge process. The GCD profiles of the full cell at a rate of 0.1 C (1 C = 100 mA g^{-1}) from 2nd to 4th cycle are shown in Figure 15.12d. The cycle performance of full cell at 1 C is shown in Figure 15.12e, which delivered a 92.4% capacity retention after 100 cycles. Furthermore, the pouch cell could power an array of LEDs even completely bended (inset of Figure 15.12d), showing the commercial potential of the $Na_2Ti_3O_7//VOPO_4$ full batteries.

To construct a high-performance sodium-ion full battery, many factors need to be taken into consideration, including specific capacity, working potential, cost, and so on. To further reduce the cost, Li and coworkers [32] constructed a $CNT/FeO_x//NaFeFe(CN)_6$ full cell with iron-based redox couples. In this design, both electrodes relied on the earth abundant and environmentally friendly element–iron. The charge and discharge profiles of the CNT/FeO_x anode

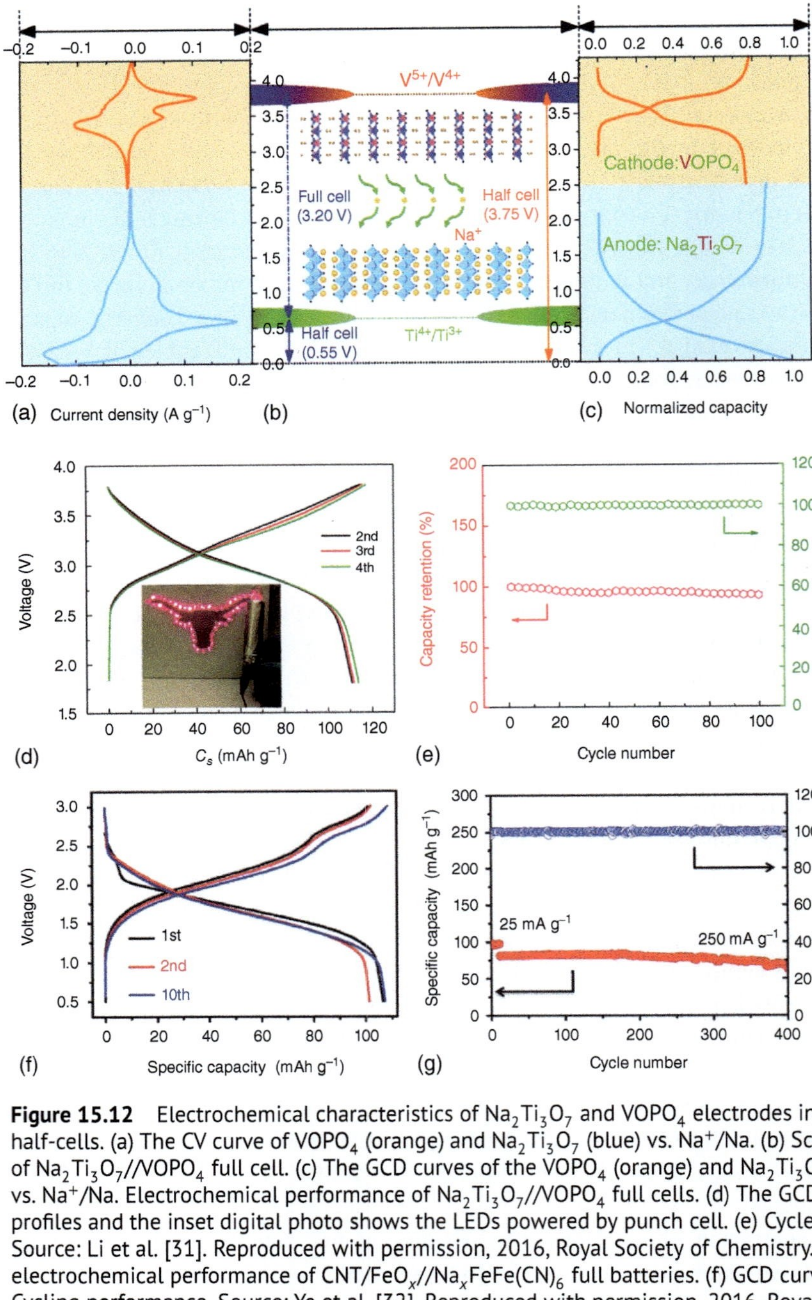

Figure 15.12 Electrochemical characteristics of $Na_2Ti_3O_7$ and $VOPO_4$ electrodes in half-cells. (a) The CV curve of $VOPO_4$ (orange) and $Na_2Ti_3O_7$ (blue) vs. Na^+/Na. (b) Schematic of $Na_2Ti_3O_7$//$VOPO_4$ full cell. (c) The GCD curves of the $VOPO_4$ (orange) and $Na_2Ti_3O_7$ (blue) vs. Na^+/Na. Electrochemical performance of $Na_2Ti_3O_7$//$VOPO_4$ full cells. (d) The GCD profiles and the inset digital photo shows the LEDs powered by punch cell. (e) Cycle tests. Source: Li et al. [31]. Reproduced with permission, 2016, Royal Society of Chemistry. The electrochemical performance of CNT/FeO_x//$Na_xFeFe(CN)_6$ full batteries. (f) GCD curves. (g) Cycling performance. Source: Ye et al. [32]. Reproduced with permission, 2016, Royal Society of Chemistry.

showed a long slope with a large irreversible capacity at the current density of 50 mA g^{-1} but it could be stabilized at around 400 mAh g^{-1} after 100 cycles. The charge and discharge profiles of the $Na_xFeFe(CN)_6$ in half-cells exhibited a long plateau at around 3.0 V, which delivered a discharge capacity of ~120 mAh g^{-1} at current density of 25 mA g^{-1} with almost no capacity loss after 200 cycles. The all-iron reversible redox sodium-ion full cell were fabricated by combining $Na_xFeFe(CN)_6$ as cathode and CNT/FeO_x as anode. The charge and discharge profiles of CNT/FeO_x//$Na_xFeFe(CN)_6$ at 25 mA g^{-1} are shown in Figure 15.12f, which delivered a specific capacity of 100 mAh g^{-1} (based on cathode) in the voltage window of 0.5–3.0 V. When cycled at 250 mA g^{-1}, the full battery still exhibited good performance for over 400 cycles (Figure 15.12g).

In recent years, sodium metal has been widely applied as the anode material because of the large theoretical capacity. However, there are still unbridgeable chasms (poor processability and safety issues) before industrial application. As well known, Na metal tends to stick to each other or other materials during the storage or processing procedure. Zhu and coworkers [33] proposed a MXene ($Ti_3C_2T_x$)-modified carbon-cloth ($Ti_3C_2T_x$-CC) as a stable and flexible host for metallic Na anodes to guarantee the uniform deposition. The digital photo and SEM image of Na-$Ti_3C_2T_x$-CC composite are presented in Figure 15.13a and the corresponding capacity was ~730 mAh g^{-1}. The Na-$Ti_3C_2T_x$-CC electrodes demonstrated long cycle life, suggesting stable stripping/plating processes. The full cells were constructed by combining Na-$Ti_3C_2T_x$-CC as anode and $Na_3V_2(PO_4)_3$ as cathode. The rate performance of Na-$Ti_3C_2T_x$-CC//$Na_3V_2(PO_4)_3$ full cell in the voltage window of 2.0–4.3 V is presented in Figure 15.13b, which delivered almost the theoretical discharge capacity at 50 mA g^{-1} and preserved a discharge capacity of 80 mAh g^{-1} at 1000 mA g^{-1}. This work also showed the Na-$Ti_3C_2T_x$-CC foil could demonstrate stable performance under different bending angles during the cycling tests at the current density of 500 mA g^{-1} (Figure 15.13c).

Besides of the above-mentioned electrode materials, organic electrode materials with redox centers have also been investigated widely. However, organic electrodes are facing two major problems, that is, electrode dissolution and low conductivity. As shown in Figure 15.13d, Guo and coworkers [34] prepared a Juglone/rGO composite electrode through a facile assembly method. The Juglone/rGO films on a copper substrate were tested in half-cells in the potential window of 0.01–2.5 V vs. Na^+/Na. A pair of redox peaks were observed at potentials of 0.6 and 1.5 V. The Juglone/rGO delivered a discharge capacity of 325 mAh g^{-1} with a smooth CV curve. The cycling performance of the Juglone/rGO anode was tested at 0.1 A g^{-1} which maintained a discharge capacity of 280 mAh g^{-1} with ~99% CE after 100 cycles. The sodium-ion full cell was assembled with Juglone/rGO as anode and $Na_3V_2(PO_4)_3$ as cathode, and the CV curves of full cell are presented in Figure 15.13e, indicating good electrochemical irreversibility during charge and discharge process. The cycle performance of full cell is displayed in Figure 15.13f, which delivered a capacity of 80 mAh g^{-1} at 0.1 A g^{-1} after 100 cycles. Additionally, the assembled pouch cell was capable of powering an electric fan (inset of Figure 15.13f).

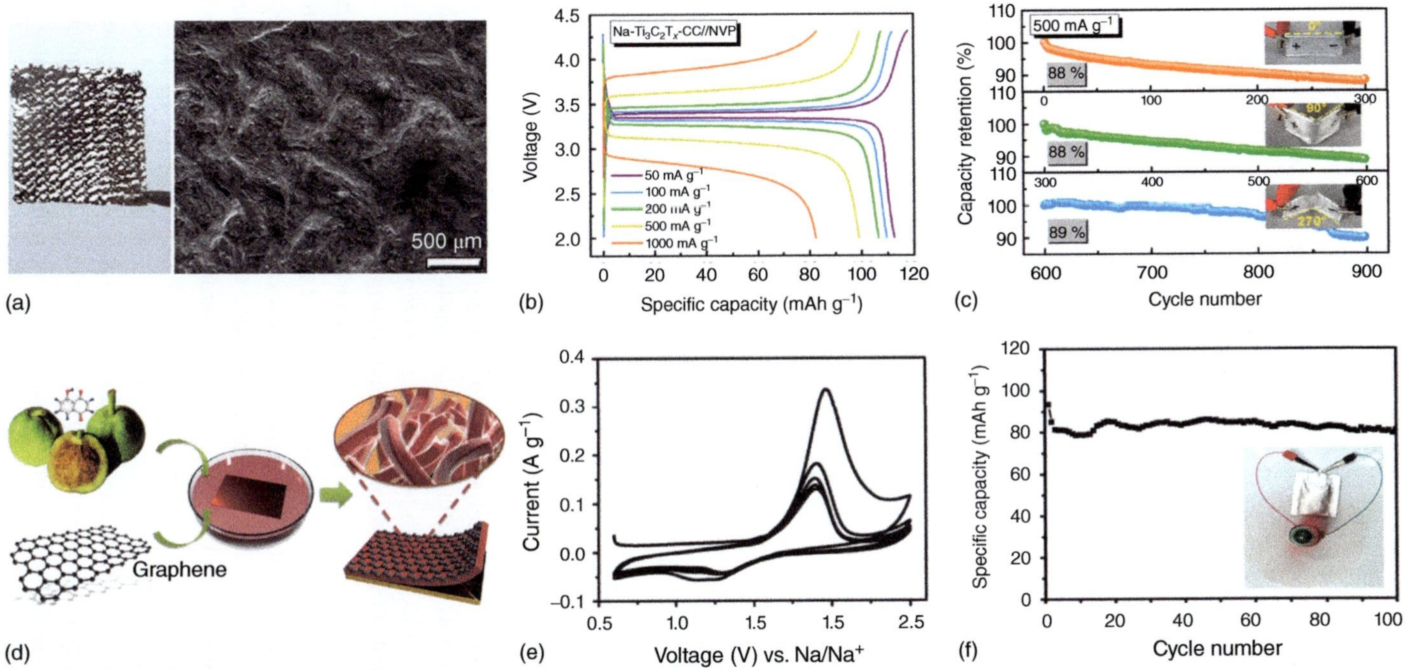

Figure 15.13 (a) Digital photos and SEM image of Na-$Ti_3C_2T_x$-CC. Electrochemical performance of Na-$Ti_3C_2T_x$-CC//NVP. (b) GCD curves. (c) Cycling stability at different angle of bend. Source: Fang et al. [33]. Reproduced with permission, 2020, American Chemical Society. (d) Schematic of fabrication process of Juglone/rGO. Electrochemical performance of Juglone/rGO//$Na_3V_2(PO_4)_3$/C full cells. (e) CV curves, and (f) cycle performance (inset: digital photo of a pouch cell powering an electric fan). Source: Wang et al. [34]. Reproduced with permission, 2015, Wiley–VCH.

15.4 Solid-state Sodium-Ion Full Batteries

From the previous two sections, we can see that the aqueous-based and organic-based full batteries have been widely studied due to the high ionic conductivity and good wettability of the electrolyte, but their applications have been limited by the poor thermal stability and the risk of flammability and leakage. As the sodium-ion solid-state battery has a stable electrolyte system, which compensates for the shortcomings of the aqueous-based and organic-based batteries. And the stable electrolyte system can be coupled with the high-potential cathode and the metal sodium anode to obtain higher energy density, which broadened its potential application. Just as shown in Table 15.3, in this section, we will introduce the quasi-solid-state batteries and the all-solid-state batteries. In addition, we also discuss in detail the current challenges and key prospects of the sodium-ion solid batteries.

15.4.1 Quasi-Solid-State Sodium-Ion Full Batteries

The quasi-solid-state battery system shows many advantages including lower risk, preferable mechanical performance, fewer side reactions, and wider electrochemical window than that of aqueous-based and organic-based SIBs, making it a better choice to serve as the next-generation energy storage device. Compared with all-solid-state batteries, it has better flexibility.

Kim et al. [35] reported for the first time the use of $NaFePO_4$ cathode, HC anode, and a HSE (ceramic NASICON ($Na_3Zr_2Si_2PO_{12}$) (70 wt%), polymer polyvinylidene fluoride-hexafluoropropylene (PVDF-HFP) (15 wt%), and a small amount of liquid (15 wt%, 1 M sodium triflate in triglyme)) to assemble a full battery. It showed a capacity of 120 mAh g^{-1} at 0.2 C with the average voltage of around 2.6 V, and after 200 cycles at 0.2 C, a good capacity retention of 96% was also achieved (Figure 15.14a,b). With the current density increased from 0.5 to 1C, the capacities were 103.1 and 75.4 mAh g^{-1}, respectively. It was distinguished from the ether electrolyte with advantages of high decomposition voltage of up to 5 V and high CE in the first cycle, making it promising energy storage appliance for the next generation. However, this battery is restricted by the poor interfacial performance. Goodenough and coworkers [36] also designed a sodium-ion rechargeable battery with an stibium anode, a $Na_3V_2(PO_4)_3$ cathode, and a low-budget composite gel-polymer electrolyte with cross-linked poly(methyl methacrylate). For this full battery, the capacities of 106.8 mAh g^{-1} at 0.1 C and 61.1 mAh g^{-1} at 10 C were delivered, respectively, with an average voltage of 2.7 V (Figure 15.14c). After 100 charge–discharge cycles, this cell retained a capacity of 86.3 mAh g^{-1} with the CE close to 100% at 1 C (Figure 15.14d), indicating a good cycle performance. On account of the good wettability of gel-polymer electrolyte, this quasi-solid battery delivered small polarization after 100 cycles. It is worth noting that, this gel-polymer electrolyte battery also worked well at 80 °C.

Table 15.3 The representative solid-state sodium-ion full batteries.

Type		Full batteries	Average working potential (V)	Specific capacity ($mAh\,g^{-1}$)	Cycle performance (cycle)	Energy density ($Wh\,kg^{-1}$)
Quasi-solid-state sodium-ion full battery		$HC/HSE/NaFePO_4$	2.6	120 (0.2 C)	96% (200)	—
		Sb/poly(methyl methacrylate)/$Na_3V_2(PO_4)_3$	2.7	86.3 (1 C)	99% (100)	—
		MoS_2/polysulfonamide/$Na_3V_2(PO_4)_3$	2	87 (0.5 C)	84% (1000)	—
		$Na/GPE/Na_3V_2(PO_4)_3$	3.4	71.4 (5 C)	81.8% (10 000)	—
		HC/gel-polymer/$Na_3V_2(PO_4)_2O_2F$	3.8	78.1 (10 C)	90% (500)	460
		$Na/IL/SE/Na_3V_2(PO_4)_3$	3.4	90 (10 C)	99% (10 000)	—
All-solid-state sodium-ion full battery	Polymer-based electrolyte	$Na/NaFSI/PEO/Na_{0.67}Ni_{0.33}Mn_{0.67}O_2$	3.3	70 (0.2 C)	93% (50)	—
		$Na/SPE/Na_3V_2(PO_4)_3$-F	3.4	84 (0.2 C)	87.5% (1000)	400
	Ceramic-based electrolyte	$Na_2Ti_3O_7$-$La_{0.8}Sr_{0.2}MnO_3$/Na-β″-Al_2O_3/ P2-$Na_{2/3}[Fe_{1/2}Mn_{1/2}]O_2$	2.5	80 (0.05 C)	87% (100)	—
		Na/Na-β″-Al_2O_3/$Na_{0.66}Ni_{0.33}Mn_{0.67}O_2$	3.4	60 (6 C)	90% (10 000)	—
		Na/CPMEA/NASICON/$NaTi_2(PO_4)_3$	2.1	102 (0.2 C)	98% (70)	—
		$Na_3V_2(PO_4)_3/Na_5YSi_4O_{12}/Na_3V_2(PO_4)_2O_2F$	2.1	79 (1 C)	89% (50)	—
		$Na/Na_3Zr_2(Si_2PO_{12})$-PCE/$Na_3V_2(PO_4)_3$	3.4	88 (1 C)	98% (100)	—
		$Na_2Ti_3O_7/Na_{3.1}Sn_{0.1}P_{0.9}S_4/Na_{2+2\delta}Fe_{2-\delta}(SO_4)_3$	2	109 (2 C)	80% (100)	—
		$Na_{15}Sn_4/Na_3PS_4/Na_4C_6O_6$	2.6	107 (0.2 C)	76% (400)	—
	Composite-based electrolyte	$Na/HSE/Na_3V_2(PO_4)_3$-C	3.4	81 (0.5 C)	85% (175)	—
	New-type electrolyte	$Na/Na_2(B_{12}H_{12})_{0.5}(B_{10}H_{10})_{0.5}/NaCrO_2$	3	68 (0.2 C)	85% (250)	—

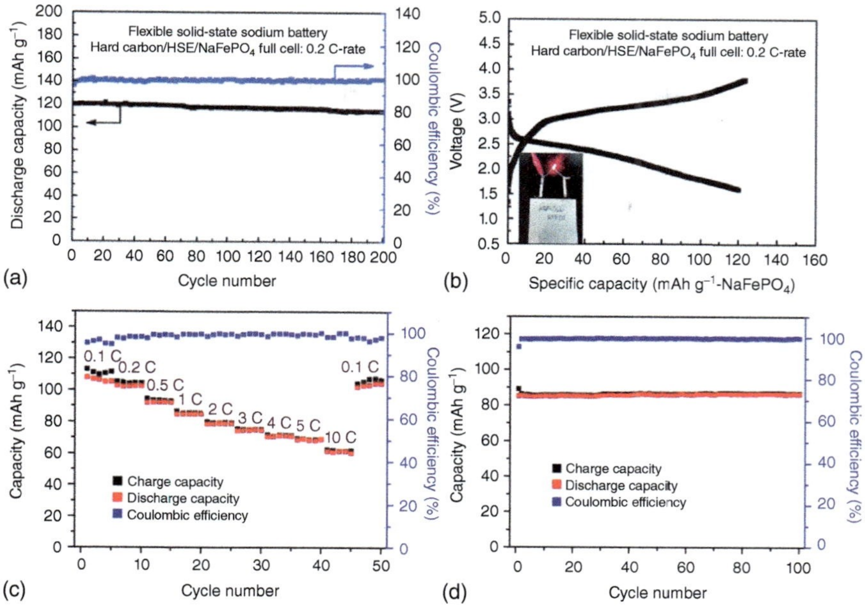

Figure 15.14 The electrochemical performance of carbon//HSE//$NaFePO_4$. (a) Cycle performance and CE. (b) Charge–discharge profile. Source: Kim et al. [35]. Reproduced with permission, 2015, Royal Society of Chemistry. The electrochemical performance of Sb//cross-linked PMMA//$Na_3V_2(PO_4)_3$. (c) Rate performance. (d) Cycling performance. Source: Gao et al. [36]. Reproduced with permission, 2016, WILEY-VCH.

Generally, compared to liquid electrolytes, quasi-solid electrolytes can withstand higher potentials which promise higher energy density. Cui and coworkers [37] fabricated a SIB with polysulfonamide electrolyte, $Na_3V_2(PO_4)_3$ cathode, and layered MoS_2 anode (Figure 15.15a). This composite-polymer-electrolyte-based sodium-ion full battery had excellent rate capability, with specific capacities of 91, 88, 83, 81, 79, and 74 mAh g^{-1} at the rates of 0.2, 0.5, 1, 2, 4, and 6 C, respectively. This cell delivered a reversible capacity of up to 70 mAh g^{-1} even when the rate increased to 10 C (Figure 15.15b). After 1000 cycles at 0.5 C, a reversible capacity of 87 mAh g^{-1} was delivered along with a high-capacity retention of 84% (Figure 15.15c). In addition, this polymer electrolyte showed higher ionic conductivity, more stable electrochemical window and better mechanical property. As a result, the SIB based on PPDE-SPE will open up a new direction for exploring the practical application of SIBs based on polymer electrolytes. Zheng et al. [38] obtained a phosphonate-based porous cross-linked gel-polymer electrolyte by using in situ thermal polymerization method and coupled it with $Na_3V_2(PO_4)_3$ and metallic Na as cathode and anode, respectively. The Na//GPE//$Na_3V_2(PO_4)_3$ cell yielded a high reversible capacity of 110.7 mAh g^{-1} at 1 C rate with the average voltage of 3.4 V and possessed reversible capacities with slight decay from 117.8 to 105 mAh g^{-1} when current rates increased from 0.1 to 2 C. Even at the high rate of 5 C, a relatively high capacity of 71.4 mAh g^{-1} can be still achieved, indicating the good rate performance. Moreover, it showed a

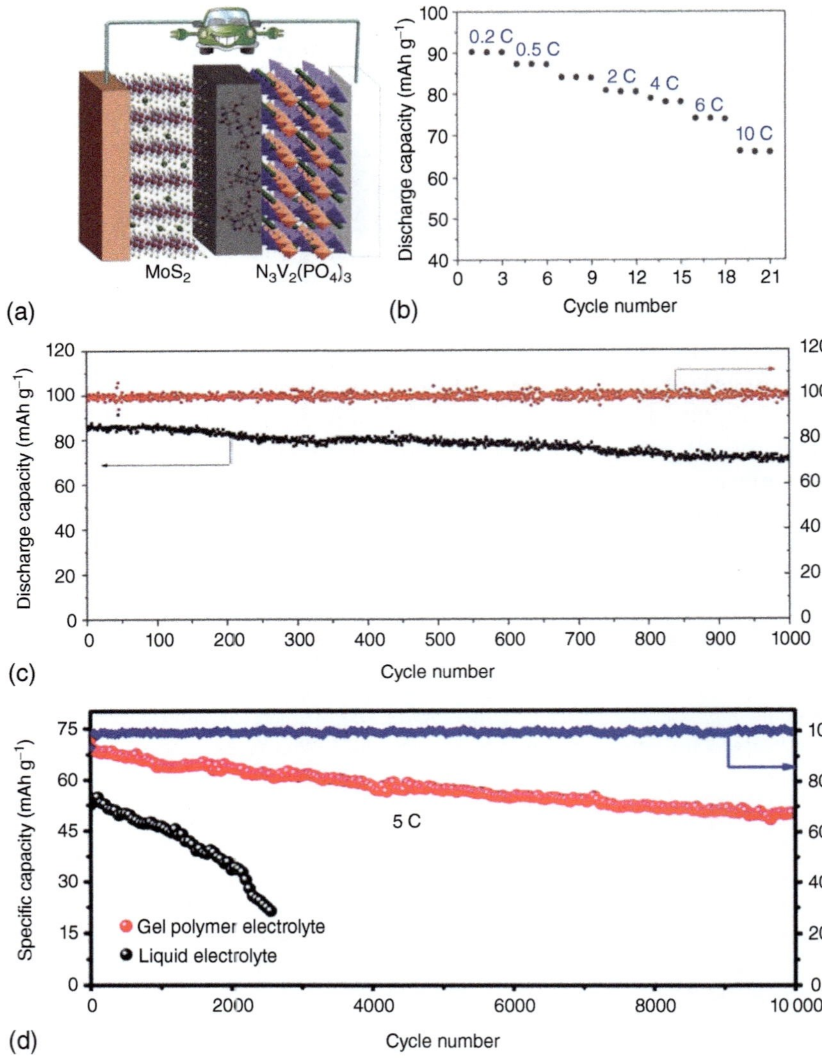

Figure 15.15 The electrochemical performance of flexible $MoS_2//Na_3V_2(PO_4)_3$ battery. (a) Schematic illustration for the fabrication of the monolithic sodium-ion full cell. (b) Rate capability. (c) Cycling performance and EC at the rate of 0.5 C. Source: Zhang et al. [37]. Reproduced with permission, 2017, Wiley–VCH. (d) The cycling performance of $Na//GPE//Na_3V_2(PO_4)_3$ batteries. Source: Zheng et al. [38]. Reproduced with permission, 2018, Royal Society of Chemistry.

superior cycle performance under the rate of 5 C, e.g. showing a high-capacity retention of 81.8% after 4500 cycles and 69.2% even after 10 000 cycles (Figure 15.15d). When the temperature increased to 60 °C, this battery can still maintain the prominent cycling performance in contrast to the battery with liquid electrolyte.

Wu and coworkers [39] designed a high-performance solid-state cell by using hard carbon as anode and high-potential $Na_3V_2(PO_4)_2O_2F$ as cathode, combined with a sodium-ion-conducting gel-polymer membrane as electrolyte. This cell realized an average voltage of 3.8 V and a high energy density of 460 Wh kg^{-1}. Moreover, the fabricated full battery also exhibited a superior rate performance with capacities of about 121 mAh g^{-1} at 0.1 C and 78.1 mAh g^{-1} at 10 C, accompanying an excellent cycle performance with 90% capacity retention for 500 cycles (Figure 15.16a). Hu and coworkers [40] reported an ultralong cycle life solid-state sodium battery by a self-forming composite electrolyte with $Na_3V_2(PO_4)_3$ cathode and Na anode, and introduced some ionic liquid (e.g. nonflammable and nonvolatile ionic liquid of *N*-methyl-*N*-propyl piperidinium-bis(fluorosulfonyl) imide (PP13FSI)) to the interface between cathode and electrolyte. It showed specific discharge capacities of 113, 112, 109, 106, 103, 97, 91, and 86 mAh g^{-1} with the rate increased from 0.2 to 10 C, respectively (Figure 15.16b). After 10 000 cycles at 10 C at room temperature, a specific capacity of about 90 mAh g^{-1} was maintained with nearly no capacity decay (Figure 15.16c). These superior performances were closely associated with the self-formed NASICON system by introducing La^{3+} ions to induce several new phases of $Na_3La(PO_4)_2$, La_2O_3, and $LaPO_4$, which improved the bulk and grain boundaries conductivity simultaneously. In addition, a few nonflammable and nonvolatile ionic liquid at the interface between the cathode and electrolyte further lowered the interface resistance.

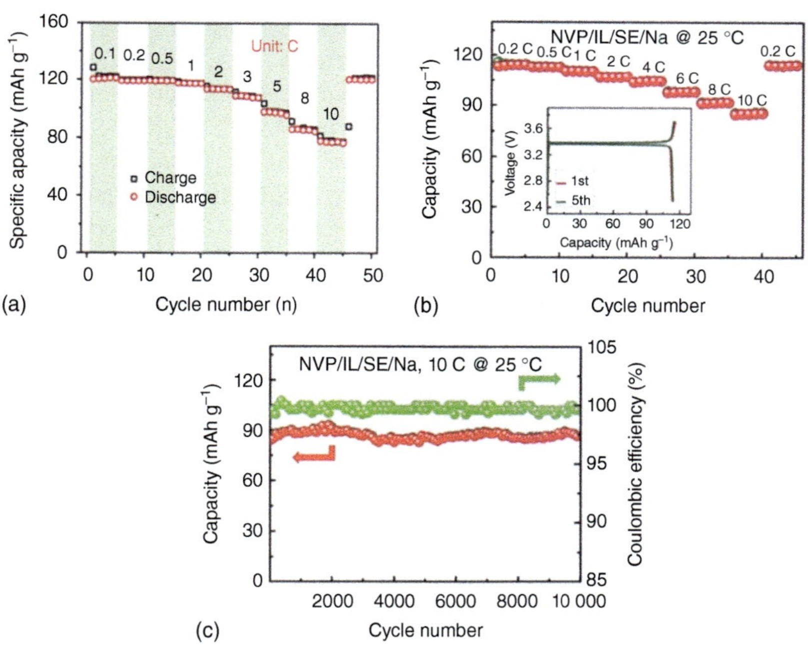

Figure 15.16 The rate capability of the CC//NVPOF full cells. Source: Guo et al. [39]. Reproduced with permission, 2018, American Chemical Society. The electrochemical performance of $Na_3V_2(PO_4)_3$//IL/SFCE//Na solid-state batteries. (b) Rate performance. (c) Cycling performance and CE. Source: Zhang et al. [40]. Reproduced with permission, 2016, WILEY-VCH.

15.4.2 All-Solid-state Sodium-Ion Full Batteries (ASSSIFBs)

Other than the above three types of batteries, the all-solid-state batteries are promising in achieving high energy density due to their prominent mechanical strength, good thermal stability, and wide voltage window. In this section, we will introduce polymer-based electrolyte, ceramic-based electrolyte, composite-based electrolyte, and borate-based electrolyte for all-solid-state batteries.

15.4.2.1 Polymer-Electrolyte-based ASSSIFBs

Compared with ceramic-based all-solid-state batteries, polymer-based all-solid-state batteries have attracted greater attention due to their good flexibility and interface wettability. They are widely used in flexible battery design. But due to their relatively narrow voltage windows, sluggish kinetics, poor chemical, and electrochemical stability at the electrolyte and electrode interface, further exploration is still needed.

Hu and coworkers [41] reported a full battery with NaFSI/PEO-blended polymer as electrolyte and $Na_{0.67}Ni_{0.33}Mn_{0.67}O_2$ as the cathode and metallic sodium as anode, displaying excellent interfacial stability and good cycling performance. At an average operating voltage of about 3.3 V, this polymer-based battery delivered a capacity of 70 mAh g^{-1} at 0.2 C under 80 °C with nearly no polarization, which is similar to the liquid-electrolyte-based cells. After 50 cycles at 80 °C, it maintained a Coulombic efficiency of nearly 100%, and good capacity retention of 93%, owning to good wettability of the interface between P2-layered NNM cathode material and flexible polymer electrolyte (Figure 15.17a). Pan and coworkers [42] reported a high-performance solid-state SIB with nanofiber structured $Na_3V_2(PO_4)_3$ as cathode and a PEO-based electrolyte (SPE), which exhibited excellent cycling performance and reversible capacity even under high rates (Figure 15.17b). The NVP-S obtained a specific capacity of 101.5 mAh g^{-1} at 0.2 C with the working voltage of 3.4 V, while NVP-F achieved a higher capacity of (107.5 mAh g^{-1}) with a lower overpotential of 45 mV (Figure 15.17c). Even when the rate increased to 2 C, the overpotential of NVP-F just slightly increased to 90 mV with a capacity of 96.8 mAh g^{-1} retained. However, under the same circumstance, for the NVP-S, the overpotential sharply increased to 170 mV with the capacity decreased to 59.6 mAh g^{-1}. As the rates increased from 0.2 to 2 C, the reversible capacities slightly declined from 108 to 97 mAh g^{-1} (Figure 15.17d). Moreover, NVP-F yielded a capacity of 84.0 mAh g^{-1} after 1000 cycles at 0.2 C, corresponding to a capacity retention of 87.5% and a near 100% Coulombic efficiency.

15.4.2.2 Ceramic-Electrolyte-based ASSSIFBs

In-depth research on high-temperature SIBs and sodium-metal chloride battery (ZEBRA) rechargeable batteries using Na-β''-Al_2O_3 electrolyte kicks off the prelude to develop all-solid-state SIBs. However, this kind of batteries usually need to work under high temperature over 250 °C, which is far beyond the melting point of Na, restricting the practical application and hampering the safety of the batteries.

In order to improve the safety and applicability of the battery, Huang and coworkers [43] fabricated a new battery consisting of Na-β''-Al_2O_3 electrolyte membrane, $Na_2Ti_3O_7$-$La_{0.3}Sr_{0.2}MnO_3$ composite anode, and P_2-$Na_{2/3}[Fe_{1/2}Mn_{1/2}]O_2$

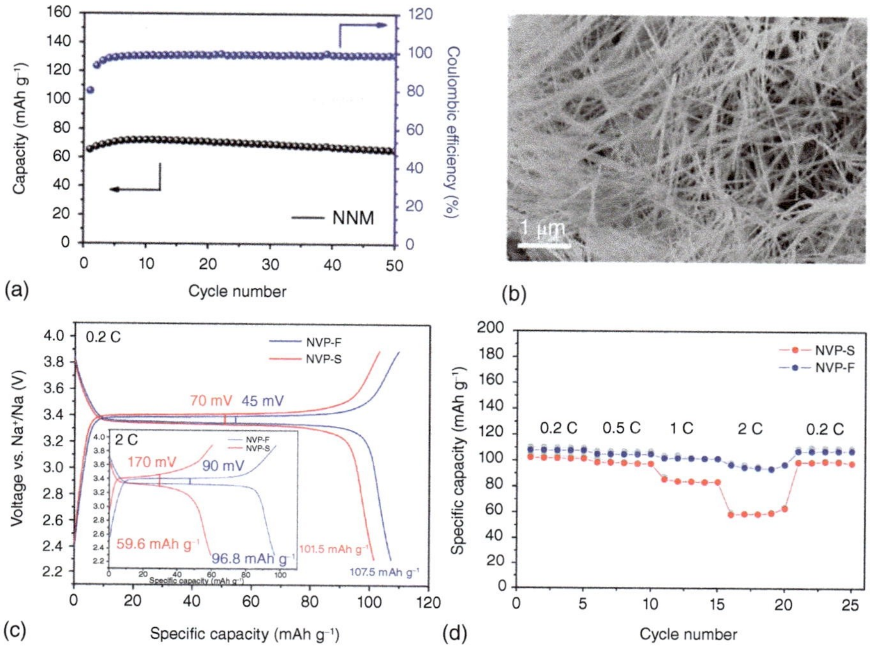

Figure 15.17 (a) The cycling performance of Na//NaFSI/PEO//NNM. Source: Qi et al. [41]. Reproduced with permission, 2016, Wiley-VCH. (b) SEM image of NVP-F, and (c, d) the electrochemical performance of Na//PEO// NVP-F or Na//PEO// NVP-S: (c) charge–discharge performance of NVP-F and NVP-S, and (d) rate performances of NVP-F and NVP-S. Source: Gao et al. [42]. Reproduced with permission, 2017, Royal Society of Chemistry.

cathode. It overcame the disadvantage associated with the metallic Na and achieved a remarkable specific capacity of 152 mAh g^{-1} at 350 °C. Hu and coworkers [44] improved the wettability of the interface using a toothpaste-like electrode to assemble an all-solid-state SIB. With increased rates from 0.1 to 8 C, this battery could deliver specific capacities from 80 to 51 mAh g^{-1}, respectively (Figure 15.18a). When operated in the voltage range of 2.5 to 3.8 V for 10000 cycles at 6 C, it showed a capacity retention of 90% (Figure 15.18b). As above, this newly designed solid-state sodium-ion battery (SS-SIBs) yielded excellent rate capability, remarkable stability and high reversibility at 70 °C, confirming the importance of optimizing the wettability at the interface to realize high-performance SS-SIBs.

Similar to the Na-β″-alumina, NASICON-type electrolytes also show a high ionic conductivity but poor interfacial contact with the electrodes. In order to overcome this issue, Goodenough and coworkers [45] fabricated a Na//$NaTi_2(PO_4)_3$ cell with a cross-linked poly(ethylene glycol) methyl ether acrylate/Na super ionic conductor (CPMEA/NASICON) double-layer electrolyte. This cell showed a good cycling stability and high Coulombic efficiency at 65 °C, due to the tactic intercalation of a dry polymer film at the interface. Dendrite formation and growth were suppressed with the increased interfacial wettability. The Na//$NaTi_2(PO_4)_3$ battery with sandwich electrolyte possessed an average voltage of 2.1 V, and delivered a steady capacity

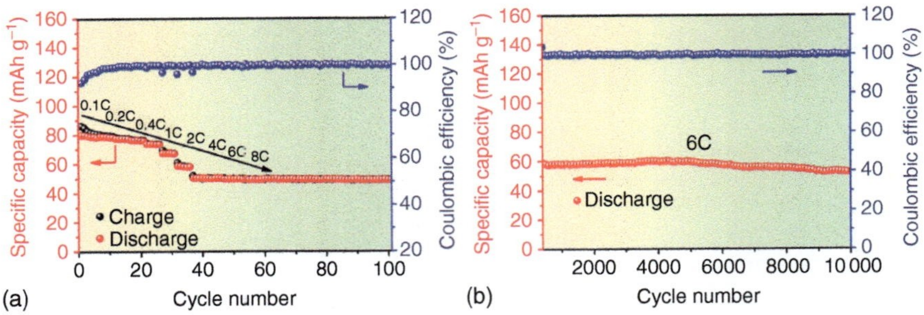

Figure 15.18 The electrochemical performance of Na//Na-β″-Al_2O_3//$Na_{0.66}Ni_{0.33}Mn_{0.67}O_2$ battery at 70 °C. (a) Cycling performance and the corresponding CE at different current densities. (b) Cycling performance at the rate of 6 C during discharge process and the corresponding CE. Source: Liu et al. [44]. Reproduced with permission, 2016, American Chemical Society.

of around 102 mAh g^{-1} with the Coulombic efficiency of about 99.7 ± 0.3% after 70 cycles at 0.2 C under 65 °C, indicating good efficiency of the Na/CPMEA/NASICON interfaces in maintaining good stability and suppressing dendrite formation during the charge–discharge process. Luo and coworkers [46] discovered a full-NASICON system battery by combining $Na_3V_2(PO_4)_3$ as anode and $Na_3V_2(PO_4)_2O_2F$ as cathode with $Na_5YSi_4O_{12}$ as electrolyte, which are all NASICON-structured materials. The full-NASICON systematic cell showed two voltage plateaus at about 2.4 and 1.9 V and displayed a specific capacity of 101 mAh g^{-1}. Importantly, the cell exhibited low risks as no flammable organic liquid electrolyte was used. At the rate of 1 C, the capacity slightly declined from 85 to 79 mAh g^{-1} after 50 cycles with a capacity retention of 89%, indicative of a good rate and cycle performance (Figure 15.19a). This full-NASICON battery may provide a new design strategy for all-solid-state sodium batteries. Goodenough and coworkers [47] reported an all-solid-state battery with good cycle stability and high rate performance by introducing a plastic–crystal electrolyte interlayer to enhance the interfacial wettability between the solid electrolyte and cathode. This battery was assembled with Na anode and $Na_3V_2(PO_4)_3$ cathode, and the NASICON-based electrolyte ($N{\equiv}C(CH_2)_2{\equiv}CN + NaClO_4[20:1]$) as interlayer. At the rate of 0.1 C, the specific capacity reached 112 mAh g^{-1} with a low polarization of about 30 mV at 50 °C (Figure 15.19b). It also possessed steady cycle performance, which achieved 98% capacity retention after 100 cycles at 1 C rate (Figure 15.19c). Furthermore, capacities of 89 and 77 mAh g^{-1} were achieved at the rates of 1 and 5 C, respectively. It is an innovation to fabricate safe, low-budget all-solid-state sodium battery (ASSSB) by introducing the plastic–crystal electrolyte, which provides a new direction to develop chemically compatible and high-voltage battery.

The ionic conductivity of solid electrolytes has always been a bottleneck restricting solid-state batteries. In order to solve this problem, the Na_3PS_4 system have received extensive attention due to their higher ionic conductivity and a moderate mechanical strength compared to the previous two solid-state electrolytes (Na-β″-alumina, and NASICON-type electrolytes). But its interfacial stability

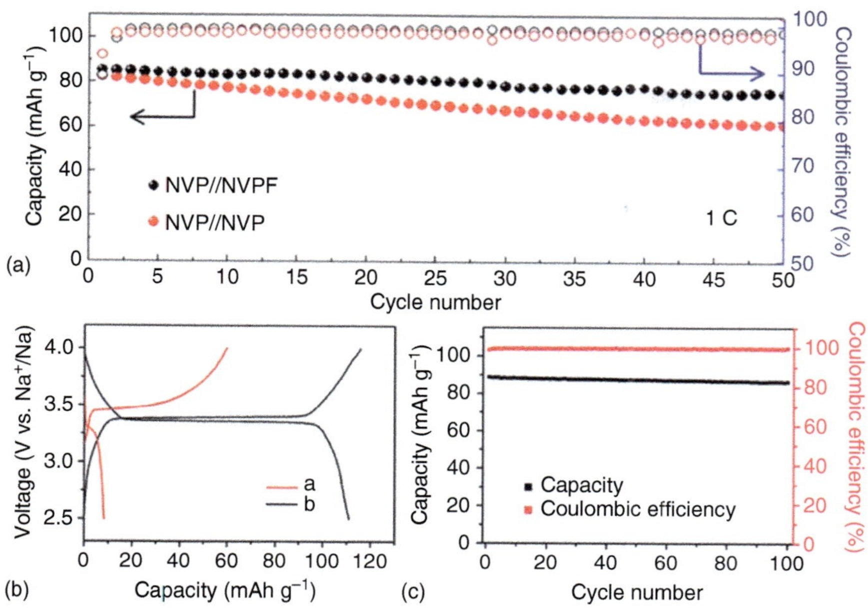

Figure 15.19 (a) Cycling stability of the NVP//NVPF and NVP//NVP full cells. Source: Sun et al. [46]. Reproduced with permission, 2019, American Chemical Society. The performance of Na//$Na_3Zr_2(Si_2PO_{12})$/PCE//NVP battery. (b) The charge–discharge performance in the initial cycle. (c) Cycle performance and the corresponding CE. Source: Gao et al. [47]. Reproduced with permission, 2017, Wiley-VCH.

is poor, as Na_3P and Na_2S can be easily formed at the interface or it can react with water to generate toxic H_2S gas, thus restricting its development. Hayashi et al. [48] developed a cubic Na_3PS_4 glass–ceramic electrolyte with high sodium-ion conductivity. The $Na_{15}Sn_4$ anode and $NaCrO_2$ cathode were then employed to coat on the electrolyte to assemble an all-solid-state battery. $Na_{15}Sn_4$//$NaCrO_2$ cells with the glass–ceramic electrolyte had an average voltage of 2.5 V and retained a capacity of about 60 mAh g^{-1} after 15 cycles at 0.013 mA cm^{-2}. But the detailed interaction mechanism between the glass–ceramic electrolyte and $NaCrO_2$ active material needs to be further explored to clarify the high stability of the glass–ceramic electrolyte under such a highly oxidative state. Jung and coworkers [49] fabricated a sodium superionic conductor Na_3SbS_4 by a solution method. This tetragonal Na_3SbS_4 electrolyte held a remarkable conductivity of 1.1 mS cm^{-1}, which can be facilely processed in large scale by MeOH or water and also performs well in dry air, serving as a potential sodium superionic conductor. With the Na_3SbS_4 as electrolyte, an ASSSB using $NaCrO_2$ as cathode and Na–Sn as anode was assembled. The full cell of $NaCrO_2$ (coated)//Na_3SbS_4//Na-Sn demonstrated much improved electrochemical performance. It could be stably operated in the voltage range of 1.2–4.0 V with a high discharge capacity of 108 mAh g^{-1}, which is comparable to that of the liquid electrolyte cells. Adams and coworkers [50] reported a high-rate all-solid-state SIB, i.e. $Na_{2+2\delta}Fe_{2-\delta}(SO_4)_3/Na_{3+x}M_xP_{1-x}S_4/Na_2Ti_3O_7$ (M = Ge^{4+}, Ti^{4+}, Sn^{4+}, X = 0, 0.1) (Figure 15.20a). Within the voltage ranging from 1.5 to

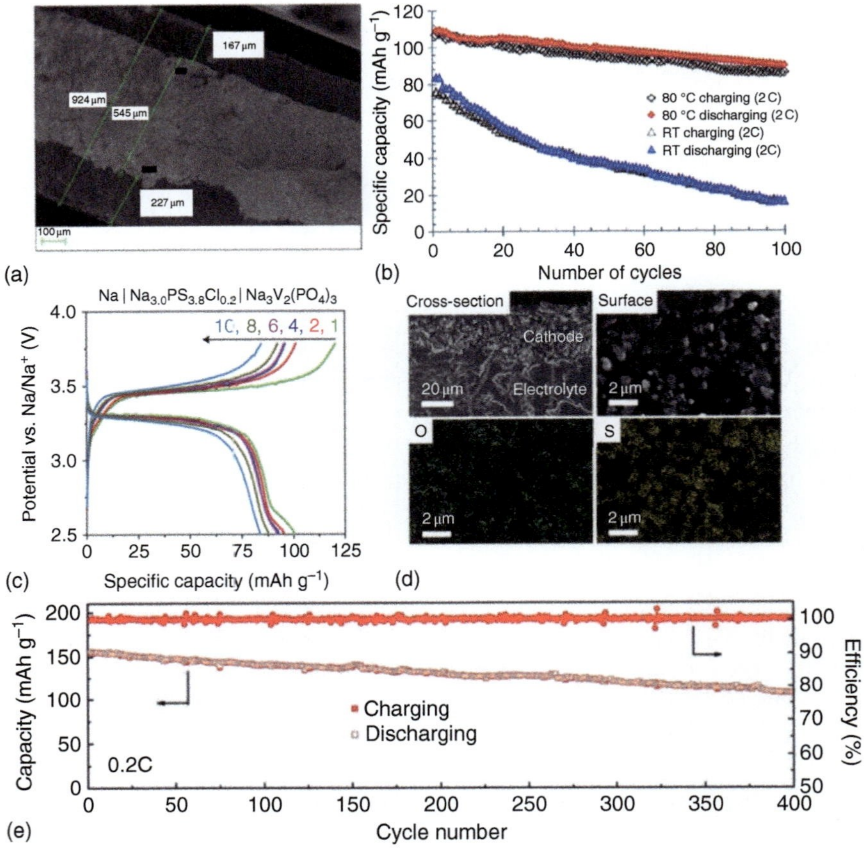

Figure 15.20 (a) Cross section of SEM image, and (b) cycling performance of $Na_{2+2\delta}Fe_{2-\delta}(SO_4)_3/Na_{3.1}Sn_{0.1}P_{0.9}S_4/Na_2Ti_3O_7$ battery. Source: Rao et al. [50]. Reproduced with permission, 2017, Royal Society of Chemistry. (c) The charge/discharge performance of Na/t-$Na_{3.0}PS_{3.8}Cl_{0.2}/Na_3V_2(PO_4)_3$. Source: Feng et al. [51]. Reproduced with permission, 2019, Wiley-VCH. (d) SEM images of $Na_4C_6O_6$ cathode/Na_3PS_4electrolyte viewed from cross section (left top) and cathode surface (right top) and the corresponding EDX mapping of O and S (bottom). (e) Cycling performance and CE of $Na_{15}Sn_4/Na_3PS_4/Na_4C_6O_6$. Source: Chi et al. [52]. Reproduced with permission, 2018, Wiley-VCH.

4.0 V, the cells with $Na_{3.1}Sn_{0.1}P_{0.9}S_4$ electrolyte exhibited good charge–discharge performance at room temperature. At the rate of 0.1 C, the initial capacity was attained to 113 mAh g^{-1}, which approaches the theoretical capacity. As the rate increased to 2 C, a specific capacity of 83 mAh g^{-1} in the initial discharge process could be obtained at room temperature, and when the temperature increased to 80 °C, a capacity of 109 mAh g^{-1} could be achieved with 80% capacity retention after100 cycles (Figure 15.20b). Even at the temperature beyond 250 °C, all components of these batteries remained stable with a low expansion of less than 3% caused by temperature increasing. Hu and coworkers [51] designed a full cell using $Na_{3.0}PS_{3.8}Cl_{0.2}$ electrolyte with Na anode and $Na_3V_2(PO_4)_3$ cathode. At the first discharge–charge cycle, it showed a reversible capacity of around 100 mAh g^{-1}

at room temperature. No obvious capacity decay was observed in the first several cycles, e.g. the capacity changed from 100 (first cycle) to 82 mAh g^{-1} (10th cycle) at 10 mA g^{-1} (Figure 15.20c). It also possessed a high average voltage of about 3.3 V. However, the unstable interfacial during cycling caused a dramatic capacity decline in the long cycles. On account of the undesirable side reaction at the interface, which creates several insulating products, the interfacial resistance increased greatly from 80 Ω at the beginning to approximately 1300 Ω after cycling. Thus, Na was replaced by Na–Sn alloy to improve surface stability to relieve the detrimental side reactions. Yao and coworkers [52] used a tailored organic electrode material of $Na_4C_6O_6$ as cathode to match sulfide electrolyte of Na_3PS_4 and anode of metallic $Na_{15}Sn_4$, obtaining a stable and high-capacity all-solid-state sodium batteries (Figure 15.20d). The battery yielded a reversible capacity of 182 mAh g^{-1} (90% of the theoretical capacity) with a 2.6 V upper cut-off potential and 99% Coulombic efficiency, suggesting the good stability. Moreover, when cycled under 60 °C at 0.2 C for 400 times, a capacity of 107 mAh g^{-1} could still be obtained (Figure 15.20e).

15.4.2.3 Composite-Electrolyte-based ASSSIFBs

For the sole polymer electrolyte, it possesses good flexibility and wettability. However, its conductivity is low, and it is easy to be pierced to cause short circuit, leading to the poor safety and cycling performance. Furthermore, owing to its narrow voltage window and poor chemical and electrochemical stability, polymer electrolyte-based battery normally possesses poor performance. Similarly, single inorganic ceramic electrolyte with high bulk conductivity and strength lacks good interface contact and grain boundary conductivity, resulting its lower overall conductivity. Thus, the ceramic-based electrolyte cells display inferior power/energy density and cycling performance. Nevertheless, the synthetic composite solid electrolyte can complement each other and improve the overall performance, which broadens the further application.

Liang and coworkers [53] reported an hybrid solid electrolyte (HSE)-based ASSSIB with a PVDF-HFP polymer mixed with $Na_3Zr_2Si_2PO_{12}$ (NASICON) as electrolyte. Metallic Na and $Na_3V_2(PO_4)_3/C$ were utilized as anode and cathode, respectively. This battery showed an average voltage of 3.4 V and delivered a reversible capacity of 98 mAh g^{-1} at 0.2 C (Figure 15.21a), which equaled to 83.05% of the theoretical capacity of $Na_3V_2(PO_4)_3/C$. Reversible capacities of 93, 95, 95, and 89 mAh g^{-1} were achieved at the different rates of 0.1, 0.2, 0.5, and 1 C, respectively. When returned to 0.1 C, it also delivered a specific capacity of 95 mAh g^{-1} (Figure 15.21b). Furthermore, around 85% capacity retention was maintained after 175 cycles at 0.5 C, indicating good cycle performance (Figure 15.21c). Na//HSE//$Na_3V_2(PO_4)_3/C$ solid-state batteries also exhibited small battery polarization, e.g. lower than 200 mV at 0.2 C, displaying excellent electrochemical stability and great interface wettability between electrode and flexible hybrid electrolyte. Mixing $Na_3Zr_2Si_2PO_{12}$ particles with polymer alleviated Na dendrite formation, meanwhile, accelerated the Na ion diffusion. This hybrid electrolyte with excellent flexibility and high ion conductivity may provide a new strategy to achieve outstanding all-solid-state sodium battery.

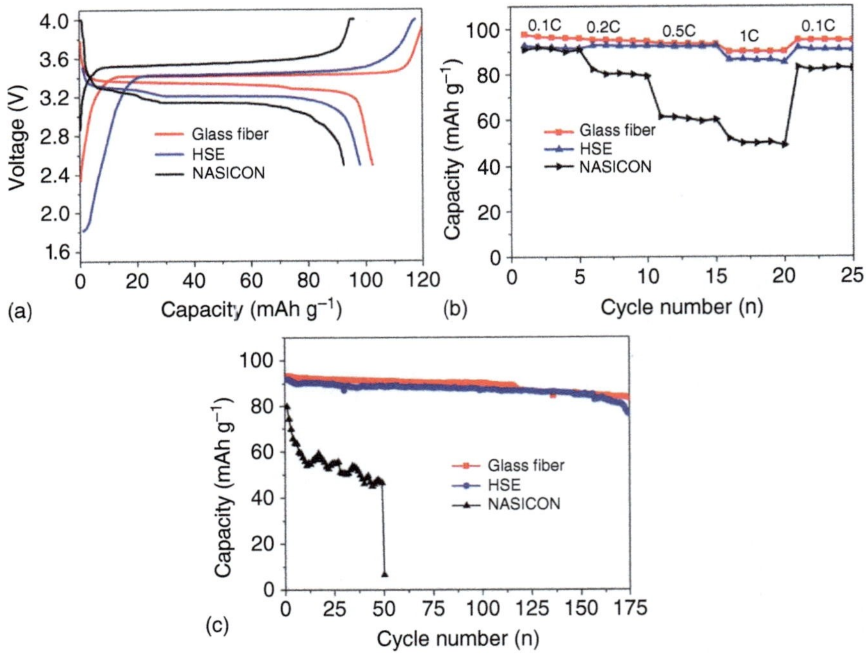

Figure 15.21 The electrochemical performance of the Na//HSE//$Na_3V_2(PO_4)_3$/C with different electrolytes. (a) The first charge–discharge curves. (b) Rate performances. (c) Cycling performance. Source: Cheng et al. [53]. Reproduced with permission, 2020, IOP Publishing.

15.4.2.4 New Types of ASSSIFBs

In recent years, a new type of all-solid-state cell called borohydride-based electrolyte battery has drawn much attention due to the high ionic conductivity of up to around 7×10^{-2} S cm^{-1}. Besides, it is light weight and shows high ductility and good compatibility with Na metals, which is promising for the high-performance all-solid-state sodium battery. However, several problems still need to be addressed. For the existing borohydride SSEs, the ionic conductivity is not the main factor that restrains the performance of the full battery. Instead, the contact between electrolyte and electrode determines the performance of the cells. Thus, the challenge has shifted to battery integration, i.e. to optimize the electrochemical stability and mechanical properties of electrodes and SSE simultaneously. It is essential to understand these two issues and conquer the integration restrictions of ASSBs to fully realize the potential of borohydride SSEs. To date, many strategies have been developed.

Remhof and coworkers [54] assembled a closo-borate electrolyte-based all-solid-battery by coupling $Na_2(B_{12}H_{12})_{0.5}(B_{10}H_{10})_{0.5}$ electrolyte with a metallic Na anode and $NaCrO_2$ cathode (Figure 15.22a). This cell realized a stable voltage of 3 V, exhibiting good solid–solid interfacial contact between cathode and electrolyte by coating the $Na_2(B_{12}H_{12})_{0.5}(B_{10}H_{10})_{0.5}$ electrolyte (i.e. accounting for 10 wt% of the total mass of cathode mixture) onto the cathode particles. As shown in Figure 15.22b, the battery had a work voltage of 3 V and could show a reversible

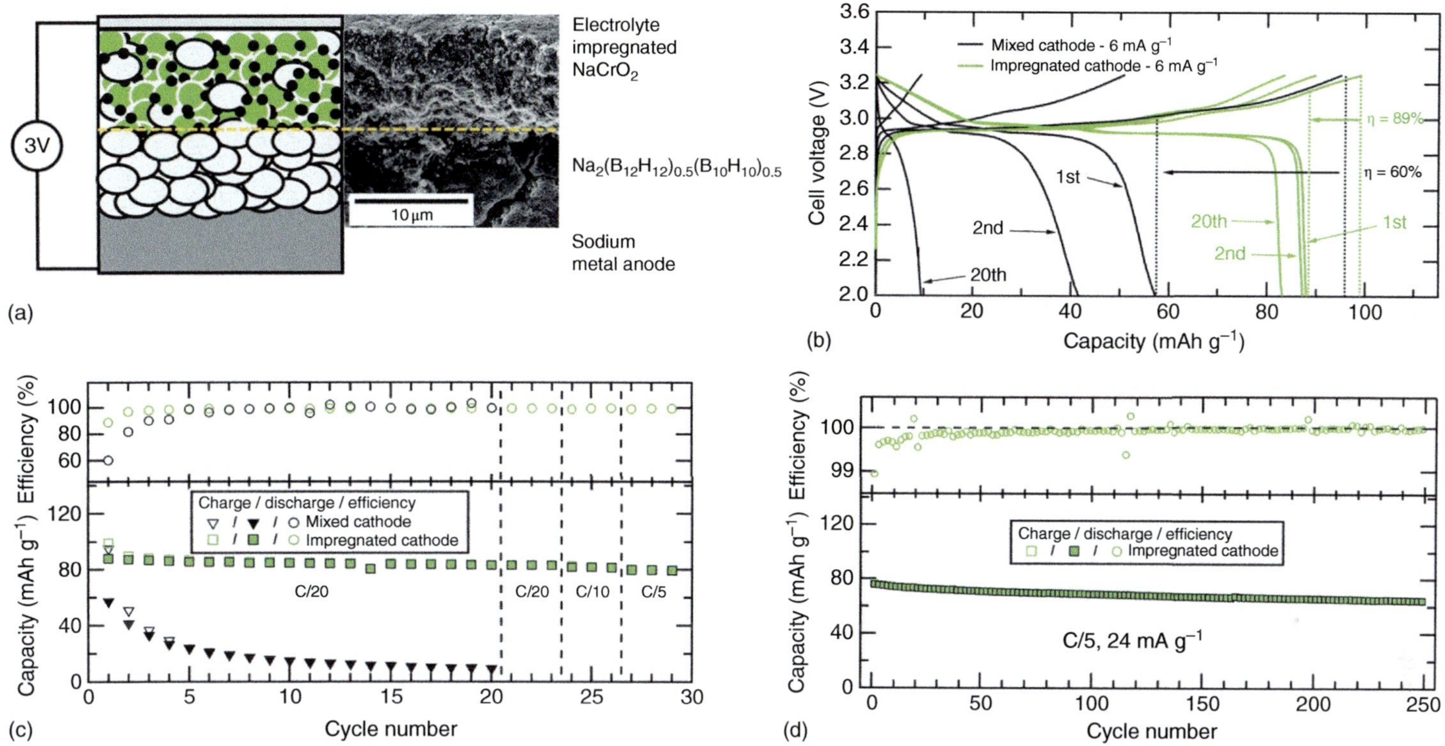

Figure 15.22 The electrochemical performance of $Na//Na_2(B_{12}H_{12})_{0.5}(B_{10}H_{10})_{0.5}//NaCrO_2$ battery. (a) Schematic and SEM cross section of the batteries. (b) Charge–discharge performance for different cycles with mixed cathode and impregnated cathode. (c) Cycling performance and the corresponding CE for the same cells at different rates. (d) Cycling performance of a cell combined with the impregnated cathode. Source: Duchêne et al. [54]. Reproduced with permission, 2017, Royal Society of Chemistry.

capacity of 85 and 80 mAh g^{-1} at C/20 and C/5, respectively. And after 20 cycles at C/20, more than 90% of the capacity was retained. When the rate increased to C/5 for 250 cycles, it could also achieve steady cycling performance with 85% capacity retention at 60 °C (Figure 15.22c,d). In addition to the conductivity improvement, interface optimization represents a novel direction in designing all-solid-state batteries.

In summary, compared with aqueous- and organic-based liquid SIBs, quasi-solid and all-solid-state batteries have good chemical and electrochemical stability, as well as better mechanical strength to prevent dendrites formation, and lead to better safety and durability of the battery. Simultaneously, a wider and stable voltage window can fully exert the performance of high-potential cathode to realize a battery system with high energy density. However, the low conductivity and untailored interface result in poor contact and undesirable side reaction, which remains bottleneck issue restricting the development of solid-state batteries. Sodium-ion solid-state batteries without properly engineered interfaces normally show low density/power density and poor cycling stability at room temperature. The interface optimization is thus the key to achieve high kinetic performance.

Some effective strategies to address the above issues are developed. For example, it is preferable to introduce interlayer to improve the interface contact or suppress the side reaction by the addition of ionic liquids, inserting polymer, and in situ/ex situ heat treatment. To realize high energy and power densities, more attention should be paid to metallic Na anode, and the interface between Na and electrolyte associated with the active Na and undesirable volume change during cycling. Tactics such as design of 3D frameworks and Na alloys can alleviate the volume change and facilitate the in situ formation of artificial thin SEI layer on sodium anode surface.

Overall, as the research on solid-state batteries is still in the early stages, much effort still needs to be made to push forward their practical applications.

References

1 Bin, D., Wang, F., Tamirat, A.G. et al. (2018). *Adv. Energy Mater.* 8: 1703008.
2 Whitacre, J.F., Tevar, A., and Sharma, S. (2010). *Electrochem. Commun.* 12: 463.
3 Whitacre, J.F., Wiley, T., Shanbhag, S. et al. (2012). *J. Power Sources* 213: 255.
4 Cao, X. and Yang, Y. (2018). *Mater. Lett.* 231: 183.
5 Ke, L., Dong, J., Lin, B. et al. (2017). *Nanoscale* 9: 4183.
6 Guo, Z., Zhao, Y., Ding, Y. et al. (2017). *Chem* 3: 348.
7 Liu, Y., Qiao, Y., Lou, X. et al. (2016). *ACS Appl. Mater. Inter.* 8: 14564.
8 Zhou, G., Li, L., Wang, D.W. et al. (2015). *Adv. Mater.* 27: 641.
9 Wang, Y., Mu, L., Liu, J. et al. (2015). *Adv. Energy Mater.* 5: 1501005.
10 Wu, X., Cao, Y., Ai, X. et al. (2013). *Electrochem. Commun.* 31: 145.
11 Wu, X.Y., Sun, M.Y., Shen, Y.F. et al. (2014). *ChemSusChem* 7: 407.
12 Pasta, M., Wessells, C.D., Liu, N. et al. (2014). *Nat. Commun.* 5: 3007.
13 Zhang, Q., Liao, C., Zhai, T., and Li, H. (2016). *Electrochim. Acta* 196: 470.

14 Kumar, P.R., Jung, Y.H., Lim, C.H., and Kim, D.K. (2015). *J. Mater. Chem. A* 3: 6271.
15 Wang, H., Zhang, T., Chen, C. et al. (2017). *Nano Res.* 11: 490.
16 Shen, Y., Han, X., Cai, T. et al. (2020). *RSC Adv.* 10: 25496.
17 Gao, H. and Goodenough, J.B. (2016). *Angew. Chem. Int. Ed.* 55: 12768.
18 Kim, D.J., Jung, Y.H., Bharathi, K.K. et al. (2014). *Adv. Energy Mater.* 4: 1400133.
19 Dong, X., Chen, L., Liu, J. et al. (2016). *Sci. Adv.*: e1501038.
20 Gu, T., Zhou, M., Liu, M. et al. (2016). *RSC Adv.* 6: 53319.
21 Long, H., Zeng, W., Wang, H. et al. (2018). *Adv. Sci.* 5: 1700634.
22 Hou, Z., Zhang, X., Ao, H. et al. (2019). *Mater. Today Energy* 14: 100337.
23 Ren, W., Zhu, Z., An, Q., and Mai, L. (2017). *Small* 13: 1604181.
24 Tang, W., Song, X., Du, Y. et al. (2016). *J. Mater. Chem. A* 4: 4882.
25 Yu, C.-Y., Park, J.-S., Jung, H.-G. et al. (2015). *Energy Environ. Sci.* 8: 2019.
26 Zhou, Y.-N., Wang, P.-F., Niu, Y.-B. et al. (2019). *Nano Energy* 55: 143.
27 Rui, X., Sun, W., Wu, C. et al. (2015). *Adv. Mater.* 27: 6670.
28 Sun, R., Wei, Q., Sheng, J. et al. (2017). *Nano Energy* 35: 396.
29 Song, M., Wang, C., Du, D. et al. (2019). *Sci. China Chem.* 62: 616.
30 Guo, J.-Z., Wang, P.-F., Wu, X.-L. et al. (2017). *Adv. Mater.* 29: 1701968.
31 Li, H., Peng, L., Zhu, Y. et al. (2016). *Energy Environ. Sci.* 9: 3399.
32 Ye, H., Wang, Y., Zhao, F. et al. (2016). *J. Mater. Chem. A* 4: 1754.
33 Fang, Y., Lian, R., Li, H. et al. (2020). *ACS Nano* 14: 8744.
34 Wang, H., Hu, P., Yang, J. et al. (2015). *Adv. Mater.* 27: 2348.
35 Kim, J.-K., Lim, Y.J., Kim, H. et al. (2015). *Energy Environ. Sci.* 8: 3589.
36 Gao, H., Zhou, W., Park, K., and Goodenough, J.B. (2016). *Adv. Energy Mater.* 6: 1600467.
37 Zhang, J., Wen, H., Yue, L. et al. (2017). *Small* 13: 1601530.
38 Zheng, J., Zhao, Y., Feng, X. et al. (2018). *J. Mater. Chem. A* 6: 6559.
39 Guo, J.Z., Yang, A.B., Gu, Z.Y. et al. (2018). *ACS Appl. Mater. Inter.* 10: 17903.
40 Zhang, Z., Zhang, Q., Shi, J. et al. (2017). *Adv. Energy Mater.* 7: 1601196.
41 Qi, X., Ma, Q., Liu, L. et al. (2016). *ChemElectroChem* 3: 1741.
42 Gao, R., Tan, R., Han, L. et al. (2017). *J. Mater. Chem. A* 5: 5273.
43 Wei, T., Gong, Y., Zhao, X., and Huang, K. (2014). *Adv. Funct. Mater.* 24: 5380.
44 Liu, L., Qi, X., Ma, Q. et al. (2016). *ACS Appl. Mater. Inter.* 8: 32631.
45 Zhou, W., Li, Y., Xin, S., and Goodenough, J.B. (2017). *ACS Cent. Sci.* 3: 52.
46 Sun, H.B., Guo, J.Z., Zhang, Y. et al. (2019). *ACS Appl. Mater. Interfaces* 11: 24192.
47 Gao, H., Xue, L., Xin, S. et al. (2017). *Angew. Chem. Int. Ed.* 56: 5541.
48 Hayashi, A., Noi, K., Tanibata, N. et al. (2014). *J. Power Sources* 258: 420.
49 Banerjee, A., Park, K.H., Heo, J.W. et al. (2016). *Angew. Chem. Int. Ed.* 55: 9634.
50 Rao, R.P., Chen, H., Wong, L.L., and Adams, S. (2017). *J. Mater. Chem. A* 5: 3377.
51 Feng, X., Chien, P.-H., Zhu, Z. et al. (2019). *Adv. Funct. Mater.* 29: 1807951.
52 Chi, X., Liang, Y., Hao, F. et al. (2018). *Angew. Chem. Int. Ed.* 57: 2630.
53 Cheng, M., Qu, T., Zi, J. et al. (2020). *Nanotechnology* 31: 425401.
54 Duchêne, L., Kühnel, R.S., Stilp, E. et al. (2017). *Energy Environ. Sci.* 10: 2609.

16

Perspectives for Sodium-Ion Batteries

Effective electrochemical energy storage and conversion technologies, such as rechargeable batteries, have catalyzed the information and communications industry. As one of the most promising next-generation rechargeable batteries, sodium-ion batteries (SIBs) have attracted great attention due to the abundant sodium resources and low prices. With the urgent need for large-scale energy storage, the development of SIBs is opening a new era full of opportunities and challenges, and suitable electrode materials are undoubtedly the key. A successful SIB should have the characteristics of low cost, reliable safety, and long cycle life. The design of active materials is critical to increase the energy density of SIBs. Although great progresses have been achieved in Chapters 3–10, the low capacity and unsuitable potential of current electrode materials are the main factors that limit the increase in the energy density of SIBs. In order to realize their large-scale application, ideal active materials for SIBs should have the following features: (i) the ability to hold a large amount of Na ions (high specific capacity); (ii) high redox potential for positive electrode materials and low redox potential for negative electrode materials; (iii) high resistance against structural integrity during cycling (long life); (iv) large diffusion coefficient for Na^+ ions and high electronic conductivity (good rate performance); (v) high chemical and thermal stability and good compatibility with electrolytes (good security); (vi) readily accessible and abundant source materials on the earth; and (vii) safe to handle.

Cathode Materials. Layered oxides are the cathode materials closest to large-scale commercialization for SIBs as they boast advantages of low cost and simple synthesis process. P-type-layered oxides present excellent rate performance, but their capacities are limited due to the low sodium content. The vacancies in their structure result in the high Coulombic efficiency, e.g., greater than 100% in the first cycle, when assembled in a half-cell with the metal sodium as anode. The Coulombic efficiency of a full battery is lower and additional supplement of sodium ions is required. On the other hand, due to the large ionic radius of Na^+ ions, the Na-intercalated layered oxide material typically undergoes complex phase transition, demonstrating undesirable structural evolutions during repeated charging and discharging. The deep insertion and de-insertion of Na ions will inevitably lead to large expansion and contraction of the crystal lattice volume, as well as the irreversible phase transitions, resulting in severe damage to the structure. The current strategies for stabilizing

Sodium-Ion Batteries: Energy Storage Materials and Technologies, First Edition. Yan Yu.

anion redox and enhancing the cycle life of such materials mainly include bulk doping to delay the irreversible phase transition of the material during the extraction and insertion of sodium ions, and surface coating to reduce side reactions at the liquid–solid surface. To overcome these challenges, joint efforts from academia and industry are in urgent demand. And these are still far from enough. To solve these problems fundamentally, it is necessary for the academic community to reconsider the design and optimization of this type of material. In addition, it is worthwhile to develop new layered oxides as cathode materials for SIBs. The targets can be met through two strategies. The first one is to delay irreversible structural loss. The other one is to increase the capacity and voltage. Using anions to participate in charge compensation at high potential redox is an effective strategy to increase the energy density of SIBs, but at present, the anion redox reaction is unstable, and the material still suffers from the attenuation of voltage and capacity during cycling.

Compared with layered transition metal oxides that usually come with larger volume changes during the extraction or insertion of sodium ions, the characteristics of zero-strain or low-strain polyanionic electrode materials are suitable for the long life and high safety of practical cathode materials for SIBs. However, these materials face problems such as low electronic conductivity, low working potential, and low capacity. In order to deal with these inherent problems of polyanionic compounds, some strategies have been adopted to enhance the electrochemical performance: (i) design of three-dimensional conductive framework; (ii) the introduction of cation ions that participate in charge compensation to increase the working potential and Na-ion diffusion coefficient; (iii) controlled synthesis of specific shapes to optimize the structure of the material. Though some progress has been made based on the above-mentioned strategies, there is still large room for the further improvement. Future developments can focus on the following aspects: (i) effective strategies to enhance the electronic conductivity; (ii) design new polyanionic cathode material with high potential and high capacity; and (iii) further improvement of ionic conductivity and chemical reaction kinetics through doping or other modification strategies.

For the Prussian blue-type cathode materials, the inherently low specific capacity makes them unsuitable for applications that require high specific energy. But it may be more suitable for energy storage systems with low energy density that emphasize long cycle life. Fe is the basic element for the Prussian blue cathodes. The Prussian blue cathode based on Fe has low cost, good cycling performance, and satisfactory rate performance. Manganese-based Prussian blue has also attracted wide attention due to its higher working voltage, which is helpful in boosting the energy density of the battery. Introduction of nickel and cobalt could further improve the electrochemical performance, but the higher cost of these metals needs to be considered in the future large-scale application process. From the perspective of raw material cost, iron-based Prussian blue is still the preferable choice. Because Prussian blues are often synthesized in liquid phase, water molecules are often involved in the structure, which is fatal to the battery life. In addition, as a new type of cathode material, it requires further exploration of the reaction mechanism of the Prussian blues, including phase changes, water in the structure, and active sodium storage sites during electrochemical cycling to provide better theoretical guidance. Intensive efforts are needed to reduce water molecules and vacancies to prepare high-quality Prussian blue cathodes. In addition to composition design, metal substitution, dehydration treatment, and modification with conductive materials are important development directions for enhancing the electrochemical performance in the future. For large-scale applications, we also need to consider the increased costs involved in the additional dehydration and optimization processes, wastewater treatment, and recycling strategies.

The organic composite is another type of attractive cathode material for SIBs due to its low cost and high specific capacity. However, the application is hindered by the low conductivity and high solubility. Carbon incorporation, anchoring, encapsulation, polymerization, salt formation, and electrolyte solidification will still be the main strategies for the modification of organic composites. The academic community has devoted great efforts to design new organic composites with high capacity and long life. When synthesizing organic compounds, the environmental pollution of toxic waste should be taken into account.

Overall, the existing cathode materials for SIBs need to be redesigned. In the design process, advanced calculations and prediction methods such as material genomes can be used. At the same time, as a new battery system, SIB still has many remaining unclear mechanism problems to be explored. For cathode materials, (i) the relationship between crystal structure, local structure, and Na storage behaviors is still not fully understood; (ii) the source and capacity of charge compensation need further exploration; (iii) the relationship between performance and reaction anisotropy needs to be further studied; and (iv) the thermodynamics and kinetics of electrode material should be figured out. Advanced characterization techniques play a key role in understanding these scientific points in SIBs. In situ/ex situ synchrotron radiation and neutron diffraction have been widely used to explore the reaction mechanism of the cathode materials, more advanced instruments need to be developed. In addition, in large-scale production, suitable synthesis equipment and simple preparation procedure are demanded. The above-mentioned problems all require the joint development of academia and industry, multidisciplinary cooperation from chemistry, physics, materials science, engineering, environmental science, and other relative fields to realize the commercialization of SIBs.

Electrolytes. The electrolyte as a key component plays a critical role in the whole battery system. The excellent compatibility with active electrodes (both cathodes and anodes) would contribute to develop more advanced SIBs. In view of the various types of electrolytes, we need to identify the requirements of SIBs and identify the proper proposals to achieve the specific properties.

(i) For organic liquid electrolytes in SIBs, sufficient ionic conductivity, low viscosity under the working temperature, and high stability in electrochemical and chemical behaviors should be essential. Moreover, high thermostability and low cost would be also favorable. However, it is almost impossible for a single solvent to fulfill all these requirements. Though SIB is a newly burgeoning system with relatively deficient basic research, it can inherit much knowledge from the lithium-ion batteries (LIB) system due to similar working principles.

For example, superior electrochemical performance can be realized via the optimum blend of various non-aqueous solvents and sodium salts for SIBs. Among those, carbonate-based electrolytes are the most common ones in SIBs due to their high chemical/electrochemical stability and competent ability in dissolving salts. And numerous effective additives can also benefit the electrolyte design. Besides, ether solvents can induce uniform solid electrode interface (SEI) with decent thickness and composition on sodium metal anode. Although plenty of organic liquid electrolyte formulas/systems have been explored in recent years, deep understandings and investigations are still needed.

Firstly, the cognition of transport properties at a deeper level should be gained. The sodium salts with different anions display distinctive dissolving capacity and ion conductivity in electrolyte with the same solvent. This effect needs to be systematically explored with more valid research both in experiments and calculations. In addition, the diffusion of the ion–solvent complex in the bulk liquid phase and further charge transfer behaviors across the electrolyte/electrode interphase is another noteworthy issue. With the advanced computational–experimental approach, their thermodynamic and kinetic factors resulted from the solvent effect can also be investigated.

The second is the chemistry essence of electrochemical stability for electrolyte systems. The desirable electrolyte is supposed to possess a large distance between the critical energy level (HOMO and LUMO). As such, the voltage window could be extended with improved electrochemical stability. Hence, some types of solvents including sulfones, nitriles, and fluorine-based electrolytes might be worth exploring due to their large electrochemical working window. Besides, in view of the general applicability of carbonate-based solvents, the functional transformation on these ester molecules is advisable. Appropriate introduction of more electronegative elements (such as N, F, and S) can lower the HOMO level via the withdrawing properties toward electrons. Thus, the functioned carbonates can enhance their stability against oxidation to adapt to the high-voltage cathode materials. More importantly, fundamental understandings of the mechanism of the SEI formation in organic electrolytes are relatively inadequate. The existence of an interfacial layer can effectively isolate the liquid electrolytes and active electrodes and enhance the compatibility between those materials, and this layer highly depends on the components in organic electrolytes like solvent, salts, and additives. Therefore, to guide the formation of the proper SEI layer, the mechanism of this layer should be comprehensively studied. For example, ion de-solvation across the electric double layer and interfacial transport mechanism in the interface need to be identified at the atomic level. Moreover, appropriate additives with multifunctional groups are critical to match with various electrodes in SIBs. Those additives preferable in forming the passive layer can inhibit the side reactions and stabilize the interfacial regions between the electrode and the liquid. Meanwhile, there is no need to sacrifice the cycling performance with the functional additives. With the modulation of the whole liquid electrolyte, long life SIBs with high performance can be achieved. Besides, changing the concentration of sodium salts in the same system can also make some difference. Super-concentrated electrolytes show enhanced electrochemical stability due to the unique solvation structure, but they also show some defects including high viscosity and high cost. Recently, electrolytes with low concentration demonstrate huge advantages in cycling performance and cost savings. However, further research on the actual mechanism is needed as well as the verification of pouch-type batteries.

Another critical factor needs to be concerned for organic liquid electrolytes is the safety characteristics. Since the current SIBs mainly rely on highly flammable organic electrolytes, the security situation is overall grim and needs further improvement. Firstly, the overall thermal stability of the electrolytes is a vital research project. By sorting out each individual component in electrolytes, the thermal stability can be analyzed in detail by thermoanalysis techniques. At the meantime, the thermal stability among the electrodes and electrolytes would be more difficult to identify. More advanced thermal analysis instruments would contribute to understand the real thermal behaviors of various materials. Then the experience gained will provide feedback to the further modulation on the electrolyte formula. Furthermore, some flame-retardant additives such as phosphazene can be introduced to the electrolytes to enhance the safety, while the electrochemical performance might be partly affected.

(ii) For ionic liquid electrolytes, they can serve as a medium-short-term supplement for organic solvents because of their combined properties such as high electrochemical stability, wide temperature range, high safety, and low volatility. However, their high viscosity and low ionic conductivity are the main constraints toward the practical application. In future research, more cations and anions should be investigated to gain the optimum formula that is not limited to imidazolium-based ionic liquids. In addition, the salt concentration in ionic liquid (IL)-based electrolyte is also an important parameter that determines the bulk ionic conductivity and overall viscosity. The proper concentration of sodium salt needs to be identified, which induces adequate transference numbers and improves the interfacial properties. Another route to enhance salt solubility in the IL-based electrolyte is by directly introducing functional groups in complex cation, such as ether-functionalized groups adhering to a quaternary ammonium cation. With the good sodium chelating ability of the ether functional group, the solvating properties of the ILs can be remarkably boosted. However, their relatively high cost would be the obstacle in meeting the demand for larger scale application in SIBs.

(iii) For gel polymer electrolytes, they have shown advantageous properties such as good compatibility with electrodes and excellent flexibility, which make them popular for diversified applications. Compared with liquid electrolytes, gel polymer electrolytes avoid liquid leakage to enhance their safety. Moreover, their ionic conductivity can exceed that of solid polymer electrolytes due to the existence of solvents or liquid plasticizers. In comparison with glass-ceramic-based electrolytes, they have shown better mechanical performance with superior flexibility and processability. To further design more effective gel polymer materials in battery applications, the ion diffusion mechanism needs to be deeply investigated. According to their ion transfer mechanism and preferable pathway, the proper design principle can be obtained from the molecular level. One effective method is the precise introduction with useful functional groups on the polymer molecules. Hence, with the controllable pathway or the guided diffusion of Na ions, the corresponding ionic conductivity would be significantly improved to sustain the high-rate cycling.

(iv) For solid electrolytes, they can radically circumvent the serious safety issues derived from flammable liquid organics. Moreover, excellent mechanical properties make them more attractive in the next-generation Na-based batteries. But their low ionic conductivity in bulk phase and high interfacial contact resistance are still the challenging bottlenecks toward the practical applications. Hence, there are some directions for advancing the solid electrolytes in SIBs.

More work should be focused on developing novel electrolytes to enable high-performance SIBs. The first criteria of novel solid electrolytes should be the adequate ionic conductivity comparable to that of the liquid. Meanwhile, high-throughput screening is a proper method to select the appropriate materials with desirable properties. After systematic calculations, the electrochemical properties and ion diffusion paths can be theoretically clarified. Then experimental results can work out the optimum electrolyte formula. Besides, the simple and energy-saving synthetic technology, such as chemical synthesis, should be developed.

The high impedance across the electrode/electrolyte interface is another challenge. Many solid electrolytes suffer from the mismatch between the active electrodes and electrolytes, leading to the high interface contact resistance. More intuitive situ detection tools and computational calculations are required to characterize the actual physical/chemical process on this interface. With a more detailed understanding, the more effective strategies including coating and protective layers can be presented. To further alleviate the high impedance, many improved processing technologies in all-solid-state batteries need to be developed. For example, combining the electrolyte and active materials through solution treatment can remarkably reduce interface resistance. Other wetting materials like organic polymer can serve as a buffer layer and effectively improve the physical contact across the interface. Moreover, the short circuit behaviors in sodium-based solid-state electrolytes as well as the corresponding mechanism should be paid more attention. This insight can provide more valuable information on the subsequent design and modifications toward safer solid electrolytes.

High fabrication cost is a more realistic obstacle for further large-scale applications. When initiating the major deployment of solid electrolytes, the large-area ceramic electrolytes would be strived to develop with the inevitable cost increase. Fortunately, many manufacturing infrastructures and emerging manufacturing workshops for all-solid-electrolyte LIBs are comparatively mature, which can provide some reference for the development of practical SIBs.

Overall, the electrolyte is a crucial component in electrochemical storage devices, whose development greatly affects the practical process of low-cost SIBs. In this sense, extensive efforts are necessary to develop effective composition formulas and other novel-type electrolyte materials for available large-scale applications. Moreover, more deep insights and effective design principles need to be summarized to provide timely feedback and deliver the significant closed-loop.

Anode Materials. So far, the research on anode materials for SIBs has achieved phased achievements, and a variety of promising anode materials have been developed, including carbon-based materials, sodium titanium phosphates, metal oxides and sulfides, metal alloys, phosphorus and phosphides, as well as sodium metal, etc. Although many of them typically deliver considerable electrochemical performance (high capacity, stable cycling performance, or high reversible capacity), however, for the successful application of SIBs, the unit energy cost and overall performance of next-generation SIBs should be comparable or even superior to that of commercial LIBs, so the revolution of anode materials with high performance cannot stop.

Based on the working mechanism, there are three main types of anode materials for SIBs: intercalation (carbon materials and titanium-based oxides, etc.), alloys (Sn, Sb, P, etc.), and conversion types (MO_x, MS_x, MF_x, etc.). Intercalated anodes typically demonstrate better cycle stability, but lower specific capacity. Although alloys and conversion anodes deliver higher theoretical capacity, they will display larger volume expansion during cycling, disrupting the integrity of the electrode structure, and seriously affecting the cycle stability. In addition, the poor conductivity of the conversion anode material is also a fatal defect. Therefore, it is necessary to make targeted improvements according to the different characteristic and electrochemical reaction mechanism of different anode materials.

Graphite-like carbon materials are not suitable for anode of SIBs due to insufficient interlayer spacing, while hard carbon and soft carbon materials are gradually developed and matured, but their specific capacities of sodium storage are still low. The specific capacity of hard or soft carbon can be improved by nanostructure design and heteroatom doping. Besides, hard carbon typically possesses higher reversible capacity, while soft carbon has better rate capability, which is due to the discrepancy in graphitization degree. So, there is an interesting strategy to improve the electrochemical performance of carbon materials, that is, mixing precursors of soft and hard carbon. It is necessary to construct a carbon material with a unique micro/nanoarchitecture that includes a shorter sodium ion diffusion pathway and abundant reversible sodium storage active sites, to overcome the low initial Coulombic efficiency and insufficient rate capability of existing carbon materials. For practical applications, an initial Coulombic efficiency higher than 90% is needed. However, there has been rare report about the carbon materials with an initial Coulombic efficiency exceeding 80%, and most of them are below 70%. From the perspective of commercialization, carbon molecular sieve as a new type of negative electrode material for SIBs possesses the advantages including high capacity, high Coulombic efficiency, and long cycle life. At the same time, it also can be prepared on a large scale and has been commercialized. Its similarity with graphite directly promotes its practical application in LIBs, although the Na storage mechanism of carbon sieves has not yet been clarified. On the other hand, sodium titanium phosphates can deliver a relatively high voltage plateau. When used as an anode in a full battery, it needs to couple with a high voltage cathode so as to achieve high energy output, which poses higher requirements for cathode materials. For sodium titanium phosphate, although the structure is stable with good cycling performance, its theoretical capacity is low. Element doping or composite with high specific capacity materials are two viable strategies to further improve this material.

At present, the research on alloy anodes for SIBs is mainly focused on Sn and Sb, and more alloy materials with excellent electrochemical properties need to be discovered. Metal alloys, metal oxides, and sulfides can store sodium ion by alloying reaction or multi-electron conversion reaction, respectively, which deliver high sodium storage specific capacity. However, a fatal problem that they will suffer from is the huge volume expansion, e.g., typically over 50% during cycling, and the low initial Coulombic efficiency, e.g., less than 80%, which leads to poor cycle performance. While the volume change of commercial batteries should be less than 30%, the volume expansion problem can be restrained and the comprehensive battery performance can be improved by nanocrystallization, alloying, and combination with carbon. In addition, poor electrical conductivity is also a major problem facing metal oxides, sulfides, and phosphates. Therefore, delicate design of these anode materials is needed. The most common method is to fabricate composites with unique nanoarchitectures. However, it is worth to mention that a considerable number of inactive components would be introduced during the formation of nanocomposites, which significantly reduces the energy density of active materials. Other than increasing the specific capacity of electrode materials, we still need to increase the overall energy density of SIBs. Therefore, it is necessary optimize the nanocomposite architecture to achieve high energy density.

Similar to other anodes, phosphorus and phosphide anodes also face challenges such as unclear electrochemical mechanism, poor conductivity, and volume expansion, but it is worth to note that in most phosphorus-based anodes, the phosphorus content is very low. Generally, the active material should account for 70–90% of the weight of the entire anode; however, the phosphorus-based anodes often include a large ratio of the inactive conductive component. It may be possible to increase the proportion of phosphorus in the composite material (>70%) by constructing a free-standing anode free of inactive components, which is a prerequisite to achieve high energy density.

In view of the fundamental difference between Li and Na, further theoretical and experimental studies should be carried out to reveal the dendrite growth behavior, nucleation, and SEI formation mechanism. When Na metal anodes are paired with different electrolytes or cathodes, they will exhibit different working mechanisms, so specific modification strategies are needed. If Na metal anode is not incorporated into the high-performance full battery, its protection will be meaningless. Therefore, researchers must consider the full battery system when designing the protection strategies of Na metal anode.

As stated above, the current anode materials work via different sodiation mechanisms, such as intercalation, conversion, alloying, or combined reactions, and their failure mode is also different. Nevertheless, the sodiation reaction as well as failure mechanism of these anode materials has not been fully revealed or still under controversy. Therefore, more advanced characterization tools, especially in situ techniques, are needed to get in-depth and thorough insights to further optimize their electrochemical performance. For example, although the electrochemical performance of the anode material largely depends on the interface chemistry such as nanostructure and composition of SEI layer, the interface chemistry between the anode materials and the electrolytes is not yet well understood. Practically, it is unreasonable to talk about the impact of electrolyte on electrochemical performance without considering the type of anode materials. Besides, at present, most high-performance anodes as reported are tested with low mass loading, but for practical application, the mass loading needs to reach at least 5 $mg\,cm^{-1}$. The low tap density of nanomaterials will significantly decrease the volume capacity of SIBs. In order to effectively preserve the volumetric capacity of nanostructured anode materials, it is necessary to design more reasonable nanostructures. On the other hand, building 3D free-standing flexible anode may be another way to reduce the cost and increase the energy density of current SIBs. Heterostructured negative electrodes composed of two or more types of materials require more attention to achieve high rate, high energy density, or high-safety anodes, because the single-phase anode materials typically cannot meet all the performance requirements at the same time. In addition, most of the synthesis schemes to fabricate nanostructured anodes suffer from low yield and tedious steps, which makes it difficult to obtain the expected nanostructures in large scale. Scalable methods with lower cost are highly desirable for large-scale commercial application. The ultimate solution to realize high-performance SIBs may be the combination of various modification strategies, but there is still a long way to achieve the large-scale industrial production, which cannot be separated from the continuous experimental and theoretical study.

In the past 40 years, the research on LIB anodes has been relatively matured, regardless of advanced characterization or electrochemical performance optimization, which have provided effective guidance for the development of SIBs. Research on anodes of SIBs should be regarded as a long-term campaign. With constant attention to this field, people will achieve more fundamental achievements. For the future commercialization of SIBs, it is a critical task to explore cheap and high-performance anode materials suitable for industrial production.

In general, as a low-cost and high-performance system, SIBs show huge potential along with great challenges. As a new type of battery system, the reaction mechanisms still need continuous exploration. It will need to optimize the cathode, electrolyte, anode, and other aspects, as well as the cooperation of the system. We believe that in the near future, with joint efforts from different fields, many scientific problems facing current SIBs will be gradually solved, and large-scale commercialization of SIBs is also foreseeable.

Index

Sodium-Ion Batteries: Energy Storage Materials and Technologies, First Edition. Yan Yu.

C

n

q

r

s

x

y

z